Principles of Spread-Spectrum Communication Systems

Don Torrieri

Principles of Spread-Spectrum Communication Systems

Fourth Edition

Springer

Don Torrieri
Silver Spring, MD, USA

Additional material to this book can be downloaded from http://www.springer.com/in/book/9783319705682.

ISBN 978-3-030-09970-1 ISBN 978-3-319-70569-9 (eBook)
https://doi.org/10.1007/978-3-319-70569-9

Softcover re-print of the Hardcover 4th edition 2018

Printed on acid-free paper

This Springer imprint is published by the registered company Springer International Publishing AG part of Springer Nature.
The registered company address is: Gewerbestrasse 11, 6330 Cham, Switzerland

To my family

Preface

The continuing vitality of spread-spectrum communication systems and my desire to expand the scope of the content motivated me to undertake this fourth edition of *Principles of Spread-Spectrum Communication Systems*. This edition is intended to enable readers to understand the current state-of-the-art spread-spectrum communication systems. This edition includes new or enhanced sections on code tracking, the normalized LMS algorithm, frequency-hopping diversity, direct-sequence multicode and multiple-input multiple-output systems, interference cancelers, optimal frequency-hopping patterns, complementary codes, Laplace transforms and characteristic functions, orthonormal functions, and Hermitian positive-definite matrices. The remainder of the material has been thoroughly revised to improve the presentation.

This book provides a comprehensive and intensive examination of spread-spectrum communication systems that is suitable for graduate students, practicing engineers, and researchers with a solid background in the theory of digital communication. As the title indicates, this book stresses principles rather than specific current or planned systems, which are described in many less advanced books. The principal goal of this book is to provide a concise, lucid explanation of the fundamentals of spread-spectrum systems; with an emphasis on theoretical principles and methods of mathematical analysis that will facilitate future research. The choice of specific topics to include was tempered by my judgment of their practical significance and interest to both researchers and system designers. The book contains many improved derivations of the classical theory and presents the latest research results, bringing the reader to the frontier of the field. The analytical methods and subsystem descriptions are applicable to a wide variety of communication systems. Problems at the end of each chapter are intended to assist readers in consolidating their knowledge and to provide practice of analytical techniques. The listed references are those I recommend for further study and as sources of additional references.

A *spread-spectrum signal* is a signal with extra modulation that expands the signal bandwidth greatly beyond what is required by the underlying channel code and modulation. Spread-spectrum communication systems are useful for suppressing interference and jamming, making it difficult to detect and process secure communications, accommodating fading and multipath channels, and providing a multiple-access capability without requiring synchronization across the entire network. The most practical and dominant spread-spectrum systems are *direct-sequence* and *frequency-hopping* systems.

There is no fundamental theoretical barrier to the effectiveness of spread-spectrum communications. This remarkable fact is not immediately apparent because the increased bandwidth of a spread-spectrum signal necessitates a receive filter that passes more noise power to the demodulator. However, when any signal and white Gaussian noise are applied to a filter matched to the signal, the sampled filter output has a signal-to-noise ratio that depends solely on the energy-to-noise-density ratio. Thus, the bandwidth of the input signal is irrelevant, and spread-spectrum signals have no inherent limitations.

Chapter 1 reviews fundamental results of coding and modulation theory that are essential to a full understanding of spread-spectrum systems. In this chapter, coding and modulation theory are used to derive the required receiver computations and the error probabilities of the decoded information bits. *Channel codes*, which are also called *error-correction* or *error-control* codes, are vital in fully exploiting the potential capabilities of spread-spectrum systems. Although direct-sequence systems can greatly suppress interference, practical systems require channel codes to limit the effects of the residual interference and channel impairments, such as fading. Frequency-hopping systems are designed to avoid interference, but the possibility of hopping into an unfavorable spectral region usually requires a channel code to maintain adequate performance.

Chapter 2 presents the fundamentals of direct-sequence systems. This chapter describes basic spreading sequences and waveforms and provides a detailed analysis of how the direct-sequence receiver suppresses various forms of interference. *Direct-sequence modulation* entails the direct addition of a high-rate spreading sequence and a lower-rate data sequence, resulting in a transmitted signal with a relatively wide bandwidth. The removal of the spreading sequence in the receiver causes a contraction of the bandwidth that can be exploited by applying appropriate filtering to remove a large portion of the interference.

Chapter 3 covers the fundamentals of frequency-hopping systems. *Frequency hopping* is the periodic changing of the carrier frequency of a transmitted signal. This time-varying characteristic potentially endows a communication system with great strength against interference. Whereas a direct-sequence system relies on spectral spreading, spectral despreading, and filtering to suppress interference, the basic mechanism of interference suppression in a frequency-hopping system is that of avoidance. When the avoidance fails, it is only temporary because of the periodic changing of the carrier frequency. The impact of the interference is further mitigated by the pervasive use of channel codes, which are more essential for frequency-hopping systems than for direct-sequence systems. The basic concepts, spectral and performance aspects, and coding and modulation issues are presented. The effects of partial-band interference and multitone jamming are examined, and the most important issues in the design of frequency synthesizers are described.

The methods of code synchronization for both direct-sequence and frequency-hopping systems are presented in Chapter 4. A spread-spectrum receiver requires *code synchronization* to generate a spreading sequence or frequency-hopping pattern that is synchronized with the received sequence or pattern. After code synchronization, the received and receiver-generated chips or dwell intervals must precisely or nearly coincide. Any misalignment causes the signal amplitude at the demodulator output to fall in accordance with the autocorrelation or partial autocorrelation function. A practical implementation of code synchronization is greatly facilitated by dividing synchronization into two operations: acquisition and tracking. *Code acquisition* provides coarse synchronization by limiting the possible timing offsets of the receiver-generated chips or dwell intervals to a finite number of quantized candidates. Code acquisition is almost always the dominant design issue and most expensive component of a complete spread-spectrum system. Following code acquisition, *code tracking* is activated to provide fine synchronization by which synchronization errors are further reduced or at least maintained within certain bounds. *Symbol synchronization*, which is needed to provide timing pulses for symbol detection to the decoder, is derived from the code-synchronization system. Although the use of precision clocks in both the transmitter and the receiver limit the timing uncertainty of sequences or patterns in the receiver, clock drifts, range uncertainty, and the Doppler shift may cause synchronization problems.

Adaptive filters and adaptive arrays have numerous applications as components of communication systems. Chapter 5 covers those adaptive filters and adaptive arrays that are amenable to exploiting the special spectral characteristics of spread-spectrum signals to enable interference suppression beyond that inherent in the despreading or dehopping. Adaptive filters for the rejection of narrowband interference or primarily for the rejection of wideband interference are presented. The least-mean-square (LMS), normalized LMS, and Frost algorithms are derived, and conditions for the convergence of their mean weight vectors are determined. Adaptive arrays for both direct-sequence systems and frequency-hopping systems are described and shown to potentially provide a very high degree of interference suppression.

Chapter 6 provides a general description of the most important aspects of fading and the role of diversity methods in counteracting it. *Fading* is the variation in received signal strength due to changes in the physical characteristics of the propagation medium, which alter the interaction of multipath components of the transmitted signal. The principal means of counteracting fading are *diversity methods*, which are based on the exploitation of the latent redundancy in two or more independently fading copies of the same signal. The basic concept of diversity is that even if some copies are degraded, there is a high probability that others will not be. Both direct-sequence and frequency-hopping signals are shown to provide diversity. The rake demodulator, which is of central importance in most direct-sequence systems, is shown to be capable of exploiting undesired multipath signals rather than simply attempting to reject them. The multicarrier direct-sequence system and frequency-domain equalization are shown to be alternative methods of advantageously processing multipath signals.

Multiple access is the ability of many users to communicate with each other while sharing a common transmission medium. Wireless multiple-access communications are facilitated if the transmitted signals are orthogonal or separable in some sense. Signals may be separated in time (*time-division multiple access* or TDMA), frequency (*frequency-division multiple access* or FDMA), or code (*code-division multiple access* or CDMA).

Chapter 7 presents the general characteristics of *direct-sequence* CDMA (DS-CDMA) and *frequency-hopping* CDMA (FH-CDMA) systems. The use of spread-spectrum modulation in CDMA allows the simultaneous transmission of signals from multiple users in the same frequency band. All signals use the entire allocated spectrum, but the spreading sequences or frequency-hopping patterns differ. Information theory indicates that in an isolated cell, CDMA systems achieve the same spectral efficiency as TDMA or FDMA systems only if optimal multiuser detection is used. However, even with single-user detection, CDMA has advantages for mobile communication networks because it eliminates the need for frequency and time-slot coordination, allows carrier-frequency reuse in adjacent cells, imposes no sharp upper bound on the number of users, and provides resistance to interference and interception. The vast potential and practical difficulties of spread-spectrum multiuser detectors, such as optimal, decorrelating, minimum mean-square error, or adaptive detectors, are described and assessed. The tradeoffs and design issues of direct-sequence multiple-input multiple-output with spatial multiplexing or beamforming are determined.

The impact of multiple-access interference on mobile ad hoc and cellular networks with DS-CDMA and FH-CDMA systems are analyzed in Chapter 8. Phenomena and issues that become prominent in mobile networks using a spread spectrum include exclusion zones, guard zones, power control, rate control, network policies, sectorization, and the selection of various spread-spectrum parameters. The outage probability, which is the fundamental network performance metric, is derived for both ad hoc and cellular networks and both DS-CDMA and FH-CDMA systems. Acquisition and synchronization methods that are needed within a cellular DS-CDMA network are addressed.

Chapter 9 examines the role of iterative channel estimation in the design of advanced spread-spectrum systems. The estimation of channel parameters, such as the fading amplitude and the power spectral density of the interference and noise, is essential to the effective use of soft-decision decoding. Channel estimation may be implemented by the transmission of pilot signals that are processed by the receiver, but pilot signals entail overhead costs, such as the loss of data throughput. Deriving maximum-likelihood channel estimates directly from the received data symbols is often prohibitively difficult. There is an effective alternative when turbo or low-density parity-check codes are used. The expectation-maximization algorithm, which is derived and explained, provides an iterative approximate solution to the maximum-likelihood equations and is inherently compatible with iterative demodulation and decoding. Two examples of advanced spread-spectrum systems that apply iterative channel estimation, demodulation, and decoding are described and analyzed. These systems provide good illustrations of the calculations required in the design of advanced systems.

The ability to detect the presence of spread-spectrum signals is often required by cognitive radio, ultra-wideband, and military systems. Chapter 10 presents an analysis of the detection of spread-spectrum signals when the spreading sequence or the frequency-hopping pattern is unknown and cannot be accurately estimated by the detector. Thus, the detector cannot mimic the intended receiver, and alternative procedures are required. The goal is limited in that only detection is sought, not demodulation or decoding. Nevertheless, detection theory leads to impractical devices for the detection of spread-spectrum signals. An alternative procedure is to use a radiometer or energy detector, which relies solely on energy measurements to determine the presence of unknown signals. The radiometer has applications not only as a detector of spread-spectrum signals, but also as a general sensing method in cognitive radio and ultra-wideband systems.

Eight appendices contain important mathematical details about Gaussian processes and the central limit theorem, the moment-generating function and the Laplace transform, the Fourier transform and the characteristic function, deterministic and random signal characteristics, probability distribution functions, orthonormal functions and parameter estimation, Hermitian positive-definite matrices, and special functions.

In writing this book, I have relied heavily on notes and documents prepared and the perspectives gained during my work at the US Army Research Laboratory. I am thankful to my colleagues Matthew Valenti and Hyuck Kwon for their thorough reviews of the original manuscript. I am grateful to my wife, Nancy, who provided me not only with her usual unwavering support but also with extensive editorial assistance.

Silver Spring, MD, USA Don Torrieri

Contents

Chapter 1
Channel Codes and Modulation

This chapter reviews the fundamental results of coding and modulation theory that are essential to a full understanding of spread-spectrum systems. *Channel codes*, which are also called *error-correction* or *error-control* codes, are vital in fully exploiting the potential capabilities of spread-spectrum communication systems. Although direct-sequence systems greatly suppress interference, practical systems require channel codes to limit the effects of the residual interference and channel impairments, such as fading. Frequency-hopping systems are designed to avoid interference, but the possibility of hopping into an unfavorable spectral region usually requires a channel code to maintain adequate performance. In this chapter, coding and modulation theory are used to derive the required receiver computations and the error probabilities of the decoded information bits.

1.1 Block Codes

A *channel code* for forward error control or error correction [7, 57, 69] is a set of *codewords* that are used to improve communication reliability. An (n, k) *block code* uses a codeword of n code symbols to represent k information symbols. Each symbol is selected from an alphabet of q symbols that belong to the Galois Field GF(q), and there are q^k codewords. If $q = 2^m$, then a q-ary symbol may be represented by m bits, and a nonbinary codeword of n symbols may be mapped into an (mn, mk) binary codeword. A block encoder can be implemented by using logic elements or memory to map a k-symbol information word into an n-symbol codeword.

A block code of length n over GF(q) is called a *linear block code* if its q^k codewords form a k-dimensional subspace of the vector space of sequences with n symbols. Thus, the vector sum of two codewords or the vector difference between them is a codeword. Since a linear block code is a subspace of a vector space, it

D. Torrieri, *Principles of Spread-Spectrum Communication Systems*,
https://doi.org/10.1007/978-3-319-70569-9_1

must contain the additive identity. Thus, the all-zero sequence is always a codeword in any linear block code. Since nearly all practical block codes are linear, henceforth block codes are assumed to be linear.

The number of symbol positions in which the symbol of one sequence differs from the corresponding symbol of another equal-length sequence is called the *Hamming distance* between the sequences. The minimum Hamming distance between any two codewords of a code is called the *minimum distance* of the code.

The *Hamming weight* of a codeword is the number of nonzero symbols in a codeword. For binary block codes, the Hamming weight is the number of ones in a codeword. For any linear block code, the vector difference between two codewords is another codeword with a weight equal to the distance between the two original codewords. By subtracting the codeword **c** to all the codewords, we find that the set of Hamming distances from any codeword **c** is the same as the set of Hamming distances from the all-zero codeword. Consequently, *the minimum Hamming distance of a code is equal to the minimum Hamming weight of the nonzero codewords.*

Let **m** denote a row vector of k information symbols and **c** denote a row vector of n codeword symbols. Let **G** denote a $k \times n$ *generator matrix*, each row of which is a basis vector of the subspace of codewords. A linear block code computes

$$\mathbf{c} = \mathbf{m}\mathbf{G} \tag{1.1}$$

to generate a codeword.

The *orthogonal complement* of the row space of **G** is an $(n-k)$-dimensional subspace of the n-dimensional vector space such that each of its linearly independent vectors is orthogonal to the row space of **G**, and hence to the codewords. An $(n-k) \times n$ *parity-check matrix* **H** has row vectors that span the orthogonal complement. Therefore,

$$\mathbf{G}\mathbf{H}^T = \mathbf{0}. \tag{1.2}$$

A *systematic block code* is a code in which the information symbols appear unchanged in the codeword, which also has additional parity symbols. Thus, a systematic codeword can be expressed in the form $\mathbf{c} = [\mathbf{m}\ \mathbf{p}]$, where **p** is the row vector of $n-k$ parity symbols, and the generator matrix has the form

$$\mathbf{G} = [\mathbf{I}_k \quad -\mathbf{P}] \tag{1.3}$$

where $\mathbf{I}_k$ is the $k \times k$ identity matrix and **P** is a $k \times (n-k)$ matrix. This equation and (1.2) indicate that the parity check matrix for a linear block code is

$$\mathbf{H} = \left[\mathbf{P}^T \ \mathbf{I}_{n-k}\right]. \tag{1.4}$$

In terms of performance, every linear code is equivalent to a systematic linear code that is easier to implement. Therefore, systematic block codes are the standard choice and are assumed henceforth. Substituting (1.3) into (1.1), we obtain

$$\mathbf{c} = [\mathbf{m} \ \ \mathbf{mP}] \tag{1.5}$$

which indicates the dependence of the parity symbols on the information symbols. For a binary block code, which uses an alphabet of symbols 0 and 1, the parity bits of a codeword are modulo-2 sums of information bits.

After the waveform representing a codeword is received and demodulated, the decoder uses the demodulator output to determine the information symbols corresponding to the codeword. If the demodulator produces a sequence of discrete symbols and the decoding is based on these symbols, the demodulator is said to make *hard decisions*. Conversely, if the demodulator produces analog or multilevel quantized samples of the waveform, the demodulator is said to make *soft decisions*. The advantage of soft decisions is that reliability or quality information is provided to the decoder, which can use this information to improve its performance.

When the demodulator makes hard decisions, the demodulator output symbols are called *channel symbols,* and the output sequence is called the *received sequence*, or the *received word*. Hard decisions imply that the overall channel between the encoder output and the decoder input is the classical binary symmetric channel. A decoder that processes the received word is called a *hard-decision decoder*. If the channel-symbol error probability is less than one-half, then the maximum-likelihood criterion implies that the correct codeword is the one that is the smallest Hamming distance from the received word. A *complete decoder*is a hard-decision decoder that implements the maximum-likelihood criterion. An *incomplete decoder* does not attempt to correct all the words it receives.

The n-dimensional vector space of sequences is conceptually represented as a three-dimensional space in Figure 1.1. Each codeword occupies the center of a *decoding sphere* with radius t in Hamming distance, where t is a positive integer.

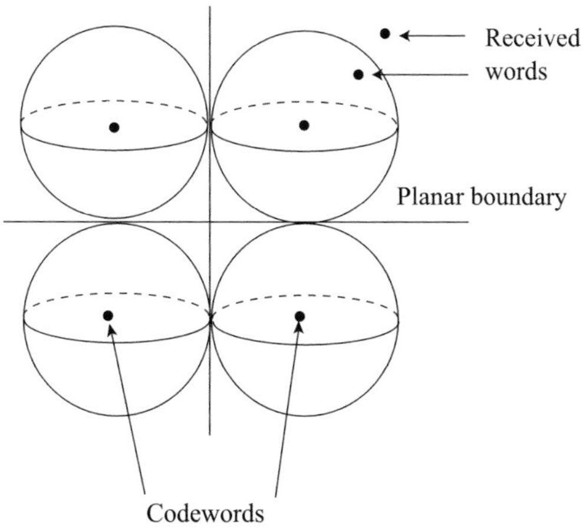

Fig. 1.1 Conceptual representation of n-dimensional vector space of sequences

A complete decoder has decision regions defined by planar boundaries surrounding each codeword. A *bounded-distance decoder* is an incomplete decoder that attempts to correct symbol errors in a received word if it lies within one of the decoding spheres. Since unambiguous decoding requires none of the spheres to intersect, the maximum number of random errors that can be corrected by a bounded-distance decoder is

$$t = \lfloor (d_m - 1)/2 \rfloor \tag{1.6}$$

where d_m is the minimum Hamming distance between codewords, and $\lfloor x \rfloor$ denotes the largest integer less than or equal to x. When more than t errors occur, the received word may lie within a decoding sphere surrounding an incorrect codeword or it may lie in the interstices (regions) outside the decoding spheres. If the received word lies within a decoding sphere, the decoder selects the incorrect codeword at the center of the sphere and produces an output word of information symbols with undetected errors. If the received word lies in the interstices, the decoder cannot correct the errors, but recognizes their existence.

Since there are $\binom{n}{i}(q-1)^i$ words at exactly distance i from the center of the sphere, the number of words in a decoding sphere of radius t is

$$V = \sum_{i=0}^{t} \binom{n}{i} (q-1)^i. \tag{1.7}$$

Since a block code has q^k codewords, $q^k V$ words are enclosed in spheres. The number of possible received words is $q^n \geq q^k V$, which yields

$$q^{n-k} \geq \sum_{i=0}^{t} \binom{n}{i} (q-1)^i. \tag{1.8}$$

This inequality implies an upper bound on t and hence d_m. The upper bound on d_m is called the *Hamming bound*.

A *cyclic code* is a linear block code in which a cyclic shift of the symbols of a codeword produces another codeword. This characteristic allows the implementation of encoders and decoders that use linear feedback shift registers. Relatively simple encoding and hard-decision decoding techniques are known for cyclic codes belonging to the class of *Bose-Chaudhuri-Hocquenghem* (BCH) *codes*. A BCH code has a length that is a divisor of $q^m - 1$, where $m \geq 2$, and is designed to have an error-correction capability of $t = \lfloor (\delta - 1)/2 \rfloor$, where δ is the *design distance*. Although the minimum distance may exceed the design distance, the standard BCH decoding algorithms cannot correct more than t errors. The parameters (n, k, t) for binary BCH codes with $7 \leq n \leq 127$ are listed in Table 1.1.

A *perfect code* is a block code such that every n-symbol sequence is at a maximum distance of t from some n-symbol codeword, and the sets of all sequences at distance t or less from each codeword are disjoint. Thus, the Hamming bound is

Table 1.1 Binary BCH codes

n	k	t	D_p	n	k	t	D_p	n	k	t	D_p
7	4	1	1	63	45	3	0.1592	127	92	5	0.0077
7	1	3	1	63	39	4	0.0380	127	85	6	0.0012
15	11	1	1	63	36	5	0.0571	127	78	7	1.68×10^{-4}
15	7	2	0.4727	63	30	6	0.0088	127	71	9	2.66×10^{-4}
15	5	3	0.5625	63	24	7	0.0011	127	64	10	2.48×10^{-5}
15	1	7	1	63	18	10	0.0044	127	57	11	2.08×10^{-6}
31	26	1	1	63	16	11	0.0055	127	50	13	1.42×10^{-6}
31	21	2	0.4854	63	10	13	0.0015	127	43	14	9.11×10^{-8}
31	16	3	0.1523	63	7	15	0.0024	127	36	15	5.42×10^{-9}
31	11	5	0.1968	63	1	31	1	127	29	21	2.01×10^{-6}
31	6	7	0.1065	127	120	1	1	127	22	23	3.56×10^{-7}
31	1	15	1	127	113	2	0.4962	127	15	27	7.75×10^{-7}
63	57	1	1	127	106	3	0.1628	127	8	31	8.10×10^{-7}
63	51	2	0.4924	127	99	4	0.0398	127	1	63	1

Table 1.2 Code words of Hamming (7,4) code

0000000	0001011	0010110	0011101
0100111	0101100	0110001	0111010
1000101	1001110	1010011	1011000
1100010	1101001	1110100	1111111

satisfied with equality, and a complete decoder of a perfect code is also a bounded-distance decoder. The only perfect codes are the binary repetition codes of odd length, the Hamming codes, the binary (23,12) Golay code, and the ternary (11,6) Golay code.

Repetition codes represent each information bit by n binary code symbols. When n is odd, the $(n, 1)$ repetition code is a perfect code with $d_m = n$ and $t = (n-1)/2$. A hard-decision decoder makes a decision based on the state of most of the demodulated symbols. Although repetition codes are not efficient for the additive white Gaussian noise (AWGN) channel, they can improve the system performance for fading channels if the number of repetitions is properly chosen. An (n, k) *Hamming code* is a perfect BCH code with $d_m = 3$ and $n = (q^{n-k}-1)/(q-1)$. Since $t = 1$, a Hamming code is capable of correcting all single errors. Binary Hamming codes with $n \leq 127$ are found in Table 1.1. The 16 codewords of a (7,4) Hamming code are listed in Table 1.2. The first four bits of each codeword are the information bits. The perfect (23,12) *Golay code* is a binary cyclic code with $d_m = 7$ and $t = 3$. The perfect (11,6) Golay code is a ternary cyclic code with $d_m = 5$ and $t = 2$.

Any (n, k) linear block code with an odd value of d_m can be converted into an $(n+1, k)$ *extended code* by adding a parity symbol. The advantage of the extended code stems from the fact that the minimum distance of the block code is increased by one, which improves the performance, but the decoding complexity and code rate are usually changed insignificantly. The (24,12) *extended Golay code* is formed by adding an overall parity symbol to the (23,12) Golay code, thereby increasing the minimum distance to $d_m = 8$. As a result, some received sequences with four errors can be corrected with a complete decoder. The (24,12) code is often preferable to the (23,12) code because the *code rate*, which is defined as the ratio k/n for a binary code, is exactly one-half, which simplifies the system timing.

Some systematic codewords have only one nonzero information symbol. Since there are at most $n-k$ parity symbols, these codewords have Hamming weights that cannot exceed $n - k + 1$. Since the minimum distance of the code is equal to the minimum codeword weight,

$$d_m \leq n - k + 1. \tag{1.9}$$

This upper bound is called the *Singleton bound*. A linear block code with a minimum distance equal to the Singleton bound is called a *maximum-distance separable code*.

Nonbinary block codes can accommodate high data rates efficiently because decoding operations are performed at the symbol rate rather than the higher information-bit rate. A *Reed-Solomon code* is a nonbinary BCH code and a maximum-distance separable code with $n = q - 1$ and $d_m = n - k + 1$. For convenience in implementation, q is usually chosen so that $q = 2^m$, where m is the number of bits per symbol. Thus, $n = 2^m - 1$ and the code provides correction of 2^m-ary symbols. Most Reed-Solomon decoders are bounded-distance decoders with $t = \lfloor (d_m - 1)/2 \rfloor$.

The most important single determinant of the code performance is its *weight distribution,* which is a list or function that gives the number of codewords with each possible weight. The weight distributions of the Golay codes are listed in Table 1.3. Analytical expressions for the weight distribution are known in a few cases. Let A_l denote the number of codewords with weight l. For a binary Hamming code, each A_l can be determined from the weight-enumerator polynomial

Table 1.3 Weight distributions of Golay codes

	Number of codewords	
Weight	(23,12)	(24,12)
0	1	1
7	253	0
8	506	759
11	1288	0
12	1288	2576
15	506	0
16	253	759
23	1	0
24	0	1

$$A(x) = \sum_{l=0}^{n} A_l x^l = \frac{1}{n+1}[(1+x)^n + n(1+x)^{(n-1)/2}(1-x)^{(n+1)/2}]. \quad (1.10)$$

For example, the (7,4) Hamming code gives $A(x) = \frac{1}{8}[(1+x)^7+7(1+x)^3(1-x)^4] = 1 + 7x^3 + 7x^4 + x^7$, which yields $A_0 = 1, A_3 = 7, A_4 = 7, A_7 = 1$, and $A_l = 0$, otherwise. For a maximum-distance separable code, $A_0 = 1$ and

$$A_l = \binom{n}{l}(q-1)\sum_{i=0}^{l-d_m}(-1)^i\binom{l-1}{i}q^{l-i-d_m}, \quad d_m \le l \le n. \quad (1.11)$$

The weight distribution of other codes can be determined by examining all valid codewords if the number of codewords is not too large for a computation.

Hard-Decision Decoders

There are two types of bounded-distance decoders: erasing decoders and reproducing decoders. They both produce errors when a received word falls within an incorrect decoding sphere, which is called an *undetected error*. They differ only in their actions following the detection of uncorrectable errors in a received word, which is called a *decoding failure*. An *erasing decoder* discards the received word after a decoding failure and may initiate an automatic retransmission request. For a systematic block code, a *reproducing decoder* reproduces the information symbols of the received word as its output after a decoding failure.

Let P_s denote the *channel-symbol error probability*, which is the probability of error in a demodulated code symbol. We assume that the channel-symbol errors are statistically independent and identically distributed, which is an accurate model for systems with appropriate symbol interleaving (Section 1.4). Let P_w denote the *word error probability*, which is the probability that a decoder does not produce the correct information symbols of a codeword because of an undetected error or decoding failure. There are $\binom{n}{i}$ distinct ways in which i errors may occur among n channel symbols. Since a received sequence may have more than t errors but no information-symbol errors, a reproducing decoder that corrects t or few errors has

$$P_w \le \sum_{i=t+1}^{n}\binom{n}{i}P_s^i(1-P_s)^{n-i}. \quad (1.12)$$

For an erasing decoder, (1.12) becomes an equality if erased words are considered word errors.

For error correction with reproducing decoders, t is given by (1.6) because it is pointless to make the decoding spheres smaller than the maximum allowed by the code. However, if a block code is used for both error correction and error detection, an erasing decoder is often designed with t fewer than the maximum. If a block code is used exclusively for error detection, then $t = 0$.

A complete decoder correctly decodes even if the number of symbol errors exceeds t, provided that the received word is closest in Hamming distance to the correct codeword. When a received sequence is equidistant from two or more codewords, a complete decoder selects one of them according to some arbitrary rule. Thus, the word error probability for a complete decoder satisfies (1.12).

The word error probability is a performance measure that is important, primarily in applications for which only a decoded word completely without symbol errors is acceptable. When the utility of a decoded word degrades in proportion to the number of information bits that are in error, the *information-bit error probability* is frequently used as a performance measure. To evaluate it for block codes that may be nonbinary, we first examine the information-symbol error probability.

Let $\mathcal{P}(\nu)$ denote the probability of an error in information symbol ν at the decoder output. To avoid assuming that $\mathcal{P}(\nu)$ is independent of ν, the *information-symbol error probability* is defined as the average error probability of the information symbols:

$$P_{is} = \frac{1}{k}\sum_{\nu=1}^{k}\mathcal{P}(\nu). \tag{1.13}$$

The random variables Z_ν, $\nu = 1, 2, \ldots, k$, are defined so that $Z_\nu = 1$ if information symbol ν is in error and $Z_\nu = 0$ if it is correct. Let $E[\cdot]$ denote the expected value. The expected number of information-symbol errors is

$$E[I] = E\left[\sum_{\nu=1}^{k} Z_\nu\right] = \sum_{\nu=1}^{k} E[Z_\nu] = \sum_{\nu=1}^{k}\mathcal{P}(\nu) = kP_{is} \tag{1.14}$$

which implies that the *information-symbol error rate*, which is defined as $E[I]/k$, *is equal to the information-symbol error probability*. Similarly, we find that the decoded-symbol error probability, which is defined as *the average error probability of all the symbols, is equal to the decoded-symbol error rate*.

Consider an erasing bounded-distance decoder, which may produce an error in an information symbol only if there is an undetected error. As shown previously, the set of Hamming distances from a specific codeword to the other codewords is the same for all specific codewords of a linear block code. Therefore, it is legitimate to assume for convenience in evaluating P_{is} that the all-zero codeword was transmitted. If channel-symbol errors in a received word are statistically independent and occur with the same probability P_s, then the probability of a specific set of i erroneous symbols among the n codeword symbols is

$$P_e(i) = \left(\frac{P_s}{q-1}\right)^i (1 - P_s)^{n-i}. \tag{1.15}$$

For an undetected error to occur at the output of a bounded-distance decoder, the number of channel-symbol errors must exceed t, and the received word must lie within an incorrect decoding sphere of radius t. Consider an incorrect decoding sphere of radius t associated with a codeword of weight l, where $d_m \leq l \leq n$. If $l - t \leq i \leq l + t$, let $N(l, i)$ denote the number of sequences in the set $S(i, l)$ of sequences with Hamming weight i that lie within this decoding sphere. If a received word with i channel-symbol errors matches one of the sequences in $S(i, l)$, then an incorrect codeword with Hamming weight l is selected, and the decoder-symbol and information-symbol error probabilities are l/n. Therefore, (1.15) implies that the *information-symbol error probability for an erasing bounded-distance decoder* is

$$P_{is} = \sum_{i=t+1}^{n} \sum_{l=\max(i-t,d_m)}^{\min(i+t,n)} A_l N(l,i) P_e(i) \frac{l}{n}$$

$$= \sum_{i=t+1}^{n} \left(\frac{P_s}{q-1}\right)^i (1-P_s)^{n-i} \sum_{l=\max(i-t,d_m)}^{\min(i+t,n)} A_l N(l,i) \frac{l}{n}. \tag{1.16}$$

Consider sequences of weight i that are at distance s from a particular codeword of weight l, where $|l - i| \leq s \leq t$ so that the sequences are within the decoding sphere of the codeword. By counting these sequences and then summing over the allowed values of s, we can determine $N(l, i)$. The counting is done by considering changes in the symbols of this codeword that can produce one of these sequences.

Let ν denote the number of nonzero codeword symbols that are changed to zeros, α the number of codeword zeros that are changed to any of the $(q - 1)$ nonzero symbols in the alphabet, and β the number of nonzero codeword symbols that are changed to any of the other $(q-2)$ nonzero symbols. For a sequence at distance s to result, it is necessary that $0 \leq \nu \leq s$. The number of sequences that can be obtained by changing any ν of the l nonzero symbols to zeros is $\binom{l}{\nu}$, where $\binom{b}{a} = 0$ if $a > b$. For a specified value of ν, it is necessary that $\alpha = \nu + i - l$ to ensure a sequence of weight i. The number of sequences that result from changing any α of the $n - l$ zeros to nonzero symbols is $\binom{n-l}{\alpha}(q-1)^\alpha$. For a specified value of ν and hence α, it is necessary that $\beta = s - \nu - \alpha = s + l - i - 2\nu$ to ensure a sequence at distance s. The number of sequences that result from changing β of the $l - \nu$ remaining nonzero components is $\binom{l-\nu}{\beta}(q-2)^\beta$, where $0^x = 0$ if $x \neq 0$ and $0^0 = 1$. Summing over the allowed values of s and ν, we obtain

$$N(l,i) = \sum_{s=|l-i|}^{t} \sum_{\nu=0}^{s} \binom{l}{\nu}\binom{n-l}{\nu+i-l}\binom{l-\nu}{s+l-i-2\nu} \times (q-1)^{\nu+i-l}(q-2)^{s+l-i-2\nu}. \tag{1.17}$$

Equations (1.16) and (1.17) allow the exact calculation of P_{is}.

When $q = 2$, the only term in the inner summation of (1.17) that is nonzero, has the index $\nu = (s + l - i)/2$, provided that this index is an integer and $0 \leq (s + l - i)/2 \leq s$. Using this result, we find that for binary codes,

$$N(l, i) = \sum_{s=|l-i|}^{t} \binom{n-l}{\frac{s+i-l}{2}} \binom{l}{\frac{s+l-i}{2}} , \quad q = 2 \tag{1.18}$$

where $\binom{m}{n} = 0$ when n is not a nonnegative integer.

The number of sequences of weight i that lie in the interstices outside the decoding spheres is

$$L(i) = (q-1)^i \binom{n}{i} - \sum_{l=\max(i-t,d_m)}^{\min(i+t,n)} A_l N(l, i) , \quad i \geq t + 1 \tag{1.19}$$

where the first term is the total number of sequences of weight i, and the second term is the number of sequences of weight i that lie within incorrect decoding spheres. When i channel-symbol errors in the received word cause a decoding failure, the decoded symbols in the output of a reproducing decoder contain i errors, and the probability of an information-symbol error is i/n. Therefore, (1.16) implies that the *information-symbol error rate for a reproducing bounded-distance decoder* is

$$P_{is} = \sum_{i=t+1}^{n} \left(\frac{P_s}{q-1} \right)^i (1 - P_s)^{n-i} \left[\sum_{l=\max(i-t,d_m)}^{\min(i+t,n)} A_l N(l, i) \frac{l}{n} + L(i) \frac{i}{n} \right] . \tag{1.20}$$

Two major problems still arise in calculating P_{is} from (1.16) or (1.20). The computational complexity may be prohibitive when n and q are large, and the weight distribution is unknown for many block codes. To derive simple approximations for reproducing decoders, we consider the packing densities of block codes.

The *packing density* is defined as the ratio of the number of words in the q^k decoding spheres to the total number of sequences of length n. From (1.7), it follows that the packing density is

$$D_p = \frac{q^k}{q^n} \sum_{i=0}^{t} \binom{n}{i} (q-1)^i . \tag{1.21}$$

For perfect codes, $D_p = 1$. If $D_p > 0.5$, undetected errors tend to occur more often than decoding failures, and the code is considered *tightly packed*. If $D_p < 0.1$, decoding failures predominate, and the code is considered *loosely packed*. The packing densities of binary BCH codes are listed in Table 1.1. The perfect BCH codes and the (15, 5, 3) BCH code are tightly packed. If $n \geq 63$ and $t \geq 4$, the BCH codes are loosely packed.

Consider the transmission of the all-zero codeword of n symbols, independent channel-symbol errors, and tightly packed codes. If a received word has i channel-symbol errors at the decoder input and $d_m \leq i \leq n$, then a bounded-distance decoder usually chooses a codeword with Hamming weight i. However, there is no codeword with a Hamming weight between 0 and d_m. Therefore, if a received word has i channel-symbol errors and $t+1 \leq i \leq d_m$, then a reproducing bounded-distance decoder usually chooses a codeword with Hamming weight d_m. Therefore, the identity $\binom{n}{i}\frac{i}{n} = \binom{n-1}{i-1}$ indicates that P_{is} for reproducing bounded-distance decoders of tightly packed codes is approximated by

$$P_{is} \approx \sum_{i=t+1}^{d_m} \frac{d_m}{n}\binom{n}{i} P_s^i (1-P_s)^{n-i} + \sum_{i=d_m+1}^{n} \binom{n-1}{i-1} P_s^i (1-P_s)^{n-i}. \tag{1.22}$$

The virtues of this approximation are its simplicity and lack of dependence on the code weight distribution. Let P_{df} and P_{ud} denote the probability of a decoding failure and the probability of an undetected error, respectively. Computations for specific codes indicate that the accuracy of (1.22) tends to increase with P_{ud}/P_{df}. The right-hand side of (1.22) gives an approximate upper bound on P_{is} for complete decoders because some received sequences with $t+1$ or more errors can be corrected and hence produce no information-symbol errors.

For loosely packed codes, the first term on the right side of (1.19) is much larger than the second term. Using this result in (1.20) indicates that P_{is} for reproducing bounded-distance decoders of loosely packed codes is approximated by

$$P_{is} \approx \sum_{i=t+1}^{n} \binom{n-1}{i-1} P_s^i (1-P_s)^{n-i}. \tag{1.23}$$

The virtues of this approximation are its simplicity and independence from the code weight distribution. The approximation is accurate when decoding failures are the predominant error mechanism. For cyclic Reed-Solomon codes, numerical examples indicate that the exact P_{is} and the approximation are quite close for all values of P_s when $t \geq 3$, a result that is not surprising in view of the paucity of sequences in the decoding spheres for a Reed-Solomon code with $t \geq 3$.

A symbol is said to be erased when the demodulator, after deciding that a symbol is unreliable, instructs the decoder to ignore that symbol during the decoding. *Symbol erasures* are used to strengthen hard-decision decoding. If a code has a minimum distance d_m and a received word is assigned ϵ symbol erasures, then all codewords differ in at least $d_m - \epsilon$ of the unerased symbols. Hence, ν errors can be corrected if $2\nu + 1 \leq d_m - \epsilon$. If d_m or more erasures are assigned, a decoding failure occurs. Let P_e denote the probability of an erasure. For independent symbol errors and erasures, the probability that a received sequence has i errors and ϵ erasures is $P_s^i P_e^{\epsilon}(1-P_s-P_e)^{n-i-\epsilon}$. Therefore, for a bounded-distance *errors-and-erasures decoder* of an (n, k) block code,

$$P_w \leq \sum_{\epsilon=0}^{n} \sum_{i=i_0}^{n-\epsilon} \binom{n}{\epsilon} \binom{n-\epsilon}{i} P_s^i P_e^\epsilon (1 - P_s - P_e)^{n-i-\epsilon} ,$$
$$i_0 = \max(0, \lceil (d_m - \epsilon)/2 \rceil) \tag{1.24}$$

where $\lceil x \rceil$ denotes the smallest integer greater than or equal to x. For the AWGN channel, decoding with optimal erasures provides an insignificant performance improvement relative to hard-decision decoding, but erasures are often effective against fading or sporadic interference. Codes for which *errors-and-erasures decoding* is most useful are those with relatively large minimum distances, such as Reed-Solomon codes.

Soft-Decision Decoders

A classical soft-decision decoder uses demodulator output samples to associate a number called the *codeword metric* with each possible codeword, decides that the codeword with the largest metric is the transmitted codeword, and then produces the corresponding information bits as the decoder output. Let $\mathbf{y}$ denote the n-dimensional vector of noisy output samples y_i, $i = 1, 2, \ldots, n$, produced by a demodulator that receives a sequence of n code symbols representing k information symbols. Let $\mathbf{x}_c$ denote the cth codeword vector with symbols x_{ci}, $i = 1, 2, \ldots, n$. Let $f(\mathbf{y}|\mathbf{x}_c)$ denote the *likelihood function*, which is the conditional density function of $\mathbf{y}$ given that $\mathbf{x}_c$ was transmitted. A soft-decision decoder may use the likelihood function as the codeword metric, or any monotonically increasing function of $f(\mathbf{y}|\mathbf{x}_c)$ may serve as the metric. A convenient choice is often proportional to the natural logarithm of $f(\mathbf{y}|\mathbf{x}_c)$, which is called the *log-likelihood function* and is denoted by $\ln f(\mathbf{y}|\mathbf{x}_c)$. For statistically independent demodulator outputs, the log-likelihood function for each of the q^k possible codewords of the (n, k) block code is

$$\ln f(\mathbf{y}|\mathbf{x}_c) = \sum_{i=1}^{n} \ln f(y_i|x_{ci}) \, , \quad c = 1, 2, \ldots, q^k \tag{1.25}$$

where $f(y_i|x_{ci})$ is the conditional density function of y_i given the value of x_{ci}.

A fundamental property of a probability, called *countable subadditivity*, is that the probability of a finite or countable union of events B_n, $n = 1, 2, \ldots$, satisfies

$$P[\cup_n B_n] \leq \sum_n P[B_n]. \tag{1.26}$$

In communication theory, a bound obtained from this inequality is called a *union bound*.

To determine upper bounds on P_w and P_{is} for linear block codes, it suffices to assume that the *all-zero codeword was transmitted*. Let $P_2(l)$ denote the probability that the metric for an incorrect codeword at Hamming distance l from the correct codeword, and hence with Hamming weight l, exceeds the metric for the correct codeword. The union bound and the relation between weights and distances imply that P_w for soft-decision decoding satisfies

$$P_w \leq \sum_{l=d_m}^{n} A_l P_2(l). \tag{1.27}$$

Let $B_{lm\nu}$ denote the event that the mth incorrect codeword with Hamming weight l has a larger metric than the correct codeword and has a nonzero symbol for its νth information symbol. The information-symbol error probability, which is defined by (1.13), is

$$P_{is} = \frac{1}{k} \sum_{\nu=1}^{k} P\left(\cup_l \cup_m B_{lm\nu}\right)$$

$$\leq \frac{1}{k} \sum_{l=d_m}^{n} \sum_{\nu=1}^{k} \sum_{m=1}^{A_l} P\left(B_{lm\nu}\right). \tag{1.28}$$

Let $\delta_{lm\nu} = 1$ if the mth incorrect codeword with Hamming weight l has a nonzero symbol for its νth information symbol, and $\delta_{lm\nu} = 0$, otherwise. Then $P\left(B_{lm\nu}\right) = \delta_{lm\nu} P_2(l)$, and substitution into (1.28) yields

$$P_{is} \leq \sum_{l=d_m}^{n} \frac{\beta_l}{k} P_2(l) \tag{1.29}$$

where

$$\beta_l = \sum_{\nu=1}^{k} \sum_{m=1}^{A_l} \delta_{lm\nu} \tag{1.30}$$

denotes the total information-symbol weight of the codewords of weight l.

To determine β_l for any cyclic (n, k) code, consider the set S_l of A_l codewords of weight l. The total weight of all the codewords in S_l is $A_T = lA_l$. Let α and β denote any two fixed positions in the codewords. By definition, any cyclic shift of a codeword produces another codeword of the same weight. Therefore, for every codeword in S_l that has a zero in α, there is some codeword in S_l that results from a cyclic shift of that codeword and has a zero in β. Thus, among the codewords of S_l, the total weight of all the symbols in a fixed position is the same regardless of the position and is equal to A_T/n. The total weight of all the information symbols in S_l is $\beta_l = kA_T/n = klA_l/n$. Therefore,

$$P_{is} \leq \sum_{l=d_m}^{n} \frac{l}{n} A_l P_2(l). \tag{1.31}$$

This upper bound depends on $P_2(l)$, $d_m \leq l \leq n$, which depends on the modulation system.

1.2 Modulations and Code Metrics

A basic operation performed in demodulators is *matched filtering*. A filter is said to be *matched* to a signal $x(t)$ that is zero outside the interval $[0, T]$ if the impulse response of the filter is $h(t) = x^*(T - t)$, where the asterisk denotes complex conjugation. When $x(t)$ is applied to a filter matched to it, the filter output is

$$\begin{aligned} y(t) &= \int_{-\infty}^{\infty} x(u)h(t-u)du = \int_{-\infty}^{\infty} x(u)x^*(u+T-t)du \\ &= \int_{\max(t-T,0)}^{\min(t,T)} x(u)x^*(u+T-t)du. \end{aligned} \tag{1.32}$$

If the matched-filter output is sampled at $t = T$, then

$$y(T) = \int_0^T |x(u)|^2 \, du \tag{1.33}$$

which is equal to the energy of the signal.

Pulse Amplitude Modulation

Consider *pulse amplitude modulation* (PAM), which includes q-ary *quadrature amplitude modulation* (QAM) and *phase-shift keying (PSK)*. One of the q^k codewords of an (n, k) block code is transmitted over the AWGN channel. For symbol i of codeword c, the received signal is

$$\begin{aligned} r_i(t) = \text{Re}\left[\alpha_i \sqrt{2\mathcal{E}_s} x_{ci} \psi_s \left[t - (i-1)T_s\right] e^{j(2\pi f_c t + \theta_i)}\right] + n(t) \\ (i-1)T_s \leq t \leq iT_s, \; i = 1, 2, \ldots, n \end{aligned} \tag{1.34}$$

where $j = \sqrt{-1}$, α_i is the fading amplitude, $\mathcal{E}_s$ is the average symbol energy when $\alpha_i = 1$, T_s is the symbol duration, f_c is the carrier frequency, x_{ci} is a complex number representing a point in the signal constellation corresponding to symbol

i of codeword c, $\psi_s(t)$ is the real-valued symbol waveform, θ_i is the carrier phase, and $n(t)$ is zero-mean Gaussian noise. The fading amplitude is a real-valued positive attenuation that may vary from symbol to symbol. The symbol waveform $\psi_s(t)$ is assumed to be largely confined to a single symbol interval to avoid intersymbol interference and has unit energy in a symbol interval:

$$\int_0^{T_s} \psi_s^2(t)dt = 1. \tag{1.35}$$

The complex number x_{ci} is any of the q complex numbers $x_{ci}(k)$, $k = 1, 2, \ldots, q$, in the signal constellation, which is normalized so that

$$\frac{1}{q}\sum_{m=1}^{q} |x_{ci}(m)|^2 = 1. \tag{1.36}$$

We assume that the spectrum of $\psi_s(t)$ is negligible unless $|f| < f_c$. The average symbol energy is defined as

$$\frac{1}{q}\sum_{m=1}^{q}\int_{(i-1)T_s}^{iT_s} \left\{\mathrm{Re}\left[\sqrt{2\mathcal{E}_s}x_{ci}(k)\,\psi_s\left[t-(i-1)T_s\right]e^{j(2\pi f_c t+\theta_i)}\right]\right\}^2 dt \tag{1.37}$$

which equals $\mathcal{E}_s$, as verified by expanding the right-hand side in terms of integrals and then using the spectral assumption to eliminate negligible integrals.

A frequency translation or *downconversion* to baseband is followed by matched filtering. Assuming ideal frequency synchronization in the receiver, the downconversion is represented by the multiplication of the received signal by $\sqrt{2}exp\left(-j2\pi f_c t - j\phi_i\right)$, where ϕ_i is a phase introduced during the downconversion, and the factor $\sqrt{2}$ has been inserted for mathematical convenience. The downconversion is physically realized as an in-phase and quadrature decomposition. After downconversion, the signal is applied to a filter matched to $\psi_s(t)$ and sampled. After discarding a negligible integral, we find that the matched-filter output samples are

$$y_i = \alpha_i\sqrt{\mathcal{E}_s}x_{ci}e^{j(\theta_i-\phi_i)} + n_i, \quad i = 1, 2, \ldots, n \tag{1.38}$$

where

$$n_i = \sqrt{2}\int_{(i-1)T_s}^{iT_s} n(t)\psi_{si}(t)\,e^{-2\pi f_c t+\phi_i}dt \tag{1.39}$$

and $\psi_{si}(t) = \psi_s\left[t-(i-1)T_s\right]$. To allow for time-varying interference that can be modeled as zero-mean Gaussian noise with a time-varying power spectrum, we generalize the AWGN channel. For the *time-varying AWGN channel*, the autocorrelation of the zero-mean Gaussian noise process is modeled as

$$E[n(t)n(t+\tau)] = \frac{N_{0i}}{2}\delta(\tau), \;\; (i-1)T_s \le t \le iT_s, \;\; i = 1, 2, \ldots, n \tag{1.40}$$

where $N_{0i}/2$ is the two-sided power spectral density (PSD) of $n(t)$ during the interval $(i-1)T_s \le t \le iT_s$, and $\delta(\tau)$ is the Dirac delta function.

Let the superscript T denote the transpose of a vector or matrix. For the time-varying AWGN channel, the approximating Riemann sums of the $2n$ real and imaginary parts of the $\{n_i\}$ are sums of independent zero-mean Gaussian random variables. According to Appendix A.1 and Theorem A1, the $2n$ components of the $\{n_i\}$ are *jointly zero-mean Gaussian random variables*, and the $n \times 1$ random noise vector $\mathbf{n} = [n_1 \; n_2 \; \ldots \; n_n]^T$ is a zero-mean, complex Gaussian random vector.

Equations (1.39) and (1.40) imply that

$$\begin{aligned} E\left[n_i n_l\right] &= 2\int_{(i-1)T_s}^{iT_s} \psi_{si}(t) e^{-j(2\pi f_c t+\phi_i)} dt \\ &\times \int_{(l-1)T_s}^{lT_s} E[n(t)n(t_1)]\psi_{si}(t_1)e^{-j(2\pi f_c t+\phi_i)} dt_1 \\ &= \delta_{il} N_{0i} \int_{(i-1)T_s}^{iT_s} \psi_{si}^2(t) e^{-j(4\pi f_c t+2\phi_i)} dt \end{aligned} \tag{1.41}$$

where $\delta_{il} = 0, \; i \neq l$, and $\delta_{il} = 1, \; i = l$. The remaining integral is proportional to the Fourier transform of $\psi_{si}^2(t)$ at frequency $2f_c$. Since the spectrum of $\psi_{si}(t)$ is negligible unless $|f| < f_c$, this integral is zero, and

$$E\left[n_i n_l\right] = 0, \;\; 1 \le i, l \le n \tag{1.42}$$

which implies that

$$E\left[\mathbf{n}\mathbf{n}^T\right] = \mathbf{0}. \tag{1.43}$$

A zero-mean, complex random vector $\mathbf{n}$ that satisfies this equation is said to have *circular symmetry*. An alternative derivation of the circular symmetry of noise following a downconversion and matched filtering is given in Appendix D.3.

Similar calculations indicate that

$$E\left[\mathbf{n}\mathbf{n}^H\right] = N_{0i}\mathbf{I} \tag{1.44}$$

where the superscript H denotes the conjugate transpose of a vector or matrix, and $\mathbf{I}$ denotes the identity matrix. Therefore, as shown in Appendix A.1, the 2ν real and imaginary components of the $\{n_i\}$ are all independent Gaussian random variables with the same variance $N_{0i}/2$.

The density function of a complex *Gaussian random variable* is defined as the joint density function of its real and imaginary components. The density function of n_i with independent, identically distributed components is

$$f(n_i) = \frac{1}{\pi N_{0i}} \exp\left(-\frac{|n_i|^2}{N_{0i}}\right), \ i = 1, 2, \ldots, n. \tag{1.45}$$

Therefore, the conditional density function of y_i given the value of x_{ci} is

$$f(y_i|x_{ci}) = \frac{1}{\pi N_{0i}} \exp\left(-\frac{\left|y_i - \alpha_i\sqrt{\mathcal{E}_s}x_{ci}e^{j(\theta_i-\phi_i)}\right|^2}{N_{0i}}\right), \ i = 1, 2, \ldots, n. \tag{1.46}$$

Let $\mathbf{y} = [y_1 \ y_2 \ \ldots \ y_n]^T$ denote the vector of matched-filter samples, and let $\mathbf{x}_c = [x_{c1} \ x_{c2} \ \ldots x_{cn}]^T$ denote the vector of codeword symbols for codeword c. Equations (1.25) and (1.46) imply that the log-likelihood function for the codeword is

$$\ln f(\mathbf{y}|\mathbf{x}_c) = -\frac{1}{2}\sum_{i=1}^{n}\log(\pi N_{0i}) - \sum_{i=1}^{n}\frac{\left|y_i - \alpha_i\sqrt{\mathcal{E}_s}x_{ci}e^{j(\theta_i-\phi_i)}\right|^2}{N_{0i}}. \tag{1.47}$$

For coherent demodulation, the receiver is synchronized with the carrier phase so that $\phi_i = \theta_i$. Since the first sum in (1.47) is independent of the codeword c, it may be discarded in metrics derived from the log-likelihood function. For the AWGN channel, the $\{N_{0i}\}$ are all equal and each $\alpha_i = 1$. Thus, these factors are irrelevant to the decision-making and may be discarded. As a result, a suitable codeword metric for coherent PAM and the AWGN channel is

$$U(c) = -\left\|\mathbf{y} - \sqrt{\mathcal{E}_s}\mathbf{x}_c\right\|^2 \tag{1.48}$$

where $\|\cdot\|$ denotes the Euclidean norm of a vector. This equation indicates that the optimal decision is to choose the signal constellation vector $\sqrt{\mathcal{E}_s}\mathbf{x}_c$ that has the minimum Euclidean distance from the received vector $\mathbf{y}$. A simplification of the computation of the codeword metric results from expanding the Euclidean distance and then discarding irrelevant terms. Thus, the *codeword metric for coherent PAM, an (n, k) block code, and the AWGN channel* is

$$U(c) = \sum_{i=1}^{n}[2\sqrt{\mathcal{E}_s}\,\mathrm{Re}\left(x_{ci}^{*}y_i\right) - \mathcal{E}_s\,|x_{ci}|^2], \quad c = 1, 2, \ldots, q^k \tag{1.49}$$

which requires the receiver to accurately estimate $\mathcal{E}_s$. Similarly, the *codeword metric for coherent PAM and the time-varying AWGN channel* is

$$U(c) = \sum_{i=1}^{n} \frac{[2\alpha_i \sqrt{\mathcal{E}_s} \operatorname{Re}\left(x_{ci}^* y_i\right) - \mathcal{E}_s |x_{ci}|^2]}{N_{0i}}, \quad c = 1, 2, \ldots, q^k \tag{1.50}$$

which requires the receiver to extract *channel-state information* that leads to accurate estimates of $\sqrt{\mathcal{E}_s}\alpha_i/N_{0i}$ and $\mathcal{E}_s/N_{0i}$, $i = 1, 2, \ldots, n$.

Consider coherent q-ary PSK and the time-varying AWGN channel for which $|x_{ci}| = 1$. After discarding irrelevant terms and factors in (1.50), we obtain the *codeword metric for q-ary PSK and the time-varying AWGN channel*:

$$U(c) = \sum_{i=1}^{n} \frac{\alpha_i \operatorname{Re}\left(x_{ci}^* y_i\right)}{N_{0i}}, \quad c = 1, 2, \ldots, q^k \tag{1.51}$$

which requires channel-state information about $\alpha_i/N_{0i}, i = 1, 2, \ldots, n$.

For binary PSK (BPSK), $x_{ci} = +1$ when binary symbol i is a 1 and $x_{ci} = -1$ when binary symbol i is a 0. Therefore, the *codeword metric for BPSK* and the time-varying AWGN channel is

$$U(c) = \sum_{i=1}^{n} \frac{\alpha_i x_{ci} y_{ri}}{N_{0i}}, \quad c = 1, 2, \ldots, 2^k \tag{1.52}$$

where $y_{ri} = \operatorname{Re}(y_i)$. Equation (1.46) implies that

$$f\left(y_{ri}|x_{ci}\right) = \frac{1}{\sqrt{\pi N_{0i}}} \exp\left[-\frac{\left(y_{ri} - \alpha_i \sqrt{\mathcal{E}_s} x_{ci}\right)^2}{N_{0i}}\right], \; i = 1, 2, \ldots, n. \tag{1.53}$$

Let r denote the *code rate*, which is the ratio of information bits to transmitted channel symbols. For (n, k) block codes with $m = \log_2 q$ information bits per symbol, $r = mk/n$. Thus, the energy per received channel symbol $\mathcal{E}_s$ is related to the energy per information bit $\mathcal{E}_b$ by

$$\mathcal{E}_s = r\mathcal{E}_b = \frac{mk}{n}\mathcal{E}_b. \tag{1.54}$$

Consider BPSK and the AWGN channel in which the $\{N_{0i}\}$ are all equal and each $\alpha_i = 1$. The codeword metric is

$$U(c) = \sum_{i=1}^{n} x_{ci} y_{ri}, \quad c = 1, 2, \ldots, 2^k. \tag{1.55}$$

After reordering the samples $\{y_{ri}\}$, the difference between the metrics for the correct codeword and an incorrect one at Hamming distance l may be expressed as

$$D(l)=\sum_{i=1}^{l}(x_{1i}-x_{2i})y_{Ri}=2\sum_{i=1}^{l}x_{1i}y_{ri} \tag{1.56}$$

where the sum includes only the l terms that differ, x_{1i} refers to the correct codeword, x_{2i} refers to the incorrect codeword, and $x_{2i}=-x_{1i}$. Since each of its terms is independent and each y_{ri} has a Gaussian distribution, $D(l)$ has a Gaussian distribution with mean $l\sqrt{\mathcal{E}_s}$ and variance $lN_0/2$. Since $P_2(l)$ is the probability that $D(l)<0$ and $\mathcal{E}_s=r\mathcal{E}_b$, a straightforward calculation yields

$$P_2(l)=Q\left(\sqrt{\frac{2lr\mathcal{E}_b}{N_0}}\right) \tag{1.57}$$

where $r=k/n$ is the code rate, the Gaussian Q-function is defined as

$$Q(x)=\frac{1}{\sqrt{2\pi}}\int_x^{\infty}\exp\left(-\frac{y^2}{2}\right)dy=\frac{1}{2}\operatorname{erf}c\left(\frac{x}{\sqrt{2}}\right) \tag{1.58}$$

and $erfc(\cdot)$ is the complementary error function. Equation (1.57) may be used in (1.31) to calculate P_{is} for a block code.

Optimal soft-decision decoding cannot be efficiently implemented except for very short block codes, primarily because the number of codewords for which the metrics must be computed is prohibitively large, but approximate maximum-likelihood decoding algorithms are available. The *Chase algorithm* generates a small set of candidate codewords that almost always include the codeword with the largest metric. Test patterns are generated by first making hard decisions on each of the received symbols to determine a received word of channel symbols. Assuming coherent PAM, let $\sqrt{\mathcal{E}_s}\mathbf{x}_c^d$ denote the constellation vector corresponding to the channel symbols. A *reliability measure* for the ith channel symbol is

$$M_r(i)=\left|y_i-\sqrt{\mathcal{E}_s}x_{ci}^d\right|. \tag{1.59}$$

Using this reliability measure, the Chase algorithm alters the least reliable symbols in the received word and generates a set of test patterns. Hard-decision decoding of each test pattern and the discarding of decoding failures generate the candidate codewords. The decoder selects the candidate codeword with the largest metric and declares it to be the transmitted codeword.

The quantization of soft-decision information to more than two levels requires analog-to-digital conversion of the demodulator output samples. Since the optimal location of the levels is a function of the signal, thermal noise, and interference powers, automatic gain control is often necessary. For the AWGN channel, an eight-level quantization represented by three bits and a uniform spacing between threshold levels is found to cause a loss of no more than a few tenths of a decibel relative to what could theoretically be achieved with unquantized analog voltages or infinitely fine quantization.

The results for soft-decision decoding can be used to derive channel-symbol error probabilities when hard decisions are made. When a single BPSK symbol is considered and hence $l = 1$, (1.57) indicates that the channel-bit or channel-symbol error probability for the AWGN channel is

$$P_b = P_s = Q\left(\sqrt{\frac{2r\mathcal{E}_b}{N_0}}\right). \tag{1.60}$$

Consider the detection of a single quadriphase-shift keying (QPSK) symbol transmitted over the AWGN channel. The codeword metric for a single symbol is called the *symbol metric*. The constellation symbols are $x_c = (\pm 1 \pm j)/\sqrt{2}$. Since $n = 1$ and $k = 2$, the symbol metric of (1.51) becomes

$$U(c) = \text{Re}\left(x_c^* y\right), \quad c = 1, 2, 3, 4. \tag{1.61}$$

Without loss of generality because of the constellation symmetry, we assume that the transmitted symbol is $x_c = (1 + j)/\sqrt{2}$. A symbol error occurs if y does not lie in the first quadrant. Since $\text{Re}(y)$ and $\text{Im}(y)$ are independent, the channel-symbol error probability is

$$P_s = 1 - P[\text{Re}(y) > 0]\, P[\text{Im}(y) > 0]. \tag{1.62}$$

Since $\text{Re}(y)$ and $\text{Im}(y)$ have Gaussian distributions, an evaluation using $\mathcal{E}_s = 2r\mathcal{E}_b$ yields

$$\begin{aligned} P_s &= 2Q\left[\sqrt{\frac{2r\mathcal{E}_b}{N_0}}\right] - Q^2\left[\sqrt{\frac{2r\mathcal{E}_b}{N_0}}\right] \\ &\simeq 2Q\left[\sqrt{\frac{2r\mathcal{E}_b}{N_0}}\right], \quad \frac{2r\mathcal{E}_b}{N_0} >> 1. \end{aligned} \tag{1.63}$$

If the alphabets of the code symbols and the transmitted symbols differ, then the q-ary code symbols may be mapped into q_1-ary transmitted symbols. Typically, $q = 2^\nu$, $q_1 = 2^{\nu_1}$, $\nu/\nu_1 \geq 1$, and ν/ν_1 is a positive integer. Under these conditions, there are ν/ν_1 received symbols per code symbol. For hard-decision decoding, if any of these received symbols is demodulated incorrectly, the corresponding code symbol is incorrect. If the demodulator errors are independent and the ν/ν_1 demodulated received symbols constitute a channel symbol that is one of the possible code symbols, then the channel-symbol error probability is

$$P_s = 1 - (1 - P_{dr})^{\nu/\nu_1} \tag{1.64}$$

where P_{dr} is the error probability of a demodulated received symbol. A common application is to map nonbinary code symbols into binary channel symbols ($\nu_1 = 1$). For coherent BPSK, (1.60) and (1.64) imply that

$$P_s = 1 - \left[1 - Q\left(\sqrt{\frac{2r\mathcal{E}_b}{N_0}}\right)\right]^{\nu}. \tag{1.65}$$

A *constellation labeling* is the mapping of m bits to the $q = 2^m$ two-dimensional or complex-valued constellation points representing the possible symbols. When a PAM signal is transmitted over the AWGN channel and hard-decision symbol decoding is used, the relation between P_s and the bit error probability P_b for a channel bit depends on the constellation labeling. A *Gray labeling* or *Gray coding* labels adjacent symbols that are closest in Euclidean distance with the same bits except for one, thereby minimizing the number of bit errors that occur if an adjacent symbol of a received symbol is assigned the highest likelihood or largest metric by the decoder. Since the most likely erroneous symbol selection is the adjacent symbol, the channel-bit error probability is

$$P_b \simeq \frac{1}{m} P_s \tag{1.66}$$

for Gray labeling. For QPSK symbols with Gray labeling, (1.66), (1.63), and (1.60) indicate that P_b is approximately the same as it is for BPSK. Thus, there is not much of a loss in transmitting QPSK symbols as two BPSK symbols transmitted over orthogonal carriers, which is usually done in practice.

Orthogonal Modulation

An orthogonal modulation system transmits one of a set of orthogonal signals for each codeword symbol. Consider the transmission of a codeword of n symbols using delayed versions of q-ary orthogonal complex-valued symbol waveforms: $\sqrt{\mathcal{E}_s}s_1(t)$, $\sqrt{\mathcal{E}_s}s_2(t)$, ..., $\sqrt{\mathcal{E}_s}s_q(t)$. The receiver requires q matched filters, each implemented as a pair of baseband matched filters. The $nq \times 1$ observation vector is $\mathbf{y} = [\mathbf{y}_1\ \mathbf{y}_2 \ldots\ \mathbf{y}_q]^T$, where each $\mathbf{y}_l$ is a $1 \times n$ row vector of matched-filter output samples for filter l with components $y_{l,i}$, $i = 1, 2, \ldots, n$. Suppose that symbol i of codeword c uses orthogonal waveform $s_{w_{ci}}(t - iT_s)$, where T_s is the code-symbol duration. For the time-varying AWGN channel, the received signal for symbol i can be expressed as

$$r_i(t) = \text{Re}\left[\alpha_i \sqrt{2\mathcal{E}_s} s_{w_{ci}}\left[t - (i-1)T_s\right] e^{j(2\pi f_c t + \theta_i)}\right] + n(t)$$
$$(i-1)T_s \le t \le iT_s,\ i = 1, 2, \ldots, n \tag{1.67}$$

where α_i is the fading amplitude, and $n(t)$ is the zero-mean time-varying white Gaussian noise with PSD equal to $N_{0i}/2$. Since the symbol energy for all the waveforms is $\mathcal{E}_s$,

$$\int_0^{T_s} |s_l(t)|^2 dt = 1\ , \quad l = 1, 2, \ldots, q. \tag{1.68}$$

The orthogonality of symbol waveforms implies that

$$\int_0^{T_s} s_r(t)s_l^*(t)dt = 0\ , \quad r \neq l. \tag{1.69}$$

A frequency translation or *downconversion* to baseband is followed by matched filtering. Matched-filter l, which is matched to $s_l(t)$, produces the output samples

$$y_{l,i} = \sqrt{2}\int_{(i-1)T_s}^{iT_s} r_i(t)e^{-j(2\pi f_c t+\phi_i)}s_l^*\left[t-(i-1)T_s\right]dt,$$
$$i = 1, 2, \ldots, \nu\ , \quad l = 1, 2, \ldots, q \tag{1.70}$$

where the factor $\sqrt{2}$ has been inserted for mathematical convenience. The substitution of (1.67) into (1.70), (1.69), and the assumption that each of the $\{s_l(t)\}$ has a spectrum confined to $|f| < f_c$ yields

$$y_{l,i} = \alpha_i\sqrt{\mathcal{E}_s}e^{j(\theta_i-\phi_i)}\delta_{l,w_{ci}} + n_{l,i} \tag{1.71}$$

where $\delta_{l,w_{ci}} = 1$ if $l = w_{ci}$, and $\delta_{l,w_{ci}} = 0$ otherwise, and

$$n_{l,i} = \sqrt{2}\int_{(i-1)T_s}^{iT_s} n(t)e^{-j(2\pi f_c t+\phi_i)}s_l^*\left[t-(i-1)T_s\right]dt. \tag{1.72}$$

The $nq \times 1$ noise vector is $\mathbf{n} = [\mathbf{n}_1\ \mathbf{n}_2 \ldots\ \mathbf{n}_q]^T$, where each $\mathbf{n}_l$ is an $1 \times n$ row vector of noise outputs for matched filter l with components $n_{l,i},\ i = 1, 2, \ldots, n$. Since (1.72) has the same form as (1.39), $\mathbf{n}$ is a zero-mean, complex Gaussian random vector with

$$E\left[\mathbf{n}\mathbf{n}^H\right] = N_{0i}\mathbf{I} \tag{1.73}$$

and circular symmetry:

$$E\left[\mathbf{n}\mathbf{n}^T\right] = \mathbf{0}. \tag{1.74}$$

Therefore, as shown in Appendix A.1, the $2nq$ real and imaginary components of the $\{n_{l,i}\}$ are all independent zero-mean Gaussian random variables with the same variance $N_{0i}/2$.

Since the density function of each $n_{l,i}$ is defined as the joint density function of its real and imaginary parts, the conditional density function of y_{li} given $\psi_i = \theta_i - \phi_i$ and w_{ci} is

$$f(y_{l,i}\,|w_{ci}, \psi_i) = \frac{1}{\pi N_{0i}} \exp\left(-\frac{\left|y_{l,i} - \alpha_i\sqrt{\mathcal{E}_s}e^{j\psi_i}\delta_{l,w_{ci}}\right|^2}{N_{0i}}\right),$$

$$i = 1, 2, \ldots, n, \quad l = 1, 2, \ldots, q. \tag{1.75}$$

The likelihood function of the $qn \times 1$ observation vector $\mathbf{y}$, which has components equal to the $\{y_{li}\}$, is the product of the qn density functions specified by (1.75):

$$f(\mathbf{y}\,|\mathbf{w}_c, \boldsymbol{\psi}) = \prod_{i=1}^{n}\left[\left(\frac{1}{\pi N_{0i}}\right)^q \exp\left(-\frac{\alpha_i^2\mathcal{E}_s - 2\alpha_i\sqrt{\mathcal{E}_s}\,\mathrm{Re}\left(y^*_{w_{ci}}e^{j\psi_i}\right)}{N_{0i}} - \sum_{l=1}^{q}\frac{|y_{l,i}|^2}{N_{0i}}\right)\right] \tag{1.76}$$

where $\boldsymbol{\psi}$ and $\mathbf{w}_c$ are the n-dimensional vectors that have the $\{\psi_i\}$ and $\{w_{ci}\}$ as components, respectively.

For *coherent* signals, the $\{\theta_i\}$ are tracked by the phase synchronization system, and thus ideally $\theta_i = \phi_i$ and $\psi_i = 0$. Forming the log-likelihood function and eliminating irrelevant terms that are independent of c, we obtain the *codeword metric for coherent orthogonal signals, an (n, k) block code, and the time-varying AWGN channel*:

$$U(c) = \sum_{i=1}^{n}\frac{\alpha_i\mathrm{Re}(V_{ci})}{N_{0i}}, \quad c = 1, 2, \ldots, q^k \tag{1.77}$$

where V_{ci} is the sampled output i of the filter matched to the signal representing symbol i of codeword c. The maximum-likelihood decoder finds the value of c for which $U(c)$ is largest. If this value is c_0, the decoder decides that codeword c_0 was transmitted. A problem with this metric is that each α_i/N_{0i} value must be known or estimated. If it is known that $\alpha_i = 1$ and each $N_{0i} = N_0$, then the *codeword metric for coherent orthogonal signals and the AWGN channel* is

$$U(c) = \sum_{i=1}^{\nu}\mathrm{Re}\left(V_{ci}\right), \quad c = 1, 2, \ldots, q^k \tag{1.78}$$

and the common value N_0 does not need to be known to apply this metric.

For *noncoherent* signals, we assume that each $\psi_i = \theta_i - \phi_i$ is independent and uniformly distributed over $[0, 2\pi)$, which preserves the independence of the $\{y_{li}\}$. Expanding the argument of the exponential function in (1.76), expressing $y_{w_{ci}}$ in polar form, and using (H.16) of Appendix H.3 to integrate over each ψ_i, we obtain the likelihood function of the observation vector $\mathbf{y}$:

$$f(\mathbf{y}\,|\mathbf{w}_c) = \prod_{i=1}^{n}\left[\left(\frac{1}{\pi N_{0i}}\right)^q \exp\left(-\frac{\alpha_i \mathcal{E}_s}{N_{0i}} - \sum_{l=1}^{q}\frac{|y_{l,i}|^2}{N_{0i}}\right) I_0\left(\frac{2\alpha_i\sqrt{\mathcal{E}_s}\,|y_{w_{ci}}|}{N_{0i}}\right)\right] \tag{1.79}$$

where $I_0(\cdot)$ is the modified Bessel function of the first kind and order zero. Let $R_{ci} = |y_{w_{ci}}|$ denote the magnitude of the output produced by the filter matched to $s_{w_{ci}}(t)$, the signal representing symbol i of codeword c. We form the log-likelihood function and eliminate terms and factors that do not depend on the codeword, thereby obtaining the *codeword metric for noncoherent orthogonal signals and the time-varying AWGN channel*:

$$U(c) = \sum_{i=1}^{n} \ln I_0\left(\frac{2\alpha_i\sqrt{\mathcal{E}_s}R_{ci}}{N_{0i}}\right), \quad c = 1, 2, \ldots, q^k \tag{1.80}$$

which requires each $\alpha_i\sqrt{\mathcal{E}_s}/N_{0i}$ value to be known or estimated. If $\alpha_i = 1$ and each $N_{0i} = N_0$, then the *codeword metric for noncoherent orthogonal signals and the AWGN channel* is

$$U(c) = \sum_{i=1}^{n} \ln I_0\left(\frac{2\sqrt{\mathcal{E}_s}R_{ci}}{N_0}\right), \quad c = 1, 2, \ldots, q^k \tag{1.81}$$

and $\sqrt{\mathcal{E}_s}/N_0$ must be known to apply this metric.

It is desirable to have an approximation of (1.81) to reduce the computational requirements. Comparing the series representation in (H.14) to that of $\exp\left(x^2/4\right)$, it follows that

$$I_0(x) \le \exp\left(\frac{x^2}{4}\right). \tag{1.82}$$

From the integral representation in (H.15), we obtain

$$I_0(x) \le \exp(|\,x\,|). \tag{1.83}$$

The upper bound in (1.82) is tighter for $0 \le x < 4$, while the upper bound in (1.83) is tighter for $4 < x < \infty$. If we assume that $\sqrt{\mathcal{E}_s}R_{ci}/N_0$ is often less than 2, then the approximation of $I_0(x)$ by $\exp(x^2/4)$ is reasonable. Dropping irrelevant constants, we obtain the *square-law metric*

$$U(c) = \sum_{i=1}^{n} R_{ci}^2, \quad c = 1, 2, \ldots, q^k \tag{1.84}$$

which does not depend on knowing $\sqrt{\mathcal{E}_s}/N_0$.

To determine the symbol metric for the AWGN channel, we set $n = k = 1$ and $c = l$ and drop the unnecessary subscripts in (1.78) and (1.81). We find that the symbol metric is $Re(V_l)$ for coherent orthogonal signals and $ln I_0(2\sqrt{\mathcal{E}_s}R_l/N_0)$ for noncoherent orthogonal signals, where the index l ranges over the symbol alphabet. Since the latter function increases monotonically with $R_l = |y_l|$, optimal symbol metrics or decision variables for noncoherent orthogonal signals are R_l for $l = 1, 2, \ldots, q$.

For the noncoherent detection of a single symbol, (1.75) implies that the conditional density function of matched-filter output y_l given $\psi = \theta - \phi$ and the transmission of $s_1(t)$ is

$$f(y_l\,|s_1(t), \psi) = \frac{1}{\pi N_0}\exp\left(-\frac{\left|y_l - \sqrt{\mathcal{E}_s}e^{j\psi}\delta_{l,1}\right|^2}{N_0}\right), \quad l = 1, 2, \ldots, q. \tag{1.85}$$

Therefore, the joint density function of $R_{l1} = \text{Re}\,(y_l)$ and $R_{l2} = \text{Im}\,(y_l)$ given the transmission of $s_1(t)$ and the value of ψ is

$$g_1(r_{l1}, r_{l2}) = \frac{1}{\pi N_0}\exp\left[-\frac{(r_{l1} - \sqrt{\mathcal{E}_s}\delta_{l,1}\cos\psi)^2 + (r_{l2} - \sqrt{\mathcal{E}_s}\delta_{l,1}\sin\psi)^2}{N_0}\right]. \tag{1.86}$$

Define R_l and Φ_l implicitly by $R_{l1} = R_l\cos\Phi_l$ and $R_{l2} = R_l\sin\Phi_l$. Transforming variables, we find that the joint density function of R_l and Φ_l is

$$g_2(r, \phi) = \frac{r}{\pi N_0}\exp\left[-\frac{r^2 - 2\sqrt{\mathcal{E}_s}\delta_{l,1}r\cos\phi\cos\psi - 2\sqrt{\mathcal{E}_s}\delta_{l,1}r\sin\phi\sin\psi + \mathcal{E}_s\delta_{l,1}}{N_0}\right],$$
$$r \geq 0, \quad |\phi| \leq \pi. \tag{1.87}$$

The density function of R_l is obtained by integration of (1.87) over ϕ. Using trigonometry and (H.16), we obtain the density function for R_l, $l = 1, 2, \ldots, q$:

$$f_1(r) = \frac{2r}{N_0}\exp\left(-\frac{r^2 + \mathcal{E}_s}{N_0}\right) I_0\left(\frac{2\sqrt{\mathcal{E}_s}r}{N_0}\right) u(r) \tag{1.88}$$

$$f_l(r) = \frac{2r}{N_0}\exp\left(-\frac{r^2}{N_0}\right) u(r), \quad l = 2, \ldots, q \tag{1.89}$$

where $u(r)$ is the *unit step function* defined as $u(r) = 1$ if $r \geq 0$, and $u(r) = 0$ if $r < 0$.

The statistical independence of the $\{y_l\}$ implies that the random variables $\{R_l\}$ are independent. A symbol error occurs when $s_1(t)$ is transmitted if R_1 is not the largest of the $\{R_l\}$. Since the $\{R_l\}$ are identically distributed for $l = 2, \cdots, q$, the probability of a symbol error when $s_1(t)$ is transmitted is

$$P_s = 1 - \int_0^\infty \left[\int_0^r f_2(y)dy\right]^{q-1} f_1(r)dr. \tag{1.90}$$

Evaluating the inner integral yields

$$\int_0^r f_2(y)dy = 1 - \exp\left(-\frac{r^2}{N_0}\right). \tag{1.91}$$

Expressing the $(q-1)$th power of this result as a binomial expansion and then substituting it into (1.90), the remaining integration may be done by using

$$\int_0^\infty r \exp\left(-\frac{r^2}{2b^2}\right) I_0\left(\frac{r\sqrt{\lambda}}{b^2}\right) dr = b^2 \exp\left(\frac{\lambda}{2b^2}\right) \tag{1.92}$$

which follows from the fact that the density function in (1.88) must integrate to unity. The final result is the symbol error probability for noncoherent q-ary orthogonal symbols over the AWGN channel:

$$P_s = \sum_{i=1}^{q-1} \frac{(-1)^{i+1}}{i+1} \binom{q-1}{i} \exp\left[-\frac{i\mathcal{E}_s}{(i+1)N_0}\right]. \tag{1.93}$$

When $q = 2$, this equation reduces to the classical formula for binary orthogonal symbols:

$$P_s = \frac{1}{2}\exp\left(-\frac{\mathcal{E}_s}{2N_0}\right). \tag{1.94}$$

Orthogonal signals are *q-ary symmetric* insofar as an incorrectly decoded symbol is equally likely to be any of the remaining $q-1$ symbols in the alphabet. Consider a linear (n, k) block code and q-ary orthogonal signals. Among the $q-1$ incorrect symbols, a given bit is incorrect in $q/2$ instances. Therefore, the bit error rate is

$$P_b = \frac{q}{2(q-1)} P_{is} \tag{1.95}$$

which reduces to $P_b = P_{is}$ when $q = 2$.

Detection of FSK Symbols

For noncoherent orthogonal frequency-shift keying (FSK), a pair of baseband filters are matched to each unit-energy waveform

$$s_l(t) = \exp(j2\pi f_l t)/\sqrt{T_s}, \quad 0 \le t \le T_s, \quad l = 1, 2, \ldots, q. \tag{1.96}$$

The orthogonality condition (1.69) is satisfied if the adjacent frequencies are separated by k/T_s, where k is a nonzero integer. If $r(t)$ is the received signal, a parallel set of matched filters and magnitude detectors provide the symbol metrics

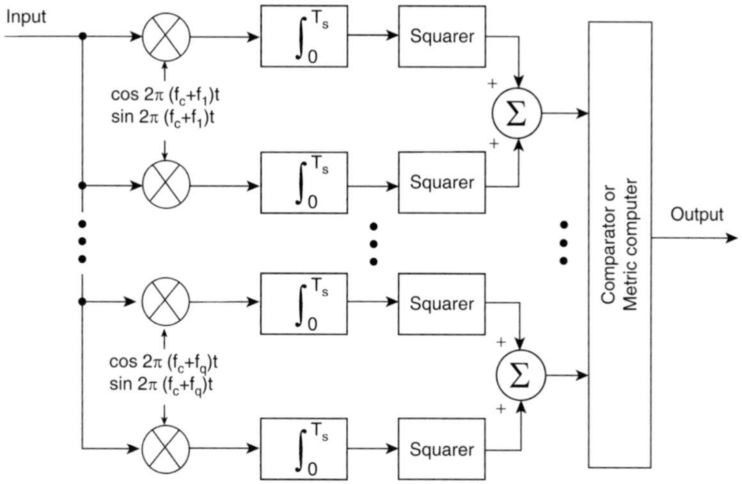

Fig. 1.2 Noncoherent orthogonal FSK demodulator with baseband matched filters

$$R_l = \left| \int_0^{T_s} r(t) e^{-j2\pi f_c t} e^{-j2\pi f_l t} dt \right| \tag{1.97}$$

where an irrelevant constant has been omitted. Expanding R_l^2, we obtain

$$R_l^2 = R_{lc}^2 + R_{ls}^2 \tag{1.98}$$

$$R_{lc} = \int_0^{T_s} r(t) \cos\left[2\pi(f_c + f_l)t\right] dt \tag{1.99}$$

$$R_{ls} = \int_0^{T_s} r(t) \sin\left[2\pi(f_c + f_l)t\right] dt. \tag{1.100}$$

These equations imply the demodulator structure depicted in Figure 1.2. If hard-decision decoding is used, then for each symbol a comparator decides what symbol was transmitted by observing which comparator input is the largest. Although R_l could be computed, the use of R_l^2 in the comparisons is simpler and entails no loss in performance. The comparator decisions are applied to the decoder. If soft-decision decoding is used, then a metric computer transfers its outputs to the decoder.

To derive an alternative implementation, we observe that when the received waveform $r(t) = A\cos[2\pi(f_c + f_l)t + \theta]$, $0 \leq t \leq T_s$ is applied to a filter with impulse response $\cos 2\pi(f_c + f_l)(T_s - t)$, $0 \leq t \leq T_s$, the filter output at time t is

$$\begin{aligned} y_l(t) &= \int_0^t r(\tau) \cos\left[2\pi(f_c + f_l)(\tau - t + T_s)\right] d\tau \\ &= \left\{ \int_0^t r(\tau) \cos\left[2\pi(f_c + f_l)\tau\right] d\tau \right\} \cos\left[2\pi(f_c + f_l)(t - T_s)\right] \end{aligned}$$

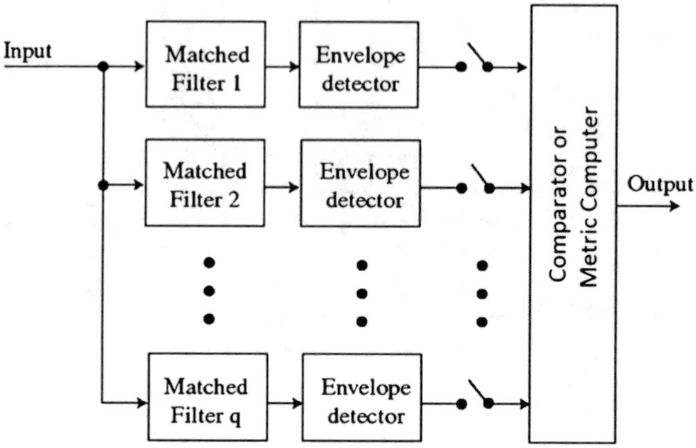

Fig. 1.3 Noncoherent orthogonal FSK demodulator with matched filters and envelope detectors

$$+\left\{\int_0^t r(\tau)\sin\left[2\pi(f_c+f_l)\tau\right]d\tau\right\}\sin\left[2\pi(f_c+f_l)(t-T_s)\right]$$
$$=R_l(t)\cos\left[2\pi(f_c+f_l)(t-T_s)+\phi(t)\right]\,,\quad 0\le t\le T_s \tag{1.101}$$

where the *envelope* of $y_l(t)$ is

$$R_l(t)=\left\{\left[\int_0^t r(\tau)\cos\left[2\pi(f_c+f_l)\tau\right]d\tau\right]^2+\left[\int_0^t r(\tau)\sin\left[2\pi\left(f_c+f_l\right)\tau\right]d\tau\right]^2\right\}^{1/2}. \tag{1.102}$$

The envelope is extracted by an *envelope detector* and sampled to produce $R_l(T_s)=R_l$, which is given by (1.98) to (1.100). Thus, we obtain the demodulator structure depicted in Figure 1.3. A practical envelope detector consists of a peak detector followed by a lowpass filter.

Performance Examples

The *coding gain* of one code compared with another one is the reduction in the value of E_b/N_0 required to produce a specified information-bit or information-symbol error probability. Calculations for specific communication systems and codes operating over the AWGN channel have shown that an optimal soft-decision decoder provides a coding gain of approximately 2 dB relative to a hard-decision decoder. However, soft-decision decoders are much more complex to implement and may be too slow for the processing of high rates of information. For a given level of implementation complexity, hard-decision decoders can accommodate much longer block codes, thereby at least partially overcoming the inherent advantage of soft-

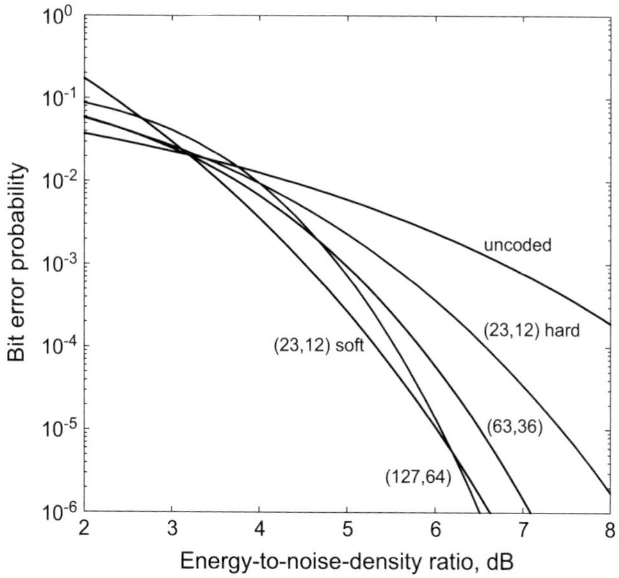

Fig. 1.4 Information-bit error probability for binary block (n, k) codes and coherent BPSK

decision decoders. In practice, soft-decision decoding other than errors-and-erasures decoding is seldom used with block codes of length greater than 50.

Figure 1.4 depicts the information-bit error probability $P_b = P_{is}$ versus $\mathcal{E}_b/N_0$ for various binary block codes with coherent BPSK over the AWGN channel. Equation (1.22) is used to compute P_b for the (23,12) Golay code with hard decisions. Since the packing density D_p is small for these codes, (1.23) is used for hard-decision decoding of the (63,36) BCH code, which corrects $t = 5$ errors, and the (127,64) BCH code, which corrects $t = 10$ errors. Equation (1.60) is used for P_s. Inequality (1.31), Table 1.2, and (1.57) are used to compute the upper bound on $P_b = P_{is}$ for the (23,12) Golay code with optimal soft decisions. The graphs illustrate the power of the soft-decision decoding. For the (23,12) Golay code, soft-decision decoding provides an approximately 2-dB coding gain for $P_b = 10^{-5}$ relative to hard-decision decoding. Only when $P_b < 10^{-5}$ does the (127,64) BCH code begin to outperform the (23,12) Golay code with soft decisions. If $\mathcal{E}_b/N_0 \leq$ 3 dB, an uncoded system with coherent BPSK provides a lower P_b than a similar system that uses one of the block codes of the figure.

Figure 1.5 illustrates the performance of loosely packed Reed-Solomon codes with hard-decision decoding over the AWGN channel as a function of $\mathcal{E}_b/N_0$. Equation (1.23) is used to compute the approximate information-bit error probabilities for binary channel symbols with coherent BPSK and nonbinary channel symbols with noncoherent orthogonal FSK. For the nonbinary channel symbols, (1.93) is applicable, and (1.95) is correct or provides a good approximation. For the binary channel symbols, (1.65) is used. For the chosen values of n, the best performance

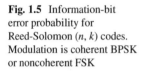

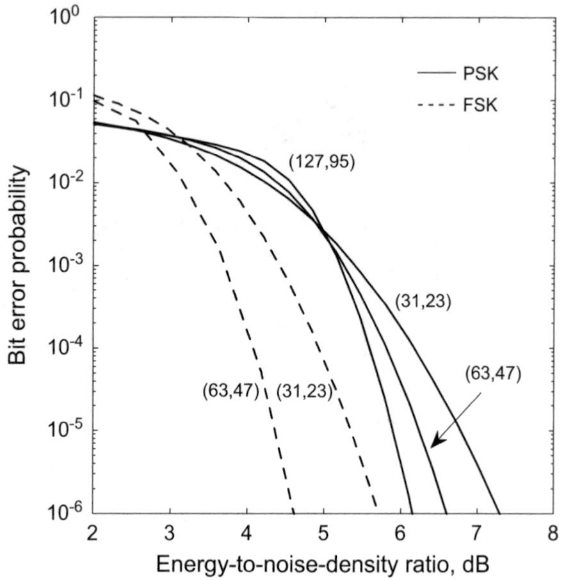

Fig. 1.5 Information-bit error probability for Reed-Solomon (n, k) codes. Modulation is coherent BPSK or noncoherent FSK

at $P_b = 10^{-5}$ is obtained if the code rate is $k/n \approx 3/4$. Further gains result from increasing n and hence the complexity of the implementation.

Although the figure indicates the performance advantage of Reed-Solomon codes with q-ary orthogonal FSK, there is a major bandwidth penalty. Let B denote the bandwidth required for an uncoded BPSK signal. If the same data rate is accommodated by using uncoded binary FSK (BFSK), the required bandwidth for demodulation with envelope detectors is approximately $2B$. For uncoded orthogonal FSK using $q = 2^m$ frequencies, the required bandwidth is $2^m B/m$ because each symbol represents m bits. If a Reed-Solomon (n, k) code is used with FSK, the required bandwidth becomes $2^m nB/mk$.

1.3 Convolutional Codes and Trellis Codes

In contrast to a block codeword, a convolutional codeword represents an entire message of indefinite length. A convolutional encoder over the binary field $GF(2)$ uses shift registers of bistable memory elements to convert each input of k information bits into an output of n code bits, each of which is the modulo-2 sum of both current and previous information bits. A convolutional code is *linear* because the modulo-2 sums imply that the superposition property applies to the input–output relations and that the all-zero codeword is a member of the code. The *constraint length K* of a convolutional code is the maximum number of sets of n output bits that

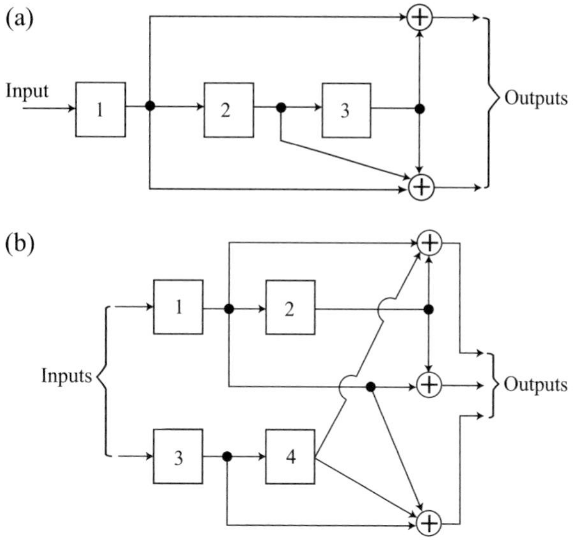

Fig. 1.6 Encoders of nonsystematic convolutional codes with (**a**) $K = 3$ and rate $= 1/2$ and (**b**) $K = 2$ and rate $= 2/3$

can be affected by an input bit. A convolutional code is *systematic* if the information bits appear unaltered in each codeword; otherwise, it is nonsystematic.

A nonsystematic convolutional encoder with two memory stages, $k = 1, n = 2$, and $K = 3$ is shown in Figure 1.6 (a). Information bits enter the shift register in response to clock pulses. After each clock pulse, the most recent information bit becomes the content and output of the first memory stage, the previous contents of stages are shifted to the right, the previous content of the final stage is shifted out of the register, and a new bit appears at the input of the first memory stage. The outputs of the modulo-2 adders (exclusive-OR gates) provide two code bits. The impulse responses of the encoder are the two output streams of K bits in response to an input 1 bit followed by $K - 1$ input 0 bits. The *generators* of the code bits are the vectors $\mathbf{g}_1 = [1\,0\,1]$ and $\mathbf{g}_2 = [1\,1\,1]$, which indicate the impulse responses at the two outputs starting from the left-hand side. In octal form, the three bits of the two generator vectors are represented by (5, 7).

The encoder of a nonsystematic convolutional code with four memory stages, $k = 2, n = 3$, and $K = 2$ is shown in Figure 1.6 (b). Its generators, which represent the impulse responses at the three outputs due to the upper input bit followed by the lower input bit, are $\mathbf{g}_1 = [1\,1\,0\,1]$, $\mathbf{g}_2 = [1\,1\,0\,0]$, and $\mathbf{g}_3 = [1\,0\,1\,1]$. By octally representing groups of 3 bits starting from the right-hand side and inserting zeros when fewer than 3 bits remain, we obtain the octal forms of the generators, which are (15, 14, 13) for this encoder.

Polynomials allow a compact description of the input and output sequences of an encoder. A *polynomial* over the binary field $GF(2)$ has the form

$$f(x) = f_0 + f_1 x + f_2 x^2 + \cdots + f_n x^n \tag{1.103}$$

where the coefficients $f_0, f_1, \cdots, f_n$ are elements of $GF(2)$ and the symbol x is an indeterminate introduced into the calculations for convenience. The *degree* of a polynomial is the largest power of x with a nonzero coefficient.

The *sum* of a polynomial $f(x)$ of degree n_1 and a polynomial $g(x)$ of degree n_2 is another polynomial over $GF(2)$ defined as

$$f(x) + g(x) = \sum_{i=0}^{\max(n_1,n_2)} (f_i \oplus g_i)\, x^i \tag{1.104}$$

where $max(n_1, n_2)$ denotes the larger of n_1 and n_2, and $\oplus$ denotes modulo-2 addition. An example is

$$(1 + x^2 + x^3) + (1 + x^2 + x^4) = x^3 + x^4. \tag{1.105}$$

The *product* of a polynomial $f(x)$ of degree n_1 and a polynomial $g(x)$ of degree n_2 is another polynomial over $GF(2)$ defined as

$$f(x)g(x) = \sum_{i=0}^{n_1+n_2} \left(\sum_{k=0}^{i} f_k g_{i-k} \right) x^i \tag{1.106}$$

where the inner addition is modulo-2. For example,

$$(1 + x^2 + x^3)(1 + x^2 + x^4) = 1 + x^3 + x^5 + x^6 + x^7. \tag{1.107}$$

It is easily verified that associative, commutative, and distributive laws apply to polynomial addition and multiplication.

The input sequence $m_0, m_1, m_2, \ldots$ is represented by the input polynomial $m(x) = m_0 + m_1 x + m_2 x^2 + \ldots$, and similarly the output polynomial $c(x) = c_0 + c_1 x + c_2 x^2 + \ldots$ represents an output stream. The *transfer function* $g(x) = g_0 + g_1 x + \ldots + g_{K-1} x^{K-1}$ represents the K bits of an impulse response. When a single input sequence $\mathbf{m}$ is applied to an encoder, an encoder output sequence $\mathbf{c}$ is called the convolution of $\mathbf{m}$ and the impulse response $\mathbf{g}$ and may be represented by a polynomial multiplication so that $c(x) = m(x)\, g(x)$. In general, if there are k input sequences represented by the vector $\mathbf{m}(x) = [m_1(x)\ m_2(x)\ \ldots\ m_k(x)]$ and n encoder output sequences represented by the vector $\mathbf{c}(x) = [c_1(x)\ c_2(x)\ \ldots\ c_n(x)]$, then

$$\mathbf{c}(x) = \mathbf{m}(x)\, \mathbf{G}(x) \tag{1.108}$$

where $\mathbf{G}(x) = [\mathbf{g}_1(x)\ \mathbf{g}_2(x)\ \ldots\ \mathbf{g}_n(x)]$ is the $k \times n$ *generator matrix* with n columns of k generator polynomials.

In Figure 1.6 (a), $m(x)$ is the input polynomial, $\mathbf{c}(x) = [c_1(x)\ c_2(x)]$ is the output vector, and $\mathbf{G}(x) = [g_1(x)\ g_2(x)]$ is the generator matrix with transfer functions $g_1(x) = 1 + x^2$ and $g_2(x) = 1 + x + x^2$. In Figure 1.6 (b), $\mathbf{m}(x) = [m_1(x)\ m_2(x)]$, $\mathbf{c}(x) = [c_1(x)\ c_2(x)\ c_3(x)]$, and

$$\mathbf{G}(x) = \begin{bmatrix} 1+x & 1+x & 1 \\ x & 0 & 1+x \end{bmatrix}. \tag{1.109}$$

Since k bits exit from the shift register as k new bits enter it, only the contents of the $(K-1)k$ memory stages prior to the arrival of new bits affect the subsequent output bits of a convolutional encoder. Therefore, the contents of these $(K-1)k$ stages define the *state* of the encoder. The initial state of a feedforward encoder, which has no feedback connections, is generally the *zero state in which the contents of all stages are zeros.* After the message sequence has been encoded $(K-1)k$ zeros must be inserted into the feedforward encoder to complete and terminate the codeword. If the number of message bits is much higher than $(K-1)k$, these terminal zeros have a negligible effect and the *code rate* is well-approximated by $r = k/n$. However, the need for the terminal zeros renders the convolutional codes unsuitable for short messages. For example, if 12 information bits are to be transmitted, the (23, 12) Golay code provides a better performance than convolutional codes.

A *recursive systematic convolutional code* uses feedback and has a generator matrix with at least one rational function. A recursive systematic convolutional code with $K = 4$ and rate $= 1/2$ is generated by the encoder shown in Figure 1.7. Let $m(x)$ and $m_1(x)$ denote the input polynomial and the output polynomial of the first adder, respectively. Then the output polynomial of memory-stage n is $m_1(x)x^n$, $n = 1, 2, 3$. The diagram indicates that

$$m_1(x) = m(x) + m_1(x)x^2 + m_1(x)x^3 \tag{1.110}$$

which implies that $m_1(x)\left(1 + x^2 + x^3\right) = m(x)$. The output polynomials are $c_1(x) = m(x)$ and $c_2(x) = m_1(x)\left(1 + x + x^3\right)$. Therefore, the output vector $\mathbf{c}(x) = [c_1(x)\ c_2(x)]$ is given by (1.108) with $\mathbf{G}(x) = [1\ \ G_2(x)]$ and

$$G_2(x) = \frac{\left(1 + x + x^3\right)}{\left(1 + x^2 + x^3\right)} \tag{1.111}$$

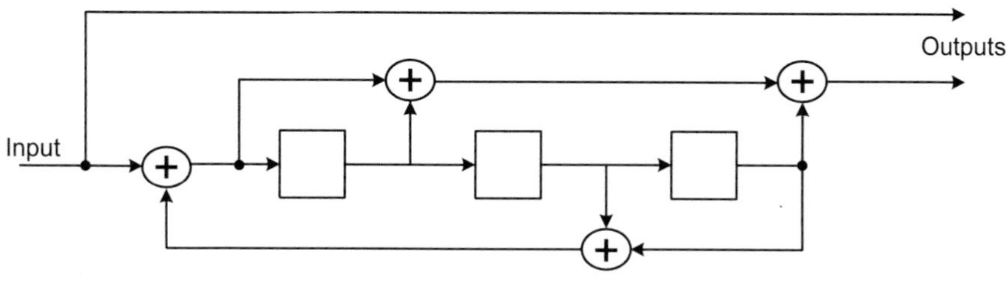

Fig. 1.7 Encoder of recursive systematic convolutional code with $K = 4$ and rate = 1/2

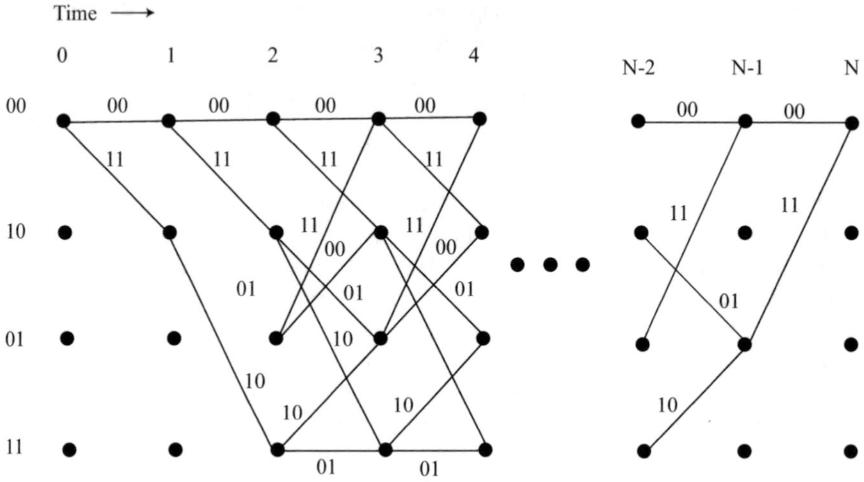

Fig. 1.8 Trellis diagram corresponding to encoder of Figure 1.7

which may be expressed as a polynomial after long division.

To bring a recursive encoder back to the zero state after a codeword transmission, consecutive feedback bits are inserted as input bits to the leftmost adder until the encoder returns to the zero state. In the encoder of Figure 1.7, the zero state is restored after three clock pulses.

A *trellis diagram* displays the possible progression of the states of a finite-state machine, such as the encoder of a convolutional code. A trellis diagram corresponding to the encoder of Figure 1.6 (a) is shown in Figure 1.8. Each of the nodes in a column of a trellis diagram represents the state of the encoder at a specific time prior to a clock pulse. The first bit of a state represents the content of the first memory stage, while the second bit represents the content of the second memory stage. Branches connecting nodes represent possible changes of state. Each branch is labeled with the output bits or symbols produced following a clock pulse and the formation of a new encoder state. In this example, the first bit of a branch label refers to the upper output of the encoder. The upper branch leaving a node corresponds to a 0 input bit, while the lower branch corresponds to a 1. Every path from left to right through the trellis represents a possible codeword. If the encoder begins in the zero state, not all the other states can be reached until the initial contents have been shifted out. The trellis diagram then becomes identical from column to column until the final $(K-1)k$ input bits force the encoder back to the zero state.

Each branch of the trellis is associated with a *branch metric*, and the metric of a codeword is defined as the sum of the branch metrics for the path associated with the codeword. A maximum-likelihood decoder selects the codeword with the largest metric. The branch metrics are determined by the modulation, code, and communication channel. For example, if BPSK signals are transmitted over the AWGN channel, then (1.52) indicates that the metric of branch i and code c is

$\alpha_i x_{ci} y_{ri}$, where α_i is the fading amplitude for branch i, $x_{ci} = \pm 1$, $y_{ri} = \mathrm{Re}(y_i)$, and y_i is the received sample corresponding to branch i.

The *Viterbi decoder* implements maximum-likelihood decoding efficiently by sequentially eliminating many of the possible paths. At any node, only the partial path reaching that node with the largest partial metric is retained, because any partial path stemming from the node adds the same branch metrics to all paths that merge at that node.

Since the decoding complexity grows exponentially with constraint length, Viterbi decoders are limited to use with convolutional codes of short constraint lengths. A Viterbi decoder for a rate-1/2, $K = 7$ convolutional code has approximately the same complexity as a Reed-Solomon (31,15) decoder. If the constraint length is increased to $K = 9$, the complexity of the Viterbi decoder increases by a factor of approximately 4.

The suboptimal *sequential decoder* of convolutional codes does not invariably provide maximum-likelihood decisions, but its implementation complexity only weakly depends on the constraint length. Thus, very low error probabilities can be attained by using long constraint lengths. The number of computations needed to decode a frame of data is fixed for the Viterbi decoder, but is a random variable for the sequential decoder. When strong interference is present, the excessive computational demands and consequent memory overflows of sequential decoding usually result in a higher bit error probability than for Viterbi decoding and a much longer decoding delay. Thus, Viterbi decoding predominates in communication systems.

To bound bit error probability for the Viterbi decoder of a convolutional code, we use the fact that the distribution of either Hamming or Euclidean distances from a codeword to the other codewords is invariant to the choice of a reference codeword. Consequently, whether the demodulator makes hard or soft decisions, the assumption that the all-zero sequence is transmitted entails no loss of generality in the derivation of the error probability.

Although the encoder follows the all-zero path through the trellis, the decoder in the receiver essentially observes successive columns in the trellis, eliminating possible paths and thereby sometimes introducing errors at each node. The decoder may retain an incorrect path that merges at node ν with the correct path, thereby eliminating the correct path and introducing errors that occurred over the unmerged segment of the incorrect retained path.

Let $E[N_e(\nu)]$ denote the expected value of the number of errors introduced at node ν. As shown in Section 1.1, the information-bit error probability P_b equals the *information-bit error rate*, which is defined as the ratio of the expected number of information-bit errors to the number of information bits applied to the convolutional encoder. If there are N branches in a complete path, then

$$P_b = \frac{1}{kN} \sum_{\nu=1}^{N} E[N_e(\nu)]. \tag{1.112}$$

Let $B_\nu(l,i)$ denote the event that among the paths merging at node ν, the one with the largest metric has a Hamming weight l and i incorrect information bits over its unmerged segment. Let d_f denote the *minimum free distance*, which is the minimum Hamming distance between any two codewords that can be generated by the encoder. Then,

$$E\left[N_e(\nu)\right] = \sum_{i=1}^{I_\nu}\sum_{l=d_f}^{D_\nu} E\left[N_e(\nu)|B_\nu(l,i)\right] P\left[B_\nu(l,i)\right] \tag{1.113}$$

when $E[N_e(\nu)|B_\nu(l,i)]$ is the conditional expectation of $N_e(\nu)$ given event $B_\nu(l,i)$, $P[B_\nu(l,i)]$ is the probability of this event, and I_ν and D_ν are the maximum values of i and l, respectively, that are consistent with the position of node ν in the trellis. When $B_\nu(l,i)$ occurs, i bit errors are introduced into the decoded bits; thus,

$$E[N_e(\nu)|B_\nu(l,i)] = i. \tag{1.114}$$

Consider the paths that would merge with the correct path at node ν if the trellis were infinitely long. Let $a(l,i)$ denote the number of such paths that have a Hamming weight l and i incorrect information symbols over the unmerged segments of the path before it merges with the correct all-zero path. Thus, these paths are at a Hamming distance of l from the correct all-zero path. The union bound, and the fact that the number of paths in an infinite trellis exceeds the number in a finite trellis, implies that

$$P\left[B_\nu(l,i)\right] \le a(l,i)P_2(l) \tag{1.115}$$

where $P_2(l)$ is the probability that the correct path segment has a smaller metric than an unmerged path segment that differs in l code symbols.

The *information-weight spectrum* or *distribution* is defined as

$$B(l) = \sum_{i=1}^{\infty} ia(l,i), \quad l \ge d_f. \tag{1.116}$$

Substituting (1.113) to (1.115) into (1.112), extending the two summations to ∞, and then using (1.116), we obtain

$$P_b \le \frac{1}{k}\sum_{l=d_f}^{\infty} B(l)P_2(l). \tag{1.117}$$

When the demodulator makes hard decisions and a correct path segment is compared with an incorrect one, correct decoding results if the number of symbol errors in the demodulator output is fewer than half the number of symbols in which the two segments differ. If the number of symbol errors is exactly half the number of differing symbols, then either of the two segments is chosen with equal probability. If independent symbol errors occur with probability P_s, then

$$P_2(l) = \begin{cases} \sum\limits_{i=(l+1)/2}^{l} \binom{l}{i} P_s^i (1-P_s)^{l-i}, & l \text{ is odd} \\ \sum\limits_{i=l/2+1}^{l} \binom{l}{i} P_s^i (1-P_s)^{l-i} + \frac{1}{2}\binom{l}{l/2} \left[P_s(1-P_s)\right]^{l/2}, & l \text{ is even.} \end{cases} \tag{1.118}$$

for hard-decision decoding.

Soft-decision decoding typically provides a 2-dB power saving at $P_b = 10^{-5}$ compared with hard-decision decoding for communications over the AWGN channel. The additional implementation complexity of soft-decision decoding is minor, and the loss due to even three-bit quantization is usually only 0.2–0.3 dB. Consequently, soft-decision decoding is highly preferable to hard-decision decoding. For coherent BPSK signals over an AWGN channel and soft decisions, (1.57) and (1.117) yield

$$P_b \leq \frac{1}{k} \sum_{l=d_f}^{\infty} B(l) Q\left(\sqrt{\frac{2lr\mathcal{E}_b}{N_0}}\right). \tag{1.119}$$

Among the convolutional codes of a given code rate and constraint length, the one giving the smallest upper bound in (1.119) can sometimes be determined by a complete computer search. The codes with the largest value of d_f are selected, and the *catastrophic codes*, for which a finite number of demodulated symbol errors can cause an infinite number of decoded information-bit errors, are eliminated. All remaining codes that do not have the minimum information-weight spectrum $B(d_f)$ are eliminated. If more than one code remains, codes are eliminated on the basis of the minimal values of $B(d_f + i)$, $i \geq 1$, until one code remains.

Convolutional codes with these favorable distance properties have been determined [14] for codes with $k = 1$ and rates 1/2, 1/3, and 1/4. For these codes and constraint lengths up to 12, Tables 1.4, 1.5 and 1.6 list the corresponding values of d_f and $B(d_f + i)$, $i = 0, 1, \ldots, 7$. Also listed in octal form are the generator sequences that determine which shift-register stages feed the modulo-2 adders associated with each code bit. For example, the best $K = 3$, rate-1/2 code in Table 1.4 has generators 5 and 7, which specify the connections illustrated in Figure 1.6 (a).

Approximate upper bounds on P_b for rate-1/2, rate-1/3, and rate-1/4 convolutional codes with coherent BPSK, soft-decision decoding, and infinitely fine quantization are depicted in Figures 1.9, 1.10, and 1.11. The graphs are computed by using (1.119), $k = 1$, and Tables 1.4, 1.5, and 1.6 in (1.117) and then truncating the series after seven terms. This truncation gives a tight upper bound on P_b for $P_b \leq 10^{-2}$. However, the truncation may exclude significant contributions to the upper bound when $P_b > 10^{-2}$, and the bound itself becomes looser as P_b increases. The figures indicate that the code performance improves with increases in the

Table 1.4 Parameter values of rate-1/2 convolutional codes with favorable distance properties

			$B(d_f + i)$ for $i = 0, 1, \ldots, 6$						
K	d_f	Generators	0	1	2	3	4	5	6
3	5	5, 7	1	4	12	32	80	192	448
4	6	15, 17	2	7	18	49	130	333	836
5	7	23, 35	4	12	20	72	225	500	1324
6	8	53, 75	2	36	32	62	332	701	2342
7	10	133, 171	36	0	211	0	1404	0	11,633
8	10	247, 371	2	22	60	148	340	1008	2642
9	12	561, 763	33	0	281	0	2179	0	15,035
10	12	1131, 1537	2	21	100	186	474	1419	3542
11	14	2473, 3217	56	0	656	0	3708	0	27,518
12	15	4325, 6747	66	98	220	788	2083	5424	13,771

Table 1.5 Parameter values of rate-1/3 convolutional codes with favorable distance properties

			$B(d_f + i)$ for $i = 0, 1, \ldots, 6$						
K	d_f	Generators	0	1	2	3	4	5	6
3	8	5, 7, 7	3	0	15	0	58	0	201
4	10	13, 15, 17	6	0	6	0	58	0	118
5	12	25, 33, 37	12	0	12	0	56	0	320
6	13	47, 53, 75	1	8	26	20	19	62	86
7	15	117, 127, 155	7	8	22	44	22	94	219
8	16	225, 331, 367	1	0	24	0	113	0	287
9	18	575, 673, 727	2	10	50	37	92	92	274
10	20	1167, 1375, 1545	6	16	72	68	170	162	340
11	22	2325, 2731, 3747	17	0	122	0	345	0	1102
12	24	5745, 6471, 7553	43	0	162	0	507	0	1420

Table 1.6 Parameter values of rate-1/4 convolutional codes with favorable distance properties

			$B(d_f + i)$ for $i = 0, 1, \ldots, 6$						
K	d_f	Generators	0	1	2	3	4	5	6
3	10	5, 5, 7, 7	1	0	4	0	12	0	32
4	13	13, 13, 15, 17	4	2	0	10	3	16	34
5	16	25, 27, 33, 37	8	0	7	0	17	0	60
6	18	45, 53, 67, 77	5	0	19	0	14	0	70
7	20	117, 127, 155, 171	3	0	17	0	32	0	66
8	22	257, 311, 337, 355	2	4	4	24	22	33	44
9	24	533, 575, 647, 711	1	0	15	0	56	0	69
10	27	1173, 1325, 1467, 1751	7	10	0	28	54	58	54

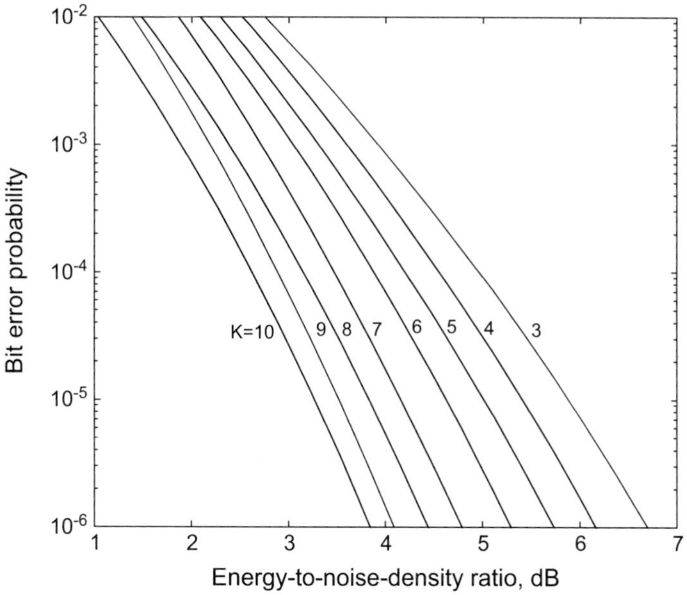

Fig. 1.9 Information-bit error probability for rate-1/2 convolutional codes with different constraint lengths and coherent BPSK

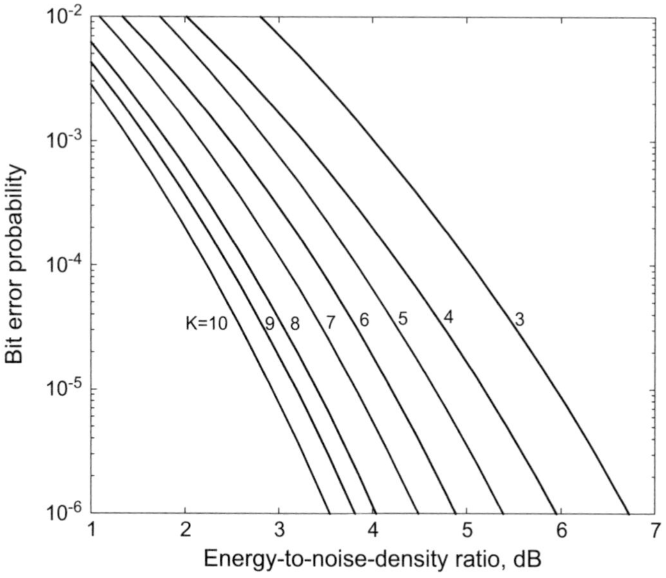

Fig. 1.10 Information-bit error probability for rate-1/3 convolutional codes with different constraint lengths and coherent BPSK

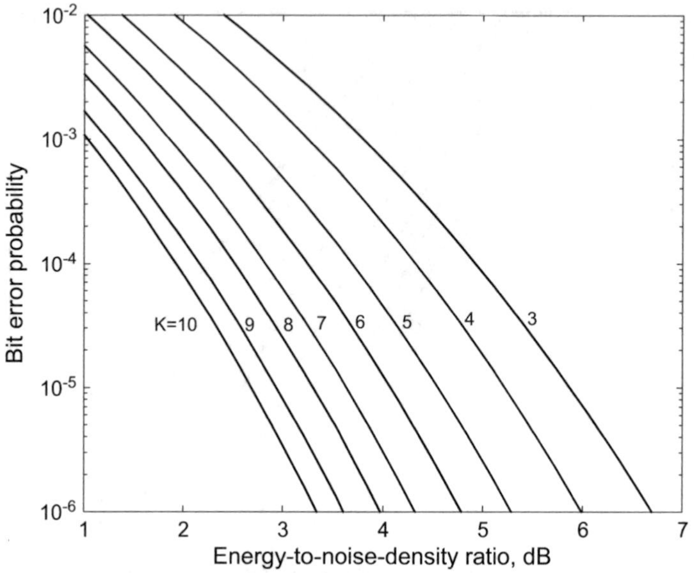

Fig. 1.11 Information-bit error probability for rate-1/4 convolutional codes with different constraint lengths and coherent BPSK

constraint length and decreases in the code rate if $K \geq 4$. The decoder complexity is almost exclusively dependent on K because there are 2^{K-1} encoder states. However, as the code rate decreases, more bandwidth is required, and bit synchronization becomes more challenging because of a reduced energy per symbol.

For convolutional codes of rate $1/n$, two trellis branches enter each state. For higher-rate codes with k information bits per branch, 2^k trellis branches enter each state, and the computational complexity may be large. This complexity can be avoided by using *punctured convolutional codes*. These codes are generated by periodically deleting bits from one or more output streams of an encoder for an unpunctured rate-$1/n$ code. For a punctured code with a period p, the encoder generates and writes p sets of n bits into a buffer. From these np bits, $p + \nu$ bits provide transmitted code bits, where $1 \leq \nu < (n-1)p$. Thus, a punctured convolutional code has rate

$$r = \frac{p}{p+\nu} \ , \quad 1 \leq \nu < (n-1)p. \tag{1.120}$$

The decoder of a punctured code may use the same decoder and trellis as the parent code, but uses only the metrics of the unpunctured bits as it proceeds through the trellis.

The pattern of puncturing is concisely described by an $n \times p$ puncturing matrix $\mathbf{P}$ in which each column specifies which encoder output bits are transmitted. Matrix element P_{ij} is set equal to 1 if code-bit i is transmitted during epoch j of the puncturing period p; otherwise, $P_{ij} = 0$. For most code rates, there are punctured codes with the largest minimum free distance of any convolutional code with that code rate. Punctured convolutional codes enable the efficient implementation of a variable-rate error-control system with a single encoder and decoder. However, the periodic character of the trellis of a punctured code requires the decoder to acquire frame synchronization.

Coded nonbinary sequences can be produced by converting the outputs of a binary convolutional encoder into a single nonbinary symbol, but this procedure does not optimize the nonbinary code's Hamming distance properties. Better nonbinary codes are possible, but performance is not as good as that of the nonbinary Reed-Solomon codes with the same transmission bandwidth.

In principle, $B(l)$ can be determined from the *generating function*, $T(D, I)$, which can be derived for some convolutional codes by treating the state diagram as a signal flow graph. The generating function is a polynomial in D and I of the form

$$T(D,I) = \sum_{i=1}^{\infty} \sum_{l=d_f}^{\infty} a(l,i) D^l I^i \tag{1.121}$$

where $a(l, i)$ denotes the number of distinct paths that have Hamming weight l and i incorrect information symbols before merging with the correct path. The derivative at $I = 1$ is

$$\left. \frac{\partial T(D,I)}{\partial I} \right|_{I=1} = \sum_{i=1}^{\infty} \sum_{l=d_f}^{\infty} i a(l,i) D^l = \sum_{l=d_f}^{\infty} B(l) D^l. \tag{1.122}$$

Thus, the bound on P_b given by (1.117) is determined by substituting $P_2(l)$ in place of D^l in the polynomial expansion of the derivative of $T(D, I)$ and multiplying the result by $1/k$. In many applications, it is possible to establish an inequality of the form

$$P_2(l) \leq \alpha Z^l \tag{1.123}$$

where α and Z are independent of l. It then follows from (1.117), (1.122), and (1.123) that

$$P_b \leq \left(\frac{\alpha}{k} \right) \left. \frac{\partial T(D,I)}{\partial I} \right|_{I=1, D=Z}. \tag{1.124}$$

For soft-decision decoding and coherent BPSK, $P_2(l)$ is given by (1.119). Inequality (H.25) of Appendix H.4 indicates that

$$Q(\sqrt{\nu + \beta}) \leq \exp\left(-\frac{\beta}{2} \right) Q(\sqrt{\nu}), \quad \nu \geq 0, \;\; \beta \geq 0. \tag{1.125}$$

Substituting this inequality into (1.119) with the appropriate choices for ν and β gives

$$P_2(l) \leq Q\left(\sqrt{\frac{2d_f r\mathcal{E}_b}{N_0}}\right) \exp\left[-(l-d_f)r\mathcal{E}_b/N_0\right]. \quad (1.126)$$

Thus, the upper bound on $P_2(l)$ may be expressed in the form given by (1.123) with

$$\alpha = Q\left(\sqrt{\frac{2d_f r\mathcal{E}_b}{N_0}}\right) \exp(d_f r\mathcal{E}_b/N_0) \quad (1.127)$$

$$Z = \exp(-r\mathcal{E}_b/N_0). \quad (1.128)$$

For other channels, codes, and modulations, an upper bound on $P_2(l)$ in the form given by (1.123) can often be derived from the Chernoff bound.

Chernoff Bound

The Chernoff bound is an upper bound on the probability that a random variable equals or exceeds a constant. The usefulness of the Chernoff bound stems from the fact that it is often much more easily evaluated than the probability it bounds. The *moment-generating function* of the random variable X with distribution function $F(x)$ is defined as (Appendix B.1)

$$M(s) = E\left[e^{sX}\right] = \int_{-\infty}^{\infty} \exp(sx)dF(x) \quad (1.129)$$

for all real-valued s for which the integral is finite. For all nonnegative s, the probability that $X \geq 0$ is

$$P[X \geq 0] = \int_0^{\infty} dF(x) \leq \int_0^{\infty} \exp(sx)dF(x). \quad (1.130)$$

Since the moment-generating function is finite in a neighborhood of $s = 0$,

$$P[X \geq 0] \leq M(s), \quad 0 \leq s < s_1 \quad (1.131)$$

where $s_1 > 0$ is the upper limit of an open interval in which $M(s)$ is defined. To make this bound as tight as possible, we choose the value of s that minimizes $M(s)$. Therefore,

$$P[X \geq 0] \leq \min_{0 \leq s < s_1} M(s) \quad (1.132)$$

where the upper bound is called the *Chernoff bound*. From (1.132) and (1.129), we obtain the generalization

$$P\,[X \geq b] \leq \min_{0 \leq s < s_1} M(s) \exp(-sb). \tag{1.133}$$

Further analysis requires some facts about convex functions. A real-valued function f over an interval is a *convex function* if $f\,[\alpha x + (1-\alpha)\,y] \leq \alpha f\,(x) + (1-\alpha) f\,(y)$ for all x and y in the interval and $\alpha \in [0,1]$. A sufficient condition for f to be a convex function is for it to have a nonnegative second derivative (see Section 7.4). A local minimum of a convex function over an interval is a global minimum, and similarly for a local maximum [19].

Since the moment-generating function is finite in a neighborhood of $s = 0$, we may differentiate under the integral sign in (1.129) to obtain the derivative of $M(s)$, as shown in Appendix B.1. The result is

$$M'(s) = \int_{-\infty}^{\infty} x \exp(sx) dF(x) \tag{1.134}$$

which implies that $M'(0) = E[X]$. Differentiating (1.134) gives the second derivative

$$M''(s) = \int_{-\infty}^{\infty} x^2 \exp(sx) dF(x) \tag{1.135}$$

which implies that $M''(s) \geq 0$. Consequently, $M(s)$ is a convex function over its interval of definition. Consider a random variable is such that

$$E(X) < 0\,, \quad P[X > 0] > 0. \tag{1.136}$$

The first inequality implies that $M'(0) < 0$, and the second inequality implies that $M(s) \to \infty$ as $s \to \infty$. Thus, since $M(0) = 1$, the convex function $M(s)$ has a minimum value that is less than unity at some positive $s = s_m$. We conclude that (1.136) is sufficient to ensure that the Chernoff bound is less than unity, which is required if this bound is to be useful.

In some applications, X has an asymmetric density function $f(x)$ such that

$$f(-x) \geq f(x)\,, \quad x \geq 0. \tag{1.137}$$

Under this condition, the Chernoff bound can be tightened. For $s \in [0, s_1)$, (1.129) and (1.137) yield

$$\begin{aligned} M(s) &= \int_0^{\infty} \exp(sx) f(x) dx + \int_{-\infty}^{0} \exp(sx) f(x) dx \\ &\geq \int_0^{\infty} [\exp(sx) + \exp(-sx)] f(x) dx = \int_0^{\infty} 2\cosh(sx) f(x) dx \end{aligned}$$

$$\geq 2\int_{0}^{\infty} f(x)dx = 2P[X \geq 0]. \tag{1.138}$$

This inequality implies that

$$P\left[X \geq 0\right] \leq \frac{1}{2} \min_{0 \leq s < s_1} M(s). \tag{1.139}$$

If X is such that both (1.136) and (1.137) are true, then the Chernoff bound of (1.139) is less than $1/2$.

In soft-decision decoding, the encoded sequence or codeword with the largest associated metric is converted into the decoded output. Let $U(\nu)$ denote the value of the metric associated with sequence ν of length L. Consider additive metrics with the form

$$U(\nu) = \sum_{i=1}^{L} m(\nu, i) \tag{1.140}$$

where $m(\nu, i)$ is the *symbol metric* associated with symbol i of the encoded sequence. Let $\nu = 1$ label the correct sequence and $\nu = 2$ label an incorrect one. Let $P_2(l)$ denote the probability that the metric for an incorrect codeword at distance l from the correct codeword exceeds the metric for the correct codeword. By suitably relabeling the l symbol metrics that may differ for the two sequences, we obtain

$$\begin{aligned} P_2(l) &\leq P\left[U(2) \geq U(1)\right] \\ &= P\left[\sum_{i=1}^{l} [m(2,i) - m(1,i)] \geq 0\right] \end{aligned} \tag{1.141}$$

where the inequality results because $U(2) = U(1)$ does not necessarily cause an error if it occurs. In all practical cases, (1.136) is satisfied for the random variable $X = U(2) - U(1)$. Therefore, the Chernoff bound implies that

$$P_2(l) \leq \alpha \min_{0<s<s_1} E\left[\exp\left\{s \sum_{i=1}^{l} [m(2,i) - m(1,i)]\right\}\right] \tag{1.142}$$

where s_1 is the upper limit of the interval over which the expected value is defined. Depending on which version of the Chernoff bound is valid, either $\alpha = 1$ or $\alpha = 1/2$.

If $m(2,i) - m(1,i)$, $i = 1, 2, \ldots, l$, are independent, identically distributed random variables,

$$P_2(l) \leq \alpha \min_{0<s<s_1} \{E\left[\exp\{s\left[m(2,i) - m(1,i)\right]\}\right]\}^l . \tag{1.143}$$

Since the power function $(\cdot)^l$ is monotonically increasing, the minimization in this equation can be taken within the power function. Therefore, if we define

$$Z = \min_{0<s<s_1} E\left[\exp\left\{s\left[m(2,i) - m(1,i)\right]\right\}\right] \tag{1.144}$$

then

$$P_2(l) \le \alpha Z^l. \tag{1.145}$$

This bound is usually much simpler to compute than the exact $P_2(l)$.

Since $X = U(2) - U(1)$ is the sum of l independent, identically distributed random variables with a finite mean and variance, the central limit theorem (Corollary 1, Appendix A.2) implies that the distribution of $X = U(2) - U(1)$ approximates the Gaussian distribution when l is large. Thus, for large enough l, (1.137) is satisfied when $E[X] < 0$, and we can set $\alpha = 1/2$ in (1.145). For small l, (1.137) may be difficult to establish mathematically, but is often intuitively clear; if not, setting $\alpha = 1$ in (1.145) is always valid.

Trellis-Coded Modulation

A *coded modulation system* is one that integrates the modulation and channel-coding processes. The encoder generates code bits that are grouped into q-ary code symbols that are applied to a q-ary modulator. The use of q-ary symbols increases the bandwidth efficiency. However, if the number of signal constellation points associated with the modulation remains fixed when the channel coding is integrated, the additional code parity bits cause an increase in the bandwidth of the communication system.

To add a channel code to a communication system while avoiding a bandwidth expansion, the number of signal constellation points must be increased. For example, if a rate-2/3 code is added to a system using QPSK, then the bandwidth is preserved if the modulation is changed to eight-phase PSK (8-PSK). Since each symbol of the latter modulation represents 3/2 as many bits as a QPSK symbol, the channel-symbol rate is unchanged. The problem is that the change from QPSK to the more compact 8-PSK constellation causes an increase in the probability of channel-symbol error that cancels most of the decrease due to the encoding. This problem is avoided by integrating coding into a *trellis-coded modulation* system.

The encoder for trellis-coded modulation has the form shown in Figure 1.12. For $k > 1$, each input of k information bits is divided into two groups and encoded as $k + 1$ bits. One group of k_1 bits is applied to a convolutional encoder with $k_1 + 1$ output bits. The other group of $k_2 = k - k_1$ bits remains uncoded. The $k_1 + 1$ output bits of the convolutional encoder select one of 2^{k_1+1} possible subsets of the points in a constellation of 2^{k+1} points. The k_2 uncoded bits select one of 2^{k_2} points

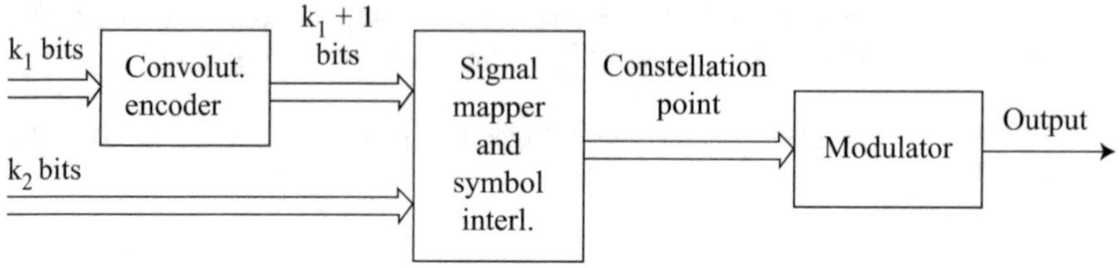

Fig. 1.12 Encoder for trellis-coded modulation

Fig. 1.13 The constellation of 8-PSK symbols partitioned into four subsets

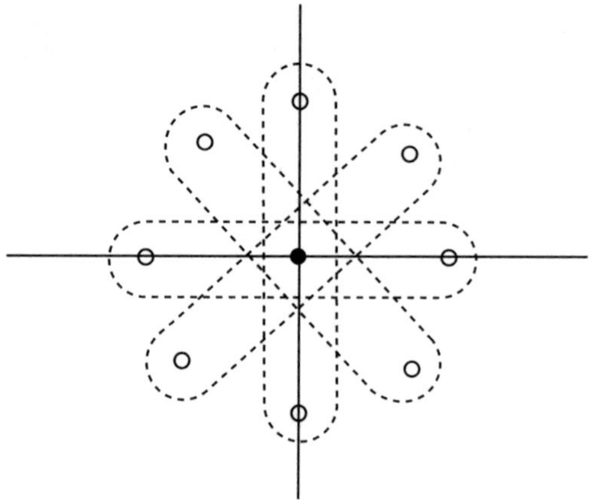

in the chosen subset. If $k_2 = 0$, there are no uncoded bits and the convolutional encoder output bits select the constellation point. After symbol interleaving of the constellation points, they are applied to the modulator. Each constellation point is a complex number representing an amplitude and phase. The process of using code bits and uncoded bits to select or label constellation points is called *set partitioning*.

For example, suppose that $k = 2,\ k_1 = k_2 = 1$ in the encoder of Figure 1.12, and an 8-PSK modulator produces an output from a constellation of eight points. Each of the four subsets that may be selected by the two convolutional-code bits comprises two antipodal points in the 8-PSK constellation, as shown in Figure 1.13. If the convolutional encoder has the form of Figure 1.6 (a), then the trellis of Figure 1.8 illustrates the state transitions of both the underlying convolutional code and the trellis code. The presence of the single uncoded bit implies that each transition between states in the trellis corresponds to two different transitions and two different phases of the transmitted 8-PSK waveform.

In general, there are 2^{k_2} parallel transitions between every pair of states in the trellis. Often, the dominant error events consist of mistaking one of these parallel transitions for the correct one. If the symbols corresponding to parallel transitions are separated by large Euclidean distances, and the constellation subsets associated with transitions are suitably chosen, then the trellis-coded modulation with soft-decision Viterbi decoding can yield a substantial coding gain. This gain usually ranges from 4 to 6 dB, depending on the number of states and hence the complexity of the implementation.

For coherent BPSK signals over an AWGN channel and soft decisions, (1.119) gives an upper bound on the probability of information-bit error. The minimum Euclidean distance between a correct trellis-code path and an incorrect one is called the *free Euclidean distance* and is denoted by $d_f\sqrt{\mathcal{E}_s}$, where $\mathcal{E}_s = r\mathcal{E}_b$ is the energy per received symbol. When $\mathcal{E}_s/N_0$ is high, the free Euclidean distance largely determines P_b.

1.4 Interleavers

An *interleaver* is a device that permutes the order of a sequence of symbols. A *deinterleaver* is the corresponding device that restores the original order of the sequence. A major application is the interleaving of symbols transmitted over a communication channel subject to fading or interference. After deinterleaving at the receiver, a burst of channel-symbol errors or corrupted symbols is dispersed over a number of codewords or constraint lengths, thereby facilitating the removal of the errors by the decoding. Ideally, the interleaving and deinterleaving ensure that the decoder encounters statistically independent symbol decisions or metrics, as it would if the channel were memoryless.

A *block interleaver* performs identical permutations on successive blocks of symbols. As illustrated in Figure 1.14, mn successive input symbols are stored in a random-access memory (RAM) as an array of m rows and n columns. The input sequence is written into the interleaver in successive rows, but successive columns are read to produce the interleaved sequence. Thus, if the input sequence is numbered $1, 2, \ldots, n, n+1, \ldots, mn$, the interleaved sequence is $1, n+1, 2n+1, \ldots, 2, n+2, \ldots, mn$. For continuous interleaving, two RAM arrays are needed. Symbols are written into one array while previous symbols are read from the other. In the deinterleaver, symbols are stored by column in one array, while previous symbols are read by rows from another. Consequently, a delay of $2mn$ symbol periods must be accommodated, and synchronization is required at the deinterleaver.

When channel symbols are interleaved, the parameter n equals or exceeds the block codeword length or a few constraint lengths of a convolutional code. Consequently, if a burst of m or fewer consecutive symbol errors occurs and there are no other errors, then each block codeword or constraint length, after deinterleaving, has one error at the most, which can be eliminated by the error-correcting code. Similarly, a block code that can correct t errors is capable of correcting a single burst of errors spanning as many as mt symbols. Since fading can cause correlated errors, it is necessary that mT_s exceed the channel coherence time. Interleaving effectiveness can be thwarted by slow fading that cannot be accommodated without large buffers that cause an unacceptable delay.

A *pseudorandom interleaver* permutes each block of symbols pseudorandomly. Pseudorandom interleavers may be applied to channel symbols, but their main

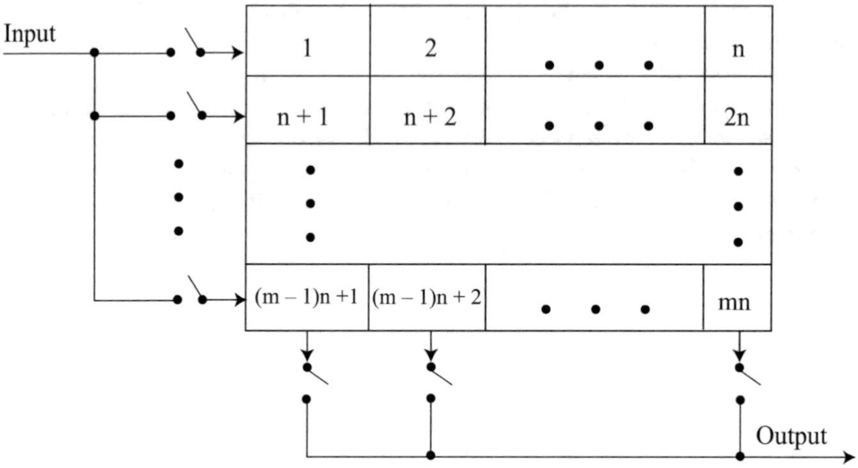

Fig. 1.14 Block interleaver

application is as critical elements in encoders of turbo codes and codes that use iterative decoding (Sections 1.5 and 1.6). The desired permutation may be stored in a read-only memory (ROM) as a sequence of addresses or permutation indices. Each block of symbols is written sequentially into a RAM array and then interleaved by reading them in the order dictated by the contents of the ROM.

If the interleaver is large, it is often preferable to generate the permutation indices by an algorithm rather than by storing them in a ROM. If the interleaver size is $N = mn = 2^{\nu} - 1$, then a linear feedback shift register with ν stages that produces a maximal-length sequence can be used. The binary outputs of the shift-register stages constitute the *state* of the register. The state specifies the index from 1 to N that defines a specific interleaved symbol. The shift register generates all N states and indices periodically.

An *S-random interleaver* is a pseudorandom interleaver that constrains the minimum interleaving distance so that error bursts of length S or less are dispersed. A tentative permutation index is randomly selected from among N uniformly distributed indices and then compared with the S previously selected indices, where $1 \leq S < N$. If the tentative index does not differ sufficiently from the S previous ones, then it is discarded and replaced by a new tentative index. If it does differ enough, then the tentative index becomes the next selected index. This procedure continues until all N pseudorandom indices are selected.

A *convolutional interleaver* consists of a bank of shift registers of successively increasing length [57]. Convolutional interleavers and their corresponding deinterleavers confer an efficiency advantage in that the ratio of the length of the smallest error burst that exceeds the correction capability of a block code to the interleaver memory is higher than the ratio for comparable block interleavers. However, convolutional interleavers do not possess the inherent simplicity and compatibility with block structures that block, pseudorandom, and S-random interleavers have.

1.5 Classical Concatenated Codes

Classical concatenated codes are serially concatenated codes. Multiple codes could be used, but no significant advantages of having more than two codes have been found. Two serially concatenated codes are called the outer code and the inner code, and the encoder has the form depicted in Figure 1.15 (a). The outer encoder produces binary or nonbinary symbols. After interleaving, these symbols are converted, if necessary, into symbols that are encoded by the inner encoder. The concatenated-code symbols at the output of the inner encoder are applied to the modulator.

In the receiver, the demodulator makes hard decisions and sends concatenated-code symbols to the decoder. As depicted in Figure 1.15 (b), the inner decoder removes the inner code and groups the output symbols into outer-code symbols. Symbol deinterleaving disperses the outer-code symbol errors. Consequently, the outer decoder is able to correct most symbol errors originating in the inner-decoder output. The concatenated code has a rate

$$r = r_1 r_0 \tag{1.146}$$

where r_1 is the inner-code rate and r_0 is the outer-code rate. The bandwidth required by a concatenated code is B/r, where B is the uncoded bandwidth.

The dominant and most powerful concatenated code of this type comprises a binary convolutional inner code and a Reed-Solomon outer code. At the output of a convolutional inner decoder using the Viterbi algorithm, the channel-bit errors occur over spans with an average length that depends on the value of $\mathcal{E}_b/N_0$. The deinterleaver is designed to ensure that Reed-Solomon symbols formed from bits in the same typical error span do not belong to the same Reed-Solomon codeword.

Let $m = \log_2 q$ denote the number of bits in a Reed-Solomon code symbol. In the worst case, the inner decoder produces channel-bit errors that are separated enough that each one causes a separate channel-symbol error at the input to the Reed-Solomon decoder. Since there are m times as many bits as symbols, the symbol error probability P_{s1} is upper-bounded by m times the bit error probability at the inner-decoder output. Since P_{s1} is no smaller than it would be if each set of m bit errors caused a single symbol error, P_{s1} is lower-bounded by the bit error probability

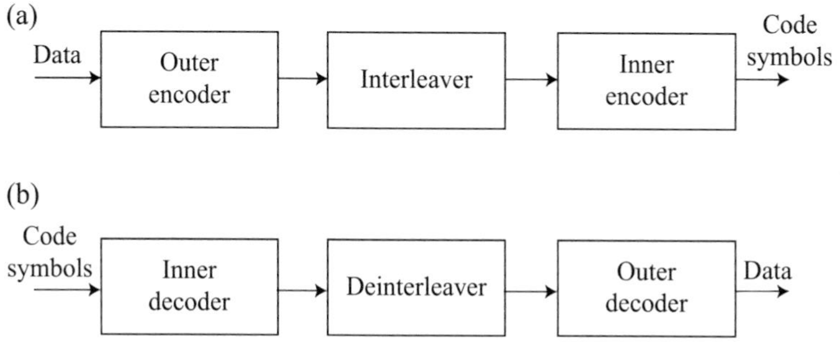

Fig. 1.15 Structure of classical (**a**) encoder and (**b**) decoder for serially concatenated code

at the inner-decoder output. Thus, for binary convolutional inner codes,

$$\frac{1}{k}\sum_{l=d_f}^{\infty} B(l)P_2(l) \leq P_{s1} \leq \frac{\log_2 q}{k}\sum_{l=d_f}^{\infty} B(l)P_2(l) \tag{1.147}$$

where $P_2(l)$ is given by (1.145) and (1.144).

Assuming that the deinterleaving ensures independent symbol errors at the outer-decoder input, and that the Reed-Solomon code is loosely packed, (1.23) and (1.95) imply that

$$P_b \approx \frac{q}{2(q-1)} \sum_{i=t+1}^{n} \binom{n-1}{i-1} P_{s1}^{i}(1-P_{s1})^{n-i}. \tag{1.148}$$

For coherent BPSK modulation with soft decisions, $P_2(l)$ is given by (1.119); if hard decisions are made, (1.118) applies.

Figure 1.16 depicts the upper bound on the bit error probability for concatenated codes, the AWGN channel, coherent BPSK, soft demodulator decisions, an inner binary convolutional code with $k = 1$, $K = 7$, rate-$1/2$, and $d_f = 10$, and various Reed-Solomon outer codes. Equation (1.148), the upper bound in (1.147), and the information-weight spectrum in Table 1.4 are used.

Product Codes

A *product code* is a serially concatenated code that is constructed from a particular type of multidimensional array and linear block codes. A set of $k = k_1k_2$ information bits are placed into a $k_2 \times k_1$ array. Each of the k_2 rows is encoded as a codeword of an (n_1, k_1) code and n_1 columns are formed. Each of these columns is encoded as an (n_2, k_2) code and n_2 rows are formed. The resulting $n_2 \times n_1$ array, which is depicted in Figure 1.17, defines a codeword of a product code with $n = n_1n_2$ code bits and code rate

$$r = \frac{k_1k_2}{n_1n_2}. \tag{1.149}$$

The row codewords serve as inner codewords, and the column codewords serve as outer codewords. The dual bits in the lower right corner of the figure serve as parity bits for both the inner and outer codewords.

Let d_{m1} and d_{m2} denote the minimum Hamming distances of the row and column codes, respectively. For a nonzero product codeword, every nonzero row in the array must have a weight of at least d_{m1}, and there must be at least d_{m2} nonzero rows. Thus, the minimum Hamming distance of the product code, which is equal to the minimum Hamming weight of a nonzero codeword, is at least $d_{m1}d_{m2}$. Let $\mathbf{c}_1$ and $\mathbf{c}_2$ denote row and column minimum-weight codewords, respectively. One of the valid product codewords is defined by an array in which all columns corresponding to

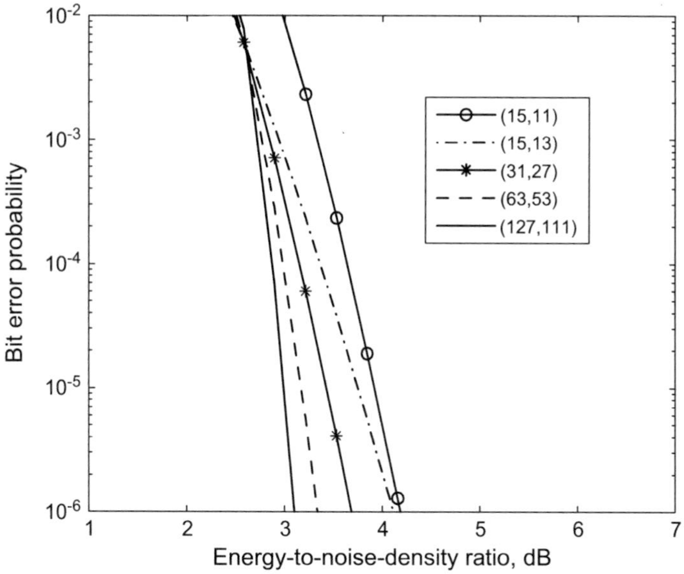

Fig. 1.16 Information-bit error probability for concatenated codes with inner convolutional code ($K = 7$, rate-1/2), various Reed-Solomon (n, k) outer codes, and coherent BPSK

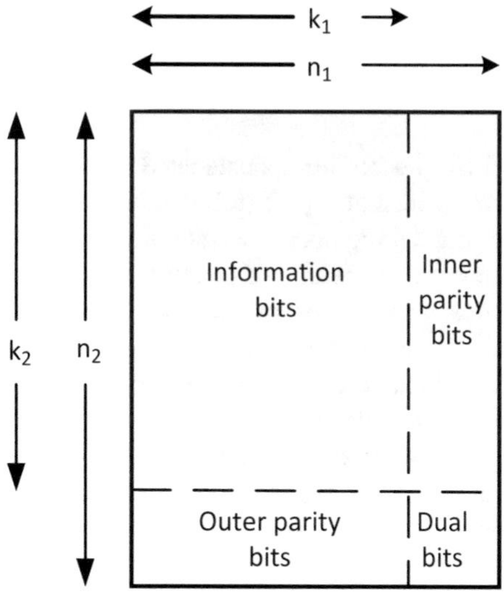

Fig. 1.17 Codeword structure for turbo product code

zeros in $\mathbf{c}_1$ are zeros and all columns corresponding to ones in $\mathbf{c}_1$ are the same as $\mathbf{c}_2$. Therefore, a product codeword of weight $d_{m1}d_{m2}$ exists, and the product code has a minimum distance

$$d_m = d_{m1}d_{m2}. \tag{1.150}$$

The $n_2 \times n_1$ array of code bits may be transmitted row by row or column by column. Hard-decision demodulator decisions are done sequentially on an $n_2 \times n_1$ array of received code bits. Successive rows (or columns) of codewords are decoded, and code-bit errors are corrected. Any residual errors are then corrected during the decoding of successive columns (or rows) of codewords.

Let t_1 and t_2 denote the error-correcting capability of the row and column codes, respectively. Incorrect decoding of the row codewords requires there to be at least $t_1 + 1$ errors in at least one row codeword. For the column decoder to fail to correct the residual errors, there must be at least $t_2 + 1$ row codewords that have $t_1 + 1$ or more errors, and the errors must occur in certain array positions. Thus, the number of errors that are always correctable is

$$t = (t_1 + 1)(t_2 + 1) - 1 \tag{1.151}$$

which is roughly half of what (1.6) guarantees for classical block codes with the same minimum distance d_m. Moreover, although not all patterns with more than t errors are correctable, many of them are.

1.6 Turbo Codes

Turbo codes are concatenated codes that use iterative decoding [30]. There are two principal types: turbo codes with parallel concatenated codes, which are the original turbo codes, and serially concatenated turbo codes, which are related to the classical concatenated codes. Each of the concatenated codes is called a *component code*. Since three or more component codes have not been found to provide a significant performance advantage, the use of two component codes is predominant. The iterative decoding requires both component codes to be systematic and of the same type; that is, both convolutional or both block.

The turbo decoder processes the sampled demodulator output streams. Although soft-decision decoders for block and convolutional codes, such as the Viterbi decoder, minimize the probability that a received codeword or an entire received sequence is in error, a turbo decoder is designed to minimize the error probability of each information bit. A turbo decoder comprises separate component decoders for each component code, which is theoretically suboptimal but crucial in reducing the complexity of the decoder. Each component decoder uses a version of the *BCJR* (Bahl, Cocke, Jelinek, and Raviv) *algorithm*, which is also known as the *maximum a posteriori (MAP) algorithm*.

BCJR Decoding Algorithm

The *BCJR algorithm* is a decoding algorithm that uses a priori information to enable the generation of *soft* or numerical estimates of the a posteriori probabilities of the bits and the states in a trellis. The provision of soft output information for each information bit after each iteration enables the iterative decoding of turbo codes. The BCJR algorithm exploits the Markov properties of a convolutional code or other code that can be described in terms of a trellis structure.

Let the random vector $\mathbf{Y} = [\mathbf{Y}_1, \mathbf{Y}_2, \ldots, \mathbf{Y}_N]$ denote the sequence of N observation symbols produced by the demodulator and used to calculate inputs to the BCJR algorithm. Let the sequence of vectors $\mathbf{y} = [\mathbf{y}_1, \mathbf{y}_2, \ldots, \mathbf{y}_N]$ denote an outcome or particular observation of the random vector $\mathbf{Y}$. Associated with $\mathbf{Y}_k$ is the transmitted symbol $\mathbf{S}_k$ that represents a group of bits or a constellation point of a code symbol. Each symbol $\mathbf{S}_k, k = 1, 2, \ldots, N$, has a form determined by the modulation and coding. If binary code bits are transmitted using BPSK, then each $\mathbf{S}_k$ is the vector corresponding to an information bit and its associated parity bits. If q-ary symbols are transmitted using PAM, then $\mathbf{S}_k$ may be a complex number representing a constellation point.

Let $b_l(k)$ denote the lth bit represented by $\mathbf{S}_k$. The BCJR algorithm computes the log-likelihood ratio (LLR) of the a posteriori probability that this bit is 1 or 0 given that $\mathbf{Y} = \mathbf{y}$. Accordingly, the *bit LLR* is

$$\Lambda_{k,l} = \ln\left[\frac{P[b_l(k) = 1|\mathbf{y}]}{P[b_l(k) = 0|\mathbf{y}]}\right] \tag{1.152}$$

where $P[b_l(k) = b|\mathbf{y}]$ is a compact notation for the probability that $b_l(k) = b$ given that $\mathbf{Y} = \mathbf{y}$. Since the a posteriori probabilities are related by $P[b_l(k) = 1|\mathbf{y}] = 1 - P[b_l(k) = 0|\mathbf{y}]$, $\Lambda_{k,l}$ completely characterizes the a posteriori probabilities. The decoder using the BCJR algorithm observes $\mathbf{y}$ and computes $\Lambda_{k,l}$.

The *state of the encoder* of a trellis code is the set of stored bits that can influence subsequent code bits. Let ψ_k denote the discrete random variable that specifies the state of the encoder at discrete-time k. The transmission of symbol $\mathbf{S}_k$ and bit $b_l(k)$ is associated with a state transition from $\psi_k = s'$ to $\psi_{k+1} = s$. Applying the definition of a conditional probability, algebraic cancelation, and the mutually exclusive nature of state transitions to (1.152) yields

$$\begin{aligned}\Lambda_{k,l} &= \ln\left[\frac{P[b_l(k) = 1, \mathbf{y}]}{P[b_l(k) = 0, \mathbf{y}]}\right] \\ &= \ln\left[\frac{\sum\limits_{s',s:b_l(k)=1} f(\psi_k = s', \psi_{k+1} = s, \mathbf{y})}{\sum\limits_{s',s:b_l(k)=0} f(\psi_k = s', \psi_{k+1} = s, \mathbf{y})}\right]\end{aligned} \tag{1.153}$$

where $f(\cdot)$ is a generic joint probability and density function, and the summations are over all states consistent with $b_l(k) = 1$ and $b_l(k) = 0$, respectively.

The observed vector sequence $\mathbf{y}$ is decomposed into three separate sets of observations: $\mathbf{y}_k^- = \{\mathbf{y}_l, l < k\}$ represents the observations prior to time k, $\mathbf{y}_k$ is the current observation, and $\mathbf{y}_k^+ = \{\mathbf{y}_l, l > k\}$ represents the observations that occur after time k. We define two joint probability and density functions and a density function:

$$\alpha_k(s') = f(\psi_k = s', \mathbf{y}_k^-) \tag{1.154}$$

$$\gamma_k(s', s) = f(\psi_{k+1} = s, \mathbf{y}_k \mid \psi_k = s') \tag{1.155}$$

$$\beta_{k+1}(s) = f(\mathbf{y}_k^+ \mid \psi_{k+1} = s). \tag{1.156}$$

After conditioning on the event $\psi_{k+1} = s$, $\mathbf{y}_k^+$ is independent of $\mathbf{y}_k$, $\mathbf{y}_k^-$, and the event that $\psi_k = s'$. It follows that

$$\begin{aligned} f(\psi_k = s', \psi_{k+1} = s, \mathbf{y}) &= f(\mathbf{y}_k^+ \mid \psi_{k+1} = s) f(\psi_k = s', \psi_{k+1} = s, \mathbf{y}_k, \mathbf{y}_k^-) \\ &= \beta_{k+1}(s) f(\psi_{k+1} = s, \mathbf{y}_k \mid \psi_k = s', \mathbf{y}_k^-)\, \alpha_k(s'). \end{aligned} \tag{1.157}$$

After conditioning on the event $\psi_k = s'$, the event $(\psi_{k+1} = s, \mathbf{y}_k)$ is independent of $\mathbf{y}_k^-$. Therefore,

$$f(\psi_k = s', \psi_{k+1} = s, \mathbf{y}) = \alpha_k(s')\, \gamma_k(s', s)\, \beta_{k+1}(s). \tag{1.158}$$

Substitution of this equation into (1.153) yields the bit LLR:

$$\Lambda_{k,l} = \ln \left[\frac{\sum\limits_{s',s:b_l(k)=1} \alpha_k(s')\, \gamma_k(s', s)\, \beta_{k+1}(s)}{\sum\limits_{s',s:b_l(k)=0} \alpha_k(s')\, \gamma_k(s', s)\, \beta_{k+1}(s)} \right]. \tag{1.159}$$

The density function $\gamma_k(s', s)$ serves as a *branch metric* in a trellis diagram of the states. Let $b_1(k)$ denote the information bit associated with $\mathbf{S}_k$. There is a unique information bit $b(s', s)$ that is associated with the transition from $\psi_k = s'$ to $\psi_{k+1} = s$. Let $P[b_1(k) = b(s', s)]$ denote the a priori probability that $b_1(k) = b(s', s)$. Then

$$P(\psi_{k+1} = s | \psi_k = s') = P[b_1(k) = b(s', s)]. \tag{1.160}$$

Therefore, the *branch metric* is

$$\gamma_k(s', s) = f(\mathbf{y}_k | \psi_{k+1} = s, \psi_k = s')\, P[b_1(k) = b(s', s)] \tag{1.161}$$

where the conditional density function $f(\mathbf{y}_k|\psi_{k+1} = s, \psi_k = s')$ can be calculated from knowledge of the modulation, demodulation, and communication channel. Equation (1.161) explicitly shows how the a priori information $P[b_1(k) = b(s', s)]$ is used by the BCJR algorithm.

To compute the bit LLRs from the branch metrics, the BCJR algorithm first computes $\alpha_{k+1}(\nu)$ and $\beta_k(\nu)$ recursively. The *forward recursion* for $\alpha_{k+1}(\nu)$ can be derived from its definition. If $\{0, 1, \ldots, Q-1\}$ denotes the set of Q states, then the theorem of total probability implies that

$$\begin{aligned}\alpha_{k+1}(\nu) &= \sum_{\mu=0}^{Q-1} f\left(\psi_{k+1} = \nu, \mathbf{y}_k^-, \mathbf{y}_k, \psi_k = \mu\right)\\ &= \sum_{\mu=0}^{Q-1} f\left(\psi_{k+1} = \nu, \mathbf{y}_k|\mathbf{y}_k^-, \psi_k = \mu\right)\alpha_k(\mu). \end{aligned} \tag{1.162}$$

After conditioning on the event $\psi_k = \mu$, the event $(\psi_{k+1} = \nu, \mathbf{y}_k)$ is independent of $\mathbf{y}_k^-$. Therefore,

$$\alpha_{k+1}(\nu) = \sum_{\mu=0}^{Q-1} \alpha_k(\mu)\gamma_k(\mu, \nu). \tag{1.163}$$

The *backward recursion* for $\beta_{k+1}(s)$ can be derived in a similar manner. We have

$$\begin{aligned}\beta_k(\nu) &= \sum_{\mu=0}^{Q-1} f\left(\mathbf{y}_k^+, \mathbf{y}_k, \psi_{k+1} = \mu|\psi_k = \nu\right)\\ &= \sum_{\mu=0}^{Q-1} f\left(\mathbf{y}_k^+|\mathbf{y}_k, \psi_{k+1} = \mu, \psi_k = \nu\right)\gamma_k(\nu, \mu). \end{aligned} \tag{1.164}$$

After conditioning on the event $\psi_{k+1} = \mu$, the event $\mathbf{y}_k^+$ is independent of the event $(\mathbf{y}_k, \psi_k = \nu)$. Therefore,

$$\beta_k(\nu) = \sum_{\mu=0}^{Q-1} \gamma_k(\nu, \mu)\beta_{k+1}(\mu). \tag{1.165}$$

Assuming that the encoder begins in the zero state and ends in the zero state at the time $k = L$, recursions for $\alpha_{k+1}(\nu)$ and $\beta_k(\nu)$ are initialized according to

$$\alpha_0(x) = \delta(x), \quad \beta_L(x) = \delta(x) \tag{1.166}$$

where $\delta(x) = 1$ if $x = 0$, and $\delta(x) = 0$ otherwise.

Example. A systematic rate-1/3 binary convolutional code has its code symbols transmitted as BPSK symbols over the AWGN channel. Let $\mathbf{x}_k = [x_{k,1}\ x_{k,2}\ x_{k3}]$ and $\mathbf{y}_k = [y_{k,1}\ y_{k,2}\ y_{k,3}]$ denote the transmitted antipodal and observed code symbols, respectively, associated with the transition from state $\psi_k = s'$ to state $\psi_{k+1} = s$. The antipodal mapping $\{0 \to +1,\ 1 \to -1\}$ implies that $x_{k,l} = 1 - 2b_l(k)$, $l = 1, 2, 3$, where $b_l(k)$ is a binary code-symbol 1 or 0. The transmitted information bit is $b_1(k) = b(s', s)$. Equation (1.53) implies that

$$f\left(\mathbf{y}_k|\psi_{k+1} = s, \psi_k = s'\right) = \frac{1}{(\pi N_0)^{3/2}} \exp\left[-\frac{1}{N_0}\sum_{l=1}^{3}(y_{k,l} - \sqrt{\mathcal{E}_s}x_{k,l})^2\right]. \tag{1.167}$$

This equation is used to evaluate the branch metric of (1.161), which is required by the BCJR algorithm. □

The generic name for a version of the BCJR algorithm or an approximation of it is *soft-in soft-out (SISO) algorithm*. The *log-MAP algorithm* is an SISO algorithm that transforms the BCJR algorithm into the logarithmic domain, thereby simplifying operations and reducing numerical problems while causing no performance degradation. The log-MAP algorithm requires both a forward and a backward recursion through the code trellis. Since the log-MAP algorithm also requires additional memory and calculations, it is roughly four times as complex as the standard Viterbi algorithm.

The log-MAP algorithms expedites computations by using the *max-star function*, which is defined as

$$\max_i{}^*\{x_i\} = \ln\left(\sum_i e^{x_i}\right). \tag{1.168}$$

By separately considering $x > y$ and $x < y$, we verify that for two variables

$$\max{}^*(x, y) = \max(x, y) + \ln\left(1 + e^{-|x-y|}\right) \tag{1.169}$$

whereas for more than two variables, the calculation can be performed recursively. For example, for three variables,

$$\max{}^*(x, y, z) = \max{}^*[x, \max{}^*(y, z)] \tag{1.170}$$

which is verified by applying (1.169) to the right-hand side of (1.170). The numerical values of the second term on the right-hand side of (1.169) can be stored in a table accessed by the log-MAP algorithm.

The log-MAP algorithm computes

$$\overline{\alpha}_k\left(s'\right) = \ln\left(\alpha_k\left(s'\right)\right),\ \overline{\beta}_{k+1}(s) = \ln\left(\beta_{k+1}(s)\right),\ \overline{\gamma}_k\left(s', s\right) = \ln\left(\gamma_k\left(s', s\right)\right). \tag{1.171}$$

From (1.163) and (1.165), it follows that the forward and backward recursions are

$$\overline{\alpha}_{k+1}(s) = \max_{s'}{}^{*}\{\overline{\alpha}_k(s') + \overline{\gamma}_k(s', s)\} \tag{1.172}$$

$$\overline{\beta}_k(s') = \max_{s}{}^{*}\{\overline{\beta}_{k+1}(s) + \overline{\gamma}_k(s', s)\} \tag{1.173}$$

respectively. From (1.159), we obtain the LLR:

$$\begin{aligned}\Lambda_{k,l} = & \max_{s',s:b_k=1}{}^{*}\{\overline{\alpha}_k(s') + \overline{\gamma}_k(s', s) + \overline{\beta}_{k+1}(s)\} \\ & - \max_{s',s:b_k=0}{}^{*}\{\overline{\alpha}_k(s') + \overline{\gamma}_k(s', s) + \overline{\beta}_{k+1}(s)\}.\end{aligned} \tag{1.174}$$

If the encoder begins and terminates in the zero state, then the recursions are initialized by $\overline{\alpha}_0(s) = \overline{\beta}_L(s) = 0$ if $s = 0$; $\overline{\alpha}_0(s) = \overline{\beta}_L(s) = -\infty$, otherwise.

The *max-log-MAP algorithm* and the *soft-output Viterbi algorithm (SOVA)* are SISO algorithms that reduce the complexity of the log-MAP algorithm at the cost of some performance degradation. The max-log-MAP algorithm uses the approximation $\max{}^*(x.y) \simeq \max(x, y)$ to reduce its complexity to roughly two thirds that of the log-MAP algorithm. The SOVA algorithm is similar to the Viterbi algorithm except that it uses a posteriori probabilities in computing branch metrics, and it produces a soft output, indicating the reliability of its decisions. The SOVA algorithm is roughly one third as complex as the log-MAP algorithm, but is less accurate. The BCJR, log-MAP, max-log-MAP, and SOVA algorithms have complexities that increase linearly with the number of states of the component codes.

Turbo Codes with Parallel Concatenated Codes

As shown in Figure 1.18, the encoder of a turbo code with two parallel concatenated codes uses two parallel component encoders. Component encoder 1 receives the information bits and generates information bits and associated parity bits, while

Fig. 1.18 Encoder of turbo code

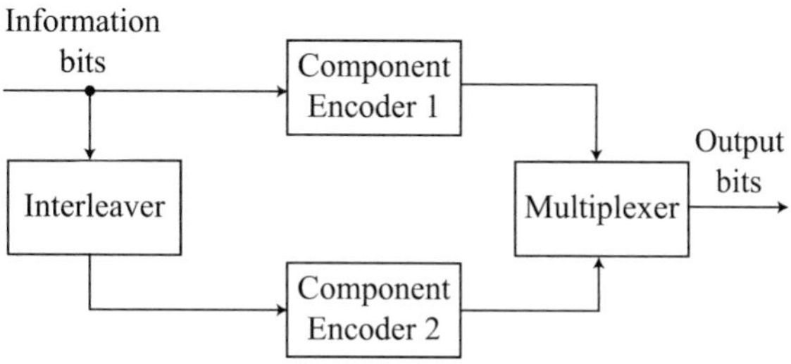

component encoder 2 receives interleaved information bits and generates parity bits associated with the interleaved information bits. Within this architecture, the component codes are usually identical and might be convolutional codes, block codes, or trellis-coded modulation. Multiple concatenated codes could be used, but the use of two codes is predominant and assumed subsequently.

The multiplexer output of the turbo encoder in Figure 1.18 comprises both the information bits received from component encoder 1 and the parity bits produced by both encoders. If the multiplexer punctures the parity streams, higher-rate codes can be generated. Although it requires frame synchronization in the decoder, the puncturing may serve as a convenient means of adapting the code rate to the channel conditions. The purpose of the interleaver, which may be a block or pseudorandom interleaver, is to permute the input bits of encoder 2 so that it is unlikely that both component codewords have a low weight even if the input word has a low weight. As a result, the turbo codewords tend to have large weights, which increases the Hamming distances between codewords.

Convolutional Turbo Codes

A *convolutional turbo code* uses convolutional codes as its component codes. *Recursive systematic convolutional encoders* are used in the component encoders because they map most low-weight information sequences into higher-weight coded sequences. Although the number of low-weight codewords is greatly reduced, some remain, and there is little or no change in the minimum free distance. Thus, a turbo code has very few low-weight codewords, regardless of whether its minimum free distance is large. When large interleavers are used, the weight distribution of the codewords of a turbo code resembles that of a random code.

Terminating tail bits are inserted into both component convolutional codes so that the turbo trellis terminates in the zero state, and the turbo code can be treated as a block code. Recursive encoders require nonzero tail bits that are functions of the preceding nonsystematic output bits.

A rate-1/2 turbo code can be produced from rate-1/2 systematic convolutional component codes by alternate puncturing prior to transmission of the even parity bits generated by encoder 1 and the odd parity bits generated by encoder 2. An even information bit has an associated parity bit of component code 1 transmitted. However, because of the interleaving that precedes encoder 2, an odd information bit may have neither its associated parity bit of component code 1 nor that of component code 2 transmitted. Some odd information bits may have both associated parity bits transmitted, although not successively because of the interleaving. Since some information bits have no associated parity bits transmitted, the decoder is less likely to be able to correct errors in those information bits. A convenient means of avoiding this problem, and ensuring that exactly one associated parity bit is transmitted for each information bit, is to use a block interleaver with an odd number of rows and an odd number of columns. If bits are written into the interleaver array in successive rows, but successive columns are read, then odd and even information

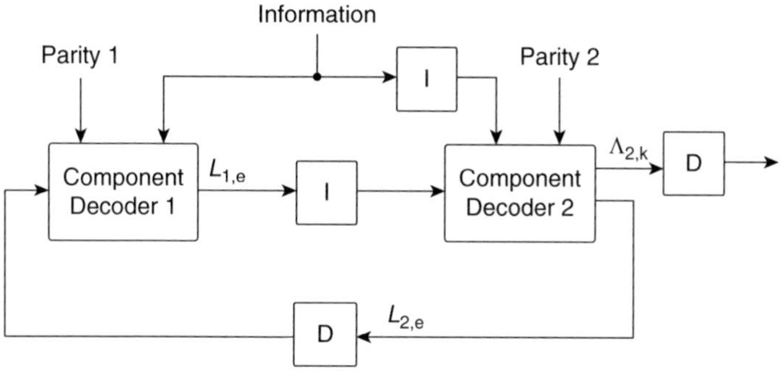

Fig. 1.19 Decoder of turbo code. *I* interleaver; *D* deinterleaver

bits alternate at the input of encoder 2, thereby ensuring that all information bits have an associated parity bit that is transmitted once. Simulation results confirm that this procedure, which is called *odd–even separation*, improves the system performance when puncturing and block interleavers are used.

The interleaver size is equal to the block length or frame length of the codes. The number of low-weight or minimum-distance codewords tends to be inversely proportional to the interleaver size. With a large block length and interleaver and a sufficient number of decoder iterations, the performance of the convolutional turbo code can approach within less than 1 dB of the information-theoretic limit. However, as the block length increases, so does the *system latency*, which is the delay between the input and final output.

The basic structure of a turbo decoder is illustrated in Figure 1.19. Component decoder 1 is fed by demodulator outputs due to the information bits and the parity bits generated by encoder 1. Component decoder 2 is fed by interleaved demodulator outputs due to the information bits and the parity bits generated by encoder 2. Let $\mathbf{Y} = [\mathbf{Y}_1, \mathbf{Y}_2, \ldots, \mathbf{Y}_N]$ denote the sequence of demodulator output vectors associated with the successive information bits $b_1, b_2, \ldots, b_N$. Let $\mathbf{Z}_1 = [\mathbf{Z}_{11}, \mathbf{Z}_{12}, \ldots, \mathbf{Z}_{1N}]$ and $\mathbf{Z}_2 = [\mathbf{Z}_{21}, \mathbf{Z}_{22}, \ldots, \mathbf{Z}_{2N}]$ denote the sequences of demodulator output vectors associated with the parity bits of encoders 1 and 2, respectively. If the encoder uses puncturing, then some vectors corresponding to parity bits are absent. For each information bit b_k, the BCJR algorithms of decoders 1 and 2 compute estimates of the *log-likelihood ratios* (LLRs) of the a posteriori probabilities that this bit is 1 or 0 given by

$$\Lambda_{1,k} = \ln\left[\frac{P(b_k = 1|\mathbf{y}_k, \mathbf{z}_1)}{P(b_k = 0|\mathbf{y}_k, \mathbf{z}_1)}\right], \quad \Lambda_{2,k} = \ln\left[\frac{P(b_k = 1|\mathbf{y}_k, \mathbf{z}_2)}{P(b_k = 0|\mathbf{y}_k, \mathbf{z}_2)}\right] \tag{1.175}$$

respectively, where $P(b_k = b|\mathbf{y}_k, \mathbf{z}_\delta)$ is a compact notation for $P(b_k = 1|\mathbf{Y}_k = \mathbf{y}_k, \mathbf{Z}_\delta = \mathbf{z}_\delta)$. The LLRs of the information bits are iteratively updated in the two component decoders by passing information between them and using the BCJR algorithm in each of them.

The key to the iterative processing of the turbo decoder is the decomposition of $\Lambda_{\delta,k}$, $\delta = 1, 2$. From the definition of a conditional probability, (1.175) may be expressed in terms of density functions as

$$\Lambda_{\delta,k} = \ln\left[\frac{f(b_k = 1, \mathbf{y}_k, \mathbf{z}_\delta)}{f(b_k = 0, \mathbf{y}_k, \mathbf{z}_\delta)}\right], \quad \delta = 1, 2. \tag{1.176}$$

Given $b_k = b$, $\mathbf{Z}_\delta$ is independent of $\mathbf{Y}_k$. Therefore,

$$f(b_k = b, \mathbf{y}_k, \mathbf{z}_\delta) = f(\mathbf{z}_\delta|b_k = b)f(\mathbf{y}_k|b_k = b)P(b_k = b), \quad b = 1, 2. \tag{1.177}$$

Substituting this equation into (1.176) and decomposing the results, we obtain

$$\Lambda_{\delta,k} = L_a(b_k) + L_c(\mathbf{y}_k) + L_{\delta,e}(b_k), \quad \delta = 1, 2 \tag{1.178}$$

where the a priori *LLR* is

$$L_a(b_k) = \ln\left[\frac{P(b_k = 1)}{P(b_k = 0)}\right] \tag{1.179}$$

the *extrinsic information is*

$$L_{\delta,e}(b_k) = \ln\left[\frac{f(\mathbf{z}_\delta|b_k = 1)}{f(\mathbf{z}_\delta|b_k = 0)}\right], \quad \delta = 1, 2 \tag{1.180}$$

and the *channel LLR*, which represents information about b_k provided by the demodulator, is

$$L_c(\mathbf{y}_k) = \ln\left[\frac{f(\mathbf{y}_k|b_k = 1)}{f(\mathbf{y}_k|b_k = 0)}\right] \tag{1.181}$$

where $f(\mathbf{y}_k|b_k = b)$, $b = 0, 1$, is the conditional density function of $\mathbf{Y}_k$ given that $b_k = b$. Since $P(b_k = 1) + P(b_k = 0) = 1$, (1.179) implies that

$$P(b_k = b) = \frac{\exp[bL_a(b_k)]}{1 + \exp[L_a(b_k)]}, \quad b = 0, 1. \tag{1.182}$$

Applying the theorem of total probability, the channel LLR may be expressed as

$$L_c(\mathbf{y}_k) = \ln\left[\frac{\sum\limits_{\mathbf{s}\in D_k^1} f(\mathbf{y}_k|\mathbf{s})P(\mathbf{S}_k = \mathbf{s})}{\sum\limits_{\mathbf{s}\in D_k^0} f(\mathbf{y}_k|\mathbf{s})P(\mathbf{S}_k = \mathbf{s})}\right] \tag{1.183}$$

where $P(\mathbf{S}_k = \mathbf{s})$ is the probability that the kth transmitted symbol was $\mathbf{s}$, $f(\mathbf{y}_k|\,\mathbf{s})$ is the conditional density function of $\mathbf{y}_k$ given that $\mathbf{S}_k = \mathbf{s}$, and D_k^b denotes the set of all symbols such that $b_k = b$. If a priori information is known about the symbol probabilities, then it is used to determine $P(\mathbf{S}_k = \mathbf{s})$. If no a priori information is available, then $P(\mathbf{S}_k = \mathbf{s})$ is assumed to be uniformly distributed over the symbol constellation.

In each decoder δ, $\Lambda_{\delta,k}$ is computed by the BCJR algorithm. Equation (1.178) indicates that the extrinsic information, which is passed between the two decoders, can be computed as

$$L_{\delta,e}(b_k) = \Lambda_{\delta,k} - L_a(b_k) - L_c(\mathbf{y}_k), \quad \delta = 1, 2. \tag{1.184}$$

Except for the first iteration of component decoder 1, the a priori LLR $L_a(b_k)$ for one decoder is set equal to the extrinsic information calculated by the other decoder at the end of its previous iteration. The BCJR algorithm of each decoder uses (1.182) to calculate the branch metrics and bit LLRs.

The extrinsic information $L_{\delta,e}(b_k)$ is calculated by using (1.184) after $\Lambda_{\delta,k}$ has been computed by the BCJR algorithm. Since the extrinsic information depends primarily on different parity symbols and not on the a priori information provided by the other decoder, the extrinsic information provides additional information to the decoder to which it is transferred. As indicated in Figure 1.19, appropriate interleaving or deinterleaving is required to ensure that the extrinsic information $L_{1,e}(b_k)$ or $L_{2,e}(b_k)$ is applied to each component decoder in the correct sequence.

Let $B\{\mathbf{u}, \mathbf{v}, E\}$ denote the function calculated by the BCJR algorithm during a single iteration of a component decoder, where $\mathbf{u}$ is an input vector, $\mathbf{v}$ is an input sequence of vectors, and E is the a priori or extrinsic information applied to the component decoder. Let $I[\cdot]$ denote the interleave operation, $D[\cdot]$ the deinterleave operation, and a numerical superscript (n) the nth iteration. For each information bit and each iteration $n \geq 1$, the turbo decoder calculates the following functions:

$$\Lambda_{1,k}^{(n)} = B\{\mathbf{y}_k, \mathbf{z}_1, D[L_{2,e}^{(n-1)}(b_k)]\} \tag{1.185}$$

$$L_{1,e}^{(n)}(b_k) = \Lambda_{1,k}^{(n)} - L_c(\mathbf{y}_k) - D[L_{2,e}^{(n-1)}(b_k)] \tag{1.186}$$

$$\Lambda_{2,k}^{(n)} = B\{I[\mathbf{y}_k], \mathbf{z}_2, I[L_{1,e}^{(n)}(b_k)]\} \tag{1.187}$$

$$L_{2,e}^{(n)}(b_k) = \Lambda_{2,k}^{(n)} - L_c(\mathbf{y}_k) - I[L_{1,e}^{(n)}(b_k)] \tag{1.188}$$

where $D[L_{2,e}^{(0)}] = L_c(\mathbf{y}_k)$ is the a priori bit LLR used during the first iteration of component decoder 1.

When the iterative process terminates after N iterations, the LLR $\Lambda_{2,k}^{(N)}$ from component decoder 2 is deinterleaved and then applied to a device that makes a hard decision. Let u_k denote an information bit after the antipodal mapping

$\{0 \rightarrow +1,\ 1 \rightarrow -1\}$ so that $u_k = 1 - 2b_k$. Then the decision for bit k is

$$\widehat{u_k} = sgn\{D[\Lambda_{2,k}^{(N)}(b_k)]\},\ \ \widehat{b_k} = (1 - \widehat{u_k})/2 \tag{1.189}$$

where the *signum function* is defined as $sgn(x) = 1,\ x \geq 0$, and $sgn(x) = -1,\ x < 0$. Performance improves with the number of iterations N, but simulation results indicate that typically little is gained beyond 4–12 iterations.

The calculation of $f(\mathbf{y}_k|\mathbf{s})$ in (1.183) by the demodulator depends on the modulation. For the AWGN channel and coherent q-ary PAM, we apply (1.46) to the reception of a single symbol k with $\phi_k = \theta_k$. Discarding irrelevant factors that are common to the numerator and denominator in (1.183), we obtain the *normalized density function for coherent q-ary PAM:*

$$f\left(\mathbf{y}_k\left|\mathbf{s}\right.\right) = f\left(y_k\left|x_s\right.\right) = \exp\left(\frac{2\alpha\sqrt{\mathcal{E}_s}\mathrm{Re}\left(x_s^* y_k\right) - \alpha^2\mathcal{E}_s\left|x_s\right|^2}{N_0}\right) \tag{1.190}$$

where y_k is the complex number representing received symbol $\mathbf{y}_k$, x_s is the complex number representing the point in the signal constellation associated with the transmitted symbol $\mathbf{s}$, and amplitude α is inserted to allow for fading. Similarly, the *normalized density function for coherent q-ary PSK is*

$$f\left(\mathbf{y}_k\left|\mathbf{s}\right.\right) = \exp\left(\frac{2\alpha\sqrt{\mathcal{E}_s}\mathrm{Re}\left(x_s^* y_k\right)}{N_0}\right). \tag{1.191}$$

For the AWGN channel and *orthogonal* modulation, application of (1.76) to the reception of a single symbol yields the conditional density of $\mathbf{y}_k$, the q-dimensional received vector with $y_{l,k}$ as component l :

$$f\left(\mathbf{y}_k\left|\mathbf{s},\psi\right.\right) = \left(\frac{1}{\pi N_0}\right)^q \exp\left(-\frac{\alpha^2\mathcal{E}_s - 2\alpha\sqrt{\mathcal{E}_s}\,\mathrm{Re}\left(y_{\nu,k}^* e^{j\psi}\right)}{N_0} - \sum_{l=1}^{q}\frac{\left|y_{l,k}\right|^2}{N_0}\right) \tag{1.192}$$

where ν is the index of the matched filter that is matched to symbol $\mathbf{s}$, ψ is the phase difference, and N_0 is the noise-power spectral density associated with the received symbol. The amplitude α is inserted to allow for fading. For *coherent* orthogonal modulation, an ideal phase synchronization system provides $\psi = 0$. Discarding irrelevant factors that are common to the numerator and denominator in (1.183), we obtain the *normalized density function for coherent q-ary orthogonal modulation*:

$$f\left(\mathbf{y}_k\left|\mathbf{s}\right.\right) = \exp\left(\frac{2\alpha\sqrt{\mathcal{E}_s}\mathrm{Re}\left(y_{\nu,k}\right)}{N_0}\right). \tag{1.193}$$

For *noncoherent* q-ary orthogonal modulation, we assume that ψ is uniformly distributed over $[0, 2\pi)$. Applying (1.79) to the reception of a single symbol and discarding irrelevant factors that are common to the numerator and denominator in (1.215), we obtain the *normalized density function for noncoherent q-ary orthogonal modulation*:

$$f\left(\mathbf{y}_k\left|\mathbf{s}\right.\right) = I_0\left(\frac{2\alpha\sqrt{\mathcal{E}_s}\left|y_{\nu,k}\right|}{N_0}\right). \tag{1.194}$$

For coherent q-ary PAM, (1.183) with a uniformly distributed a priori symbol probability, (1.190), and (1.168) yield

$$L_c(\mathbf{y}_k) = \max_{\mathbf{s}\in D_k^1}{}^{*}\left\{\frac{2\alpha\sqrt{\mathcal{E}_s}\mathrm{Re}\left(x_s^* y_k\right) - \alpha^2\mathcal{E}_s\left|x_s\right|^2}{N_0}\right\} - \max_{\mathbf{s}\in D_k^0}{}^{*}\left\{\frac{2\alpha\sqrt{\mathcal{E}_s}\mathrm{Re}\left(x_s^* y_k\right) - \alpha^2\mathcal{E}_s\left|x_s\right|^2}{N_0}\right\} \tag{1.195}$$

which is computationally efficient. Similarly, for coherent q-ary orthogonal modulation with a uniformly distributed a priori symbol probability,

$$L_c(\mathbf{y}_k) = \max_{\mathbf{s}\in D_k^1}{}^{*}\left\{\frac{2\alpha\sqrt{\mathcal{E}_s}\mathrm{Re}\left(y_{\nu,k}\right)}{N_0}\right\} - \max_{\mathbf{s}\in D_k^0}{}^{*}\left\{\frac{2\alpha\sqrt{\mathcal{E}_s}\mathrm{Re}\left(y_{\nu,k}\right)}{N_0}\right\}. \tag{1.196}$$

For coherent BPSK over the AWGN channel, let $x_s = 2b_k - 1$ be the mapping of information bit b_k into its constellation symbol. Since the exponential function is a monotonically increasing function of its argument,

$$L_c(y_k) = 4\alpha\frac{\sqrt{\mathcal{E}_s}}{N_0}y_k \quad \text{(BPSK)}. \tag{1.197}$$

When the modulation is BFSK, $b_k = 0$ is transmitted as signal 0, $b_k = 1$ is transmitted as signal 1, and $\mathbf{y}_k = [y_{0,k}y_{1,k}]^T$ is the two-dimensional vector comprising the sampled outputs of filters matched to the two signals due to the transmission of information bit b_k. For coherent demodulation, we obtain

$$L_c(\mathbf{y}_k) = \frac{2\alpha\sqrt{\mathcal{E}_s}\,\mathrm{Re}\left(y_{1,k} - y_{0,k}\right)}{N_0} \quad \text{(coherent BFSK)}. \tag{1.198}$$

Similarly, for noncoherent BFSK demodulation with a uniformly distributed a priori symbol probability,

$$L_c(\mathbf{y}_k) = \ln\left[\frac{I_0\left(\frac{2\alpha\sqrt{\mathcal{E}_s}\left|y_{1,k}\right|}{N_0}\right)}{I_0\left(\frac{2\alpha\sqrt{\mathcal{E}_s}\left|y_{0,k}\right|}{N_0}\right)}\right] \quad \text{(noncoherent BFSK)}. \tag{1.199}$$

The preceding equations indicate that both q-ary PAM and orthogonal modulation require channel-state information about $\alpha\sqrt{\mathcal{E}_s}/N_0$.

For identical component decoders and typically eight algorithm iterations, the overall complexity of a turbo decoder using the log-MAP algorithm is roughly 64 times that of a Viterbi decoder for one of the component codes. The complexity of the decoder increases, while the performance improves as the constraint length K of each component code increases. The complexity of a turbo decoder using eight iterations and component convolutional codes with $K = 3$ is approximately the same as that of a Viterbi decoder for a convolutional code with $K = 9$. The max-log-MAP algorithm is roughly two-thirds as complex as the log-MAP algorithm and typically degrades the bit error probability of the turbo code by 0.1–0.2 dB at $P_b = 10^{-4}$. The SOVA algorithm is roughly one-third as complex as the log-MAP algorithm and typically degrades the bit error probability of the turbo code by 0.5–1.0 dB at $P_b = 10^{-4}$.

Parallel Block Turbo Codes

A *parallel block turbo code* uses two linear block codes as its component codes. Linear block codes can be represented by a trellis diagram [57], which makes them amenable to soft-decision decoding by the BCJR algorithm or an SISO algorithm. To limit the decoding complexity, high-rate binary BCH codes are generally used as the component codes, and the turbo code is called a *turbo BCH code*. The encoder of a block turbo code has the form of Figure 1.18. Puncturing is not used because it causes a significant performance degradation. Suppose that the component block codes are binary systematic (n_1, k_1) and (n_2, k_2) codes, respectively. Encoder 1 converts each set of k_1 information bits into n_1 codeword bits that are passed to the multiplexer. Simultaneously, $k_2 k_1$ information bits are written successively row-by-row into the interleaver as k_1 columns and k_2 rows. Encoder 2 converts each column of k_2 interleaver bits into a codeword of n_2 bits, but passes only the $n_2 - k_2$ parity bits of each codeword to the multiplexer. Consequently, information bits are transmitted only once, and the code rate of the block turbo code is

$$r = \frac{k_1 k_2}{k_2 n_1 + (n_2 - k_2) k_1}. \tag{1.200}$$

If the two block codes are identical, then $r = k/(2n - k)$.

Consider component codes with minimum Hamming distances d_{m1} and d_{m2} respectively. We can define a valid nonzero block-turbo codeword with all zeros in the first k_2 rows and k_1 columns of information bits except for one bit that defines a nonzero row and a nonzero column. The weight of the nonzero row is at least d_{m1}, and the weight of the nonzero column is at least d_{m2}. Since the row and column have one bit in common, the weight of this defined block-turbo codeword is at least $d_{m1} + d_{m2} - 1$, and a codeword with this weight can be generated. Therefore, the minimum distance of the block turbo code is

$$d_m = d_{m1} + d_{m2} - 1. \tag{1.201}$$

The decoder of a block turbo code has the form of Figure 1.19, and only slight modifications of the BCJR or SISO decoding algorithms are required. Long, high-rate turbo BCH codes approach the Shannon limit in performance, but their complexities are greater than those of convolutional turbo codes of comparable performance.

Turbo Trellis-Coded Modulation

Turbo trellis-coded modulation (TTCM), which produces a nonbinary bandwidth-efficient modulation, is obtained by using identical trellis codes as the component codes in a turbo code. The encoder has the form illustrated in Figure 1.20. The code rate and hence the required bandwidth of the trellis code are preserved by the TTCM encoder because it alternately selects constellation points or complex symbols generated by the two parallel component encoders. To ensure that all information bits, which constitute the encoder inputs, are transmitted only once and that the parity bits are provided alternately by the two component encoders, the symbol interleaver transfers symbols in odd positions to odd positions and symbols in even positions to even positions, where each symbol is a group of bits. After the complex symbols are produced by signal mapper 2, the symbol deinterleaver restores the original ordering. The selector passes the odd-numbered complex symbols from mapper 1 and the even-numbered complex symbols from mapper 2. To ensure that the code rate of the trellis code is maintained, the parity symbols of the two encoders are alternately punctured. The channel interleaver permutes the selected complex symbols prior to the modulation. The TTCM decoder uses a symbol-based SISO algorithm analogous to that used by convolutional-turbo decoders. TTCM can provide a performance close to the theoretical limit for the AWGN channel, but its implementation complexity is much greater than that of conventional trellis-coded modulation.

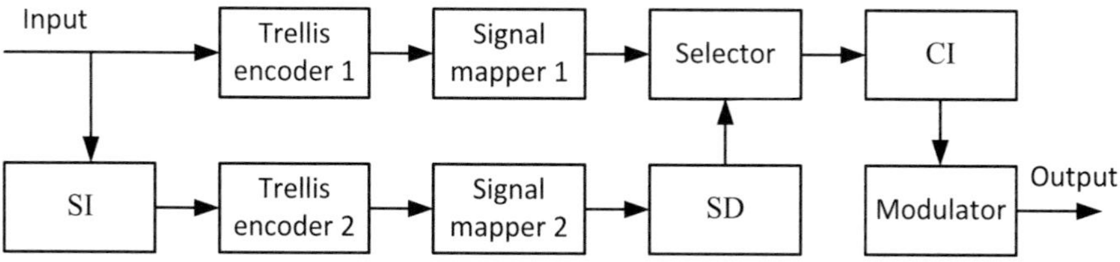

Fig. 1.20 Encoder for turbo trellis-coded modulation. *SI* symbol interleaver. *SD* symbol deinterleaver. *CI* channel interleaver

Serially Concatenated Turbo Codes

Serially concatenated turbo codes differ from classical ones in their use of large interleavers and method of iterative decoding. Both the inner and outer codes must be amenable to efficient decoding by the BCJR algorithm or an SISO algorithm. Thus, the codes are either binary systematic block codes or binary systematic convolutional codes. The encoder for a serially concatenated turbo code has the form of Figure 1.15 (a). The outer encoder generates a codeword of n_1 bits for every k_1 information bits and fills the block interleaver row-by-row with n_1 outer codewords. Since the interleaver columns are read by the inner encoder to provide the channel symbols, there is interleaving of the outer-code symbols prior to the inner encoding. After the interleaving, each set of n_1 bits is converted by the inner encoder into n_2 bits. Thus, the overall code rate of the serially concatenated code is $k_1n_1/n_2n_1 = k_1/n_2$. If the component codes are block codes, then an outer (n_1, k_1) code and an inner (n_1, n_2) code are used.

A functional block diagram of an iterative decoder for a serially concatenated code is illustrated in Figure 1.21. For each inner codeword, the inner-decoder input comprises the demodulator outputs corresponding to the n_2 bits. For each iteration, the inner decoder computes the LLRs for the n_1 systematic bits. After a deinterleaving, these LLRs provide extrinsic information about the n_1 code bits of the outer code. The outer decoder then computes the LLRs for all its code bits. After an interleaving, these LLRs provide extrinsic information about the n_1 systematic bits of the inner code. The final output of the iterative decoder comprises the k_1n_1 information bits of a concatenated codeword of n_1n_2 bits.

In the absence of interleaving, severely corrupted inner codewords cannot provide significant information for decoding of the outer codewords during the iterative decoding. Thus, the interleaver is an essential part of the encoder of a serially concatenated turbo code. The deinterleaving greatly diminishes the possibility that corrupted inner codewords can undermine the iterative process in the turbo decoder.

Turbo Product Codes

A *turbo product code* is a product code (Section 1.5) that uses iterative decoding of the component codewords. The primary SISO algorithm uses a version of the Chase

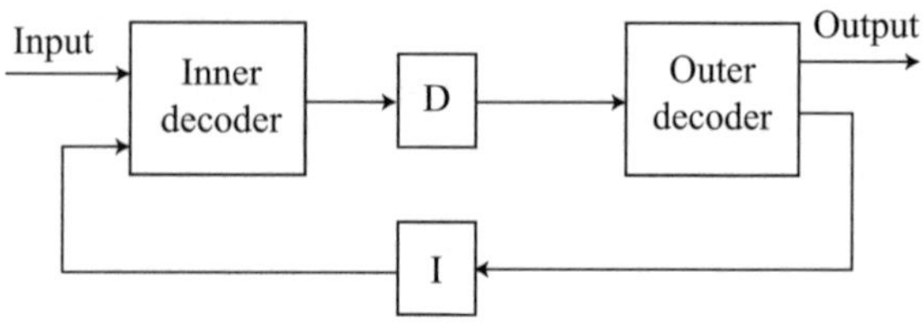

Fig. 1.21 Iterative decoder for serially concatenated code. *D* deinterleaver; *I* interleaver

algorithm (Section 1.2). The resulting limited search for the maximum-likelihood codewords of the component codes rather than an exhaustive search enables a practical implementation with only a small loss in potential performance.

The encoder generates an $n_2 \times n_1$ array of code bits, which is illustrated in Figure 1.17. Each of the k_2 rows in the array is a codeword of an (n_1, k_1) code, and each of the n_1 columns is the codeword of an (n_2, k_2) code. The array defines a codeword of a product code with $n = n_1 n_2$ code bits representing $k_1 k_2$ information bits. Each transmitted codeword is received as an $n_2 \times n_1$ array $\mathbf{R}$ of sampled matched-filter outputs. The array $\mathbf{R}$ is partitioned into row and column vectors that are separately decoded by the SISO algorithm.

During the first half-iteration, both hard decisions and bit metrics for the row vectors are provided by the demodulator. In the second half-iteration and subsequent iterations, extrinsic information is used in processing either the column vectors or the row vectors. The array $\mathbf{R}$ provides a *bit-metric vector* $\mathbf{y}$ of matched-filter output samples y_i, $1 \leq i \leq n_1 n_2$. During each half-iteration, the bit-metric vector is updated by using extrinsic information. Then hard decisions and the reliability metric defined by (1.59) are used to detect the r least reliable bits in the array, and 2^r different test patterns are generated by allowing the least reliable bits to have either of their two possible values. For each test pattern, hard-decision decoding of the component codewords is performed, and 2^r candidate turbo codewords for the $n_2 \times n_1$ array are produced. The candidate codeword that is closest in Euclidean distance to the bit-metric vector is selected and stored in the array as the vector $\mathbf{c}_1$. Similarly, the candidate codeword that is second closest in Euclidean distance to the bit-metric vector is selected and stored in the array as the vector $\mathbf{c}_2$.

For the nth half-iteration, where $n \geq 2$, the turbo decoder updates its bit-metric vector with [59, 75]

$$\overline{\mathbf{y}}(n) = \mathbf{y} + \alpha(n)\,\mathbf{w}(n), \quad n \geq 2 \tag{1.202}$$

where $\alpha(n)$ is a scaling factor that can be determined experimentally, and $\mathbf{w}(n)$ is extrinsic information computed as

$$\mathbf{w}(n+1) = \widetilde{\mathbf{y}}(n) - \overline{\mathbf{y}}(n), \quad n \geq 1 \tag{1.203}$$

where

$$\widetilde{\mathbf{y}}(n) = \left[\frac{\|\overline{\mathbf{y}}(n) - \mathbf{c}_2\|^2 - \|\overline{\mathbf{y}}(n) - \mathbf{c}_1\|^2}{4}\right]\mathbf{c}_1. \tag{1.204}$$

The iterative decoding process terminates when $\overline{\mathbf{y}}(n)$ remains unchanged for two half-iterations.

Turbo product codes are in competition with other turbo codes with respect to performance and computational complexity. Advantages of turbo product codes are the capability of high coding gains at high code rates without resorting to puncturing

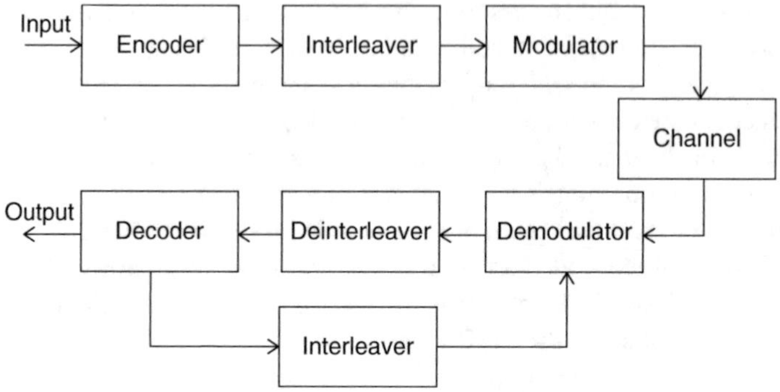

Fig. 1.22 Iterative demodulation and decoding with transmitter and receiver

and the inherent error-detection capability without the need for additional parity bits. Many modified turbo product codes have been proposed or used and provide various potential advantages and disadvantages.

1.7 Iterative Demodulation and Decoding

The concept of performing iterative computations using two decoders can be extended to a demodulator and decoder by designing the demodulator to exploit a priori information provided by the decoder, which itself receives a priori information from the demodulator and may be internally iterative [111]. The major components of a communication system with iterative decoding and demodulation are shown in Figure 1.22. The code can be a block, convolutional, turbo, or low-density parity-check (LDPC) code. In the transmitter, message bits are encoded, bit-interleaved or symbol-interleaved, and then applied to the modulator. Prior to the modulation, the modulator applies a *constellation labeling map* or *labeling map, which* is the mapping of a bit pattern to each symbol or point in a signal-set constellation. Each set of $m = \log_2 q$ consecutive bits in the input $\mathbf{b} = \{b_1, \ldots, b_m\} \in [0, 1]^m$ is mapped into a q-ary symbol $\mathbf{s} = \mu(\mathbf{b})$, where $\mu(\mathbf{b})$ is the labeling map, and the set of constellation symbols has cardinality q.

In the receiver, the demodulator partitions the received signal into a sequence of received symbols. A demapper within the demodulator processes each received symbol to produce a vector of bit LLRs. This vector provides extrinsic information that is deinterleaved and passed to the decoder, which generates a vector of bit LLRs that are interleaved and passed to the demapper. The demapper and decoder exchange extrinsic information until bit decisions are made by the decoder after a specified number of iterations.

When the demodulator receives a q-ary symbol $\mathbf{S}_k$, it computes an LLR for each of the $m = \log_2 q$ code bits constituting the symbol. Let $b_l(k)$ denote the

lth bit associated with q-ary symbol $\mathbf{S}_k$, and the LLR for bit l produced by the demodulator is

$$\Lambda_{k,l} = \ln\left[\frac{P\left[b_l\left(k\right) = 1|\mathbf{y}_k\right]}{P\left[b_l\left(k\right) = 0|\mathbf{y}_k\right]}\right] \tag{1.205}$$

where $\mathbf{y}_k$ is the observation corresponding to $\mathbf{S}_k$. Let $\mathbf{s}$ denote a particular realization of symbol $\mathbf{S}_k$ representing m bits including bit b_k. From the theorem of total probability and Bayes' rule,

$$\begin{aligned} P\left[b_l\left(k\right) = b|\mathbf{y}_k\right] &= \sum_{\mathbf{s}\in D} P\left[b_l\left(k\right) = b, \mathbf{S}_k = \mathbf{s}|\mathbf{y}_k\right] \\ &= \sum_{\mathbf{s}\in D} f(\mathbf{y}_k \mid b_l\left(k\right) = b, \mathbf{S}_k = \mathbf{s})P\left[b_l\left(k\right) = b, \mathbf{S}_k = \mathbf{s}\right]/f\left(\mathbf{y}_k\right) \end{aligned} \tag{1.206}$$

where $b = 1$ or 0, $P\left(\cdot\right)$ is a generic probability function, $f\left(\cdot\right)$ is a generic density function or joint probability and density function, and the summation is over the set D of q possible symbols. Let D_l^b denote the set of all symbols such that $b_l = b$. Then

$$P\left[b_l\left(k\right) = b, \mathbf{S}_k = \mathbf{s}\right] = \begin{cases} 0, & \mathbf{s} \notin D_l^b \\ P\left(\mathbf{S}_k = \mathbf{s}\right), & \mathbf{s} \in D_l^b \end{cases} \tag{1.207}$$

and if $\mathbf{s} \in D_l^b$, then

$$f(\mathbf{y}_k \mid b_l\left(k\right) = b, \mathbf{S}_k = \mathbf{s}) = f(\mathbf{y}_k|\mathbf{S}_\mathbf{k} = \mathbf{s}) = f(\mathbf{y}_k|\ \mathbf{s}). \tag{1.208}$$

Using these results in (1.206) before substituting it into (1.205) yields the bit LLR

$$\Lambda_{k,l} = \ln\left[\frac{\sum\limits_{\mathbf{s}\in D_l^1} f(\mathbf{y}_k|\mathbf{s})P\left(\mathbf{S}_k = \mathbf{s}\right)}{\sum\limits_{\mathbf{s}\in D_l^0} f(\mathbf{y}_k|\mathbf{s})P\left(\mathbf{S}_k = \mathbf{s}\right)}\right]. \tag{1.209}$$

During the first iteration of the iterative demodulation and decoding, the known a priori symbol probability $P\left(\mathbf{S}_k = \mathbf{s}\right)$ is used, or each symbol is assumed to be uniformly distributed over the symbol constellation. After the demodulator output is passed to the decoder, the decoder feeds back a posteriori probabilities that become the a priori probabilities of the demodulator. During the second and subsequent iterations, the assumption of statistically independent code bits implies that

$$P\left(\mathbf{S}_k = \mathbf{s}\right) = \prod_{l=1}^{m} P[b_l\left(k\right) = b_l\left(\mathbf{s}\right)] \tag{1.210}$$

where $b_l(\mathbf{s})$ is the value of bit l of symbol $\mathbf{s}$.

Since $b_l(\mathbf{s}) = b$ when $\mathbf{s} \in D_l^b$, a factor $P[b_l(k) = b]$ appears in (1.210). Therefore, when (1.210) is substituted into (1.209), the latter can be decomposed as

$$\Lambda_{k,l} = v_{k,l} + z_{k,l} \tag{1.211}$$

where the *LLR* for bit l of symbol k is

$$v_{k,l} = \ln\left[\frac{P[b_l(k) = 1]}{P[b_l(k) = 0]}\right] \tag{1.212}$$

the *extrinsic LLR* for bit l of symbol k is

$$z_{k,l} = \ln\left[\frac{\sum_{\mathbf{s}\in D_l^1} f(\mathbf{y}_k|\mathbf{s}) \prod_{i=1,i\neq l}^{m} P[b_i(k) = b_i(\mathbf{s})]}{\sum_{\mathbf{s}\in D_l^0} f(\mathbf{y}_k|\mathbf{s}) \prod_{i=1,i\neq l}^{m} P[b_i(k) = b_i(\mathbf{s})]}\right] \tag{1.213}$$

and the products are omitted in (1.213) if the symbols are binary.

The *LLR* $v_{k,l}$ for each bit l of each symbol k is calculated by the decoder and fed back to the demodulator. Since $P[b_l(k) = 1] + P[b_l(k) = 0] = 1$, it can be verified by substituting $b_l(\mathbf{s}) = 1$ and then $b_l(\mathbf{s}) = 0$ and using (1.212) that

$$P[b_l(k) = b_l(\mathbf{s})] = \frac{\exp[b_l(\mathbf{s})\, v_{k,l}]}{1 + \exp(v_{k,l})}. \tag{1.214}$$

The substitution of (1.214) into (1.213) and a cancelation yields

$$z_{k,l} = \ln\left[\frac{\sum_{\mathbf{s}\in D_l^1} f(\mathbf{y}_k|\mathbf{s}) \prod_{i=1,i\neq l}^{m} \exp[b_i(\mathbf{s})\, v_{k,i}]}{\sum_{\mathbf{s}\in D_l^0} f(\mathbf{y}_k|\mathbf{s}) \prod_{i=1,i\neq l}^{m} \exp[b_i(\mathbf{s})\, v_{k,i}]}\right]. \tag{1.215}$$

For the iterative demodulation and decoding, the demodulator transfers the $\{z_{k,l}\}$ to the decoder during the second and subsequent iterations. The decoder then uses the $\{z_{k,l}\}$ as bit LLRs in the computation of branch metrics in the BCJR algorithm and to compute the $\{v_{k,i}\}$, which are fed back to the demodulator. Each receiver iteration includes a demodulator iteration followed by one or more decoder iterations. After the final receiver iteration, the final decoded bits are the hard decisions based on the final demodulator or decoder LLRs.

The calculation of $f(\mathbf{y}_k|\mathbf{s})$ in (1.215) by the demodulator depends on the modulation. For the AWGN channel and coherent q-ary PAM, coherent orthogonal modulation, and noncoherent orthogonal modulation, we apply (1.190), (1.193), and (1.194), respectively. For coherent PAM, (1.215), (1.190), and (1.168) yield

$$\begin{aligned} z_{k,l} = &\max_{\mathbf{s}\in D_1^1}{}^* \left\{ \frac{2\alpha\sqrt{\mathcal{E}_s}\mathrm{Re}\left(x_s^* y_l\right) - \alpha^2 \mathcal{E}_s \left|x_s\right|^2}{N_0} + \sum_{i=1, i\neq l}^{m} b_i\left(\mathbf{s}\right) v_{k,i} \right\} \\ &- \max_{\mathbf{s}\in D_l^0}{}^* \left\{ \frac{2\alpha\sqrt{\mathcal{E}_s}\mathrm{Re}\left(x_s^* y_l\right) - \alpha^2 \mathcal{E}_s \left|x_s\right|^2}{N_0} + \sum_{i=1, i\neq l}^{m} b_i\left(\mathbf{s}\right) v_{k,i} \right\} \end{aligned} \quad (1.216)$$

which is computationally efficient. Similarly, for coherent orthogonal modulation,

$$\begin{aligned} z_{k,l} = &\max_{\mathbf{s}\in D_l^1}{}^* \left\{ \frac{2\alpha\sqrt{\mathcal{E}_s}\mathrm{Re}\left(y_{\nu,l}\right)}{N_0} + \sum_{i=1, i\neq l}^{m} b_i\left(\mathbf{s}\right) v_{k,i} \right\} \\ &- \max_{\mathbf{s}\in D_l^0}{}^* \left\{ \frac{2\alpha\sqrt{\mathcal{E}_s}\mathrm{Re}\left(y_{\nu,l}\right)}{N_0} + \sum_{i=1, i\neq l}^{m} b_i\left(\mathbf{s}\right) v_{k,i} \right\}. \end{aligned} \quad (1.217)$$

Bit-Interleaved Coded Modulation

In a communication system with *symbol-interleaved coded modulation,* an encoder generates code bits that are grouped into nonbinary code symbols that are interleaved over a depth exceeding the channel coherence time to provide time diversity (Section 6.2). The q-ary code symbols are then applied to a q-ary modulator that provides bandwidth efficiency. A more pragmatic approach is to interleave the code bits over a depth exceeding the channel coherence time prior to forming the q-ary code symbols and performing q-ary modulation. The resulting coded modulation, which is known as *bit-interleaved coded modulation* (BICM) [13], increases time diversity, thereby providing improved performance over a fading channel. Although true maximum-likelihood decoding of BICM requires joint demodulation and decoding, the BICM demodulator separately generates bit metrics that are applied to the decoder. BICM has become a standard method for transmitting over fading channels, forming the basis of most cellular, satellite, and wireless networking systems. However, the reduced minimum Euclidean distance due to the bit interleaving degrades the performance of BICM over the AWGN channel.

In the transmitter of a system with BICM, encoded bits are interleaved and then placed into a $1 \times N_d$ vector $\mathbf{d}$ with elements $d_i \in \{1, 2, \ldots, q\}$, each of which represents $m = \log_2 q$ bits. The vector $\mathbf{d}$ generates the modulated signal, which passes through an AWGN or fading channel. The received signal passes through a bank of q matched filters. The output of each matched filter is sampled at the symbol rate to produce a sequence of complex numbers. Assuming that symbol synchronization exists, the complex samples are used to compute an $m \times N_d$ matrix $\mathbf{Z}$ of demodulator bit LLRs. After the deinterleaving of $\mathbf{Z}$, the ordered bit LLRs provide a priori information to the decoder, which generates the final bit decisions. The demodulator bit LLRs are given by (1.209), where $f(\mathbf{y}_i|\mathbf{s})$ depends on the modulation. For the AWGN channel and coherent q-ary PAM, coherent orthogonal modulation, and noncoherent orthogonal modulation, we apply (1.190), (1.193), and (1.194), respectively.

Bit-interleaved coded modulation with iterative decoding and demodulation (BICM-ID) uses BICM, bit LLRs generated by both the demodulator and the decoder, and iterative demodulation and decoding [111]. For binary modulations, BICM-ID and BICM are identical. However, when used with a convolutional, turbo, or LDPC code and a nonbinary modulation with a two-dimensional constellation, such as QAM or q-ary PSK, BICM-ID can offer superior bit-error-rate performance relative to BICM with any symbol labeling. The use of BICM-ID maintains the advantage of BICM over fading channels while minimizing any performance degradation experienced by BICM over the AWGN channel. The disadvantage of BICM-ID relative to BICM is the much greater computational complexity of BICM-ID.

As described above, the BICM-ID demodulator computes an $m \times N_d$ matrix $\mathbf{Z}$ of bit LLRs that are deinterleaved and then provided as a priori information to the decoder. The bit metrics computed by the decoder are interleaved and then fed back to the demodulator as a priori information in the form of an $m \times N_d$ matrix $\mathbf{V}$ of decoder LLRs, which provides the $\{v_{k,i}\}$ of (1.215) in the received sequence order. The iterative process then proceeds as previously described.

The *constellation labeling* has a major impact on both BICM and BICM-ID. *Gray labeling* (Section 1.2) is not always possible. With binary or nonbinary orthogonal modulation, Gray labeling does not exist as all neighbors are equidistant. When Gray labeling exists, it provides the optimal performance for BICM, but usually not for BICM-ID [112].

Compared with trellis-coded modulation, BICM-ID has a small free Euclidean distance, but the decoder feedback exploits the time diversity and greatly mitigates this disadvantage. As a result, BICM-ID with BPSK modulation and convolutional codes outperform systems with trellis-coded modulation or turbo trellis-coded modulation of similar computational complexity over both the AWGN and Rayleigh fading channels [48].

Simulation Examples

The performance examples in this section, which are plots of the bit error probability as a function of $\mathcal{E}_b/N_0$, are generated by Monte Carlo simulations. We assume here and elsewhere in this book that the front-end and lowpass filters of both the transmitter and receiver are perfect.

CDMA2000 is a family of communication standards for the wireless transmission of data. Figures 1.23, 1.24, 1.25, 1.26 and 1.27 illustrate the performance of a CDMA2000 system that uses a rate-1/2 turbo code with parallel rate-1/2 concatenated codes. The turbo encoder has the form shown in Figure 1.18. Each component code is a recursive systematic convolutional code that has the configuration shown in Figure 1.7. The puncturing matrix for each pair of input information bits is

$$\mathbf{P} = \begin{bmatrix} 1 & 1 \\ 1 & 0 \\ 0 & 0 \\ 0 & 1 \end{bmatrix} \tag{1.218}$$

where each column defines which output bits of the two component encoders are transmitted following each input bit. Each set of $\log_2 q$ bits is encoded as a q-ary channel symbol that is transmitted by using orthogonal q-ary FSK (q-FSK). Channel-state information is assumed to be available in the receiver, and noncoherent demodulation is used. The codeword size is $k = 1530$ information bits, and the AWGN or the Rayleigh fading (see Chapter 6) channel is assumed. Ideal bit interleaving is assumed for binary symbols when BICM is used. When BICM is not used, ideal symbol-interleaved coded modulation (SICM) is assumed. Ideal interleaving is implemented in the simulation for the Rayleigh channel by having each modulated symbol multiplied by an independent fading coefficient.

Figure 1.23 illustrates the improvement in the bit error probability of a system that uses 4-FSK and BICM-ID over the Rayleigh channel as the number of decoding iterations increases. The improvement exhibits rapidly diminishing returns beyond four iterations and is insignificant beyond ten iterations. The diminishing returns indicate that the adaptive curtailment of iterations based on some measure of the subsequent potential gain is desirable to reduce system latency, which is one of the principal limitations of turbo codes. Several potentially effective methods for stopping iterations are available [15].

Figure 1.24 compares the bit error probabilities of noncoherent systems using SICM, BICM, and BICM-ID with 4-FSK over the Rayleigh channel, while Figure 1.25 makes the same comparisons for communications over the AWGN channel. In both figures, it is observed that BICM provides a small improvement relative to SICM, but BICM-ID provides a more substantial improvement. Figure 1.26 illustrates the degradation that occurs when the turbo decoder of the system with BICM-ID uses the max-log-MAP algorithm instead of the log-MAP algorithm for communications over the Rayleigh channel. The degradation for the AWGN channel is similar.

As the alphabet size increases, the performance improves at the cost of larger signal bandwidths. For example, Figure 1.27 compares the bit error probabilities over the Rayleigh channel of a BICM-ID system with 4-FSK and several systems

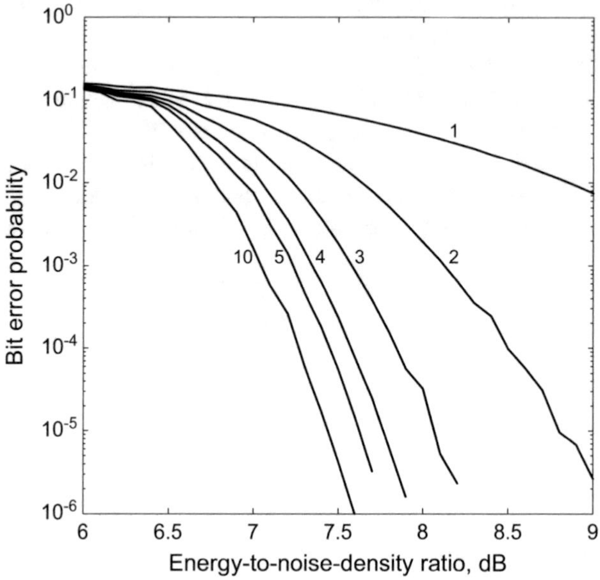

Fig. 1.23 Performance of turbo code with 4-FSK and BICM-ID over the Rayleigh channel as the number of decoding iterations varies

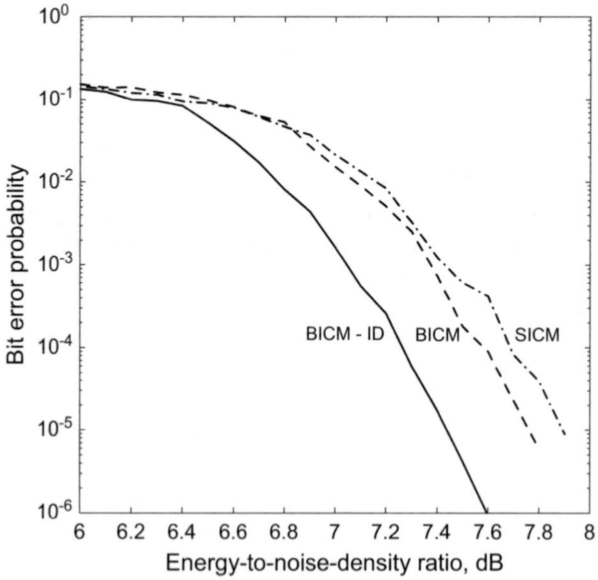

Fig. 1.24 Performance of turbo code with 4-FSK over the Rayleigh channel for SICM, BICM, and BICM-ID

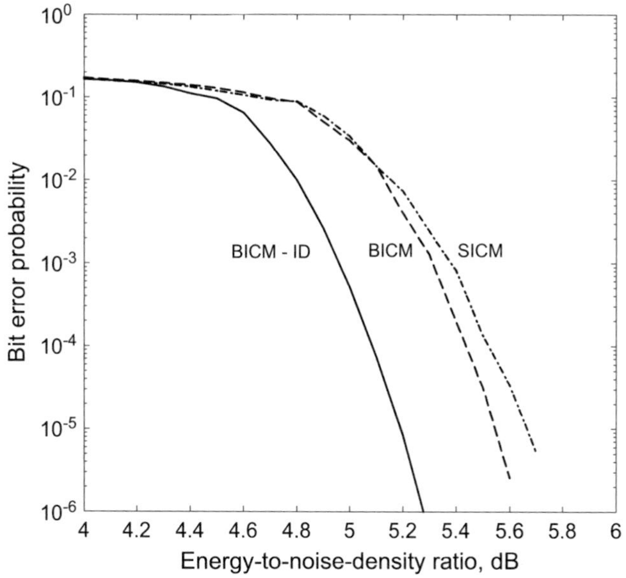

Fig. 1.25 Performance of turbo code with 4-FSK over the AWGN channel for SICM, BICM, and BICM-ID

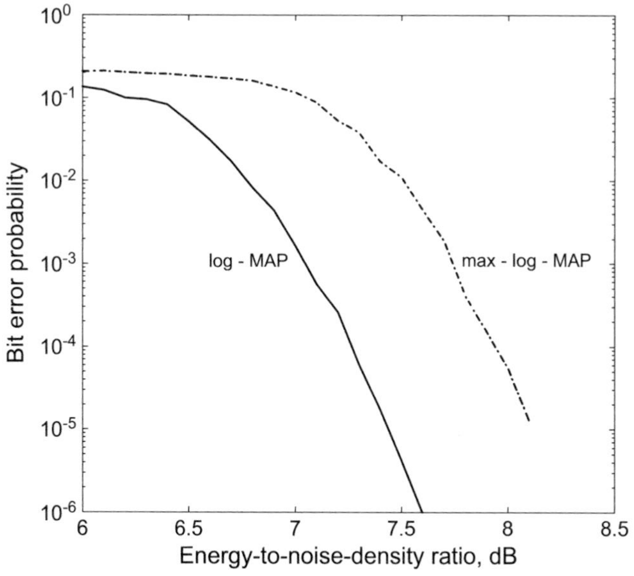

Fig. 1.26 Performance of turbo code with 4-FSK and BICM-ID over the Rayleigh channel for log-MAP and max-log-MAP algorithms

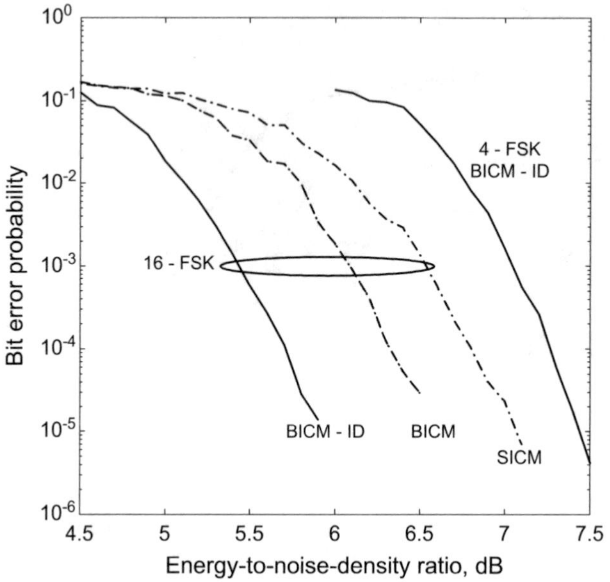

Fig. 1.27 Performance of turbo code with 16-FSK over the Rayleigh channel for SICM, BICM, and BICM-ID and comparison with 4-FSK and BICM-ID

with 16-FSK. It is observed that the 16-FSK system with BICM-ID provides an improvement of roughly 1.5 dB at a bit error probability of 10^{-5} relative to the 4-FSK system with BICM-ID, but the bandwidth requirement is increased by a factor of 4. A code-rate reduction also provides improved performance, but an increased signal bandwidth.

Plots of the bit error probability for systems with iterative decoding generally exhibit a *waterfall* region, which is characterized by a rapid decrease in the bit error probability as $\mathcal{E}_b/N_0$ increases, and an *error-floor* region, in which the bit error probability decreases much more slowly. A hypothetical plot illustrating these regions is shown in Figure 1.28. A low error floor may be important for radio-relay communication, space–ground communication, compressed-data transfer, optical transmission, or when an automatic-repeat request is not feasible because of the variable delays. Figures 1.23, 1.24, 1.25, 1.26 and 1.27 do not exhibit error floors because bit error probabilities lower than 10^{-6} are not displayed. A Gray labeling map tends to provide an early onset of the waterfall region, but the error floor, which for the AWGN channel is determined by the minimum Euclidean distance of the symbol set, is lower for other labeling maps.

The potentially high system latency, the system complexity, and sometimes the error floor are the primary disadvantages of turbo codes.

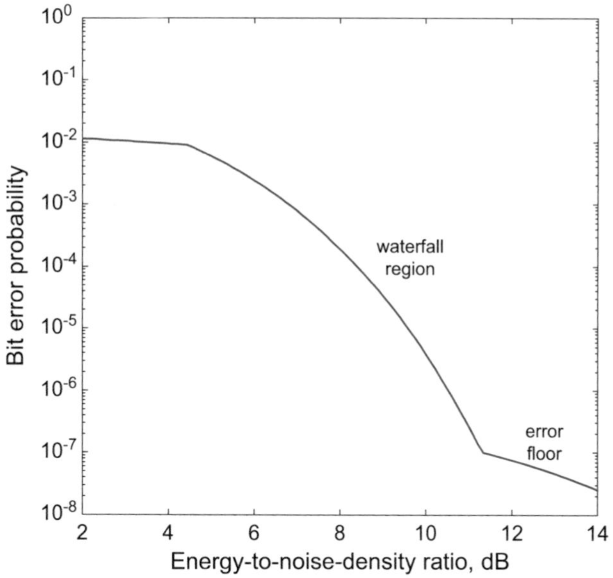

Fig. 1.28 Illustration of regions of bit error probability plot

1.8 Low-Density Parity-Check Codes

Low-density parity-check (LDPC) *codes* [10, 80] are linear block codes specified by an $(n-k) \times n$ parity-check matrix $\mathbf{H}$ that is sparsely populated with nonzero elements. Nonbinary LDPC codes provide excellent performance when the codewords are short and allow a direct combination with high-order modulations. However, binary codes are henceforth assumed because of their predominance in applications. A *regular* LDPC code has the same number of ones in each column and the same number of ones in each row of $\mathbf{H}$; otherwise, the LDPC code is *irregular*. Irregular LDPC codes are competitive with turbo codes in terms of the performance obtained for a given level of implementation or computational complexity.

Since (1.1) and (1.2) imply that

$$\mathbf{Hc}^T = \mathbf{0} \tag{1.219}$$

each of the $n-k$ rows of $\mathbf{H}$ specifies a parity-check equation that must be satisfied. A *Tanner graph* is a bipartite graph that represents the parity-check matrix and equations as two sets of nodes. One set of n nodes, called the *variable nodes*, represents the codeword symbols. Another set of $n-k$ nodes, called the *check nodes*, represents the parity-check equations. An edge connects variable-node i to check-node l if component $H_{li} = 1$. The $(7, 4)$ Hamming code has the parity-check matrix

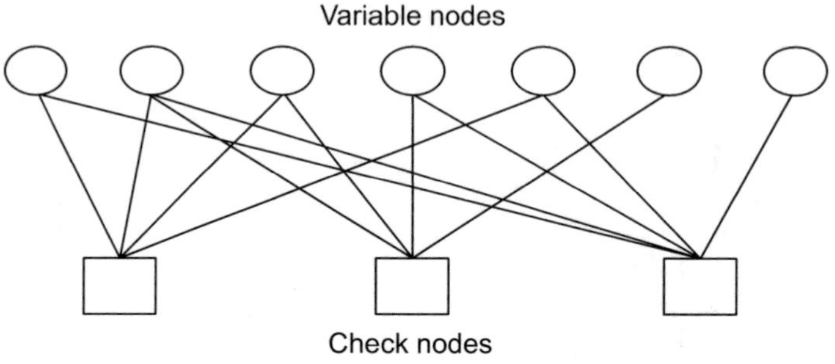

Fig. 1.29 Tanner graph for (7, 4) Hamming code

$$\mathbf{H} = \begin{bmatrix} 1\,1\,1\,0\,1\,0\,0 \\ 0\,1\,1\,1\,0\,1\,0 \\ 1\,1\,0\,1\,1\,0\,1 \end{bmatrix} \tag{1.220}$$

and the associated Tanner graph is shown in Figure 1.29.

If the ones in a large **H** are approximately randomly distributed, then any bit that is corrupted or subjected to a deep fade is likely to be applied to a check node that also receives more reliable information from other bits. The sparseness of ones in **H** reduces the probability that a set of corrupted bits are applied to the same check nodes.

A *cycle* of a graph is a sequence of distinct edges that start and terminate at the same node. The length of the shortest cycle in a graph is called its *girth*. The Tanner graph of the $(7, 4)$ Hamming code has a girth equal to 4, which is the minimum possible length of a cycle. The most effective LDPC codes have girths exceeding 4 because a low girth corresponds to a limited amount of independent information exchange among some variable and check nodes.

A well-designed LDPC code does not require an interleaver following the encoder because interleaving is equivalent to the permutation of the columns of the parity-check matrix **H**. Since a deinterleaver is not required prior to its decoder, an LDPC code usually has less latency than a turbo code of similar complexity.

The soft-decision decoding algorithm for LDPC codes is called the *sum-product*, *message-passing*, or *belief-propagation* algorithm. The first name refers to the main computations required. The second name refers to the fact that if each node is regarded as a processor, then the algorithm can be interpreted as the iterative passing of information messages between a set of variable nodes and a set of check nodes in a Tanner graph. The third name emphasizes that the messages are measures of the credibility of the most recent computations. The sparseness of the parity-check matrix facilitates LDPC decoding.

The basis of the sum-product algorithm is a *parity-check LLR* λ_s that is a measure of the likelihood that a parity-check equation is satisfied and can be expressed as a function of the LLRs of the bits. Let $b_1, b_2, \ldots, b_n$ denote n codeword bits with

LLRs defined as

$$\lambda_i = \ln\left[\frac{P(b_i = 1)}{P(b_i = 0)}\right], \quad i = 1, 2, \ldots, n. \tag{1.221}$$

The parity-check equation for the set of these bits is the modulo-2 sum $s = \sum_{i=1}^{n} b_i$, which is equal to zero or one. It may be verified by considering an even number of ones among the n bits and then an odd number of ones that

$$s = \frac{1}{2}\left[1 - \prod_{i=1}^{n}(1 - 2b_i)\right]. \tag{1.222}$$

The parity-check LLR associated with the sum s is defined as

$$\lambda_s = \ln\left[\frac{P(s = 1)}{P(s = 0)}\right] \tag{1.223}$$

which indicates that $\lambda_s \to -\infty$ if $P(s = 0) \to 1$, and $\lambda_s \to \infty$ if $P(s = 0) \to 0$.

As $s = 1$ or $s = 0$, $E[s] = P(s = 1)$. Equation (1.221) implies that $E[b_i] = P(b_i = 1) = \exp(\lambda_i) / [1 + \exp(\lambda_i)]$. Therefore, (1.222) and the independence of the bits implies that

$$\begin{aligned} P(s = 1) = E[s] &= \frac{1}{2}\left[1 - \prod_{i=1}^{n}(1 - 2E[b_i])\right] \\ &= \frac{1}{2}\left[1 - \prod_{i=1}^{n}\left(1 - \frac{2e^{\lambda_i}}{1 + e^{\lambda_i}}\right)\right]. \end{aligned} \tag{1.224}$$

Using algebra and the definition $tanh(x) = (e^x - e^{-x}) / (e^x + e^{-x})$, we obtain

$$P(s = 1) = \frac{1}{2}\left[1 - \prod_{i=1}^{n}\tanh\left(-\frac{\lambda_i}{2}\right)\right]. \tag{1.225}$$

Combining (1.223) and the fact that $P(s = 1) + P(s = 0) = 1$, we obtain $P(s = 1) = \exp(\lambda_s) / [1 + \exp(\lambda_s)]$. From this equation, (1.225), the definition of $tanh(x)$, and an inversion, we obtain

$$\lambda_s = -2\tanh^{-1}\left[\prod_{i=1}^{n}\tanh\left(-\frac{\lambda_i}{2}\right)\right] \tag{1.226}$$

which relates the parity-check LLR λ_s to the LLRs of the individual bits.

Prior to the first iteration of the sum-product algorithm, variable-node i uses the matched-filter output vector $\mathbf{y}_i$ associated with code-bit b_i to compute an a posteriori

LLR. Using Bayes' rule and assuming equal a priori bit probabilities, we obtain the *channel LLR*:

$$\lambda_i^{(0)} = \ln\left[\frac{P(b_i = 1|\mathbf{y}_i)}{P(b_i = 0|\mathbf{y}_i)}\right] = \ln\left[\frac{P(\mathbf{y}_i|b_i = 1)}{P(\mathbf{y}_i|b_i = 0)}\right]$$
$$= L_c(\mathbf{y}_i) \tag{1.227}$$

where $L_c(\mathbf{y}_i)$ is given by (1.197), (1.198), or (1.199) for BPSK, coherent BFSK, and noncoherent BFSK respectively. Thus, the calculation of $\lambda_i^{(0)}$ requires *channel-state information* that enables the estimation of $\alpha_i\sqrt{\mathcal{E}_s}/N_0$. During the first iteration, the channel LLR is the message sent to each check node connected to variable-node i in the Tanner graph. Each check node receives LLRs from *adjacent nodes*, which are those variable nodes corresponding to bits that contribute to the check node's parity-check equation.

During iteration ν, check-node l combines its input LLRs to produce output LLRs that constitute the messages sent to each of its adjacent variable nodes in subsequent iterations. After receiving message $\nu - 1, \nu = 1, 2 \ldots, \nu_{\max}$, check-node l updates the LLR that is subsequently sent to variable-node i by slightly modifying the parity-check LLR of (1.226):

$$\mu_{l,i}^{(\nu)} = -2\tanh^{-1}\left[\prod_{m=N_l/i}\tanh\left(-\frac{\lambda_m^{(\nu-1)} - \mu_{l,m}^{(\nu-1)}}{2}\right)\right], \quad \nu \geq 1 \tag{1.228}$$

where N_l/i is the set of variable nodes adjacent to check-node l, but excluding variable-node i, and $\mu_{l,i}^{(0)} = 0$ for all i and l. The exclusion is to prevent redundant information originating in node i from recycling back to it, thereby causing instability. The value passed to variable-node m in the preceding iteration is subtracted from $\lambda_m^{(\nu-1)}$ to reduce the correlation with previous iterations. If $\mu_{l,i}^{(\nu)} \to -\infty$, then variable-node i learns that it is highly likely that the parity-check equation of check-node l is satisfied.

After receiving message $\nu \geq 1$, variable-node i updates its LLR as

$$\lambda_i^{(\nu)} = \lambda_i^{(0)} + \sum_{l\in M_i}\mu_{l,i}^{(\nu)}, \quad \nu \geq 1 \tag{1.229}$$

where M_i is the set of check nodes adjacent to variable-node i. The algorithm can terminate when all $n - k$ of the parity-check equations of some codewords are satisfied or after a specified number of iterations. If the algorithm terminates after ν_0 iterations, then the LDPC decoder sets $b_i = 1$ if $\lambda_i^{(\nu_0)} > 0$, and $b_i = 0$ otherwise.

Low-density parity-check codes are often characterized by the degree distributions of the nodes in their Tanner graphs. The *degree* of a node is defined as the number of edges emanating from it. The *degree distribution of the variable nodes* is defined as the polynomial

$$v(x) = \sum_{i=2}^{d_v} n_i x^{i-1} \tag{1.230}$$

where n_i denotes the fraction of variable nodes with degree i, and d_v the maximum degree or number of edges connected to a variable node. The *degree distribution of the check nodes* is defined as the polynomial

$$\chi(x) = \sum_{i=2}^{d_c} \chi_i x^{i-1} \tag{1.231}$$

where χ_i denotes the fraction of check nodes with degree i, and d_c the maximum degree or number of edges connected to a check node. Theoretically, optimal degree distributions for infinitely long codes can be used as a starting point in the design of finite LDPC codes.

Structured LDPC Codes

The sparse parity-check matrix of LDPC codes enables decoding with a complexity that increases linearly with the codeword or block length. However, the corresponding generator matrix of unstructured or pseudorandom LDPC codes is generally not sparse. Since the encoding requires the matrix multiplication indicated in (1.1), the encoding complexity increases quadratically with the codeword length. To reduce this complexity and the encoding latency, *structured LDPC codes* are often used, although the additional structure may make it difficult to match the outstanding error-correction capabilities of unstructured LDPC codes. Two of the most practical classes of structured LDPC codes, which offer rapid encoding, efficient decoding, and excellent performance, are protograph codes and irregular repeat-accumulate codes.

The *protograph LDPC codes* are a class of LDPC codes that are constructed by first repeatedly duplicating a protograph consisting of the Tanner graph of an LDPC code with a small number of nodes. The protograph may have parallel edges of different types. The *derived graph* representing the protograph code is constructed by combining the duplicates and then permuting edges that belong to the same type. Many protograph codes increase the code rate by having *punctured variable nodes*, which are potential variable nodes that represent untransmitted code symbols [24].

The *repeat-accumulate codes* are a class of structured LDPC codes that have an encoding complexity that increases linearly with the block length. This class may be considered a subclass of the protograph LDPC codes. A repeat-accumulate code is a serially concatenated code that may be decoded as either an LDPC code or a turbo code. An outer repetition encoder repeats each of k information bits n times to form a block of kn bits that are passed through an interleaver. Each interleaver output bit is successively applied to an inner encoder that is a rate-1 recursive convolutional code functioning as an accumulator. The inner-encoder output is the modulo-2 sum of its input and its previous output. A limitation of the repeat-accumulate code, whether it is systematic or not, is that its code rate cannot exceed 1/2.

The *irregular repeat-accumulate* (IRA) *codes* are generalizations of repeat-accumulate codes that retain the linear complexity of encoding and decoding, but are not limited in their code rates, are more flexible in their design options, and can provide a much better performance. The IRA code repeats each information bit a variable number of times. The repeated bits are interleaved, and then a number of them are combined and applied as successive inputs to an accumulator. A systematic (n, k) IRA encoder generates the codeword $\mathbf{c} = [\mathbf{m}\ \mathbf{p}]$, where $\mathbf{m}$ is the row vector of k information bits, and $\mathbf{p}$ is the row vector of $n - k$ accumulator outputs.

The $(n - k) \times n$ parity-check matrix for a systematic IRA code has the form

$$\mathbf{H} = [\mathbf{H}_1\ \ \mathbf{H}_2] \tag{1.232}$$

where $\mathbf{H}_1$ is an $(n - k) \times k$ sparse matrix, and $\mathbf{H}_2$ is the $(n - k) \times (n - k)$ sparse matrix with the dual-diagonal form

$$\mathbf{H}_2 = \begin{bmatrix} 1\,0\,0 & & \\ 1\,1\,0 & & \\ 0\,1\,1 & & \\ & \cdot & \\ & & \cdot \\ & & 1\,1\,0 \\ & & 0\,1\,1 \end{bmatrix}. \tag{1.233}$$

The generator matrix corresponding to $\mathbf{H}$ is

$$\mathbf{G} = \left[\mathbf{I}\ \ \mathbf{H}_1^T\mathbf{H}_2^{-T}\right] \tag{1.234}$$

which satisfies (1.2). An array inversion and transpose of (1.233) yield an upper triangular matrix:

$$\mathbf{H}_2^{-T} = \begin{bmatrix} 1\,1\,1 & & \\ 0\,1\,1 & & \\ 0\,0\,1 & & \\ & \cdot & \\ & & \cdot \\ & & 1\,1 \\ & & 0\,1 \end{bmatrix}. \tag{1.235}$$

The encoder produces the codeword

$$\mathbf{c} = \left[\mathbf{m}\ \ \mathbf{m}\mathbf{H}_1^T\mathbf{H}_2^{-T}\right]. \tag{1.236}$$

The form of $\mathbf{H}_2^{-T}$ indicates that its premultiplication by the $1 \times (n-k)$ row vector $\mathbf{m}\mathbf{H}_1^T$ may be implemented as an accumulator that generates successive elements of the $1 \times (n-k)$ row vector $\mathbf{m}\mathbf{H}_1^T\mathbf{H}_2^{-T}$. Since $\mathbf{H}_1^T$ is sparse, far fewer encoder computations are required than for unstructured LDPC codes.

LDPC Performance

In selecting an LDPC code, there is usually a tradeoff between a favorable waterfall region and a low error floor. The bit error probability of an LDPC code in the waterfall region is mostly dependent on the girth because short cycles prevent nodes from extracting a substantial amount of independent parity-check information. The bit error probability in the error-floor region is mostly dependent on the minimum Euclidean distance of the symbol set, which may be decreased if the girth is too large. Well-designed irregular LDPC codes can provide lower bit error probabilities than regular codes of the same size in the waterfall region. The price is increased complexity of implementation and a raised error floor primarily because of the large girths. For example, an increase in some column weights tends to decrease the girth, but may increase the minimum Euclidean distance. LDPC codes often provide lower error floors than turbo codes of similar complexity.

WiMAX is a telecommunications standard designed to provide broadband access. Figure 1.30 illustrates the performance over the Rayleigh channel of a WiMAX system with coherent 16-QAM, rate-1/2 pseudorandom LDPC codes, and perfect channel-state information. The codeword lengths are 2304, 4608, and 9216 bits. The steepness of the waterfall region increases rapidly with increases in the codeword length.

1.9 Problems

1 (a) Use (1.17) to show that $N(d_m, d_m - t) = \binom{d_m}{t}$. Can the same result be derived directly? (b) Use (1.18) to derive $N(l, i)$ for Hamming codes. Consider the cases $l = i,\ i+1,\ i-1$, and $i-2$ separately.

2 (a) Use (1.19) with $d_m = 2t + 1$ to derive an upper bound on A_{d_m}. (b) Explain why this upper bound becomes an equality for perfect codes. (c) Show that $A_3 = n(n-1)/6$ for Hamming codes. (d) Show that for perfect codes as $P_s \to 0$, both the exact equation (1.20) and the approximation (1.22) give the same expression for P_{is}.

3 Use erasures to show that a Reed-Solomon codeword with k information symbols can be recovered from any k correct symbols.

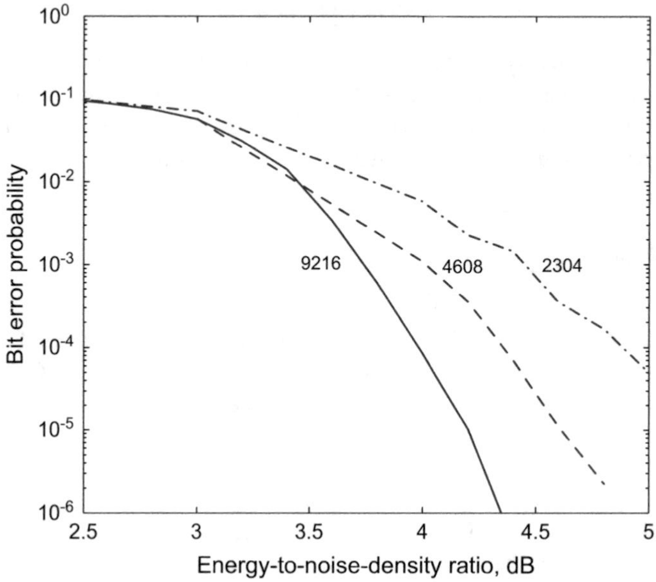

Fig. 1.30 Performance of LDPC code with 16-QAM over the Rayleigh channel for codewords of length 2304, 4608, and 9216

4 Suppose that a binary (7,4) Hamming code is used for coherent BPSK communications over the AWGN channel with no fading. The received output samples are -0.4, 1.0, 1.0, 1.0, 1.0, 1.0, and 0.4. Use the table of (7,4) Hamming codewords to find the decision made when the codeword metric (1.55) is used.

5 Prove that the word error probability for BPSK and a block code with soft-decision decoding satisfies

$$P_w \leq (q^k - 1)Q\left(\sqrt{\frac{2d_m r\mathcal{E}_b}{N_0}}\right).$$

6 Use (1.57) and (1.31) to show that the coding gain or power advantage of the BPSK system with a binary block code and maximum-likelihood decoding is roughly $d_m r$ relative to no code when $\mathcal{E}_b/N_0 \rightarrow \infty$.

7 (a) Show that $P[X \geq b] \geq 1 - \min_{0\leq s}[M(-s)e^{sb}]$, where $M(s)$ is the moment-generating function of X. (b) Derive the Chernoff bound on $P[X \geq b]$ for a standard Gaussian random variable with zero mean and unit variance.

8 The Chernoff bound can be applied to hard-decision decoding, which can be regarded as a special case of soft-decision decoding with the following symbol metric. If symbol i of a candidate binary sequence ν agrees with the corresponding symbol detected at the demodulator output, then $m(\nu, i) = 1$; otherwise $m(\nu, i) = 0$. Prove that

$$P_2(l) \leq [4P_s(1 - P_s)]^{l/2}.$$

This upper bound is not always tight but has great generality since no specific assumptions have been made about the modulation or coding.

9 Consider a system that uses coherent BPSK and a convolutional code over the AWGN channel. (a) What is the coding gain of a binary system with soft decisions, $K = 7$, and $r = 1/2$ relative to an uncoded system for large $\mathcal{E}_b/N_0$? (b) Use the approximation

$$Q(x) \approx \frac{1}{\sqrt{2\pi}x} \exp(-\frac{x^2}{2}), \quad x > 0$$

to show that as $\mathcal{E}_b/N_0 \to \infty$, soft-decision decoding of a binary convolutional code has slightly less than a 3 dB coding gain relative to hard-decision decoding.

10 A concatenated code comprises an inner binary $(2^m, m)$ *Hadamard* block code and an outer (n, k) Reed-Solomon code. The outer encoder maps every set of m bits into one Reed-Solomon symbol, and every set of k symbols is encoded as an n-symbol codeword. After symbol interleaving, the inner encoder maps every Reed-Solomon symbol into 2^m bits. After the interleaving of these bits, they are transmitted using a binary modulation. (a) Describe the removal of the encoding by the inner and outer decoders. (b) What is the value of n as a function of m ? (c) What are the block length and code rate of the concatenated code?

11 Consider the example in Section 1.6 for a systematic rate-1/3 binary convolutional code with its code symbols transmitted as BPSK symbols over the AWGN channel. Define the a priori LLR of information bit $b_1(k)$ as

$$L_k(b_1) = \ln \left[\frac{P(b_1(k) = 1)}{P(b_1(k) = 0)} \right].$$

Show that each branch metric may be economically represented by

$$\gamma_k(s', s) \sim \exp \left[\frac{2\sqrt{\mathcal{E}_s}}{N_0} \sum_{l=1}^{3} y_{k,l} x_{k,l} + b(s', s) L_k(b_1) \right]$$

where $x_{k,1} = 1 - 2b_{k,1} = 1 - 2b(s', s)$.

12 Consider the transmission of information bits over the binary symmetric channel with bit error probability p. A transmitted bit b_k is received as y_k. Show that the channel LLR is

$$L_c(y_k) = (-1)^{y_k} \log\left(\frac{p}{1-p}\right).$$

13 If N_0 is unknown and may be significantly different from symbol to symbol, a potential procedure is to replace the channel LLR of (1.181) with the *generalized channel LLR:*

$$L_g(\mathbf{y}_k) = \ln\left[\frac{f(\mathbf{y}_k|b_k = 1, N_0 = N_1)}{f(\mathbf{y}_k|b_k = 0, N_0 = N_2)}\right]$$

where N_1 and N_2 are maximum-likelihood estimates of N_0 obtained from $f(\mathbf{y}_k|b_k = 1)$ and $f(\mathbf{y}_k|b_k = 0)$, respectively. For BPSK with no coding, derive the estimators for N_1 and N_2 from (1.191). Then calculate the corresponding $L_g(y_k)$ in terms of α and $\mathcal{E}_s$. What practical difficulty is encountered if attempts are made to use this LLR?

14 (a) Verify (1.170). (b) Use (1.215) to show that BICM-ID and BICM are identical for binary modulations.

15 Consider an LDPC decoder. (a) If check-node l receives inputs only from variable-nodes α and β, what are the LLRs $\mu_{l,\alpha}^{(1)}$ and $\mu_{l,\beta}^{(1)}$? (b) If variable-node α then receives an input only from check-node l, what is $\lambda_{\alpha}^{(1)}$? Observe how the LLR of one variable node becomes part of the LLR of another variable node in this case.

Chapter 2
Direct-Sequence Systems

A *spread-spectrum signal* is a signal with extra modulation that expands the signal bandwidth greatly beyond what is required by the underlying channel code and modulation. Spread-spectrum communication systems are useful for suppressing interference, making secure communications difficult to detect and process, accommodating fading and multipath channels, and providing multiple-access capability. Spread-spectrum signals cause relatively minor interference to other systems operating in the same spectral band. The most practical and dominant spread-spectrum systems are *direct-sequence* and *frequency-hopping* systems.

There is no fundamental theoretical barrier to the effectiveness of spread-spectrum communications. This remarkable fact is not immediately apparent since the increased bandwidth of a spread-spectrum signal necessitates a receive filter that passes more noise power to the demodulator. However, when any signal and white Gaussian noise are applied to a filter matched to the signal, the sampled filter output has a signal-to-noise ratio that depends solely on the energy-to-noise-density ratio. Thus, the bandwidth of the input signal is irrelevant, and spread-spectrum signals have no inherent limitations.

After the information bits are mapped into code symbols, *direct-sequence modulation* entails the direct addition of a high-rate spreading sequence to the lower-rate code-symbol sequence, resulting in a transmitted signal with a relatively wide bandwidth. The removal of the spreading sequence in the receiver causes a contraction of the bandwidth that can be exploited by applying appropriate filtering to remove a large portion of the interference. This chapter describes basic spreading sequences and waveforms and provides a detailed analysis of how the direct-sequence receiver suppresses various forms of interference.

D. Torrieri, *Principles of Spread-Spectrum Communication Systems*,
https://doi.org/10.1007/978-3-319-70569-9_2

2.1 Definitions and Concepts

A *direct-sequence signal* is a spread-spectrum signal generated by the direct mixing of the data with a spreading waveform before the final signal modulation. Ideally, a direct-sequence signal with binary phase-shift keying (BPSK) or differential PSK (DPSK) data modulation can be represented by

$$s(t) = Ad(t)p(t)\cos(2\pi f_c t + \theta) \tag{2.1}$$

where A is the signal amplitude, $d(t)$ is the data modulation, $p(t)$ is the spreading waveform, f_c is the carrier frequency, and θ is the phase at $t = 0$. The data modulation is a sequence of nonoverlapping rectangular pulses of duration T_s, each of which has an amplitude $d_i = +1$ if the associated data symbol is a 1 and $d_i = -1$ if it is a 0 (alternatively, the mapping could be $1 \to -1$ and $0 \to +1$). The *spreading waveform* has the form

$$p(t) = \sum_{i=-\infty}^{\infty} p_i \psi(t - iT_c) \tag{2.2}$$

where each p_i equals $+1$ or -1 and represents one *chip* of the *spreading sequence*. The *chip waveform* $\psi(t)$ is designed to limit interchip interference in the receiver and is mostly confined to the time interval $[0, T_c]$. Figure 2.1 depicts an example of $d(t)$ and $p(t)$ for a rectangular chip waveform.

Message privacy is provided by a direct-sequence system if a transmitted message cannot be recovered without knowledge of the spreading sequence. Although message secrecy can be protected by cryptography, message privacy provides protection even if cryptography is not used. If the data-symbol and chip transitions do not coincide, then it is theoretically possible to separate the data symbols from the chips by detecting and analyzing the transitions. To ensure message privacy, which is assumed henceforth, the data-symbol transitions must coincide with the chip transitions. Another reason for common transitions is the simplification of the receiver implementation. The common transitions imply that the number of chips per data symbol is a positive integer. If W is the bandwidth of $p(t)$ and B is the bandwidth of $d(t)$, the spreading due to $p(t)$ ensures that $s(t)$ has a bandwidth $W >> B$.

Figure 2.2 is a functional or conceptual block diagram of the basic operation of a direct-sequence system with BPSK or DPSK. To provide common transitions, data symbols and chips, which are represented by digital sequences of 0's and 1's, are synchronized by the same clock and then modulo-2 added. The adder output is converted according to $0 \to -1$ and $1 \to +1$ before the chip-waveform modulation shown in Figure 2.2 (a). After upconversion, the modulated signal is transmitted. As indicated in Figure 2.2 (b), the received signal is filtered and then multiplied by a synchronized local replica of $p(t)$. If $\psi(t)$ is rectangular with unit amplitude, then

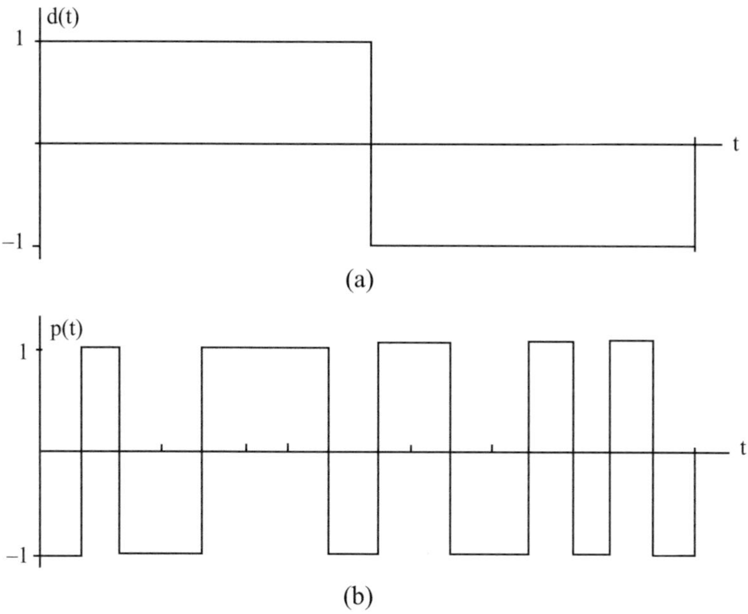

Fig. 2.1 Examples of (**a**) data modulation and (**b**) spreading waveform

$p(t) = \pm 1$ and $p^2(t) = 1$. Therefore, if the filtered signal is given by (2.1), the multiplication yields the *despread signal*

$$s_1(t) = p(t)s(t) = Ad(t)\cos(2\pi f_c t + \theta) \tag{2.3}$$

at the input of the BPSK or DPSK demodulator. A standard demodulator extracts the data symbols or provides the decoder with symbol metrics.

Figure 2.3 (a) is a qualitative depiction of the relative spectra of the desired signal and narrowband interference at the output of the wideband filter. Multiplication of the received signal by the spreading waveform, which is called *despreading*, produces the spectra of Figure 2.3 (b) at the demodulator input. The signal bandwidth is reduced to B, while the interference energy is spread over a bandwidth exceeding W. Since the filtering action of the demodulator then removes most of the interference spectrum that does not overlap the signal spectrum, most of the original interference energy is eliminated.

The *spreading factor or processing gain* is defined as the positive integer

$$G = \frac{T_s}{T_c} \tag{2.4}$$

which equals the number of chips in a symbol interval. An approximate measure of the interference suppression capability is given by the ratio W/B. Whatever the precise definition of a bandwidth is, W and B are proportional to $1/T_c$ and $1/T_s$,

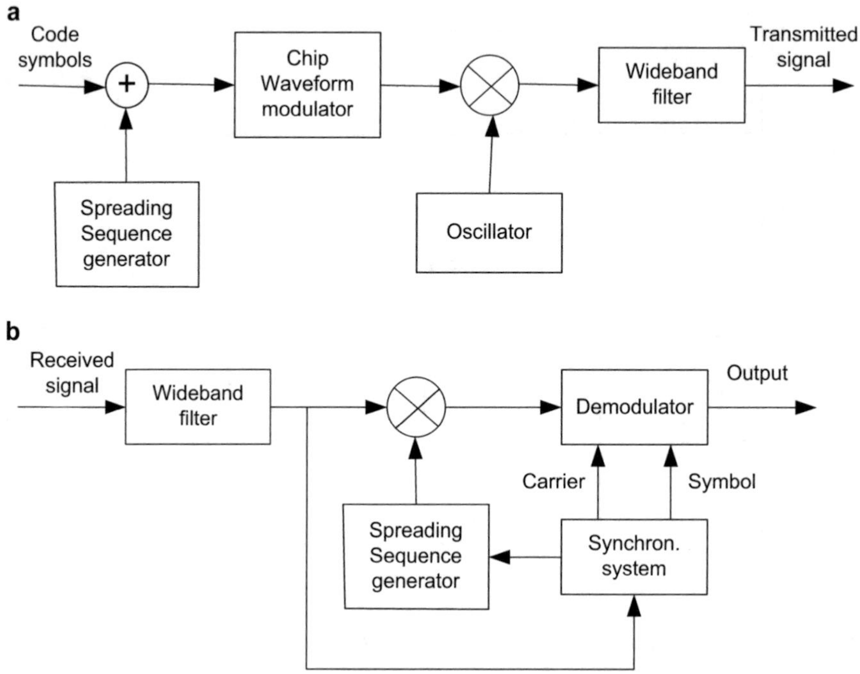

Fig. 2.2 Functional block diagram of a direct-sequence system with BPSK or DPSK: (**a**) transmitter and (**b**) receiver

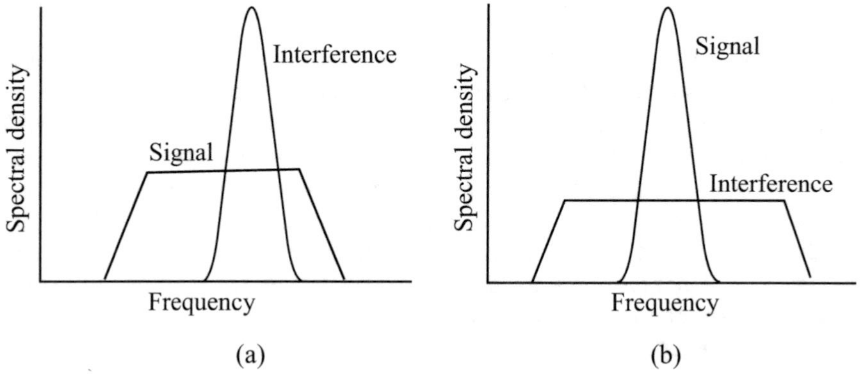

Fig. 2.3 Spectra of the desired signal and interference: (**a**) wideband-filter output and (**b**) demodulator input

respectively, with the same proportionality constant. Therefore, $W/B = T_s/T_c$, and hence the spreading factor is a measure of the interference suppression illustrated in Figure 2.3. Since its spectrum is unchanged by the despreading, white Gaussian noise is not suppressed by a direct-sequence system.

The spectrum of the spreading waveform $p(t)$ is largely determined by the chip waveform $\psi(t)$, which is designed to cause negligible *interchip interference* among the matched-filter output samples in the receiver. If the bandwidth of $\psi(t)$ is large enough, then the energy in $\psi(t)$ is mainly concentrated within a chip interval of duration T_c. However, in a practical system, a wideband filter in the transmitter is used to limit the out-of-band radiation. This filter and the propagation channel disperse the chip waveform so that the received chip waveform is no longer completely confined to $[0, T_c]$. To prevent significant interchip interference in the receiver, the filtered chip waveform must be designed so that the Nyquist criterion for no interchip interference is approximately satisfied. A rectangular chip waveform is ideal in the sense that it causes no interchip interference and has a minimal peak-to-average power ratio, but it can only be approximated in practice.

2.2 Spreading Sequences

A direct-sequence receiver computes the correlation between the received spreading sequence and a stored replica of the spreading sequence it is expecting. The correlation should be high when the receiver is synchronized with the received sequence, and low when it is not. Thus, it is critical that the spreading sequence has suitable autocorrelation properties. Synchronization issues are explained in Chapter 4.

To assess what is desirable in an autocorrelation and its associated power spectral density, we first examine the random binary sequence.

Random Binary Sequence

The *autocorrelation* of a stochastic process $y(t)$ is defined as

$$R_y(t, \tau) = E[y(t)y(t+\tau)]. \tag{2.5}$$

A stochastic process is *wide-sense stationary* (Appendix D.2) if its mean $m_y(t)$ is constant and $R_y(t, \tau)$ is a function of τ alone. Thus, the autocorrelation of a wide-sense-stationary process may be denoted by $R_y(\tau)$.

A *cyclostationary process* $y(t)$ is a process that has a mean and autocorrelation with the same period T. Therefore,

$$m_y(t+T) = m_y(t) \tag{2.6}$$

$$R_y(t+T, \tau) = R_y(t, \tau). \tag{2.7}$$

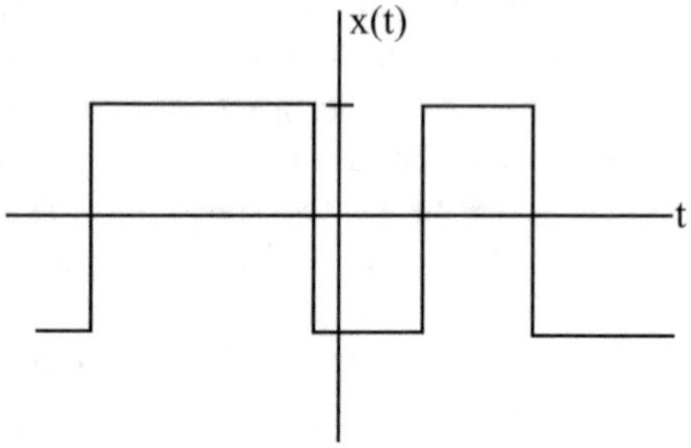

Fig. 2.4 Sample function of a random binary sequence

A deterministic function $y(t)$ with period T is a cyclostationary process with $R_y(t,\tau) = y(t)y(t+\tau)$.

A *random binary sequence* $x(t)$ is a stochastic process that consists of independent, identically distributed symbols, each of duration T. Each symbol takes the value $+1$ with probability $1/2$ or the value -1 with probability $1/2$. The first symbol transition or start of a new symbol after $t = 0$ is a random variable uniformly distributed over the half-open interval $(0, T]$. Therefore, $E[x(t)] = 0$ for all t, and

$$P\left[x\left(t\right) = i\right] = \frac{1}{2}, \quad i = +1, -1. \tag{2.8}$$

A sample function of a *random binary sequence* $x(t)$ is illustrated in Figure 2.4.

From the definition of an expected value and the two possibilities, it follows that the random binary sequence has autocorrelation

$$\begin{aligned} R_x(t,\tau) &= P[x(t+\tau) = x(t)] - P[x(t+\tau) \neq x(t)] \\ &= 1 - 2P[x(t+\tau) \neq x(t)]. \end{aligned} \tag{2.9}$$

This equation and the constant symbol duration T indicate that (2.7) is satisfied, and hence the random binary sequence is cyclostationary.

The random binary sequence is also wide-sense stationary. Consider any half-open interval $I(T)$ of length T such that $I_\tau \in I(T)$. Exactly one transition occurs in $I(T)$. Since the first transition for $t > 0$ is assumed to be uniformly distributed over $(0, T]$, the probability that a transition in $I(T)$ occurs in I_τ is $|\tau|/T$. If a transition occurs in I_τ, then $x(t)$ and $x(t+\tau)$ are independent, and hence differ with probability $1/2$; otherwise, $x(t) = x(t+\tau)$. Consequently, $P[x(t+\tau) \neq x(t)] = |\tau|/2T$ if $|\tau| \leq T$. If $|\tau| > T$, $P[x(t+\tau) \neq x(t)] = 1/2$. Substitution of this preceding result into (2.9) confirms the wide-sense stationarity of $x(t)$ and gives the *autocorrelation of the random binary sequence*:

$$R_x(\tau) = \Lambda\left(\frac{\tau}{T}\right) \tag{2.10}$$

where the *triangular function* is defined by

$$\Lambda(t) = \begin{cases} 1 - |t|, & |t| \leq 1 \\ 0, & |t| > 1. \end{cases} \tag{2.11}$$

The *power spectral density (PSD) of a stationary stochastic process* is the Fourier transform (Appendix C.1) of the autocorrelation. An elementary integration gives the PSD of the random binary sequence:

$$\begin{aligned} S_x(f) &= \int_{-\infty}^{\infty} \Lambda\left(\frac{t}{T}\right) \exp(-j2\pi f t)\, dt \\ &= T \text{sinc}^2 fT \end{aligned} \tag{2.12}$$

where $j = \sqrt{-1}$ and the *sinc function* is

$$\text{sinc}(x) = \begin{cases} (\sin \pi x)/\pi x, & x \neq 0 \\ 1, & x = 0. \end{cases} \tag{2.13}$$

The autocorrelation of a cyclostationary process that is not wide-sense stationary is not a function of the relative delay τ alone, and hence the PSD is not defined. The *periodic autocorrelation* or *average autocorrelation* of a cyclostationary process $y(t)$ with period T is defined as

$$\overline{R}_y(\tau) = \frac{1}{T} \int_c^{c+T} R_y(t, \tau) dt \tag{2.14}$$

where c is an arbitrary constant. The *average PSD* $\overline{S}_y(f)$ *of a* cyclostationary process is the Fourier transform of the average autocorrelation.

The assumption about the random location of the first symbol transition of the random binary sequence is artificial in many applications. For the cyclostationary binary process $x(t)$, this assumption is not made. If c is chosen as the start of a symbol interval, then the substitution of (2.9) into (2.14) and an evaluation yields

$$\overline{R}_x(\tau) = \Lambda\left(\frac{\tau}{T}\right) \tag{2.15}$$

which indicates that the average autocorrelation of the cyclostationary binary process is identical to the autocorrelation of the wide-sense-stationary random binary sequence. Therefore, the average PSD of the cyclostationary binary process is identical to that of the wide-sense-stationary random binary sequence.

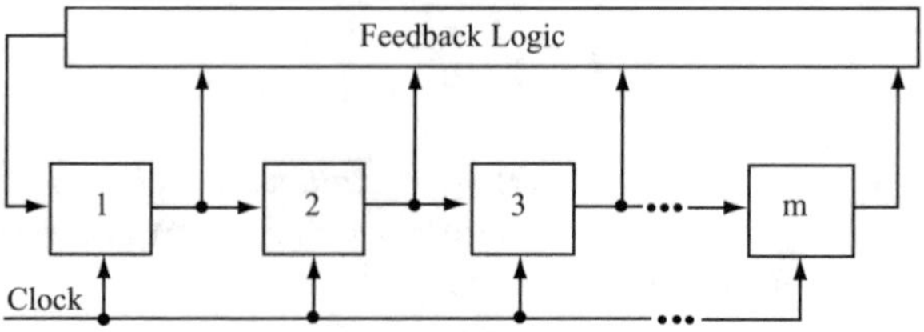

Fig. 2.5 General feedback shift register with m stages

Shift-Register Sequences

Although it has a favorable autocorrelation, a random binary sequence is impractical as a spreading sequence because the latter must be known at both the transmitter and receiver. Practical spreading sequences are periodic binary sequences. A *shift-register sequence* is a periodic binary sequence generated by the output of a feedback shift register or by combining the outputs of feedback shift registers. A *feedback shift register*, which is diagrammed in Figure 2.5, consists of consecutive two-state memory or storage stages and feedback logic. Binary sequences drawn from the alphabet $\{0,1\}$ are shifted through the shift register in response to clock pulses. The *contents* of the stages, which are identical to their outputs, are logically combined to produce the input to the first stage. The initial contents of the stages and the feedback logic determine the successive contents of the stages. If the feedback logic consists entirely of modulo-2 adders (exclusive-OR gates), the feedback shift register and its generated sequence are called *linear.*

Figure 2.6 (a) illustrates a linear feedback shift register with three stages and an output sequence extracted from the final stage. The input to the first stage is the modulo-2 sum of the contents of the second and third stages. After each clock pulse, the contents of the first two stages are shifted to the right, and the input to the first stage becomes its content. If the initial contents of the shift-register stages are 0 0 1, the subsequent contents after successive shifts are listed in Figure 2.6 (b). Since the shift register returns to its initial state after seven shifts, the shift-register sequence extracted from the final stage has a period of 7 bits.

The *state* of an m-stage shift register after clock pulse i is the vector

$$\mathbf{S}(i) = [s_1(i)\ s_2(i) \ldots s_m(i)], \quad i \geq 0 \tag{2.16}$$

where $s_l(i)$ denotes the content of stage l after clock pulse i, and $\mathbf{S}(0)$ is the initial state. The definition of a shift register implies that

$$s_l(i) = s_{l-k}(i-k), \quad i \geq k \geq 0, \quad k \leq l \leq m \tag{2.17}$$

where $s_0(i)$ denotes the input to stage 1 after clock pulse i. If a_i denotes the state of bit i of the shift-register sequence, then $a_i = s_m(i)$. The state of a feedback shift

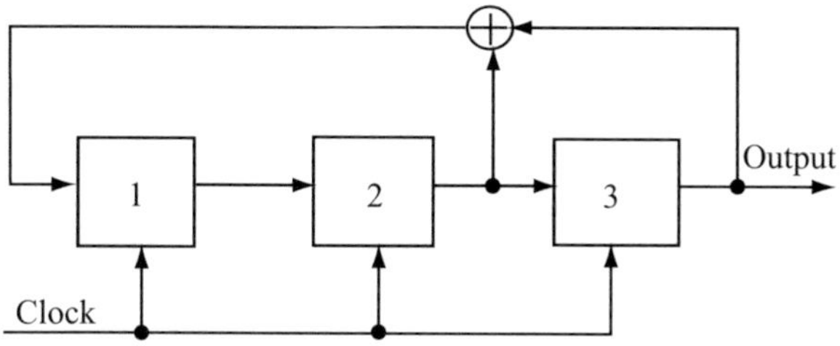

(a)

Shift	Contents		
	Stage 1	Stage 2	Stage 3
Initial	0	0	1
1	1	0	0
2	0	1	0
3	1	0	1
4	1	1	0
5	1	1	1
6	0	1	1
7	0	0	1

(b)

Fig. 2.6 (**a**) Three-stage linear feedback shift register and (**b**) contents after successive shifts

register uniquely determines the subsequent sequence of states and the shift-register sequence.

The *period of a shift-register sequence* $\{a_i\}$ is defined as the smallest positive integer N for which $a_{i+N} = a_i,\ i \geq 0$. Since the number of distinct states of an m-stage nonlinear feedback shift register is 2^m, the sequence of states and the shift-register sequence have period $N \leq 2^m$.

The input to stage 1 of a linear feedback shift register is

$$s_0(i) = \sum_{k=1}^{m} c_k s_k(i), \quad i \geq 0 \tag{2.18}$$

where the additions are modulo-2, and the feedback coefficient c_k equals either 0 or 1, depending on whether the output of stage k feeds a modulo-2 adder. An m-stage shift register is defined to have $c_m = 1$; otherwise, the final state would not contribute to the generation of the output sequence, but would only provide a one-shift delay. For example, Figure 2.6 gives $c_1 = 0, c_2 = c_3 = 1$, and $s_0(i) = s_2(i) \oplus s_3(i)$, where $\oplus$ denotes modulo-2 addition. A general representation of a linear feedback shift register is shown in Figure 2.7 (a). If $c_k = 1$, the corresponding switch is closed; if $c_k = 0$, it is open.

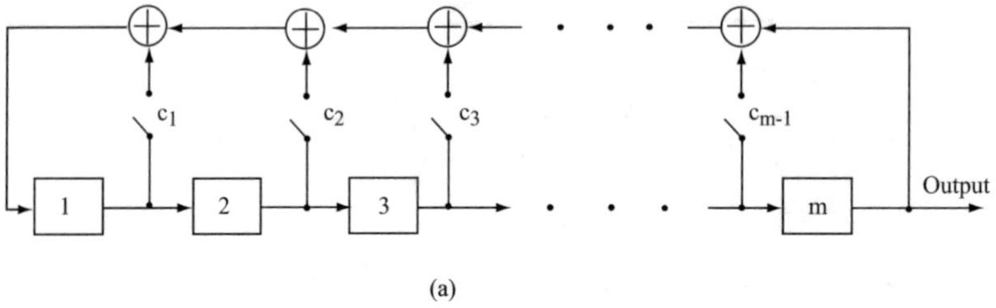

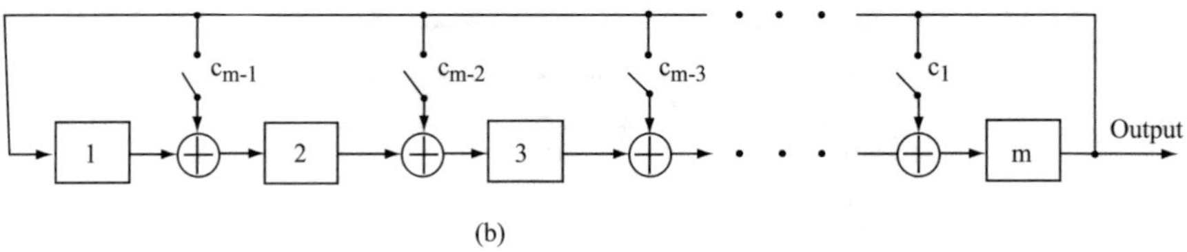

Fig. 2.7 Linear feedback shift register: (**a**) standard representation and (**b**) high-speed form

Since the shift-register sequence bit $a_i = s_m(i)$, (2.17) and (2.18) imply that for $i \geq m$,

$$a_i = s_0(i-m) = \sum_{k=1}^{m} c_k s_k(i-m) = \sum_{k=1}^{m} c_k s_m(i-k) \tag{2.19}$$

which indicates that each bit satisfies a *linear recurrence relation*:

$$a_i = \sum_{k=1}^{m} c_k a_{i-k}, \quad i \geq m. \tag{2.20}$$

The first m bits of the shift-register sequence are determined solely by the initial state:

$$a_i = s_{m-i}(0), \quad 0 \leq i \leq m-1. \tag{2.21}$$

Figure 2.7 (a) is not the fastest configuration to generate a particular shift-register sequence. Figure 2.7 (b) illustrates an implementation that allows higher-speed operation. The diagram indicates that

$$s_l(i) = s_{l-1}(i-1) \oplus c_{m-l+1} s_m(i-1), \quad i \geq 1, \quad 2 \leq l \leq m \tag{2.22}$$

$$s_1(i) = s_m(i-1) \qquad i \geq 1. \tag{2.23}$$

Repeated application of (2.22) implies that

$$\begin{aligned}
s_m(i) &= s_{m-1}(i-1) \oplus c_1 s_m(i-1) \ , \qquad i \geq 1 \\
s_{m-1}(i-1) &= s_{m-2}(i-2) \oplus c_2 s_m(i-2) \ , \qquad i \geq 2 \\
&\vdots \\
s_2(i-m+2) &= s_1(i-m+1) \oplus c_{m-1} s_m(i-m+1) \ , \qquad i \geq m-1.
\end{aligned} \tag{2.24}$$

The addition of these $m-1$ equations yields

$$s_m(i) = s_1(i-m+1) \oplus \sum_{k=1}^{m-1} c_k s_m(i-k), \quad i \geq m-1. \tag{2.25}$$

Substituting (2.23) and then $a_i = s_m(i)$ into (2.25), we obtain

$$a_i = a_{i-m} \oplus \sum_{k=1}^{m-1} c_k a_{i-k}, \quad i \geq m. \tag{2.26}$$

Since $c_m = 1$, (2.26) is the same as (2.20). Thus, the two implementations can produce the same shift-register sequence indefinitely if the first m bits coincide. However, they require different initial states and have different sequences of states.

Successive substitutions into the first equation of sequence (2.24) yield

$$s_m(i) = s_{m-i}(0) \oplus \sum_{k=1}^{i} c_k s_m(i-k) \ , \qquad 1 \leq i \leq m-1. \tag{2.27}$$

Substituting $a_i = s_m(i)$, $a_{i-k} = s_m(i-k)$, and $j = m-i$ into (2.27) and then using binary arithmetic, we obtain

$$s_l(0) = a_{m-l} \oplus \sum_{k=1}^{m-l} c_k a_{m-l-k} \ , \qquad 1 \leq l \leq m. \tag{2.28}$$

If $a_0, a_1, \ldots a_{m-1}$ are specified, then (2.28) gives the corresponding initial state of the high-speed shift register.

The sum of binary sequence $\mathbf{a} = (a_0, a_1, \cdots)$ and binary sequence $\mathbf{b} = (b_0, b_1, \cdots)$ is defined to be the binary sequence $\mathbf{d} = \mathbf{a} \oplus \mathbf{b}$, each bit of which is

$$d_i = a_i \oplus b_i \ , \quad i \geq 0. \tag{2.29}$$

Consider sequences **a** and **b**, that are generated by the same linear feedback shift register but may differ because the initial states may be different. For the sequence $\mathbf{d} = \mathbf{a} \oplus \mathbf{b}$, (2.29), (2.20), and the associative and distributive laws of binary fields imply that

$$d_i = \sum_{k=1}^{m} c_k a_{i-k} \oplus \sum_{k=1}^{m} c_k b_{i-k} = \sum_{k=1}^{m} (c_k a_{i-k} \oplus c_k b_{i-k})$$
$$= \sum_{k=1}^{m} c_k (a_{i-k} \oplus b_{i-k}) = \sum_{k=1}^{m} c_k d_{i-k}. \quad (2.30)$$

Since the linear recurrence relation is identical, **d** can be generated by the same linear feedback logic as **a** and **b**. Thus, if **a** and **b** are two output sequences of a linear feedback shift register, then $\mathbf{a} \oplus \mathbf{b}$ is also an output sequence.

Maximal Sequences

If a linear feedback shift register reached the zero state with all its contents equal to 0 at some time, it would always remain in the zero state, and the output sequence would subsequently be all 0's. Since a linear m-stage feedback shift register has exactly $2^m - 1$ nonzero states, the period of its output sequence cannot exceed $2^m - 1$. A nonzero sequence of period $2^m - 1$ generated by a linear feedback shift register is called a *maximal* or *maximal-length sequence*. If a linear feedback shift register generates a maximal sequence, then all of its nonzero output sequences are maximal, regardless of the initial states.

Out of 2^m possible states of a linear feedback shift register, the content of the last stage, which is a bit of the shift-register sequence, is a 0 in 2^{m-1} states. Among the nonzero states, this bit is a 0 in $2^{m-1} - 1$ states. Therefore, in one period of a maximal sequence, the number of 0's is exactly $2^{m-1} - 1$, while the number of 1's is exactly 2^{m-1}.

Let $\mathbf{a}(l) = (a_l, a_{l+1}, \ldots)$ denote a maximal sequence that is shifted by l bits relative to sequence $\mathbf{a}(0)$. If $l \neq 0$, modulo $2^m - 1$, then $\mathbf{a}(0) \oplus \mathbf{a}(l)$ is not the sequence of all 0's. Since $\mathbf{a}(0) \oplus \mathbf{a}(l)$ is generated by the same shift register as $\mathbf{a}(0)$, it must be a maximal sequence and hence some cyclic shift of $\mathbf{a}(0)$. We conclude that the modulo-2 sum of a maximal sequence and a cyclic shift of itself by l digits, where $l \neq 0$, modulo $2^m - 1$, produces another cyclic shift of the original sequence; that is,

$$\mathbf{a}(0) \oplus \mathbf{a}(l) = \mathbf{a}(k), \quad l \neq 0 \text{ (modulo } 2^m - 1). \quad (2.31)$$

In contrast, a nonmaximal linear sequence $\mathbf{a}(0) \oplus \mathbf{a}(l)$ is not necessarily a cyclic shift of $\mathbf{a}(0)$ and may not even have the same period. As an example, consider the linear feedback shift register depicted in Figure 2.8. The possible state transitions depend on the initial state. If the initial state is 0 1 0, then the second state diagram indicates that there are two possible states, and hence the shift-register sequence has a period of two. The shift-register sequence is $\mathbf{a}(0) = (0, 1, 0, 1, 0, 1, \ldots)$, which implies that $\mathbf{a}(1) = (1, 0, 1, 0, 1, 0, \ldots)$. Therefore,

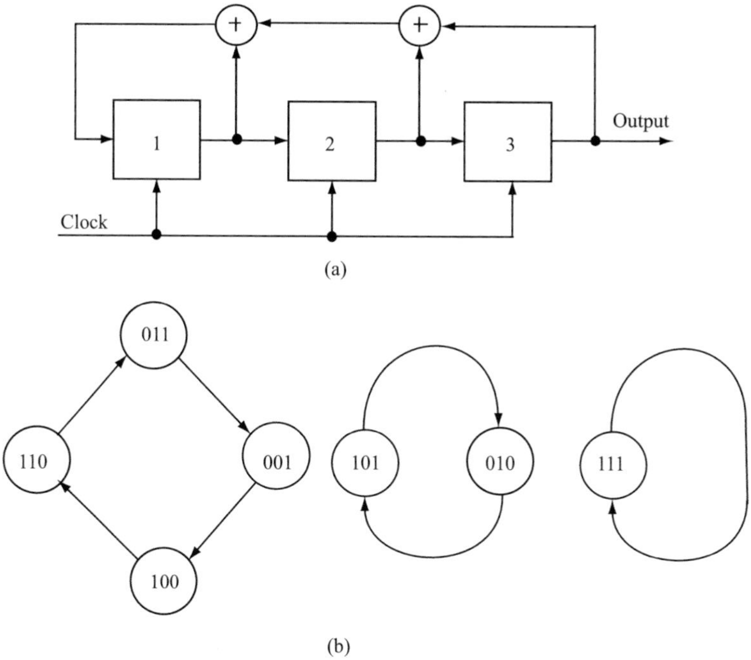

Fig. 2.8 (**a**) Nonmaximal linear feedback shift register and (**b**) state diagrams

$\mathbf{a}(0) \oplus \mathbf{a}(1) = (1, 1, 1, 1, 1, 1, \ldots)$, which indicates that there is no value of k for which (2.31) is satisfied.

Maximal sequences have many properties that make them difficult to distinguish from random sequences. As an example, suppose that one observes i bits within a maximal sequence of period $2^m - 1$ bits, and $i \leq m$. The i bits are part of a state sequence of m bits that constitute a state of the shift register that generated the maximal sequence. If not all of the i bits are 0's, the $m - i$ unobserved bits in the m-bit state sequence may be any of 2^{m-i} possible sequences. Since there are $2^m - 1$ possible m-bit state sequences, the relative frequency of a particular sequence of i observed bits is $2^{m-i}/\,(2^m - 1)$ if the bits are not all 0's. If i 0's are observed, then the unobserved bits cannot all be 0's because the m-bit state sequence constitutes a state of the maximal-sequence generator. Thus, the relative frequency of i observed 0's is $(2^{m-i} - 1) \,/\, (2^m - 1)$. Both of these ratios approach 2^{-i} as $m \to \infty$ which is what would happen if the m-bit sequence were random.

Characteristic Polynomials

Polynomials over the binary field $GF(2)$ (Section 1.3) allow a compact description of the dependence of the shift-register sequence of a linear feedback shift register on its feedback coefficients and initial state. The *characteristic polynomial*

$$f(x) = 1 + \sum_{i=1}^{m} c_i x^i \tag{2.32}$$

defines a linear feedback shift register of m stages, with feedback coefficients c_i, $i = 1, 2, \ldots, m$. The coefficient $c_m = 1$ assumes that stage m contributes to the generation of the shift-register sequence. The *generating function* associated with the shift-register sequence is defined as

$$G(x) = \sum_{i=0}^{\infty} a_i x^i. \tag{2.33}$$

Substitution of (2.20) into this equation yields

$$\begin{aligned} G(x) &= \sum_{i=0}^{m-1} a_i x^i + \sum_{i=m}^{\infty} \sum_{k=1}^{m} c_k a_{i-k} x^i \\ &= \sum_{i=0}^{m-1} a_i x^i + \sum_{k=1}^{m} c_k x^k \sum_{i=m}^{\infty} a_{i-k} x^{i-k} \\ &= \sum_{i=0}^{m-1} a_i x^i + \sum_{k=1}^{m} c_k x^k \left[G(x) + \sum_{i=0}^{m-k-1} a_i x^i \right]. \end{aligned} \tag{2.34}$$

Combining this equation with (2.32), and defining $c_0 = 1$, we obtain

$$\begin{aligned} G(x)f(x) &= \sum_{i=0}^{m-1} a_i x^i + \sum_{k=1}^{m} c_k x^k \left(\sum_{i=0}^{m-k-1} a_i x^i \right) \\ &= \sum_{k=0}^{m-1} c_k x^k \left(\sum_{i=0}^{m-k-1} a_i x^i \right) = \sum_{k=0}^{m-1} \sum_{i=0}^{m-k-1} c_k a_i x^{k+i} \\ &= \sum_{k=0}^{m-1} \sum_{l=k}^{m-1} c_k a_{l-k} x^l = \sum_{l=0}^{m-1} \sum_{k=0}^{l} c_k a_{l-k} x^l \end{aligned} \tag{2.35}$$

which implies that

$$G(x) = \frac{\sum_{i=0}^{m-1} x^i \left(\sum_{k=0}^{i} c_k a_{i-k} \right)}{f(x)}, \qquad c_0 = 1. \tag{2.36}$$

Thus, the generating function of the shift-register sequence generated by a linear feedback shift register with characteristic polynomial $f(x)$ may be expressed in the form

$$G(x) = \frac{\phi(x)}{f(x)} \tag{2.37}$$

where the degree of $\phi(x)$ is less than the degree of $f(x)$. The shift-register sequence is said to be *generated* by $f(x)$. Equation (2.36) explicitly shows that the shift-register sequence is completely determined by the feedback coefficients c_k, $k = 1, 2, \ldots, m$, and the initial state $a_i = s_{m-i}(0)$, $i = 0, 1, \ldots, m-1$.

In Figure 2.6, the feedback coefficients are $c_1 = 0$, $c_2 = 1$, and $c_3 = 1$, and the initial state gives $a_0 = 1$, $a_1 = 0$, and $a_2 = 0$. Therefore,

$$G(x) = \frac{1 + x^2}{1 + x^2 + x^3}. \tag{2.38}$$

Performing the long polynomial division according to the rules of binary arithmetic yields $1 + x^3 + x^5 + x^6 + x^7 + x^{10} + \ldots$, which implies the shift-register sequence listed in the figure.

A polynomial $b(x)$ is *divisible* by the polynomial $p(x)$ if there is a polynomial $h(x)$ such that $b(x) = h(x)p(x)$. A polynomial $p(x)$ over $GF(2)$ of degree m is called *irreducible* if $p(x)$ is not divisible by any polynomial over $GF(2)$ of a degree less than m but greater than zero. If $p(x)$ is irreducible over $GF(2)$, then $p(0) \neq 0$, because otherwise x would divide $p(x)$. If $p(x)$ has an even number of terms, then $p(1) = 0$. Therefore, the fundamental theorem of algebra over $GF(2)$ implies that $x + 1$ divides $p(x)$, and thus *an irreducible polynomial over* $GF(2)$ *must have an odd number of terms*. However, since $1 + x + x^5 = (1 + x^2 + x^3)(1 + x + x^2)$, some polynomials with an odd number of terms are not irreducible.

If a shift-register sequence $\{a_i\}$ is periodic with period n, then its generating function $G(x) = \phi(x)/f(x)$ may be expressed as

$$\begin{aligned} G(x) &= g(x) + x^n g(x) + x^{2n} g(x) + \cdots = g(x) \sum_{i=0}^{\infty} x^{in} \\ &= \frac{g(x)}{1 + x^n} \end{aligned} \tag{2.39}$$

where $g(x)$ is a polynomial of degree $n - 1$ and represents the first period of the shift-register sequence. Therefore,

$$g(x) = \frac{\phi(x)(1+x^n)}{f(x)} \tag{2.40}$$

and $f(x)$ divides $\phi(x)(1+x^n)$. If $\phi(x) = 1$, then the generating function is $G(x) = 1/f(x)$. Therefore, (2.40) implies that $1/f(x)$ *has a period* n *or less only if* $f(x)$ *divides* $1 + x^n$.

Conversely, if the characteristic polynomial $f(x)$ divides $1 + x^n$, then $f(x)h(x) = 1 + x^n$ for some polynomial $h(x)$. Therefore,

$$G(x) = \frac{\phi(x)}{f(x)} = \frac{\phi(x)h(x)}{1+x^n} = \phi(x)h(x)\sum_{i=0}^{\infty} x^{in} \tag{2.41}$$

which implies that *a sequence with period* n *or less is generated if* $f(x)$ *divides* $1+x^n$. It follows from the two preceding results that *if* $f(x)$ *divides* $1+x^N$ *but does not divide* $1+x^n$ *for* $n < N$, *then* $1/f(x)$ *has period* N.

The nonmaximal linear feedback shift register of Figure 2.8 has characteristic polynomial $1+x+x^2+x^3$, which divides $1+x^4$. Sequences of periods 4, 2, and 1 are generated.

A polynomial $f(x)$ over $GF(2)$ of degree m is called *primitive* if the smallest positive integer n for which the polynomial divides $1 + x^n$ is $n = 2^m - 1$. Thus, $1/f(x)$ has period $2^m - 1$. Suppose that a primitive characteristic polynomial $f(x)$ of positive degree m is not irreducible and can be factored so that $f(x) = f_1(x)f_2(x)$, where $f_1(x)$ is of positive degree m_1 and $f_2(x)$ is of positive degree $m - m_1$. A partial-fraction expansion of the generating function $G(x) = 1/f(x)$ yields

$$\frac{1}{f(x)} = \frac{a(x)}{f_1(x)} + \frac{b(x)}{f_2(x)}. \tag{2.42}$$

Since $f_1(x)$ and $f_2(x)$ can serve as characteristic polynomials, the period of the first term in the expansion cannot exceed $2^{m_1} - 1$, whereas the period of the second term cannot exceed $2^{m-m_1} - 1$. Therefore, the period of $1/f(x)$ cannot exceed $(2^{m_1} - 1)(2^{m-m_1} - 1) \leq 2^m - 3$. However, the definition of a primitive polynomial implies that the period of $1/f(x)$ is $2^m - 1$, and hence we have a contradiction. Thus, *a primitive characteristic polynomial must be irreducible.*

Theorem *A characteristic polynomial of degree* m *generates a maximal sequence of period* $2^m - 1$ *if and only if it is a primitive polynomial.*

Proof To prove sufficiency, we observe that if $f(x)$ is a primitive characteristic polynomial, then the generating function $1/f(x)$ has period $2^m - 1$, which indicates that a maximal sequence is generated. If a sequence of smaller period could be generated, then the irreducible $f(x)$ would have to divide $1 + x^{n_1}$ for $n_1 < n$, which contradicts the assumption of a primitive polynomial. To prove necessity, we observe that if the characteristic polynomial $f(x)$ generates the maximal sequence with period $n = 2^m - 1$, then $f(x)$ cannot divide $1 + x^{n_1}$, $n_1 < n$, because a sequence

Table 2.1 Primitive polynomials

Degree	Primitive	Degree	Primitive	Degree	Primitive
2	7	7	103	8	534
3	51		122	9	1201
	31		163	10	1102
4	13		112	11	5004
	32		172	12	32101
5	15		543	13	33002
	54		523	14	30214
	57		532	15	300001
	37		573	16	310012
	76		302	17	110004
	75		323	18	1020001
6	141		313	19	7400002
	551		352	20	1100004
	301		742	21	50000001
	361		763	22	30000002
	331		712	23	14000004
	741		753	24	702000001
			772	25	110000002

with a smaller period would result, and such a sequence cannot be generated by the maximal sequence generator defined by $f(x)$. Since $f(x)$ does divide $1 + x^n$, it must be a primitive polynomial. □

Primitive polynomials have been tabulated and may be generated by recursively producing polynomials and evaluating whether they are primitive by using them as characteristic polynomials [57]. Those that generate maximal sequences are primitive. Primitive polynomials for which $m \leq 7$ and one of those of minimal coefficient weight for $8 \leq m \leq 25$ are listed in Table 2.1 as octal numbers in increasing order (e.g., $51 \leftrightarrow 1\,0\,1\,1\,0\,0 \leftrightarrow 1 + x^2 + x^3$). For any positive integer m, the number of different primitive polynomials of degree m over $GF(2)$ is [57]

$$\lambda(m) = \frac{\phi_e\left(2^m - 1\right)}{m} \tag{2.43}$$

where the *Euler function* $\phi_e(n)$ is the number of positive integers that are less than and relatively prime to the positive integer n. If n is a prime number, $\phi_e(n) = n - 1$. In general,

$$\phi_e(n) = n \prod_{i=1}^{k} \frac{\nu_i - 1}{\nu_i} \leq n - 1 \tag{2.44}$$

where $\nu_1, \nu_2, \ldots, \nu_k$ are the prime integers that divide n. Thus, $\lambda(6) = \phi_e(63)/6 = 6$ and $\lambda(13) = \phi_e(8191)/13 = 630$.

Autocorrelations and Power Spectra

A deterministic function $x(t)$ with period T is a cyclostationary process with autocorrelation $R_x(t, \tau) = x(t)x(t + \tau)$. To determine the average autocorrelation of a periodic spreading waveform $p(t)$, we first determine the average autocorrelation of a periodic binary sequence. Prior to the application of a binary sequence to the modulation waveform, a binary sequence $\mathbf{a}$ with components $a_i \in GF(2)$ is mapped into a binary antipodal sequence $\mathbf{p}$ with components $p_i \in \{-1, +1\}$ by means of the transformation

$$p_i = (-1)^{a_i+1}, \quad i \geq 0 \tag{2.45}$$

or, alternatively, $p_i = (-1)^{a_i}$. The *periodic autocorrelation* of a periodic binary sequence $\mathbf{a}$ with period N is defined as the periodic autocorrelation of the corresponding binary antipodal sequence $\mathbf{p}$:

$$\begin{aligned}\theta_p(k) &= \frac{1}{N}\sum_{i=n}^{n+N-1} p_i p_{i+k}, \quad n = 0, 1, \ldots \\ &= \frac{1}{N}\sum_{i=0}^{N-1} p_i p_{i+k}\end{aligned} \tag{2.46}$$

which has period N because $\theta_p(k + N) = \theta_p(k)$, where k is an integer. Substitution of (2.45) into (2.46) yields

$$\theta_p(k) = \frac{1}{N}\sum_{i=0}^{N-1}(-1)^{a_i+a_{i+k}} = \frac{1}{N}\sum_{i=0}^{N-1}(-1)^{a_i \oplus a_{i+k}} = \frac{A_k - D_k}{N} \tag{2.47}$$

where A_k denotes the number of agreements in the corresponding bits of $\mathbf{a}$ and its shifted version $\mathbf{a}(k)$, and D_k denotes the number of disagreements. Equivalently, A_k is the number of 0's in one period of $\mathbf{a} \oplus \mathbf{a}(k)$, and $D_k = N - A_k$ is the number of 1's.

We derive the average autocorrelation of the spreading waveform $p(t)$ assuming an ideal periodic spreading waveform of infinite extent and the rectangular chip waveform $\psi(t) = w(t, T_c)$, where

$$w(t, T) = \begin{cases} 1, & 0 \leq t < T \\ 0, & \text{otherwise.} \end{cases} \tag{2.48}$$

If a spreading sequence has period N, then $p(t)$ has period $T = NT_c$. If a chip boundary is aligned with $t = 0$, then (2.2), (2.48), and (2.14) with $c = 0$ give the average autocorrelation of $p(t)$:

$$\overline{R}_p(\tau) = \frac{1}{NT_c} \sum_{i=0}^{N-1} p_i \sum_{l=-\infty}^{\infty} p_l \int_0^{NT_c} w(t - iT_c, T_c) w(t - lT_c + \tau, T_c)\, dt. \tag{2.49}$$

Any delay can be expressed in the form $\tau = kT_c + \epsilon$, where k is an integer and $0 \leq \epsilon < T_c$. Substituting this form into (2.49), we observe that the integrand is nonzero only if $l = i + k$ or $l = i + k + 1$. Therefore,

$$\overline{R}_p(kT_c + \epsilon) = \frac{1}{NT_c} \sum_{i=0}^{N-1} p_i p_{i+k} \int_0^{NT_c} w(t - iT_c, T_c) w(t - iT_c + \epsilon, T_c)\, dt$$
$$+ \frac{1}{NT_c} \sum_{i=0}^{N-1} p_i p_{i+k+1} \int_0^{NT_c} w(t - iT_c, T_c) w(t - iT_c + \epsilon - T_c, T_c)\, dt. \tag{2.50}$$

Using (2.46) and (2.48) in (2.50), we obtain

$$\overline{R}_p(kT_c + \epsilon) = \left(1 - \frac{\epsilon}{T_c}\right) \theta_p(k) + \frac{\epsilon}{T_c} \theta_p(k+1) \tag{2.51}$$

where $\theta_p(k)$ is given by (2.47).

For a maximal sequence with $N = 2^m - 1$, (2.47) and (2.31) imply that

$$\theta_p(k) = \begin{cases} 1, & k = 0 \text{ (modulo } N) \\ -\frac{1}{N}, & k \neq 0 \text{ (modulo } N). \end{cases} \tag{2.52}$$

The substitution of (2.52) into (2.51) yields $\overline{R}_p(\tau)$ over one period:

$$\overline{R}_p(\tau) = \frac{N+1}{N} \Lambda\left(\frac{\tau}{T_c}\right) - \frac{1}{N}, \quad |\tau| \leq NT_c/2. \tag{2.53}$$

Since it has period NT_c, the *average autocorrelation of the maximal-sequence waveform* can be compactly expressed as

$$\overline{R}_p(\tau) = -\frac{1}{N} + \frac{N+1}{N} \sum_{i=-\infty}^{\infty} \Lambda\left(\frac{\tau - iNT_c}{T_c}\right). \tag{2.54}$$

Over one period, this autocorrelation resembles that of a random binary sequence, which is given by (2.10) with $T = T_c$. Both autocorrelations are shown in Figure 2.9.

Since the infinite series in (2.54) is a periodic function of τ, it can be expressed as a complex exponential Fourier series. Applying (2.12), we obtain

$$\overline{R}_p(\tau) = -\frac{1}{N} + \frac{N+1}{N^2} \sum_{i=-\infty}^{\infty} \text{sinc}^2\left(\frac{i}{N}\right) \exp\left(\frac{j2\pi i\tau}{NT_c}\right). \tag{2.55}$$

Since the Fourier transform of a complex exponential is a delta function, the *average PSD of the maximal-sequence waveform* is

$$S_p(f) = \frac{1}{N^2}\delta(f) + \frac{N+1}{N^2} \sum_{i=-\infty,\, i\neq 0}^{\infty} \text{sinc}^2\left(\frac{i}{N}\right)\delta\left(f - \frac{i}{NT_c}\right) \tag{2.56}$$

where $\delta(\cdot)$ is the Dirac delta function. This function, which consists of an infinite series of delta functions, is depicted in Figure 2.10.

For the direct-sequence signal

$$s(t) = Ad(t)p(t)\cos(2\pi f_c t + \theta) \tag{2.57}$$

the data modulation $d(t)$ is modeled as a random binary sequence with autocorrelation given by (2.10), the chip waveform is rectangular, and θ is modeled as a random variable uniformly distributed over $[0, 2\pi)$ and statistically independent of $d(t)$. We obtain the autocorrelation

$$R_s(t,\tau) = \frac{A^2}{2}p(t)p(t+\tau)\Lambda\left(\frac{\tau}{T_s}\right)\cos 2\pi f_c \tau \tag{2.58}$$

where $p(t)$ is the periodic spreading waveform with period T_s, and $\Lambda(\cdot)$ is the triangular function defined by (2.11). Since its mean is zero and $R_s(t+T_s,\tau) = R_s(t,\tau)$, $s(t)$ is a cyclostationary process with period T_s. Substituting (2.58) into (2.14), we obtain the *average autocorrelation of a direct-sequence signal*:

$$\bar{R}_s(\tau) = \frac{A^2}{2}\bar{R}_p(\tau)\,\Lambda\left(\frac{\tau}{T_s}\right)\cos 2\pi f_c \tau \tag{2.59}$$

where $\overline{R}_p(\tau)$ is the average autocorrelation of $p(t)$ and is given by (2.54) for a maximal-sequence waveform. If the spreading factor is $G = N$, the chip duration is $T_c = T_s/G$, (2.59), (2.56), (2.12), and the application of the convolution theorem (Appendix C.1) twice give the *average PSD of a direct-sequence signal with a maximal spreading sequence:*

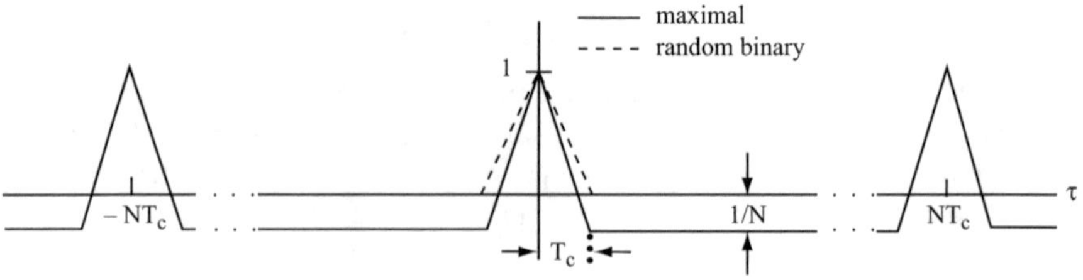

Fig. 2.9 Autocorrelations of maximal sequence and random binary sequence

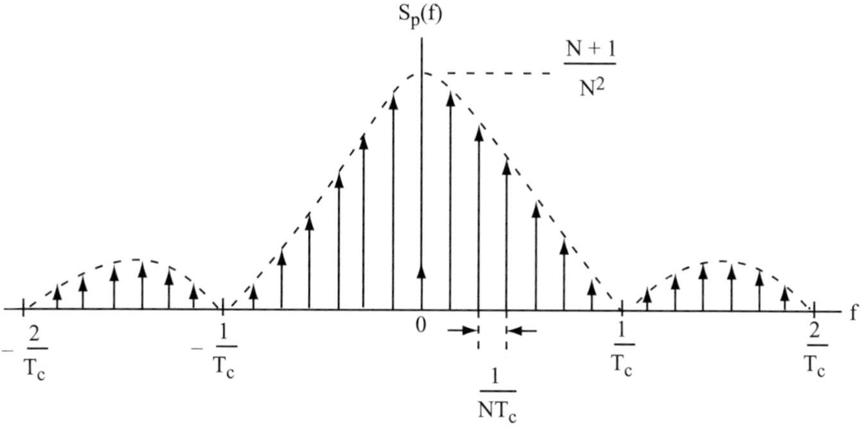

Fig. 2.10 Power spectral density of maximal sequence

$$\bar{S}_s(f) = \frac{A^2}{4}\left[S_{s1}(f - f_c) + S_{s1}(f + f_c)\right] \tag{2.60}$$

where the lowpass equivalent PSD is

$$S_{s1}(f) = \frac{T_c}{G}\text{sinc}^2(fGT_c) + \frac{G+1}{G}T_c \sum_{i=-\infty,\, i\neq 0}^{\infty} \text{sinc}^2\left(\frac{i}{G}\right)\text{sinc}^2(fGT_c - i). \tag{2.61}$$

A *pseudonoise* or *pseudorandom sequence* is a periodic binary sequence with a nearly even balance of 0's and 1's and an autocorrelation that roughly resembles, over one period, the autocorrelation of a random binary sequence. Maximal sequences are pseudonoise sequences with autocorrelations that facilitate code synchronization in the receiver (Chapter 4). However, maximal sequences do not have favorable cross-correlations and are inadequate when direct-sequence systems are part of a network with multiple-access interference. Consequently, maximal sequences are rarely used as spreading sequences in networks of direct-sequence systems, but they provide building blocks for the construction of other sequences that are inherently resistant to multiple-access interference (Chapter 7).

2.3 Long Nonlinear Sequences

A *long sequence* or *long code* is a spreading sequence with a period that is much longer than the data-symbol duration and may even exceed the message duration. A *short sequence* or *short code* is a spreading sequence with a period that is equal to or less than the data-symbol duration. Since short sequences are susceptible to interception, mathematical cryptanalysis, and hence regeneration, long spreading sequences and programmable sequence generators are needed for communications with a high level of security. However, if a modest level of security is acceptable, short or moderate-length spreading sequences are preferable for rapid acquisition, burst communications, multiple-access communications, and multiuser detection. Suitable short spreading sequences, notably the Hadamard, Gold, and Kasami sequences, are covered in Chapter 7.

Many direct-sequence systems use a spreading sequence that is a composite of two spreading sequences. The composite sequence is formed by the modulo-2 addition of the two component sequences. One of the component sequences, which is called the *scrambling sequence*, is a long sequence that provides security, whereas the other sequence is shorter and suppresses multiple-access interference from other direct-sequence systems. In a receiver, the component spreading sequences are despread successively.

Susceptibility to Cryptanalysis

The algebraic structure of linear feedback shift registers makes them susceptible to cryptanalysis. Let

$$\mathbf{c} = [c_1 \; c_2 \ldots c_m]^T \tag{2.62}$$

denote the column vector of the m feedback coefficients of an m-stage linear feedback shift register, where T denotes the transpose of a matrix or vector. The column vector of m successive sequence bits produced by the shift register starting at bit i is

$$\mathbf{a}_i = [a_i \; a_{i+1} \; \ldots a_{i+m-1}]^T . \tag{2.63}$$

Let $\mathbf{A}(i)$ denote the $m \times m$ matrix with columns consisting of the $\mathbf{a}_k$ vectors for $i \leq k \leq i+m-1$:

$$\mathbf{A}(i) = \begin{bmatrix} a_{i+m-1} & a_{i+m-2} & \cdots a_i \\ a_{i+m} & a_{i+m-1} & \cdots a_{i+1} \\ \vdots & \vdots & \vdots \\ a_{i+2m-2} & a_{i+2m-3} & \cdots a_{i+m-1} \end{bmatrix} . \tag{2.64}$$

The linear recurrence relation (2.20) indicates that the shift-register sequence and feedback coefficients are related by

$$\mathbf{a}_{i+m} = \mathbf{A}(i)\mathbf{c}, \quad i \geq 0. \tag{2.65}$$

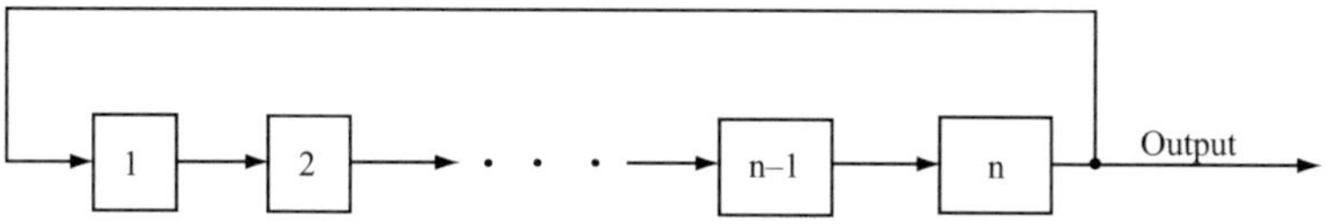

Fig. 2.11 Linear generator of binary sequence with period n

If $2m$ consecutive shift-register sequence bits are known, then $\mathbf{A}(i)$ and $\mathbf{a}_{i+m}$ are completely known for some i. If $\mathbf{A}(i)$ is invertible, then the feedback coefficients can be computed from

$$\mathbf{c} = \mathbf{A}^{-1}(i)\mathbf{a}_{i+m} \ , \quad i \geq 0. \tag{2.66}$$

A shift-register sequence is completely determined by the feedback coefficients and any state vector. Since any m successive sequence bits determine a state vector, $2m$ successive bits provide enough information to reproduce the shift-register sequence unless $\mathbf{A}(i)$ is not invertible. In that case, one or more additional bits are required.

Nonlinear Generators

If a binary sequence has period n, it can always be generated by an n-stage linear feedback shift register by connecting the output of the last stage to the input of the first stage and inserting n consecutive bits of the sequence into the shift-register, as illustrated in Figure 2.11. The polynomial associated with one period of the binary sequence is

$$g(x) = \sum_{i=0}^{n-1} a_i x^i. \tag{2.67}$$

Let $gcd(g(x), 1 + x^n)$ denote the greatest common polynomial divisor of the polynomials $g(x)$ and $1 + x^n$. Then (2.39) implies that the generating function of the sequence may be expressed as

$$G(x) = \frac{g(x)/gcd\,(g(x), 1 + x^n)}{(1 + x^n)\,/gcd\,(g(x), 1 + x^n)}. \tag{2.68}$$

If $gcd(g(x), 1 + x^n) \neq 1$, the degree of the denominator of $G(x)$ is less than n. Therefore, the sequence represented by $G(x)$ can be generated by a linear feedback shift register with fewer stages than n and with the characteristic polynomial given by the denominator. The appropriate initial state can be determined from the coefficients of the numerator.

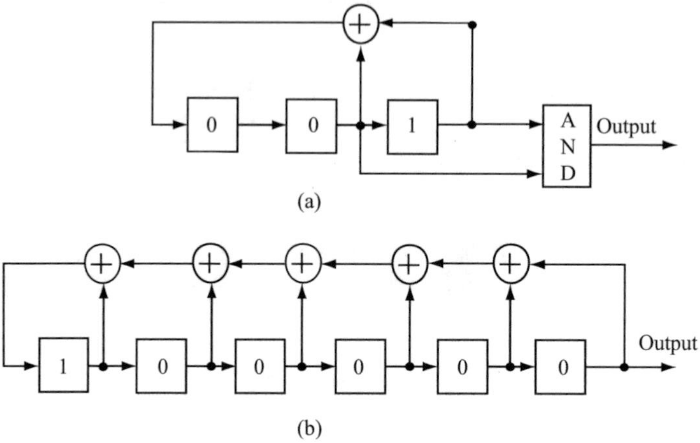

Fig. 2.12 (**a**) Nonlinear generator and (**b**) its linear equivalent

The *linear equivalent* of the generator of a sequence is the linear shift register with the fewest stages that produces the sequence. The number of stages in the linear equivalent is called the *linear span* or *linear complexity* of the sequence. If the linear span is equal to m, then (2.66) determines the linear equivalent after the observation of $2m$ consecutive sequence bits.

Security improves as the period of a shift-register sequence increases, but there are practical limits to the number of shift-register stages. To produce shift-register sequences with an adequate linear span for high security, the feedback logic in Figure 2.5 may be nonlinear. Among other alternatives, the outputs of two or more linear feedback shift registers or several outputs of shift-register stages may be applied to a nonlinear device to produce the final shift-register sequence. Nonlinear generators with relatively few shift-register stages can produce sequences with an enormous linear span.

As an example, Figure 2.12 (a) depicts a nonlinear generator that preserves the linear feedback shift register but performs a nonlinear operation on the outputs of the stages. Two stages have their outputs applied to an AND gate to produce the shift-register sequence. The initial contents of the shift-register stages are indicated by the enclosed binary numbers. Since the linear generator produces a maximal sequence of length 7, the shift-register sequence has period 7. The first period of the sequence is (0 0 0 0 0 1 1), from which the linear equivalent with the initial contents shown in Figure 2.12 (b) is derived by evaluating (2.68).

Although a large linear span is necessary for the cryptographic integrity of a sequence, it is not necessarily sufficient because other statistical characteristics, such as a nearly even distribution of 1's and 0's, are required. For example, a long sequence of many 0's followed by a single 1 has a linear span equal to the length of the sequence, but the sequence is very weak. The generator of Figure 2.12 (a)

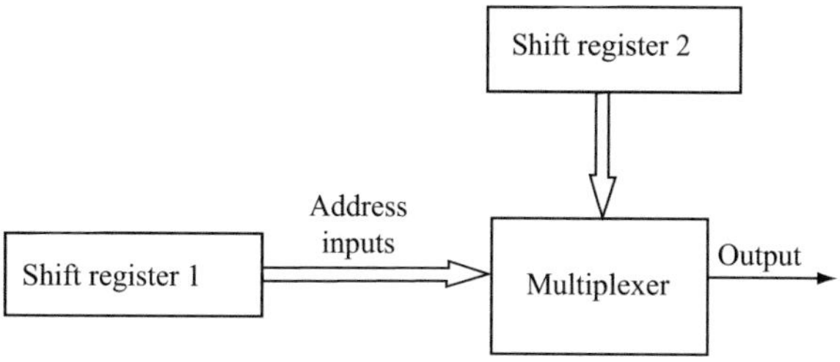

Fig. 2.13 Nonlinear generator that uses a multiplexer

produces a relatively large number of 0's because the AND gate produces a 1 only if both of its inputs are 1's.

A *spread-spectrum key* is a set of parameters that determines the generation of a specific spreading sequence. The initial states of the shift registers, the feedback connections, and which stages are accessed for other purposes may be determined by the spread-spectrum key that is the ultimate source of the security of the spreading sequences. The basic architecture or configuration of a sequence generator cannot be kept secret indefinitely; thus, the key does not contain this type of information. For example, a key for the nonlinear generator of Figure 2.12 (a) comprises the six bits 001011, the first three of which specify the initial contents, and the second three of which specify the feedback connections. A key for the linear generator of Figure 2.12 (b) comprises the 12 bits 100000111111. A change in the key of a generator usually results in a major change in its shift-register sequence.

As another example, a nonlinear generator that uses a multiplexer is shown in Figure 2.13. The outputs of various stages of feedback shift register 1 are applied to the multiplexer, which interprets the binary number determined by these outputs as an address. The multiplexer uses this address to select one of the stages of feedback shift register 2. The selected stage provides the multiplexer output and hence one bit of the output sequence. Suppose that shift register 1 has m stages and shift register 2 has n stages. If h stages of shift register 1, where $h < m$, are applied to the multiplexer, then the address is one of the numbers $0, 1, \ldots, 2^h - 1$. Therefore, if $n \geq 2^h$, each address specifies a distinct stage of shift register 2. A key of $2m+2n+h$ bits specifies the initial states of the two registers, the feedback connections, and which stages are used for addressing. The feedback of both shift registers may be nonlinear, but the structures of the nonlinearities are not specified by the key.

2.4 Systems with BPSK Modulation

A received direct-sequence signal with coherent BPSK modulation and ideal carrier synchronization can be represented by (2.1) with $\theta = 0$ to reflect the absence of phase uncertainty. The direct-sequence signal is

$$s(t) = \sqrt{2\mathcal{E}_s} d(t) p(t) \cos 2\pi f_c t \tag{2.69}$$

where $\sqrt{2\mathcal{E}_s}$ is the signal amplitude, $d(t)$ is the data modulation, $p(t)$ is the spreading waveform, and f_c is the carrier frequency. The data modulation is a sequence of nonoverlapping rectangular pulses, each of which has an amplitude equal to +1 or −1. Each pulse of $d(t)$ represents a data symbol or code bit and has a duration of T_s. The spreading waveform is given by (2.2), and the spreading factor, which is a positive integer, is given by (2.4). Assuming that the chip waveform $\psi(t)$ is well-approximated by a waveform of duration T_c, it is convenient, and entails no loss of generality, to normalize the energy of $\psi(t)$ according to

$$\int_0^{T_c} \psi^2(t)dt = \frac{1}{G}. \tag{2.70}$$

Using this equation and assuming that $f_c >> 1/T_c$ so that an integral over a double-frequency term is negligible, an integration of $s(t)$ over a symbol interval indicates that $\mathcal{E}_s$ is the energy per symbol.

A normalized *rectangular chip waveform* has

$$\psi(t) = \begin{cases} \frac{1}{\sqrt{T_s}}, & 0 \leq t < T_c \\ 0, & \text{otherwise} \end{cases} \tag{2.71}$$

and a normalized *sinusoidal chip waveform* has

$$\psi(t) = \begin{cases} \sqrt{\frac{2}{T_s}} \sin\left(\frac{\pi}{T_c}t\right), & 0 \leq t \leq T_c \\ 0, & \text{otherwise.} \end{cases} \tag{2.72}$$

A practical direct-sequence system differs from the functional diagram of Figure 2.2. The transmitter needs practical devices, such as a power amplifier and a filter that limits out-of-band radiation. In the receiver, the radio-frequency front end includes devices for wideband filtering and automatic gain control. These devices are assumed to have a negligible effect on the operation of the demodulator in the subsequent analysis. Thus, the front-end circuitry is omitted from Figure 2.14, which shows the optimal demodulator in the form of a correlator for the detection of a single symbol of a direct-sequence signal with BPSK in the presence of white Gaussian noise. This correlator is more practical and flexible for digital processing than the alternative shown in Figure 2.2. It is a suboptimal but reasonable approach against non-Gaussian interference. Matched-filter implementation is not practical for a long sequence that extends over many data symbols. Consequently, long sequences are processed as periodically changing short sequences, and the matched filter for a symbol replaces one for the entire signal.

In the correlator of Figure 2.14, the output of a synchronization system is applied to a mixer that removes the carrier frequency of the received signal and thereby provides a baseband input to the chip-matched filter. The output of the chip-matched filter is applied to the analog-to-digital converter (ADC) that uses the synchronized chip-rate clock to produce chip-rate samples. These samples are

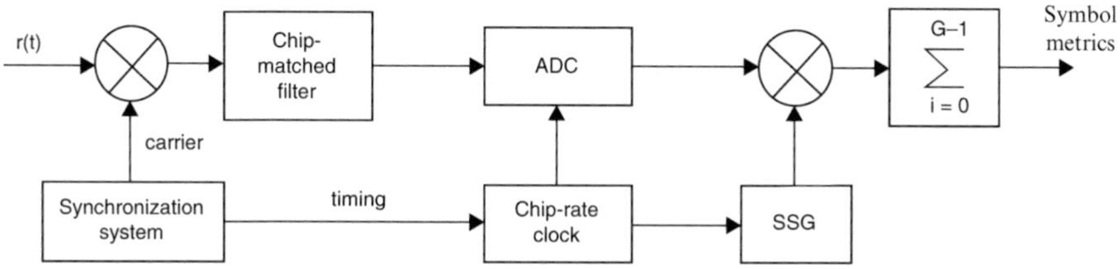

Fig. 2.14 Basic elements of the correlator for direct-sequence signal with coherent BPSK

applied to a despreader consisting of a mixer followed by an accumulator. The samples are multiplied by the synchronized chips generated by the spreading-sequence generator (SSG). Then, G successive products are added in an accumulator to produce the symbol metrics at the symbol rate. The symbol metrics are applied to a soft-decision decoder, or hard decisions are made on successive symbol metrics to produce a symbol sequence that is applied to a hard-decision decoder.

The sequence generator, multiplier, and adder comprise a discrete-time matched filter that is matched to each G-bit sequence of the spreading sequence. The matched filter has a fixed impulse response for short spreading sequences and has a time-varying impulse response for long spreading sequences. Since the chip waveform is of short duration, the response of this matched filter to a G-bit sequence is insignificantly affected by previous G-bit sequences. Thus, if the multipath delay spread (Section 6.3) is less than the data-symbol duration T_s and the spreading factor G is adequate, then the *intersymbol interference* is negligible, as explained in Section 6.10. *The lack of significant intersymbol interference is an important advantage of direct-sequence communications* and is always assumed in this chapter.

In the subsequent analysis, perfect phase, frequency, sequence, and symbol synchronization are assumed. Although higher sampling rates may be advantageous in practical systems, chip-rate sampling suffices in principle and is assumed in the analysis. The received signal is

$$r(t) = s(t) + i(t) + n(t) \tag{2.73}$$

where $i(t)$ is the interference, and $n(t)$ denotes the zero-mean, white Gaussian noise. The output of the chip-matched filter, which is matched to the chip waveform, is sampled at the chip rate to provide G samples per data symbol. We assume that the Nyquist criterion is approximately satisfied so that the interchip interference is negligible. If $d(t) = d_0$ over $[0, T_s]$ and $f_c >> 1/T_c$, then (2.69) to (2.73) and Figure 2.14 indicate that the ith received chip associated with this data symbol is

$$Z_i = \sqrt{2}\int_{iT_c}^{(i+1)T_c} r(t)\psi(t - iT_c)\cos 2\pi f_c t\, dt = S_i + J_i + N_{si}, \quad 0 \le i \le G-1 \tag{2.74}$$

where the $\sqrt{2}$ is introduced for mathematical convenience, and

$$S_i = \sqrt{2}\int_{iT_c}^{(i+1)T_c} s(t)\psi(t - iT_c)\cos 2\pi f_c t\, dt = d_0\frac{\sqrt{\mathcal{E}_s}}{G} \tag{2.75}$$

$$J_i = \sqrt{2}\int_{iT_c}^{(i+1)T_c} i(t)\psi(t - iT_c)\cos 2\pi f_c t\, dt \tag{2.76}$$

$$N_{si} = \sqrt{2}\int_{iT_c}^{(i+1)T_c} n(t)\psi(t - iT_c)\cos 2\pi f_c t\, dt. \tag{2.77}$$

The symbol metric, which is proportional to the despread symbol in the absence of interference and noise, is

$$V = \sum_{i=0}^{G-1} p_i Z_i = d_0\sqrt{\mathcal{E}_s} + V_1 + V_2 \tag{2.78}$$

where

$$V_1 = \sum_{i=0}^{G-1} p_i J_i \tag{2.79}$$

$$V_2 = \sum_{i=0}^{G-1} p_i N_{si}. \tag{2.80}$$

The white Gaussian noise has autocorrelation

$$R_n(\tau) = \frac{N_0}{2}\delta(\tau) \tag{2.81}$$

where $N_0/2$ is the two-sided noise PSD. As explained in Section 1.2, the $\{N_{si}\}$ are real-valued, zero-mean, independent, identically distributed Gaussian random variables. Since V_2 is a linear combination of these independent random variables, V_2 is a Gaussian random variable. Assuming that $f_c >> 1/T_c$, (2.77), (2.80), and (2.81) yield

$$E[V_2] = 0, \quad var\,(V_2) = \frac{N_0}{2}. \tag{2.82}$$

It is natural and analytically desirable to model a long spreading sequence as a random binary sequence. This model does not seem to obscure important exploitable characteristics of long sequences and is a reasonable approximation even for short sequences in networks with asynchronous communications. A random binary sequence consists of statistically independent symbols, each of which takes the value +1 with probability 1/2 or the value −1 with probability 1/2. Thus, $E[p_i] = E[p(t)] = 0$. It then follows from (2.79) that $E[V_1] = 0$. Since p_i and p_k are independent for $i \neq k$,

$$E[p_i p_k] = 0, \quad i \neq k \tag{2.83}$$

and hence

$$E[V_1] = 0, \;\; var\,(V_1) = \sum_{i=0}^{G-1} E\left[J_i^2\right]. \tag{2.84}$$

Therefore, the mean and variance of the symbol metric are

$$E[V] = d_0\sqrt{\mathcal{E}_s}, \;\; var\,(V) = \frac{N_0}{2} + var\,(V_1)\,. \tag{2.85}$$

Consider hard decisions based on the symbol metrics. If $d_0 = +1$ represents the logic symbol 1 and $d_0 = -1$ represents the logic symbol 0, then the decision device produces the symbol 1 if $V > 0$ and the symbol 0 if $V < 0$. An error occurs if $V < 0$ when $d_0 = +1$ or if $V > 0$ when $d_0 = -1$. The probability that $V = 0$ is zero. When $d_0 = +1$, $E[V] = |E[V]|$, and when $d_0 = -1$, $E[V] = -|E[V]|$. Therefore, if V has a Gaussian distribution, then a straightforward evaluation indicates that the symbol-error probability is

$$P_s = Q\left[\frac{|E[V]|}{\sqrt{var\,(V)}}\right] \tag{2.86}$$

where $Q(x)$ is defined by (1.58).

Tone Interference at Carrier Frequency

For tone interference with the same carrier frequency as the desired signal, a nearly exact, closed-form equation for the symbol error probability can be derived. The tone interference has the form

$$i\,(t) = \sqrt{2I}\cos\left(2\pi f_c t + \phi\right) \tag{2.87}$$

where I is the average power and ϕ is the phase relative to the desired signal. Assuming that $f_c >> 1/T_c$, then (2.76), (2.79), (2.87), and a change of variables give

$$V_1 = \sqrt{I}\cos\phi\sum_{i=0}^{G-1} p_i\int_0^{T_c}\psi\,(t)\,dt. \tag{2.88}$$

Let k_1 denote the number of chips in $[0, T_s]$ for which $p_i = +1$; the number for which $p_i = -1$ is $G - k_1$. Equations (2.88), (2.71), and (2.72) yield

$$V_1 = \sqrt{\frac{I\kappa}{T_s}} T_c (2k_1 - G) \cos\phi \tag{2.89}$$

where κ depends on the chip waveform, and

$$\kappa = \begin{cases} 1, & \textit{rectangular chip} \\ \frac{8}{\pi^2}, & \textit{sinusoidal chip}. \end{cases} \tag{2.90}$$

These equations indicate that the use of sinusoidal chip waveforms instead of rectangular ones effectively reduces the interference power by a factor $8/\pi^2$ if $V_1 \neq 0$. Thus, the advantage of sinusoidal chip waveforms is 0.91 dB against tone interference at the carrier frequency. Equation (2.89) indicates that tone interference at the carrier frequency would be completely rejected if $k_1 = G/2$ in every symbol interval.

In the random-binary-sequence model, p_i is equally likely to be $+1$ or -1. Therefore, the conditional symbol error probability given the value of ϕ is

$$P_s(\phi) = \sum_{k_1=0}^{G} \binom{G}{k_1} \left(\frac{1}{2}\right)^G \left[\frac{1}{2} P_s(\phi, k_1, +1) + \frac{1}{2} P_s(\phi, k_1, -1)\right] \tag{2.91}$$

where $P_s(\phi, k_1, d_0)$ is the conditional symbol error probability given the values of ϕ, k_1, and d_0. Under these conditions, V_1 is a constant, and hence V has a Gaussian distribution. Equations (2.78) and (2.89) imply that the conditional expected value of V is

$$E[V|\phi, k_1, d_0] = d_0\sqrt{\mathcal{E}_s} + \sqrt{\frac{I\kappa}{T_s}} T_c (2k_1 - G)\cos\phi. \tag{2.92}$$

The conditional variance of V is equal to the variance of V_2, which is given by (2.82). Using (2.86) to evaluate $P_s(\phi, k_1, +1)$ and $P_s(\phi, k_1, -1)$ separately and then consolidating the results yields

$$P_s(\phi, k_1, d_0) = Q\left[\sqrt{\frac{2\mathcal{E}_s}{N_0}} + d_0\sqrt{\frac{2IT_c\kappa}{GN_0}} (2k_1 - G)\cos\phi\right]. \tag{2.93}$$

Assuming that ϕ is uniformly distributed over $[0, 2\pi)$ during each symbol interval and exploiting the periodicity of $\cos\phi$, we obtain the symbol error probability

$$P_s = \frac{1}{\pi}\int_0^{\pi} P_s(\phi)\, d\phi \tag{2.94}$$

where $P_s(\phi)$ is given by (2.91) and (2.93).

General Tone Interference

To simplify the preceding equations for P_s and to examine the effects of tone interference with a carrier frequency different from the desired frequency, a Gaussian approximation is used. Consider interference due to a single tone of the form

$$i(t) = \sqrt{2I}\cos(2\pi f_1 t + \theta_1) \tag{2.95}$$

where I, f_1, and θ_1 are the average power, frequency, and phase angle of the interference signal at the receiver, respectively. The frequency f_1 is assumed to be close enough to the desired frequency f_c that the tone is undisturbed by the initial wideband filtering that precedes the correlator. If $f_1 + f_c >> f_d = f_1 - f_c$ so that a term involving $f_1 + f_c$ is negligible, then (2.95), (2.76), and a change of variable yield

$$J_i = \sqrt{I}\int_0^{T_c} \psi(t)\cos(2\pi f_d t + \theta_1 + i2\pi f_d T_c)dt. \tag{2.96}$$

Substitution into (2.84) gives

$$var(V_1) = I\sum_{i=0}^{G-1}\left[\int_0^{T_c} \psi(t)\cos(2\pi f_d t + \theta_1 + i2\pi f_d T_c)dt\right]^2. \tag{2.97}$$

For a rectangular chip waveform, evaluation of the integral and trigonometry yields

$$var(V_1) = \frac{IT_c}{G}\text{sinc}^2(f_d T_c)\sum_{i=0}^{G-1}[\cos(i2\pi f_d T_c + \theta_2)]^2 \tag{2.98}$$

where

$$\theta_2 = \theta_1 + \pi f_d T_c. \tag{2.99}$$

Expanding the squared cosine, we obtain

$$var(V_1) = \frac{IT_c}{2G}\text{sinc}^2(f_d T_c)\left[G + \sum_{i=0}^{G-1}\cos(i4\pi f_d T_c + 2\theta_2)\right]. \tag{2.100}$$

To evaluate the inner summation, we use the identity

$$\sum_{\nu=0}^{n-1}\cos(a + \nu b) = \cos\left(a + \frac{n-1}{2}b\right)\frac{\sin(nb/2)}{\sin(b/2)} \tag{2.101}$$

which is proved by using mathematical induction and trigonometric identities. Evaluation and simplification yield

$$var(V_1) = \frac{IT_c}{2}\text{sinc}^2\left(f_d T_c\right)\left[1 + \frac{\text{sinc}\left(2f_d T_s\right)}{\text{sinc}\left(2f_d T_c\right)}\cos 2\phi\right] \tag{2.102}$$

where

$$\phi = \theta_2 + \pi f_d\left(T_s - T_c\right) = \theta_1 + \pi f_d T_s. \tag{2.103}$$

We now make the plausible assumption that the conditional distribution of V_1 given the value of ϕ is approximately Gaussian if $G >> 1$. The independence of the thermal noise and the interference then implies that the conditional distribution of V is approximately Gaussian with mean and variance given by (2.85). Using (2.102) and (2.86), we find that the conditional symbol error probability for rectangular chip waveforms is approximated by

$$P_s\left(\phi\right) = Q\left[\sqrt{\frac{2\mathcal{E}_s}{N_{0e}(\phi)}}\right] \tag{2.104}$$

where

$$N_{0e}(\phi) = N_0 + IT_c\text{sinc}^2\left(f_d T_c\right)\left[1 + \frac{\text{sinc}\left(2f_d GT_c\right)}{\text{sinc}\left(2f_d T_c\right)}\cos 2\phi\right] \tag{2.105}$$

and $N_{0e}(\phi)/2$ can be interpreted as the *equivalent two-sided PSD* of the interference and noise, given the value of ϕ. For sinusoidal chip waveforms, the substitution of (2.72) into (2.97), and then the use of trigonometric identities, simple integrations, and (2.101) yields (2.104) with

$$N_{0e}(\phi) = N_0 + IT_c\left(\frac{8}{\pi^2}\right)\left(\frac{\cos \pi f_d T_c}{1 - 4f_d^2 T_c^2}\right)^2\left[1 + \frac{\text{sinc}\left(2f_d GT_c\right)}{\text{sinc}\left(2f_d T_c\right)}\cos 2\phi\right]. \tag{2.106}$$

Equations (2.105) and (2.106) indicate that a sinusoidal chip waveform provides a $\pi^2/8 = 0.91$ dB advantage relative to a rectangular chip waveform when $f_d = 0$, but this advantage decreases as $|f_d|$ increases and ultimately disappears.

If θ_1 in (2.103) is modeled as a random variable that is uniformly distributed over $[0, 2\pi)$ during each symbol interval, then the modulo-2π character of $\cos 2\phi$ in (2.105) implies that its distribution is unchanged if ϕ is assumed to be uniformly distributed over $[0, 2\pi)$. The symbol error probability, which is obtained by averaging $P_s(\phi)$ over the range of ϕ, is

$$P_s = \frac{2}{\pi}\int_0^{\pi/2} Q\left[\sqrt{\frac{2\mathcal{E}_s}{N_{0e}\left(\phi\right)}}\right] d\phi \tag{2.107}$$

where the fact that $\cos 2\phi$ takes all its possible values over $[0, \pi/2]$ has been used to shorten the integration interval.

Figure 2.15 depicts the symbol error probability as a function of the *despread signal-to-interference ratio*, $\mathcal{E}_s/IT_c$, for one tone-interference signal, rectangular chip waveforms, $f_d = 0, G = 50 = 17$ dB, and $\mathcal{E}_s/N_0 = 14$ and 20 dB. One pair of graphs are computed using the approximate model of (2.105) and (2.107), whereas the other pair are derived from the nearly exact model of (2.91), (2.93), and (2.94) with $\kappa = 1$. For the nearly exact model, P_s depends not only on $\mathcal{E}_s/IT_c$, but also on G. A comparison of the two graphs indicates that the error introduced by the Gaussian approximation is on the order of or less than 0.1 dB when $P_s \geq 10^{-6}$. This example and others provide evidence that the Gaussian approximation introduces insignificant error if $G \geq 50$ and practical values for the other parameters are assumed.

Figure 2.16 uses the approximate model to plot P_s versus the normalized frequency offset $f_d T_c$ for rectangular and sinusoidal chip waveforms, $G = 17$ dB, $\mathcal{E}_s/N_0 = 14$ dB, and $\mathcal{E}_s/IT_c = 10$ dB. There is a sharp decrease in the symbol error probability even for small values of the frequency offset. The performance advantage of sinusoidal chip waveforms is apparent, but their realization or that of Nyquist chip waveforms in a transmitted BPSK waveform is difficult because of the distortion introduced by a nonlinear power amplifier in the transmitter when the signal does not have a constant amplitude.

Gaussian Interference

Gaussian interference is interference $i(t)$ that approximates a zero-mean, Gaussian process (Appendix A.1) for which the $\{J_i\}$ are not only Gaussian but also independent random variables. Assuming that $\psi(t)$ is rectangular,

$$E\left[J_i^2\right] = \frac{1}{T_s}\int_0^{T_c}\int_0^{T_c} R\left(t_1 - t_2\right)\cos\left(2\pi f_c t_1\right)\cos\left(2\pi f_c t_2\right) dt_1 dt_2 \tag{2.108}$$

where $R(t)$ is the autocorrelation of $i(t)$. After a trigonometric expansion, we change variables by using $\tau = t_1 - t_2$ and $s = t_1 + t_2$. Since the Jacobian of this transformation is 2, we obtain

$$\begin{aligned} E\left[J_i^2\right] &= \frac{1}{2T_s}\int_{-T_c}^{T_c} R(\tau)d\tau\left[\int_{|\tau|}^{2T_c-|\tau|}(\cos 2\pi f_c\tau + \cos 2\pi f_c s)ds\right] \\ &\simeq \frac{1}{G}\int_{-T_c}^{T_c} R(\tau)\Lambda\left(\frac{\tau}{T_c}\right)\cos 2\pi f_c\tau \, d\tau \end{aligned} \tag{2.109}$$

where the approximation is valid if $2\pi f_c T_c >> 1$.

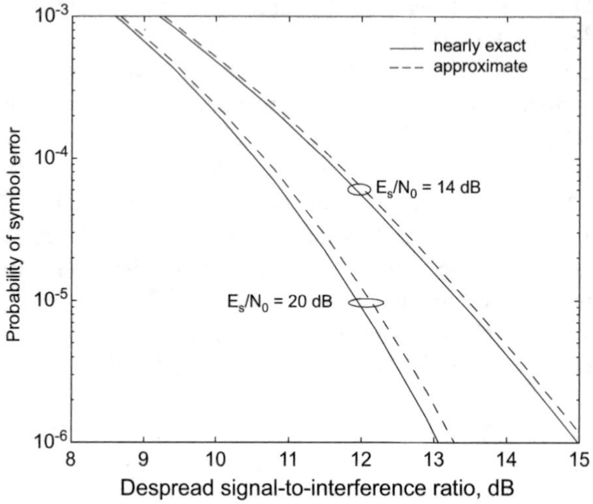

Fig. 2.15 Symbol error probability of the direct-sequence system with BPSK and tone interference at carrier frequency and $G = 17\,\text{dB}$

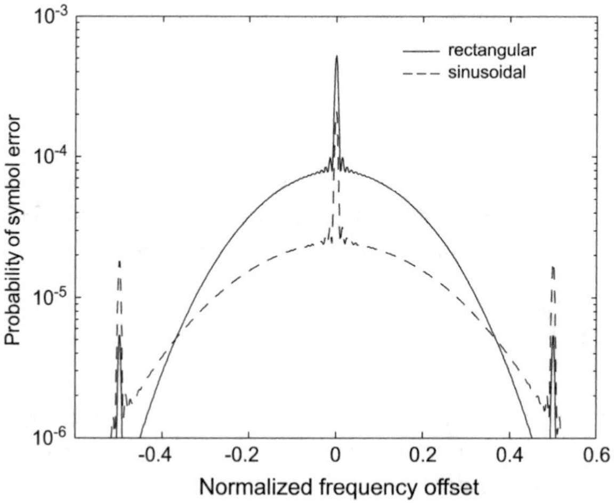

Fig. 2.16 Symbol error probability for direct-sequence system with BPSK, rectangular and sinusoidal chip waveforms, $G = 17\,\text{dB}$, $\mathcal{E}_s/N_0 = 14\,\text{dB}$, and $\mathcal{E}_s/IT_c = 10\,\text{dB}$ in the presence of tone interference

Since $E[J_i^2]$ does not depend on the index i, V_1 is the sum of G independent, identically distributed Gaussian random variables. Therefore, V_1 has a Gaussian distribution function, and (2.84) gives

$$var(V_1) = GE\left[J_i^2\right] \approx \int_{-\infty}^{\infty} R(\tau)\Lambda\left(\frac{\tau}{T_c}\right)\cos 2\pi f_c\tau\, d\tau \tag{2.110}$$

where the integration limits are extended to $\pm\infty$ because the integrand is truncated. Since $R(\tau)\Lambda\left(\frac{\tau}{T_c}\right)$ is an even function, the cosine function may be replaced by a complex exponential. Then, the convolution theorem and the Fourier transform of $\Lambda(t)$ yield the alternative form

$$var(V_1) \approx T_c \int_{-\infty}^{\infty} S(f)\,\text{sinc}^2\left[(f - f_c)\,T_c\right] df \tag{2.111}$$

where $S(f)$ is the PSD of the interference after passage through the initial wideband filter of the receiver.

Since the independence of the thermal noise and the interference implies that $V_1 + V_2$ is the sum of independent Gaussian random variables, V has a Gaussian distribution function. The mean and variance of V are given by (2.85) with $var(V_1)$ given by (2.111). Thus, (2.86) yields the symbol error probability:

$$P_s = Q\left(\sqrt{\frac{2\mathcal{E}_s}{N_{0e}}}\right) \tag{2.112}$$

where

$$N_{0e} \approx N_0 + 2T_c \int_{-\infty}^{\infty} S(f)\,\text{sinc}^2\left[(f - f_c)\,T_c\right] df. \tag{2.113}$$

Suppose that the PSD of the interference is

$$S(f) = \begin{cases} \frac{I}{2W_1}, & |f - f_1| \le \frac{W_1}{2}, \quad |f + f_1| \le \frac{W_1}{2} \\ 0, & \text{otherwise.} \end{cases} \tag{2.114}$$

If $f_c >> 1/T_c$, the integration over negative frequencies in (2.113) is negligible and

$$N_{0e} \approx N_0 + \frac{IT_c}{W_1} \int_{f_1 - W_1/2}^{f_1 + W_1/2} \text{sinc}^2\left[(f - f_c)T_c\right] df. \tag{2.115}$$

which indicates that $f_1 \approx f_c$ increases the impact of the interference power.

Since the integrand in (2.115) is upper-bounded by unity, $N_{0e} < N_0 + IT_c$. Equation (2.112) yields

$$P_s < Q\left(\sqrt{\frac{2\mathcal{E}_s}{N_0 + IT_c}}\right). \tag{2.116}$$

A plot of this upper bound with the parameter values of Figure 2.15 indicates that roughly 2 dB more interference power is required for worst-case Gaussian interference to degrade P_s as much as tone interference at the carrier frequency.

Complex Binary Spreading Sequences

A *complex binary spreading sequence* is a sequence comprising two binary sequences that serve as real and imaginary components of the spreading sequence. Although they confer no performance advantage, spreading by complex binary sequences is sometimes used to enable a reduction in the peak-to-average power ratio of the transmitted signal.

Consider a complex binary spreading sequence represented by the row vector $\mathbf{p} = \mathbf{p}_1 + j\mathbf{p}_2$, where $\mathbf{p}_1$ and $\mathbf{p}_2$ are binary sequences that have component or chip values $\pm 1/\sqrt{2}$, and $j = \sqrt{-1}$. Similarly, a complex binary data sequence is represented by $\mathbf{d} = \mathbf{d}_1 + j\mathbf{d}_2$, where $\mathbf{d}_1$ and $\mathbf{d}_2$ are binary sequences that have bit values $\pm 1/\sqrt{2}$. If the spreading factor is G, there is one data bit for every G chips. As shown in Figure 2.17, the product of the complex multiplication of bit $d_i = d_{1i} + jd_{2i}$ by chip $p_k = p_{1k} + jp_{2k}$ is $y_k = d_i p_k = y_{1k} + jy_{2k}$, where y_k is a chip of the transmitted sequence, and

$$y_{1k} = d_{1i}p_{1k} - d_{2i}p_{2k}, \quad y_{2k} = d_{2i}p_{1k} + d_{1i}p_{2k}. \tag{2.117}$$

If the waveforms $d(t)$ and $p(t)$ are associated with the sequences $\mathbf{d}$ and $\mathbf{p}$, respectively, then the transmitted signal may be represented as

$$s(t) = \text{Re}\left\{Ad(t)p(t)e^{j2\pi f_c t}\right\} \tag{2.118}$$

where Re$\{x\}$ denotes the real part of x, and A is the amplitude.

The receiver extracts a signal proportional to $y(t) = d(t)p(t)$. Since the complex symbols have unity magnitude, the despreading entails the complex multiplication of $y(t)$ by the conjugate sequence $p^*(t)$ to produce $d(t)|p(t)|^2 = d(t)$ if the chip waveforms are rectangular. A representation of the receiver in terms of complex variables is illustrated in Figure 2.18. The real and imaginary parts of the downconverted signal are applied to chip-matched filers and sampled. For G chips per data bit, the summation of G multiplications by the complex spreading sequence produces a symbol metric. An actual implementation of this receiver uses the quadrature downconverter described in the next section.

Complex sequences ensure balanced power in the in-phase and quadrature branches of the transmitter, which limits the peak-to-average power fluctuations. Suppose that different bit rates or quality-of-service requirements make it desirable for the data sequences $d_1(t)$ and $d_2(t)$ to have unequal amplitudes d_1 and d_2, respectively. If the symbols of $d_1(t)$ and $d_2(t)$ are zero-mean, antipodal, and

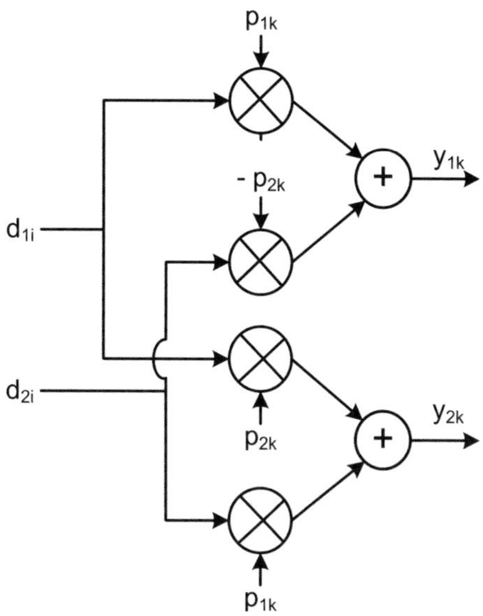

Fig. 2.17 Product of complex binary spreading chip k and complex binary data bit i

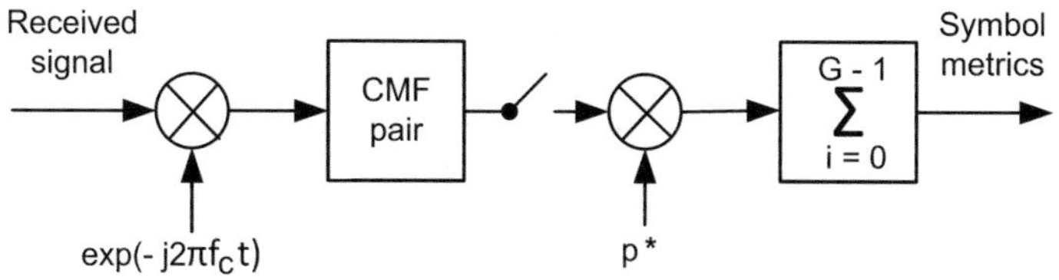

Fig. 2.18 Receiver for direct-sequence system with complex binary spreading sequences. *CMF* chip-matched filter

independent, the chip waveforms are rectangular, and $p_1^2(t) = p_2^2(t) = 1/2$, then $E[y_1^2(t)] = E[y_2^2(t)] = (d_1^2 + d_2^2)/2$. This result indicates that the power in the in-phase and quadrature components of $y(t) = p(t)d(t)$ are equal, despite any disparity between d_1^2 and d_2^2. Complex-valued quaternary sequences are considered in Section 7.2.

2.5 Quaternary Systems

A received *quaternary direct-sequence signal* with perfect phase synchronization and a chip waveform of duration T_c can be represented by

$$s(t) = \sqrt{\mathcal{E}_s}d_1(t)p_1(t)\cos 2\pi f_c t + \sqrt{\mathcal{E}_s}d_2(t+t_0)p_2(t+t_0)\sin 2\pi f_c t \tag{2.119}$$

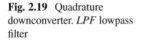

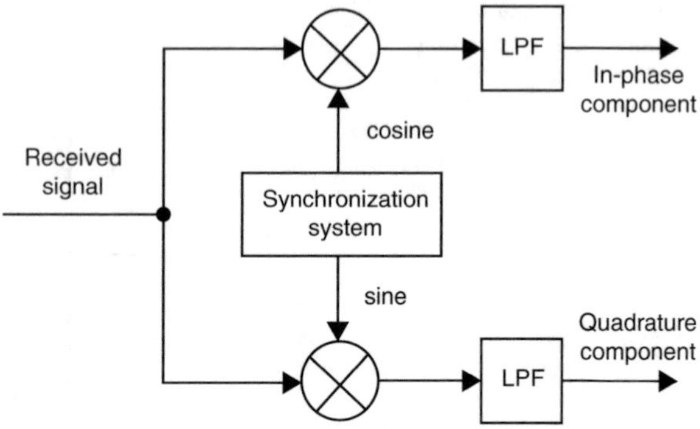

Fig. 2.19 Quadrature downconverter. *LPF* lowpass filter

where two spreading waveforms, $p_1(t)$ and $p_2(t)$, and two data signals, $d_1(t)$ and $d_2(t)$, are used with two quadrature carriers, and t_0 is the relative delay between the in-phase and quadrature components of the signal. For a *quadriphase direct-sequence system*, which uses quadriphase-shift keying (QPSK), $t_0 = 0$. For a direct-sequence system with offset QPSK (OQPSK) or minimum-shift keying (MSK), $t_0 = T_c/2$. For OQPSK, the chip waveforms are rectangular; for MSK, which can be represented as two offset components, the chips are sinusoidal. The OQPSK waveform lacks the 180° phase transitions of QPSK, which limits spectral regrowth after filtering and nonlinear amplification. However, in the absence of filtering to limit sidelobes, the PSDs of OQPSK and QPSK are the same. MSK may be used to limit the spectral sidelobes of the direct-sequence signal, which may interfere with other signals.

Let T_s denote the duration of the data symbols or code bits before the generation of (2.119), and let $T_{s1} = 2T_s$ denote the duration of each of the binary channel-symbol components, which are transmitted in pairs. Of the available desired-signal power, half is in each of the two components of (2.119). Since $T_{s1} = 2T_s$, the energy per binary channel-symbol component is $\mathcal{E}_s$, the same as for a direct-sequence system with BPSK.

Let T_c denote the common duration of the chips of the spreading waveforms $p_1(t)$ and $p_2(t)$, each of which has the form of (2.2). The spreading factor of each binary channel-symbol component is the positive integer $G_1 = T_{s1}/T_c$. This spreading factor is twice that of a BPSK system with the same values of T_c and T_s. However, we cannot expect any great improvement in the suppression of interference. In both binary and quaternary systems, the despreading distributes the interference over a similar spectral band largely determined by T_c and the chip waveform. The chip waveform has normalized energy according to (2.70).

After passing through the receiver front end, the received signal $r(t)$ is applied to a *quadrature downconverter* that produces in-phase and quadrature components near baseband, as illustrated in Figure 2.19. The pair of mixers have inputs from a phase synchronization system that generates sinusoidal signals at frequency f_c. The mixer outputs are passed through lowpass filters (LPFs) to remove the double-frequency components. The filter outputs are in-phase and quadrature components

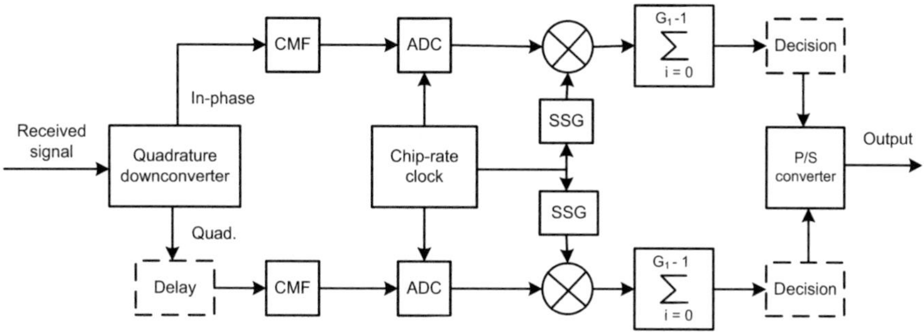

Fig. 2.20 Receiver for direct-sequence signal with dual quaternary modulation. *CMF* chip-matched filter, *SSG* spreading-sequence generator, *P/S* parallel-to-serial, *ADC* analog-to-digital converter. Delay = 0 for QPSK; delay = $T_c/2$ for OQPSK and MSK

proportional to filtered versions of $r(t)\cos 2\pi f_c t$ and $r(t)\sin 2\pi f_c t$ respectively. If $r(t)$ is bandlimited and the output signals are set equal to the real and negative imaginary parts of a complex-valued signal, then the latter represents the complex envelope of $r(t)$, as shown in Appendix D.1.

We assume that the timing synchronization is perfect and that the Nyquist criterion is approximately satisfied so that the interchip interference is negligible. The lowpass filters in the quadrature downconverter are sufficiently wideband compared with the chip-matched filters that their ultimate effect on the interference and noise is negligible. Because of the perfect phase synchronization, crosstalk terms are double-frequency terms that are blocked by the lowpass filters in the quadrature downconverter.

Dual Quaternary System

Consider the *classical* or *dual* quaternary system in which $d_1(t)$ and $d_2(t)$ are independent. As illustrated in Figure 2.20, the in-phase and quadrature output signals of the quadrature downconverter are applied to chip-matched filters with outputs that are sampled at the chip rate by analog-to-digital converters (ADCs). The ADC outputs are the demodulated signals, which are then despread. The decision devices are present when hard-decision decoding is used. Output symbols from the two decision devices are applied to a parallel-to-serial (P/S) converter, the output of which is applied to the decoder. When soft-decision decoding is used, the decision devices are absent, and the symbol metrics are directly applied to the P/S converter and then the decoder.

If the received signal is given by (2.119), then the upper symbol metric applied to the decision device at the end of a symbol interval during which $d_1(t) = d_{10}$ is

$$V = d_{10}\sqrt{2\mathcal{E}_s} + V_1 + V_2 \tag{2.120}$$

where

$$V_1 = \sum_{i=0}^{G_1-1} p_{1i}J_i, \; V_2 = \sum_{i=0}^{G_1-1} p_{1i}N_{si} \tag{2.121}$$

and J_i and N_{si} are given by (2.76) and (2.77), respectively. Similarly, the lower symbol metric at the end of a channel-symbol interval during which $d_2(t) = d_{20}$ is

$$U = d_{20}\sqrt{2\mathcal{E}_s} + U_1 + U_2 \tag{2.122}$$

where

$$U_1 = \sum_{i=0}^{G_1-1} p_{2i}J_i', \; U_2 = \sum_{i=0}^{G_1-1} p_{2i}N_i' \tag{2.123}$$

$$J_i' = \sqrt{2}\int_{iT_c}^{(i+1)T_c} i(t)\psi\,(t - iT_c)\sin 2\pi f_c t\; dt \tag{2.124}$$

$$N_i' = \sqrt{2}\int_{iT_c}^{(i+1)T_c} n(t)\psi\,(t - iT_c)\sin 2\pi f_c t\; dt. \tag{2.125}$$

Of the available desired-signal power S, half is in each of the two components of (2.119). Since $T_{s1} = 2T_s$, the energy per channel-symbol component is $\mathcal{E}_s = ST_s$, the same as for a direct-sequence system with BPSK, and

$$E[V] = d_{10}\sqrt{2\mathcal{E}_s}, \quad E(U) = d_{20}\sqrt{2\mathcal{E}_s}. \tag{2.126}$$

For the reasons stated in Section 2.4, both V_2 and U_2 are real-valued, zero-mean, identically distributed Gaussian random variables. Since $n(t)$ is white noise with PSD $N_0/2$, we find that

$$var\,(V_2) = var\,(U_2) = N_0. \tag{2.127}$$

Consider the general tone-interference model. We assume that both V_1 and U_1 have distributions that are approximately Gaussian. Averaging the error probabilities for the two parallel symbol streams of the dual quaternary system, we obtain the conditional symbol error probability:

$$P_s(\phi) = \frac{1}{2}Q\left[\sqrt{\frac{2\mathcal{E}_s}{N_{0e}^{(0)}(\phi)}}\right] + \frac{1}{2}Q\left[\sqrt{\frac{2\mathcal{E}_s}{N_{0e}^{(1)}(\phi)}}\right] \tag{2.128}$$

where $N_{0e}^{(0)}(\phi)$ and $N_{0e}^{(1)}(\phi)$ arise from the upper and lower branches of Figure 2.20 respectively. The equivalent two-sided PSD $N_{0e}^{(0)}(\phi)$ is given by (2.105) and (2.106), with symbol duration $T_{s1} = 2T_s$ in place of T_s. To calculate $N_{0e}^{(1)}(\phi)$, we use (2.95), a trigonometric expansion, and a change of integration variable in (2.124). If $f_1 + f_c >> f_d = f_1 - f_c$ so that a term involving $f_1 + f_c$ is negligible, we obtain

$$J_i' = -\sqrt{I}\int_0^{T_c} \psi(t)\sin(2\pi f_d t + \theta_1 + i2\pi f_d T_c)dt \tag{2.129}$$

which is the same as (2.96) with θ_1 replaced by $\theta_1 + \pi/2$. Therefore, $N_{0e}^{(1)}(\phi)$ is given by (2.105) and (2.106) with $T_s \rightarrow T_{s1} = G_1 T_c$ and $\phi \rightarrow \phi + \pi/2$. For rectangular chip waveforms (QPSK and OQPSK signals),

$$N_{0e}^{(l)}(\phi) = N_0 + IT_c \text{sinc}^2(f_d T_c)\left[1 + \frac{\text{sinc}(2f_d G_1 T_c)}{\text{sinc}(2f_d T_c)}\cos(2\phi + l\pi)\right] \tag{2.130}$$

and for sinusoidal chip waveforms,

$$N_{0e}^{(l)}(\phi) = N_0 + IT_c\left(\frac{8}{\pi^2}\right)\left(\frac{\cos\pi f_d T_c}{1 - 4f_d^2 T_c^2}\right)^2\left[1 + \frac{\text{sinc}(2f_d G_1 T_c)}{\text{sinc}(2f_d T_c)}\cos(2\phi + l\pi)\right] \tag{2.131}$$

where

$$\phi = \theta_1 + 2\pi f_d T_s,\ l = 0, 1. \tag{2.132}$$

These equations indicate that $P_s(\phi)$ for a quaternary direct-sequence system and the worst value of ϕ is usually lower than $P_s(\phi)$ for a binary direct-sequence system with the same chip waveform and the worst value of ϕ. The symbol error probability is determined by integrating $P_s(\phi)$ over the distribution of ϕ during a symbol interval. For a uniform distribution, the two integrals are equal. Using the periodicity of $\cos 2\phi$ to shorten the integration interval, we obtain

$$P_s = \frac{2}{\pi}\int_0^{\pi/2} Q\left[\sqrt{\frac{2\mathcal{E}_s}{N_{0e}^{(0)}(\phi)}}\right]d\phi. \tag{2.133}$$

The quaternary system provides a slight advantage relative to the binary system against tone interference. Both systems provide the same P_s when $f_d = 0$ and nearly the same P_s when $f_d > 1/T_s$. Figure 2.21 illustrates P_s versus the normalized frequency offset $f_d T_c$ for quaternary and binary systems, $G = 17$ dB, $\mathcal{E}_s/N_0 = 14$ dB, and $\mathcal{E}_s/IT_c = 10$ dB.

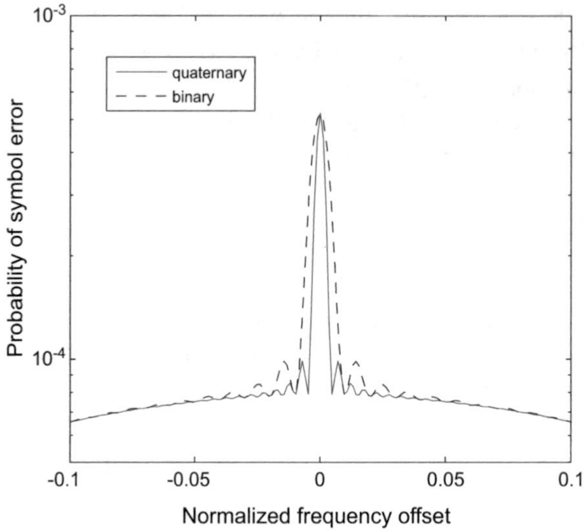

Fig. 2.21 Symbol error probability for quaternary and binary direct-sequence systems with $G =$ 17 dB, $\mathcal{E}_s/N_0 = 14$ dB, and $G\mathcal{E}_s/IT_s = 10$ dB in the presence of tone interference

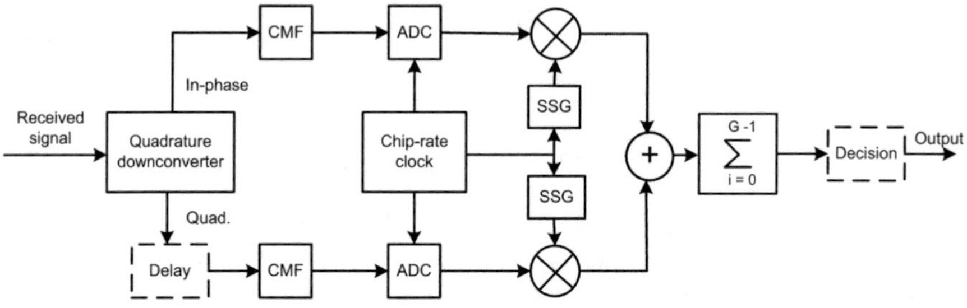

Fig. 2.22 Receiver for direct-sequence signal with balanced quaternary modulation (delay = 0 for QPSK and delay = $T_c/2$ for OQPSK and MSK); *CMF* chip-matched filter; *ADC* analog-to-digital converter, *SSG* spreading-sequence generator

Balanced Quaternary System

In a *balanced quaternary system,* the same data symbols are carried by both the in-phase and quadrature components, which implies that the received direct-sequence signal has the form given by (2.119) with $d_1(t) = d_2(t) = d(t)$. Thus, although the spreading is performed by quadrature carriers, the data modulation may be regarded as BPSK. A receiver for this system is shown in Figure 2.22.

The synchronization system is assumed to operate perfectly so that the crosstalk terms are negligible. The duration of both a data symbol and a channel symbol is T_s, the chip duration is T_c, and the spreading factor for each quadrature component is $G = T_s/T_c$. If the data symbol is $d_{10} = d_{20} = d_0$, then the symbol metric at the input to the decision device is

$$V = d_0\sqrt{2\mathcal{E}_s} + \sum_{i=0}^{G-1} p_{1i}J_i + \sum_{i=0}^{G-1} p_{2i}J_i' + \sum_{i=0}^{G-1} p_{1i}N_i + \sum_{i=0}^{G-1} p_{2i}N_i'. \tag{2.134}$$

If $p_1(t)$ and $p_2(t)$ are approximated by independent random binary sequences, then the last four terms of (2.134) are zero-mean uncorrelated random variables. Therefore, the variance of V is equal to the sum of the variances of these four random variables, and

$$E[V] = d_0\sqrt{2\mathcal{E}_s}\,. \tag{2.135}$$

Straightforward evaluations verify that both types of quaternary signals provide the same symbol-error probability against Gaussian interference as direct-sequence signals with BPSK, which is defined by (2.112) with N_{0e} given by (2.113) or (2.115) and upper bounded by $N_0 + IT_c$.

A *balanced QPSK system* is a balanced quaternary system with $t_0 = 0$. Assume that a balanced QPSK system receives tone interference. For a rectangular chip waveform, variance calculations of the interference terms in (2.134) similar to those leading to (2.102) indicate that

$$\begin{aligned} var(V) &= N_0 + \frac{1}{2}IT_c\text{sinc}^2\,(f_dT_c)\left[1 + \frac{\text{sinc}\,(2f_dT_s)}{\text{sinc}\,(2f_dT_c)}\cos 2\phi\right] \\ &+ \frac{1}{2}IT_c\text{sinc}^2\,(f_dT_c)\left[1 - \frac{\text{sinc}\,(2f_dT_s)}{\text{sinc}\,(2f_dT_c)}\cos 2\phi\right] \\ &= N_0 + IT_c\text{sinc}^2\,(f_dT_c)\,. \end{aligned} \tag{2.136}$$

Thus, $P_s(\phi)$ is independent of ϕ, and (2.86) yields

$$P_s = P_s\,(\phi) = Q\left(\sqrt{\frac{2\mathcal{E}_s}{N_{0e}}}\right) \tag{2.137}$$

where for rectangular chip waveforms,

$$N_{0e} = N_0 + IT_c\text{sinc}^2\,(f_dT_c)\,. \tag{2.138}$$

Similarly, for sinusoidal chip waveforms,

$$N_{0e} = N_0 + IT_c\left(\frac{8}{\pi^2}\right)\left(\frac{\cos \pi f_dT_c}{1 - 4f_d^2T_c^2}\right)^2. \tag{2.139}$$

If $f_d = 0$ and the interference is given by (2.87), a nearly exact model similar to that in Section 2.4 implies that the conditional symbol error probability is

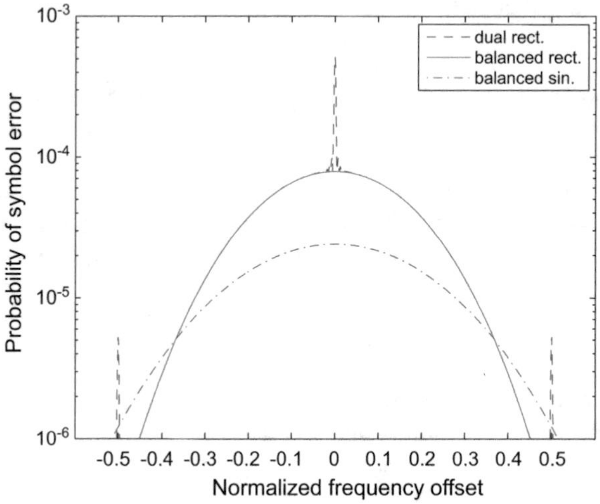

Fig. 2.23 Symbol error probability for direct-sequence systems with balanced QPSK and dual quaternary modulations, rectangular and sinusoidal chip waveforms, $G = 17\,\text{dB}$, $\mathcal{E}_s/N_0 = 14\,\text{dB}$, and $G\mathcal{E}_s/IT_s = 10\,\text{dB}$ in the presence of tone interference

$$P_s(\phi) = \sum_{k_1=0}^{G}\sum_{k_2=0}^{G}\binom{G}{k_1}\binom{G}{k_2}\left(\frac{1}{2}\right)^{2G}\left[\frac{1}{2}P_s(\phi, k_1, k_2, +1) + \frac{1}{2}P_s(\phi, k_1, k_2, -1)\right] \tag{2.140}$$

where k_1 and k_2 are the number of chips in a symbol for which $p_1(t) = +1$ and $p_2(t) = +1$, respectively, and $P_s(\phi, k_1, k_2, d_0)$ is the conditional symbol error probability, given the values of ϕ, k_1, and k_2, and that $d(t) = d_0$.

Substitution of (2.87) into (2.76) and (2.124) gives $J_i = \sqrt{I}\cos\phi$ and $J_i' = -\sqrt{I}\sin\phi$. A derivation analogous to that of (2.93) yields

$$P_s(\phi, k_1, k_2, d_0) = Q\left\{\sqrt{\frac{2\mathcal{E}_s}{N_0}} + d_0\sqrt{\frac{IT_c\kappa}{GN_0}}\left[(2k_1 - G)\cos\phi - (2k_2 - G)\sin\phi\right]\right\}. \tag{2.141}$$

If ϕ is uniformly distributed over $[0, 2\pi)$ during a symbol interval, then

$$P_s = \frac{1}{2\pi}\int_0^{2\pi} P_s(\phi)d\phi. \tag{2.142}$$

Numerical comparisons of the nearly exact model with the approximate results given by (2.137) for $f_d = 0$ indicate that the approximate results typically introduce an insignificant error if $G \geq 50$.

Figure 2.23 illustrates the performance advantage of the balanced QPSK system of Figure 2.22 against tone interference when $f_d < 1/T_s$. Equations (2.128) to (2.133) and (2.137) to (2.139) are used for the dual quaternary and the balanced

QPSK systems, respectively, and $G_1 = 2G$, $G = 17$ dB, $\mathcal{E}_s/N_0 = 14$ dB, and $\mathcal{E}_s/IT_c = 10$ dB. The normalized frequency offset is $f_d T_c$. The advantage of the balanced QPSK system when f_d is small exists because a tone at the carrier frequency cannot have a phase that causes desired-signal cancelation simultaneously in both receiver branches.

Systems with Channel Codes

A major benefit of direct-sequence spread spectrum is that the despreading and filtering in the receiver tend to whiten the interference PSD over the code-symbol passband. If the spreading sequence is modeled as a random binary sequence, then (2.79), (2.121), and (2.123) indicate that the terms of V_1 and U_1 are uncorrelated. However, they are not necessarily independent, and hence the central limit theorem is not applicable. Therefore, V_1 and U_1 do not necessarily have Gaussian distribution functions. If the distribution functions of V_1 and U_1 do approximate Gaussian distributions, then the net effect of the interference after the despreading is similar to what it would have been if the interference and noise were white Gaussian noise with the equivalent two-sided PSD $N_{0e}/2$ or $N_{0e}(\phi)/2$. In this section and the previous one, these PSDs have been derived for tone interference and narrowband Gaussian interference. Chapter 9 presents an analysis and simulation of a direct-sequence system with an LDPC channel code and Gaussian interference. If the distribution functions of V_1 and U_1 do not approximate Gaussian distributions, then the bit error probability must be determined by a simulation.

2.6 Pulsed Interference and Decoding Metrics

Pulsed interference occurs sporadically for brief periods. Whether it is generated unintentionally or by an opponent, pulsed interference can cause a substantial increase in the bit error rate of a communication system relative to the rate caused by continuous interference with the same average power. Pulsed interference may be produced in a receiver by a signal with a variable center frequency that sweeps over a frequency range that intersects or includes the receiver passband.

Consider a direct-sequence system with BPSK that operates in the presence of pulsed interference. Let μ denote either the pulse duty cycle, which is the ratio of the pulse duration to the repetition period, or the probability of pulse occurrence if the pulses occur randomly. During a pulse, the interference is modeled as white Gaussian interference with two-sided PSD $I_0/2\mu$, where $I_0/2$ is the equivalent PSD of continuous interference ($\mu = 1$). Let $N_{0e}/2$ denote the equivalent two-sided noise PSD. In the absence of a pulse, $N_{0e} = N_0$, whereas in the presence of a pulse,

$$N_{0e} = N_0 + I_0/\mu. \tag{2.143}$$

If the interference pulse duration approximately equals or exceeds the channel-symbol duration, then (2.112) implies that the probability of an error in a binary code symbol is

$$P_s \simeq \mu Q\left(\sqrt{\frac{2\mathcal{E}_s}{N_0 + I_0/\mu}}\right) + (1-\mu)\, Q\left(\sqrt{\frac{2\mathcal{E}_s}{N_0}}\right), \quad 0 \le \mu \le 1. \tag{2.144}$$

If $\mathcal{E}_s >> N_0$ and $I_0 >> N_0$, calculus gives the value of μ that maximizes P_s:

$$\mu_0 \simeq \begin{cases} 0.7\left(\frac{\mathcal{E}_s}{I_0}\right)^{-1}, & \frac{\mathcal{E}_s}{I_0} > 0.7 \\ 1\,, & \frac{\mathcal{E}_s}{I_0} \le 0.7. \end{cases} \tag{2.145}$$

Thus, worst-case pulsed interference is more damaging than continuous interference if $\mathcal{E}_s/I_0 > 0.7$.

By substituting $\mu = \mu_0$ into (2.144), we obtain an approximate expression for the worst-case P_s when $I_0 \gg N_0$:

$$P_s \simeq \begin{cases} 0.083\left(\frac{\mathcal{E}_s}{I_0}\right)^{-1}, & \frac{\mathcal{E}_s}{I_0} > 0.7 \\ Q\left(\sqrt{\frac{2\mathcal{E}_s}{I_0}}\right), & \frac{\mathcal{E}_s}{I_0} \le 0.7 \end{cases} \tag{2.146}$$

which indicates that the worst-case P_s varies inversely, rather than exponentially, with $\mathcal{E}_s/I_0$ if this ratio is large enough. To restore a nearly exponential dependence on $\mathcal{E}_s/I_0$, a channel code and symbol interleaving are necessary.

Decoding metrics that are effective against white Gaussian noise are not necessarily effective against worst-case pulsed interference. We examine the performance of five different metrics against pulsed interference when the direct-sequence system uses BPSK, ideal symbol interleaving, a binary convolutional code, and Viterbi decoding [97]. The results are the same when either dual or balanced QPSK is the modulation.

Let $B(l)$ denote the total information weight of the paths at Hamming distance l from the correct path over an unmerged segment in the trellis diagram of the convolutional code. Let $P_2(l)$ denote the probability of an error in comparing the correct path segment with a particular path segment that differs in l symbols. According to (1.117) with $k = 1$, the information-bit error rate is upper-bounded by

$$P_b \le \sum_{l=d_f}^{\infty} B(l) P_2(l) \tag{2.147}$$

where d_f is the minimum free distance. If r is the code rate, $\mathcal{E}_b$ is the energy per information bit, T_b is the bit duration, and G_u is the spreading factor of the uncoded system, then

$$\mathcal{E}_s = r\mathcal{E}_b, \quad T_s = rT_b, \quad G = rG_u. \tag{2.148}$$

The decrease in the spreading factor is compensated by the coding gain. An upper bound on P_b for worst-case pulsed interference is obtained by maximizing the right-hand side of (2.147) with respect to μ, where $0 \leq \mu \leq 1$. The maximizing value of μ, which depends on the decoding metric, is not necessarily equal to the actual worst-case μ because a bound rather than an equality is maximized. However, the discrepancy is small when the bound is tight.

The simplest practical decoding metric to implement is provided by applying the input sequence to a hard-decision decoder. Assuming that the deinterleaving ensures the independence of symbol errors, (1.118) indicates that hard-decision decoding gives

$$P_2(l) = \begin{cases} \sum_{i=(l+1)/2}^{l} \binom{l}{i} P_s^i (1 - P_s)^{l-i}, & l \text{ is odd} \\ \sum_{i=l/2+1}^{l} \binom{l}{i} P_s^i (1 - P_s)^{l-i} + \frac{1}{2} \binom{l}{l/2} [P_s (1 - P_s)]^{l/2}, & l \text{ is even.} \end{cases} \tag{2.149}$$

Since $\mu = \mu_0$ approximately maximizes P_s, it also approximately maximizes the upper bound on P_b for hard-decision decoding given by (2.146) to (2.149).

Figure 2.24 depicts the upper bound on P_b as a function of $\mathcal{E}_b/I_0$ for worst-case pulsed interference, $\mathcal{E}_b/N_0 = 20\,\text{dB}$, and binary convolutional codes with several constraint lengths and rates. Tables 1.4 and 1.5 for $B(l)$ are used, and the series in (2.147) is truncated after the first seven terms. This truncation gives reliable results only if $P_b \leq 10^{-3}$ because the series converges very slowly. However, the truncation error is partially offset by the error incurred by the use of the union bound because the latter error is in the opposite direction. Figure 2.24 indicates the significant advantage of raising the constraint length K and reducing r at the cost of increased implementation complexity and synchronization requirements respectively.

Let $N_{0i}/2$ denote the equivalent noise PSD due to interference and noise in output sample y_i of a coherent BPSK demodulator. For convenience, y_i is assumed to have the form of the right-hand side of (2.78) normalized by multiplying the latter by $\sqrt{2/T_s}$. Thus, y_i has variance $N_{0i}/2$. Given that code symbol i of sequence k has value x_{ki}, the conditional density function of y_i is determined from the Gaussian character of the interference and noise. For a sequence of L code symbols, the density function is

$$f(y_i|x_{ki}) = \frac{1}{\sqrt{\pi N_{0i}}} \exp\left[-\frac{(y_i - x_{ki})^2}{N_{0i}}\right], \quad i = 1, 2, \ldots, L. \tag{2.150}$$

From the log-likelihood function and the statistical independence of the samples, it follows that when the values of $N_{01}, N_{02}, \ldots, N_{0L}$ are known, the *maximum-likelihood metric* for optimal soft-decision decoding of the sequence of L code symbols is

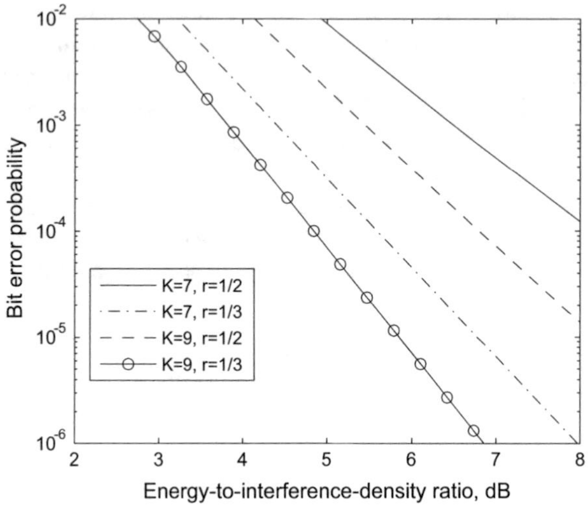

Fig. 2.24 Worst-case performance against pulsed interference of the direct-sequence system with BPSK, convolutional codes of constraint length K and rate r, $\mathcal{E}_b/N_0 = 20\,\text{dB}$, and hard decisions

$$U(k) = \sum_{i=1}^{L} \frac{x_{ki} y_i}{N_{0i}}. \tag{2.151}$$

This metric weights each output sample y_i according to the level of the equivalent noise. Since each y_i is assumed to be an independent Gaussian random variable, $U(k)$ is a Gaussian random variable.

Let $k = 1$ label the correct sequence, and let $k = 2$ label an incorrect one at distance l. If the quantization of the sample values is infinitely fine, the probability that $U(2) = U(1)$ is zero. Therefore, the probability of an error in comparing a correct sequence with an incorrect one that differs in l symbols, $P_2(l)$, is equal to the probability that $M_0 = U(2) - U(1) > 0$. The symbols that are the same in both sequences are irrelevant to the calculation of $P_2(l)$ and are subsequently ignored. Since either $x_{2i} = x_{1i}$ or $x_{2i} = -x_{1i}$, after reordering the terms, we have

$$M_0 = \sum_{i=1}^{L} \frac{(x_{2i} - x_{1i})\, y_i}{N_{0i}} = -\sum_{i=1}^{l} \frac{2x_{1i} y_i}{N_{0i}} \tag{2.152}$$

where l is the number of disagreements between the sequences $\{x_{2i}\}$ and $\{x_{1i}\}$. Let $P_2(l|\nu)$ denote the conditional probability that $M_0 > 0$ given that an interference pulse occurs during ν out of l differing symbols and does not occur during $l - \nu$ symbols. Since M_0 is a Gaussian random variable, we obtain

$$P_2(l|\nu) = Q\left(\frac{-E\,[M_0|\nu]}{\sqrt{var\,[M_0|\nu]}}\right) \tag{2.153}$$

where $E[M_0|\nu]$ is the conditional mean of M_0 and $var[M_0|\nu]$ is the conditional variance of M_0. Because of the interleaving, the probability that a symbol is interfered with is statistically independent of the rest of the sequence and equals μ. Evaluating $P_2(l)$ and substituting into (2.147) yields

$$P_b \le \sum_{l=d_f}^{\infty} B(l) \sum_{\nu=0}^{l} \binom{l}{\nu} \mu^{\nu}(1-\mu)^{l-\nu} P_2(l/\nu). \tag{2.154}$$

When an interference pulse occurs, $N_{0i} = N_0 + I_0/\mu$; otherwise, $N_{0i} = N_0$. Reordering the symbols for notational simplicity and observing that $x_{2i} = -x_{1i}$, $x_{1i}^2 = \mathcal{E}_s$, and $E[y_i] = x_{1i}$, we obtain

$$\begin{aligned} E[M_0|\nu] &= \sum_{i=1}^{\nu} \frac{(x_{2i}-x_{1i})\,E\,[y_i]}{N_0 + I_0/\mu} + \sum_{i=\nu+1}^{l} \frac{(x_{2i}-x_{1i})\,E\,[y_i]}{N_0} \\ &= \sum_{i=1}^{\nu} \frac{-2\mathcal{E}_s}{N_0 + I_0/\mu} + \sum_{i=\nu+1}^{l} \frac{-2\mathcal{E}_s}{N_0} \\ &= -2\mathcal{E}_s \left[\frac{\nu}{N_0 + I_0/\mu} + \frac{l-\nu}{N_0} \right]. \end{aligned} \tag{2.155}$$

Using the statistical independence of the samples and observing that $var[y_i] = N_{0i}/2$, we similarly find that

$$var\,[M_0|\nu] = 2\mathcal{E}_s \left[\frac{\nu}{N_0 + I_0/\mu} + \frac{l-\nu}{N_0} \right]. \tag{2.156}$$

Substituting (2.155) and (2.156) into (2.153), we obtain

$$P_2[l|\nu] = Q\left\{ \sqrt{\frac{2\mathcal{E}_s}{N_0}} \left[l - \nu \left(1 + \frac{\mu N_0}{I_0} \right)^{-1} \right]^{1/2} \right\}. \tag{2.157}$$

The substitution of this equation into (2.154) gives the upper bound on P_b for the maximum-likelihood metric.

The upper bound on P_b versus $\mathcal{E}_b/I_0$ for worst-case pulsed interference, $\mathcal{E}_b/N_0 = 20$ dB, and several binary convolutional codes is shown in Figure 2.25. Although the worst value of μ varies with $\mathcal{E}_b/I_0$, it is found that worst-case pulsed interference causes very little degradation relative to continuous interference. When $K = 9$ and $r = 1/2$, the *maximum-likelihood metric* provides a performance that is more than 4 dB superior at $P_b = 10^{-5}$ to that provided by hard-decision decoding; when $K = 9$ and $r = 1/3$, the advantage is approximately 2.5 dB. However, the implementation of the maximum-likelihood metric entails knowledge not only of the presence of interference, but also its PSD. Estimates of the N_{0i} might be based

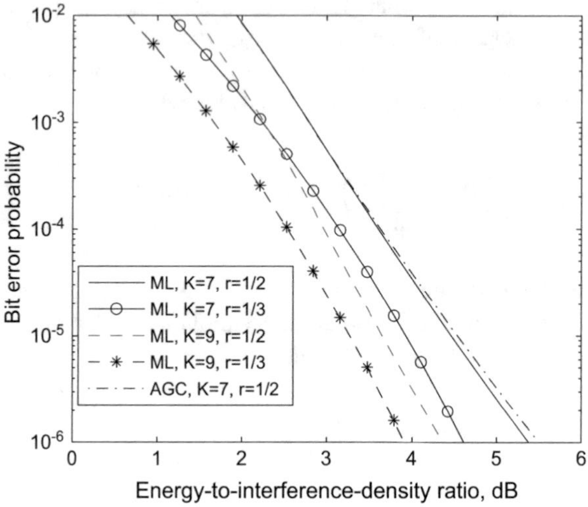

Fig. 2.25 Worst-case performance against pulsed interference of the direct-sequence system with convolutional codes of constraint length K and rate r, $\mathcal{E}_b/N_0 = 20\,\text{dB}$, and the maximum-likelihood and *automatic gain control* metrics

on power measurements in adjacent frequency bands only if the interference PSD is fairly uniform over the desired-signal and adjacent bands. Any measurement of the power within the desired-signal band is contaminated by the presence of the desired signal, the average power of which is usually unknown a priori because of the fading. Since iterative estimation of the N_{0i} and decoding is costly in terms of system latency and complexity, we examine another approach.

Consider an automatic gain control (AGC) device that measures the average power at the demodulator output before sampling and then weights the sampled demodulator output y_i in proportion to the inverse of the measured power to form the *automatic gain control metric*. The average power during channel-symbol i is $N_{0i}B + \mathcal{E}_s/T_s$, where B is the equivalent bandwidth of the demodulator and T_s is the channel-symbol duration. If the power measurement is perfect and $BT_s \approx 1$, then the *AGC metric* is

$$U(k) = \sum_{i=1}^{L} \frac{x_{ki}y_i}{N_{0i} + \mathcal{E}_s} \tag{2.158}$$

which is a Gaussian random variable. For this metric,

$$\begin{aligned} E[M_0|\nu] &= -\sum_{i=1}^{\nu} \frac{2\mathcal{E}_s}{N_0 + I_0/\mu + \mathcal{E}_s} - \sum_{i=\nu+1}^{l} \frac{2\mathcal{E}_s}{N_0 + \mathcal{E}_s} \\ &= -2\mathcal{E}_s\left[\frac{\nu}{N_0 + I_0/\mu + \mathcal{E}_s} + \frac{l-\nu}{N_0 + \mathcal{E}_s}\right] \end{aligned} \tag{2.159}$$

where ν is the number of symbols affected by interference pulses. Similarly, since $var[y_i] = N_{0i}/2$,

$$var[M_0|\nu] = 2\mathcal{E}_s \left[\frac{\nu\,(N_0 + I_0/\mu)}{(N_0 + I_0/\mu + \mathcal{E}_s)^2} + \frac{(l-\nu)\,N_0}{(N_0 + \mathcal{E}_s)^2} \right]. \tag{2.160}$$

Substitution of these equations into (2.153) yields

$$P_2(l/\nu) = Q\left\{ \sqrt{\frac{2\mathcal{E}_s}{N_0}} \frac{l\,(N_0 + \mathcal{E}_s + I_0/\mu) - \nu I_0/\mu}{\left[l\,(N_0 + \mathcal{E}_s + I_0/\mu)^2 - \nu\left(N_0 + I_0/\mu - \mathcal{E}_s^2/N_0\right) I_0/\mu \right]^{1/2}} \right\}. \tag{2.161}$$

This equation and (2.154) give the upper bound on P_b for the AGC metric.

The upper bound on P_b versus $\mathcal{E}_b/I_0$ for worst-case pulsed interference, the AGC metric, the rate-1/2 binary convolutional code with $K = 7$, and $\mathcal{E}_b/N_0 = 20$ dB is plotted in Figure 2.25. The figure indicates that the potential performance of the automatic gain control metric is nearly as good as that of the maximum-likelihood metric.

The energy $N_{0i}BT_s + \mathcal{E}_s$ may be measured using a *radiometer*, which is a device that determines the energy at its input. An ideal radiometer (Section 10.2) provides an unbiased estimate of the energy received during a symbol interval. The radiometer outputs are accurate estimates only if the standard deviation of the output is much lower than its expected value. This criterion and theoretical results for $BT_s = 1$ indicate that the energy measurements over a symbol interval are unreliable if $\mathcal{E}_s/N_{0i} \leq 10$ dB during interference pulses. Thus, the potential performance of the AGC metric is expected to be significantly degraded in practice unless each interference pulse extends over many channel symbols and its energy is measured over the corresponding interval.

The maximum-likelihood metric for continuous interference (N_{0i} is constant for all i) is the *white-noise metric*:

$$U(k) = \sum_{i=1}^{L} x_{ki} y_i \tag{2.162}$$

which is much simpler to implement than the AGC metric. For the white-noise metric, calculations similar to those above yield

$$P_2(l|\nu) = Q\left[\sqrt{\frac{2\mathcal{E}_s}{N_0}}\, l \left(l + \nu \frac{I_0}{\mu N_0} \right)^{-1/2} \right]. \tag{2.163}$$

This equation and (2.154) give the upper bound on P_b for the white-noise metric. Figure 2.26 illustrates the upper bound on P_b versus $\mathcal{E}_b/I_0$ for $K = 7$, $r =$

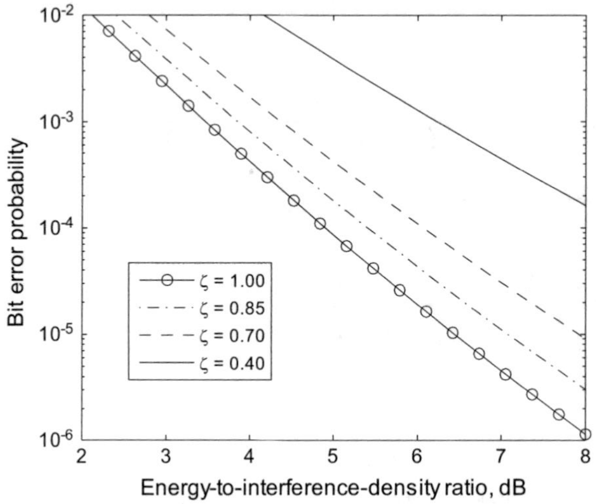

Fig. 2.26 Performance against pulsed interference of the direct-sequence system with convolutional code and the white-noise metric, $K = 7$, $r = 1/2$, and $\mathcal{E}_b/N_0 = 20\,\text{dB}$

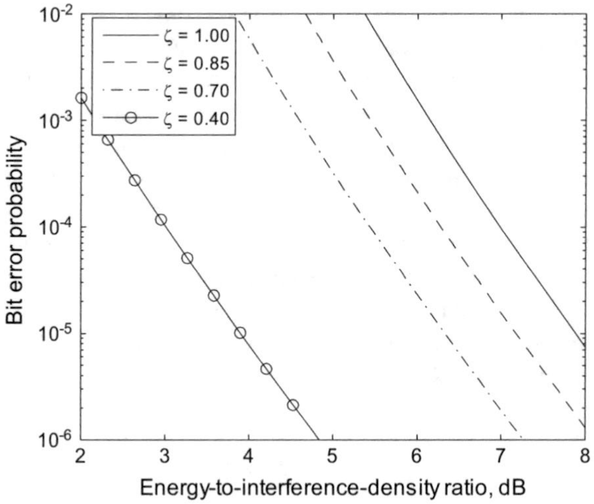

Fig. 2.27 Performance against pulsed interference of the direct-sequence system with convolutional code and the erasures metric, $K = 7$, $r = 1/2$, and $\mathcal{E}_b/N_0 = 20\,\text{dB}$

$1/2$, $\mathcal{E}_b/N_0 = 20\,\text{dB}$, and several values of $\zeta = \mu/\mu_0$. The figure demonstrates the vulnerability of soft-decision decoding with the white-noise metric to short high-power pulses if interference power is conserved. The high values of P_b for $\zeta < 1$ are due to the domination of the metric by a few degraded symbol metrics.

Consider a coherent BPSK demodulator that erases its output and hence a received symbol whenever an interference pulse occurs. The presence of the pulse might be detected by examining a sequence of the demodulator outputs and determining which ones have inordinately large magnitudes compared with the others. Alternatively, the demodulator might decide that a pulse has occurred if an output has a magnitude that exceeds a known upper bound for the desired signal. Consider an ideal demodulator that unerringly detects the pulses and erases the corresponding received symbols. Following the deinterleaving of the demodulated symbols, the decoder processes symbols that have a probability of being erased equal to μ. The unerased symbols are decoded by using the white-noise metric. The erasing of ν symbols causes two sequences that differ in l symbols to be compared on the basis of $l - \nu$ symbols where $0 \leq \nu \leq l$. As a result, the *erasures metric* provides

$$P_2(l|\nu) = Q\left[\sqrt{\frac{2\mathcal{E}_s}{N_0}(l-\nu)}\right]. \tag{2.164}$$

The substitution of this equation into (2.154) gives the upper bound on P_b for errors-and-erasures decoding.

The upper bound on P_b is illustrated in Figure 2.27 for $K = 7$, $r = 1/2$, $\mathcal{E}_b/N_0 = 20$ dB, and several values of $\zeta = \mu/\mu_0$. In this example, erasures provide no advantage over the white-noise metric in reducing the required $\mathcal{E}_b/I_0$ for $P_b = 10^{-5}$ if $\zeta > 0.85$, but are increasingly useful as ζ decreases.

Consider an ideal demodulator that unerringly activates erasures only when μ is small enough that the erasures are more effective than the white-noise metric. When this condition does not occur, the white-noise metric is used. The upper bound on P_b for this *ideal erasures metric*, worst-case pulsed interference, $\mathcal{E}_b/N_0 = 20$ dB, and several binary convolutional codes is illustrated in Figure 2.28. The required $\mathcal{E}_b/I_0$ at $P_b = 10^{-5}$ is roughly 2 dB less than for worst-case hard-decision decoding. However, a practical demodulator sometimes erroneously makes erasures or fails to erase, and the performance advantage of the ideal erasures metric may be much more modest.

2.7 Noncoherent Systems

A noncoherent direct-sequence system avoids the need for phase synchronization with the carrier in the receiver. The noncoherent receiver produces ADC outputs in the manner described in Section 2.5. The ADC outputs are chip-rate in-phase and quadrature sequences that are applied to a metric generator, as shown in Figure 2.29. For hard-decision decoding, the symbol metrics produced by the metric generator are compared to make a symbol decision every symbol period. For soft-decision decoding, the symbol metrics are directly applied to the decoder.

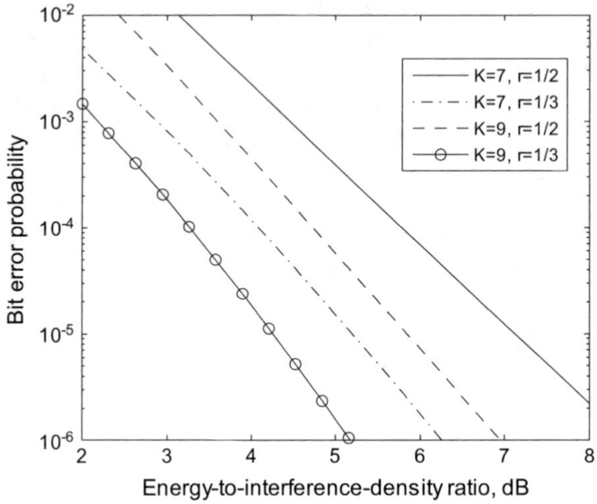

Fig. 2.28 Worst-case performance against pulsed interference of the direct-sequence system with convolutional codes of constraint length K and rate r, ideal erasures metric, and $\mathcal{E}_b/N_0 = 20\,\text{dB}$

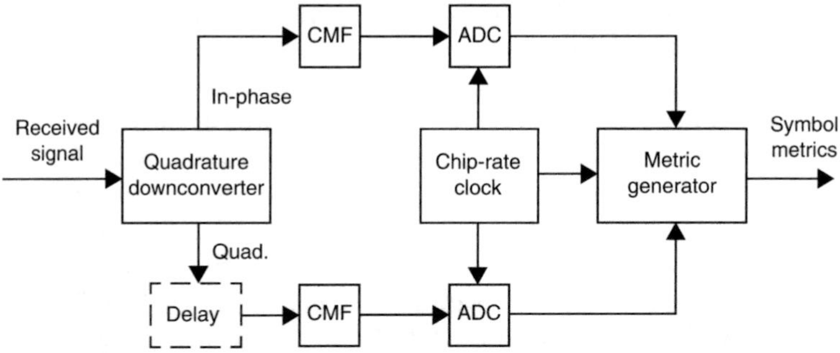

Fig. 2.29 Receiver for noncoherent direct-sequence system

A direct-sequence system with q-ary code-shift keying (CSK) encodes each group of m bits as one of $q = 2^m$ nonbinary symbols that are represented by orthogonal binary code sequences. Each code sequence has length G and serves as a spreading sequence. For orthogonality of the q code or spreading sequences, the sequences must have length $G \geq q$.

In the metric generator for noncoherent detection, each received sequence is applied to q parallel matched filters that are matched to the orthogonal code sequences, as shown in Figure 2.30. The matched-filter outputs are sampled at the symbol rate. Since the detection is noncoherent, corresponding sampled matched-filter outputs are each squared and then combined to produce symbol metrics.

As in the coherent receivers of Sections 2.4 and 2.5, each matched filter in Figure 2.30 may be implemented by multiplying an in-phase or quadrature received

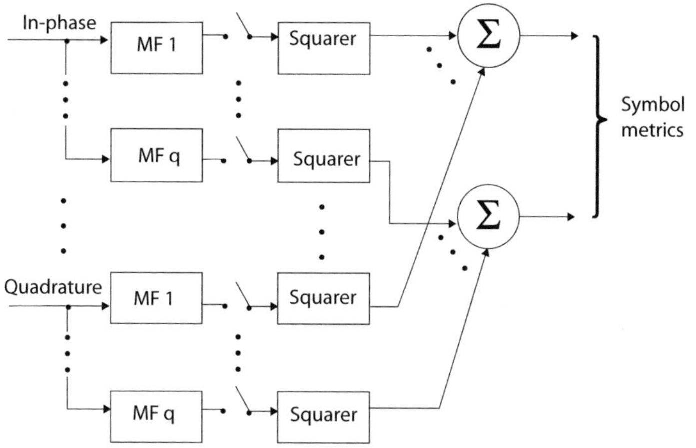

Fig. 2.30 Metric generator for noncoherent detection of q orthogonal code or spreading sequences

sequence by the corresponding orthogonal code or spreading sequence, provided that the receiver-generated sequence is synchronized with the received sequence (Sections 4.1-4.6). *Symbol or timing synchronization*, which is needed to provide timing pulses to the decoder, is derived from the receiver-generated sequence.

The received *direct-sequence signal* with q-ary CSK for symbol k of duration T_s is represented as

$$s_k(t) = \sqrt{\mathcal{E}_s} p_k(t) \cos(2\pi f_c t + \phi) + \sqrt{\mathcal{E}_s} p_k(t + t_0) \sin(2\pi f_c t + \phi), \quad 0 \leq t \leq T_s \tag{2.165}$$

where $\mathcal{E}_s$ is the energy per q-ary symbol, t_0 is the relative delay between the in-phase and quadrature components of the signal, ϕ is the received phase, the code or spreading waveform is

$$p_k(t) = \sum_{i=-0}^{G-1} p_{ki} \psi(t - iT_c), \tag{2.166}$$

and p_{ki} is chip i of spreading-sequence k. The chip duration is T_c, and the energy of the chip waveform satisfies (2.70). The total received signal during reception of symbol k is

$$r(t) = s_k(t) + i(t) + n(t) \tag{2.167}$$

where $i(t)$ is the interference, and $n(t)$ is the additive white Gaussian noise. An evaluation similar to that in Section 2.5 indicates that the in-phase sequence applied to the metric generator of Figure 2.29 is

$$I_i = \frac{\sqrt{\mathcal{E}_s/2}}{G} p_{ki} (\cos\phi + \sin\phi) + J_i + N_{si}, \quad i = 0, 1, \ldots, G-1 \tag{2.168}$$

where J_i and N_{si} are defined by (2.76) and (2.77), respectively. Similarly, the quadrature sequence applied to the metric generator is

$$Q_i = \frac{\sqrt{\mathcal{E}_s/2}}{G} p_{ki} (\cos\phi - \sin\phi) + J_i^{'} + N_{si}^{'}, \ i = 0, 1, \ldots, G-1 \tag{2.169}$$

where $J_i^{'}$ and $N_{si}^{'}$ are defined by (2.124) and (2.125), respectively.

In the metric generator of Figure 2.30, the sampled outputs of the in-phase and quadrature matched filters that are matched to transmitted symbol k are

$$I_{sk} = \sqrt{\mathcal{E}_s/2} (\cos\phi + \sin\phi) + \sum_{i=0}^{G-1} p_{ki} (J_i + N_{si}) \tag{2.170}$$

$$Q_{sk} = \sqrt{\mathcal{E}_s/2} (\cos\phi - \sin\phi) + \sum_{i=0}^{G-1} p_{ki} \left(J_i^{'} + N_{si}^{'} \right) \tag{2.171}$$

respectively. Using the orthogonality of the spreading sequences, we find that the sampled outputs of the in-phase and quadrature matched filters that are matched to symbol $l \neq k$ are

$$I_{sl} = \sum_{i=0}^{G-1} p_{li} (J_i + N_{si}), \quad Q_{sl} = \sum_{i=0}^{G-1} p_{li} \left(J_i^{'} + N_{si}^{'} \right), \quad l \neq k \tag{2.172}$$

respectively. We assume that the interference approximates the Gaussian interference of Section 2.4. Then

$$var(I_{sl}) = var(Q_{sl}) = N_{0e}/2, \quad l = 1, 2, \ldots q \tag{2.173}$$

where N_{0e} is given by (2.115). We make the plausible assumption that the $\{I_{sk}, Q_{sk}\}$ are approximately independent of the $\{I_{sl}, Q_{sl}\}$, $l \neq k$, in the sense that any dependence is negligible.

For hard-decision decoding, the symbol metrics are compared and a symbol decision is made. As shown in Appendix E.1, after the squaring and combining operations, each symbol metric $R_l = I_{sl}^2 + Q_{sl}^2$, $l \neq k$, has a central chi-squared density with two degrees of freedom and variance $\sigma^2 = N_{0e}/2$. The symbol metric R_k associated with transmitted symbol k has a chi-squared density with two degrees of freedom, variance $\sigma^2 = N_{0e}/2$, and noncentral parameter $\lambda = \mathcal{E}_s$. Using the probability density functions (E.10) and (E.15), a derivation paralleling that of (1.93) leads to the symbol error probability:

$$P_s = \sum_{i=1}^{q-1} \frac{(-1)^{i+1}}{i+1} \binom{q-1}{i} \exp\left[-\frac{i\mathcal{E}_s}{(i+1)N_{0e}}\right]. \tag{2.174}$$

For the same N_{0e} and $\mathcal{E}_s = m\mathcal{E}_b$, a comparison of (2.174) with (2.112) indicates that the bit error probability as a function of $\mathcal{E}_b/N_{0e}$ of the direct-sequence system with noncoherent binary CSK is approximately 4 dB worse than that of the system with coherent BPSK. This difference arises because binary CSK uses orthogonal rather than antipodal signals. A much more complicated system with coherent binary CSK would only recover roughly 1 dB of the disparity. The performance of a direct-sequence system with noncoherent 8-ary CSK in the presence of wideband Gaussian interference and the same N_{0e} is slightly better than that of a direct-sequence system with coherent BPSK. However, eight matched filters are required, which offsets the advantage that phase synchronization is not required.

In a system that uses a single binary CSK sequence and a minimum amount of hardware, the symbol 1 is signified by the transmission of the sequence, whereas the symbol 0 is signified by the absence of a transmission. Decisions are made after comparing the envelope-detector output with a threshold. One problem with this system is that the optimal threshold is a function of the amplitude of the received signal, which must somehow be estimated. Another problem is the degraded performance of the symbol synchronizer when many consecutive zeros are transmitted. Thus, a system with two binary CSK sequences is much more practical.

A direct-sequence system with DPSK signifies the symbol 1 by the transmission of a spreading sequence without any change in the carrier phase; the symbol 0 is signified by the transmission of the same sequence after a phase shift of π radians in the carrier phase. Thus, the symbol sequence is determined by the phase shifts between consecutive spreading sequences. In the metric generator, chip-rate in-phase and quadrature sequences are multiplied by the spreading sequence to produce symbol-rate despread sequences, as illustrated in Figure 2.31. These sequences are multiplied by previous despread sequences in mixers, the outputs of which are added to generate the symbol metrics. Since the symbol information is embedded in the phase shifts, pairs of symbol metrics are used in the decoding. An analysis of this system for hard decisions and wideband Gaussian interference indicates that it is more than 2 dB superior to the system with binary coherent CSK. However, the system with DPSK is more sensitive to Doppler shifts and is more than 1 dB inferior to a system with coherent BPSK.

2.8 Despreading with Bandpass Matched Filters

A matched filter can be implemented at baseband as a digital filter. Alternatively, bandpass matched filtering can be implemented by analog devices. Despreading short spreading sequences with bandpass matched filters provides pulses that can be

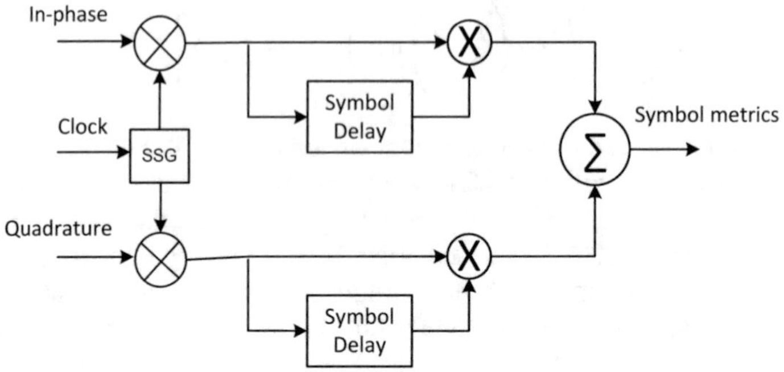

Fig. 2.31 Metric generator for a direct-sequence system with differential phase-shift keying

used for code synchronization and as the basis for producing the symbol metrics of simple direct-sequence receivers, which are described in this section.

The spreading waveform for a short spreading sequence may be expressed as

$$p(t) = \sum_{i=-\infty}^{\infty} p_1(t - iT) \tag{2.175}$$

where $p_1(t)$ is one period of the spreading waveform and T is its period. If the short spreading sequence has length N, then

$$p_1(t) = \begin{cases} \sum_{i=0}^{N-1} p_i \psi(t - iT_c), & 0 \le t \le T \\ 0, & otherwise \end{cases} \tag{2.176}$$

where $\psi(t)$ is the chip waveform, $p_i = \pm 1$, and $T = NT_c$.

As explained in Section 1.2, a filter is said to be *matched* to a signal $x(t)$, $0 \le t \le T$, if the impulse response of the filter is $h(t) = x(T - t)$. Consider a *bandpass-matched filter* that is matched to

$$x(t) = \begin{cases} p_1(t)\cos(2\pi f_c t + \theta_1), & 0 \le t \le T \\ 0, & otherwise \end{cases} \tag{2.177}$$

where $p_1(t)$ is one period of a spreading waveform, and f_c is the desired carrier frequency. We evaluate the filter response to the received signal corresponding to a single data symbol:

$$s(t) = \begin{cases} 2Ap_1(t - t_0)\cos(2\pi f_1 t + \theta), & t_0 \le t \le t_0 + T \\ 0, & otherwise \end{cases} \tag{2.178}$$

where t_0 is a measure of the unknown arrival time, the polarity of A is determined by the data symbol, and f_1 is the received carrier frequency, which differs from f_c

because of oscillator instabilities and the Doppler shift. If $f_c >> 1/T$, the matched-filter output is

$$\begin{aligned} y_s(t) &= \int_{\max(t-T,t_0)}^{\min(t,t_0+T)} s(u)p_1(u+T-t)\cos\left[2\pi f_c(u+T-t)+\theta_1\right] du \\ &= A\int_{\max(t-T,t_0)}^{\min(t,t_0+T)} p_1\left(u-t_0\right)p_1(u-t+T)\cos\left(2\pi f_d u + 2\pi f_c t + \theta_2\right) du \end{aligned} \tag{2.179}$$

where $\theta_2 = \theta - \theta_1 - 2\pi f_c T$ is the phase mismatch and $f_d = f_1 - f_c$.

If $f_d << 1/T$, the carrier-frequency error is inconsequential, and

$$y_s(t) \approx A_s(t)\cos\left(2\pi f_c t + \theta_3\right), \quad t_0 \le t \le t_0 + 2T \tag{2.180}$$

where $\theta_3 = \theta_2 + 2\pi f_d t_0$ and

$$A_s(t) = A\int_{\max(t-T,t_0)}^{\min(t,t_0+T)} p_1\left(u-t_0\right)p_1(u-t+T)du. \tag{2.181}$$

In the absence of noise, the matched-filter output $y_s(t)$ is a sinusoidal pulse of duration $2T$ with a polarity determined by A. Equation (2.181) indicates that the peak magnitude, which occurs at the ideal sampling time $t = t_0 + T$, equals $|A|T$. However, if $f_d > 0.1/T$, then (2.179) is not well-approximated by (2.180), and the matched-filter output is significantly degraded.

The response of the matched filter to the interference and noise, denoted by $N(t) = i(t) + n(t)$, may be expressed as

$$\begin{aligned} y_n(t) &= \int_{t-T}^{t} N(u)p_1(u+T-t)\cos\left[2\pi f_c(u+T-t)+\theta_1\right] du \\ &= N_1(t)\cos\left(2\pi f_c t + \theta_2\right) + N_2(t)\sin\left(2\pi f_c t + \theta_2\right) \end{aligned} \tag{2.182}$$

where

$$N_1(t) = \int_{t-T}^{t} N(u)p_1(u+T-t)\cos\left(2\pi f_c u + \theta\right) du \tag{2.183}$$

$$N_2(t) = \int_{t-T}^{t} N(u)p_1(u+T-t)\sin\left(2\pi f_c u + \theta\right) du. \tag{2.184}$$

These equations exhibit the spreading and filtering of the interference spectrum.

Assuming that $f_d t_0 << 1$, $\theta_3 \cong \theta_2$ and the envelope of the matched-filter output $y(t) = y_s(t) + y_n(t)$ is

$$E(t) = \left\{[A_s(t) + N_1(t)]^2 + N_2^2(t)\right\}^{1/2}. \quad (2.185)$$

At the ideal sampling time $t = t_0 + T$, $A_s(t_0 + T) = AT$. If

$$|A_s(t) + N_1(t)| >> |N_2(t)| \quad (2.186)$$

then (2.185) implies that

$$E(t_0 + T) \approx |AT + N_1(t_0 + T)|. \quad (2.187)$$

Since $A_s(t) \leq AT$, (2.180), (2.182), and (2.186) imply that

$$y(t) \leq |A_s(t) + N_1(t)| + |N_2(t)| \approx |AT + N_1(t)|. \quad (2.188)$$

A comparison of this approximate upper bound with (2.187) indicates that there is relatively little degradation in using an envelope detector after the matched filter rather than directly detecting the peak magnitude of the matched-filter output, which is much more difficult.

Surface Acoustic Wave Filters

Figure 2.32 illustrates the basic form of a *surface acoustic wave* (SAW) *transversal filter*, which is a passive matched filter that essentially stores a replica of the underlying spreading sequence and waits for the received sequence to align itself

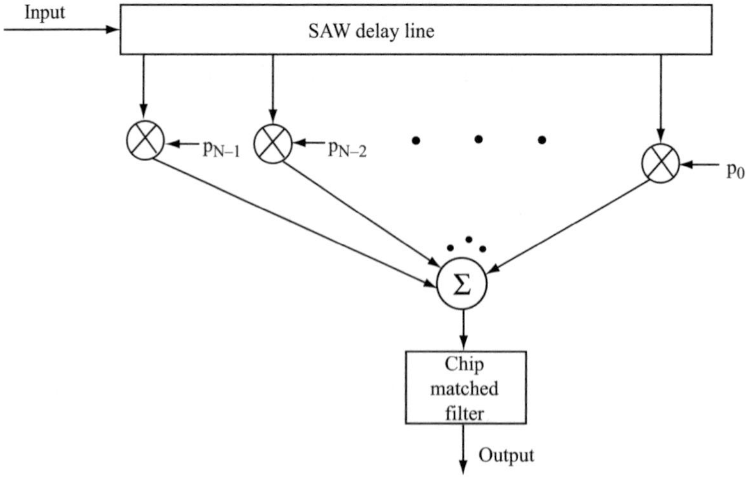

Fig. 2.32 Matched filter that uses a surface acoustic wave (SAW) transversal filter. Output is $y_s(t) + y_n(t)$

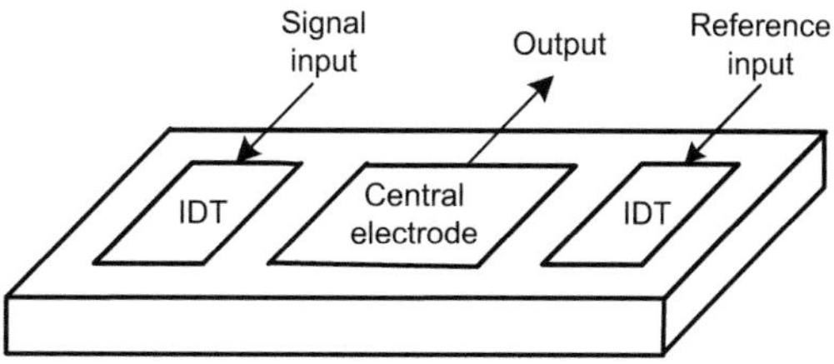

Fig. 2.33 SAW convolver. *IDT* interdigital transducer

with the replica. The SAW delay line consists primarily of a piezoelectric substrate, which serves as the acoustic propagation medium, and interdigital transducers, which serve as the taps and the input transducer. The transversal filter is matched to one period of the spreading waveform, the propagation delay between taps is T_c, and f_cT_c is an integer. The chip-matched filter following the summer is matched to $\psi(t)\cos(2\pi f_ct+\theta)$. It is easily verified that the impulse response of the transversal filter is that of a filter matched to $p_1(t)\cos(2\pi f_ct+\theta)$.

An *active matched filter* can be implemented as a *SAW convolver* [58], which is depicted in Figure 2.33. The received signal and a reference signal are applied to separate interdigital transducers that generate acoustic waves at opposite ends of a substrate. The reference signal is a recirculating, time-reversed replica of the spreading waveform. The acoustic waves travel in opposite directions with speed v, and the acoustic terminations suppress reflections. The received-signal wave is launched at position $x = 0$ and the reference wave at $x = L$. The received-signal wave travels to the right in the substrate and has the form

$$F(t,x) = f\left(t-\frac{x}{v}\right)\cos\left[2\pi f_c\left(t-\frac{x}{v}\right)+\theta\right] \tag{2.189}$$

where $f(t)$ is the modulation at position $x = 0$. The reference wave travels to the left and has the form

$$G(t,x) = g\left(t+\frac{x-L}{v}\right)\cos\left[2\pi f_c\left(t+\frac{x-L}{v}\right)+\theta_1\right] \tag{2.190}$$

where $g(t)$ is the modulation at position $x = L$. Both $f(t)$ and $g(t)$ are assumed to have bandwidths much smaller than f_c.

Beam compressors, which consist of thin metallic strips, focus the acoustic energy to increase the convolver's efficiency. When the acoustic waves overlap beneath the central electrode, a nonlinear piezoelectric effect causes a surface charge distribution that is spatially integrated by the electrode. The primary component of the convolver output is proportional to

$$y(t) = \int_0^L [F(t,x)+G(t,x)]^2 dx. \tag{2.191}$$

Substituting (2.189) and (2.190) into (2.191) and using trigonometry, we find that $y(t)$ is the sum of a number of terms, some of which are negligible if $f_cL/v >>$

1. Others are slowly varying and are easily blocked by a filter. The most useful component of the convolver output is

$$y_s(t) = \left[\int_0^L f\left(t - \frac{x}{\upsilon}\right) g\left(t + \frac{x - L}{\upsilon}\right) dx\right] \cos\left(4\pi f_c t + \theta_2\right) \tag{2.192}$$

where $\theta_2 = \theta + \theta_1 - 2\pi f_c L/\upsilon$. Changing variables, we find that the amplitude of the output is

$$A_s(t) = \int_{t-L/\upsilon}^{t} f(y)g(2t - y - L/\upsilon)dy \tag{2.193}$$

where the factor $2t$ results from the counterpropagation of the two acoustic waves.

Suppose that an acquisition pulse is a single period of the spreading waveform. Then, $f(t) = Ap_1(t - t_0)$ and $g(t) = p(T - t)$, where t_0 is the uncertainty in the arrival time of an acquisition pulse relative to the launching of the reference signal at $x = L$. The periodicity of $g(t)$ allows the time origin to be selected so that $0 \le t_0 \le T$. Equations (2.193) and (2.175) and a change of variables yield

$$A_s(t) = A \sum_{i=-\infty}^{\infty} \int_{t-t_0-L/\upsilon}^{t-t_0} p_1(y)p_1\left(y + iT + t_0 - 2t + L/\upsilon\right) dy. \tag{2.194}$$

Since $p_1(t) = 0$ unless $0 \le t < T$, $A_s(t) = 0$ unless $t_0 < t < t_0 + L/\upsilon + T$.

For every positive integer k, let

$$\tau_k = \frac{kT + t_0 + L/\upsilon}{2}, \qquad k \ge 1. \tag{2.195}$$

Only one term in (2.194) can be nonzero when $t = \tau_k$, and

$$A_s\left(\tau_k\right) = A \int_{\tau_k-t_0-L/\upsilon}^{\tau_k-t_0} p_1^2(y)dy. \tag{2.196}$$

The maximum possible magnitude of $A_s(\tau_k)$ is produced if $\tau_k - t_0 \ge T$ and $\tau_k - t_0 - L/\upsilon \le 0$; that is, if

$$t_0 + T \le \tau_k \le t_0 + \frac{L}{\upsilon}. \tag{2.197}$$

Since (2.195) indicates that $\tau_{k+1} - \tau_k = T/2$, there is some τ_k that satisfies (2.197) if

$$L \ge \frac{3}{2}\upsilon T. \tag{2.198}$$

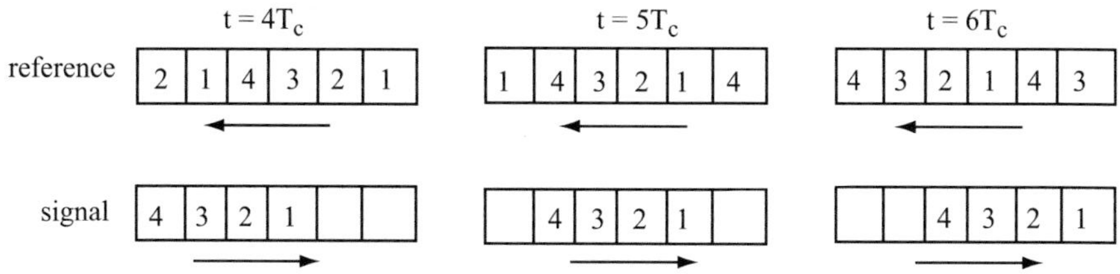

Fig. 2.34 Chip configurations within the convolver at time instants $t = 4T_c$, $5T_c$, and $6T_c$ when $t_0 = 0$, $L/\upsilon = T_c$, and $T = 4T_c$

Thus, if L is large enough, then there is some k such that $A_s(\tau_k) = AT$, and the envelope of the convolver output at $t = \tau_k$ has the maximum possible magnitude. If $L = 3\upsilon T/2$ and $t_0 \neq T/2$, only one peak value occurs in response to the single received pulse.

As an example, let $t_0 = 0$, $L/\upsilon = 6T_c$, and $T = 4T_c$. The chips propagating in the convolver for three separate time instants $t = 4T_c$, $5T_c$, and $6T_c$ are illustrated in Figure 2.34. The top diagrams refer to the counterpropagating periodic reference signal, whereas the bottom diagrams refer to the single received pulse of four chips. The chips are numbered consecutively. The received pulse is completely contained within the convolver during $4T_c \leq t \leq 6T_c$. The maximum magnitude of the output occurs at time $t = 5T_c$, which is the instant of perfect alignment of the reference signal and the received chips.

Multipath-Resistant Coherent System

The coherent demodulation of a direct-sequence signal requires the generation of a phase-coherent synchronization signal with the correct carrier frequency in the receiver. Prior to the despreading, the signal-to-noise ratio (SNR) may be too low for the received signal to serve as the input to a phase-locked loop that produces a synchronization signal. An inexpensive method of generating a synchronization signal is to use a *recirculation loop* , which is a loop designed to reinforce a periodic input signal by positive feedback.

As illustrated in Figure 2.35, the feedback elements are an attenuator of gain K and a delay line with a delay $\widehat{T}_s$ that approximates a symbol duration T_s. The basic concept behind this architecture is that successive signal pulses are coherently added while the interference and noise are noncoherently added, thereby producing an output pulse with an improved SNR.

The periodic input consists of N symbol pulses such that

$$s_0(t) = \sum_{i=0}^{N} g\left(t - iT_s\right) \tag{2.199}$$

Fig. 2.35 Recirculation loop

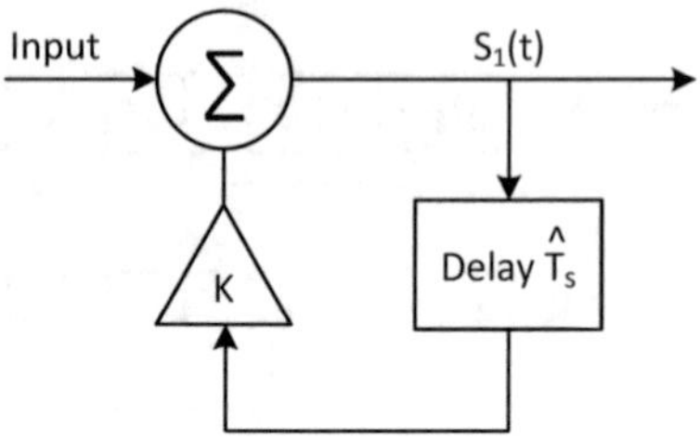

where $g(t) = 0$ for $t < 0$ or $t \geq T_s$. The figure indicates that the loop output is

$$s_1(t) = s_0(t) + Ks_1\left(t-\widehat{T}_s\right) \tag{2.200}$$

where $s_1(t) = 0$ for $t < 0$. Substitution of this equation into itself yields

$$s_1(t) = s_0(t) + Ks_0\left(t-\widehat{T}_s\right) + K^2 s_1\left(t - 2\widehat{T}_s\right). \tag{2.201}$$

Repeating this substitution process n times leads to

$$s_1(t) = \sum_{m=0}^{n} K^m s_0\left(t - m\widehat{T}_s\right) + K^{n+1} s_1\left[t - (n+1)\widehat{T}_s\right] \tag{2.202}$$

which indicates that $s_1(t)$ increases with n if $K \geq 1$ and enough input pulses are available. To prevent an eventual loop malfunction, $K < 1$ is a design requirement that is assumed henceforth.

During the interval $[n\widehat{T}_s, (n+1)\widehat{T}_s)$, n or fewer recirculations of the symbols have occurred. Since $s_1(t) = 0$ for $t < 0$, the substitution of (2.199) into (2.202) yields

$$s_1(t) = \sum_{m=0}^{n}\sum_{i=0}^{N} K^m g\left(t - m\widehat{T}_s - iT_s\right), \qquad n\widehat{T}_s \leq t < (n+1)\widehat{T}_s. \tag{2.203}$$

This equation indicates that if $\widehat{T}_s$ is not exactly equal to T_s, then the pulses do not add coherently, and may combine destructively. However, as $K < 1$, the effect of a particular pulse decreases as m increases and is eventually negligible. The delay $\widehat{T}_s$ is designed to match T_s. Suppose that the design error is small enough that

$$N\left|\widehat{T}_s - T_s\right| << T_s. \tag{2.204}$$

If $n \leq N$, then $t - m\widehat{T}_s - iT_s = t - (m+i)T_s - m(\widehat{T}_s - T_s) \approx t - (m+i)T_s$. Thus, as $g(t)$ is time-limited, only the term in (2.203) with $i = n - m$ contributes appreciably to the output. Therefore,

$$s_1(t) \approx \sum_{m=0}^{n} K^m g\left[t - nT_s - m\left(\widehat{T}_s - T_s\right)\right], \qquad n\widehat{T}_s \le t < (n+1)\widehat{T}_s. \tag{2.205}$$

Consider an input pulse of the form

$$g(t) = A(t)\cos 2\pi f_c t, \qquad 0 \le t < \min\left(T_s, \widehat{T}_s\right). \tag{2.206}$$

Assume that the amplitude $A(t)$ varies slowly enough that

$$A\left[t - nT_s - m\left(\widehat{T}_s - T_s\right)\right] \approx A\left(t - nT_s\right), \qquad 0 \le m \le n \tag{2.207}$$

and that the design error is small enough that

$$nf_c\left|\widehat{T}_s - T_s\right| << 1. \tag{2.208}$$

Then (2.205) to (2.208) yield

$$\begin{aligned} s_1(t) &\approx g\left(t - nT_s\right)\sum_{m=0}^{n} K^m \\ &= g\left(t - nT_s\right)\left(\frac{1 - K^{n+1}}{1 - K}\right), \qquad n\widehat{T}_s \le t < (n+1)\widehat{T}_s. \end{aligned} \tag{2.209}$$

If S is the average power in an input pulse, then (2.209) indicates that the average power in an output pulse during the interval $n\widehat{T}_s \le t < (n+1)\widehat{T}_s$ is approximately

$$S_n = \left(\frac{1 - K^{n+1}}{1 - K}\right)^2 S, \qquad K < 1. \tag{2.210}$$

If $\widehat{T}_s$ is large enough that the recirculated noise is uncorrelated with the input noise, which has average power σ^2, then the output noise power after n recirculations is

$$\begin{aligned} \sigma_n^2 &= \sigma^2 \sum_{m=0}^{n} \left(K^2\right)^m \\ &= \sigma^2\left(\frac{1 - K^{2n+2}}{1 - K^2}\right), \qquad K < 1. \end{aligned} \tag{2.211}$$

The improvement in the SNR due to the presence of the recirculation loop is

$$\begin{aligned} I\left(n, K\right) &= \frac{S_n/\sigma_n^2}{S/\sigma^2} = \frac{\left(1 - K^{n+1}\right)\left(1 + K\right)}{\left(1 + K^{n+1}\right)\left(1 - K\right)} \\ &\le \frac{1 + K}{1 - K}, \quad K < 1. \end{aligned} \tag{2.212}$$

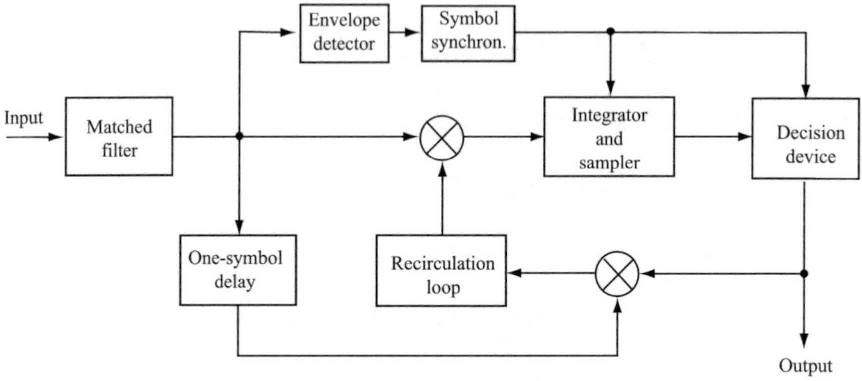

Fig. 2.36 Coherent decision-directed demodulator

If K^n is small, the upper bound on $I(n, K)$ is nearly attained after n recirculations. However, the upper bound on n for the validity of (2.208) decreases as the loop phase error $2\pi f_c|\widehat{T}_s - T_s|$ increases, and it may be necessary to decrease K to maximize $I(n, K)$ as the loop phase error increases.

Figure 2.36 illustrates a *coherent decision-directed demodulator* for a direct-sequence signal with BPSK and the same carrier phase at the beginning of each symbol. The bandpass matched filter removes the spreading waveform and produces compressed sinusoidal pulses, as indicated by (2.180) and (2.181), when A is bipolar. A compressed pulse due to a direct-path signal may be followed by one or more compressed pulses due to multipath signals, as illustrated conceptually in Figure 2.37 (a) for pulses corresponding to the transmitted symbols 101. Each compressed pulse is delayed by one symbol and then mixed with the demodulator's output symbol. If this symbol is correct, it coincides with the same data symbol that is modulated onto the compressed pulse. Consequently, the mixer removes the data modulation and produces a phase-coherent reference pulse that is independent of the data symbol, as illustrated in Figure 2.37 (b), where the two middle pulses are inverted in phase relative to the corresponding pulses in Figure 2.37 (a). The reference pulses are amplified by a recirculation loop. The loop output and the matched-filter output are applied to a mixer that produces the baseband integrator input illustrated in Figure 2.37 (c). The length of the integration interval is equal to a symbol duration. The sampling times, which occur at the boundaries of the integration intervals, are determined by the synchronized pulses produced by the symbol synchronizer. The sampled integrator output is applied to a decision device that produces the data output. Since multipath components are coherently integrated, the demodulator provides improved performance in a fading environment.

Even if the desired-signal multipath components are absent, the coherent decision-directed receiver potentially suppresses interference approximately as much as the correlator of Figure 2.14. The decision-directed receiver is much simpler to implement because code acquisition and tracking systems are

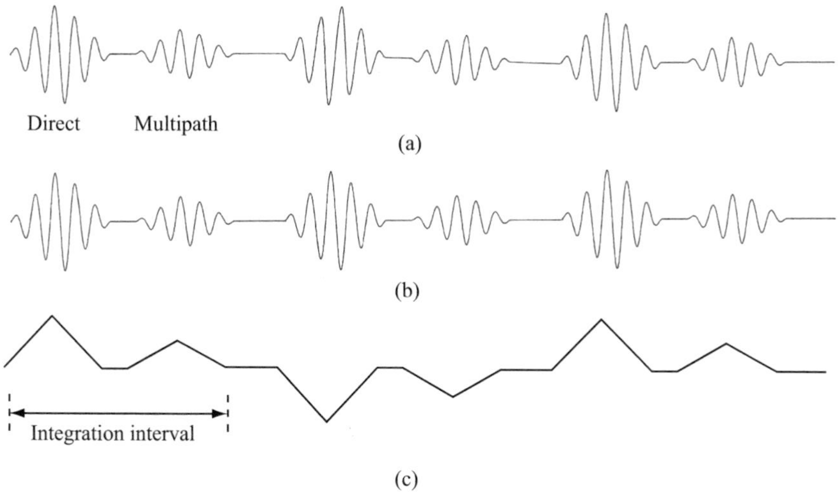

Fig. 2.37 Conceptual waveforms of the demodulator: (**a**) matched-filter output, (**b**) recirculation loop input or output, and (**c**) baseband integrator input

unnecessary, but it requires a short spreading sequence and an accurate recirculation loop. More efficient exploitation of multipath components is possible with rake combining (Section 6.10).

2.9 Problems

1 The characteristic polynomial associated with a linear feedback shift register is $f(x) = 1 + x^2 + x^3 + x^5 + x^6$. The initial state is $a_0 = a_1 = 0,\ a_2 = a_3 = a_4 = a_5 = 1$. Use polynomial long division to determine the first nine bits of the output sequence.

2 If the characteristic polynomial associated with a linear feedback shift register is $1 + x^m$, what is the linear recurrence relation? Write the generating function associated with the output sequence. Use polynomial long division to determine the possible periods of the output sequence.

3 Prove by exhaustive search that the polynomial $f(x) = 1 + x^2 + x^3$ is primitive.

4 Derive (2.55) and (2.56) using the steps specified in the text.

5 Derive the characteristic polynomial of the linear equivalent of Figure 2.12 (a). Verify the structure of Figure 2.12 (b) and derive the initial contents indicated in the figure.

6 This problem illustrates the limitations of an approximate model in an extreme case. Suppose that tone interference at the carrier frequency is coherent with a BPSK direct-sequence signal so that $\phi = 0$ in (2.87). Assume that $N_0 \to 0$ and $\mathcal{E}_s > GIT_c\kappa$. Show that $P_s \to 0$. Show that the general tone-interference model of Section 2.4 leads to a nonzero approximate expression for P_s.

7 Derive (2.109) using the steps specified in the text.

8 Derive the expression for $E[V|\phi, k_1, k_2, d_0]$ that leads to (2.141).

9 Consider a direct-sequence system with BPSK and hard decisions, a required $P_s = 10^{-5}$, and $N_0 = 0$. How much additional power is required against worst-case pulsed interference beyond that required against continuous interference? Use $Q(\sqrt{20}) = 10^{-5}$.

10 Assuming that $\mathcal{E}_s \gg N_0$ and $I_0 \gg N_0$, (2.112) implies that the probability of an error in a binary code symbol is

$$P_s \simeq \mu Q\left(\sqrt{\frac{2\mathcal{E}_s\mu}{I_0}}\right), \quad 0 \leq \mu \leq 1.$$

Use this equation and the fact $Q(1.2) = 0.115$ to verify (2.145).

11 What are the values of $E[M_0|\nu]$ and $var[M_0|\nu]$ for the white-noise metric?

12 For a direct-sequence system with binary DPSK and hard decisions, $P_s = \frac{1}{2}\exp(-\mathcal{E}_s/N_0)$ in the presence of white Gaussian noise. Derive the worst-case duty cycle and P_s for strong pulsed interference when the PSD of continuous interference is $I_0/2 >> N_0/2$. Show that DPSK has a more than 3 dB disadvantage relative to BPSK against worst-case pulsed interference when $\mathcal{E}_s/I_0$ is large.

13 Expand (2.182) to determine the degradation in $A_s(t_0 + T)$ when $f_d \neq 0$ and the chip waveform is rectangular.

14 Evaluate the impulse response of a transversal filter with the form of Figure 2.32. Show that this impulse response is equal to that of a filter matched to $p_1(t)cos(2\pi f_c t + \theta)$ if $f_c T_c$ is an integer.

15 Consider a convolver for which $L/v = nT$ for some positive integer n and $g(t) = p(T - t)$, where $p(t)$ is the periodic spreading waveform of period T. The received signal is $f(t) = Ap(t - t_0)$, where A is a positive constant. Express $A_s(t)$ as a function of $\overline{R}_p(\cdot)$, the periodic autocorrelation of the spreading waveform. How might this result be applied to acquisition?

Chapter 3
Frequency-Hopping Systems

Frequency hopping is the periodic changing of the carrier frequency of a transmitted signal. This time-varying characteristic potentially endows a communication system with great strength against interference. Whereas a direct-sequence system relies on spectral spreading, spectral despreading, and filtering to suppress interference, the basic mechanism of interference suppression in a frequency-hopping system is avoidance. When avoidance fails, it is only temporary because of the periodic changing of the carrier frequency. The impact of the interference is further mitigated by the pervasive use of channel codes, which are more essential for frequency-hopping systems than for direct-sequence systems. The basic concepts, spectral and performance aspects, and coding and modulation issues of frequency-hopping systems are presented in this chapter. The effects of partial-band interference and multitone jamming are examined, and the most important issues in the design of frequency synthesizers are described.

3.1 Concepts and Characteristics

The sequence of carrier frequencies transmitted by a frequency-hopping system is called the *frequency-hopping pattern,* and the set of M possible carrier frequencies $\{f_1,f_2,\ldots,f_M\}$ is the *hopset*. The rate at which the carrier frequency changes is the *hop rate*, and frequency hopping occurs over a frequency band called the *hopping band*, which includes *M frequency channels*. Each frequency channel is defined as a spectral region that includes a single carrier frequency of the hopset as its center frequency and has a bandwidth B large enough to include most of the power in a signal pulse with a specific carrier frequency. Figure 3.1 illustrates the frequency channels associated with a particular frequency-hopping pattern. The time interval between hops is called the *hop interval*. The length of the hop interval is the *hop duration* and is denoted as T_h. The hopping band has *hopping bandwidth* $W \geq MB$.

D. Torrieri, *Principles of Spread-Spectrum Communication Systems*,
https://doi.org/10.1007/978-3-319-70569-9_3

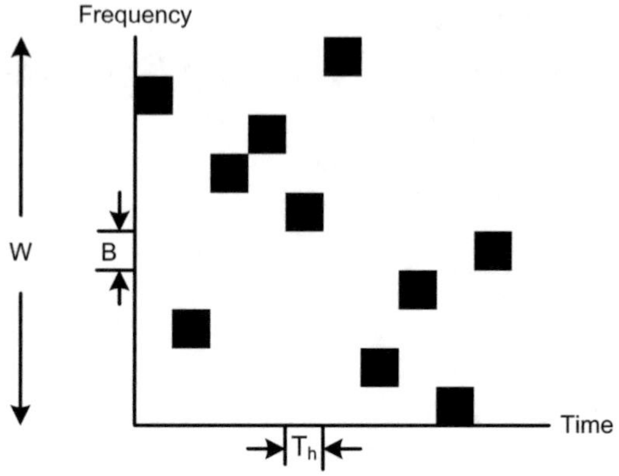

Fig. 3.1 Frequency-hopping pattern

Figure 3.2 (a) depicts the general form of a frequency-hopping transmitter. Pattern-control bits, which are the output bits of a *pattern generator*, change at the hop rate so that a frequency synthesizer produces a frequency-hopping pattern. The data-modulated signal is mixed with the frequency-hopping pattern to produce the frequency-hopping signal. If the data modulation is some form of angle modulation $\phi(t)$, then the received signal for the ith hop is

$$s(t) = \sqrt{2\mathcal{E}_s/T_s}\cos\left[2\pi f_{ci}t + \phi(t) + \phi_i\right], \quad (i-1)T_h \leq t \leq iT_h \tag{3.1}$$

where $\mathcal{E}_s$ is the energy per symbol, T_s is the symbol duration, f_{ci} is the carrier frequency for the ith hop, and ϕ_i is a random phase angle for the ith hop.

The frequency-hopping pattern produced by the receiver synthesizer of Figure 3.2 (b) is synchronized with the pattern produced by the transmitter, but is offset by a fixed intermediate frequency, which may be zero. The mixing operation removes the frequency-hopping pattern from the received signal and is hence called *dehopping*. The mixer output is applied to a bandpass filter that excludes double-frequency components and power that originated outside the appropriate frequency channel and produces the data-modulated *dehopped signal*, which has the form of (3.1) with f_{ci} for all hops replaced by the common intermediate frequency.

Although it provides no advantage against white noise, frequency hopping enables signals to hop out of frequency channels with interference or slow frequency-selective fading. To fully escape from the effects of narrowband interference signals, disjoint frequency channels are necessary. The disjoint channels may be contiguous or have unused spectral regions between them. Some spectral regions with steady interference or a susceptibility to fading may be omitted from the hopset, a process called *spectral notching*.

To ensure that a frequency-hopping pattern is difficult to reproduce or dehop by an opponent, the pattern should be pseudorandom with a large period and an approximately uniform distribution over the frequency channels. The pattern generator is a nonlinear sequence generator that maps each generator state to the

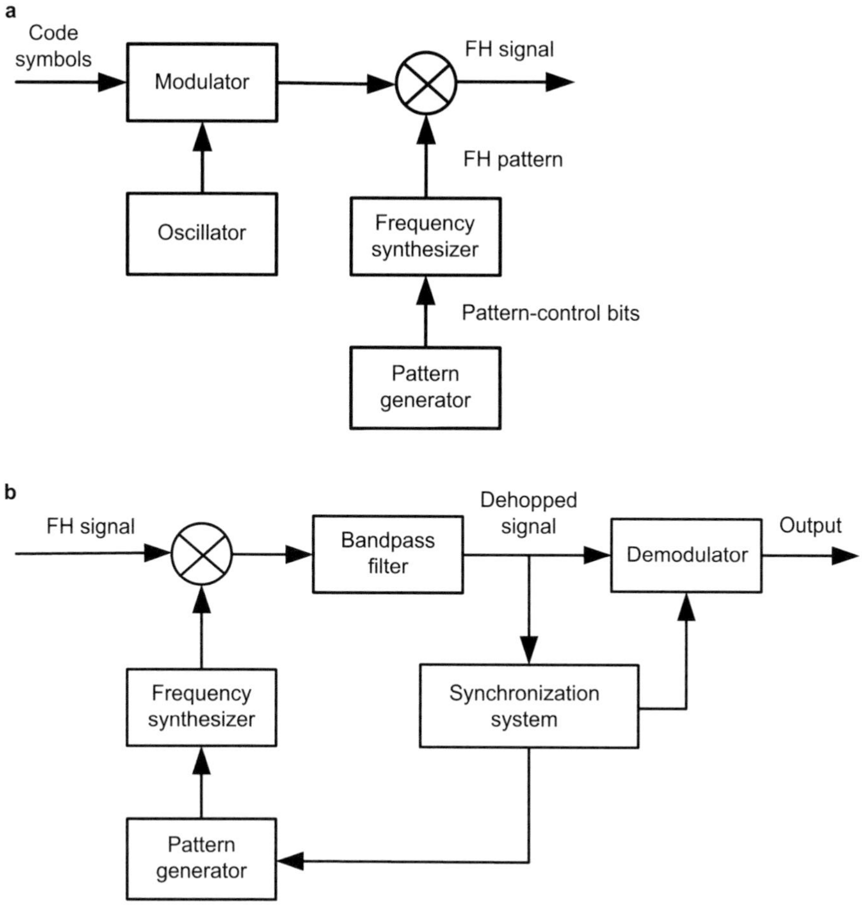

Fig. 3.2 General form of frequency-hopping system: (**a**) transmitter and (**b**) receiver

pattern-control bits that specify a frequency. The *linear span* or *linear complexity* of a nonlinear sequence is the number of stages of the shortest linear feedback shift register that can generate the sequence or the successive generator states. A large linear span inhibits the reconstruction of a frequency-hopping pattern from a short segment of it. More is required of frequency-hopping patterns to alleviate multiple-access interference when similar frequency-hopping systems are part of a network (Section 7.4).

An architecture that enhances the *transmission security* of a frequency-hopping system is shown in Figure 3.3. The structure or algorithm of the pattern generator is determined by a set of pattern-control bits that comprise the spread-spectrum key and the time-of-day (TOD). The *spread-spectrum key*, which is the ultimate source of security, is a set of bits that are changed infrequently. The spread-spectrum key may be generated by combining secret bits with two sets of address bits that

Fig. 3.3 Secure method of synthesizer control

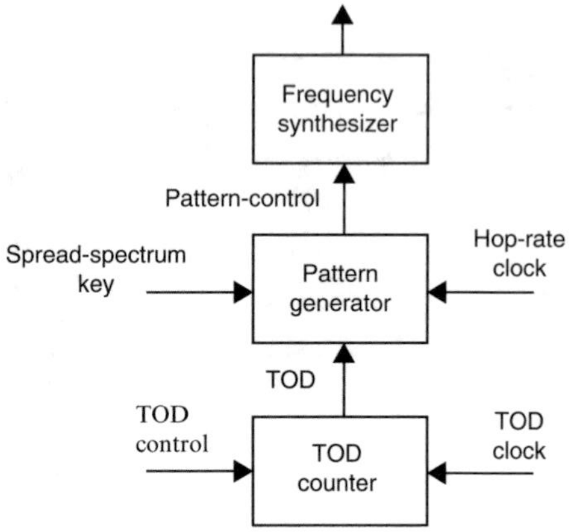

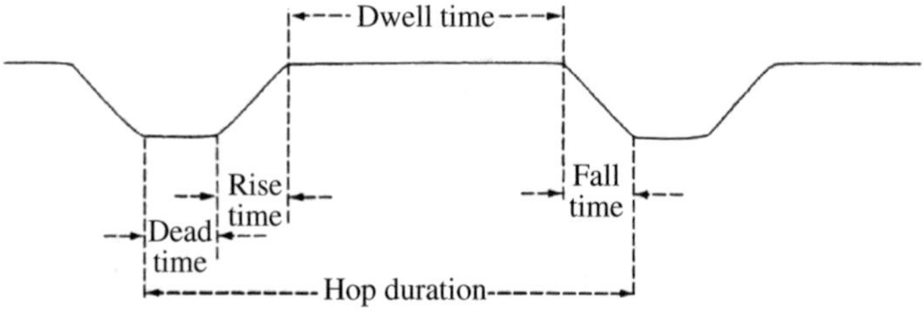

Fig. 3.4 Time durations and amplitude changes of a frequency-hopping pulse

identify both the transmitting device and the receiving device at the other end of a communication link. The TOD is a set of bits that are derived from the stages of the TOD counter and change with every transition of the TOD clock. For example, the spread-spectrum key might change daily, whereas the TOD might change every second. The purpose of the TOD is to vary the pattern-generator algorithm without constantly changing the spread-spectrum key. In effect, the pattern-generator algorithm is controlled by a time-varying key. The hop-rate clock, which regulates when the changes occur in the pattern generator, operates at a much higher rate than the TOD clock. In a receiver, the hop-rate clock is produced by the synchronization system. In both a transmitter and a receiver, the TOD clock may be derived from the hop-rate clock. The TOD control bits initiate or reset the TOD counter to a desired state.

A frequency-hopping pulse with a fixed carrier frequency occurs during a portion of the hop interval called the *dwell interval*. As illustrated in Figure 3.4, the *dwell time* is the duration of the dwell interval during which the channel symbols are transmitted and the peak amplitude occurs. The hop duration T_h is equal to the sum of the dwell time T_d and the switching time T_{sw}. The *switching time* is equal to the *dead time*, which is the duration of the interval when no signal is present, plus the rise and fall times of a pulse. Even if the switching time is insignificant in the transmitted signal, it is more substantial in the dehopped signal in the receiver because of the imperfect synchronization of received and receiver-generated waveforms. The nonzero switching time, which may include an intentional *guard time*, decreases the transmitted symbol duration T_s. If T_{so} is the symbol duration in the absence of frequency hopping, then $T_s = T_{so}(T_d/T_h)$. The reduction in symbol duration expands the transmitted spectrum and thereby reduces the number of frequency channels within a fixed hopping band. Since the receiver filtering ensures that rise and fall times of pulses have durations on the order of a symbol duration, $T_{sw} > T_s$ in practical systems.

Frequency hopping may be classified as fast or slow. *Fast frequency hopping* occurs if there is more than one hop for each information symbol. *Slow frequency hopping* ensues if one or more information symbols are transmitted in the time interval between frequency hops. Although these definitions do not refer to the absolute hop rate, fast frequency hopping is an option only if a hop rate that exceeds the information-symbol rate can be implemented. Slow frequency hopping is preferable because the transmitted waveform is much more spectrally compact (see Section 3.4) and the overhead cost of the switching time is greatly reduced.

To obtain the full advantage of block or convolutional channel codes in a slow frequency-hopping system, the code symbols should be interleaved in such a way that the symbols of a block codeword or the symbols within a few free distances in a convolutional code fade independently. In frequency-hopping systems operating over a frequency-selective fading channel, the realization of this independence requires certain constraints among the system parameter values (Section 6.7).

Frequency-selective fading and Doppler shifts make it difficult to maintain phase coherence from hop to hop between frequency synthesizers in the transmitter and the receiver. Furthermore, the time-varying delay between the frequency changes of the received signal and those of the synthesizer output in the receiver causes the phase shift in the dehopped signal to differ for each hop interval. Thus, frequency-hopping systems use noncoherent or differentially coherent demodulators unless a pilot signal is available, the hop duration is very long, or elaborate iterative phase estimation (perhaps as part of turbo decoding) is used.

In military applications, the ability of frequency-hopping systems to avoid interference is potentially neutralized by a *repeater jammer* (also known as a *follower jammer*), which is a device that intercepts a signal, processes it, and then transmits jamming at the same center frequency. To be effective against a frequency-hopping system, the jamming energy must reach the victim receiver before it hops to a new carrier frequency. Thus, the hop rate is the critical factor in protecting a system against a repeater jammer. Required hop rates and the limitations of repeater jamming are analyzed in [98].

3.2 Frequency Hopping with Orthogonal CPFSK

In a slow frequency-hopping system with frequency-shift keying as its data modulation, the implementation of phase continuity from symbol to symbol within a hop dwell interval prevents excessive spectral splatter outside a frequency channel (Section 3.3). Thus, the data modulation is called continuous-phase frequency-shift keying (CPFSK). In a frequency-hopping system with CPFSK (FH/CPFSK system), one of a set S_q of q CPFSK frequencies is selected to offset the carrier frequency for each transmitted symbol within each hop dwell interval. The general transmitter of an FH/CPFSK system is illustrated in Figure 3.5 (a). The pattern-generator output bits, which define a carrier frequency, and the digital symbols, which define a frequency in S_q, are combined to determine the frequency generated by the synthesizer during a symbol interval. In the standard design, the q subchannels are contiguous, and each set of q subchannels constitutes a frequency channel within the hopping band.

An orthogonal FH/CPFSK system adds frequency hopping to orthogonal CPFSK signals (Section 1.2). Figure 3.5 (b) depicts the main elements of a noncoherent orthogonal FH/CPFSK receiver. The frequency-hopping carrier frequency is removed, and the resulting orthogonal CPFSK signal is applied to the demodulator

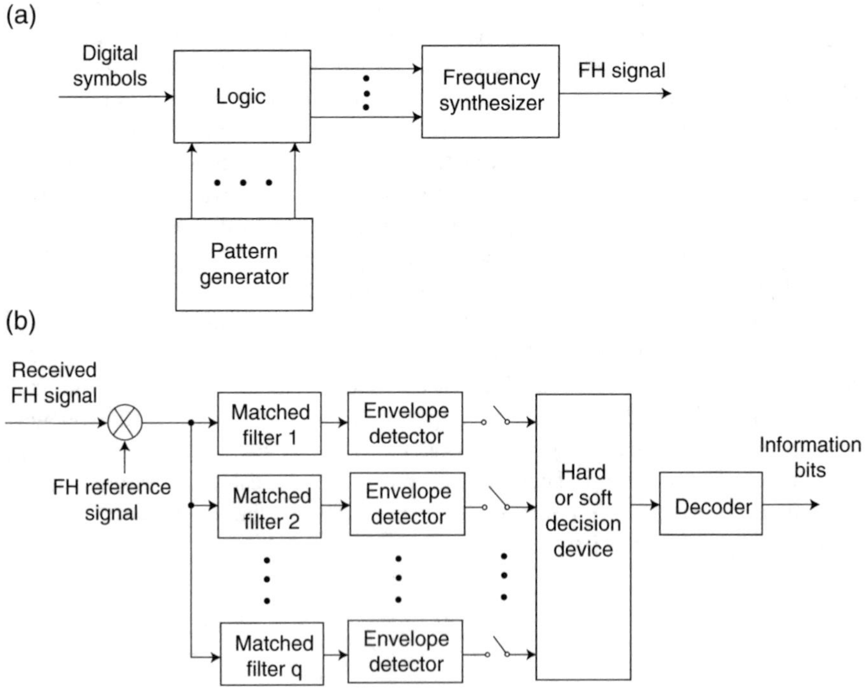

Fig. 3.5 Orthogonal FH/CPFSK (**a**) transmitter and (**b**) noncoherent receiver

and decoder. Each matched filter in the bank of matched filters corresponds to a CPFSK subchannel.

Illustrative System

To illustrate some basic issues of frequency-hopping communications, we consider an orthogonal FH/CPFSK system that uses a repetition code as its channel code, suboptimal metrics, and the receiver of Figure 3.5 (b). The system has significant deficiencies, such as the very weak repetition code, but is amenable to an approximate mathematical analysis that provides insight into the design issues of a frequency-hopping system. Each information symbol is transmitted as a codeword of n identical code symbols. The interference is modeled as wideband Gaussian noise uniformly distributed over part of the hopping band. Along with perfect dehopping and negligible switching times, either slow frequency hopping with ideal interleaving over many hop intervals or fast frequency hopping is assumed. Both ensure the independence of code-symbol errors.

The difficulty of implementing the maximum-likelihood metric (1.80) leads to consideration of the suboptimal metrics. The square-law metric defined by (1.84), which performs *linear square-law combining,* has the advantage that no *channel-state information* or *side information* providing information about the reliability of code symbols is required for its computation. However, this metric is unsuitable against strong partial-band interference because a single large symbol metric can corrupt the codeword metric.

A dimensionless metric that is computationally simpler than (1.80) but retains the division by N_{0i} is the *nonlinear square-law metric*:

$$U(l) = \sum_{i=1}^{n} \frac{R_{li}^2}{N_{0i}}, \quad l = 1, 2, \ldots, q \tag{3.2}$$

where R_{li} is the sample value of the envelope-detector output that is associated with code symbol i of candidate information-symbol l, and $N_{0i}/2$ is the two-sided power spectral density (PSD) of the interference and noise over all the CPFSK subchannels during code symbol i. The advantage of this metric is that the impact of a large R_{li} is alleviated by a large N_{0i}, which must be known. The subsequent analysis is for the nonlinear square-law metric.

The union bound (1.26) implies that the information-symbol error probability satisfies

$$P_{is} \leq (q-1)P_2 \tag{3.3}$$

where P_2 is the probability of an error in comparing the metric associated with the transmitted information symbol with the metric associated with an alternative one. We assume that there are enough frequency channels that n distinct carrier frequencies are used for the n code symbols. Since the CPFSK tones are orthogonal, the code-symbol metrics $\{R_{li}^2\}$ are independent and identically distributed for all values of l and i (Section 1.2). Therefore, the Chernoff bound given by (1.145) and (1.144) with $\alpha = 1/2$ yields

$$P_2 \leq \frac{1}{2}Z^n \tag{3.4}$$

$$Z = \min_{0<s<s_1} E\left[\exp\left\{\frac{s}{N_1}\left(R_2^2 - R_1^2\right)\right\}\right] \tag{3.5}$$

where R_1 is the sampled output of an envelope detector when the desired signal is present at the input of the associated matched filter, R_2 is the output from a different matched filter when the desired signal is absent at its input, and $N_1/2$ is the two-sided PSD of the interference and noise that is assumed to cover all the CPFSK subchannels during a code symbol and may vary from symbol to symbol. Since the CPFSK tones are orthogonal, and hence q-ary symmetric, (1.95), (3.3), and (3.4) give an upper bound on the information-bit error probability:

$$P_b \leq \frac{q}{4}Z^n. \tag{3.6}$$

The received signal has the form

$$r(t) = \sqrt{2\mathcal{E}_s/T_s}\cos\left[2\pi\left(f_c + f_1\right)t + \theta\right] + n_1(t), \quad 0 \leq t \leq T_s$$

where $\mathcal{E}_s$ is the energy per code symbol, T_s is a code-symbol duration, and $n_1(t)$ is a zero-mean Gaussian interference-and-noise process. Assuming that $n_1(t)$ is white,

$$E[n_1(t)n_1(t+\tau)] = \frac{N_1}{2}\delta(\tau). \tag{3.7}$$

As in Section 1.5, R_l^2 may be expressed as

$$R_l^2 = R_{lc}^2 + R_{ls}^2\,, \quad l = 1, 2. \tag{3.8}$$

Using the orthogonality of the symbol waveforms and (3.7) and assuming that $f_c + f_l >> 1/T_s$ in (1.99) and (1.100), we verify the independence of the Gaussian random variables R_{1c}, R_{1s}, R_{2c}, and R_{2s}. We obtain the moments

$$E[R_{1c}] = \sqrt{\mathcal{E}_s T_s/2}\cos\theta\ , \quad E[R_{1s}] = \sqrt{\mathcal{E}_s T_s/2}\sin\theta \tag{3.9}$$

$$E[R_{2c}] = E[R_{2s}] = 0 \tag{3.10}$$

$$\mathrm{var}(R_{lc}) = \mathrm{var}(R_{ls}) = N_1 T_s/4\ , \quad l = 1, 2. \tag{3.11}$$

For a Gaussian random variable X with mean m and variance σ^2, a straightforward calculation yields

$$E[\exp(aX^2)] = \frac{1}{\sqrt{1-2a\sigma^2}}\exp\left(\frac{am^2}{1-2a\sigma^2}\right)\ , \quad a < \frac{1}{2\sigma^2} \tag{3.12}$$

which can be used to partially evaluate the expectation in (3.5). After conditioning on N_1, (3.8) to (3.12) and the substitution of $\lambda = sT_s/2N_1$ give

$$Z = \min_{0<\lambda<1} E\left[\frac{1}{1-\lambda^2}\exp\left(-\frac{\lambda\mathcal{E}_s/N_1}{1+\lambda}\right)\right] \tag{3.13}$$

where the remaining expectation is with respect to the statistics of N_1.

To simplify the analysis, we assume that the thermal noise is negligible. When a repetition symbol encounters no interference, $N_1 = 0$; when it does, $N_1 = I_{t0}/\mu$, where μ is the fraction of the hopping band with interference, and I_{t0} is the PSD that would exist if the interference power were uniformly spread over the entire hopping band. We define

$$\gamma = \frac{(\log_2 q)\,\mathcal{E}_b}{I_{t0}} \tag{3.14}$$

where $\log_2 q$ is the number of bits per information symbol, and $\mathcal{E}_b$ is the energy per information bit. Since μ is the probability that interference is encountered by a received code symbol and $\mathcal{E}_s = (\log_2 q)\,\mathcal{E}_b/n$, (3.13) becomes

$$Z = \min_{0<\lambda<1}\left[\frac{\mu}{1-\lambda^2}\exp\left(-\frac{\lambda\mu\gamma/n}{1+\lambda}\right)\right]. \tag{3.15}$$

Using calculus, we find that

$$Z = \frac{\mu}{1-\lambda_1^2}\exp\left(-\frac{\lambda_1\mu\gamma/n}{1+\lambda_1}\right) \tag{3.16}$$

where

$$\lambda_1 = -\left(\frac{1}{2}+\frac{\mu\gamma}{4n}\right)+\left[\left(\frac{1}{2}+\frac{\mu\gamma}{4n}\right)^2+\frac{\mu\gamma}{2n}\right]^{1/2}. \tag{3.17}$$

Substituting (3.16) and (3.14) into (3.6), we obtain

$$P_b \le \frac{q}{4}\left(\frac{\mu}{1-\lambda_1^2}\right)^n \exp\left(-\frac{\lambda_1\mu\gamma}{1+\lambda_1}\right). \tag{3.18}$$

Suppose that the interference is worst-case partial-band interference, which occurs when the interference power is distributed in the most damaging way. An approximate upper bound on P_b is obtained by maximizing the right-hand side of (3.18) with respect to μ, where $0 \le \mu \le 1$. Calculus yields the maximizing value of μ:

$$\mu_0 = \min\left(\frac{3n}{\gamma}, 1\right). \tag{3.19}$$

Since μ_0 is obtained by maximizing a bound rather than an equality, it is not necessarily equal to the actual worst-case μ. Setting $\mu = \mu_0$ in (3.18), we obtain the approximate upper bound on P_b for worst-case partial-band interference:

$$P_b \le \begin{cases} \frac{q}{4}\left(\frac{4n}{e\gamma}\right)^n, & n \le \gamma/3 \\ \frac{q}{4}(1-\lambda_0^2)^{-n}\exp\left(-\frac{\lambda_0\gamma}{1+\lambda_0}\right), & n > \gamma/3 \end{cases} \tag{3.20}$$

where

$$\lambda_0 = -\left(\frac{1}{2}+\frac{\gamma}{4n}\right)+\left[\left(\frac{1}{2}+\frac{\gamma}{4n}\right)^2+\frac{\gamma}{2n}\right]^{1/2}. \tag{3.21}$$

Further analysis requires some facts about convex functions. A real-valued function f over an interval is a *convex function* if $f[\alpha x + (1-\alpha)y] \le \alpha f(x) + (1-\alpha)f(y)$ for all x and y in the interval and $\alpha \in [0, 1]$. A sufficient condition for f to be a convex function is for it to have a nonnegative second derivative (see Section 7.3). A local minimum of a convex function over an interval is a global minimum, and similarly a local maximum is a global maximum [19].

If the value of γ is known, then the number of repetitions can be chosen to minimize the upper bound on P_b for worst-case partial-band interference. We treat n as a continuous variable such that $n \ge 1$ and let n_0 denote the minimizing value of n. A calculation indicates that the derivative with respect to n of the second line on the right-hand side of (3.20) is positive. Therefore, if $\gamma/3 < 1$ so that the second line is applicable for $n \ge 1$, then $n_0 = 1$. If $\gamma/3 > 1$, the upper bound on P_b is minimized by minimizing the strictly convex function $f(n) = (q/4)(4n/e\gamma)^n$ over the compact set $[1, \gamma/3]$. Since $f(n)$ is strictly convex, a stationary point of $f(n)$ represents a global minimum. Further calculation yields

$$n_0 = \max\left(\frac{\gamma}{4}, 1\right). \tag{3.22}$$

Since n must be an integer, the optimal number of repetitions against worst-case partial-band interference is approximately $\lfloor n_0 \rfloor$, where $\lfloor x \rfloor$ denotes the largest integer less than or equal to x. Equation (3.22) indicates that the optimal number of repetitions increases with γ.

The approximate upper bound on P_b for worst-case partial-band interference and the optimal number of repetitions is obtained by substituting the preceding results for n_0 and (3.14) into (3.20), which gives

$$P_b \leq \begin{cases} \frac{q}{4}(1-\lambda_0^2)^{-1}\exp\left(-\frac{\lambda_0 \mathcal{E}_b/I_{t0}}{1+\lambda_0}\right), & \frac{\mathcal{E}_b}{I_{t0}} < 3\log_2 q \\ \frac{q}{e(\log_2 q)}\left(\frac{\mathcal{E}_b}{I_{t0}}\right)^{-1}, & 3\log_2 q \leq \frac{\mathcal{E}_b}{I_{t0}} < 4\log_2 q \\ \frac{q}{4}\exp\left(-\frac{\mathcal{E}_b/I_{t0}}{4(\log_2 q)}\right), & \frac{\mathcal{E}_b}{I_{t0}} \geq 4\log_2 q \end{cases} \tag{3.23}$$

which indicates that the upper bound on P_b decreases exponentially with $\mathcal{E}_b/I_{t0}$ only if the optimal number of repetitions is chosen and $\mathcal{E}_b/I_{t0} \geq 4\log_2 q$. Figure 3.6 illustrates the upper bound on P_b as a function of $\mathcal{E}_b/I_{t0}$ for $q = 2$, 4, and 8 and both

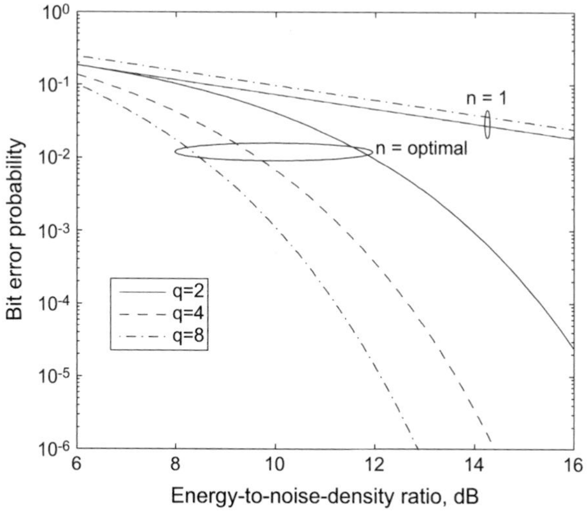

Fig. 3.6 Upper bound on the bit error probability of the frequency-hopping system in the presence of worst-case partial-band interference for $q = 2$, 4, and 8 and both the optimal number of repetitions and $n = 1$

the optimal number of repetitions and $n = 1$. The figure indicates that the nonlinear square-law metric and optimal repetitions sharply limit the performance degradation caused by worst-case partial-band interference relative to full-band interference with the same power. For example, setting $N_0 \to I_{t0}$ in (1.94) and $q = 2$ in (3.23) and then comparing the equations, we find that this degradation is approximately 3 dB for binary CPFSK.

Substituting (3.22) into (3.19), we obtain

$$\mu_0 = \begin{cases} 1, & \frac{\mathcal{E}_b}{I_{t0}} < 3\log_2 q \\ \frac{3}{\log_2 q}\left(\frac{\mathcal{E}_b}{I_{t0}}\right)^{-1} & 3\log_2 q \le \frac{\mathcal{E}_b}{I_{t0}} < 4\log_2 q \\ \frac{3}{4}, & \frac{\mathcal{E}_b}{I_{t0}} \ge 4\log_2 q \end{cases} \tag{3.24}$$

which indicates that if the optimal number of repetitions is used, then the worst-case interference covers three-fourths or more of the hopping band.

For frequency hopping with binary CPFSK and the nonlinear square-law metric, a more precise derivation [44] that does not use the Chernoff bound confirms that (3.23) provides an approximate upper bound on the information-bit error rate caused by worst-case partial-band interference when the thermal noise is negligible, although the optimal number of repetitions is much smaller than is indicated by (3.22). Thus, the appropriate weighting of terms in the nonlinear square-law metric prevents the domination by a single corrupted symbol metric and limits the inherent *noncoherent combining loss* resulting from the fragmentation of the symbol energy.

The implementation of the nonlinear square-law metric requires the measurement of the interference power. An iterative method of power estimation based on the expectation-maximization algorithm (Section 9.1) provides approximate maximum-likelihood estimates, but a stronger code than the repetition code is required. Other strategies are to revert to either hard-decision decoding or to the square-law metric with clipping or soft-limiting of each envelope-detector output R_{li}. Both strategies prevent a single corrupted sample from undermining the codeword detection. Although clipping is potentially more effective than hard-decision decoding, its implementation requires an accurate measurement of the signal power for properly setting the clipping level. Instead of the square-law metric with clipping, an easily implemented suboptimal metric has the form

$$U(l) = \sum_{i=1}^{n} f\left(\frac{R_{li}^2}{\sum_{j=1}^{n} R_{lj}^2}\right), \quad l = 1, 2, \ldots, q \tag{3.25}$$

for some monotonically increasing function $f(\cdot)$. Although this metric does not require the estimation of each N_{0i}, it is sensitive to each N_{0i}.

Multitone Jamming

Orthogonal CPFSK tones produce negligible responses in the incorrect subchannels if the frequency separation between tones is k/T_s, where k is a nonzero integer, and T_s denotes the symbol duration. To maximize the hopset size when the CPFSK subchannels are contiguous, $k = 1$ is selected. Consequently, the bandwidth of a frequency channel for slow frequency hopping with many symbols per dwell interval and negligible switching time is

$$B \approx \frac{q}{T_s} = \frac{q}{T_b \log_2 q} \tag{3.26}$$

where T_b is the duration of a bit, and the factor $\log_2 q$ accounts for the increase in symbol duration when a nonbinary modulation is used. If the hopping band has bandwidth W and uniformly separated contiguous frequency channels of bandwidth B are assigned, the hopset size is

$$M = \left\lfloor \frac{W}{B} \right\rfloor \tag{3.27}$$

and thus

$$M = \left\lfloor \frac{WT_b \log_2 q}{q} \right\rfloor. \tag{3.28}$$

When the CPFSK subchannels are contiguous, it is not advantageous to a jammer to transmit the jamming in all the subchannels of an CPFSK set because only a single subchannel needs to be jammed to cause a symbol error. A sophisticated jammer with knowledge of the spectral locations of the CPFSK sets can cause increased system degradation by placing one jamming tone or narrowband jamming signal in every CPFSK set, which is called *multitone jamming*.

To assess the impact of this sophisticated multitone jamming on hard-decision decoding in the receiver of Figure 3.5 (b), we assume that thermal noise is absent and that each jamming tone coincides with one CPFSK tone in a frequency channel encompassing q orthogonal CPFSK tones. Whether a jamming tone coincides with the transmitted CPFSK tone or an incorrect one, there will be no symbol error if the desired-signal power S exceeds the jamming power. Thus, if I_t is the total available jamming power, then the jammer can maximize symbol errors by placing tones with power levels slightly above S whenever possible in approximately J frequency channels such that

$$J = \begin{cases} 1, & I_t < S \\ \left\lfloor \frac{I_t}{S} \right\rfloor, & S \le I_t < MS \\ M, & MS \le I_t. \end{cases} \tag{3.29}$$

If a transmitted tone enters a jammed frequency channel and $I_t \geq S$, then with probability $(q-1)/q$, the jamming tone does not coincide with the transmitted tone and causes a symbol error after a hard decision is made. If the jamming tone does coincide with the correct tone, it does not cause a symbol error in the absence of thermal noise. Since J/M is the probability with which a frequency channel is jammed, and no error occurs if $I_t < S$, the symbol error probability is

$$P_s = \begin{cases} 0, & I_t < S \\ \frac{J}{M}\left(\frac{q-1}{q}\right), & I_t \geq S. \end{cases} \tag{3.30}$$

Substitution of (3.28) and (3.29) into (3.30) and the approximation $\lfloor x \rfloor \approx x$ yield

$$P_s = \begin{cases} \frac{q-1}{q}, & \frac{\mathcal{E}_b}{I_{t0}} < \frac{q}{\log_2 q} \\ \left(\frac{q-1}{\log_2 q}\right)\left(\frac{\mathcal{E}_b}{I_{t0}}\right)^{-1}, & \frac{q}{\log_2 q} \leq \frac{\mathcal{E}_b}{I_{t0}} \leq WT_b \\ 0, & \frac{\mathcal{E}_b}{I_{t0}} > WT_b \end{cases} \tag{3.31}$$

where $\mathcal{E}_b = ST_b$ denotes the energy per bit, and $I_{t0} = I_t/W$ denotes the PSD of the interference power that would exist if it were uniformly spread over the hopping band. This equation exhibits an inverse linear dependence of P_s on $\mathcal{E}_b/I_{t0}$, which indicates that the jamming has an impact that is qualitatively similar to that of Rayleigh fading and to what is observed in Figure 3.6 for worst-case partial-band interference and $n = 1$. The symbol error probability increases with q, which is the opposite of what is observed for worst-case partial-band interference when n is optimized. The reason for this increase in P_s is the increase in the bandwidth of each frequency channel as q increases, which provides a larger target for multitone jamming. Thus, binary CPFSK is advantageous in relation to this sophisticated multitone jamming.

3.3 Frequency Hopping with DPSK and CPM

In a network of frequency-hopping systems and a fixed hopping bandwidth, it is highly desirable to choose a spectrally compact data modulation so that the hopset is large and hence the number of collisions among the frequency-hopping signals is kept low. A spectrally compact modulation also helps to ensure that the bandwidth of a frequency channel is less than the coherence bandwidth (Section 6.3) so that equalization in the receiver is not necessary.

The limiting of spectral splatter is another desirable characteristic of data modulation for frequency hopping. *Spectral splatter* is the interference produced in frequency channels other than that being used by a frequency-hopping pulse. It is caused by the time-limited nature of transmitted pulses. The degree to which spectral splatter causes errors depends primarily on the separation F_s between carrier frequencies and the percentage of the signal power included in a frequency channel. In practice, this percentage must be at least 90% to avoid signal distortion and is often more than 95%. Usually, only pulses in adjacent channels produce a significant amount of spectral splatter in a frequency channel.

The *adjacent splatter ratio* K_s is the ratio of the power due to spectral splatter from an adjacent channel to the corresponding power that arrives at the receiver in that channel. For example, if B is the bandwidth of a frequency channel that includes 97% of the signal power and $F_s \geq B$, then no more than 1.5% of the power from a transmitted pulse can enter an adjacent channel on one side of the frequency channel used by the pulse; therefore, $K_s \leq 0.015/0.97 = 0.0155$. A given maximum value of K_s can be reduced by an increase in F_s, but eventually the number of frequency channels M must be reduced if the hopping bandwidth is fixed. As a result, the rate at which users hop into the same channel increases. This increase may cancel any improvement due to the reduction of the spectral splatter. The opposite procedure (reducing F_s and B so that more frequency channels become available) increases not only the spectral splatter but also signal distortion and intersymbol interference, so the amount of useful reduction is limited.

To avoid spectral spreading due to amplifier nonlinearity, it is desirable for the data modulation to have a constant amplitude, as it is often impossible to implement a filter with the appropriate bandwidth and center frequency for spectral shaping of a signal after it emerges from the final power amplifier. Since they produce constant amplitudes and do not require coherent demodulation, a good modulation candidate is differential phase-shift keying (DPSK) or some form of spectrally compact continuous-phase modulation.

FH/DPSK

A DPSK demodulator compares the phases of two successive received symbols. If the magnitude of the phase difference is does not exceed $\pi/2$, then the demodulator decides that a 1 was transmitted; otherwise, it decides that a 0 was transmitted.

Consider multitone jamming of an FH/DPSK system with negligible thermal noise. Each tone is assumed to have a frequency identical to the center frequency of one of the frequency channels. The composite signal, consisting of the transmitted signal and the jamming tone, has a constant phase over two successive received symbols in the same hop dwell interval if a 1 was transmitted and the thermal noise is absent; thus, the demodulator correctly detects the 1.

Suppose that a 0 was transmitted, and that the desired signal is $\sqrt{2S}\cos 2\pi f_c t$ during the first symbol interval, where S is the average power and f_c is the carrier frequency of the frequency-hopping signal during the dwell interval. When a jamming tone is present, trigonometric identities indicate that the composite signal during the first symbol interval may be expressed as

$$\sqrt{2S}\cos 2\pi f_c t + \sqrt{2I}\cos(2\pi f_c t + \theta) = \sqrt{2S + 2I_t + 4\sqrt{SI}\cos\theta}\cos(2\pi f_c t + \phi_1) \tag{3.32}$$

where I is the average power of the jamming tone, θ is the phase of the tone relative to the phase of the transmitted signal, and ϕ_1 is the phase of the composite signal:

$$\phi_1 = \tan^{-1}\left(\frac{\sqrt{I}\sin\theta}{\sqrt{S} + \sqrt{I}\cos\theta}\right). \tag{3.33}$$

Since the desired signal during the second symbol is $-\sqrt{2S}\cos 2\pi f_c t$, the composite signal during the second symbol interval is

$$-\sqrt{2S}\cos 2\pi f_c t + \sqrt{2I}\cos(2\pi f_c t + \theta) = \sqrt{2S + 2I_t + 4\sqrt{SI}\cos\theta}\cos(2\pi f_c t + \phi_2) \tag{3.34}$$

where

$$\phi_2 = \tan^{-1}\left(\frac{\sqrt{I}\sin\theta}{-\sqrt{S} + \sqrt{I}\cos\theta}\right). \tag{3.35}$$

Using trigonometry, it is found that

$$\phi_2 - \phi_1 = \cos^{-1}\left[\frac{I - S}{\sqrt{S^2 + I^2 + 2SI(1 - 2\cos^2\theta)}}\right]. \tag{3.36}$$

If $I \geq S$, then $|\phi_2 - \phi_1| \leq \pi/2$, so the demodulator incorrectly decides that a 1 was transmitted. If $I < S$, no mistake is made. Thus, multitone jamming with total power I_t is most damaging when J frequency channels given by (3.29) are jammed and each tone has power $I = I_t/J$. If the information bits 0 and 1 are equally likely, then the symbol error probability given that a frequency channel is jammed with $I \geq S$ is $P_s = 1/2$, the probability that a 0 was transmitted. Therefore, $P_s = J/2M$ if $I_t \geq S$, and $P_s = 0$, otherwise. Using (3.27) and (3.29) with $S = \mathcal{E}_b/T_b$, $I_t = I_{t0}W$, and $\lfloor x \rfloor \approx x$, we obtain the symbol error probability for FH/DPSK and multitone jamming:

$$P_s \approx \begin{cases} \frac{1}{2}, & \frac{\mathcal{E}_b}{I_{t0}} < BT_b \\ \frac{1}{2}BT_b\left(\frac{\mathcal{E}_b}{I_{t0}}\right)^{-1}, & BT_b \leq \frac{\mathcal{E}_b}{I_{t0}} \leq WT_b \\ 0, & \frac{\mathcal{E}_b}{I_{t0}} > WT_b \end{cases} \tag{3.37}$$

which indicates that the multitone jamming has a much stronger impact on P_s than white Gaussian noise with the same power.

FH/CPM

We define the *frequency pulse* $g(t)$ as a piece-wise continuous function that vanishes outside the interval $[0, LT_s]$; that is,

$$g(t) = 0\ , \quad t < 0\ , \quad t > LT_s \tag{3.38}$$

where L is a positive integer and T_s is the symbol duration. The function is normalized so that

$$\int_0^{LT_s} g(x)dx = \frac{1}{2}. \tag{3.39}$$

The *phase response* is defined as the continuous function

$$\phi\,(t) = \begin{cases} 0, & t < 0 \\ \int_0^t g(x)dx, & 0 \leq t \leq LT_s \\ 1/2, & t > LT_s. \end{cases} \tag{3.40}$$

The general form of a signal with *continuous-phase modulation (CPM)* is

$$s(t) = A\cos[2\pi f_c t + \phi(t, \boldsymbol{\alpha})] \tag{3.41}$$

where A is the amplitude, f_c is the carrier frequency, and $\phi(t, \boldsymbol{\alpha})$ is the *phase function* that carries the message. The phase function has the form

$$\phi(t, \boldsymbol{\alpha}) = 2\pi h \sum_{i=-\infty}^{\infty} \alpha_i \phi(t - iT_s) \tag{3.42}$$

where h is a constant called the *deviation ratio* or *modulation index*, and the vector $\boldsymbol{\alpha}$ is a sequence of q-ary channel symbols. Each symbol α_i takes one of q values; if q is even, the values are $\pm 1, \pm 3, \ldots, \pm(q-1)$. The phase function is continuous and (3.42) indicates that the phase in any specified symbol interval depends on the previous symbols.

Since $g(t)$ is a piece-wise continuous function, $\phi(t, \boldsymbol{\alpha})$ is differentiable. The frequency function of the CPM signal, which is proportional to the derivative of $\phi(t, \boldsymbol{\alpha})$, is

$$\frac{1}{2\pi}\phi'(t, \boldsymbol{\alpha}) = h \sum_{i=-\infty}^{n} \alpha_i g(t - iT_s), \quad nT_s \leq t \leq (n+1)\,T_s. \tag{3.43}$$

If $L = 1$, the continuous-phase modulation is called a *full-response modulation*; if $L > 1$, it is called a *partial-response modulation*, and each frequency pulse extends over two or more symbol intervals. The normalization condition for a full-response modulation implies that the phase change over a symbol interval is equal to $h\pi\alpha_i$.

Continuous-phase frequency-shift keying (CPFSK) is a full-response subclass of CPM for which the instantaneous frequency is constant over each symbol interval. Because of the normalization, a CPFSK frequency pulse is given by

$$g(t) = \begin{cases} \frac{1}{2T_s}, & 0 \leq t \leq T_s \\ 0, & \text{otherwise} \end{cases} \tag{3.44}$$

and its phase response is

$$\phi(t) = \begin{cases} 0, & t < 0 \\ \frac{t}{2T_s}, & 0 \leq t \leq T_s \\ \frac{1}{2}, & t > T_s. \end{cases} \tag{3.45}$$

A CPFSK signal shifts among frequencies separated by $f_d = h/T_s$. The substitution of (3.45) into (3.42) indicates that

$$\phi(t, \boldsymbol{\alpha}) = \pi h \sum_{i=-\infty}^{n-1} \alpha_i + \frac{\pi h \alpha_n}{T_s}(t - nT_s), \quad nT_s \leq t \leq (n+1)\,T_s. \tag{3.46}$$

The main difference between CPFSK and FSK is that h can have any positive value for CPFSK but is relegated to integer values for FSK so that the tones are orthogonal to each other. Both modulations may be detected with matched filters, envelope detectors, and frequency discriminators. Although CPFSK explicitly requires phase continuity and FSK does not, FSK is almost always implemented with phase continuity to avoid the generation of spectral splatter, and hence is equivalent to CPFSK with $h = 1$. *Minimum-shift keying (MSK)* is defined as binary CPFSK with $h = 1/2$, and hence the two frequencies are separated by $f_d = 1/2T_s$.

With *multisymbol noncoherent detection* [110], CPFSK systems can provide a better symbol error probability than coherent BPSK systems without multisymbol detection. For r-symbol detection, the optimal receiver correlates the received waveform over all possible r-symbol patterns before making a decision. The drawback is the considerable implementation complexity of multisymbol detection, even for three-symbol detection.

Many communication signals are modeled as bandpass signals having the form

$$s(t) = \text{Re}\,[s_l(t) \exp(j2\pi f_c t)] \tag{3.47}$$

where $j = \sqrt{-1}$, and $s_l(t)$ is a complex-valued function. Multiplication of $s(t)$ by $2\exp(-j2\pi f_c t)$ and lowpass filtering produces $s_l(t)$, which is called the *complex envelope* or *equivalent lowpass waveform* of $s(t)$. We consider quadrature modulations that have the form

$$s_l(t) = \frac{A[d_1(t) + jd_2(t)]\exp(j\theta)}{\sqrt{2}} \tag{3.48}$$

where $d_1(t)$ and $d_2(t)$ are data modulations, and θ is the random phase uniformly distributed over $[0, 2\pi)$. Then $s(t)$ may be expressed as

$$s(t) = \frac{A}{\sqrt{2}}d_1(t)\cos(2\pi f_c t + \theta) + \frac{A}{\sqrt{2}}d_2(t)\sin(2\pi f_c t + \theta) \tag{3.49}$$

where the $\sqrt{2}$ has been inserted because the power is divided between the quadrature components.

We consider data modulations with the idealized form

$$d_i(t) = \sum_{k=-\infty}^{\infty} a_{ik}\psi(t - kT_s - T_0 - t_i), \quad i = 1, 2 \tag{3.50}$$

where $\{a_{ik}\}$ is a sequence of independent, identically distributed random variables, $a_{ik} = +1$ with probability 1/2 and $a_{ik} = -1$ with probability 1/2, $\psi(t)$ is a pulse waveform, T_s is the pulse duration and symbol duration, t_i is the relative pulse offset, and T_0 is an independent random variable that is uniformly distributed over the interval $[0, T_s)$ and reflects the arbitrariness of the origin of the coordinate system. Equation (3.50) describes an infinite stream of data symbols, which is an approximation that serves to simplify the evaluations of the autocorrelations and PSDs. As shown in Section 3.4, a finite set of symbols has a less compact spectrum.

Since a_{ik} is independent of a_{in} when $n \neq k$, it follows that $E[a_{ik}a_{in}] = 0, n \neq k$. Therefore, the autocorrelation of the real-valued $d_i(t)$ is

$$\begin{aligned} R_{di}(\tau) &= E[d_i(t)d_i(t+\tau)] \\ &= \sum_{k=-\infty}^{\infty} E[\psi(t - kT_s - T_0 - t_i)\psi(t - kT_s - T_0 - t_i + \tau)]. \end{aligned} \tag{3.51}$$

Expressing the expected value as an integral over the range of T_0 and changing variables, we obtain

$$\begin{aligned} R_{di}(\tau) &= \sum_{k=-\infty}^{\infty} \frac{1}{T_s}\int_{t-kT_s-T_s-t_i}^{t-kT_s-t_i} \psi(x)\psi(x+\tau)dx \\ &= \frac{1}{T_s}\int_{-\infty}^{\infty} \psi(x)\psi(x+\tau)dx, \quad i = 1, 2. \end{aligned} \tag{3.52}$$

This equation indicates that $d_1(t)$ and $d_2(t)$ are wide-sense-stationary processes with the same autocorrelation. The autocorrelation of the complex-valued $s_l(t)$ is

$$R_l(\tau) = E[s_l(t)s_l^*(t+\tau)]. \tag{3.53}$$

The independence of $d_1(t)$ and $d_2(t)$ imply that

$$R_l(\tau) = \frac{A^2}{2}R_{d1}(\tau) + \frac{A^2}{2}R_{d2}(\tau). \tag{3.54}$$

The two-sided PSD $S_l(f)$ of the complex envelope $s_l(t)$ is the Fourier transform (Appendix C.1) of $R_l(\tau)$. The Fourier transform of $\psi(t)$ is

$$G(f) = \int_{-\infty}^{\infty} \psi(t)e^{-j2\pi ft}dt. \tag{3.55}$$

From (3.54), (3.52), and the convolution theorem (Appendix C.1), we obtain the PSD

$$S_l(f) = A^2\frac{|G(f)|^2}{T_s}. \tag{3.56}$$

In a quadriphase-shift keying (QPSK) signal, $d_1(t)$ and $d_2(t)$ are modeled as independent random binary sequences with $t_1 = t_2 = 0$ and symbol duration $T_s = 2T_b$ in (3.50), where T_b is a bit duration. If $\psi(t)$ is rectangular with unit amplitude over $[0, 2T_b]$, then (3.56), (3.55), and an elementary integration indicate that the *PSD for QPSK* is

$$S_l(f) = 2A^2T_b\, sinc^2(2T_bf) \tag{3.57}$$

which is the same as the PSD for BPSK.

A binary minimum-shift-keying (MSK) signal with the same component amplitude can be represented by (3.49) and (3.50) with $t_1 = 0$, $t_2 = -\pi/2$, $T_s = T_b$, and

$$\psi(t) = \sqrt{2}\sin\left(\frac{\pi t}{2T_b}\right), \quad 0 \le t < 2T_b. \tag{3.58}$$

A straightforward evaluation of $G(f)$ using trigonometry and trigonometric integrals gives the *PSD of MSK:*

$$S_l(f) = \frac{16A^2T_b}{\pi^2}\left[\frac{\cos(2\pi T_bf)}{16T_b^2f^2 - 1}\right]^2. \tag{3.59}$$

A measure of the spectral compactness of a signal is provided by the *fractional in-band* power $F_{ib}(b)$ defined as the fraction of power for $f \in [-b, b]$. Thus,

$$F_{ib}(b) = \frac{\int_{-b}^{b} S_l(f)df}{\int_{-\infty}^{\infty} S_l(f)df}, \quad b \geq 0. \tag{3.60}$$

The bandwidth B of a frequency channel is determined by setting $F_{ib}(B/2)$ equal to the fraction of demodulated signal power in a frequency channel. The autocorrelations of the complex envelopes of practical signals with symbol duration T_s are functions of τ/T_s : $R_l(\tau) = f_1(\tau/T_s)$. For these autocorrelations, the PSD of the complex envelope has the form $S_l(f) = T_s f_2(fT_s)$. Therefore, B is inversely proportional to T_s. We define the *normalized bandwidth* as

$$\zeta = BT_s. \tag{3.61}$$

Usually, $F_{ib}(B/2)$ must exceed at least 0.9 to prevent significant signal distortion and performance degradation in communications over a bandlimited channel. The normalized bandwidth for which $F_{ib}(B/2) = 0.99$ is approximately 1.2 for binary MSK, but approximately 8 for BPSK.

The *fractional out-of-band* power of the complex envelope is defined as $F_{ob}(f) = 1 - F_{ib}(f)$. The adjacent-splatter ratio, which is due to out-of-band power on one side of the center frequency, has the upper bound given by

$$K_s < \frac{1}{2} F_{ob}(B/2). \tag{3.62}$$

The closed-form expressions for the PSDs of QPSK and binary MSK are used to generate Figure 3.7. The graphs depict $F_{ob}(f)$ in decibels as a function of f in units of $1/T_b$, where $T_b = T_s/\log_2 q$ for a q-ary modulation.

An even more compact spectrum than binary MSK is obtained by passing the MSK frequency pulses through a Gaussian filter with transfer function

$$H(f) = \exp\left[-\frac{(\ln 2)}{B^2} f^2\right] \tag{3.63}$$

where B_1 is the 3-dB bandwidth, which is the positive frequency such that $H(f) \geq H(0)/2$, $|f| \leq B_1$. The filter response to a unit-amplitude MSK frequency pulse is the *Gaussian MSK (GMSK)* pulse:

$$g(t) = Q\left[\frac{2\pi B_1}{\sqrt{\ln 2}}(t - \frac{T_s}{2})\right] - Q\left[\frac{2\pi B_1}{\sqrt{\ln 2}}(t + \frac{T_s}{2})\right] \tag{3.64}$$

where $T_s = T_b$. As B_1 decreases, the spectrum of a GMSK signal becomes more compact. However, each pulse is longer, and hence there is more intersymbol interference. If $B_1 T_s = 0.3$, which is specified in the Global System for Mobile (GSM) cellular communication system, the normalized bandwidth for which $F_{ib}(B/2) = 0.99$ is approximately $\zeta = 0.92$. Each pulse may be truncated for $|\ t\ | > 1.5T_s$

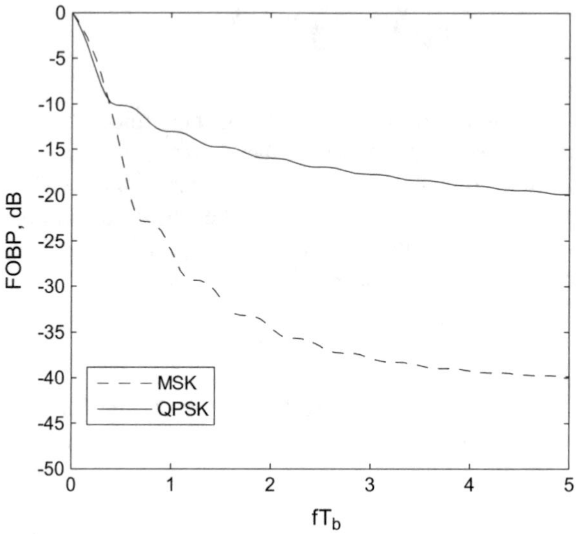

Fig. 3.7 Fractional out-of-band power for equivalent lowpass waveforms of QPSK and MSK

with little loss. The performance loss relative to MSK is approximately 0.46 dB for coherent demodulation and presumably also for noncoherent demodulation.

The *cross-correlation parameter* for two transmitted signals $s_1(t)$ and $s_2(t)$, each with energy $\mathcal{E}_s$, is defined as

$$C = \frac{1}{\mathcal{E}_s} \int_0^{T_s} s_1(t)s_2(t)dt. \tag{3.65}$$

For binary CPFSK, the two possible transmitted signals, each representing a different channel symbol, are

$$s_1(t) = \sqrt{2\mathcal{E}_s/T_s}\cos(2\pi f_1 t + \phi_1)\ , \quad s_2(t) = \sqrt{2\mathcal{E}_s/T_s}\cos(2\pi f_2 t + \phi_2). \tag{3.66}$$

The substitution of these equations into (3.65), a trigonometric expansion and discarding of an integral that is negligible if $(f_1 + f_2)T_s >> 1$, and the evaluation of the remaining integral give

$$C = \frac{1}{2\pi f_d T_s}[\sin(2\pi f_d T_s + \phi_d) - \sin\phi_d]\ , \quad f_d \neq 0 \tag{3.67}$$

where $f_d = f_1 - f_2$ and $\phi_d = \phi_1 - \phi_2$. Because of the phase synchronization in a coherent demodulator, we may take $\phi_d = 0$. Therefore, the orthogonality condition $C = 0$ is satisfied if $h = f_d T_s = k/2$, where k is any nonzero integer. The smallest value of h for which $C = 0$ is $h = 1/2$, which corresponds to MSK.

In a noncoherent demodulator, ϕ_d is a random variable that is assumed to be uniformly distributed over $[0, 2\pi)$. Equation (3.67) indicates that $E[C] = 0$ for all values of h. The variance of C is

$$\begin{aligned} var(C) &= \left(\frac{1}{2\pi f_d T_s}\right)^2 E\left[\sin^2(2\pi f_d T_s + \phi_d) + \sin^2\phi_d - 2\sin\phi_d \sin(2\pi f_d T_s + \phi_d)\right] \\ &= \left(\frac{1}{2\pi f_d T_s}\right)^2 (1 - \cos 2\pi f_d T_s) \\ &= \frac{1}{2}\text{sinc}^2 h. \end{aligned} \quad (3.68)$$

Since $var(C) \neq 0$ for $h = 1/2$, MSK does not provide orthogonal signals for noncoherent demodulation. If h is any nonzero integer, then both (3.68) and (3.67) indicate that the two CPFSK signals are orthogonal for any ϕ_d. This result justifies the previous assertion that FSK tones must be separated by $f_d = k/T_s$ to provide noncoherent orthogonal signals.

Consider the multitone jamming of an FH/CPM or FH/CPFSK system in which the thermal noise is absent and each jamming tone is randomly placed within a single frequency channel. It is reasonable to assume that a symbol error occurs with probability $(q-1)/q$ when the frequency channel contains a jamming tone with power exceeding S. The substitution of (3.27), (3.20), $\mathcal{E}_b = ST_b$, and $I_{t0} = I_t/W$ into (3.22) yield

$$P_s = \begin{cases} \frac{q-1}{q}, & \frac{\mathcal{E}_b}{I_{t0}} < BT_b \\ \left(\frac{q-1}{q}\right) BT_b \left(\frac{\mathcal{E}_b}{I_{t0}}\right)^{-1}, & BT_b \leq \frac{\mathcal{E}_b}{I_{t0}} \leq WT_b \\ 0, & \frac{\mathcal{E}_b}{I_{t0}} > WT_b \end{cases} \quad (3.69)$$

for sophisticated multitone jamming. The symbol error probability increases with B because the enlarged frequency channels present better targets for the multitone jamming.

As implied by Figure 3.7, the bandwidth requirement of DPSK with $K_s > 0.9$, which is the same as that of BPSK or QPSK and less than that of orthogonal FSK, exceeds that of MSK. Thus, if the hopping bandwidth W is fixed, the number of frequency channels available for FH/DPSK is smaller than it is for noncoherent FH/MSK. This increase in B and reduction in frequency channels offsets the intrinsic performance advantage of FH/DPSK and implies that noncoherent FH/MSK gives a lower P_s than FH/DPSK in the presence of worst-case multitone jamming, as indicated in (3.37). Alternatively, if the bandwidth of a frequency channel is fixed, an FH/DPSK signal experiences more distortion and spectral splatter than an FH/MSK signal. Any pulse shaping of the DPSK symbols alters their constant amplitude. An FH/DPSK system is more sensitive to Doppler shifts and frequency instabilities than an FH/MSK system. Another disadvantage of FH/DPSK is the usual lack of phase coherence from hop to hop, which necessitates an extra phase-reference symbol at the start of every dwell interval. This extra symbol reduces $\mathcal{E}_s$ by a factor $(N-1)/N$, where N is the number of symbols per hop or dwell interval and $N \geq 2$. Thus, DPSK is not as suitable a means of modulation as noncoherent MSK for most applications of frequency-hopping communications, and the main competition for MSK comes from other forms of CPM.

3.4 Power Spectral Density of FH/CPM

The finite extent of the dwell intervals causes an expansion of the spectral density of an FH/CPM signal relative to a CPM signal of the same duration. The reason is that an FH/CPM signal has a continuous phase over each dwell interval with N symbols, but has a phase discontinuity every $T_h = NT_s + T_{sw}$ seconds at the beginning of another dwell interval, where N is the number of symbols per dwell interval. The signal may be expressed as

$$s(t) = A \sum_{i=-\infty}^{\infty} w(t - iT_h, T_d) \cos\left[2\pi f_{ci}t + \phi(t, \boldsymbol{\alpha}) + \theta_i\right] \tag{3.70}$$

where A is the amplitude during a dwell interval, $w(t, T_d)$ as defined in (2.48) is a unit-amplitude rectangular pulse of duration $T_d = NT_s$, f_{ci} is the carrier frequency during hop-interval i, $\phi(t, \boldsymbol{\alpha})$ is the *phase function* defined by (3.42), and θ_i is the phase at the beginning of dwell-interval i.

The PSD of the complex envelope of an FH/CPM signal, which is the same as the dehopped PSD, depends on the number of symbols per dwell interval N because of the finite dwell time. To simplify the derivation of the PSD, we neglect the switching time and set $T_h = T_d = NT_s$. Let $w(t) = 1,\ 0 \leq t < NT_s$, and $w(t) = 0$, otherwise. The complex envelope of the FH/CPM signal is

$$F(t, \boldsymbol{\alpha}) = A \sum_{i=-\infty}^{\infty} w(t - iNT_s) \exp\left[j\phi(t, \boldsymbol{\alpha}) + j\theta_i\right] \tag{3.71}$$

where the $\{\theta_i\}$ are assumed to be independent and uniformly distributed over $[0, 2\pi)$. Therefore, $E\left[\exp\left(j\theta_i - j\theta_k\right)\right] = 0,\ i \neq k$, and the autocorrelation of $F(t, \boldsymbol{\alpha})$ is

$$\begin{aligned} R_f(t, t+\tau) &= E\left[F^*(t, \boldsymbol{\alpha})\, F(t+\tau, \boldsymbol{\alpha})\right] \\ &= A^2 \sum_{i=-\infty}^{\infty} w(t - iNT_s) w(t + \tau - iNT_s) R_c(t, t+\tau) \end{aligned} \tag{3.72}$$

where the asterisk denotes the complex conjugate, and the autocorrelation of the complex envelope of the underlying CPM signal is

$$R_c(t, t+\tau) = E\{\exp[j\phi(t+\tau, \boldsymbol{\alpha}) - j\phi(t, \boldsymbol{\alpha})]\}. \tag{3.73}$$

Assuming that the symbols of $\boldsymbol{\alpha}$ are independent and identically distributed, (3.42) and (3.73) imply that $R_c(t, t+\tau)$ is periodic in t with period T_s. Since the series in (3.72) is infinite, each term in the series has a duration less than NT_s. Therefore, the periodicity of $R_c(t, t+\tau)$ implies that $R_f(t, t+\tau)$ is periodic in t with period NT_s.

The average autocorrelation of $F(t, \boldsymbol{\alpha})$, found by substituting (3.72) into the definition (2.14), is

$$R_f(\tau) = \frac{A^2}{NT_s}\int_0^{NT_s} R_f(t, t+\tau)\, dt. \tag{3.74}$$

Equation (3.73) indicates that $R_c(t, t-\tau) = R_c^*(t-\tau, t)$. This equation, (3.72), and (3.74) imply that

$$R_f(-\tau) = R_f^*(\tau). \tag{3.75}$$

Since $w(t - iNT_s)w(t + \tau - iNT_s) = 0$ if $\tau \geq NT_s$ in (3.72),

$$R_f(\tau) = 0, \quad \tau \geq NT_s. \tag{3.76}$$

Thus, only $R_f(\tau)$ for $0 \leq \tau < NT_s$ remains to be evaluated.

Since $w(t - iNT_s)w(t + \tau - iNT_s) = 0$ if $\tau \geq 0$ and $t \in [NT_s - \tau, NT_s]$, the substitution of (3.72) into (3.74) yields

$$R_f(\tau) = \frac{A^2}{NT_s}\int_0^{NT_s-\tau} R_c(t, t+\tau)\, dt, \quad \tau \in [0, NT_s). \tag{3.77}$$

Let $\tau = \nu T_s + \epsilon$, where ν is a nonnegative integer, $0 \leq \nu < N$, and $0 \leq \epsilon < T_s$. Since $R_c(t, t+\tau)$ is periodic in t with period T_s, the integration interval in (3.77) can be divided into smaller intervals with similar integrals. Thus, we obtain

$$R_f(\tau) = A^2\left[\frac{N-\nu-1}{NT_s}\int_0^{T_s} R_c(t, t+\tau)\, dt + \frac{1}{NT_s}\int_0^{T_s-\epsilon} R_c(t, t+\tau)\, dt\right],$$
$$\nu = \lfloor \tau/T_s \rfloor < N, \ \ \epsilon = \tau - \nu T_s, \quad \tau = \nu T_s + \epsilon. \tag{3.78}$$

The assumption is that the symbols of $\boldsymbol{\alpha}$ are statistically independent and that the substitution of (3.42) into (3.73) yields an infinite product. The finite duration of $\phi(t)$ in (3.40) implies that if $k > (t+\tau)/T_s$ or $k < 1 - L$, then $\phi(t + \tau - kT_s) - \phi(t - kT_s) = 0$, and the corresponding factors in the infinite product are unity. Thus, we obtain

$$R_c(t, t+\tau) = \prod_{k=1-L}^{\lfloor (t+\tau)/T_s \rfloor} E\left\{\exp[j2\pi h \alpha_k \phi_d(t, \tau, k)]\right\},$$
$$t \in [0, T_s),\ \tau \in [0, NT_s) \tag{3.79}$$

where

$$\phi_d(t, \tau, k) = \phi(t + \tau - kT_s) - \phi(t - kT_s). \tag{3.80}$$

If each symbol is equally likely to have any of the q possible values, then

$$R_c(t, t+\tau) = \prod_{k=1-L}^{\lfloor (t+\tau)/T_s \rfloor} \left\{ \frac{1}{q} \sum_{l=-(q-1),odd}^{q-1} \exp[j2\pi h l \phi_d(t, \tau, k)] \right\} \tag{3.81}$$

where the sum only includes odd values of the index l. After a change of the index in the sum to $m = (l + q - 1)/2$, (3.81) can be evaluated as a geometric series. We obtain

$$R_c(t, t+\tau) = \prod_{k=1-L}^{\lfloor (t+\tau)/T_s \rfloor} \frac{1}{q} \frac{\sin[2\pi h q \phi_d(t, \tau, k)]}{\sin[2\pi h \phi_d(t, \tau, k)]},$$
$$t \in [0, T_s),\ \tau \in [0, NT_s) \tag{3.82}$$

where a factor in the product is set equal to +1 if $2h\phi_d(t, \tau, k)$ is equal to an even integer, and is set equal to -1 if $2h\phi_d(t, \tau, k)$ is equal to an odd integer.

Equation (3.82) indicates that $R_c(t, t+\tau)$ is real-valued, and then (3.77) and (3.75) indicate that $R_f(\tau)$ is a real-valued, even function. Therefore, the average PSD of the dehopped signal, which is the Fourier transform of the average autocorrelation $R_f(\tau)$, is

$$S_l(f) = 2\int_0^{NT_s} R_f(\tau)\cos(2\pi f \tau)\, d\tau. \tag{3.83}$$

The average PSD can be calculated by substituting (3.78) and (3.82) into (3.83) and then numerically evaluating the integrals, which extend over finite intervals [41]. The PSD of CPM without frequency hopping but with many data symbols is obtained by taking $N \to \infty$.

For FH/CPFSK with binary CPFSK, $L = 1, q = 2$, and analytical simplifications are computationally useful. Substitution of (3.80) into (3.82) and a trigonometric identity yield

$$R_c\left(t, t+\tau\right) = \cos\{2\pi h[\phi\left(t+\tau\right) - \phi\left(t\right)]\} \prod_{k=1}^{\lfloor (t+\tau)/T_s \rfloor} \cos\left[2\pi h \phi\left(t - kT_s + \tau\right)\right],$$
$$0 \le t < T_s, \quad \tau \in [0, NT_s) \tag{3.84}$$

where $\prod_{k=1}^{0}(\cdot) = 1$. Let $\tau = \nu T_s + \epsilon$, where ν is a nonnegative integer, $0 \le \nu < N$, and $0 \le \epsilon < T_s$. Substituting (3.45) and evaluating the product, we obtain

$$R_c\left(t, t+\tau\right) = \cos a\epsilon, \;\; 0 \le t < T_s - \epsilon, \;\; \nu = 0 \tag{3.85}$$

$$R_c\left(t, t+\tau\right) = \left(\cos \pi h\right)^{\nu-1} \cos\left(at - \pi h\right) \cos\left(at + a\epsilon\right),$$
$$0 \le t < T_s - \epsilon, \; 1 \le \nu < N \tag{3.86}$$

$$R_c\left(t, t+\tau\right) = \left(\cos \pi h\right)^{\nu} \cos\left(at - \pi h\right) \cos\left(at + a\epsilon - \pi h\right),$$
$$T_s - \epsilon \le t < T_s, \; 0 \le \nu < N \tag{3.87}$$

where

$$a = \pi h / T_s, \;\; \nu = \lfloor \tau / T_s \rfloor < N, \;\; \epsilon = \tau - \nu T_s, \;\; \tau = \nu T_s + \epsilon. \tag{3.88}$$

If $h = 1/2$, then $(\cos \pi/2)^{\nu} = 0$, $\nu \ge 1$, and $(\cos \pi/2)^{0} = 1$. Substitution of $R_c\left(t, t+\tau\right)$ into (3.78), use of trigonometric identities, and the evaluation of basic trigonometric integrals yields

$$R_f\left(\epsilon\right) = A^2 \left[1 - \frac{\left(N+1\right)\epsilon}{2NT_s}\right] \cos a\epsilon + A^2 \frac{N-1}{N2\pi h} \sin a\epsilon, \;\; \nu = 0 \tag{3.89}$$

$$R_f\left(\nu T_s + \epsilon\right) = A^2 \frac{N-\nu}{NT_s} \left(\cos \pi h\right)^{\nu-1} \left[\frac{T_s - \epsilon}{2} \cos\left(a\epsilon + \pi h\right) - \frac{T_s}{2\pi h} \sin\left(a\epsilon - \pi h\right)\right]$$
$$+ A^2 \frac{N-\nu-1}{NT_s} \left(\cos \pi h\right)^{\nu} \left[\frac{\epsilon}{2} \cos a\epsilon + \frac{T_s}{2\pi h} \sin a\epsilon\right], \;\; 1 \le \nu < N. \tag{3.90}$$

Equation (3.83) indicates that the PSD is

$$S_l(f) = 2\sum_{\nu=0}^{N-1} \int_0^{T_s} R_f(\nu T_s + \epsilon) \cos[2\pi f(\nu T_s + \epsilon)] d\epsilon. \tag{3.91}$$

Substitution of (3.89) and (3.90) into (3.91) followed by analytical or numerical integrations gives the PSDs for FH/CPFSK with binary CPFSK.

When frequency hopping is absent, $L = 1$, and there are many data symbols per hop, we take $N \to \infty$ in (3.78), (3.82), and (3.83) to enable the calculation of a closed-form expression for the PSD, but only after a very lengthy series of evaluations of trigonometric integrals and algebraic and trigonometric manipulations. For CPFSK with $L = 1$, the PSD is [69]

$$S_l(f) = A^2 \frac{T_s}{q} \sum_{n=1}^{q} \left[A_n^2(f) + \frac{2}{q} \sum_{m=1}^{q} A_n(f) A_m(f) B_{nm}(f) \right] \tag{3.92}$$

where

$$A_n(f) = \frac{\sin\left[\pi f T_s - \frac{\pi h}{2}(2n - q - 1)\right]}{\pi f T_s - \frac{\pi h}{2}(2n - q - 1)} \tag{3.93}$$

$$B_{nm}(f) = \frac{\cos(2\pi f T_s - \alpha_{nm}) - \Phi \cos \alpha_{nm}}{1 + \Phi^2 - 2\Phi \cos 2\pi f T_s} \tag{3.94}$$

$$\alpha_{nm} = \pi(n + m - q - 1), \quad \Phi = \frac{\sin q\pi h}{q \sin \pi h}. \tag{3.95}$$

If the denominator in (3.93) is zero, we set $A_n(f) = 1$. If h is an even integer, we set $\Phi = 1$; if h is an odd integer, we set $\Phi = -1$. For the special case of MSK, the substitution of $q = 2$ and $h = 1/2$ into the preceding equations and considerable mathematical simplification lead again to (3.59).

The normalized 99-% bandwidth is determined from (3.60) by setting $F_{ib}(B/2) = 0.99$ and solving for $\zeta = BT_s$. The normalized 99-% bandwidths of FH/CPFSK with deviation ratios $h = 0.5$ (MSK) and $h = 0.7$ are listed in Table 3.1 for different values of N. As N increases, the PSD becomes more compact and approaches that of CPFSK without frequency hopping. For $N \geq 64$, the frequency hopping causes little spectral spreading.

Fast frequency hopping, which corresponds to $N = 1$, entails a very large 99-% bandwidth. *This fact and the long switching times are the main reasons why slow frequency hopping is preferable to the fast form and is the predominant form of frequency hopping. Consequently, frequency hopping is always assumed to be the slow form unless explicitly stated otherwise.*

An advantage of FH/CPFSK with $h < 1$ or FH/GMSK is that it requires less bandwidth than orthogonal CPFSK ($h = 1$). The increased number of frequency

Table 3.1 Normalized bandwidth (99%) for FH/CPFSK

	Deviation ratio	
Symbols/dwell	$h = 0.5$	$h = 0.7$
1	18.844	18.688
2	9.9375	9.9688
4	5.1875	5.2656
16	1.8906	2.1250
64	1.2813	1.8750
256	1.2031	1.8125
1024	1.1875	1.7969
No hopping	1.1875	1.7813

channels due to the decreased bandwidth does not improve performance over the additive white Gaussian noise (AWGN) channel. However, the increase is advantageous against a fixed number of interference tones, optimized jamming, and multiple-access interference in a network of frequency-hopping systems (Section 9.4).

3.5 Digital Demodulation of FH/CPFSK

In principle, the frequency hopping/continuous-phase frequency-shift keying (FH/CPFSK) receiver does symbol-rate sampling of matched-filter outputs. However, the details of the actual implementation are more involved, primarily because aliasing should be avoided. A practical digital demodulation is described in this section.

The dehopped FH/CPFSK signal during a dwell interval has the form

$$s_1(t) = A\cos\left[2\pi f_1 t + \phi(t, \boldsymbol{\alpha}) + \phi_0\right] \tag{3.96}$$

where A is the amplitude, f_1 is the intermediate frequency (IF), $\phi(t, \boldsymbol{\alpha})$ is the phase function, and ϕ_0 is the initial phase. This signal is applied to the noncoherent digital demodulator illustrated in Figure 3.8. An error f_e in the estimated carrier frequency used in the dehopping leads to an f_1 that differs from the desired IF f_{IF} by the *IF offset frequency* $f_e = f_1 - f_{IF}$. The quadrature downconverter, which is shown in Figure 2.17, uses a sinusoidal signal at frequency $f_{IF} - f_o$, where f_o is the *baseband offset frequency,* and a pair of mixers to produce in-phase and quadrature components near baseband. The mixer outputs are passed through lowpass filters to remove the double-frequency components. The filter outputs are the in-phase and quadrature CPFSK signals with center frequency at $f_o + f_e$. As shown in Figure 3.8, each of these signals is sampled at the rate f_s by an analog-to-digital converter (ADC).

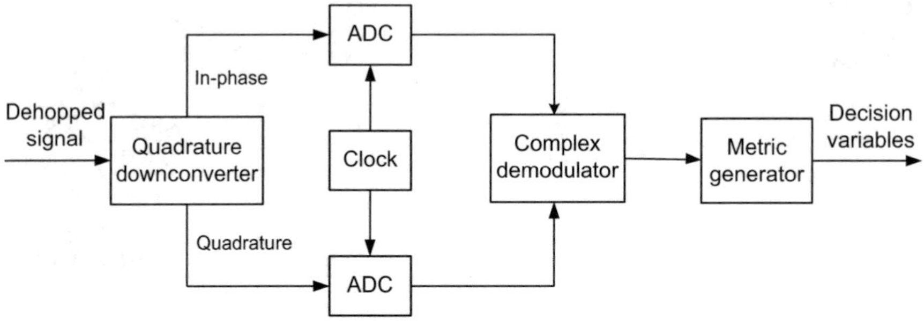

Fig. 3.8 Digital demodulator of dehopped FH/CPFSK signal

A critical choice in the design of the digital demodulator is the sampling rate of the ADCs. This rate must be large enough to prevent aliasing and to accommodate the IF offset. To simplify the demodulator implementation, it is highly desirable for the sampling rate to be an integer multiple of the symbol rate $1/T_s$. Thus, we assume a sampling rate $f_s = L/T_s$, where L is a positive integer.

To determine the appropriate sampling rate and offset frequency, we use the *sampling theorem* (Appendix D.4), which relates a continuous-time signal $x(t)$ and the discrete-time sequence $x_n = x(nT)$ in the frequency domain:

$$X(e^{j2\pi fT}) = \frac{1}{T}\sum_{i=-\infty}^{\infty} X\left(f - \frac{i}{T}\right) \tag{3.97}$$

where $X(f)$ is the Fourier transform of $x(t)$, $X(e^{j2\pi fT})$ is the *discrete-time Fourier transform (DTFT) of x_n defined as*

$$X(e^{j2\pi fT}) = \sum_{n=-\infty}^{\infty} x_n e^{-j2\pi nfT} \tag{3.98}$$

and $1/T$ is the sampling rate. The DTFT $X(e^{j2\pi fT})$ is a periodic function of frequency f with period $1/T$. If $x(t)$ is sampled at rate $1/T$, then $x(t)$ can be recovered from the samples if the transform is bandlimited so that $X(f) = 0$ for $|f| > 1/2T$. If the sampling rate is not high enough to satisfy this condition, then the terms of the sum in (3.97) overlap, which is called *aliasing,* and the samples may not correspond to a unique continuous-time signal.

The lowpass filters of the quadrature downconverter have bandwidths wide enough to accommodate both f_e and the bandwidth of $s_1(t)$. The receiver timing or symbol synchronization may be derived from the frequency-hopping pattern synchronization (Section 4.6). If the timing is correct, then the sampled ADC output due to the desired signal in the upper branch of Figure 3.8 is the sequence

$$x_n = A\cos\left[2\pi\left(f_o + f_e\right)nT_s/L + \phi(nT_s/L, \boldsymbol{\alpha}) + \phi_0\right] \tag{3.99}$$

where $\phi(nT_s/L, \boldsymbol{\alpha})$ is the sampled phase function of the CPFSK modulation, and ϕ_0 is the unknown phase. A similar sequence

$$y_n = A \sin\left[2\pi\left(f_o + f_e\right) nT_s/L + \phi(nT_s/L, \boldsymbol{\alpha}) + \phi_0\right] \quad (3.100)$$

is produced in the lower branch. Equation (3.46) indicates that during *symbol interval m*,

$$\phi(nT_s/L, \boldsymbol{\alpha}) = \frac{\pi h \alpha_m}{L}(n - Lm) + \phi_1(m), \quad Lm \le n \le Lm + L - 1 \quad (3.101)$$

where α_m denotes the symbol received during the interval,

$$\phi_1(m) = \pi h \sum_{i=-\infty}^{m-1} \alpha_i \quad (3.102)$$

and the $\{\alpha_i\}$ are previous symbols.

The Fourier transforms of the in-phase and quadrature outputs of the quadrature downconverter occupy the upper band $[f_o + f_e - B/2, f_o + f_e + B/2]$ and the lower band $[-f_o - f_e - B/2, -f_o - f_e + B/2]$, where B is the one-sided bandwidth of the dehopped FH/CPFSK signal. To avoid aliasing when the sampling rate is L/T_s, the sampling theorem requires that

$$f_o + f_e + \frac{B}{2} < \frac{L}{2T_s}. \quad (3.103)$$

To prevent the upper and lower bands from overlapping, and hence distorting the DTFTs, it is necessary that $f_o + f_e - B/2 > -f_o - f_e + B/2$. Thus, the necessary condition is

$$f_o > f_{\max} + \frac{B}{2} \quad (3.104)$$

where $f_{\max} \ge |f_e|$ is the maximum IF offset frequency that is likely to occur. Combining this inequality with (3.103), we obtain another necessary condition:

$$L > 2T_s\left(2f_{\max} + B\right). \quad (3.105)$$

Inequalities (3.104) and (3.105) indicate that both the baseband offset frequency and the sampling rate must be increased as $f_{\max}$ or B increases.

If we set

$$f_o = \frac{L}{4T_s} \quad (3.106)$$

then (3.104) is automatically satisfied when (3.105) is satisfied. Thus, if $f_{\max} < B/2$, then $L = 4BT_s$ and $f_o = B$ are satisfactory choices.

Once the ADC outputs with appropriate DTFTs are produced, the offset frequency plays no further role. It is removed by a *complex demodulator* that computes

$$\begin{aligned} z_n &= (x_n + jy_n) \exp\left[-j2\pi f_o n T_s / L - j\widehat{\phi}_1(m)\right] \\ &= A \exp\left\{ j2\pi \left[n \left(\frac{2f_e T_s + h\alpha_m}{2L} \right) - \frac{h\alpha_m m}{2} \right] + j\phi_e(m) \right\}, \\ Lm &\leq n \leq Lm + L - 1 \end{aligned} \tag{3.107}$$

where $\widehat{\phi}_1(m)$ is the estimate of $\phi_1(m)$ obtained from previous demodulated symbols, and $\phi_e(m) = \phi_0 + \phi_1(m) - \widehat{\phi}_1(m)$. As shown in Figure 3.8, this sequence and the accompanying noise are applied to a metric generator.

The *metric generator* comprises q discrete-time symbol-matched filters. The symbol-matched filter k of the metric generator, which is matched to symbol β_k, has an impulse response $g_{k,n}$ of length L:

$$g_{k,l} = \exp\left\{ j2\pi \left[\frac{h\beta_k (L-1-l)}{2L} \right] \right\}, \quad 0 \leq l \leq L-1, \ \ 1 \leq k \leq q. \tag{3.108}$$

The response of the symbol-matched filter k to z_n at discrete-time $Lm + L - 1$, which is denoted by $C_k(\alpha_m)$, is the result of a complex discrete-time convolution:

$$C_k(m) = \sum_{n=Lm}^{Lm+L-1} z_n g^*_{k,Lm+L-1-n}, \quad 1 \leq k \leq q \tag{3.109}$$

which is generated at the symbol rate. We define the function

$$D(\theta, L) = \begin{cases} \frac{\sin(\theta L/2)}{\sin(\theta/2)} & 0 < \theta < 2\pi \\ L & \theta = 0 \end{cases} \tag{3.110}$$

which is positive and monotonically decreasing over $0 \leq \theta < 2\pi$. We assume that f_e is small enough and L is large enough that

$$h(\alpha_m - \beta_k) + 2f_e T_s < 2L, \quad 1 \leq k \leq q. \tag{3.111}$$

Substituting (3.107) and (3.108) into (3.109), changing the summation index, evaluating the resulting geometric series, and then using (3.110), we obtain the q *selected matched-filter outputs* in the absence of noise:

$$C_k(m) = AD\left(\left[\frac{\pi h(\alpha_m - \beta_k)}{L} + \frac{2\pi f_e T_s}{L} \right], L \right)$$

$$\times \exp\left\{ j\pi \left[\frac{(L-1)\left(2f_e T_s + h\left(\alpha_m - \beta_k\right)\right)}{2L} + 2f_e T_s m \right] + j\phi_e\left(m\right) \right\}$$

$$1 \leq k \leq q. \tag{3.112}$$

These outputs are generated at the symbol rate.

For noncoherent detection, the q symbol metrics produced at the symbol rate are the magnitudes of the selected matched-filter outputs. Assuming that $\beta_{k_0}=\alpha_m$, the symbol metrics in the absence of noise are

$$|C_k(m)| = \begin{cases} AD\left(\frac{2\pi f_e T_s}{L}, L\right), & k = k_0 \\ AD\left(\left[\frac{\pi h(\alpha_m - \beta_k)}{L} + \frac{2\pi f_e T_s}{L}\right], L\right), & k \neq k_0 \end{cases} \tag{3.113}$$

where (3.111) ensures that $D\left(\cdot, L\right)$ is positive and monotonically decreasing. When hard decisions are made, the largest of the $|C_k(m)|$ determines the symbol decision. Equation (3.113) indicates that if (3.111) is satisfied, then as h increases, the distinctions among the symbol metrics increase, and hence the symbol metrics become increasingly favorable for either hard-decision or soft-decision decoding. However, the CPFSK bandwidth decreases, and the frequency-hopping system can accommodate more frequency channels, thereby strengthening it against multitone jamming and multiple-access interference. This issue is examined in Section 9.4.

In the absence of noise, (3.112) indicates that the sequence obtained by taking, at each discrete-time m, the matched-filter output with the largest magnitude is

$$C_{k_0}(m) = AD\left(\frac{2\pi f_e T_s}{L}, L\right) \exp\left\{ j\pi \left[\frac{(L-1) f_e T_s}{L} + 2f_e T_s m \right] + j\phi_e\left(m\right) \right\} \tag{3.114}$$

which has a complex exponential with a component that varies linearly with $f_e m$. Therefore, we can estimate the IF offset frequency f_e from the DTFT of successive values of this symbol-rate sequence [68]. Iterative decision-directed feedback can then be used to reduce f_e so that (3.111) is increasingly likely to be satisfied.

3.6 Partial-Band Interference and Channel Codes

If partial-band interference has power that is uniformly distributed over J frequency channels out of M in the hopping band, then the fraction of the hopping band with interference is

$$\mu = \frac{J}{M}. \tag{3.115}$$

The interference PSD in each of the interfered channels is $I_{t0}/2\mu$, where $I_{t0}/2$ denotes the interference PSD that would exist if the interference power were uniformly distributed over the hopping band. When the frequency-hopping signal uses a carrier frequency that lies within the spectral region occupied by the partial-band interference, this interference is modeled as AWGN, which increases the two-sided noise PSD from $N_0/2$ to $N_0/2 + I_{t0}/2\mu$. Therefore, for hard-decision decoding, the symbol error probability is

$$P_s = \mu G\left(\frac{\mathcal{E}_s}{N_0 + I_{t0}/\mu}\right) + (1-\mu)G\left(\frac{\mathcal{E}_s}{N_0}\right)$$
$$\approx \mu G\left(\frac{\mu\mathcal{E}_s}{I_{t0}}\right), \; I_{t0} >> \mu N_0 \tag{3.116}$$

where the conditional symbol error probability is a monotonically decreasing function $G(x)$ that depends on the modulation and fading.

Consider an *orthogonal FH/CPFSK system with noncoherent detection and hard decisions*. For the AWGN channel, (1.93) indicates that

$$G(x) = \sum_{i=1}^{q-1} \frac{(-1)^{i+1}}{i+1}\binom{q-1}{i}\exp\left[-\frac{ix}{(i+1)}\right] \text{ (AWGN)} \tag{3.117}$$

where q is the alphabet size of the orthogonal CPFSK symbols. For binary orthogonal CPFSK, this equation reduces to

$$G(x) = \frac{1}{2}\exp(-\frac{x}{2}) \tag{3.118}$$

and (3.116) yields

$$P_s \approx \frac{\mu}{2}\exp\left(-\frac{\mu\mathcal{E}_s}{2I_{t0}}\right), \; 0 \le \mu \le 1. \tag{3.119}$$

For the fading channel, the symbol energy may be expressed as $\mathcal{E}_s\alpha^2$, where $\mathcal{E}_s$ represents the average energy, and α is a random fading amplitude with $E[\alpha^2] = 1$. For Ricean fading, which is fully discussed in Section 6.2, the density function of α is

$$f_\alpha(r) = 2(\kappa+1)r\exp\{-\kappa - (\kappa+1)r^2\}I_0(\sqrt{\kappa(\kappa+1)}2r)u(r) \tag{3.120}$$

where κ is the Rice factor, and $u(r)$ is the unit step function. Replacing x with $x\alpha^2$ in (3.117), an integration over the density function (3.120) and the use of (1.92) yield

$$G(x) = \sum_{i=1}^{q-1}(-1)^{i+1}\binom{q-1}{i}\frac{\kappa+1}{\kappa+1+(\kappa+1+x)i}\exp\left[-\frac{\kappa x i}{\kappa+1+(\kappa+1+x)i}\right]$$
$$\text{(Ricean).} \tag{3.121}$$

For Rayleigh fading and binary orthogonal FH/CPFSK, we set $\kappa = 0$ and $q = 2$ in (3.121) to obtain

$$G(x) = \frac{1}{2+x} \text{ (Rayleigh, binary).} \tag{3.122}$$

This equation and (3.116) indicate that the worst-case value of μ in the presence of strong interference is $\mu_0 = 1$. Thus, for binary orthogonal FH/CPFSK over the Rayleigh channel, strong interference spread uniformly over the entire hopping band hinders communications more than interference concentrated over part of the band.

If a large amount of interference power is received over a small portion of the hopping band, then unless accurate channel-state information is available, soft-decision decoding metrics for the AWGN channel may be ineffective because of the possible dominance of a path or code metric by a single symbol metric (see Section 2.6 on pulsed interference). This dominance is reduced by hard decisions or the use of a practical two- or three-bit quantization of symbol metrics instead of unquantized symbol metrics.

Reed-Solomon Codes

For orthogonal FH/CPFSK with a negligible switching time and a channel code, the relation between the code-symbol duration T_s and the information-bit duration T_b is $T_s = r(\log_2 q)T_b$, where r is the code rate. Therefore, the energy per channel symbol is

$$\mathcal{E}_s = r(\log_2 q)\mathcal{E}_b. \tag{3.123}$$

The use of a Reed-Solomon code with orthogonal CPFSK is advantageous against partial-band interference for two principal reasons. First, a Reed-Solomon code is maximum-distance separable (Section 1.1) and hence accommodates many erasures. Second, the use of nonbinary orthogonal CPFSK symbols to represent code symbols allows a relatively large symbol energy, as indicated by (3.123).

Consider an orthogonal FH/CPFSK system that uses a Reed-Solomon code with no erasures in the presence of partial-band interference and Ricean fading. The demodulator comprises a parallel bank of noncoherent detectors and a device that makes hard decisions. In a frequency-hopping system, symbol interleaving within a hop dwell interval and among different dwell intervals and subsequent deinterleaving in the receiver are used to disperse errors due to the fading or interference. This dispersal facilitates the removal of the errors by the decoder. The frequency channels are assumed to be separated enough and the interleaving sufficient to ensure independent symbol errors. Orthogonal CPFSK modulation implies a q-ary symmetric channel. Therefore, for hard-decision decoding of loosely packed Reed-Solomon codes, (1.23) and (1.95) indicate that

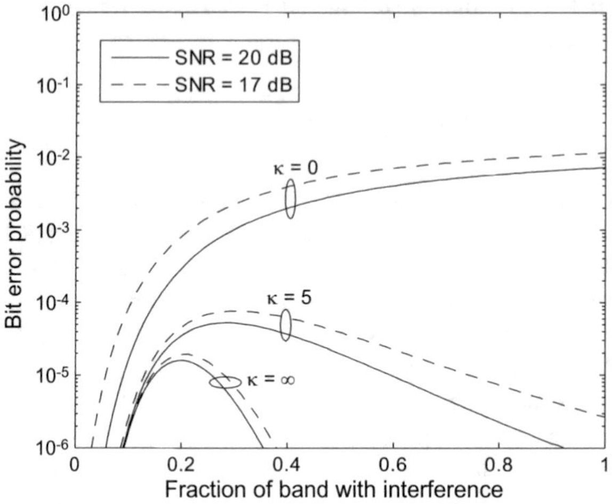

Fig. 3.9 Performance of an orthogonal FH/CPFSK system with Reed-Solomon (32,12) code, $q = 32$, no erasures, $SIR = 10\,\text{dB}$, and Ricean factor κ

$$P_b \approx \frac{q}{2(q-1)} \sum_{i=t+1}^{n} \binom{n-1}{i-1} P_s^i (1-P_s)^{n-i}. \tag{3.124}$$

Figure 3.9 shows P_b for an FH/CPFSK system with $q = 32$, $\mathcal{E}_b/I_{t0} = 10\,\text{dB}$, and an extended Reed-Solomon (32,12) code in the presence of Ricean fading. The bit signal-to-noise ratio is $SNR = \mathcal{E}_b/N_0$, and the bit signal-to-interference ratio is $SIR = \mathcal{E}_b/I_{t0}$. Equations (3.116), (3.121), (3.123), and (3.124) are applicable. For $\kappa > 0$, the graphs exhibit peaks as the fraction of the band with interference varies. These peaks indicate that the concentration of the interference power over part of the hopping band (perhaps intentionally by a jammer) is more damaging than uniformly distributed interference. Smaller peaks become sharper and occur at smaller values of μ as $\mathcal{E}_b/I_{t0}$ increases. For Rayleigh fading, which corresponds to $\kappa = 0$, peaks are absent in the figure, and full-band interference is the most damaging.

Much better performance against partial-band interference can be obtained by inserting erasures (Section 1.1) among the demodulator output symbols before the symbol deinterleaving and hard-decision decoding. The decision to erase is made independently for each code symbol. It is based on *channel-state information*, which indicates the codeword symbols that have a high probability of being incorrectly demodulated. The channel-state information must be reliable so that only degraded symbols are erased.

Channel-state information may be obtained from N_t known *pilot symbols* that are transmitted along with the data symbols in each dwell interval of a frequency-hopping signal. A *hit* is said to occur in a dwell interval if the signal encounters partial-band interference during the interval. If δ or more of the N_t pilot symbols are incorrectly demodulated, then the receiver decides that a hit has occurred, and all

N symbols in the same dwell interval are erased. Only one symbol of a codeword is erased if the interleaving ensures that only a single symbol of the codeword is in any particular dwell interval. Pilot symbols decrease the information rate, but this loss is negligible if $N_t << N$, which is assumed henceforth.

The probability of the erasure of a code symbol is

$$P_\epsilon = \mu P_{\epsilon 1} + (1-\mu) P_{\epsilon 0} \tag{3.125}$$

where $P_{\epsilon 1}$ is the erasure probability given that a hit occurred, and $P_{\epsilon 0}$ is the erasure probability given that no hit occurred. If δ or more errors among the N_t known pilot symbols causes an erasure, then

$$P_{\epsilon i} = \sum_{j=\delta}^{N_t} \binom{N_t}{j} P_{si}^j (1-P_{si})^{N_t - j}, \quad i = 0, 1 \tag{3.126}$$

where P_{s1} is the conditional channel-symbol error probability given that a hit occurred, and P_{s0} is the conditional channel-symbol error probability given that no hit occurred.

A codeword symbol error can only occur if there is no erasure. Since pilot and codeword symbol errors are statistically independent when the partial-band interference is modeled as a white Gaussian process, the probability of a codeword symbol error is

$$P_s = \mu(1-P_{\epsilon 1}) P_{s1} + (1-\mu)(1-P_{\epsilon 0}) P_{s0} \tag{3.127}$$

and the conditional channel-symbol error probabilities are

$$P_{s1} = G\left(\frac{\mathcal{E}_s}{N_0 + I_{t0}/\mu}\right), \quad P_{s0} = G\left(\frac{\mathcal{E}_s}{N_0}\right) \tag{3.128}$$

where $G(x)$ depends on the modulation and fading.

The word error probability for errors-and-erasures decoding is upper-bounded in (1.24). Since most word errors result from decoding failures, it is reasonable to assume that $P_b \approx P_w/2$. Therefore, the information-bit error probability is given by

$$P_b \approx \frac{1}{2} \sum_{j=0}^{n} \sum_{i=i_0}^{n-j} \binom{n}{j} \binom{n-j}{i} P_s^i P_\epsilon^j (1 - P_s - P_\epsilon)^{n-i-j} \tag{3.129}$$

where $i_0 = \max(0, \lceil (d_m - j)/2 \rceil)$ and $\lceil x \rceil$ denotes the smallest integer greater than or equal to x.

Figures 3.10 and 3.11 plot the bit error probability P_b given by (3.125) to (3.129) for an FH/CPFSK system with orthogonal CPFSK and errors-and-erasures decoding. In Figure 3.10, the FH/CPFSK system transmits over the AWGN channel and

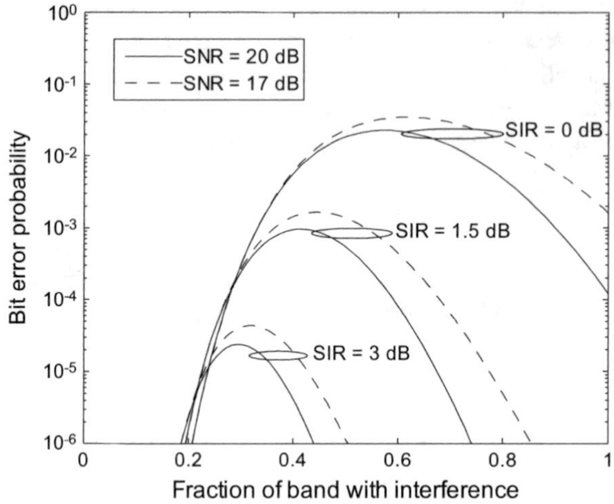

Fig. 3.10 Performance of orthogonal FH/CPFSK system over the AWGN channel with Reed-Solomon (32,12) code, $q = 32$, $N_t = 2$, and $\delta = 1$

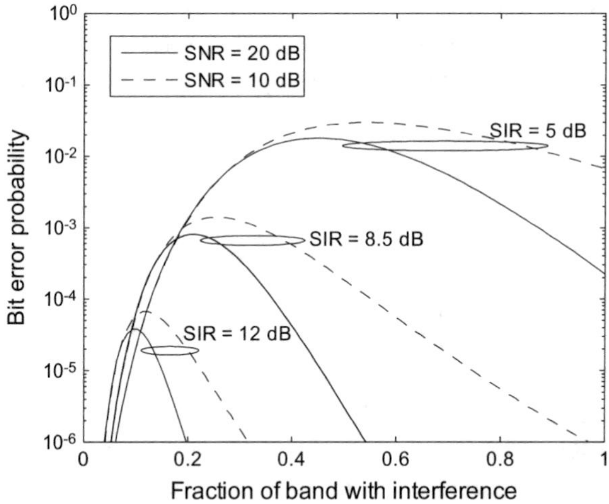

Fig. 3.11 Performance of orthogonal FH/CPFSK system over the AWGN channel with Reed-Solomon (8,3) code, $q = 8$, $N_t = 4$, and $\delta = 1$

uses $q = 32$, an extended Reed-Solomon (32,12) code, $N_t = 2$, and $\delta = 1$. A comparison of this figure with the $\kappa = \infty$ graphs of Figure 3.9 indicates that when $\mathcal{E}_b/N_0 = 20$ dB, erasures provide nearly a 7 dB improvement in the required $\mathcal{E}_b/I_{t0}$ for $P_b = 10^{-5}$. The erasures also confer strong protection against partial-band interference that is concentrated in less than 20% of the hopping band.

There are other options for generating channel-state information in addition to demodulating pilot symbols. A radiometer (Section 10.2) may be used to measure the energy in the current frequency channel, a future channel, or an adjacent channel. Erasures are inserted if the energy is inordinately large. This method does not have the overhead cost in information rate that is associated with the use of pilot symbols. Other methods include attaching a parity-check bit to each code symbol representing multiple bits to check whether the symbol was correctly received, or using the soft information provided by the inner decoder of a concatenated code.

Consider the receiver for noncoherent detection of orthogonal CPFSK signals shown in Figure 3.5 (b). The envelope-detector outputs provide the symbol metrics used in several low-complexity schemes for erasure insertion [4]. The *output threshold test* (OTT) compares the largest symbol metric with a threshold to determine whether the corresponding demodulated symbol should be erased. The *ratio threshold test* (RTT) computes the ratio of the largest symbol metric to the second largest one. This ratio is then compared with a threshold to determine an erasure. If the values of both $\mathcal{E}_b/N_0$ and $\mathcal{E}_b/I_{t0}$ are known, then optimal thresholds for the OTT, the RTT, or a hybrid method can be calculated. It is found that the OTT is resilient against fading and tends to outperform the RTT when $\mathcal{E}_b/I_{t0}$ is sufficiently low, but the opposite is true when $\mathcal{E}_b/I_{t0}$ is sufficiently high. The main disadvantage of the OTT and the RTT relative to the pilot-symbol method is the need to estimate $\mathcal{E}_b/N_0$ and either $\mathcal{E}_b/I_{t0}$ or $\mathcal{E}_b/(N_0+I_{t0})$. The joint *maximum-output ratio threshold test* (MO-RTT) uses both the maximum and the second largest of the symbol metrics. It is robust against both fading and partial-band interference.

Proposed erasure methods are based on the use of orthogonal CPFSK symbols, and their performances against partial-band interference improve as the alphabet size q increases. For a fixed hopping band, the number of frequency channels decreases as q increases, thereby making an FH/CPFSK system more vulnerable to multitone jamming or multiple-access interference (Chapter 7).

Figure 3.11 depicts P_b for an FH/CPFSK system over the AWGN channel with $q = 8$, an extended Reed-Solomon (8,3) code, $N_t = 4$, and $\delta = 1$. A comparison of Figures 3.11 and 3.10 indicates that reducing the alphabet size while preserving the code rate has increased the system sensitivity to $\mathcal{E}_b/N_0$, increased the susceptibility to interference concentrated in a small fraction of the hopping band, and raised the required $\mathcal{E}_b/I_{t0}$ for a specified P_b by 5 to 9 dB.

Another approach is to represent each nonbinary code symbol as a sequence of $\log_2 q$ consecutive binary channel symbols. Then, an FH/MSK or FH/DPSK system can be implemented to provide a large number of frequency channels and hence better protection against multiple-access interference. Equations (3.125), (3.126), (3.128), and (3.129) are applicable. However, since a code-symbol error occurs if any of its $\log_2 q$ component channel symbols is incorrect, (3.127) is replaced by

$$P_s = 1 - [1 - \mu(1 - P_{\epsilon 1})P_{s1} - (1 - \mu)(1 - P_{\epsilon 0})P_{s0}]^{\log_2 q}. \tag{3.130}$$

For the AWGN channel, (3.118) applies to MSK, whereas

$$G(x) = \frac{1}{2}\exp(-x) \tag{3.131}$$

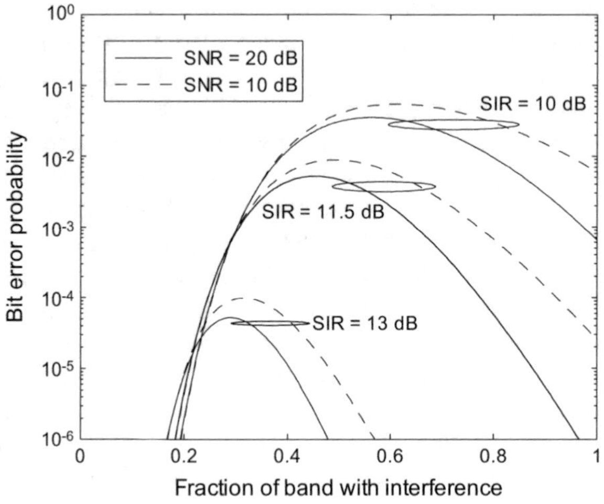

Fig. 3.12 Performance of the FH/DPSK system over the AWGN channel with Reed-Solomon (32,12) code, binary channel symbols, $N_t = 10$, and $\delta = 1$

applies to DPSK. The results for an FH/DPSK system with an extended Reed-Solomon (32,12) code, $N_t = 10$ binary pilot symbols, and $\delta = 1$ are shown in Figure 3.12. We assume that $N >> 10$ so that the loss due to the reference symbol in each dwell interval is negligible. The graphs in Figure 3.12 are similar in form to those of Figure 3.10, but the transmission of binary rather than nonbinary symbols has caused an approximately 10 dB increase in the required $\mathcal{E}_b/I_{t0}$ for a specified P_b. Figure 3.12 is applicable to orthogonal CPFSK and MSK if $\mathcal{E}_b/I_{t0}$ and $\mathcal{E}_b/N_0$ are both increased by 3 dB.

An alternative to erasures that uses binary channel symbols is an FH/DPSK system with concatenated coding (Section 1.5). Consider a concatenated code comprising a Reed-Solomon (n, k) outer code, a binary convolutional inner code, and a channel interleaver to ensure independent channel-symbol errors. After demodulation and deinterleaving in the receiver, the inner Viterbi decoder performs hard-decision decoding to limit the impact of individual symbol metrics. For the AWGN channel, the symbol error probability is given by (3.119) and (3.131). The probability of a Reed-Solomon symbol error, P_{s1}, at the output of the Viterbi decoder is upper-bounded by (1.147) and (1.118), and (1.148) then provides an upper bound on P_b. Figure 3.13 depicts this bound for an outer Reed-Solomon (31,21) code and an inner rate-1/2, $K = 7$ convolutional code. This concatenated code provides a better performance than the Reed-Solomon (32,12) code with binary channel symbols, but a much worse performance than the latter code with nonbinary channel symbols. Figures 3.10 through 3.13 indicate that a reduction in the alphabet size for channel symbols increases the system susceptibility to partial-band interference. The primary reasons are the reduced energy per channel symbol and the orthogonal CPFSK signals.

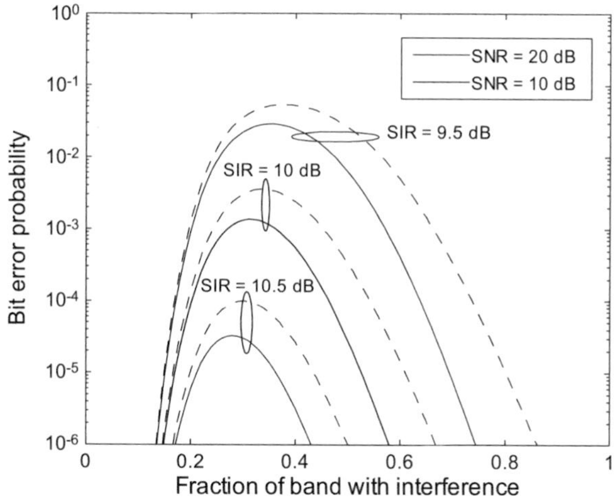

Fig. 3.13 Performance of FH/DPSK system over the AWGN channel with concatenated code, binary channel symbols, and hard decisions. Inner code is convolutional (rate = 1/2, $K = 7$) code, and outer code is Reed-Solomon (31,21) code

Trellis-Coded Modulation

Trellis-coded modulation (Section 1.3) is a combined coding and modulation method that is usually applied to coherent digital communications over bandlimited channels. Multilevel and multiphase modulations are used to enlarge the signal constellation while not expanding the bandwidth beyond what is required for the uncoded signals. Since the signal constellation is more compact, there is some modulation loss that detracts from the coding gain, but the overall gain can be substantial. Since a noncoherent demodulator is usually required for frequency-hopping communications; thus, the usual coherent trellis-coded modulations are not suitable. Instead, the trellis coding may be implemented by expanding the signal set for $q/2$-ary CPFSK to q-ary CPFSK. Although the frequency tones are uniformly spaced, they can be nonorthogonal to limit or avoid bandwidth expansion.

Trellis-coded 4-ary CPFSK is illustrated in Figure 3.14 for a system that uses a four-state, rate-1/2, convolutional code followed by a symbol mapper. The signal set partitioning, shown in Figure 3.14 (a), partitions the set of four signals or tones into two subsets, each with two tones. The partitioning doubles the frequency separation between tones from Δ Hz to 2 ΔHz. The mapping of the code bits produced by the convolutional encoder into signals is indicated. In Figure 3.14 (b), the numerical labels denote the signal assignments associated with the state transitions in the trellis for a four-state encoder. The bandwidth of the frequency channel that accommodates the four tones is $B \approx 4\Delta$.

There is a tradeoff in the choice of Δ because a small Δ allows more frequency channels and thereby limits the effect of multiple-access interference or multitone

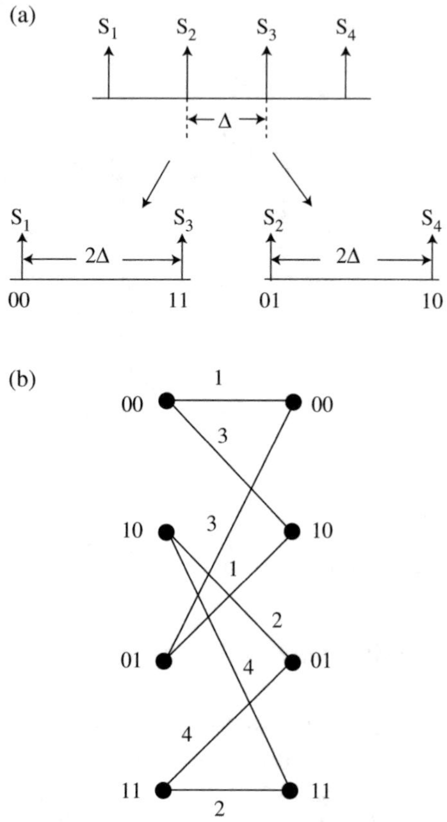

Fig. 3.14 Rate-1/2, four-state trellis-coded 4-ary FSK: (**a**) signal set partitioning and mapping of bits to signals, and (**b**) mapping of signals to state transitions

jamming, whereas a large Δ tends to improve the system performance against partial-band interference. If a trellis code uses four orthogonal tones with spacing $\Delta = 1/T_b$, where T_b is the bit duration, then $B \approx 4/T_b$. The same bandwidth results when an FH/CPFSK system uses two orthogonal tones, a rate-1/2 code, and binary channel symbols because $B \approx 2/T_s = 4/T_b$. The same bandwidth also results when a rate-1/2 binary convolutional code is used and each pair of code symbols is mapped into a 4-ary channel symbol. The performance of the four-state, trellis-coded, rate-1/2, 4-ary CPFSK frequency-hopping system [118] indicates that it is not as strong against worst-case partial-band interference as an FH/CPFSK system with a rate-1/2 convolutional code and 4-ary channel symbols or an FH/CPFSK system with a Reed-Solomon (32,16) code and errors-and-erasures decoding. Thus, trellis-coded modulation is relatively weak against partial-band interference. The advantage of trellis-coded modulation in a frequency-hopping system is its relatively low implementation complexity.

Turbo and LDPC Codes

Turbo and LDPC codes (Chapter 1) are potentially the most effective codes for suppressing partial-band interference if the system latency and computational complexity of these codes are acceptable. A turbo-coded frequency-hopping system that uses spectrally compact channel symbols also resists multiple-access interference. Accurate estimates of channel parameters, such as the variance of the interference and noise and the fading amplitude, are needed in the iterative decoding algorithms. When the channel dynamics are slower than the hop rate, all the received symbols of a dwell interval may be used in estimating the channel parameters associated with that dwell interval. After each iteration by a component decoder, its log-likelihood ratios are updated and the extrinsic information is transferred to the other component decoder. A channel estimator can convert a log-likelihood ratio transferred after a decoder iteration into a posteriori probabilities that can be used to improve the estimates of the fading attenuation and the noise variance for each dwell interval (Section 9.4). The operation of a receiver with iterative LDPC decoding and channel estimation is similar.

Known symbols may be inserted into the transmitted code symbols to facilitate the estimation, but the energy per information bit is reduced. Increasing the number of symbols per hop improves the potential estimation accuracy. However, since the reduction in the number of independent hops per information block of fixed size decreases the diversity, and hence the independence of errors, there is an upper limit on the number of symbols per hop beyond which a performance degradation occurs.

A turbo code can still provide a fairly good performance against partial-band interference, even if only the presence or absence of strong interference during each dwell interval is detected. Channel-state information about the occurrence of interference during a dwell interval can be obtained by hard-decision decoding of the output of one of the component decoders of a turbo code with parallel concatenated codes or the inner decoder of a serially concatenated turbo code. The metric for determining the occurrence of interference is the Hamming distance between the binary sequence resulting from the hard decisions and the codewords obtained by bounded-distance decoding.

3.7 Hybrid Systems

Frequency-hopping systems reject interference by avoiding it, whereas direct-sequence systems reject interference by spreading and then filtering it. Channel codes are more essential for frequency-hopping systems than for direct-sequence systems because partial-band interference is a more pervasive threat than high-power pulsed interference. When frequency-hopping and direct-sequence systems are constrained to use the same fixed bandwidth, then direct-sequence systems have an inherent advantage because they can use coherent BPSK rather than a noncoherent modulation. Coherent BPSK has an approximately 4 dB advantage relative to noncoherent MSK over the AWGN channel and an even larger advantage over fading channels. However, the potential performance advantage of direct-sequence systems is often illusory for practical reasons. A major advantage of frequency-hopping systems relative to direct-sequence systems is that it is possible to hop into noncontiguous frequency channels over a much wider band than can be occupied by a direct-sequence signal. This advantage more than compensates for the relatively inefficient noncoherent demodulation that is usually required for frequency-hopping systems. Other major advantages of frequency hopping are the possibility of excluding frequency channels with steady or frequent interference, the reduced susceptibility to the near-far problem (Section 7.7), and the relatively rapid acquisition of the frequency-hopping pattern (Section 4.7). A disadvantage of frequency hopping is that it is not amenable to transform-domain or nonlinear adaptive filtering (Section 5.3) to reject narrowband interference within a frequency channel. In practical systems, the dwell time is too short for adaptive filtering to have a significant effect.

A *hybrid frequency-hopping direct-sequence* (FH/DS) *system* is a frequency-hopping system that uses direct-sequence spreading during each dwell interval or, equivalently, a direct-sequence system in which the carrier frequency changes periodically. In the transmitter of the hybrid FH/DS system of Figure 3.15, a single

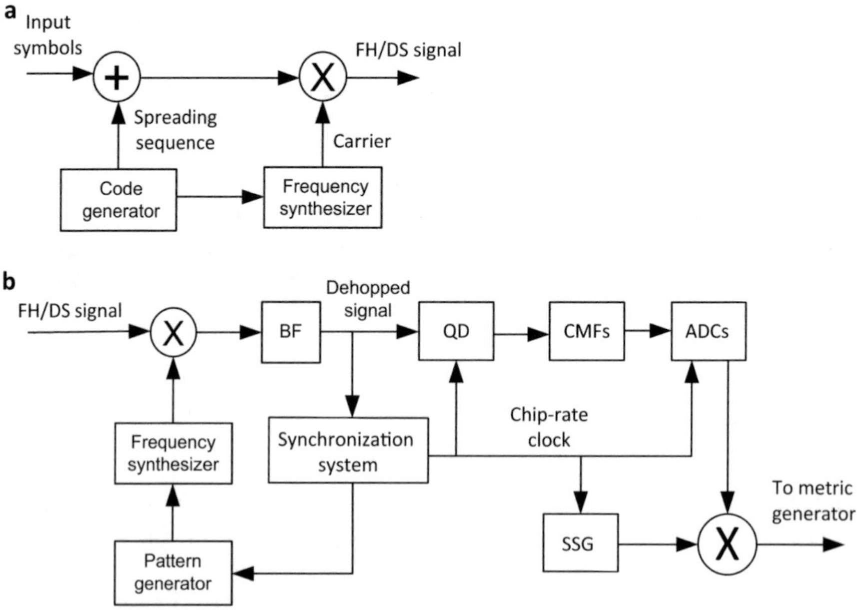

Fig. 3.15 Hybrid frequency-hopping direct-sequence system: (**a**) transmitter and (**b**) receiver. *BF* bandpass filter, *QD* quadrature downconverter, *CMFs* chip-matched filters, *ADCs* analog-to-digital converters, *SSG* spreading-sequence generator

code generator controls both the spreading sequence and the hopping pattern. The spreading sequence is added modulo-2 to the data sequence. Hops occur periodically after a fixed number of sequence chips. Because of the phase changes due to the frequency hopping, noncoherent modulation such as CSK or DPSK is usually required unless the hop rate is very low. In the noncoherent receiver, the frequency hopping and the spreading sequence of the FH/DS signal are removed in succession to produce the inputs to the metric generator. The metric generators for CSK and DPSK are shown in Figures 2.30 and 2.31, respectively.

Serial-search acquisition occurs in two stages. The first stage provides alignment of the hopping patterns, whereas the second stage over the unknown timing of the spreading sequence finishes acquisition rapidly because the timing uncertainty has been reduced by the first stage to a fraction of a hop duration.

In principle, the receiver of a hybrid FH/DS system curtails partial-band interference by both dehopping and despreading, but diminishing returns are encountered. The hopping of the FH/DS signal allows the avoidance of the interference spectrum part of the time. When the system hops into a frequency channel with interference, the interference is spread and filtered by the hybrid receiver. However, during a hop interval, interference that would be avoided by an ordinary frequency-hopping receiver is passed by the bandpass filter of a hybrid receiver because the bandwidth must be large enough to accommodate the direct-sequence signal that remains after the dehopping. This large bandwidth also limits the number of available frequency channels, which increases the susceptibility to narrowband interference and the near-far problem. Thus, hybrid FH/DS systems are seldom used, except perhaps in specialized military applications because the additional direct-sequence spreading weakens the major strengths of frequency hopping.

3.8 Frequency Synthesizers

A *frequency synthesizer* [77–79, 86] converts a standard reference frequency into a different desired frequency. In a frequency-hopping system, each hopset frequency must be synthesized. In practical applications, a hopset frequency has the form

$$f_{hi} = f_c + r_i f_r \,, \quad i = 1, 2, \ldots, M \tag{3.132}$$

where the $\{r_i\}$ are rational numbers, f_r is the reference frequency, and f_c is a frequency in the spectral band of the hopset. The *reference signal*, which is a tone at the reference frequency, is usually generated by dividing or multiplying by a positive integer the frequency of the tone produced by a stable source, such as an atomic or crystal oscillator. The use of a single reference signal ensures that any output frequency of the synthesizer has the same stability and accuracy as the reference. The three fundamental types of frequency synthesizers are the direct, digital, and indirect synthesizers. Most practical synthesizers are hybrids of these fundamental types.

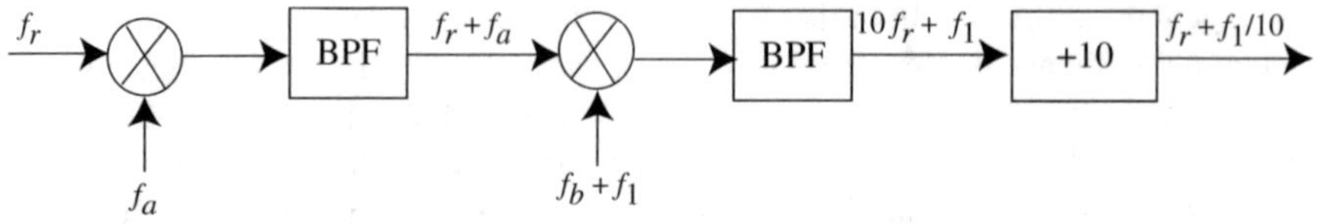

Fig. 3.16 Double-mix-divide module. *BPF* bandpass filter

Direct Frequency Synthesizer

A *direct frequency synthesizer* uses frequency multipliers and dividers, mixers, bandpass filters, and electronic switches to produce signals with the desired frequencies. Direct frequency synthesizers provide both very fine resolution and high frequencies, but often require a very large amount of hardware and do not provide a phase-continuous output after frequency changes. Although a direct synthesizer can be realized with programmable dividers and multipliers, the standard approach is to use the *double-mix-divide* (DMD) module illustrated in Figure 3.16. The reference signal at frequency f_r is mixed with a tone at the fixed frequency f_a. The bandpass filter selects the sum frequency $f_r + f_a$ produced by the mixer. A second mixing and filtering operation with a tone at $f_b + f_1$ produces the sum frequency $f_r + f_a + f_b + f_1$. If f_b is a fixed frequency such that

$$f_b = 9f_r - f_a \tag{3.133}$$

then the second bandpass filter produces the frequency $10f_r + f_1$. The divider, which reduces the frequency of its input by a factor of 10, produces the output frequency $f_r + f_1/10$. In principle, a single mixer and bandpass filter could produce this output frequency, but two mixers and bandpass filters simplify the filters. Each bandpass filter must select the sum frequency while suppressing the difference frequency and the mixer input frequencies, which may enter the filter because of mixer leakage. A bandpass filter becomes prohibitively complex and expensive as the sum frequency approaches one of these other frequencies.

Arbitrary frequency resolution is achievable by cascading enough double-mix-divide modules. Figure 3.17 illustrates a direct frequency synthesizer that provides two-digit resolution. When the synthesizer is used in a frequency-hopping system, the control bits are generated by the pattern generator, which determines the frequency-hopping pattern. Each set of control bits selects a single tone that the decade switch passes to a double-mix-divide module. The ten tones that are available to the decade switches may be produced by applying the reference frequency to appropriate frequency multipliers and dividers in the tone generator. Equation (3.133) ensures that the output frequency of the second bandpass filter in double-mix-divide module 2 is $10f_r + f_2 + f_1/10$. Thus, the final synthesizer output frequency is $f_r + f_2/10 + f_1/100$.

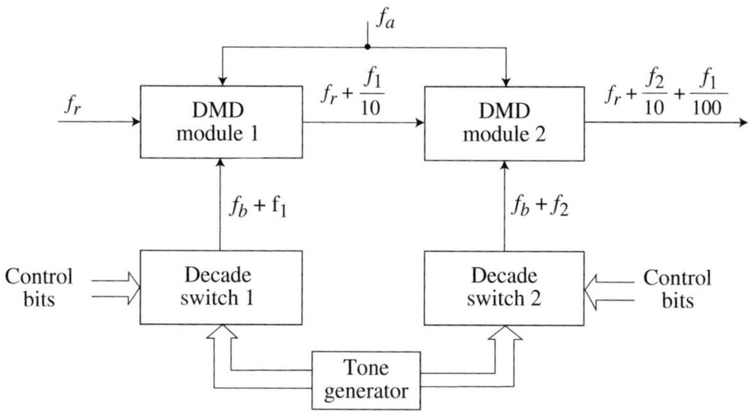

Fig. 3.17 Direct frequency synthesizer with two-digit resolution. *DMD* double-mix-divide

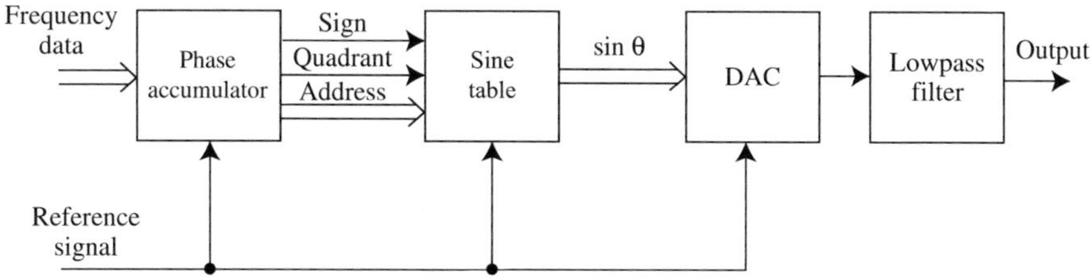

Fig. 3.18 Digital frequency synthesizer. *DAC* digital-to-analog converter

Example 3.1 It is desired to produce a 1.79 MHz tone. Let $f_r = 1$ MHz and $f_b = 5$ MHz. The ten tones provided to the decade switches are 5, 6, 7, . . ., 14 MHz so that f_1 and f_2 can range from 0 to 9 MHz. Equation (3.133) yields $f_a = 4$ MHz. If $f_1 = 7$ MHz and $f_2 = 9$ MHz, then the output frequency is 1.79 MHz. The frequencies f_a and f_b are such that the designs of the bandpass filters inside the modules are reasonably simple. □

Direct Digital Synthesizer

A *direct digital synthesizer* converts the stored sample values of a sinusoidal wave into a continuous-time sinusoidal wave with a specified frequency. The periodic and symmetric character of a sine wave implies that only sample values for the first quadrant of the sine wave need to be stored. Figure 3.18 illustrates the principal components of a digital frequency synthesizer. The *reference signal* is a sinusoidal signal with *reference frequency* f_r. The sine table stores N values of sin θ for $\theta = 0, 2\pi/N, \ldots, 2\pi(N-1)/N$. Frequency data, which is produced by the pattern control bits in a frequency-hopping system, determines the synthesized frequency by specifying a phase increment $2\pi k/N$, where k is an integer. The *phase accumulator*, which is a discrete-time integrator, converts the phase increment into successive samples of the phase by adding the increment to the content of an internal register at the rate f_r after every cycle of the reference signal. A phase sample $\theta(n) = n(2\pi k/N)$, where n denotes the reference-signal cycle, defines an address in the sine table or memory in which the values of $\sin(\theta(n))$ are stored. Each value is applied to a digital-to-analog converter (DAC), which performs a sample-and-hold operation at a sampling rate equal to the reference frequency f_r. The DAC output is applied to an anti-aliasing lowpass filter, the output of which is the desired analog signal.

Let ν denote the number of bits used to represent the $N \leq 2^\nu$ possible values of the phase accumulator contents. Since the contents are updated at the rate f_r, the longest period of distinct phase samples before repetition is N/f_r. Therefore, since the sample values of $\sin(\theta(n))$ are applied to the DAC at the rate f_r, the smallest frequency that can be generated by the direct digital synthesizer is

$$f_{\min} = \frac{f_r}{N} \tag{3.134}$$

and the phase sample is $\theta(n) = 2\pi/N$. The output frequency f_{0k} is produced when every kth stored sample value of the phase θ is applied to the DAC at the rate f_r. Thus, if the phase sample is $\theta(n) = k2\pi/N$ after every cycle of the reference signal, then

$$f_{0k} = kf_{\min}, \quad 1 \leq k \leq N \tag{3.135}$$

which implies that $f_{\min}$ is the frequency resolution.

The maximum frequency $f_{\max}$ that can be generated is produced by using only a few samples of $\sin\theta$ per period. From the Nyquist sampling theorem, it is known that $f_{\max} < f_r/2$ is required to avoid aliasing. Practical DAC and lowpass filter requirements further limit $f_{\max}$ to approximately $0.4f_r$ or less. Thus, $q \geq 2.5$ samples of $\sin\theta$ per period are used in synthesizing $f_{\max}$, and

$$f_{\max} = \frac{f_r}{q}. \tag{3.136}$$

The lowpass filter may be implemented with a linear phase across a flat passband extending slightly above $f_{\max}$. The frequencies f_r and $f_{\max}$ are limited by the speed of the DAC.

Suppose that $f_{\min}$ and $f_{\max}$ are specified minimum and maximum frequencies that must be produced by a synthesizer. Equations (3.134) and (3.136) imply that $qf_{\max}/f_{\min} = N \leq 2^\nu$, and the required number of accumulator bits is

$$\nu = \lfloor \log_2(qf_{\max}/f_{\min}) \rfloor + 1. \tag{3.137}$$

The sine table stores 2^n words, each comprising m bits, and hence has a memory requirement of $2^n m$ bits. The memory requirements of a sine table can be reduced by using trigonometric identities and hardware multipliers. Each stored word represents one possible value of sin θ in the first quadrant or, equivalently, one possible magnitude of sin θ. The input to the sine table comprises $n + 2$ parallel bits. The two most significant bits are the *sign bit* and the *quadrant bit*. The sign bit specifies the polarity of sin θ. The quadrant bit specifies whether sin θ is in the first or second quadrants or in the third or fourth quadrants. The n least significant bits of the input determine the address in which the magnitude of sin θ is stored. The address specified by the n least significant bits is appropriately modified by the quadrant bit when θ is in the second or fourth quadrants. The sign bit along with the m output bits of the sine table are applied to the DAC. The maximum number of table addresses that the phase accumulator can specify is 2^ν, but if ν input lines were applied to the sine table, the memory requirement would generally be impractically large. Since $n + 2$ bits are needed to address the sine table, the $\nu - n - 2$ least significant bits in the accumulator contents are not used to address the table.

If n is decreased, the memory requirement of the sine table is reduced, but the truncation of the potential accumulator output from ν to $n + 2$ bits causes the table output to remain constant for $\nu - n - 2$ samples of sin θ per period. Since the desired phase $\theta\,(n)$ is only approximated by $\widehat{\theta}\,(n)$, the input to the sine table, there are spurious spectral lines in the spectrum of the table output. The phase error, $\theta_e\,(n) = \theta\,(n) - \widehat{\theta}\,(n)$, is a periodic ramp with a discrete spectrum. For $\theta_e\,(n) << 1$, a trigonometric expansion indicates that

$$\sin\left(\theta\,(n)\right) \approx \sin[\widehat{\theta}\,(n)] + \theta_e\,(n)\cos[\widehat{\theta}\,(n)]. \tag{3.138}$$

The amplitudes of the spurious spectral lines are determined by the coefficients of the Fourier series of $\theta_e\,(n)$, and the largest amplitude is $2^{-(n+2)}$. The power in the largest of the spectral lines, which is often the limiting factor in applications, is

$$E_s = (2^{-(n+2)})^2 = (-6n - 12)\ dB. \tag{3.139}$$

There are several methods, each entailing an additional implementation cost, that can reduce the peak amplitudes of the spurious spectral lines. The *error feedforward method* uses an estimate of $\theta_e\,(n)$ to compute the right-hand side of (3.138), which then provides a more accurate estimate of the desired $\sin\left(\theta\,(n)\right)$. The computational requirements are a multiplication and an addition with m bits of precision at the sample rate. Other methods of reducing the peak amplitudes are based on dithering and error feedback.

Since m output bits of the sine table specify the magnitude of sin θ, the quantization error produces the worst-case *amplitude-quantization noise power*

$$E_q = (2^{-m})^2 \cong -6m\ dB. \tag{3.140}$$

Quantization noise increases the amplitudes of spurious spectral lines in the table output.

Example 3.2 A direct digital synthesizer is to be designed to cover 1 kHz to 1 MHz with $E_q \leq -45$ dB and $E_s \leq -60$ dB. According to (3.140), the use of 8-bit words in the sine table is adequate for the required quantization noise level. With $m = 8$, the table contains $2^8 = 256$ distinct words. According to (3.139), $n = 8$ is adequate for the required E_s, and hence the sine table has $n + 2 = 10$ input bits. If $2.5 \leq q \leq 4$, then because $f_{\max}/f_{\min} = 10^3$, (3.137) yields $\nu = 12$. Thus, a 12-bit phase accumulator is needed. Since $2^{12} = 4096$, we may choose $N = 4000$. If the frequency resolution and smallest frequency is to be $f_{\min} = 1$ kHz, then (3.134) indicates that $f_r = 4$ MHz is required. When the frequency $f_{\min}$ is desired, the phase increments are so small that $2^{\nu-n-2} = 4$ increments occur before a new address is specified and a new value of sin θ is produced. Thus, the 4 least-significant bits in the accumulator are not used for addressing the sine table. □

The direct digital synthesizer can be easily modified to produce a modulated output when high-speed digital data is available. For amplitude modulation, the table output is applied to a multiplier. Phase modulation may be implemented by adding the appropriate bits to the phase accumulator output. Frequency modulation entails modification of the accumulator input bits. For a quaternary modulation, the quadrature signals may be generated by separate sine and cosine tables.

A direct digital synthesizer can produce nearly instantaneous, phase-continuous frequency changes and a very fine frequency resolution despite its relatively small size, weight, and power requirements. A disadvantage is the limited maximum frequency, which restricts the bandwidth of the covered frequencies following a frequency translation of the synthesizer output. For this reason, direct digital synthesizers are sometimes used as components in hybrid synthesizers. Another disadvantage is the stringent requirement for the lowpass filter to suppress frequency spurs generated during changes in the synthesized frequency.

Indirect Frequency Synthesizers

An *indirect frequency synthesizer* uses voltage-controlled oscillators and feedback loops. Indirect synthesizers usually require less hardware than comparable direct ones, but need more time to switch from one frequency to another. Like digital synthesizers, indirect synthesizers inherently produce phase-continuous outputs after frequency changes. The principal components of a single-loop indirect synthesizer, which is similar in operation to a phase-locked loop, are depicted in Figure 3.19. The control bits, which determine the value of the modulus or divisor N, are supplied by a pattern generator. The input signal at frequency f_1 may be provided by another synthesizer. Since the feedback loop forces the frequency of the divider output, $(f_0 - f_1)/N$, to closely approximate the reference frequency f_r, the output of the voltage-controlled oscillator (VCO) is a sine wave with frequency

$$f_0 = Nf_r + f_1 \tag{3.141}$$

where N is a positive integer.

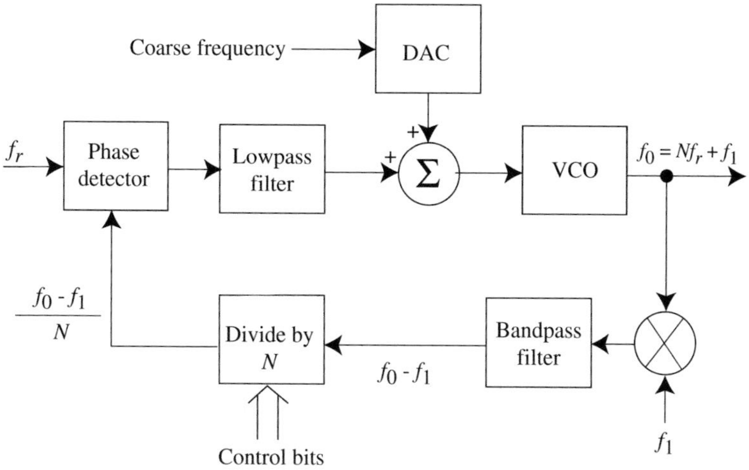

Fig. 3.19 Indirect frequency synthesizer with a single loop. *DAC* digital-to-analog converter, *VCO* voltage-controlled oscillator

Phase detectors in frequency-hopping synthesizers are usually digital devices that measure zero-crossing times rather than the phase differences measured when mixers are used. Digital phase detectors have an extended linear range, are less sensitive to input-level variations, and simplify the interface with a digital divider.

Equation (3.141) indicates that the frequency resolution of the single-loop synthesizer is f_r. It is usually necessary to have a loop bandwidth to the order of $f_r/10$ to ensure stable operation and the suppression of sidebands that are offset from f_0 by f_r. The *switching time* t_s required for changing frequencies is less than or equal to T_{sw} defined previously for frequency-hopping pulses, into which additional guard time may have been inserted. The switching time tends to be inversely proportional to the loop bandwidth and is roughly approximated by

$$t_s = \frac{25}{f_r} \tag{3.142}$$

which indicates that a low resolution and a low switching time may not be achievable by a single loop.

To decrease the switching time while maintaining the frequency resolution of a single loop, a coarse steering signal can be stored in a read only memory (ROM), converted into analog form by a DAC, and applied to the VCO (as shown in Figure 3.19) immediately after a frequency change. The steering signal reduces the frequency step that must be acquired by the loop when a hop occurs. An alternative approach is to place a fixed divider with modulus M after the loop so that the divider output frequency is $f_0 = Nf_r/M + f_1/M$. By this means, f_r can be increased without sacrificing resolution, provided that the VCO output frequency, which equals Mf_0,

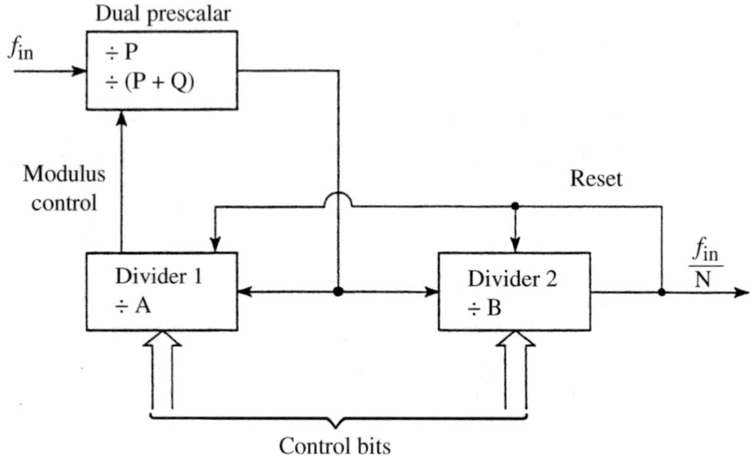

Fig. 3.20 Dual-modulus divider

is not too large for the divider in the feedback loop. To limit the transmission of spurious frequencies, it may be desirable to inhibit the transmitter output during frequency transitions.

The switching time can be dramatically reduced by using two synthesizers that alternately produce the output frequency. One synthesizer produces the output frequency while the second is being tuned to the next frequency following a command from the pattern generator. If the hop duration exceeds the switching time of each synthesizer, then the second synthesizer begins producing the next frequency before a control switch routes its output to a modulator or a mixer.

A *divider,* which is a binary counter that produces a square-wave output, counts down by one unit every time its input crosses zero. If the modulus or divisor is the positive integer N, then after N zero crossings, the divider output crosses zero and changes state. The divider then resumes counting down from N. As a result, the output frequency is equal to the input frequency divided by N. Programmable dividers have limited operating speeds that impair their ability to accommodate a high-frequency VCO output. A problem is avoided by the down-conversion of the VCO output by the mixer shown in Figure 3.19, but spurious components are introduced. Since fixed dividers can operate at much higher speeds than programmable dividers, placing a fixed divider before the programmable divider in the feedback loop could be considered. However, if the fixed divider has a modulus N_1, then the loop resolution becomes $N_1 f_r$; thus, this solution is usually unsatisfactory.

Figure 3.20 depicts a *dual-modulus divider*, which maintains a frequency resolution equal to f_r while allowing synthesizer operation at high frequencies. The principal component of the dual-modulus divider is the *dual prescalar,* which comprises two fixed dividers with divisors equal to the positive integers P and $P+Q$, respectively. Dividers 1 and 2 are programmable dividers that divide by the nonnegative integers A and B, respectively, where $B > A$. These programmable dividers are only required to accommodate a frequency f_{in}/P. The dual prescalar initially divides by the modulus $P + Q$. This modulus changes whenever a programmable divider reaches zero. After $(P + Q)A$ input transitions, divider 1 reaches zero, and the modulus control causes the dual prescalar to divide by P. Divider 2 has counted down to $B - A$. After $P(B - A)$ more input transitions, divider 2 reaches zero and causes an output transition. The two programmable dividers are then reset, and the dual prescalar reverts to division by $P+Q$. Thus, each output transition corresponds to $A(P + Q) + P(B - A) = AQ + PB$ input transitions, which implies that the dual-modulus divider has a modulus

$$N = AQ + PB\ , \quad B > A \tag{3.143}$$

and produces the output frequency f_{in}/N.

If $Q = 1$ and $P = 10$, then the dual-modulus divider is called a *10/11 divider*, and

$$N = 10B + A\ , \quad B > A \tag{3.144}$$

which can be increased in unit steps by changing A in unit steps. Since $B > A$ is required, a suitable range for A and minimum value of B are

$$0 \leq A \leq 9\ , \qquad B_{\min} = 10. \tag{3.145}$$

The relations (3.141), (3.144), and (3.145) indicate that the range of a synthesized hopset is from $f_1 + 100f_r$ to $f_1 + (10B_{\max} + 9)f_r$. Therefore, a spectral band between f_{min} and f_{max} is covered by the hopset if

$$f_1 + 100f_r \leq f_{min} \tag{3.146}$$

and

$$f_1 + (10B_{\max} + 9)f_r \geq f_{max}. \tag{3.147}$$

Example 3.3 The *Bluetooth* communication system is used to establish wireless communications among portable electronic devices. The system has a hopset of 79 carrier frequencies, its hop rate is 1600 hops per second, its hop band is between 2400 and 2483.5 MHz, and the bandwidth of each frequency channel is 1 MHz. Consider a system in which the 79 carrier frequencies are spaced 1 MHz apart from 2402 to 2480 MHz. A 10/11 divider with $f_r = 1$ MHz provides the desired increment, which is equal to the frequency resolution. Equation (3.142) indicates that $t_s = 25\,\mu$s, which indicates that 25 potential data symbols have to be omitted during each hop interval. Inequality (3.146) indicates that $f_1 = 2300$ MHz is a suitable choice. Then (3.147) is satisfied by $B_{max} = 18$. Therefore, dividers A and B require 4 and 5 control bits, respectively, to specify their potential values. If the

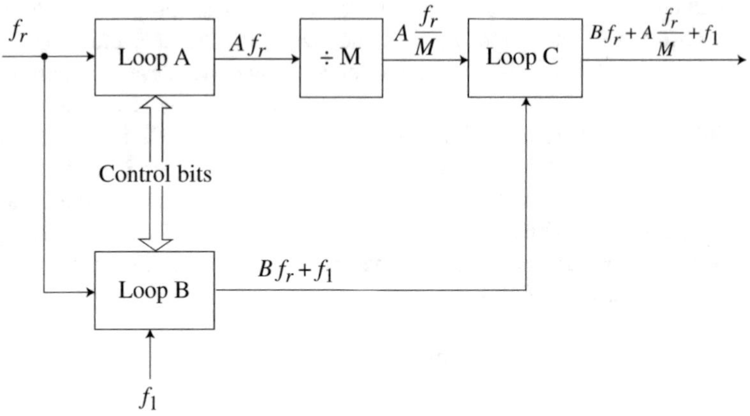

Fig. 3.21 Indirect frequency synthesizer with three loops

control bits are stored in an ROM, then each ROM location contains 9 bits. The number of ROM addresses is at least 79, the number of frequencies in the hopset. Thus, a ROM input address requires 7 bits. □

Multiple Loops

A *multiple-loop frequency synthesizer* uses two or more single-loop synthesizers to obtain both fine frequency resolution and fast switching. A three-loop frequency synthesizer is shown in Figure 3.21. Loops A and B have the form of Figure 3.19, but loop A does not have a mixer and filter in its feedback. Loop C has the mixer and filter, but lacks the divider. The reference frequency f_r is chosen to ensure that the desired switching time is realized. The divisor M is chosen so that f_r/M is equal to the desired resolution. Loop A and the divider generate increments of f_r/M, while loop B generates increments of f_r. Loop C combines the outputs of loops A and B to produce the output frequency

$$f_0 = Bf_r + A\frac{f_r}{M} + f_1 \tag{3.148}$$

where B, A, and M are positive integers because they are produced by dividers. Loop C is preferable to a mixer and bandpass filter because the filter would have to suppress a closely spaced, unwanted component when Af_r/M and Bf_r were far apart.

To ensure that each output frequency is produced by unique values of A and B, it is required that $A_{max} = A_{min} + M - 1$. To prevent degradation in the switching time, it is required that $A_{\min} > M$. Both requirements are met by choosing

$$A_{min} = M + 1, \quad A_{max} = 2M. \tag{3.149}$$

According to (3.148), a range of frequencies from f_{min} to f_{max} is covered if

$$B_{min}f_r + A_{min}\frac{f_r}{M} + f_1 \leq f_{min} \tag{3.150}$$

and

$$B_{max}f_r + A_{max}\frac{f_r}{M} + f_1 \geq f_{max}. \tag{3.151}$$

Example 3.4 Consider the Bluetooth system of Example 3.3 but with the more stringent requirement that $t_s = 2.5\,\mu$s, which only sacrifices three potential data symbols per hop interval. The single-loop synthesizer of Example 3.3 cannot provide this short switching time. The required switching time is provided by a three-loop synthesizer with $f_r = 10\,$MHz. The resolution of 1 MHz is achieved by taking $M = 10$. Equation (3.149) indicates that $A_{min} = 11$ and $A_{max} = 20$. Inequalities (3.150) and (3.151) are satisfied if $f_1 = 2300\,$MHz, $B_{min} = 9$, and $B_{max} = 16$. The maximum frequencies that must be accommodated by the dividers in loops A and B are $A_{max}f_r = 200\,$MHz and $B_{max}f_r = 160\,$MHz, respectively. Dividers A and B require 5 control bits. □

Fractional-N Synthesizer

A *fractional-N synthesizer* uses a single loop and auxiliary hardware to produce an average output frequency

$$f_0 = \left(B + \frac{A}{M}\right)f_r\,, \quad 0 \leq A \leq M - 1. \tag{3.152}$$

Although the switching time is inversely proportional to f_r, the resolution is f_r/M, which can be made arbitrarily small in principle.

The synthesis method uses a dual-modulus divider with frequency-division modulus equal to either B or $B + 1$. As shown in Figure 3.22, the modulus is controlled by a sequence of bits $\{b(n)\}$, where n is the discrete-time index. The sequence is generated at the clock rate equal to f_r, and the divider modulus is $B + b(n)$. The average value of the bit $b(n)$ is

$$\overline{b(n)} = \frac{A}{M} \tag{3.153}$$

so that the average divider modulus is $B + A/M$. The phase detector compares the arrival times of the rising edges of the frequency-divider output with those of the reference signal and produces an output that is a function of the difference in arrival times. A charge pump, which uses two current sources to charge or discharge capacitors, draws proportionate charge into the lowpass filter, the output of which

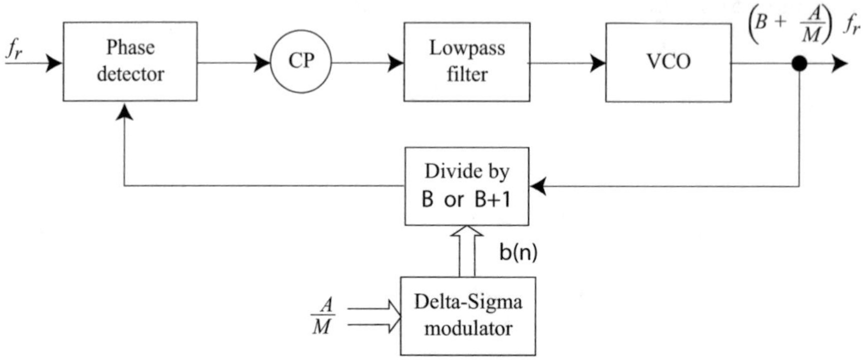

Fig. 3.22 Fractional-N frequency synthesizer. *CP* charge pump, *VCO* voltage-controlled oscillator

is the control signal of the VCO. Since the lower input to the phase detector has an average frequency of f_r, the average output frequency can be maintained at f_0.

A *delta-sigma modulator* encodes a higher-resolution digital signal into a lower-resolution one. When its input is a multiple-bit representation of A/M, the delta-sigma modulator generates the sequence $\{b(n)\}$. If this sequence were periodic, then high-level *fractional spurs*, which are harmonic frequencies that are integer multiples of f_r/M, would appear at the VCO input, thereby frequency-modulating its output signal. Fractional spurs are greatly reduced by *modulus randomization*, which randomizes $b(n)$ while maintaining (3.153). Modulus randomization is implemented by dithering or randomly changing the least significant bit of the input to the delta-sigma modulator. *Quantization noise* $q(n) = b(n) - \overline{b(n)}$ is introduced into the loop because A/M is approximated at the delta-sigma modulator output by a bit $b(n)$. To limit the effect of the quantization noise, the modulus randomization is designed such that the quantization noise has a high-pass spectrum. A lowpass loop filter can then eliminate most of the noise.

Example 3.5 Consider a fractional-N synthesizer for the Bluetooth system of Example 3.4 in which $t_s = 2.5\,\mu$s. If the output of the fractional-N synthesizer is frequency-translated by 2300 MHz, then the synthesizer itself needs to cover 102 to 180 MHz. The switching time is achieved by taking $f_r = 10$ MHz. The resolution is achieved by taking $M = 10$. Equation (3.152) indicates that the required frequencies are covered by varying B from 10 to 18 and A from 0 to 9. The integers B and A require 5 and 4 control bits, respectively. □

3.9 Problems

1 Consider FH/CPFSK with soft-decision decoding of repetition codes and large values of $\mathcal{E}_b/I_{t0}$. Suppose that the number of repetitions n is not chosen to minimize the potential impact of partial-band interference. Then, the first line of (3.20) upper-bounds P_b. Show that a nonbinary modulation with $m = \log_2 q$ bits per symbol gives a better performance than binary modulation in the presence of worst-case partial-band interference if $n > (m-1)\ln(2)/\ln(m)$.

2 Consider FH/CPFSK with soft-decision decoding of repetition codes. (a) Prove that $f(n) = (q/4)(4n/e\gamma)^n$ is a strictly convex function over the compact set $[1, \gamma/3]$. (b) Find the stationary point that gives the global minimum of $f(n)$. (c) Prove (3.22).

3 The autocorrelations of the complex envelopes of practical signals have the form $R_l(\tau) = f_1(\tau/T_s)$. (a) Prove that the power spectral density of the complex envelope has the form $S_l(f) = T_s f_2(fT_s)$. (b) Prove that if the bandwidth B of a frequency channel is determined by setting $F_{ib}(B/2) = c$, then the required B is inversely proportional to T_s.

4 For FH/CPM, use the procedures described in the text to verify (3.78) and (3.82).

5 (a) Verify (3.85)-(3.88). (b) Use the indefinite integral

$$\int \cos(ax+b)\cos(ax+b)\,dx = \frac{x}{2}\cos(b-d) + \frac{1}{4a}\sin(2ax+b+d)$$

to verify (3.89) and (3.90).

6 Derive (3.112) by following the steps described in the text.

7 Approximate $\mu = J/M$ in (3.116) by a continuous variable over the interval $[0, 1]$. What is the worst-case value of μ for binary orthogonal FH/CPFSK, noncoherent detection, hard decisions, and the AWGN channel in the presence of strong interference? What is the corresponding worst-case symbol error probability? Why does it not depend on the number of frequency channels?

8 Consider binary orthogonal FH/CPFSK, noncoherent detection, hard decisions, and the Rayleigh channel in the presence of strong interference. Show that interference spread uniformly over the entire hopping band hinders communications more than equal-power interference concentrated over part of the band.

9 This problem illustrates the importance of a channel code to a frequency-hopping system in the presence of worst-case partial-band interference. Consider binary orthogonal FH/CPFSK, noncoherent detection, hard decisions, and the AWGN channel. (a) Use the results of Problem 7 to calculate the required $\mathcal{E}_b/I_{t0}$ to obtain a bit error rate $P_b = 10^{-5}$ when no channel code is used. (b) Calculate the required $\mathcal{E}_b/I_{t0}$ for $P_b = 10^{-5}$ when a (23,12) Golay code is used. As a first step, use the first term in (1.22) to estimate the required symbol error probability. What is the coding gain?

10 It is desired to cover 198-200 MHz in 10-Hz increments using double-mix-divide modules. (a) What is the minimum number of modules required? (b) What is the range of acceptable reference frequencies? (c) Choose a reference frequency and f_b. What are the frequencies of the required tones? (d) If an upconversion by 180 MHz follows the DMD modules, what is the range of acceptable reference frequencies? Is this system more practical?

11 It is desired to cover 100-100.99 MHz in 10-kHz increments with an indirect frequency synthesizer containing a single loop and a dual-modulus divider. Let $f_1 = 0$ in Figure 3.19 and $Q = 1$ in Figure 3.20 (a) What is a suitable range of values for A? (b) What are a suitable value for P and a suitable range of values for B if it is required to minimize the highest frequency applied to the programmable dividers?

12 It is desired to cover 198-200 MHz in 10-Hz increments with a switching time equal to 2.5 ms. An indirect frequency synthesizer with three loops in the form of Figure 3.21 is used. It is desired that $B_{max} \leq 10^4$ and that f_1 is minimized. What are suitable values for the parameters f_r, M, A_{min}, A_{max}, B_{min}, B_{max}, and f_1?

13 Specify the design parameters of a fractional-N synthesizer that covers 198-200 MHz in 10-Hz increments with a switching time equal to 250 μs.

Chapter 4
Code Synchronization

A spread-spectrum receiver requires *code synchronization* to generate a spreading sequence or frequency-hopping pattern that is synchronized with the received sequence or pattern. After code synchronization, the received and receiver-generated chips or dwell intervals must precisely or nearly coincide. Any misalignment causes the signal amplitude at the demodulator output to fall in accordance with the autocorrelation or partial autocorrelation function. A practical implementation of code synchronization is greatly facilitated by dividing synchronization into two operations: acquisition and tracking. *Code acquisition* provides coarse synchronization by limiting the possible timing offsets of the receiver-generated chips or dwell intervals to a finite number of quantized candidates. Code acquisition is almost always the dominant design issue and most expensive component of a complete spread-spectrum system. Following the code acquisition, *code tracking* is activated to provide fine synchronization by which synchronization errors are further reduced or at least maintained within certain bounds. *Symbol synchronization*, which is needed to provide timing pulses for symbol detection to the decoder, is derived from the code-synchronization system. Although the use of precision clocks in both the transmitter and the receiver limits the timing uncertainty of sequences or patterns in the receiver, clock drifts, range uncertainty, and the Doppler shift may cause synchronization problems. The methods of code acquisition and code tracking for both direct-sequence and frequency-hopping systems are presented in this chapter.

4.1 Synchronization of Spreading Sequences

In nearly all applications, *code synchronization of direct-sequence signals* must precede carrier synchronization. Prior to despreading, which requires code synchronization, the signal-to-noise ratio (SNR) is unlikely to be high enough for

D. Torrieri, *Principles of Spread-Spectrum Communication Systems*,
https://doi.org/10.1007/978-3-319-70569-9_4

successful carrier tracking by a phase-locked loop. To achieve the required code synchronization and frequency synchronization, the code timing offset and the frequency offset need to be estimated.

Estimation of Code Timing and Frequency Offsets

Consider the estimation of waveform parameters [33, 47], which are components of the vector $\boldsymbol{\theta}$. The observed signal is

$$r(t) = s\,(t, \boldsymbol{\theta}) + n\,(t)\,,\ \ 0 \leq t \leq T \tag{4.1}$$

where the real-valued signal $s\,(t, \boldsymbol{\theta})$ has a known waveform except for $\boldsymbol{\theta}$, and $n\,(t)$ is zero-mean, white Gaussian noise. We assume that for all values of the waveform parameters, $s(t, \boldsymbol{\theta})$ belongs to the signal space $L^2\,[0, T]$ of complex-valued functions f such that $|f|^2$ is integrable over $[0, T]$. As shown in Appendix F.3, the *sufficient statistic for maximum-likelihood estimation* is

$$\Lambda_s\,[r(t)] = 2\int_0^T r(t)s(t, \boldsymbol{\theta})dt - \int_0^T s^2(t, \boldsymbol{\theta})dt. \tag{4.2}$$

If $s\,(t, \boldsymbol{\theta})$ depends on a random vector $\boldsymbol{\phi}$, we base the maximum-likelihood estimation on $E_{\boldsymbol{\phi}}\,[\Lambda_s\,[r(t)]]$, where $E_{\boldsymbol{\phi}}\,[\cdot]$ denotes the expected value with respect to $\boldsymbol{\phi}$. Therefore, the sufficient statistic becomes

$$\Lambda_a\,[r(t)] = E_{\boldsymbol{\phi}}\left[2\int_0^T r(t)s(t, \boldsymbol{\theta})dt - \int_0^T s^2(t, \boldsymbol{\theta})dt\right]. \tag{4.3}$$

The maximum-likelihood estimator is

$$\boldsymbol{\theta} = \arg\max_{\boldsymbol{\theta}} \Lambda_a\,[r(t)]\,. \tag{4.4}$$

For the estimation of the code timing and frequency offsets, $\boldsymbol{\theta}$=(τ,f_d). For a direct-sequence system with binary phase-shift keying (BPSK), the desired signal is

$$s(t, \boldsymbol{\theta}) = \sqrt{2S}d\,(t - \tau)\,p(t - \tau)\cos\left(2\pi f_c t + 2\pi f_d t + \phi\right),\ \ 0 \leq t \leq T \tag{4.5}$$

where S is the average power, the data modulation of code symbols is $d\,(t) = \pm 1$, $p(t)$ is the known spreading waveform, f_c is the known carrier frequency, ϕ is the received carrier phase, τ is the received code phase or timing offset, $0 \leq \tau < T$, and f_d is the frequency offset relative to the nominal carrier frequency. The frequency offset may be due to a Doppler shift or to a drift or instability in the transmitter oscillator.

The carrier phase ϕ is modeled as a random variable. The second term within the brackets (4.3) is the signal energy over the observation interval. If we assume that $\tau << T$, then this energy does not vary significantly with τ. Therefore, we drop this term and base the maximum-likelihood estimation on the sufficient statistic

$$\Lambda_a[r(t)] = E_{\boldsymbol{\phi}}\left[\int_0^T r(t)s(t,\boldsymbol{\theta})dt\right]. \tag{4.6}$$

The substitution of (4.5) into (4.6) yields

$$\Lambda_a[r(t)] = E_\phi\left\{\exp\left[\sqrt{2S}\int_0^T r(t)d\,(t-\tau)\,p(t-\tau)\cos\left(2\pi f_c t + 2\pi f_d t + \phi\right)dt\right]\right\}. \tag{4.7}$$

Acquisition Estimators

Code acquisition is the operation by which the timing of the receiver-generated spreading sequence is brought to within a fraction of a chip of the timing of the received sequence. The fundamental challenge is that the usual despreading mechanism cannot be used to reduce the bandwidth of the desired signal and suppress interference during acquisition or coarse synchronization.

We assume that $d\,(t)$ is not transmitted during acquisition, which is equivalent to setting $d\,(t) = 1$. Assuming that ϕ is uniformly distributed over $[0, 2\pi)$, a trigonometric expansion and an integration of (4.7) over ϕ using (H.16) of Appendix H.3 give

$$\Lambda_s[r(t)] = I_0\left(\sqrt{2SR\,(\tau,f_d)}\right) \tag{4.8}$$

where $I_0(\cdot)$ is the modified Bessel function of the first kind and order zero defined by (H.14), and

$$R\,(\tau,f_d) = \left[\int_0^T r(t)p(t-\tau)\cos\left(2\pi f_c t + 2\pi f_d t\right)dt\right]^2 + \left[\int_0^T r(t)p(t-\tau)\sin\left(2\pi f_c t + 2\pi f_d t\right)dt\right]^2. \tag{4.9}$$

Since $I_0(x)$ is a monotonically increasing function of x, (4.8) implies that $R(\tau,f_d)$ is a sufficient statistic for maximum-likelihood estimation. Ideally, the estimates $\widehat{\tau}$ and $\widehat{f_d}$ are determined by considering all possible values of τ and f_d, and then choosing those values that maximize (4.9):

$$\left(\widehat{\tau},\widehat{f_d}\right) = \arg\max_{(\tau,f_d)} R\,(\tau,f_d). \tag{4.10}$$

One method of implementing this joint estimation is to use a *parallel matrix* of processors, each matched to candidate pairs of quantized values of the timing and frequency offsets. The largest processor output then indicates which candidate pair is selected as the joint estimate. An alternative method of joint estimation, which is much less complex, but significantly increases the time needed to make a decision, is to serially search over the candidate pairs. The most practical approach is to separately estimate the timing and frequency offsets. Since the frequency offset may be separately estimated and tracked by standard carrier-recovery devices, the remainder of this chapter analyzes code synchronization in which only the timing offset τ is estimated.

Assuming that the instantaneous carrier frequency $f_{c1} = f_c + f_d$ is known, the sufficient statistic for the estimation of the received code phase or timing offset τ is

$$R_o(\tau) = \left[\int_0^T r(t)p(t-\tau)\cos(2\pi f_{c1}t)\,dt\right]^2 + \left[\int_0^T r(t)p(t-\tau)\sin(2\pi f_{c1}t)\,dt\right]^2 \tag{4.11}$$

and the maximum-likelihood estimate is

$$\widehat{\tau} = \arg\max_{\tau} R_o(\tau) \tag{4.12}$$

where τ is restricted to some set of possible values or uncertainty region.

The estimator defined by (4.12) essentially selects the peak value of the estimated magnitude of the autocorrelation or partial autocorrelation function of the spreading sequence. To show this, we assume that the noise is negligible so that

$$r(t) \approx s(t) = \sqrt{2S}p(t)\cos(2\pi f_{c1}t + \theta) \tag{4.13}$$

where the correct timing offset of the received signal is $\tau = 0$. Substituting this equation and (4.11) into (4.12), applying trigonometry, discarding negligible integrals, and then using the fact that $|x|^2$ is a monotonically increasing function, we obtain

$$\widehat{\tau} \approx \sqrt{\frac{S}{2}} \arg\max_{\tau} \left|\int_0^T p(t)p(t-\tau)dt\right|. \tag{4.14}$$

The integral is equal to the autocorrelation of the spreading sequence if T is equal to the period of a short spreading sequence. The integral is equal to the partial autocorrelation of the spreading sequence if T is less than the period of a long spreading sequence. Thus, if code acquisition is implemented by approximating (4.12), then *reliable acquisition depends on a low autocorrelation or partial autocorrelation function at all nonzero shifts of the spreading sequence.*

4.2 Rapid Acquisition

When the spreading sequences are relatively short, rapid acquisition is possible. For long spreading sequences, serial-search or sequential acquisition is necessary. Since the presence of the data modulation impedes acquisition, it is facilitated by the transmission of the spreading sequence without any data modulation until acquisition occurs.

Matched-Filter Acquisition

Matched-filter acquisition provides potentially rapid acquisition based on implementing (4.12) when short programmable sequences give adequate security. Successive periods of the spreading waveform are transmitted without data modulation during the time allocated to acquisition, and the matched filter is matched to one period of the spreading waveform. In an analog implementation, the envelope of the matched-filter output, which ideally comprises sharp autocorrelation spikes, is compared with one or more thresholds, one of which is close to the peak value of a spike.

The sequence length or integration time of the matched filter is limited by errors in compensating for the frequency offsets and chip-rate errors, both of which decrease the peak value or level of the spike. Since the data-symbol boundaries coincide with the beginning and end of a spreading sequence, the occurrence of a threshold crossing provides information used for both symbol or timing synchronization and code acquisition. A major application of matched-filter acquisition is for *burst communications*, which are short and infrequent communications that do not require a long spreading sequence.

A *digital matched filter* that generates $R_o(\tau)$ for the noncoherent acquisition of a binary spreading waveform is illustrated in Figure 4.1. The digital matched filter offers great flexibility, but is limited in the bandwidth it can accommodate. The received spreading waveform is decomposed into in-phase and quadrature baseband components, each of which is applied to a separate branch. In each branch, the output of a chip-matched filter is sampled at a rate $m \geq 1$ times the chip rate so that one of the samples is nearly aligned with the maximum filter output signal. The digitizers produce quantized amplitude estimates at the sampling rate. For example, a *one-bit digitizer* makes hard decisions on the received chips by observing the polarities of the sample values. Successive digitizer outputs are applied to a multidimensional shift register, which comprises several shift registers in parallel. Shifts occur at the sampling rate. The multiple-bit output of each stage of the multidimensional shift register is multiplied by stored reference weights equal to the chip values of the spreading sequence, and m consecutive reference weights

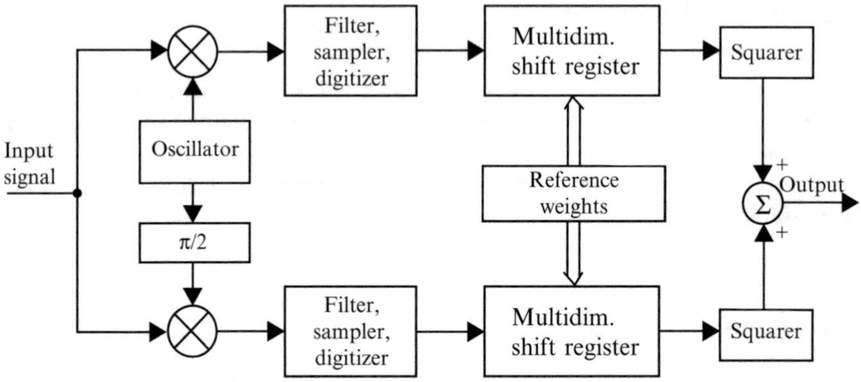

Fig. 4.1 Digital matched filter

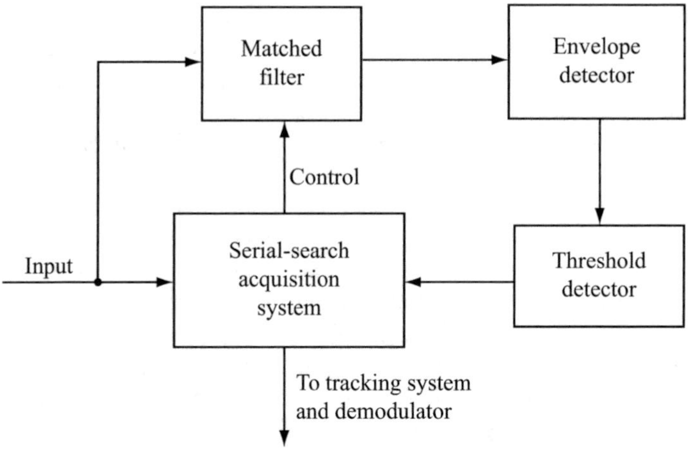

Fig. 4.2 Configuration of a serial-search acquisition system enabled by a matched filter

corresponding to each spreading-sequence chip. The sum of the products calculated by each shift register is squared, and the two squares are added to produce the final matched-filter output.

Matched-filter acquisition for continuous communications is useful when serial-search acquisition with a long sequence fails or takes too long. A short sequence for acquisition is embedded within the long sequence. The short sequence may be a subsequence of the long sequence and is stored in the programmable matched filter. Figure 4.2 depicts the configuration of a matched filter for short-sequence acquisition and a serial-search system for long-sequence acquisition. The control signal provides the short sequence that is stored or recirculated in the matched filter. The control signal activates the matched filter when it is needed and deactivates it otherwise. The short sequence is detected when the envelope of the matched-filter output crosses a threshold. The threshold-detector output starts a long-sequence generator in the serial-search system at a predetermined initial state. The long sequence is used to verify the acquisition and to despread the received direct-sequence signal. Several matched filters in parallel may be used to expedite the process.

Sequential Estimator

In a benign environment, *sequential estimation* methods [30] provide rapid acquisition when a short spreading sequence is adequate for security. The basic idea is to exploit the structure of the linear feedback shift register, which is illustrated in Figure 2.7. Successive received chips are detected and then loaded into the shift register of the receiver's code generator to establish its initial state, which then uniquely and unambiguously determines the subsequently generated spreading sequence in synchronism with the received sequence. The tracking system then ensures that the receiver's code generator maintains synchronization.

A *recursive soft sequential estimator*, which provides some protection against interference, is illustrated in Figure 4.3. Each sampled output of the chip-matched filter (Section 2.4) is applied to a recursive soft-in, soft-out (SISO) decoder similar to those used in turbo decoders (Section 1.6). The SISO decoder simultaneously receives extrinsic information from previously sampled outputs associated with earlier chips. The extrinsic information is related to the current chip, but was dispersed by the spreading-sequence generator in the transmitter. The SISO decoder processes all soft information to calculate a reliable soft output, which is a log-likelihood ratio (LLR). The LLR associated with each chip is sent to the first stage

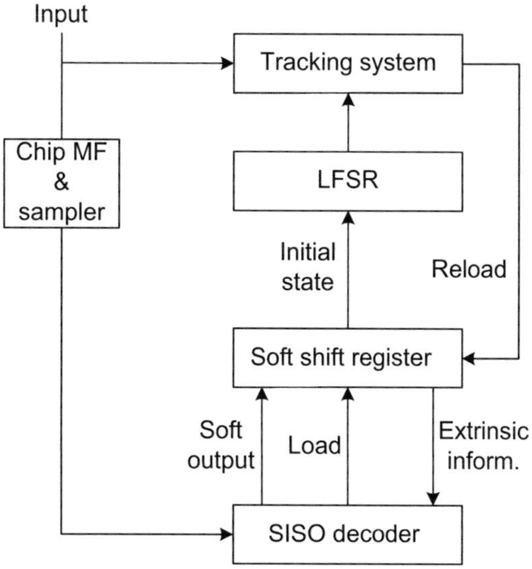

Fig. 4.3 Recursive soft sequential estimator. *LFSR* linear feedback shift register, *SISO* soft in, soft out, *MF* matched filter

of the soft shift register, which stores and shifts successive LLRs at the chip rate and provides the extrinsic information. The soft shift register stores the most recent consecutive LLRs and shifts out the oldest LLR from its final stage. If the stored LLRs have high enough values, then a load command activates hard decisions that convert the consecutive LLRs into consecutive chip values. After they are loaded into the appropriate successive stages, the chip values determine the initial state of the linear feedback shift register (LFSR). The LFSR generates the receiver's synchronized spreading sequence, which is used for despreading and is applied the tracking system. If the tracking system detects a synchronization problem, it sends a reload signal to the soft shift register, which begins a new estimation of an initial state of the LFSR.

4.3 Serial-Search Acquisition

The timing uncertainty of a local spreading sequence relative to a received spreading sequence covers a region that may be quantized into a finite number of smaller regions called *cells*. *Serial-search acquisition* is acquisition based on consecutive or serial tests of the cells until it is determined that a particular cell corresponds to the alignment of the two sequences to within a fraction of a chip. To access different cells, the receiver changes its clock rate. These changes adjust the timing offset of the *local sequence* generated by the receiver relative to the timing offset of the received sequence.

Figure 4.4 depicts the principal components of a serial-search acquisition system. The received direct-sequence signal and a local spreading sequence are applied to a noncoherent acquisition correlator that produces the statistic (4.11) associated with each cell. If the received and local spreading sequences are not aligned, the sampled correlator output is low. Therefore, the threshold is not exceeded, and the

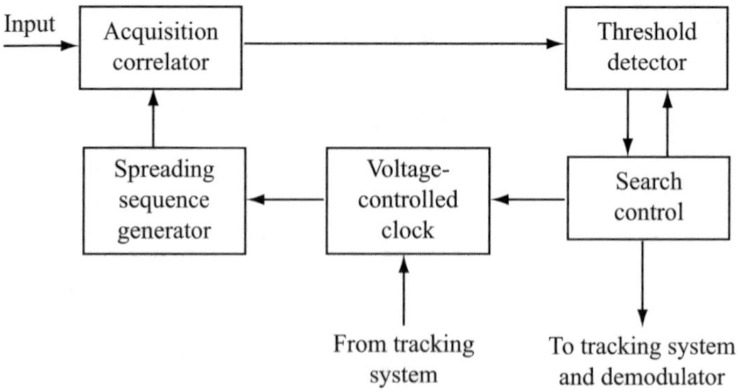

Fig. 4.4 Serial-search acquisition system

test fails. If there are one or more test failures, the cell being tested is rejected, and the phase of the local sequence is retarded or advanced, possibly by generating an extra clock pulse or by blocking one. A new cell is then tested. If the sequences are nearly aligned, the sampled correlator output is high, the threshold is exceeded, the search is stopped, and the two sequences run in parallel at some fixed phase offset. Subsequent tests verify that the correct cell has been identified. If a cell fails the verification tests, the search is resumed. If a cell passes, the two sequences are assumed to be coarsely synchronized, demodulation begins, and the tracking system is activated. The threshold-detector output continues to be monitored so that any subsequent loss of synchronization activates the serial search.

There may be several cells that potentially provide a valid acquisition. However, if none of these cells corresponds to perfect synchronization, the detected energy is reduced below its potential peak value. The *step size* is the separation between cells. If the step size is one-half of a chip, then one of the cells corresponds to an alignment within one-fourth of a chip. On average, the misalignment of this cell is one-eighth of a chip, which may cause a negligible degradation. As the step size decreases, both the average detected energy during acquisition and the number of cells to be searched increase.

The *dwell time* is the amount of time required for testing a cell and is approximately equal to the length of the integration interval in the noncoherent correlator (Section 4.4). An acquisition system is called a *single-dwell system* if a single test determines whether a cell is accepted as the correct one. If verification testing occurs before acceptance, the system is called a *multiple-dwell system*. The dwell times are either fixed or variable but bounded by some maximum value. The dwell time for the initial test of a cell is usually designed to be much shorter than the dwell times for verification tests. This approach expedites the acquisition by quickly eliminating the bulk of incorrect cells. In any serial-search system, the dwell time allotted to a test is limited by the Doppler shift, which causes the received and local chip rates to differ. As a result, an initially close alignment of the two sequences may disappear by the end of the test.

A multiple-dwell system may use a *consecutive-count strategy*, in which a failed test causes a cell to be immediately rejected, or an *up-down strategy*, in which a failed test causes a repetition of a previous test. Figures 4.5 and 4.6 depict the flow graphs of the consecutive-count and up-down strategies, respectively, that require D tests to be passed before acquisition is declared. If the threshold is not exceeded during test 1, the cell fails the test, and the next cell is tested. If it is exceeded, the cell passes the test, the search is stopped, and the system enters the *verification mode*. The same cell is tested again, but the dwell time and the threshold may be changed. Once all the verification tests have been passed, code tracking is activated, and the system enters the *lock mode*. In the lock mode, the lock detector continually verifies that code synchronization is maintained. If the lock detector decides that synchronization has been lost, *reacquisition* begins in the search mode.

The order in which the cells are tested is determined by the general search strategy. Figure 4.7 (a) depicts a *uniform search* over the q cells of the *timing uncertainty*. The broken lines represent the discontinuous transitions of the search

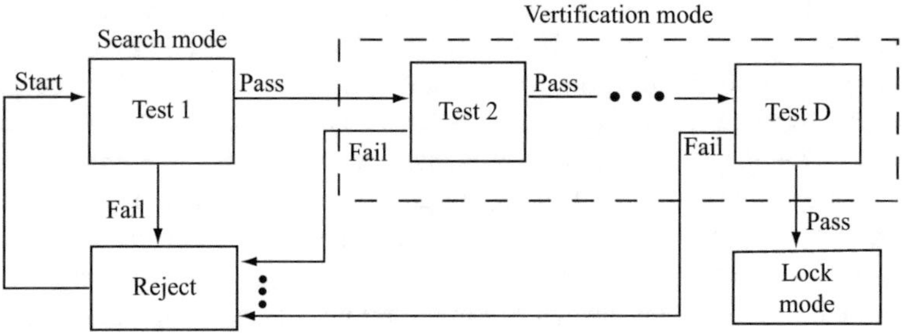

Fig. 4.5 Flow graph of a multiple-dwell system with the consecutive-count strategy

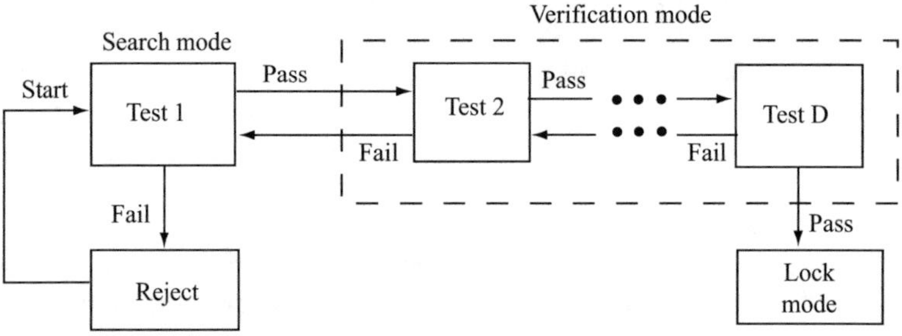

Fig. 4.6 Flow graph of multiple-dwell system with the up-down strategy

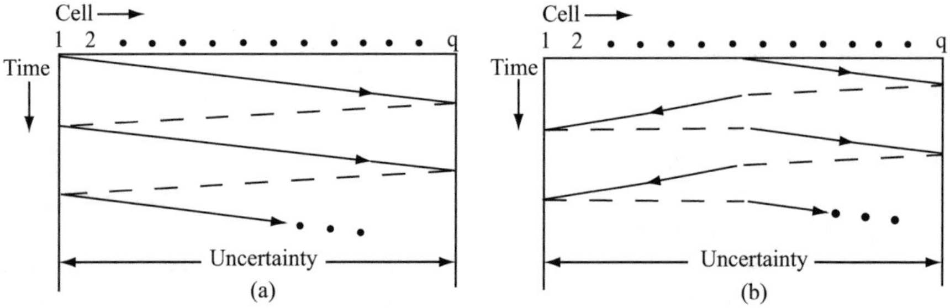

Fig. 4.7 Trajectories of search positions: (**a**) uniform search and (**b**) broken-center Z search

from the one part of the timing uncertainty to another. The *broken-center Z search*, illustrated in Figure 4.7 (b), is appropriate when a priori information makes part of the timing uncertainty more likely to contain the correct cell than the rest of the region. A priori information may be derived from the detection of a short preamble. If the sequences are synchronized with the time of day, then the receiver's estimate of the transmitter range combined with the time of day provide the a priori information.

The *acquisition time* is the amount of time required for an acquisition system to locate the correct cell and initiate the code-tracking system. To derive the statistics of the acquisition time [53], one of the q possible cells is considered the correct cell, and the other $(q-1)$ cells are incorrect. The difference in timing offsets among cells is ΔT_c, where the *step size* Δ is usually either 1 or 1/2. Let L denote the number of times the correct cell is tested before it is accepted and acquisition terminates. Let C denote the number of the correct cell and π_j denote the probability that $C = j$. Let $\nu(L, C)$ denote the number of incorrect cells tested during the acquisition process. The functional dependence is determined by the search strategy. Let $T_r(L, C)$ denote the total *rewinding time,* which is the time required for the search to move discontinuously to a different cell within the timing uncertainty. Since an incorrect cell is always ultimately rejected, there are only three types of events that occur during a serial search: the nth incorrect cell is dismissed after $T_{11}(n)$ seconds; a correct cell is falsely dismissed for the mth time after $T_{12}(m)$ seconds; or a correct cell is accepted after T_{22} seconds (where the first subscript is 1 if dismissal occurs, and 2 otherwise; the second subscript is 1 if the cell is incorrect, and 2 otherwise). Each of these decision times is a random variable. If an incorrect cell is accepted, the receiver eventually recognizes the mistake and reinitiates the search at the next cell. The wasted time expended in code tracking is a random variable called the *penalty time*. These definitions imply that the acquisition time is the random variable given by

$$T_a = \sum_{n=1}^{\nu(L,C)} T_{11}(n) + \sum_{m=1}^{L-1} T_{12}(m) + T_{22} + T_r(L, C). \tag{4.15}$$

The most important performance measures of the serial search are the mean and variance of T_a. Given that $L = i$ and $C = j$, the conditional expected value of T_a is

$$E\left[T_a|i, j\right] = \nu(i, j)\bar{T}_{11} + (i-1)\bar{T}_{12} + \bar{T}_{22} + T_r(i, j) \tag{4.16}$$

where $\bar{T}_{11}, \bar{T}_{12}$, and $\bar{T}_{22}$ are the expected values of each $T_{11}(n), T_{12}(m)$, and T_{22}, respectively. Therefore, the mean acquisition time is

$$\bar{T}_a = \bar{T}_{22} + \sum_{i=1}^{\infty} P_L(i) \sum_{j=1}^{q} \pi_j \left[\nu(i, j)\bar{T}_{11} + (i-1)\bar{T}_{12} + T_r(i, j)\right] \tag{4.17}$$

where $P_L(i)$ is the probability that $L = i$, and π_j is the probability that $C = j$. After calculating the conditional expected value of T_a^2 given that $L = i$ and $C = j$, and using the identity $\overline{x^2} = var(x) + \bar{x}^2$, we obtain

$$\begin{aligned}\overline{T_a^2} = &\sum_{i=1}^{\infty} P_L(i) \sum_{j=1}^{q} \pi_j\{[\nu(i, j)\bar{T}_{11} + (i-1)\bar{T}_{12} + \bar{T}_{22} + T_r(i, j)]^2 \\ &+ \nu(i, j)var(T_{11}) + (i-1)var(T_{12}) + var(T_{22})\}.\end{aligned} \tag{4.18}$$

The variance of T_a is

$$\sigma_a^2 = \overline{T_a^2} - \bar{T}_a^2 . \tag{4.19}$$

Assuming that the test statistics are independent and identically distributed, we obtain

$$P_L(i) = P_D\,(1 - P_D)^{i-1} \tag{4.20}$$

where P_D is the probability that the correct cell is detected when it is tested during a scan of the uncertainty region.

Application of Chebyshev's Inequality

To derive Chebyshev's inequality, consider a random variable X with distribution $F(x)$. Let $E[X] = m$ denote the expected value of X, and $P[A]$ denote the probability of event A. From elementary probability, it follows that

$$\begin{aligned} E[|X - m|^k] &= \int_{-\infty}^{\infty} |x - m|^k dF(x) \geq \int_{|x-m|\geq\alpha}^{\infty} |x - m|^k dF(x) \\ &\geq \alpha^k \int_{|x-m|\geq\alpha}^{\infty} dF(x) = \alpha^k P[|X - m| \geq \alpha]. \end{aligned} \tag{4.21}$$

Therefore,

$$P[|X - m| \geq \alpha] \leq \frac{1}{\alpha^k} E[|X - m|^k]. \tag{4.22}$$

Let $\sigma^2 = E[(X - m)^2]$ denote the variance of X. If $k = 2$, then (4.22) becomes *Chebyshev's inequality*:

$$P[|X - m| \geq \alpha] \leq \frac{\sigma^2}{\alpha^2}\,. \tag{4.23}$$

In some applications, the serial-search acquisition must be completed within a specified period of duration T_{max}. If not, the serial search is terminated, and special measures, such as the matched-filter acquisition of a short sequence, are undertaken. The probability that $T_a \leq T_{max}$ can be bounded by using Chebyshev's inequality:

$$\begin{aligned} P[T_a \leq T_{\max}] &\geq P\left[\left|T_a - \bar{T}_a\right| \leq T_{\max} - \bar{T}_a\right] \\ &\geq 1 - \frac{\sigma_a^2}{\left(T_{\max} - \bar{T}_a\right)^2}\,. \end{aligned} \tag{4.24}$$

Uniform Search with Uniform Distribution

As an important application, we consider the uniform search of Figure 4.7 (a) and a uniform a priori distribution for the location of the correct cell given by

$$\pi_j = \frac{1}{q}, \quad 1 \leq j \leq q. \tag{4.25}$$

If the cells in the figure are labeled consecutively from left to right, then

$$\nu(i,j) = (i-1)(q-1) + j - 1. \tag{4.26}$$

The rewinding time is

$$T_r(i,j) = T_r(i) = (i-1)T_r \tag{4.27}$$

where T_r is the rewinding time associated with each broken line in the figure. If the timing uncertainty covers an entire sequence period, then the cells at the two edges are actually adjacent and $T_r = 0$.

To evaluate $\bar{T}_a$ and $\overline{T_a^2}$, we substitute (4.20), (4.25), (4.26), and (4.27) into (4.17) and (4.18) and use the following identities:

$$\sum_{i=0}^{\infty} r^i = \frac{1}{1-r}, \quad \sum_{i=1}^{\infty} i r^i = \frac{r}{(1-r)^2}, \quad \sum_{i=1}^{\infty} i^2 r^i = \frac{r(1+r)}{(1-r)^3}$$

$$\sum_{i=1}^{n} i = \frac{n(n+1)}{2}, \quad \sum_{i=1}^{n} i^2 = \frac{n(n+1)(2n+1)}{6} \tag{4.28}$$

where $0 \leq |r| < 1$. Defining

$$\alpha = (q-1)\bar{T}_{11} + \bar{T}_{12} + T_r \tag{4.29}$$

we obtain

$$\bar{T}_a = (q-1)\left(\frac{2-P_D}{2P_D}\right)\bar{T}_{11} + \left(\frac{1-P_D}{P_D}\right)\left(\bar{T}_{12} + T_r\right) + \bar{T}_{22} \tag{4.30}$$

and

$$\begin{aligned}\overline{T_a^2} &= (q-1)\left(\frac{2-P_D}{2P_D}\right) var\,(T_{11}) + \left(\frac{1-P_D}{P_D}\right) var\,(T_{12}) + var\,(T_{22}) \\ &+ \frac{(2q+1)(q+1)}{6}\bar{T}_{11}^2 + \frac{\alpha^2\,(1-P_D)\,(2-P_D)}{P_D^2} + (q+1)\alpha\left(\frac{1-P_D}{P_D}\right) \\ &+ (q+1)\bar{T}_{11}\left(\bar{T}_{22} - \bar{T}_{11}\right) + 2\alpha\left(\frac{1-P_D}{P_D}\right)\left(\bar{T}_{22} - \bar{T}_{11}\right) + \left(\bar{T}_{22} - \bar{T}_{11}\right)^2.\end{aligned} \tag{4.31}$$

In most applications, the number of cells to be searched is large, and simpler *asymptotic forms* for the mean and variance of the acquisition time are applicable. As $q \to \infty$, (4.30) gives

$$\bar{T}_a \to q\left(\frac{2-P_D}{2P_D}\right)\bar{T}_{11}, \quad q \to \infty. \tag{4.32}$$

Similarly, (4.31) and (4.19) yield

$$\sigma_a^2 \to q^2\left(\frac{1}{P_D^2} - \frac{1}{P_D} + \frac{1}{12}\right)\bar{T}_{11}^2, \quad q \to \infty. \tag{4.33}$$

These equations must be modified in the presence of a large uncorrected Doppler shift. The fractional change in the received chip rate of the spreading sequence is equal to the fractional change in the carrier frequency due to the Doppler shift. If the chip rate changes from $1/T_c$ to $1/T_c + \delta$, then the average change in the code or sequence phase during the test of an incorrect cell is $\delta\bar{T}_{11}$. The change relative to the step size is $\delta\bar{T}_{11}/\Delta$. The number of cells that are actually tested in a sweep of the timing uncertainty becomes $q(1+\delta\bar{T}_{11}/\Delta)^{-1}$. Since incorrect cells predominate, the substitution of the latter quantity in place of q in (4.32) and (4.33) gives approximate asymptotic expressions for $\bar{T}_a$ and σ_a^2 when the Doppler shift is significant.

Consecutive-Count Double-Dwell System

For further specialization, consider the consecutive-count double-dwell system described by Figure 4.5 with $D = 2$. Assume that a *composite correct cell* actually subsumes two consecutive cells that have sufficiently low timing offsets that either one of these consecutive cells could be considered a correct cell. The detection probabilities of the two consecutive cells are P_a and P_b. If the test results are assumed to be statistically independent, then the probability of detection of the composite correct cell is

$$P_D = P_a + (1 - P_a)P_b. \tag{4.34}$$

Let τ_1, P_{F1}, P_{a1}, and P_{b1} denote the search-mode dwell time, false-alarm probability, and successive detection probabilities, respectively. Let τ_2, P_{F2}, P_{a2}, and P_{b2} denote the verification-mode dwell time, false-alarm probability, and successive detection probabilities, respectively. Let $\bar{T}_p$ denote the mean penalty time, which is incurred by the incorrect activation of the tracking mode. The flow graph indicates that each cell must pass two tests, and the passing probabilities are

$$P_a = P_{a1}P_{a2}, \quad P_b = P_{b1}P_{b2}. \tag{4.35}$$

If an incorrect cell is rejected, a delay of at least τ_1 occurs. If there has been a false alarm, then an additional average delay of $\tau_2 + P_{F2}\bar{T}_p$ occurs. Therefore,

$$\bar{T}_{11} = \tau_1 + P_{F1}\left(\tau_2 + P_{F2}\bar{T}_p\right). \tag{4.36}$$

Equations (4.34) to (4.36) are sufficient for the evaluation of the asymptotic values of the mean and variance given by (4.32) and (4.33).

For a more accurate evaluation of the mean acquisition time, expressions for the conditional means $\bar{T}_{22}$ and $\bar{T}_{12}$ are needed. Expressing $\bar{T}_{22}$ as the conditional expectation of the correct-cell test duration given cell detection, and enumerating the three possible durations and their conditional probabilities, we obtain

$$\begin{aligned}
\bar{T}_{22} &= \sum_{i=1}^{\infty} t_i P\left(T_{22} = t_i \,|\text{detection}\right) \\
&= \tau_1 + \tau_2 + \frac{1}{P_D}\sum_{i=1}^{\infty} t_i P\left([T_{22} = t_i + \tau_1 + \tau_2] \cap \text{detection}\right) \\
&= \tau_1 + \tau_2 + \frac{1}{P_D}\left[\tau_1\left(1 - P_{a1}\right)P_b + \left(\tau_1 + \tau_2\right)P_{a1}\left(1 - P_{a2}\right)P_b\right] \\
&= \tau_1 + \tau_2 + \tau_1\frac{\left(1 - P_a\right)P_b}{P_D} + \tau_2\frac{\left(P_{a1} - P_a\right)P_b}{P_D}.
\end{aligned} \tag{4.37}$$

Similarly,

$$\begin{aligned}
\bar{T}_{12} &= 2\tau_1 + \frac{1}{1 - P_D}\sum_{i=1}^{\infty} t_i P\left([T_{12} = t_i + 2\tau_1] \cap \text{rejection}\right) \\
&= 2\tau_1 + \tau_2\left[\frac{\left(P_{a1} - P_a\right)\left(1 - P_b\right) + \left(1 - P_a\right)\left(P_{b1} - P_b\right)}{1 - P_D}\right].
\end{aligned} \tag{4.38}$$

Single-Dwell and Matched-Filter Systems

Results for a single-dwell system are obtained by setting $P_{a2} = P_{b2} = P_{F2} = 1, \tau_2 = 0, P_a = P_{a1}, P_b = P_{b1}, P_{F1} = P_F$, and $\tau_1 = \tau_d$ in (4.36) to (4.38). We obtain

$$\bar{T}_{11} = \tau_d + P_F\bar{T}_p,\;\; \bar{T}_{22} = \tau_d\left[1 + \frac{\left(1 - P_a\right)P_b}{P_D}\right],\;\; \bar{T}_{12} = 2\tau_d. \tag{4.39}$$

Thus, (4.30) yields

$$\bar{T}_a = \frac{(q-1)\left(2 - P_D\right)\left(\tau_d + P_F\bar{T}_p\right) + 2\tau_d\left(2 - P_a\right) + 2\left(1 - P_D\right)T_r}{2P_D}. \tag{4.40}$$

Since the single-dwell system may be regarded as a special case of the double-dwell system, the latter can provide better performance by the appropriate setting of its additional parameters.

The approximate mean acquisition time for a matched filter can be derived in a similar manner. Suppose that many periods of a short spreading sequence with N chips per period are received, and the matched-filter output is sampled m times per chip. The number of cells that are tested is $q = mN$ and $T_r = 0$. Each sampled output is compared with a threshold; thus, $\tau_d = T_c/m$ is the time duration associated with a test. For $m = 1$ or 2, it is reasonable to regard two of the cells as the correct ones. These cells are effectively tested when a signal period fills or nearly fills the matched filter. Therefore, (4.34) is applicable with $P_a \approx P_b$, and (4.30) yields

$$\bar{T}_a \approx NT_c \left(\frac{2 - P_D}{2P_D} \right) (1 + mKP_F), \qquad q >> 1 \tag{4.41}$$

where $K = \bar{T}_p/T_c$. Ideally, the threshold is exceeded once per period, and each threshold crossing provides a timing marker.

Up-Down Double-Dwell System

For the up-down double-dwell system with a composite correct cell of two subsumed correct cells, the flow graph of Figure 4.6 with $D = 2$ indicates that

$$P_a = P_{a1}P_{a2} \sum_{i=0}^{\infty} [P_{a1}(1 - P_{a2})]^i = \frac{P_{a1}P_{a2}}{1 - P_{a1}(1 - P_{a2})}. \tag{4.42}$$

Similarly,

$$P_b = \frac{P_{b1}P_{b2}}{1 - P_{b1}(1 - P_{b2})} \tag{4.43}$$

and P_D is given by (4.34). If an incorrect cell passes the initial test but fails the verification test, then the cell begins the testing sequence again without any memory of the previous testing. Therefore, for an up-down double-dwell system, a recursive evaluation gives

$$\bar{T}_{11} = (1 - P_{F1})\tau_1 + P_{F1}P_{F2}\left(\tau_1 + \tau_2 + \bar{T}_p\right) + P_{F1}(1 - P_{F2})\left(\tau_1 + \tau_2 + \bar{T}_{11}\right). \tag{4.44}$$

Solving this equation yields

$$\bar{T}_{11} = \frac{\tau_1 + P_{F1}\left(\tau_2 + P_{F2}\bar{T}_p\right)}{1 - P_{F1}(1 - P_{F2})}. \tag{4.45}$$

Substitution of (4.42) to (4.45) into (4.32) to (4.34) gives the asymptotic values of the mean and variance of the acquisition time.

For a more accurate evaluation of the mean acquisition time, expressions for the conditional means $\bar{T}_{22}$ and $\bar{T}_{12}$ are needed. We begin the derivation of $\bar{T}_{22} = E\left[T_{22} \mid \text{detection}\right]$ by adding further conditioning. Let T_0 denote the additional delay when the testing of the composite correct cell restarts at the first subsumed cell, and the composite cell is ultimately detected. Let T_1 denote the additional delay when the testing of the composite correct cell restarts at the second subsumed cell, and the composite cell is ultimately detected. Then

$$\begin{aligned}\bar{T}_{22} &= E\{E\left[T_{22} \mid \text{detection}, T_0, T_1\right]\} \\ &= \tau_1 + \tau_2 + E\left\{\frac{1}{P_D}\sum_{i=1}^{\infty} t_i P\left(\left[T_{22} = t_i + \tau_1 + \tau_2\right] \cap \text{detection} \mid T_0, T_1\right)\right\} . \end{aligned} \tag{4.46}$$

From the possible durations and their conditional probabilities, we obtain

$$\bar{T}_{22} = \tau_1 + \tau_2 + \frac{1}{P_D}\left\{\begin{array}{c} P_{a1}\left(1 - P_{a2}\right) P_D E\left[T_0\right] + \left(1 - P_{a1}\right) P_{b1} P_{b2} \tau_1 \\ +\left(1 - P_{a1}\right) P_{b1}\left(1 - P_{b2}\right) P_b\left(\tau_1 + E\left[T_1\right]\right) \end{array}\right\} . \tag{4.47}$$

A recursive evaluation gives

$$\begin{aligned} E\left[T_1\right] &= \tau_1 + \tau_2 + P_{b1}\left(1 - P_{b2}\right) E\left[T_1\right] \\ &= \frac{\tau_1 + \tau_2}{1 - P_{b1} + P_b} . \end{aligned} \tag{4.48}$$

Substituting this equation and $E\left[T_0\right] = \bar{T}_{22}$ into (4.47) and solving for $\bar{T}_{22}$ yields

$$\bar{T}_{22} = \frac{1}{1 - P_{a1} + P_a}\left\{\begin{array}{c} \tau_1\left[1 + \frac{(1-P_{a1})P_b}{P_D} + \frac{(1-P_{a1})(P_{b1}-P_b)P_b}{P_D}\left(1 + \frac{1}{1-P_{b1}+P_b}\right)\right] \\ +\tau_2\left[1 + \frac{(1-P_{a1})(P_{b1}-P_b)P_b}{P_D(1-P_{b1}+P_b)}\right] \end{array}\right\} . \tag{4.49}$$

Similarly, $\bar{T}_{12}$ is determined by the recursive equation

$$\begin{aligned}\bar{T}_{12} = \tau_1 &+ \frac{\left(1 - P_{a1}\right)\left(1 - P_{a2}\right)}{1 - P_D}\tau_1 + P_{a1}\left(1 - P_{a2}\right)\left(\tau_2 + \bar{T}_{12}\right) \\ &+ \frac{\left(1 - P_{a1}\right) P_{b1}\left(1 - P_{b2}\right)\left(1 - P_b\right)}{1 - P_D}\left(\tau_1 + \tau_2 + \bar{T}'_{12}\right) \end{aligned} \tag{4.50}$$

where $\bar{T}'_{12}$ denotes the mean additional delay when the testing of the composite correct cell restarts at the second subsumed cell and the composite cell is ultimately rejected. A recursive evaluation gives

$$\begin{aligned}\bar{T}'_{12} &= \tau_1 + P_{b1}\left(1 - P_{b2}\right)\left(\tau_2 + \bar{T}'_{12}\right) \\ &= \frac{\tau_1 + \left(P_{b1} - P_b\right)\tau_2}{1 - P_{b1} + P_b} . \end{aligned} \tag{4.51}$$

Substituting this equation into (4.50) and solving $\bar{T}_{12}$ yields a

$$\bar{T}_{12} = \frac{1}{1-P_{a1}+P_a}\left\{\begin{array}{c}\tau_1\left[\begin{array}{c}1+\frac{(1-P_{a1})(1-P_{a2})}{1-P_D}\\ +\frac{(1-P_{a1})(P_{b1}-P_b)(1-P_b)}{1-P_D}(1+\frac{1}{1-P_{b1}+P_b})\end{array}\right]\\ +\tau_2\left[P_{a1}-P_a+\frac{(1-P_{a1})(P_{b1}-P_b)(1-P_b)}{1-P_D}+\frac{P_{b1}-P_b}{1-P_{b1}+P_b}\right]\end{array}\right\}. \tag{4.52}$$

Penalty Time

The *lock detector* that monitors the code synchronization in the lock mode performs tests to verify the lock condition. The time that elapses before the system incorrectly leaves the lock mode is called the *holding time*. It is desirable to have a long mean holding time and a short mean penalty time, but the realization of one of these goals tends to impede that of the other. As a simple example, suppose that each test has a fixed duration τ and that code synchronization is maintained. A single missed detection, which occurs with probability $1-P_{DL}$, causes the lock detector to assume a loss of lock and to initiate a search. Assuming the statistical independence of the lock-mode tests, the *mean holding time* is

$$\begin{aligned}\bar{T}_h &= \sum_{i=1}^{\infty} i\tau\,(1-P_{DL})\,P_{DL}^{i-1}\\ &= \frac{\tau}{1-P_{DL}}, \quad P_{DL}<1.\end{aligned} \tag{4.53}$$

This result may also be derived by recognizing that $\bar{T}_h = \tau + P_{DL}\bar{T}_h$, $P_{DL}<1$, because once the lock mode is verified, the testing of the same cell is renewed without any memory of the previous testing.

If the locally generated code phase is incorrect, the penalty time expires unless false alarms, each of which occurs with probability P_{FL}, continue to occur every τ seconds. Therefore, $\bar{T}_p = \tau + P_{FL}\bar{T}_p$, $P_{FL}<1$, which yields the *mean penalty time* for a single-dwell lock detector:

$$\bar{T}_p = \frac{\tau}{1-P_{FL}}, \quad P_{FL}<1. \tag{4.54}$$

A tradeoff between a high $\bar{T}_h$ and a low $\bar{T}_p$ exists because increasing P_{DL} tends to increase P_{FL}.

When a single test verifies the lock condition, the synchronization system is vulnerable to deep fades and pulsed interference. A preferable strategy is for the lock mode to be maintained until a number of consecutive or cumulative misses occur during a series of tests. The performance analysis is analogous to that of serial-search acquisition.

Other Search Strategies

In a *Z search*, no cell is tested more than once until all cells in the timing uncertainty have been tested. Both strategies of Figure 4.7 are Z searches. A characteristic of the Z search is that

$$\nu(i,j) = (i-1)(q-1) + \nu(1,j) \tag{4.55}$$

where $\nu(1,j)$ is the number of incorrect cells tested when $P_D = 1$ and hence $L = 1$. For simplicity, we assume that q is even. For the broken-center Z search, the search begins with cell $q/2 + 1$, and

$$\nu(1,j) = \begin{cases} j - \frac{q}{2} - 1, & j \geq \frac{q}{2} + 1 \\ q - j, & j \leq \frac{q}{2} \end{cases} \tag{4.56}$$

whereas $\nu(1,j) = j - 1$ for the uniform search. If the rewinding time is negligible, then (4.17), (4.20), and (4.55) yield

$$\bar{T}_a = \frac{1-P_D}{P_D}\left[(q-1)\bar{T}_{11} + \bar{T}_{12}\right] + \bar{T}_{22} + \bar{T}_{11}\nu(1) \tag{4.57}$$

where

$$\nu(1) = \sum_{j=1}^{q} \nu(1,j)\pi_j \tag{4.58}$$

is the average number of incorrect cells tested when $P_D = 1$.

If the correct cell number C has a uniform distribution, then $\nu(1)$ and, hence, $\bar{T}_a$ are the same for both the uniform and the broken-center Z searches. If the distribution of C is symmetrical about a pronounced central peak and $P_D \approx 1$, then a uniform search gives $\nu(1) \approx q/2$. Since a broken-center Z search usually ends almost immediately or after slightly more than $q/2$ tests,

$$\nu(1) \approx 0\left(\frac{1}{2}\right) + \frac{q}{2}\left(\frac{1}{2}\right) = \frac{q}{4} \tag{4.59}$$

which indicates that for large values of q and P_D close to unity, the broken-center Z search reduces $\bar{T}_a$ by nearly a factor of 2 relative to its value for the uniform search.

If C has a unimodal distribution with a distinct peak, then this feature can be exploited by continually retesting cells with high a priori probabilities of being the correct cell. An *expanding-window search* tests all cells within a radius R_1 from the center. If the correct cell is not found, then tests are performed on all cells within an increased radius R_2. The radius is increased successively until the boundaries of the timing uncertainty are reached. The expanding-window search then becomes

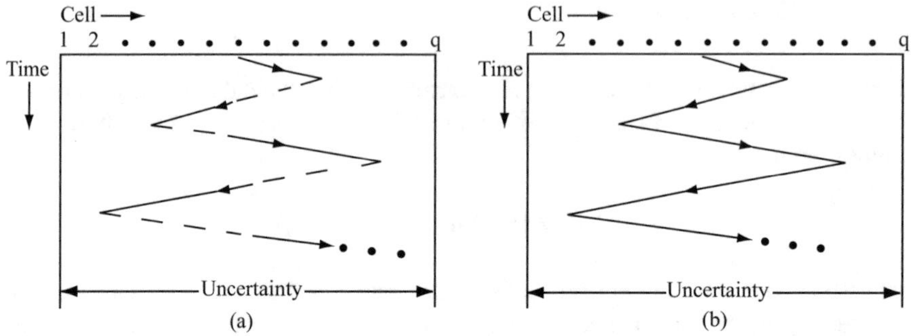

Fig. 4.8 Trajectories of expanding-window search positions: (**a**) broken-center and (**b**) continuous-center search

a Z search. If the rewinding time is negligible and C is centrally peaked, then the broken-center search of Figure 4.8 (a) is preferable to the continuous-center search of Figure 4.8 (b) because the latter retests cells before testing all the cells near the center of the timing uncertainty.

In an *equiexpanding search*, the radii have the form

$$R_n = \frac{nq}{2N}, \quad n = 1, 2, \ldots, N \tag{4.60}$$

where N is the number of sweeps before the search becomes a Z search. If the rewinding time is negligible, then it can be shown [38] that the broken-center equiexpanding-window search is optimized for $P_D \leq 0.8$ by choosing $N = 2$. For this optimized search, $\bar{T}_a$ is moderately reduced relative to its value for the broken-center Z search.

When $T_r(i,j) = 0$ and $P_D = 1$, the optimal search, which is called a *uniform alternating search*, tests the cells in order of decreasing a priori probability. For a symmetric, unimodal, centrally peaked distribution of C, this optimal search has the trajectory depicted in Figure 4.9 (a). Once all the cells in the timing uncertainty have been tested, the search repeats the same pattern. Equations (4.55) and (4.57) are applicable. If $P_D \approx 1$, $T_r(i,j) << \bar{T}_{11}$ and the distribution of C has a pronounced central peak, then $\nu(1)$ is small, and the uniform alternating search has a significant advantage over the broken-center, expanding-window search. However, computations show that this advantage dissipates as P_D decreases [53], which occurs because all cells are tested once a detection fails.

In the *nonuniform alternating search*, illustrated in Figure 4.9 (b), a uniform search is performed until a radius R_1 is reached. Then, a second uniform search is performed within a larger radius R_2. This process continues until the boundaries of the timing uncertainty are reached and the search becomes a uniform alternating search. Computations show that for a centrally peaked distribution of C, the nonuniform alternating search can result in a significant improvement over the uniform alternating search if $P_D < 0.8$, and the radii $R_n, n = 1, 2, \ldots$, are optimized [38]. However, if the radii are optimized for $P_D < 1$, then as $P_D \to 1$, the nonuniform search becomes inferior to the uniform search.

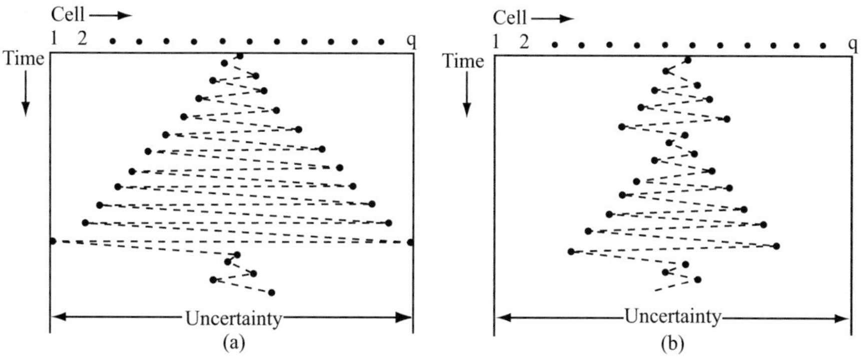

Fig. 4.9 Trajectories of alternating search positions: (**a**) uniform search and (**b**) nonuniform search

Density Function of the Acquisition Time

Let $S_n = X_1 + X_2 + \ldots + X_n$ denote the sum of n independent random variables, and let $g_i(t), 1 \leq i \leq n$, denote the density function of X_i. From elementary probability theory, it follows that the density function of S_n is equal to the n-fold convolution $g_i(t) = g_1(t) * g_2(t) * \ldots * g_{ni}(t)$, where the star $\star$ denotes the convolution operation (Appendix C.1).

The density function of T_a, which is needed to accurately calculate $P[T_a \leq T_{max}]$ and other probabilities, may be decomposed as

$$f_a(t) = P_D \sum_{i=1}^{\infty} (1 - P_D)^{i-1} \sum_{j=1}^{q} \pi_j f_a(t|i,j) \tag{4.61}$$

where $f_a(t|i,j)$ is the conditional density function of T_a given that $L = i$ and $C = j$. Let $[f(t)]^{*n}$ denote the n-fold convolution of the density function $f(t)$ with itself, $[f(t)]^{*0} = 1$, and $[f(t)]^{*1} = f(t)$. Using this notation, we obtain

$$f_a(t|i,j) = [f_{11}(t)]^{*\nu(i,j)} * [f_{12}(t)]^{*(i-1)} * [f_{22}(t)] \tag{4.62}$$

where $f_{11}(f), f_{12}(t)$, and $f_{22}(t)$ are the density functions associated with T_{11}, T_{12}, and T_{22}, respectively. If one of the decision times is a constant, then the associated density function is a delta function.

The exact evaluation of $f_a(t)$ is complicated [61], but an approximation usually suffices. Since the acquisition time conditioned on $L = i$ and $C = j$ is the sum of independent random variables, it is reasonable to approximate $f_a(t|i,j)$ by a truncated Gaussian density function with mean

$$\mu_{ij} = \nu(i,j)\bar{T}_{11} + (i-1)\bar{T}_{12} + \bar{T}_{22} + T_r(i) \tag{4.63}$$

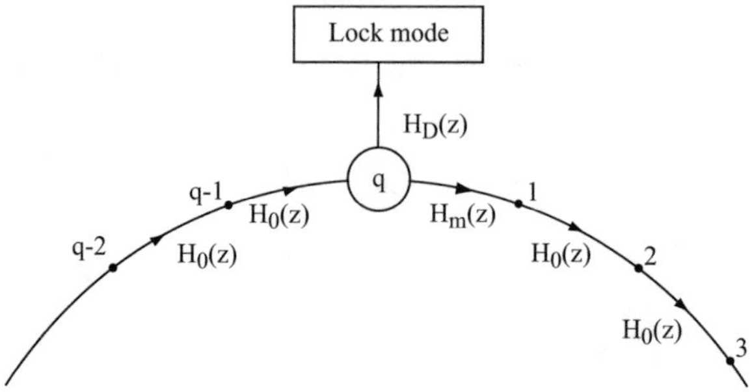

Fig. 4.10 Circular state diagram for serial-search acquisition

and variance

$$\sigma_{ij}^2 = \nu(i,j)var\,(T_{11}) + (i-1)var\,(T_{12}) + var\,(T_{22})\,. \tag{4.64}$$

The truncation is such that $f_a(t|i,j) \neq 0$ only if $0 \leq t \leq \mu_{ij} + 3\sigma_{ij}$. When P_D is large, the infinite series in (4.61) converges rapidly enough that the $f_a(t)$ can be accurately approximated by its first few terms.

Alternative Analysis

An alternative method of analyzing acquisition relies on transfer functions [66]. Each phase offset of the local code defines a *state* of the system. Of the total number of q states, $q-1$ are states that correspond to offsets (cells) that are equal to or exceed a chip duration. One state is a *collective state* that corresponds to all phase offsets that are less than a chip duration, and hence cause acquisition to be terminated and code tracking to begin. The rewinding time is assumed to be negligible. The serial-search acquisition process is represented by its *circular state diagram*, a segment of which is illustrated in Figure 4.10.

The branch labels between two states are transfer functions that contain information about the delays that may occur during the transition between the two states. Let z denote the unit-delay variable and let the power of z denote the time delay. A single-dwell system with dwell τ, false-alarm probability P_F, and constant penalty time T_p has a transfer function $H_0(z) = (1 - P_F)\,z^\tau + P_F z^{\tau+T_p}$ for all branches that do not originate in collective state q because the transition delay is τ with probability $1 - P_F$ and $\tau + T_p$ with probability P_F.

For a multiple-dwell system, $H_0(z)$ is determined by first drawing a subsidiary state diagram representing intermediate states and transitions that may occur as the system progresses from one state to the next one in the original circular state diagram. For example, Figure 4.11 illustrates the subsidiary state diagram for a consecutive-count double-dwell system with false alarms P_{F1} and P_{F2} and delays

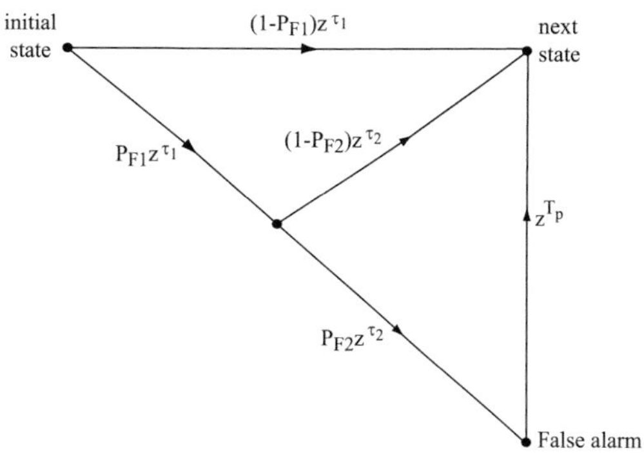

Fig. 4.11 Subsidiary state diagram for determination of $H_0(z)$ for consecutive-count double-dwell system

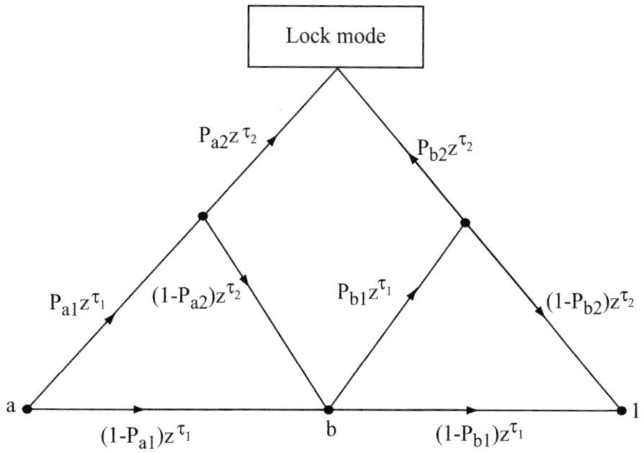

Fig. 4.12 Subsidiary state diagram for the calculation of $H_D(z)$ and $H_M(z)$ for a consecutive-count double-dwell system with a two-state collective state

τ_1 and τ_2 for the initial test and the verification test, respectively. Examination of all possible paths between the initial state and the next state indicates that

$$\begin{aligned} H_0(z) &= (1-P_{F1})\,z^{\tau_1} + P_{F1}z^{\tau_1}\left[(1-P_{F2})\,z^{\tau_2} + P_{F2}z^{\tau_2+T_P}\right] \\ &= (1-P_{F1})\,z^{\tau_1} + P_{F1}\,(1-P_{F2})\,z^{\tau_1+\tau_2} + P_{F1}P_{F2}z^{\tau_1+\tau_2+T_p}. \end{aligned} \tag{4.65}$$

Let $H_D(z)$ denote the transfer function between the collective state q and the lock mode. Let $H_M(z)$ denote the transfer function between state q and state 1 in Figure 4.10, which represents the failure to recognize code-phase offsets that are less than a chip duration. These transfer functions may be derived in the same manner as $H_0(z)$. For example, consider a consecutive-count, double-dwell system with a collective state that comprises two states. Figure 4.12 depicts the subsidiary state diagram representing intermediate states and transitions that may occur as the system progresses from state q (with subsidiary states a and b) to either the lock mode or state 1. Examination of all possible paths yields

$$\begin{aligned} H_D(z) &= P_{a1}P_{a2}z^{\tau_1+\tau_2} + P_{a1}\,(1-P_{a2})\,P_{b1}P_{b2}z^{2\tau_1+2\tau_2} \\ &\quad + (1-P_{a1})\,P_{b1}P_{b2}z^{2\tau_1+\tau_2} \end{aligned} \tag{4.66}$$

$$\begin{aligned} H_M(z) &= (1-P_{a1})(1-P_{b1})\,z^{2\tau_1} + (1-P_{a1})\,P_{b1}\,(1-P_{b2})\,z^{2\tau_1+\tau_2} \\ &+ P_{a1}\,(1-P_{a2})(1-P_{b1})\,z^{2\tau_1+\tau_2} \\ &+ P_{a1}\,(1-P_{a2})\,P_{b1}\,(1-P_{b2})\,z^{2\tau_1+2\tau_2}. \end{aligned} \tag{4.67}$$

For a single-dwell system with a collective state that comprises N states,

$$H_D(z) = P_1 z^{\tau} + \sum_{j=2}^{N} P_j \left[\prod_{i=1}^{j-1} (1-P_j) \right] z^{j\tau} \tag{4.68}$$

$$H_M(z) = \left[\prod_{j=1}^{N} (1-P_j) \right] z^{N\tau} \tag{4.69}$$

$$H_0(z) = (1-P_F)\,z^{\tau} + P_F z^{\tau+T_p} \tag{4.70}$$

where τ is the dwell time, P_F is the false-alarm probability, and P_j is the detection probability of state j within the collective state.

To calculate the statistics of the acquisition time, we seek the *generating function of the acquisition time* defined as the series

$$H(z) = \sum_{i=0}^{\infty} p_i\,(\tau_i)\,z^{\tau_i} \tag{4.71}$$

where $p_i(\tau_i)$ is the probability that the acquisition process terminates in the lock mode after τ_i seconds. Since the probability that the lock mode is reached is equal to or less than unity, $H(1) \leq 1$, and $H(z)$ converges at least for $|z| \leq 1$. If $H(z)$ is known, then a direct differentiation of (4.71) indicates that

$$\frac{dH(z)}{dz} = \sum_{i=0}^{\infty} \tau_i p_i(\tau_i) z^{\tau_i - 1}. \tag{4.72}$$

Therefore, the mean acquisition time is

$$\bar{T}_a = \sum_{i=0}^{\infty} \tau_i p_i(\tau_i) = \left. \frac{dH(z)}{dz} \right|_{z=1}. \tag{4.73}$$

Similarly, the second derivative of $H(z)$ gives

$$\left. \frac{d^2H(z)}{dz^2} \right|_{z=1} = \sum_{i=0}^{\infty} \tau_i\,(\tau_i - 1)\,p_i\,(\tau_i) = \overline{T_a^2} - \bar{T}_a. \tag{4.74}$$

Therefore, the variance of the acquisition time is

$$\sigma_a^2 = \left\{ \frac{d^2H(z)}{dz^2} + \frac{dH(z)}{dz} - \left[\frac{dH(z)}{dz}\right]^2 \right\}\Bigg|_{z=1}. \tag{4.75}$$

To derive $H(z)$, we observe that it may be expressed as

$$H(z) = \sum_{j=1}^{q} \beta_j H_j(z) \tag{4.76}$$

where β_j, $j = 1, 2, \ldots, q$, is the a priori probability that the search begins in state j, and $H_j(z)$ is the transfer function from an initial state j to the lock mode. Since the circular state diagram of Figure 4.10 may be traversed an indefinite number of times during the acquisition process,

$$\begin{aligned} H_j(z) &= H_0^{q-j}(z)H_D(z)\sum_{i=0}^{\infty}\left[H_M(z)H_0^{q-1}(z)\right]^i \\ &= \frac{H_0^{q-j}(z)H_D(z)}{1 - H_M(z)H_0^{q-1}(z)}. \end{aligned} \tag{4.77}$$

Substitution of this equation into (4.76) yields

$$H(z) = \frac{H_D(z)}{1 - H_M(z)H_0^{q-1}(z)} \sum_{j=1}^{q} \beta_j H_0^{q-j}(z). \tag{4.78}$$

For the uniform a priori distribution $\beta_j = \frac{1}{q}$, $1 \le j \le q$,

$$H(z) = \frac{H_D(z)\left[1 - H_0^q(z)\right]}{q\left[1 - H_M(z)H_0^{q-1}(z)\right]\left[1 - H_0(z)\right]}. \tag{4.79}$$

Since the progression from one state to another is inevitable until the lock mode is reached, $H_0(1) = 1$. Since $H_D(1) + H_M(1) = 1$, (4.78) and (4.73) yield

$$\bar{T}_a = \frac{1}{H_D(1)}\left\{H_D'(1) + H_M'(1) + (q-1)H_0'(1)\left[1 - \frac{H_D(1)}{2}\right]\right\} \tag{4.80}$$

where the prime indicates differentiation with respect to z. As an example, consider a single-dwell system with a two-state collective state. The evaluation of (4.80) using (4.68) to (4.70) with $N = 2$ yields (4.40) with $T_r = 0$ if we set $P_1 = P_a, P_2 = P_b$, $T_p = \bar{T}_p$, $\tau = \tau_d$, and define P_D by (4.34).

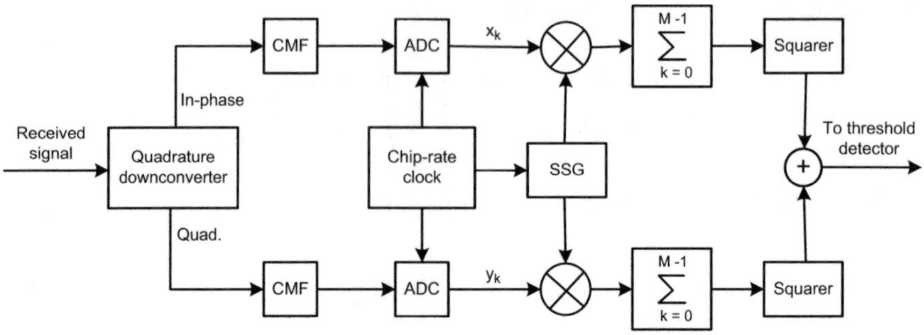

Fig. 4.13 Noncoherent acquisition correlator. *CMG* chip-matched filter, *SSG* spreading-sequence generator, *ADC* analog-to-digital converter

The presence of the multipath signals may be exploited by using a *nonconsecutive serial search*, which can be shown to provide a lower normalized mean acquisition time (NMAT) than the conventional serial search in which cells are tested serially [27, 90]. The cost is increased computational complexity.

4.4 Acquisition Correlator

An acquisition correlator must correctly identify the autocorrelation peaks and be impervious to the autocorrelation sidelobes. The *noncoherent acquisition correlator* in Figure 4.4 has the form depicted by Figure 4.13. The sequences $\{x_k\}$ and $\{y_k\}$ are obtained by in-phase and quadrature downconversion followed by chip-matched filters sampled at times $t = kT_c$. The dwell time or duration of the test interval is MT_c, where M is the number of chip-matched-filter output samples that are summed, and T_c is the chip duration. The decision variable for one test of a specific code phase is applied to a threshold detector and compared with a threshold. If the threshold is exceeded, then the particular code phase being tested is accepted as the correct one. As indicated by the figure, the decision variable is proportional to the sum of the outputs of the squares in the upper and lower branches respectively.

The acquisition correlator could be used for acquisition when the direct-sequence signal has either BPSK or QPSK modulation. For BPSK and the AWGN channel, we assume that the received signal is

$$r(t) = \sqrt{2G\mathcal{E}_c}p(t-\tau)\cos(2\pi f_c t + \theta) + n(t) \tag{4.81}$$

where $\mathcal{E}_c$ is the energy per chip, G is the spreading factor, f_c is the carrier frequency, θ is the random carrier phase, τ is the delay due to the unknown code phase, and $n(t)$ is zero-mean, white Gaussian noise with two-sided power spectral density (PSD) equal to $N_0/2$. The random carrier phase is present because the acquisition correlator processes the received signal prior to carrier synchronization. The spreading waveform has the form

$$p(t) = \sum_{i=-\infty}^{\infty} p_i \psi(t - iT_c) \tag{4.82}$$

where p_i is equal to +1 or -1 and represents one chip of a spreading sequence $\{p_i\}$. The chip waveform is normalized so that

$$\int_0^{T_c} \psi^2(t)dt = \frac{1}{G}\,. \tag{4.83}$$

With this normalization, an integration over a chip interval indicates that $\mathcal{E}_c$ is the energy per chip assuming that $f_c >> 1/T_c$ so that the integral over a double-frequency term is negligible. The data modulation $d(t)$ is omitted in (4.81) because pilot symbols with known data bits are generally transmitted during acquisition, and the modeling of $\{p_i\}$ is unaffected. We assume an ideal downconversion by the quadrature downconverter.

If the local spreading sequence produced by the spreading-sequence generator is delayed by ν chips relative to an arbitrary time origin, where ν is an integer, then the test interval begins with chip $-\nu$ of the local spreading sequence. When a cell is tested, the despreading produces the inner products

$$V_c = \sum_{k=0}^{M-1} p_{k-\nu} x_k, \quad V_s = \sum_{k=0}^{M-1} p_{k-\nu} y_k \tag{4.84}$$

where

$$x_k = \sqrt{2}\int_{kT_c}^{(k+1)T_c} r(t)\psi\,(t - kT_c)\cos 2\pi f_c t\, dt \tag{4.85}$$

$$y_k = \sqrt{2}\int_{kT_c}^{(k+1)T_c} r(t)\psi\,(t - kT_c)\sin 2\pi f_c t\, dt. \tag{4.86}$$

Since $p_k = \pm 1$, each of the inner products may be computed by either adding or subtracting each component of $\{x_k\}$ or $\{y_k\}$.

The delay τ may be expressed in the form $\tau = \nu T_c - NT_c - \epsilon T_c$, where N is an integer and $0 \le \epsilon < 1$. For a rectangular chip waveform, the preceding equations, $f_c T_c >> 1$, and the definition of chip ν yield

$$V_c = g_1(\epsilon)\cos\theta + N_c \tag{4.87}$$

$$V_s = -g_1(\epsilon)\sin\theta + N_s \tag{4.88}$$

where

$$g_1(\epsilon) = \sqrt{\frac{\mathcal{E}_c}{G}} \sum_{k=0}^{M-1} p_{k-\nu} \left[(1-\epsilon) p_{k-\nu+N} + \epsilon p_{k-\nu+N+1} \right] \tag{4.89}$$

$$N_c = \sqrt{2} \sum_{k=0}^{M-1} p_{k-\nu} \int_{kT_c}^{(k+1)T_c} n(t)\psi\left(t - kT_c\right) \cos 2\pi f_c t \, dt \tag{4.90}$$

$$N_s = \sqrt{2} \sum_{k=0}^{M-1} p_{k-\nu} \int_{kT_c}^{(k+1)T_c} n(t)\psi\left(t - kT_c\right) \sin 2\pi f_c t \, dt. \tag{4.91}$$

The alignment of the received and local spreading sequences is often close enough for acquisition if $N = -1$ or $N = 0$. If $N \neq -1, 0$, then the cell is considered incorrect.

The decision variable is

$$\begin{aligned} V &= \frac{V_c^2 + V_s^2}{N_0/G} \\ &= N_t^2 + g^2(\epsilon) + C \end{aligned} \tag{4.92}$$

where the division by N_0/G is for mathematical convenience and

$$N_t^2 = \frac{N_c^2 + N_s^2}{N_0/G}, \; g(\epsilon) = \frac{g_1(\epsilon)}{\sqrt{N_0/G}}$$

$$C = \frac{2g(\epsilon)}{\sqrt{N_0/G}} \left(N_c \cos\theta - N_s \sin\theta \right). \tag{4.93}$$

The decision variable is applied to a threshold detector with threshold $V_t \geq 0$, and acquisition is declared if $V > V_t$.

As explained in Section 1.2, the integrals in (4.90) and (4.91) are independent, zero-mean, Gaussian random variables. Since N_c and N_s are linear combinations of these random variables, N_c and N_s are independent, zero-mean Gaussian random variables. Since the noise $n(t)$ is white, we find that N_c and N_s both have a variance equal to $MN_0/2G$ if $f_c T_c >> 1$, and hence they are identically distributed. Therefore, as shown in Appendix E.2, N_t^2 has a central chi-squared distribution with two degrees of freedom,

$$E\left[N_t^2\right] = M, \; var\left(N_t^2\right) = M^2 \tag{4.94}$$

and the distribution function of N_t^2 is

$$F_n(x) = \left[1 - \exp\left(-\frac{x}{M}\right)\right] u(x) \tag{4.95}$$

where $u(x)$ is the unit step function: $u(x) = 1, x \geq 0$, and $u(x) = 0, x < 0$.

In the performance analysis, the spreading sequence $\{p_k\}$ is modeled as a zero-mean random binary sequence with independent symbols or chips, each of which is equal to $+1$ with a probability of $1/2$ and -1 with a probability of $1/2$. The spreading sequence is independent of the white Gaussian noise $n(t)$. The chip offset ϵ is modeled as an independent random variable uniformly distributed over $[0, 1]$, and the carrier phase θ is modeled as an independent random variable uniformly distributed over $[-\pi, \pi]$. Both ν and ϵ change from test to test. The models imply that

$$E[g(\epsilon)] = \begin{cases} 0, & N \neq -1, 0 \\ \frac{M}{2}\sqrt{\mathcal{R}}, & N = -1 \text{ or } 0 \end{cases} \tag{4.96}$$

$$E[g^2(\epsilon)] = \begin{cases} \frac{2M\mathcal{R}}{3}, & N \neq -1, 0 \\ \frac{M(M+1)\mathcal{R}}{3}, & N = -1 \text{ or } 0 \end{cases} \tag{4.97}$$

where

$$\mathcal{R} = \frac{\mathcal{E}_c}{N_0} . \tag{4.98}$$

The performance analysis uses the following *density-function approximation*. Let X and Y denote random variables with density functions such that $cov\,(X, Y) = 0$. Let $Z = X + Y$. If $var\,(X) >> var\,(Y)$, the density-function approximation sets the density function of Z equal to the density function of $sX + (1 - s)\,E\,[X] + E\,[Y]$, where

$$s = \sqrt{1 + \frac{var\,(Y)}{var\,(X)}} \tag{4.99}$$

and we write

$$Z \simeq sX + (1 - s)\,E\,[X] + E\,[Y] \tag{4.100}$$

as a compact representation of the approximation. The mean and variance of Z are maintained by the approximation, but the approximate density function has the same form as the density function of X, which affects the density function of Z more than that of Y. Numerical evaluations indicate that the density-function approximation is fairly good for the density functions considered subsequently if $var\,(X) > 2var\,(Y)$.

Assume that $N \neq -1, 0$ and an incorrect cell is tested. Then $g(\epsilon)$ is the sum of M independent, identically distributed, zero-mean, finite-variance random variables. We assume that $M \geq 200$ so that the central limit theorem (Corollary A1 of Appendix A.2) is applicable, which implies that the distribution function of $g(\epsilon)$ approximates a Gaussian distribution with variance

$$\sigma_1^2 = \frac{2M\mathcal{R}}{3} \tag{4.101}$$

as indicated by (4.96) and (4.97). Therefore, $g^2(\epsilon)$ has a central chi-squared distribution with one degree of freedom (Appendix C.2), and its density function is

$$f_{g2}(x) = \frac{1}{(2\pi\sigma_1^2)^{1/2}} x^{-1/2} \exp\left(-\frac{x}{2\sigma_1^2}\right) u(x), \tag{4.102}$$

and its variance is

$$var\left[g^2(\epsilon)\right] = \frac{8M^2\mathcal{R}^2}{9}. \tag{4.103}$$

From (4.94) and (4.97), we obtain

$$var(C) = \frac{4M^2\mathcal{R}}{3}. \tag{4.104}$$

Let V_0 denote the decision variable when an incorrect cell is tested. If $X = N_t^2 + g^2(\epsilon)$ and $Y = C$, then $cov(X, Y) = 0$ and $var(Y) < 0.5\, var(X)$ if $\mathcal{R} < 0.44$. The density-function approximation is applicable and indicates that

$$V_0 \simeq s_1[N_t^2 + g^2(\epsilon)] + (1 - s_1) M\left(1 + \frac{2\mathcal{R}}{3}\right), \quad \mathcal{R} < 0.44, \; M \geq 200 \tag{4.105}$$

where

$$s_1 = \sqrt{1 + \mathcal{R}\left[\frac{3}{4} + \frac{2}{3}\mathcal{R}^2\right]^{-1}}. \tag{4.106}$$

A false alarm occurs if $V_0 > V_t$ or

$$X > \frac{V_t - L}{s_1} \tag{4.107}$$

where

$$L = (1 - s_1) M\left(1 + \frac{2\mathcal{R}}{3}\right). \tag{4.108}$$

For $y \geq 0$, the distribution function of X is

$$\begin{aligned} F_1(y) &= \int_0^y F_n(y - x) f_{g2}(x)\, dx \\ &= F_{g2}(y) - \exp\left(-\frac{y}{M}\right) \int_0^y \exp\left(\frac{x}{M}\right) f_{g2}(x)\, dx \end{aligned} \tag{4.109}$$

where $F_{g2}(y)$ is the distribution function of $g^2(\epsilon)$ and (4.95) has been substituted. Substituting (4.102) into (4.109), changing variables, and using (E.20) of Appendix E.2 and (H.20) of Appendix H.4, we obtain

$$F_1(y) = 1 - 2Q\left(\sqrt{\frac{y}{\sigma_1^2}}\right) - \frac{\sigma_a}{\sigma_1}\exp\left(-\frac{y}{M}\right)\left[1 - 2Q\left(\sqrt{\frac{y}{\sigma_a^2}}\right)\right], \; y \geq 0 \quad (4.110)$$

where

$$\sigma_a^2 = \frac{\sigma_1^2 M}{M - 2\sigma_1^2}. \quad (4.111)$$

From $F_1(y)$, we obtain the false-alarm probability P_f for a test of an incorrect cell:

$$P_f \simeq \begin{cases} 1 & V_t < L \\ 1 - F_1(\frac{V_t - L}{s_1}) & V_t \geq L \end{cases}, \; \mathcal{R} < 0.44, \; M \geq 200. \quad (4.112)$$

Thus, the threshold needed to realize a specified P_f is

$$V_t \simeq s_1 F_1^{-1}\left(1 - P_f\right) + L, \; \mathcal{R} < 0.44, \; M \geq 200 \quad (4.113)$$

where $F_1^{-1}(\cdot)$ denotes the inverse of $F_1(\cdot)$.

Assume that the correct cell is tested, and hence $N = -1$ or 0. We derive the detection probability under the condition that $N = -1$. Since the detection probability is the same given that $N = 0$, the unconditional detection probability is the same. Assuming that $N = -1$,

$$\begin{aligned} g^2(\epsilon) &= \mathcal{R}\left[M\epsilon + (1-\epsilon)U\right]^2 \\ &= \mathcal{R}\left[M^2\epsilon^2 + 2M\epsilon(1-\epsilon)U + (1-\epsilon)^2 U^2\right] \end{aligned} \quad (4.114)$$

where

$$U = \sum_{k=0}^{M-1} p_{k-\nu} p_{k-\nu-1}. \quad (4.115)$$

Since the chip sequence is modeled as a random binary sequence, $E[U] = 0$ and $var(U) = M$. Therefore, if $M \geq 200$, then the density function approximation indicates that $g^2(\epsilon) \simeq \mathcal{R}\left(M^2\epsilon^2 + M/3\right)$ with negligible probability of error, and hence the decision variable for a correct cell is

$$V_1 \simeq \mathcal{R}\left(M^2\epsilon^2 + \frac{M}{3}\right) + N_t^2 + C, \; M \geq 200. \quad (4.116)$$

Equations (4.94) and (4.97) indicate that

$$var\,(C) \simeq \frac{2M^3\mathcal{R}}{3},\ M \geq 200. \tag{4.117}$$

With X equal to the first term of (4.116) and $Y = N_t^2 + C$, $cov\,(X, Y) = 0$ and the density-function approximation is applicable. The variance of X is

$$var\,(X) = \frac{4M^4\mathcal{R}^2}{45}\,. \tag{4.118}$$

Therefore, $var\,(Y) < 0.5\ var\,(X)$ if $R > 16.4/M$, and hence

$$V_1 \simeq s_2 M\mathcal{R}\left(M\epsilon^2 + \frac{1}{3}\right) + (1 - s_2)\,\frac{M^2\mathcal{R}}{3} + M\,,$$
$$\frac{16.4}{M} < \mathcal{R} < 0.44,\ M \geq 200 \tag{4.119}$$

where

$$s_2 = \left[1 + \frac{15}{2M}\mathcal{R}^{-1} + \frac{45}{4M^2}\mathcal{R}^{-2}\right]^{\frac{1}{2}}. \tag{4.120}$$

The detection probability P_d for a test of a correct cell is the probability that $V_1 > V_t$, where V_t is given by (4.113). Since ϵ is uniformly distributed over [0, 1],

$$P_d \simeq \begin{cases} 1, & V_t > L_1 \\ 1 - \sqrt{R} & L_2 \leq V_t \leq L_1 \\ 0, & V_t < L_2 \end{cases},\quad \frac{16.4}{M} < \mathcal{R} < 0.44,\ M \geq 200$$
$$R = \frac{V_t - L_1}{s_2 M^2\mathcal{R}},\ L_2 = L_1 + s_2 M^2\mathcal{R}$$
$$L_1 = M\mathcal{R}\left[\frac{s_2}{3} + \mathcal{R}^{-1} + \frac{M(1 - s_2)}{3}\right]. \tag{4.121}$$

A notable attribute of the equation for P_d is its lack of dependence on the spreading factor G. The reason for this attribute is the fact that the spectral spreading has no effect on white Gaussian noise. In the presence of other types of interference, we can expect a significant dependence.

The step size Δ of the serial search is the separation in chips between cells. When $\Delta = 1/2$, the two consecutive cells that correspond to $N = -1$ and $N = 0$ are considered the two correct cells out of the q in the timing uncertainty region. When $\Delta = 1$, it is plausible to assume that there is only one correct cell, which corresponds to either $N = -1$ or $N = 0$. Let C_u denote the number of chip durations in the timing uncertainty. The *NMAT* is defined as $\bar{T}_a/C_uT_c$. The *normalized standard deviation* (NSD) is defined as σ_a/C_uT_c. For step size $\Delta = 1$, $q = C_u$; for $\Delta = 1/2$, $q = 2C_u$.

Example 1 As an example of the application of the preceding results, consider a single-dwell system with a uniform search and a uniform a priori correct-cell location distribution. Let $\tau_d = MT_c$, where M is the number of chips per dwell time, and $\bar{T}_p = KT_c$, where $K = 10^4$ is the number of chip durations in the mean penalty time. For $\Delta = 1/2$, we assume that there are two independent correct cells with the common detection probability $P_d = P_a = P_b$. If $q >> 1$, (4.40) and (4.34) yield the NMAT:

$$NMAT = \left(\frac{2 - P_D}{2P_D}\right)\frac{q}{C_u}(M + KP_F) \tag{4.122}$$

where

$$P_D = 2P_d - P_d^2\,,\ \Delta = 1/2. \tag{4.123}$$

For $\Delta = 1$, we assume that there is one correct cell so that

$$P_D = P_d\,,\ \Delta = 1. \tag{4.124}$$

In a single-dwell system, $P_F = P_f$. Equation (4.121) relates to P_d and P_f.

Figure 4.14 shows the NMAT as a function of $\mathcal{E}_c/N_0$ for the design choices $P_f = 0.001$, $M = 200$ and $P_f = 0.01$, $M = 400$. It is observed that the relative effectiveness of these two pairs depends on $\mathcal{R} = \mathcal{E}_c/N_0$. Also shown in the figure is the minimum NMAT that is obtained if the optimal choices of P_f and M are made at each value of $\mathcal{E}_c/N_0$. The minimization of the NMAT is calculated with the constraint that $M \geq 200$. To implement the optimal choice of P_f and M at the receiver would require the accurate measurement of $\mathcal{E}_c/N_0$. The figure indicates the slight advantage of $\Delta = 1$ in a single-dwell system. From (4.33), it is found that each plot of the NSD has a shape similar to that of the corresponding NMAT plot.

The potential impact of fading is considerable. For example, suppose that $\mathcal{E}_c/N_0 = -4$ dB in the absence of fading, but 10 dB of adverse fading causes $\mathcal{E}_c/N_0 = -14$ dB during acquisition. Then, the figure indicates that the NMAT increases by a large factor relative to its value in the absence of fading. □

Example 2 Consider double-dwell systems with a uniform search, a uniform a priori correct-cell location distribution, $\Delta = 1/2$, $K = 10^4$, and two independent correct cells with $P_d = P_a = P_b$, $P_{a1} = P_{b1}$, and $P_{a2} = P_{b2}$. The dwell times are $\tau_1 = M_1T_c$ and $\tau_2 = M_2T_c$. If $q >> 1$, the NMAT is obtained from (4.32) and (4.123), where $\bar{T}_{11}$ is given by (4.36) for a consecutive-count system and (4.45) for an up-down system. Since $q/C_u = 2$, a consecutive-count system has

$$NMAT = \left(\frac{2-2P_d + P_d^2}{2P_d - P_d^2}\right)[M_1 + P_{F1}(M_2+P_{F2}K)] \tag{4.125}$$

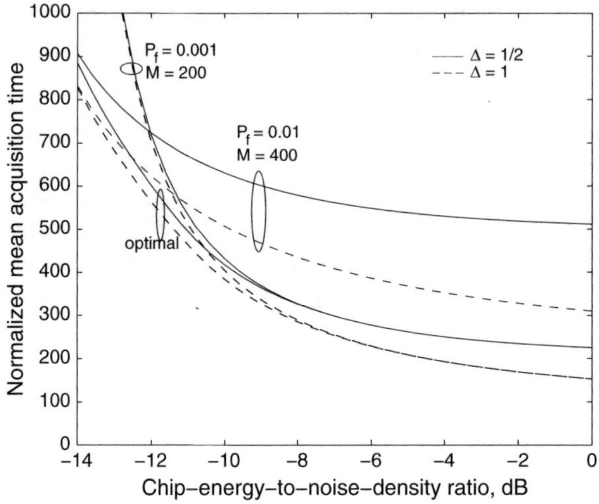

Fig. 4.14 NMAT versus $\mathcal{E}_c/N_0$ for a single-dwell system with $K = 10^4$ in the absence of fading. Three design choices of P_f and M are illustrated

and an up-down system has

$$NMAT = \left(\frac{2{-}2P_d + P_d^2}{2P_d - P_d^2}\right)\left[\frac{M_1 + P_{F1}\left(M_2 + P_{F2}K\right)}{1 - P_{F1}\left(1 - P_{F2}\right)}\right]. \tag{4.126}$$

By replacing P_d with P_{ai}, P_f with P_{Fi}, and M with M_i, the probabilities P_{ai} and P_{Fi}, $i = 1$ or 2, are related through (4.121). Equation (4.35) implies that a consecutive-count system has

$$P_d = P_{a1}P_{a2} \text{ (consecutive-count)} \tag{4.127}$$

and (4.42) and (4.43) imply that an up-down system has

$$P_d = \frac{P_{a1}P_{a2}}{1 - P_{a1}\left(1 - P_{a2}\right)} \text{ (up-down)}. \tag{4.128}$$

Figure 4.15 shows the NMAT as a function of $\mathcal{E}_c/N_0$ for double-dwell systems with the design choices $P_{F1} = 0.01$, $P_{F2} = 0.1$, $M_1 = 200$, and $M_2 = 1500$. The step size is $\Delta = 1/2$, which is found to be slightly advantageous in typical double-dwell systems. Also shown in the figure is the minimum NMAT that is obtained if the optimal choices of P_{F1}, P_{F2}, M_1, and M_2 are made at each value of $\mathcal{R} = \mathcal{E}_c/N_0$. The minimization of the NMAT is calculated with the constraints that M_1, $M_2 \geq 200$. To implement the optimal choices at the receiver would require the accurate measurement of $\mathcal{E}_c/N_0$. The figure illustrates the slight advantage of the up-down system in most practical applications. From (4.33), it is found that each plot of the

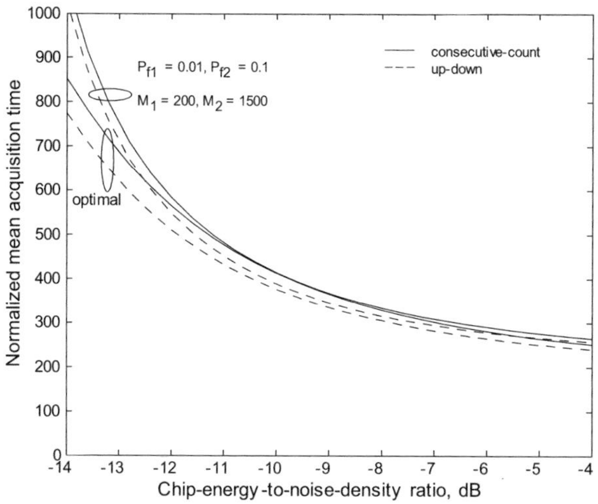

Fig. 4.15 NMAT versus $\mathcal{E}_c/N_0$ for double-dwell systems with $K = 10^4$ in the absence of fading. Step size is $\Delta = 1/2$. Two design choices P_{F1}, P_{F2}, M_1, and M_2 are illustrated

NSD has a shape similar to that of the corresponding NMAT plot. A comparison of Figure 4.15 with Figure 4.14 indicates that double-dwell systems are capable of lowering the NMAT relative to single-dwell systems. □

The NMAT may be reduced by nearly the factor η if the sequences $\{x_k\}$ and $\{y_k\}$ are applied to η parallel computations of the decision variables. In each computation, a different delay of the spreading-sequence generator output is used, and the delays are separated by multiples of the chip duration. This procedure allows a parallel search of various code phases with a moderate amount of additional hardware or software.

The threshold of (4.113), which is used to ensure a specified false-alarm rate, requires the estimation of $\mathcal{R}$. The decision variable of (4.92) requires the estimation of N_0, which may be obtained by using a radiometer (Section 10.2). When N_0 is primarily determined by rapidly varying interference power, an adaptive threshold may be set by estimating the instantaneous received power for each correlation interval prior to acquisition. As a result, the mean acquisition time is less degraded by pulsed interference [18].

The presence of either data bits or an uncompensated Doppler shift during acquisition can degrade performance. If data bits $\{d_n\}$ are present, then $p_{k-\nu}$ in (4.84) and (4.89) is replaced by $d_n p_{k-\nu}$. Although a data bit has a transition to a new value at most once every G chips, a data-bit transition in (4.89) can cause cancelations that reduce $g(\epsilon)$ when a correct cell is tested. If the residual frequency offset due to the Doppler shift is f_e, then $\theta = \theta_1 + 2\pi f_e t$ in (4.87) and (4.88). If $f_e M T_c \gtrsim 1$, then cancelations when a correct cell is tested reduce V_c and V_s and hence the decision variable. Thus, $M << 1/f_e T_c$ is required to prevent a large increase in the mean acquisition time.

An alternative to the noncoherent acquisition correlator is the *differentially coherent acquisition correlator*, which potentially provides superior performance [119]. However, the noncoherent correlator is generally preferred because it is simpler to implement and more robust.

The downlinks of cellular networks require special methods of code acquisition that differ significantly from the methods used for point-to-point communications. These methods are presented in Section 8.3.

4.5 Sequential Acquisition

The acquisition correlator of Section 4.4, which supports single-dwell or multiple-dwell serial-search acquisition systems, provides tests of fixed dwell times based on fixed numbers of chip-matched-filter output samples or observations. An alternative strategy for acquisition is *sequential acquisition* [96], which is a type of serial-search acquisition that uses only the number of samples necessary for reliable decisions. Thus, some sample sequences may allow a quick decision, whereas others may warrant using a large number of samples in the evaluation of a single cell or code phase of the spreading waveform. Sequential acquisition is based on the *sequential probability-ratio test* [47], which minimizes the average detection time for specified error probabilities when the samples are independent and identically distributed.

The principal components of a sequential acquisition system are diagrammed in Figure 4.16. To determine whether a cell passes a test, classical detection theory requires a choice between the hypothesis H_1 that the cell is correct and the hypothesis H_0 that the cell is incorrect. Let $\mathbf{V}(n) = [V_1 \, V_2 \ldots V_n]$ denote n consecutive outputs of an acquisition correlator. Let $f(\mathbf{V}(n) \,|H_i)$ denote the conditional density function of $\mathbf{V}(n)$ given hypothesis H_i, $i = 0, 1$. The sequential probability-ratio test recalculates the log-likelihood ratio

$$\Lambda\left[\mathbf{V}(n)\right] = \ln\frac{f(\mathbf{V}(n)\,|H_1)}{f(\mathbf{V}(n)\,|H_0)} \tag{4.129}$$

after each new acquisition-correlator output is produced during the testing of a cell. The log-likelihood ratio, or a function of it, is compared with both the upper and lower thresholds to determine if the test is terminated and no more outputs need to be extracted for the cell being tested. If the upper threshold is exceeded, the receiver declares acquisition, and the lock mode is entered. If $\Lambda\left[\mathbf{V}(n)\right]$ drops below the lower threshold, the test fails, and the testing of another cell begins. As long as $\Lambda\left[\mathbf{V}(n)\right]$ lies between the two thresholds, a decision is postponed, another correlator output for the same cell is observed, and $\Lambda\left[\mathbf{V}(n+1)\right]$ is computed. A truncation procedure that forces a decision after a maximum number of outputs prevents an excessive number of tests for a single cell when many ambiguous observations are encountered.

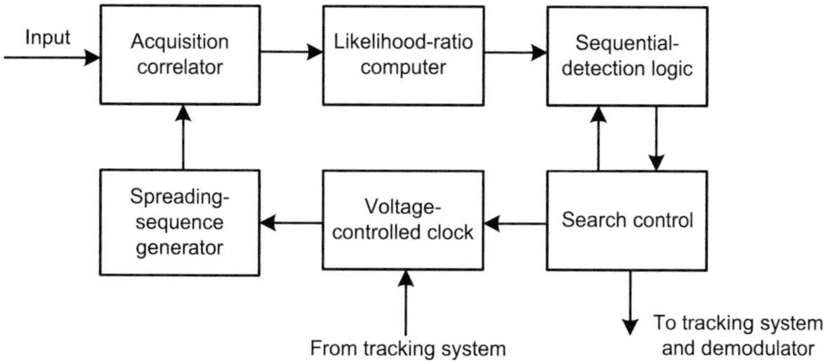

Fig. 4.16 Sequential acquisition system

Although sequential acquisition is capable of significantly reducing the mean acquisition time relative to acquisition systems that use a fixed dwell time for each decision, it presents a number of practical problems that have limited its use. Chief among them is the computational complexity of repeatedly calculating the log-likelihood ratio, particularly when the $\{V_n\}$ are not independent and identically distributed. Another problem arises when the signal strengths of the acquisition-correlator outputs for a correct cell are significantly less than expected. In that case, the sequential detector may increase the mean acquisition time relative to that required by acquisition systems with fixed dwell times.

4.6 Code Tracking

The receiver must provide code synchronization by acquisition and tracking to enable the despreading of the received signal prior to carrier synchronization and the subsequent coherent demodulation. Coherent tracking loops cannot easily accommodate the effects of data modulation. Noncoherent loops, which predominate in spread-spectrum systems, operate directly on the received signals and are insignificantly affected by the data modulation.

The data modulation $d\,(t)$ is present during the code tracking of the timing offset. Therefore, we retain $d\,(t)$ in (4.7) and obtain the sufficient statistic for code tracking of the timing offset:

$$R_o\,(\tau) = \left[\int_0^T r(t)d\,(t-\tau)\,p(t-\tau)\cos\,(2\pi f_{c1}t)\,dt\right]^2 + \left[\int_0^T r(t)d\,(t-\tau)\,p(t-\tau)\sin\,(2\pi f_{c1}t)\,dt\right]^2 \tag{4.130}$$

where f_{c1} is known. If the maximum-likelihood estimate $\hat{\tau}$ is assumed to be within the interior of its timing uncertainty region, and $R_o(\tau)$ is a differentiable function of τ, then the estimate $\hat{\tau}$ that maximizes $R_o(\tau)$ may be found by setting

$$\left.\frac{\partial R_o(\tau)}{\partial \tau}\right|_{\tau=\hat{\tau}} = 0. \tag{4.131}$$

Although $R_o(\tau)$ is not differentiable if the chip waveform is rectangular, this problem is circumvented by using a difference equation as an approximation of the derivative. Thus, for a positive δT_c, we set

$$\frac{\partial R_o(\tau)}{\partial \tau} \approx \frac{R_o(\tau + \delta T_c) - R_o(\tau - \delta T_c)}{2\delta T_c}. \tag{4.132}$$

This equation implies that the solution of (4.131) may be approximately obtained by a device that finds the $\hat{\tau}$ such that

$$R_o(\hat{\tau} + \delta T_c) - R_o(\hat{\tau} - \delta T_c) = 0. \tag{4.133}$$

To determine a practical means of solving this equation, we assume that the noise is negligible, and that the signal applied to the tracking system has the form

$$s(t) = Ad(t)p(t)\cos(2\pi f_{c1}t + \theta) \tag{4.134}$$

where the correct timing offset of the received signal is $\tau = 0$. Substituting $r(t) \approx s(t)$ into (4.130), using trigonometry, and discarding negligible terms, we obtain:

$$R_o(\hat{\tau}) \approx \frac{A^2}{4}\left[\int_0^T d(t)\,d(t-\tau)\,p(t)p(t-\hat{\tau})dt\right]^2. \tag{4.135}$$

If $\tau << T_c$ during the code tracking, then $d(t)\,d(t-\tau) = 1$ with high probability, and thus

$$R_o(\hat{\tau}) \approx \frac{A^2}{4}\left[\int_0^T p(t)p(t-\hat{\tau})dt\right]^2. \tag{4.136}$$

We model the spreading waveform $p(t)$ as a wide-sense-stationary process with autocorrelation function

$$R_p(\tau) = E[p(t)p(t+\tau)] \tag{4.137}$$

and define the interval $[0, T]$ large enough to include many chips. Then (4.135) implies that

$$R_o(\hat{\tau}) \approx \frac{A^2}{4}R_p^2(\hat{\tau}). \tag{4.138}$$

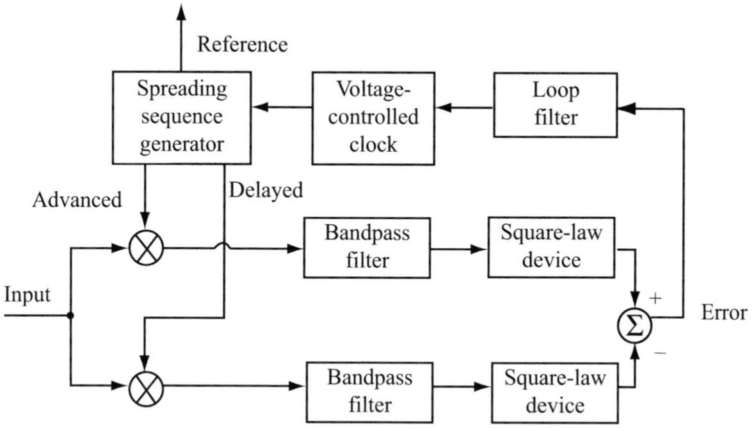

Fig. 4.17 Delay-locked loop

Using this approximation in (4.133), we find that the maximum-likelihood estimate is approximately obtained by a device that finds the $\hat{\tau}$ such that

$$R_p^2(\hat{\tau} + \delta T_c) - R_p^2(\hat{\tau} - \delta T_c) = 0. \tag{4.139}$$

Delay-Locked Loop

The noncoherent *delay-locked loop*, which resembles the early-late gates used for data-symbol synchronization [7, 69], implements an approximate computation of the difference on the left-hand side of (4.139) and then continually adjusts $\hat{\tau}$ so that this difference remains near zero. This fine adjustment establishes or maintains an accurately synchronized local spreading sequence that is used for despreading the received direct-sequence signal.

The delay-locked loop, which is diagrammed in Figure 4.17, has a spreading-sequence generator that produces three synchronous sequences, one of which is the reference sequence used for acquisition and demodulation. The other two sequences are advanced and delayed, respectively, by δT_c relative to the reference sequence. Although usually $\delta = 1/2$, other values such that $\delta \in (0, 1)$ are plausible. The advanced and delayed sequences are multiplied by the received direct-sequence signal in separate branches.

In the subsequent analysis, the processing of the direct-sequence signal and the received noise are treated separately for clarity. For the received direct-sequence signal (4.134), the desired-signal portion of the upper-branch mixer output is

$$s_{u1}(t) = Ad(t)p(t)p(t + \delta T_c - \epsilon T_c)\cos(2\pi f_c t + \theta) \tag{4.140}$$

where ϵT_c is the delay of the reference sequence relative to the received sequence, and $p(t)$ is the spreading waveform, which is modeled as a wide-sense-stationary process with autocorrelation $R_p(\tau)$. Although ϵ is a function of time because of the loop dynamics, it is slowly varying and hence is treated as a constant in the subsequent analysis. Similarly, the desired-signal portion of the lower-branch mixer output is

$$s_{l1}(t) = Ad(t)p(t)p(t - \delta T_c - \epsilon T_c)\cos(2\pi f_c t + \theta). \tag{4.141}$$

Let $H_0(f)$ denote the frequency response of a baseband filter. The filter does time-averaging over an interval short enough that $d(t)$ is not significantly distorted, but long enough that most of the spectral components of the much more rapidly varying products $p(t)p(t \pm \delta T_c - \epsilon T_c)$ are blocked except for their slowly varying time averages. We assume that these time averages are approximated by the autocorrelations $R_p(\pm\delta T_c - \epsilon T_c)$. The common frequency response (centered at f_c) of the two identical bandpass filters is

$$H_b(f) = H_0(f - f_c) + H_0(f + f_c). \tag{4.142}$$

Thus, the desired components of the upper-branch and lower-branch filter outputs are

$$s_{u2}(t) \approx Ad(t)R_p(\delta T_c - \epsilon T_c)\cos(2\pi f_c t + \theta) \tag{4.143}$$

$$s_{l2}(t) \approx Ad(t)R_p(-\delta T_c - \epsilon T_c)\cos(2\pi f_c t + \theta) \tag{4.144}$$

respectively. Both $s_{u2}(t)$ and $s_{l2}(t)$ are accompanied by residual undesired spectral components that are suppressed downstream by the loop filter.

Since $d^2(t) = 1$, the square-law devices remove the data modulation. The devices generate double-frequency components near $2f_c$, but these components are ultimately suppressed by the loop filters and thus are ignored. The signal outputs of the upper-branch and lower-branch square-law devices are

$$s_{u3}(t) \approx \frac{A^2}{2}R_p^2(\delta T_c - \epsilon T_c). \tag{4.145}$$

$$s_{l3}(t) \approx \frac{A^2}{2}R_p^2(-\delta T_c - \epsilon T_c) \tag{4.146}$$

respectively. The difference between the outputs of the two branches is the *error signal*:

$$s_e(t) \approx \frac{A^2}{2}\left[R_p^2(\delta T_c - \epsilon T_c) - R_p^2(-\delta T_c - \epsilon T_c)\right] \tag{4.147}$$

which is applied to the loop filter.

Let $H_l(f)$ denote the frequency response (centered at $f = 0$) of the loop filter, which has a significant frequency response over a band that is much narrower than the bandwidth of the bandpass filter. Because of this narrow bandwidth, the loop filter blocks not only double-frequency components but also residual undesired spectral components in $s_e(t)$ that originate in components accompanying $s_{u2}(t)$ and $s_{l2}(t)$. Since $s_e(t)$ is slowly varying (due to ϵ), the loop filter has a signal output approximately equal to $s_e(t)$. Since $R_p(\tau)$ is an even function, the error signal $s_e(t)$ is proportional to the left-hand side of (4.139) with $\hat{\tau} = \epsilon T_c$.

If $p(t)$ is modeled as a random binary sequence, then (Section 2.2)

$$R_p(\tau) = \Lambda\left(\frac{\tau}{T_c}\right) = \begin{cases} 1 - |\frac{\tau}{T_c}|, & |\tau| \leq T_c \\ 0, & |\tau| > T_c\,. \end{cases} \tag{4.148}$$

The substitution of this equation into (4.147) yields

$$s_e(t) \approx \frac{A^2}{2} S(\epsilon, \delta) \tag{4.149}$$

where $S(\epsilon, \delta)$ is the *discriminator characteristic* or *S-curve* of the tracking loop. For $\delta \geq 0$,

$$S(\epsilon, \delta) = \begin{cases} -2\left(|\delta - \epsilon| - |\delta + \epsilon| + 2\delta\epsilon\right), & |\delta - \epsilon| \leq 1, |\delta + \epsilon| \leq 1 \\ (1 - |\delta - \epsilon|)^2, & |\delta - \epsilon| \leq 1, |\delta + \epsilon| > 1 \\ -(1 - |\delta + \epsilon|)^2, & |\delta - \epsilon| > 1, |\delta + \epsilon| \leq 1 \\ 0, & \textit{otherwise} \end{cases} \tag{4.150}$$

which indicates that

$$S(-\epsilon, \delta) = -S(\epsilon, \delta),\ S(0, \delta) = 0. \tag{4.151}$$

If $0 \leq \delta \leq 1/2$,

$$S(\epsilon, \delta) = \begin{cases} 4\epsilon(1 - \delta), & 0 \leq \epsilon \leq \delta \\ 4\delta(1 - \epsilon), & \delta \leq \epsilon \leq 1 - \delta \\ 1 + (\epsilon - \delta)(\epsilon - \delta - 2), & 1 - \delta \leq \epsilon \leq 1 + \delta \\ 0, & 1 + \delta \leq \epsilon \end{cases} \tag{4.152}$$

and if $1/2 \leq \delta \leq 1$,

$$S(\epsilon, \delta) = \begin{cases} 4\epsilon(1 - \delta), & 0 \leq \epsilon \leq 1 - \delta \\ 1 + (\epsilon - \delta)(\epsilon - \delta + 2), & 1 - \delta \leq \epsilon \leq \delta \\ 1 + (\epsilon - \delta)(\epsilon - \delta - 2), & \delta \leq \epsilon \leq 1 + \delta \\ 0, & 1 + \delta \leq \epsilon\,. \end{cases} \tag{4.153}$$

Figure 4.18 illustrates the discriminator characteristic for $\delta = 1/2$.

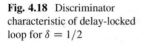

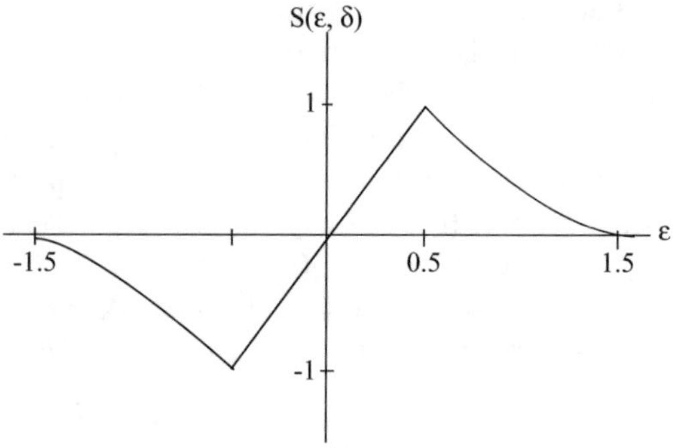

Fig. 4.18 Discriminator characteristic of delay-locked loop for $\delta = 1/2$

As shown in Figure 4.17, the filtered error signal $s_e(t)$ is applied to the voltage-controlled clock (VCC), which generates timing pulses at the clock rate for the three spreading sequences. The error signal causes the VCC to change the clock rate in such a way that the reference spreading sequence converges toward alignment with the received spreading sequence. As illustrated in Figure 4.18 for $\delta = 1/2$, $S(\epsilon, \delta)$ is positive when the reference sequence is delayed relative to the received sequence and $0 < \epsilon(t) < 1.5$. The positive error signal increases the clock rate, and hence $\epsilon(t)$ decreases. The figure indicates that $s_e(t) \to 0$ as $\epsilon(t) \to 0$. Similarly, when $-1.5 < \epsilon(t) < 0$, we find that $s_e(t) \to 0$ as $\epsilon(t) \to 0$. Thus, the delay-locked loop tracks the received code timing once the acquisition system has finished the coarse alignment.

The discriminator characteristic of a code-tracking loop has the appropriate form for tracking only over a finite range of $\epsilon(t)$. Outside that range, code tracking cannot be sustained, the synchronization system loses lock, and a reacquisition search is initiated by the lock detector. Tracking resumes once the acquisition system reduces $\epsilon(t)$ to within the range for which the discriminator characteristic leads to a reduction of $\epsilon(t)$.

To determine the noise that accompanies the error signal, we first assume that white Gaussian noise $n(t)$ with two-sided PSD $N_0/2$ enters both the upper-branch and lower-branch mixers of the delay-locked loop. Let $x(t) = 1$ if $p(t + \delta T_c - \epsilon T_c)$ and $p(t - \delta T_c - \epsilon T_c)$ are equal, and $x(t) = -1$ if they are not. At the output of the upper-branch mixer, the noise is $n_{u1}(t) = n(t)\, p(t + \delta T_c - \epsilon T_c)$. The noise output of the lower-branch mixer is $n_{l1}(t) = x(t)\, n_{u1}(t)$. Both $n_{u1}(t)$ and $n_{l1}(t)$ remain Gaussian and white with PSD $N_0/2$, but they are not always identical.

The upper-branch and lower-branch Gaussian noise outputs of the identical bandpass filters are denoted by $n_{u2}(t)$ and $n_{l2}(t)$, respectively, and $n_{l2}(t) = x(t)\, n_{u2}(t)$. The two-sided PSD of both $n_{u2}(t)$ and $n_{l2}(t)$ is $N_0 |H_b(f)|^2 / 2$. In the upper and lower branches, the squaring devices produce noise outputs $n_{u3}(t) = n_{u2}^2(t) + 2n_{u2}(t)\, s_{u2}(t)$ and $n_{l3}(t) = n_{l2}^2(t) + 2n_{l2}(t)\, s_{l2}(t)$, respectively. Provided that the delays in the two branches are identical, $n_{u2}^2(t) = n_{l2}^2(t)$. Therefore, the output noise of the subtractor, which is the input noise of the loop filter, is

$$\begin{aligned} n_4(t) &= n_{u3}(t) - n_{l3}(t) \\ &= 2n_{u2}(t)\,[s_{u2}(t) - x(t)\, s_{l2}(t)]. \end{aligned} \tag{4.154}$$

We define

$$R_1 = R_p (\delta T_c - \epsilon T_c) \tag{4.155}$$

$$R_2 = R_p (-\delta T_c - \epsilon T_c) . \tag{4.156}$$

Substituting (4.143) and (4.144) into (4.154), we obtain

$$n_4 (t) = 2Ay (t) n_{u2} (t) \cos (2\pi f_c t + \theta) \tag{4.157}$$

where

$$y (t) = R_1 d (t) - R_2 x (t) d (t) . \tag{4.158}$$

Let $S_4 (f)$ denote the PSD of $n_4 (t)$, which is obtained by taking the Fourier transform of its autocorrelation. The narrowband loop filter blocks most of the noise, except for its low-frequency components. Therefore, the average noise power that accompanies the error signal at the VCC input is

$$\mathcal{N}_e \approx 2S_4 (0) W_L \tag{4.159}$$

where the *equivalent loop bandwidth* (defined relative to the positive frequencies only) is

$$W_L = \int_0^{\infty} |H_l (f)|^2 df. \tag{4.160}$$

If we model $d(t)$ as a random binary sequence with a period T_s, then (2.12) indicates that its PSD is

$$S_d (f) = T_s \text{sinc}^2 f T_s. \tag{4.161}$$

The two-sided PSD of $n_{u2} (t)$ is $N_0 |H_b (f)|^2 /2$. The autocorrelation function of the product of two independent processes is equal to the product of the autocorrelation functions. Application of the convolution theorem (Appendix C.1) indicates that the PSD of the product of two autocorrelation functions is equal to the convolution of the PSDs of the two autocorrelation functions. Using this result and (4.142) and ignoring high-frequency components that are ultimately blocked by the narrowband loop filter, the PSD of $n_{u2}(t) \cos (2\pi f_c t + \theta)$ is

$$S_n (f) = \frac{N_0 |H_0 (f)|^2}{2}. \tag{4.162}$$

Let $S_y (f)$ and $S_x (f)$ denote the PSDs of $y (t)$ and $x (t)$, respectively. Since $p (t) d (t)$ is a random binary sequence, the PSD of $x (t) d (t)$ is equal to $S_x (f)$. We assume that $E[x(t)]$ is negligibly small compared with other contributions, so that $S_y(f) \approx R_1^2 S_d(f) + R_2^2 S_x(f)$. From the convolution of $S_y(f)$ and $S_n(f)$, we derive $S_4(0)$ and then obtain

$$\mathcal{N}_e \approx 4A^2 N_0 W_L \left(R_1^2 F_1 + R_2^2 F_2\right) \tag{4.163}$$

where

$$F_1 = \int_{-\infty}^{\infty} S_d(f) \left|H_0(f)\right|^2 df \tag{4.164}$$

$$F_2 = \int_{-\infty}^{\infty} S_x(f) \left|H_0(f)\right|^2 df \tag{4.165}$$

and $F_1 >> F_2$ because $x(t)\, d(t)$ varies much more rapidly than $d(t)$. Equations (4.149) and (4.163) indicate that the average SNR at the input of the VCC is

$$SNR \approx \frac{S^2(\epsilon, \delta)}{8N_0 W_L \left(R_1^2 F_1 + R_2^2 F_2\right)}. \tag{4.166}$$

This equation indicates that the value of δ that maximizes the average SNR is a function of ϵ.

When orthogonal short spreading sequences are used in a synchronous direct-sequence network, the multiple-access interference (Chapter 7) is zero when perfect synchronization exists, but may become significant when there is a code-phase error in the local spreading sequence. In the presence of a tracking error, the delay-locked-loop branch with the larger offset relative to the correct code phase receives relatively more noise power than the other branch. This disparity reduces the slope of the discriminator characteristic and hence degrades the tracking performance [11]. Moreover, because of the nonsymmetric character of the cross-correlations among the spreading sequences, the discriminator characteristic may be biased in one direction, which causes a tracking offset.

Tau-Dither Loop

Satisfactory performance of the noncoherent delay-locked loop depends on accurate matching of the gains, frequency responses, and delays of the two input branches. The noncoherent *tau-dither loop,* which is a lower-complexity alternative code-tracking system, is shown in Figure 4.19. The single branch rather than the two branches of the delay-locked loop resolves the issue of implementing two identical branches.

The *dither signal* $D(t)$ is a square wave that alternates between $+1$ and -1 with dither-symbol duration T_D that exceeds the data-symbol duration T_s. The dither

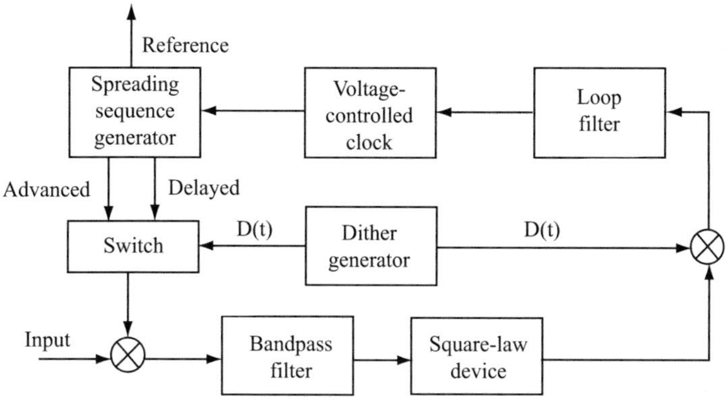

Fig. 4.19 Tau-dither loop

signal controls a switch that alternately passes an advanced or delayed version of the spreading sequence. In the absence of noise, the output of the switch can be represented by

$$s_w(t) = \left[\frac{1 + D(t)}{2}\right] p\,(t + \delta T_c - \epsilon T_c) + \left[\frac{1 - D(t)}{2}\right] p\,(t - \delta T_c - \epsilon T_c) \qquad (4.167)$$

where the two factors within brackets are orthogonal functions of time that alternate between $+1$ and 0, and only one of the factors is nonzero at any instant. The received direct-sequence signal $s(t)$ is given by (4.134). The desired part of the input applied to the bandpass filter is $s_1(t) = s(t)s_w(t)$.

The bandpass filter has a center frequency f_c and a frequency response $H_b(f)$ that has the form given by (4.142). The bandpass filter does not significantly distort $d(t)$ and $D(t)$, but blocks most of the spectral components of the much more rapidly varying products $p(t)p(t \pm \delta T_c - \epsilon T_c)$ except for their slowly varying time averages. Thus, the desired-signal portion of the filter output is

$$s_2(t) \approx Ad(t)\cos(2\pi f_c t + \theta)\left\{\left[\frac{1 + D(t)}{2}\right] R_1 + \left[\frac{1 - D(t)}{2}\right] R_2\right\} \qquad (4.168)$$

where R_1 and R_2 are defined by (4.155) and (4.156). This signal is accompanied by residual undesired spectral components that are blocked downstream by the loop filter. The signal output of the square-law device is

$$s_3(t) \approx \frac{A^2}{2}\left[\frac{1 + D(t)}{2}\right] R_1^2 + \frac{A^2}{2}\left[\frac{1 - D(t)}{2}\right] R_2^2 \qquad (4.169)$$

plus a double-frequency component that is blocked downstream by the loop filter. Since $D(t)[1 + D(t)] = 1 + D(t)$ and $D(t)[1 - D(t)] = -[1 - D(t)]$, the relevant signal input to the loop filter is

$$s_4(t) \approx \frac{A^2}{2}\left[\frac{1+D(t)}{2}\right]R_1^2 - \frac{A^2}{2}\left[\frac{1-D(t)}{2}\right]R_2^2. \tag{4.170}$$

If the slow time-variation of ϵ is ignored, then $s_4(t)$ is a rectangular wave with the same period as $D(t)$.

The loop filter, which has a frequency response $H_l(f)$ centered at $f = 0$, performs a time-averaging of its input over a time interval much longer than the period of $D(t)$. Since the loop filter has a narrow bandwidth relative to that of $D(t)$ and the bandpass filter, double-frequency components due to the squaring and residual undesired spectral components accompanying $s_2(t)$ are blocked. The desired-signal output of the loop filter is approximately equal to the average value of $s_4(t)$. Averaging the two terms of (4.170) and using (4.148), we obtain the error signal that is applied to the VCC:

$$s_5(t) \approx \frac{A^2}{4}S(\epsilon, \delta) \tag{4.171}$$

where the discriminator characteristic $S(\epsilon, \delta)$ is given by (4.151) to (4.153). The error signal regulates the clock rate of the VCC output, which causes the reference spreading sequence to align with the received spreading sequence. Thus, the tau-dither loop can track the code timing in a manner similar to that of the delay-locked loop, but with less hardware than the delay-locked loop and no need to balance the gains and delays in two branches.

The white Gaussian noise at the input to the mixer remains white Gaussian noise at its output, and the bandlimited noise $n_2(t)$ in the output of the bandpass filter remains Gaussian. The noise in the output of the square-law device is $n_3(t) = n_2^2(t) + 2n_2(t)s_2(t)$. Therefore, the noise input to the loop filter is

$$n_4(t) = D(t)n_2^2(t) + 2D(t)n_2(t)s_2(t). \tag{4.172}$$

The contribution of the first term of $n_4(t)$ to the output of the loop filter is much smaller than the contribution of the second term for two reasons. First, code tracking requires the power of $s_2(t)$ to be much greater than the average power of $n_2(t)$; hence, the first term is much smaller than the second term with a high probability. Second, the first term has more energy at high frequencies than the second term; hence, if the loop filter is chosen to have a sufficiently small bandwidth, then the first term is attenuated by the loop filter much more than the second term. Therefore, after substituting (4.168), the relevant part of the loop-filter input is

$$n_5(t) \approx Ay(t)\, n_2(t) \cos(2\pi f_c t + \theta) \tag{4.173}$$

where

$$y(t) = (R_1 - R_2)\, d(t) + (R_1 + R_2)\, d(t)\, D(t). \tag{4.174}$$

We model $d(t)$ as a random binary sequence with a PSD given by (4.161). Similarly, we model $D(t)$ as an independent random binary sequence with a PSD given by

$$S_D(f) = T_D \text{sinc}^2 fT_D. \tag{4.175}$$

Evaluating the autocorrelation function and applying the convolution theorem, we find that the PSD of $y(t)$ is

$$S_y(f) = (R_1 - R_2)^2 S_d(f) + (R_1 + R_2)^2 C(f) \tag{4.176}$$

where

$$C(f) = S_d(f) * S_D(f) \tag{4.177}$$

and the asterisk denotes a convolution. The noise $n_2(t)$ has PSD equal to $N_0 |H_b(f)|^2 /2$, where $H_b(f)$ is the frequency response of the bandpass filter. Applying the convolution theorem, using (4.142), and ignoring high-frequency components that are blocked by the narrowband loop filter, the PSD of $n_2(t) \cos(2\pi f_c t + \theta)$ is given by (4.162).

Let $S_5(f)$ denote the PSD of $n_5(t)$. The narrow loop-filter bandwidth implies that the noise power at the loop-filter output and VCC input is given by

$$\mathcal{N}_e \approx 2S_5(0) W_L \tag{4.178}$$

where W_L is given by (4.160). Calculating $S_5(0)$ by using (4.162), (4.175), and convolutions, and then substituting into (4.178), we obtain

$$\mathcal{N}_e \approx N_0 W_L A^2[(R_1 - R_2)^2 F_1 + (R_1 + R_2)^2 F_3] \tag{4.179}$$

where

$$F_3 = \int_{-\infty}^{\infty} C(f) |H_0(f)|^2 df \tag{4.180}$$

and F_1 is given by (4.164). The SNR at the VCC input is

$$SNR \approx \frac{S^2(\epsilon, \delta)}{4N_0 W_L[(R_1 - R_2)^2 F_1 + (R_1 + R_2)^2 F_3]} \tag{4.181}$$

which indicates that the value of δ that maximizes the SNR is a function of ϵ.

Assuming identical filters in the delay-locked and tau-dither loops, a comparison of (4.181) with (4.166) indicates that the ratio of the SNR at the clock input of the delay-locked loop to the SNR at the clock input of the tau-dither loop is

$$\mathcal{R} \approx \frac{(R_1 - R_2)^2 F_1 + (R_1 + R_2)^2 F_3}{2R_1^2 F_1 + 2R_2^2 F_2}. \tag{4.182}$$

For practical filters, this ratio usually exceeds unity. For example, suppose that $T_D >> T_s$, and hence $F_3 \approx F_1$. Then

$$\mathcal{R} \approx \frac{R_1^2 + R_2^2}{R_1^2 + R_2^2\,(F_2/F_1)} \tag{4.183}$$

and $\mathcal{R} \geq 1$ because $F_2 << F_1$. Thus, the choice of a code-tracking system is primarily a choice between the hardware simplicity of the tau-dither loop and the potential SNR advantage of the delay-locked loop with much more expensive hardware. More details about the tracking-loop design are found in problem 10.

4.7 Frequency-Hopping Patterns

The synchronization of the reference frequency-hopping pattern produced by the receiver synthesizer with the received pattern may be facilitated by precision clocks in both the transmitter and the receiver, feedback signals from the receiver to the transmitter, or transmitted pilot signals. However, in most applications, it is necessary or desirable for the receiver to be capable of obtaining synchronization by processing the received signal. During *acquisition*, the reference pattern is synchronized with the received pattern to within a fraction of a hop duration. The *tracking* system further reduces the synchronization error, or at least maintains it within certain bounds. For communication systems that require a strong capability to reject interference, *matched-filter acquisition* and *serial-search acquisition* are the most effective techniques. The matched filter provides rapid acquisition of short frequency-hopping patterns, but requires the simultaneous synthesis of multiple frequencies. The matched filter may also be used in the configuration of Figure 4.2 to detect short patterns embedded in much longer frequency-hopping patterns. Such a detection can be used to initialize or supplement serial-search acquisition, which is more reliable and accommodates long patterns.

Matched-Filter Acquisition

A *matched-filter acquisition system* uses one or more programmable frequency synthesizers that produce tones at frequencies $f_1, f_2, \ldots, f_N$ that are offset by a constant frequency from the consecutive frequencies of the hopping pattern for code acquisition. Each of these tones produces a downconversion of the received signal in a branch of the matched-filter acquisition system. If a tone minus the offset matches the carrier frequency of a received frequency-hopping pulse, then dehopping occurs, and the energy of the pulse is detected. A version of a matched-filter acquisition system that provides substantial protection against interference [54] is depicted in Figure 4.20. The threshold detector of branch k produces $d_k(t) = 1$ if its threshold

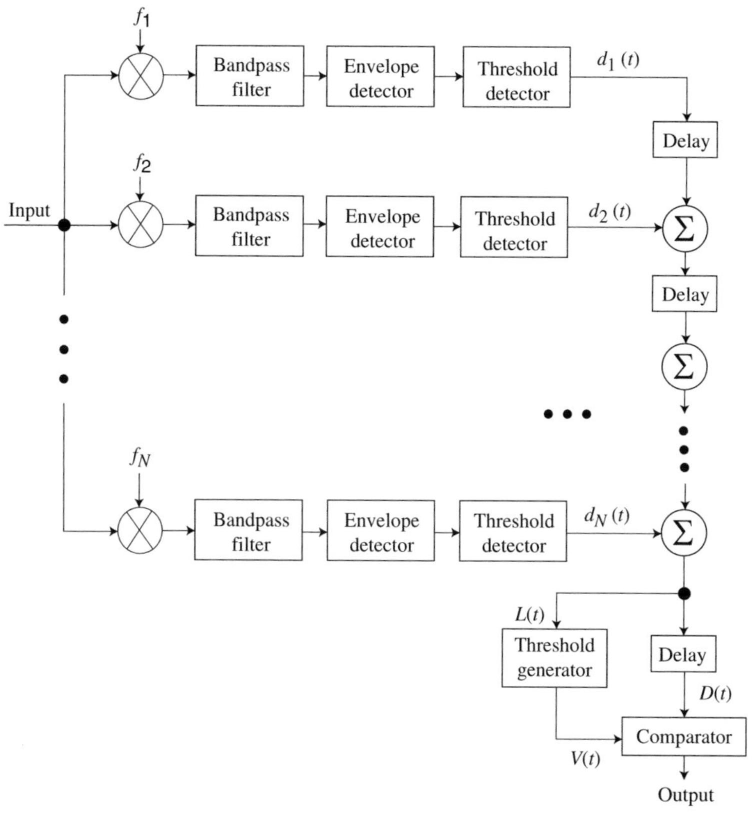

Fig. 4.20 Matched-filter acquisition system for frequency-hopping signals with protection against interference

is exceeded, which ideally occurs only if the received signal hops to a specific frequency. Otherwise, the threshold detector produces $d_k(t) = 0$. The use of binary detector outputs prevents the system from being overwhelmed by a few strong interference signals. Input $D(t)$ of the comparator is the number of frequencies in the hopping pattern that were received in succession. This discrete-valued, continuous-time function is

$$D(t) = \sum_{k=1}^{N} d_k[t - (N - k + 1)T_h] \tag{4.184}$$

where T_h is the hop duration. These waveforms are illustrated in Figure 4.21 (a) for $N = 8$. The input to the threshold generator is

$$L(t) = D(t + T_h). \tag{4.185}$$

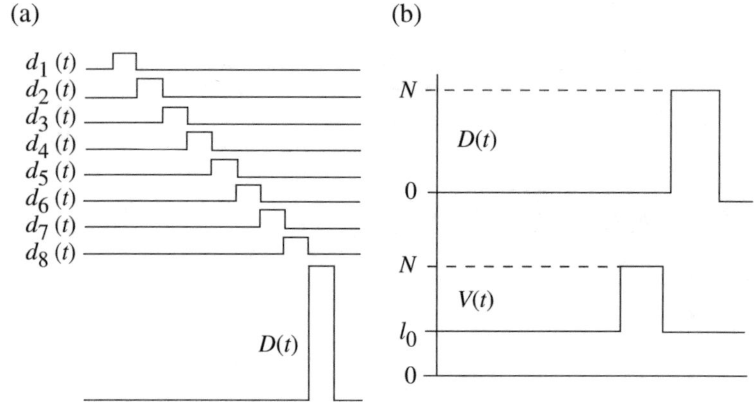

Fig. 4.21 Ideal acquisition system waveforms: (**a**) formation of $D(t)$ when $N = 8$, and (**b**) comparison of $D(t)$ and $V(t)$

Acquisition is declared when $D(t) \geq V(t)$, where $V(t)$ is an adaptive threshold that is a function of $L(t)$. An effective choice is

$$V(t) = \min[L(t) + l_0, N] \tag{4.186}$$

where the adaptation parameter l_0 is a positive integer. When acquisition is declared, a comparator output pulse is applied to a voltage-controlled clock. The clock output regulates the timing of the pattern generator shown in Figure 3.2 (b) so that the receiver-generated pattern nearly coincides with the received frequency-hopping pattern. The matched-filter acquisition system is deactivated, and the dehopped signal is applied to the demodulator.

In the absence of interference and noise, $L(t) = 0$ and $V(t) = l_0$ during the hop interval in which $D(t) = N$, as illustrated in Figure 4.21 (b). If j of the N frequency channels monitored by the matched filter receive strong, continuous interference and $j \leq N - l_0$, then $L(t) = j$ and $V(t) = j + l_0$ during this hop, and $D(t) \geq V(t)$. During other intervals, $j + l_0 \leq V(t) \leq N$, but $D(t) = j$. Therefore, $V(t) > D(t)$, and the matched filter does not declare acquisition. False alarms are prevented because $L(t)$ provides an estimate of the number of frequency channels with continuous interference.

Bandpass filters are used instead of filters matched to the acquisition tones because the appropriate sampling times are unknown. The passbands of the bandpass filters in the branches are assumed to be spectrally disjoint so that tone interference that affects one branch has a negligible effect on the other branches. If zero-mean, white Gaussian noise $n(t)$ enters the branches, then the bandpass-filter noise outputs are jointly Gaussian (Appendix A.1). The noise outputs are also statistically independent of each other if the downconversion tones are sufficiently separated in frequency. To prove this independence, let $R(\tau)$ and $S(f)$ denote the autocorrelation and PSD, respectively, of the white Gaussian noise. Let $h(t)$ denote the impulse response and let $H(f)$ denote the transfer function of each bandpass filter. Let $h_i(t) = h(t)\exp(-j2\pi f_i t)$, $i = 1, 2$, denote the impulse responses of the combined downconverter and bandpass filter in two branches, and let $H(f-f_1)$ and $H(f-f_2)$ denote the corresponding transfer functions. The cross-covariance of the jointly Gaussian, zero-mean bandpass-filter outputs is

$$\begin{aligned}
C &= E\left[\int_{-\infty}^{\infty} h_1(\tau_1)n(t-\tau_1)d\tau_1 \int_{-\infty}^{\infty} h_2(\tau_2)n(t-\tau_2)d\tau_2\right] \\
&= \int_{-\infty}^{\infty}\int_{-\infty}^{\infty} h_1(\tau_1)h_2(\tau_2)R_n(\tau_2-\tau_1)d\tau_1\, d\tau_2 \\
&= \int_{-\infty}^{\infty}\int_{-\infty}^{\infty}\int_{-\infty}^{\infty} h_1(\tau_1)h_2(\tau_2)S(f)\exp[j2\pi f(\tau_2-\tau_1)]df\, d\tau_1\, d\tau_2 \\
&= \int_{-\infty}^{\infty} S(f)H(f-f_1)H^*(f-f_2)df.
\end{aligned} \tag{4.187}$$

Since $S(f)$ is a constant, $C = 0$ if $H(f-f_1)$ and $H(f-f_2)$ are spectrally disjoint or orthogonal.

In practice, the matched filter of Figure 4.20 might operate in continuous time so that acquisition might be declared at any moment. However, for analytical simplicity, the detection and false-alarm probabilities are calculated under the assumption that there is one sample taken per hop dwell time. Suppose that when acquisition tone k is received, the signal at the bandpass-filter output in branch k of the matched filter is

$$r_k(t) = \sqrt{2S}\cos 2\pi f_0 t + \sqrt{2I}\cos(2\pi f_0 t + \phi) + n(t) \tag{4.188}$$

where f_0 is the intermediate frequency, the first term is the desired signal with average power S, the second term represents tone interference with average power I, $n(t)$ is zero-mean Gaussian interference and noise, and ϕ is the phase shift of the tone interference relative to the desired signal. The power in $n(t)$ is

$$\mathcal{N}_1 = \mathcal{N}_t + \mathcal{N}_a \tag{4.189}$$

where $\mathcal{N}_t$ is the power of the thermal noise, and $\mathcal{N}_a$ is the power of the statistically independent noise interference that affects all branches equally. According to (D.32) of Appendix D.2, the zero-mean Gaussian interference and noise have the representation

$$n(t) = n_c(t)\cos 2\pi f_0 t - n_s(t)\sin 2\pi f_0 t \tag{4.190}$$

where $n_c(t)$ and $n_s(t)$ are statistically independent zero-mean Gaussian processes with noise powers equal to $\mathcal{N}_1$. From (4.188), (4.190), and trigonometry, it follows that

$$r_k(t) = \sqrt{Z_1^2(t) + Z_2^2(t)} \cos[2\pi f_0 t + \psi(t)] \tag{4.191}$$

where

$$Z_1(t) = \sqrt{2S} + \sqrt{2I}\cos\phi + n_c(t)\ , \ \ Z_2(t) = \sqrt{2I}\sin\phi + n_s(t)$$
$$\psi(t) = \tan^{-1}\left[\frac{Z_2(t)}{Z_1(t)}\right]. \tag{4.192}$$

Since $n_c(t)$ and $n_s(t)$ are statistically independent, zero-mean Gaussian processes with the same variance, $R = \sqrt{Z_1^2(t_0) + Z_2^2(t_0)}$ at a specific sampling time t_0 has a chi-squared distribution (Appendix E.1) with two degrees of freedom and noncentral parameter

$$\lambda = 2(S + I + \sqrt{SI}\cos\phi). \tag{4.193}$$

The distribution function is given by (E.11) with $N = 2$.

We assume that the bandpass filter causes negligible distortion of R. When the acquisition tone is present, the detection probability for the threshold detector in the branch is the probability that the envelope-detector output R exceeds the threshold η. We make the pessimistic assumption that the interference tone has a frequency exactly equal to that of the acquisition tone, as indicated in (4.188). Then the conditional detection probability given the value of ϕ is

$$P_{11}(\phi) = Q_1\left(\sqrt{\frac{2S + 2I + 2\sqrt{SI}\cos\phi}{\mathcal{N}_1}}, \frac{\eta}{\sqrt{\mathcal{N}_1}}\right) \tag{4.194}$$

where $Q_1(\alpha, \beta)$ is the first-order Marcum Q-function defined by (H.26) of Appendix H.4. If ϕ is modeled as a random variable uniformly distributed over $[0, 2\pi)$, then the detection probability is

$$P_{11} = \frac{1}{\pi}\int_0^{\pi} P_{11}(\phi)d\phi \tag{4.195}$$

where the fact that $\cos\phi$ takes all its possible values over $[0, \pi]$ has been used to shorten the integration interval. In the absence of tone interference, the detection probability is

$$P_{10} = Q_1\left(\sqrt{\frac{2S}{\mathcal{N}_1}}, \frac{\eta}{\sqrt{\mathcal{N}_1}}\right). \tag{4.196}$$

If the acquisition tone is absent from a branch, the false-alarm probability when the tone interference is present or absent is

$$P_{01} = Q_1\left(\sqrt{\frac{2S}{\mathcal{N}_1}}, \frac{\eta}{\sqrt{\mathcal{N}_1}}\right), \; P_{00} = \exp\left(-\frac{\eta^2}{2\mathcal{N}_1}\right). \tag{4.197}$$

In (4.194) to (4.197), the first subscript is 1 when the acquisition tone is present and 0 otherwise, whereas the second subscript is 1 when interference is present and 0 otherwise.

Suppose that tone interference is absent, but noise interference may be present in some branches. When present, the noise interference in a branch is modeled as a zero-mean Gaussian process with power $\mathcal{N}_a$. Then the detection probability when the noise interference is present or absent is

$$P_{11} = Q_1\left(\sqrt{\frac{2S}{\mathcal{N}_a + \mathcal{N}_t}}, \frac{\eta}{\sqrt{\mathcal{N}_a + \mathcal{N}_t}}\right), \; P_{10} = Q_1\left(\sqrt{\frac{2S}{\mathcal{N}_t}}, \frac{\eta}{\sqrt{\mathcal{N}_t}}\right) \tag{4.198}$$

respectively. If the acquisition tone is absent from a branch, the false-alarm probability when the noise interference is present or absent is

$$P_{01} = \exp\left(-\frac{\eta^2}{2\mathcal{N}_a + 2\mathcal{N}_t}\right), \; P_{00} = \exp\left(-\frac{\eta^2}{2\mathcal{N}_t}\right) \tag{4.199}$$

respectively.

It is convenient to define the function

$$\beta(i, N, m, P_a, P_b) = \sum_{n=0}^{i} \binom{m}{n}\binom{N-m}{i-n} P_a^n (1-P_a)^{m-n} P_b^{i-n} (1-P_b)^{N-m-i+n} \tag{4.200}$$

where $\binom{b}{a} = 0$ if $a > b$. Given that m of the N matched-filter branches receive interference of equal power, let the index n represent the number of interfered channels with envelope-detector outputs above η. If $0 \le n \le i$, there are $\binom{m}{n}$ ways to choose n channels out of m and $\binom{N-m}{i-n}$ ways to choose $i-n$ channels with envelope-detector outputs above η from among the $N-m$ channels that are not interfered. Therefore, the conditional probability that $D(t) = i$ given that m channels receive interference is

$$P(D = i|m) = \beta(i, N, m, P_{h1}, P_{h0}), \quad h = 0, 1 \tag{4.201}$$

where $h = 1$ if the acquisition tones are present and $h = 0$ if they are not. Similarly, given that m of N acquisition channels receive interference, the conditional probability that $L(t) = l$ is

$$P(L = l|m) = \beta(l, N, m, P_{h1}, P_{h0}), \quad h = 0, 1. \tag{4.202}$$

If there are J interference signals randomly distributed among a hopset of M frequency channels, then the probability that m out of N matched-filter branches have interference is

$$P_m = \frac{\binom{N}{m}\binom{M-N}{J-m}}{\binom{M}{J}}. \tag{4.203}$$

The probability that acquisition is declared at a particular sampling time is

$$P_A = \sum_{m=0}^{\min(N,J)} P_m \sum_{l=0}^{N} P(L = l|m) \sum_{k=V(l)}^{N} P(D = k|m). \tag{4.204}$$

When the acquisition tones are received in succession, the probability of detection is determined from (4.201) to (4.204). The result is

$$P_D = \sum_{m=0}^{\min(N,J)} \frac{\binom{N}{m}\binom{M-N}{J-m}}{\binom{M}{J}} \sum_{l=0}^{N} \beta(l, N, m, P_{01}, P_{00}) \sum_{k=V(l)}^{N} \beta(k, N, m, P_{11}, P_{10}). \tag{4.205}$$

For simplicity in evaluating the probability of a false alarm, we ignore the sampling time preceding the peak value of $D(t)$ in Figure 4.21 because this probability is negligible at that time. Since the acquisition tones are absent, the probability of a false alarm is

$$P_F = \sum_{m=0}^{\min(N,J)} \frac{\binom{N}{m}\binom{M-N}{J-m}}{\binom{M}{J}} \sum_{l=0}^{N} \beta(l, N, m, P_{01}, P_{00}) \sum_{k=V(l)}^{N} \beta(k, N, m, P_{01}, P_{00}). \tag{4.206}$$

If there is no interference so that $J = 0$, then (4.205) and (4.206) reduce to

$$P_D = \sum_{l=0}^{N} \binom{N}{l} P_{00}^{l}(1 - P_{00})^{N-1} \sum_{k=V(l)}^{N} \binom{N}{k} P_{10}^{k}(1 - P_{10})^{N-k} \tag{4.207}$$

$$P_F = \sum_{l=0}^{N} \binom{N}{l} P_{00}^{l}(1 - P_{00})^{N-1} \sum_{k=V(l)}^{N} \binom{N}{k} P_{00}^{k}(1 - P_{00})^{N-k}. \tag{4.208}$$

If $D(t)$ and $L(t)$ are sampled once every hop dwell interval, then the false-alarm rate is P_F/T_h.

When tone or noise interference may be present and $S/\mathcal{N}_1$ or $S/\mathcal{N}_t$ is specified, the normalized channel threshold $\eta/\sqrt{\mathcal{N}_1}$ or $\eta/\sqrt{\mathcal{N}_t}$ and the adaptation parameter l_0 are selected to maintain a required P_F while maximizing P_D in the absence of interference. The best choice is generally $l_0 = \lfloor N/2 \rfloor$, where $\lfloor x \rfloor$ denotes the largest integer less than or equal to x.

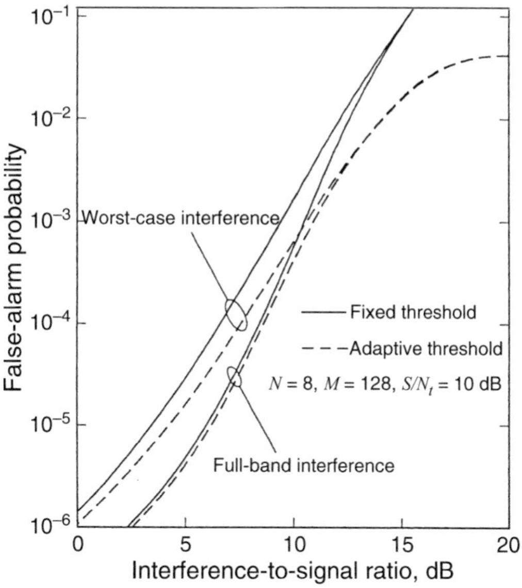

Fig. 4.22 False-alarm probability of matched-filter acquisition system for frequency-hopping signals

Example 3 Suppose that $N = 8$, $P_F = 10^{-7}$, and the SNR is $S/\mathcal{N}_1 = 10\,\text{dB}$ when an acquisition tone is received. A numerical evaluation of (4.208) then yields $\eta/\sqrt{N_1} = 3.1856$ and $l_0 = 4$ as the parameter values that maintain $P_F = 10^{-7}$ while maximizing P_D in the absence of interference. The nearly identical threshold pair $\eta/\sqrt{N_1} = 3.1896$, $l_0 = 4$ is the choice when a fixed comparator threshold $V(t) = l_0$ is used instead of the adaptive threshold of (4.186). Various other performance and design issues and the impact of frequency-hopping interference are addressed in [54]. □

Example 4 Suppose that noise interference with total power $\mathcal{N}_{tot}$ is uniformly distributed over J of the N matched-filter frequency channels so that $\mathcal{N}_a = \mathcal{N}_{tot}/J$ in each of these J frequency channels. The noise power in each of the $N - J$ other channels is $\mathcal{N}_t$. Interference tones are absent, and $N = 8$, $M = 128$, and $S/\mathcal{N}_t = 10\,\text{dB}$. To ensure that $P_F = 10^{-7}$ in the absence of interference, we set $l_0 = 4$ and $\eta/\sqrt{\mathcal{N}_t} = 3.1856$ when an adaptive comparator threshold is used, and set $l_0 = 4$ and $\eta/\sqrt{\mathcal{N}_t} = 3.1896$ when a fixed comparator threshold is used. Since P_D is relatively insensitive to J, the effect of J is assessed by examining P_F. Figure 4.22 depicts P_F as a function of $\mathcal{N}_{tot}/S$, the interference-to-signal ratio. The figure indicates that an adaptive threshold is much more resistant to partial-band interference than a fixed threshold when $\mathcal{N}_{tot}/S$ is large. When $\mathcal{N}_{tot}/S < 10\,\text{dB}$, the worst-case partial-band interference causes a considerably higher P_F than full-band interference. It is found that multitone jamming tends to produce fewer false alarms than noise interference of equal power. □

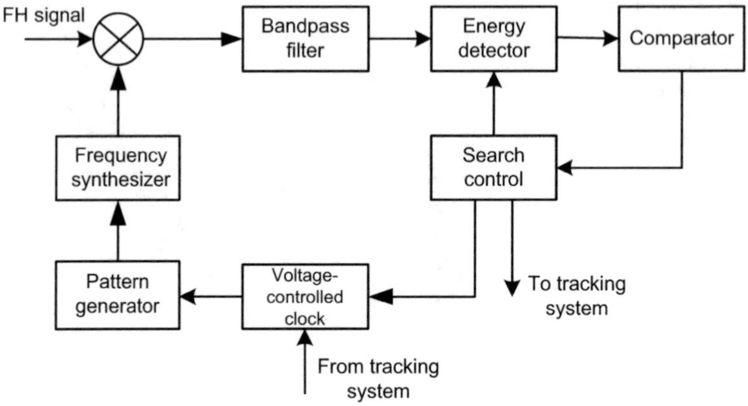

Fig. 4.23 Serial-search acquisition system for frequency-hopping signals

Serial-Search Acquisition

As illustrated by Figure 4.23, a *serial-search acquisition system* for frequency-hopping signals tests acquisition by using a locally generated frequency-hopping pattern to downconvert the received frequency-hopping pattern to a fixed intermediate frequency, and then comparing the output of a radiometer or energy detector (Section 10.2) to a threshold. If the threshold is exceeded, the test is passed; if not, the test is failed. The energy detector comprises a squarer, analog-to-digital converter, and a summer of sample values. A trial alignment of the frequency-hopping pattern synthesized by the receiver with the received pattern is called a *cell*. If a cell passes certain tests, acquisition is declared, and the tracking system is activated. When acquisition is declared, the search control system applies a constant input to a VCC that maintains the timing of the pattern generator so that the receiver-generated pattern nearly coincides with the received frequency-hopping pattern. The dehopped signal at the output of the bandpass filter is applied to the demodulator. If a cell does not pass the tests, it is rejected. A new candidate cell is produced when the search control system sends a signal to the VCC that causes it to advance or delay the reference pattern synthesized by the receiver relative to the received pattern.

A number of search techniques are illustrated in Figure 4.24, which depicts successive frequencies in the received pattern and six possible receiver-generated patterns. The small arrows indicate test times at which cells are usually rejected, and the large arrows indicate typical times at which the search-mode test is passed and subsequent verification testing begins. The step size, which is the separation in hop durations between cells, is denoted by Δ.

Techniques (a) and (b) entail inhibiting the pattern-generator clock after each unsuccessful test. Technique (c) is the same as technique (b) but extends the test duration to three hops. Technique (d) advances the reference pattern by skipping

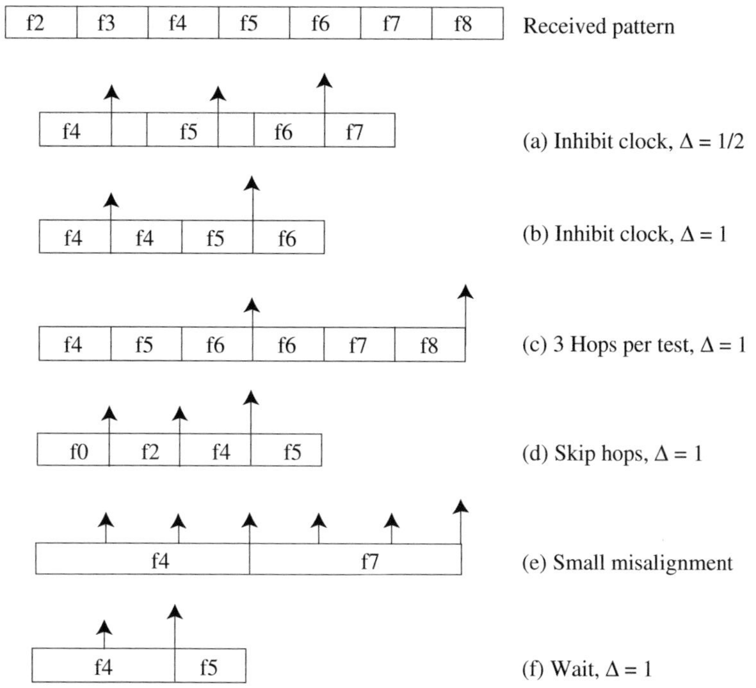

Fig. 4.24 Search techniques for acquisition of frequency-hopping signal

frequencies in the pattern after each unsuccessful test. The inhibiting or advancing of techniques (a) to (d) or an alternation of them continues until the search-mode test is passed.

The *small-misalignment technique* (e) is effective when there is a high probability that the reference and received patterns are within r hops of each other, which is usually true immediately after the tracking system loses lock. The pattern generator temporarily forces the reference pattern to remain at a frequency for $2r + 1$ hop intervals extending both before and after the interval in which the frequency would ordinarily be synthesized. If the misalignment is less than r hops relative to the central hop of the reference pattern, then the search-mode test is passed within $2r+1$ hop durations. In the figure, $r = 1$, the initial misalignment is one-half of hop duration, and we assume that the first time the reference and received frequencies coincide, detection fails, but the second time results in a detection.

The *wait technique* (f) entails waiting at a fixed reference frequency until this frequency is received. The reference frequency is determined from the estimated timing uncertainty, the key bits, and the time-of-day (TOD) bits (Section 3.1), but must be periodically shifted so that neither fading nor interference in any particular frequency channel prevents acquisition. If no acquisition verification occurs within a specified time interval, then the reference frequency must be changed. The wait technique results in a rapid search if the reference frequency can be selected so that it has an unambiguous position within the frequency-hopping pattern and is soon reached.

When the period of the frequency-hopping pattern is large, special measures may be used to reduce the timing uncertainty during an initial system acquisition. A *reduced hopset* with a short pattern period may be used temporarily to reduce the timing uncertainty and hence the acquisition time. In a network, a separate communication channel or *cueing frequency* may provide the TOD bits to subscribers.

The *synchronization-channel technique* assigns a set of dedicated synchronization frequencies and periodically selects one of them during an initial system acquisition. Prior to acquisition, the receiver waits at the selected synchronization frequency until a received signal is detected at that frequency, whereas the transmitted signal periodically hops among the dedicated synchronization frequencies. When the transmitted frequency matches the selected synchronization frequency, the demodulated and decoded data bits indicate the TOD bits of the transmitter and other information that facilitates acquisition of the timing. Once initial system acquisition is declared by the receiver and the transmitter is informed or a specified time duration expires, the transmitter begins to use the frequency-hopping pattern for communication.

The *search control system* determines the timing, the thresholds, and the logic of the tests to be conducted before acquisition is declared and the tracking system is activated. The details of the search control strategy determine the statistics of the acquisition time. The control system is usually a *multiple-dwell system* that uses an initial test with one of the search techniques to quickly eliminate improbable cells. Subsequent tests are used for verification testing of cells that pass the initial test. The multiple-dwell strategy may be a *consecutive-count strategy*, in which a failed test causes a cell to be immediately rejected, or an *up-down strategy*, in which a failed test causes a repetition of a previous test. The up-down strategy is preferable when the interference or noise level is high [73].

Since acquisition for frequency-hopping signals is analogous to acquisition for direct-sequence signals, the statistical description of acquisition given in Section 4.2 is applicable if a chip interval is interpreted as a hop dwell interval. Only the specific equations of the detection and false-alarm probabilities P_d and P_f are different. For example, consider a single-dwell system with a uniform search, a uniform a priori correct-cell location distribution, two independent correct cells with the common detection probability P_d, and q_h cells. By analogy with (4.122), the NMAT is

$$NMAT = \frac{\bar{T}_a}{C_h T_h} = \left(\frac{2-P_D}{2P_D}\right)\frac{q_h}{C_h}\left(N + K_h P_F\right) \tag{4.209}$$

where N is the number of hops per test, K_h is the number of hop durations in the mean penalty time, C_h is the number of hop durations in the timing uncertainty, $q_h >> 1$, $P_D = 2P_d - P_d^2$, and $P_F = P_f$. For step size $\Delta = 1$, $q_h/C_h = 1$; for $\Delta = 1/2$, $q_h/C_h = 2$.

Even if the detector integration is over several hop intervals, strong interference or deep fading over a single hop interval can cause a false alarm with high probability. This problem is mitigated by making a hard decision after integrating over each hop interval. After N decisions, a test for acquisition is passed or failed if the comparator threshold has been exceeded l_0 or more times out of N. Let P_{dp} and P_{da} denote the probabilities that the comparator threshold is exceeded at the end of a hop interval when the correct cell is tested and interference is present and absent respectively. Let P_d denote the probability that an acquisition test is passed when the correct cell is tested. If there is a single correct cell, then the detection probability is $P_D = P_d$; if there are two independent correct cells, then $P_D = 2P_d - P_d^2$. If J denotes the number of frequency channels with interference, and each of the N frequency channels in a test is distinct, then (4.203) gives the probability that m of the N hops encounters interference when J of the M hopset frequencies are interfered. Therefore, when a correct cell is tested, the detection probability is

$$P_d = \sum_{m=0}^{\min(N,J)} \frac{\binom{N}{m}\binom{M-N}{J-m}}{\binom{M}{J}} \sum_{l=l_0}^{N} \beta(l, N, m, P_{dp}, P_{da}) \tag{4.210}$$

where $l_0 \geq 1$. Similarly, the false-alarm probability when an incorrect cell is tested is

$$P_F = \sum_{m=0}^{\min(N,J)} \frac{\binom{N}{m}\binom{M-N}{J-m}}{\binom{M}{J}} \sum_{l=l_0}^{N} \beta(l, N, m, P_{fp}, P_{fa}) \tag{4.211}$$

where P_{fp} and P_{fa} are the probabilities that the threshold is exceeded when a single hard decision is made and interference is present or absent respectively. A suitable choice for l_0 is $\lfloor N/2 \rfloor$. Since the serial-search system of Figure 4.23 has an embedded radiometer, the performance analysis of the radiometer given in Section 10.2 can be used to obtain expressions for P_{dp}, P_{da}, P_{fp}, and P_{fa}.

Although a large step size limits the number of incorrect cells that must be tested before the correct cell is tested, it causes a loss in the average signal energy in the integrator output of Figure 4.23 when a correct cell is tested. This issue and the role of the hop dwell time T_d and the hop duration T_h are illustrated by Figure 4.25, which depicts the idealized output for a single pulse of the received and reference signals in the absence of noise. Let τ_e denote the delay of the reference pattern relative to the received pattern. Suppose that one tested cell has $\tau_e = -x$, where $0 \leq x \leq \Delta T_h$ and $0 < \delta < 1$. The next tested cell has $\tau_e = \Delta T_h - x$ following a

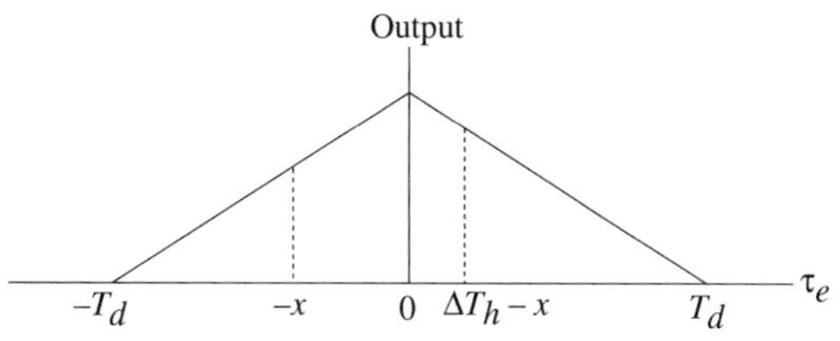

Fig. 4.25 Amplitude of integrator output as a function of the relative pattern delay of the frequency-hopping signal

cell rejection. The largest amplitude of the integrator output occurs when $|\tau_e| = y$, where

$$y = \min(x, \Delta T_h - x)\ , \quad 0 \leq x < \Delta T_h. \tag{4.212}$$

Assuming that x is uniformly distributed over $(0, \Delta T_h)$, y is uniformly distributed over $(0, \Delta T_h/2)$. Therefore,

$$E[y] = \frac{\Delta T_h}{4},\quad E[y^2] = \frac{\Delta^2 T_h^2}{12}. \tag{4.213}$$

The cell for which $|\tau_e| = y$ is the *correct cell*, or one of them. If the output function approximates the triangular shape depicted in the figure, its amplitude when $|\tau_e| = y$ is

$$A = A_{\max}\left(1 - \frac{y}{T_d}\right). \tag{4.214}$$

Therefore, when the correct cell with $|\tau_e| = y$ is tested, the average signal energy in the integrator output is attenuated by the factor

$$E\left[\left(1 - \frac{y}{T_d}\right)^2\right] = 1 - \frac{\Delta T_h}{2T_d} + \frac{\Delta^2 T_h^2}{12 T_d^2} \tag{4.215}$$

which indicates the loss due to the misalignment of patterns. For example, if $T_d = 0.9T_h$, then (4.215) indicates that the average loss is 1.26 dB when $\Delta = 1/2$; if $\Delta = 1$, then the loss is 2.62 dB. These losses should be taken into account when calculating P_{dp} and P_{da}.

The serial-search acquisition of frequency-hopping signals is faster than the acquisition of direct-sequence signals because the hop duration is much greater than a spreading-sequence chip duration for practical systems. Given the same timing uncertainty, fewer cells have to be searched to acquire frequency-hopping signals because each step covers a larger portion of the region.

Pattern Tracking

The acquisition system ensures that the receiver-synthesized frequency-hopping pattern is aligned in time with the received pattern to within a fraction of a hop duration. The pattern-tracking system must provide a fine synchronization by reducing the residual misalignment after acquisition. Although the delay-locked loop used for the code-tracking of direct-sequence signals can be adapted to frequency-hopping signals [64], the predominant form of pattern-tracking in

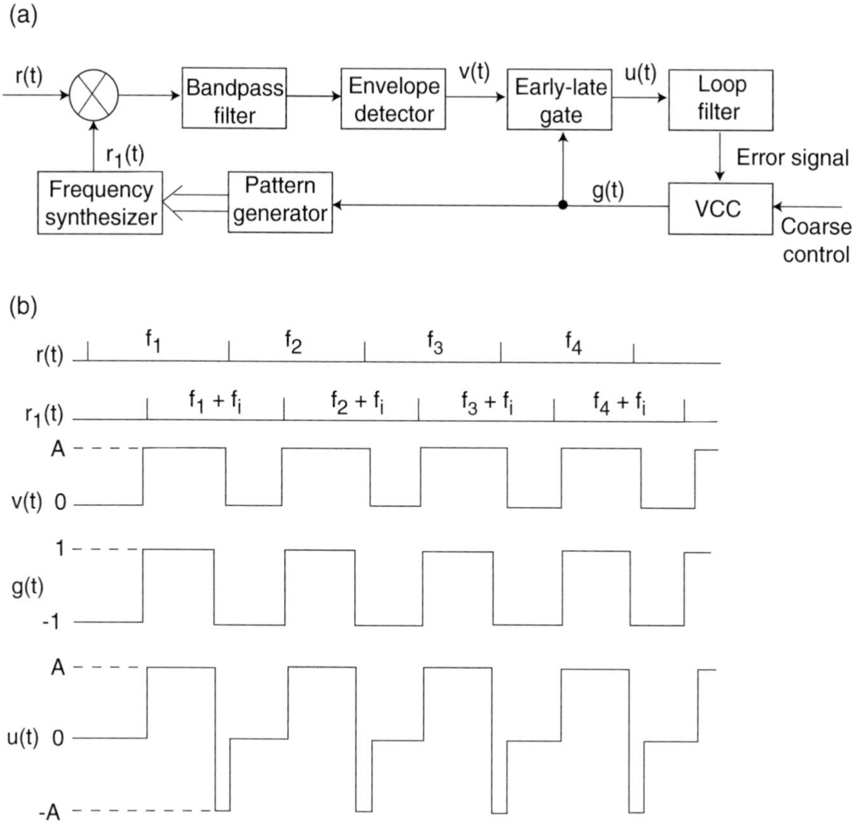

Fig. 4.26 Early-late (**a**) system and (**b**) signals

frequency-hopping systems is provided by the noncoherent *early-late system* [74], which resembles the tau-dither loop. The early-late system is shown in Figure 4.26, along with the ideal associated waveforms for a typical example in which there is a single carrier frequency during a hop dwell interval.

The bandpass filter has a center frequency equal to the intermediate frequency f_i. In the absence of noise, the bandpass filter output and hence the positive envelope-detector output $v(t)$ are significant only when the received frequency-hopping signal $r(t)$ and the receiver-generated frequency-hopping replica $r_1(t)$ are offset in frequency by f_i. As illustrated in Figure 4.26, the fraction of time that $v(t)$ is positive decreases with increases in $\epsilon\ (t)$, the normalized delay of $r_1(t)$ relative to $r(t)$. The *gating signal* $g(t)$ is a square-wave clock signal at the hop rate with transitions from -1 to $+1$ that control the frequency transitions of $r_1(t)$. The early-late gate functions as a mixer with output $u(t) = v(t)g(t)$. The narrowband loop filter produces an error signal that is approximately equal to the average value of $u(t)$ over one period of $g(t)$.

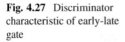

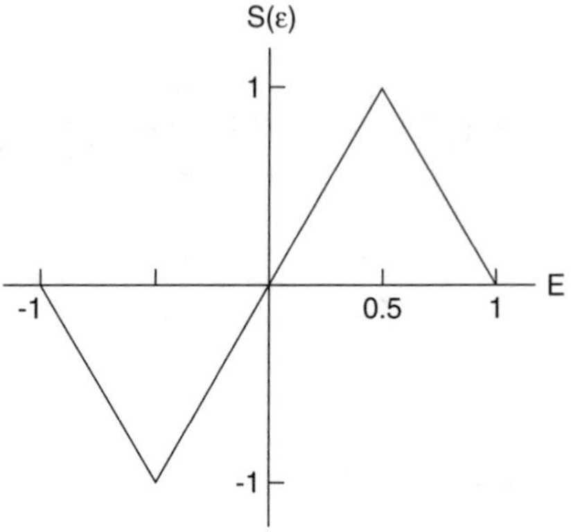

Fig. 4.27 Discriminator characteristic of early-late gate

The error signal is proportional to the discriminator characteristic $S(\epsilon)$, which is a function of $\epsilon\,(t)$. The discriminator characteristic is plotted in Figure 4.27 for an ideal loop filter. For the typical waveforms illustrated in Figure 4.26, $\epsilon\,(t)$ is positive; hence, so is $S(\epsilon)$. Therefore, the VCC increases the transition rate of the gating signal, which brings $r_1(t)$ into better time-alignment with $r(t)$.

The loop filter must be designed so that it responds to the slowly varying $u\,(t)$ and hence $\epsilon(t)$ while blocking as much noise as possible. If the pattern-tracking system loses lock and the small-misalignment test fails, then the wait technique of Figure 4.24 can be used to expedite the reacquisition.

4.8 Problems

1 Prove that for a random variable Y and a random variable X with distribution function $F(x)$, the relation $\int var(Y/x)dF(x) = var(Y)$ is not true in general. If it were, then σ_a^2 given by (4.19) could be simplified. Give a sufficient condition under which this relation is valid.

2 Consider a uniform search with a uniform a priori distribution for the location of the correct cell. (a) What is the average number of sweeps through the timing uncertainty during acquisition? (b) For a large number of cells, calculate an upper bound on $P(T_a > c\overline{T_a})$ as a function of P_D for $c > 1$. (c) For a large number of cells to be searched, show that the standard deviation of the acquisition time satisfies $\frac{\overline{T_a}}{\sqrt{3}} \leq \sigma_a \leq \overline{T_a}$.

3 Derive (4.38) for a consecutive-count double-dwell system by first expressing T_{12} as a conditional expectation and then enumerating the possible values of T_{12} and their conditional probabilities.

4 Derive (4.50) for an up-down double-dwell system by using a method similar to that used in deriving $\bar{T}_{22}$.

5 Consider a lock detector with equal detection probabilities, equal false-alarm probabilities, and equal test durations. Use recursive relations to find $\overline{T_h}$ and $\overline{T_p}$ if (a) the lock detector uses a consecutive-count double-dwell system, and (b) the lock detector maintains lock if either of two consecutive tests are passed.

6 Assume that the correct cell number C has a uniform distribution and that the rewinding time is negligible. Derive the difference between $\bar{T}_a$ for the uniform search and $\bar{T}_a$ for the broken-center Z search.

7 For the acquisition correlator, prove that N_c and N_s are statistically independent and verify (4.94).

8 (a) Assume that $M \geq 200$ and the acquisition correlator tests an incorrect cell. Show that the density-function approximation with X equal to the first two terms of (4.92) and $\mathcal{R} < 0.44$ results in (4.105). (b) Assume that $M \geq 200$ and the acquisition correlator tests the correct cell. Show that the density-function approximation with X equal to the first term of (4.116) and $\mathcal{R} > 16.4/M$ results in (4.119).

9 Derive the detection probability P_d given by (4.121) for the acquisition correlator.

10 (a) The loop filter in a tau-dither loop has a rectangular impulse response

$$h(t) = \frac{1}{T}\left[u(t) - u(t-T)\right]$$

where $u(t)$ is the unit step function. Show that this filter averages its input over a time interval of duration T. Find the frequency response of this filter. Use the definite integral

$$\int_0^\infty \frac{\sin^2 x}{x^2} dx = \frac{\pi}{2}$$

to show that the equivalent loop bandwidth is $W_L = 1/2T$. The definite integral may be evaluated by applying the second Parseval identity of Appendix C.1 to the rectangular function and its Fourier transform.

(b) Both tracking loops have $\delta = 1/2$. The bandpass filters have a corresponding baseband filter that averages its input over duration T_0 and has a frequency response denoted by $H_0(f)$. Since $\delta = 1/2, x(t)$ has the same autocorrelation and power spectral density as the spreading sequence $p(t)$. The loop parameters are selected so that

$$T_c << T_0 << \min(T_s, T_D)$$

which satisfy the basic operational requirements. Use Parseval's identity (Appendix C.1), the convolution theorem, and (2.12) to show that

$$F_1 = 1 - \frac{T_0}{3T_s}, F_2 = \frac{T_c}{T_0}\left(1 - \frac{T_c}{3T_0}\right), F_3 = 1 - \frac{T_0}{3T_s} - \frac{T_0}{3T_D} + \frac{T_0^2}{6T_sT_D}.$$

11 Compare the NMAT for a frequency-hopping system given by (4.209) with the NMAT for a direct-sequence system given by (4.122) when the penalty times and test durations for both systems are equal. Under the latter condition, it is reasonable to assume that P_D and P_F are roughly equal for both systems. With these assumptions, what is the ratio of the direct-sequence NMAT to the frequency-hopping NMAT? What is the physical reason for the advantage of frequency hopping?

12 Consider serial-search acquisition of frequency-hopping signals with $J = 0$. Reduce (4.210) to a single summation and simplify. Could the result have been derived directly?

13 Use (4.210) and (4.211) to derive P_d and P_F for serial-search acquisition of frequency-hopping signals when a single acquisition tone is used. Could the results have been derived directly?

Chapter 5
Adaptive Filters and Arrays

Adaptive filters and adaptive arrays have numerous applications as components of communication systems. This chapter covers those adaptive filters and adaptive arrays that are amenable to exploiting the special spectral characteristics of spread-spectrum signals to enable interference suppression beyond that inherent in the despreading or dehopping. Adaptive filters for the rejection of narrowband interference or primarily for the rejection of wideband interference are presented. The least-mean-square (LMS), normalized LMS, and Frost algorithms are derived and conditions for the convergence of their mean weight vectors are determined. Adaptive arrays for both direct-sequence systems and frequency-hopping systems are described and shown to potentially provide a very high degree of interference suppression.

5.1 Real and Complex Gradients

Real Gradients

Let $\mathcal{R}$ denote the set of real numbers, $\mathcal{R}^n$ denote the set of $n \times 1$ vectors with real components, and the superscript T denote the transpose. If the function $f : \mathcal{R}^n \to \mathcal{R}$ is differentiable with respect to the $n\times 1$ vector $\mathbf{x} = [x_1\ x_2\ \ldots\ x_n]^T$, then the *gradient of* f is defined as

$$\nabla_x f(\mathbf{x}) = \left[\frac{\partial f}{\partial x_1}\ \frac{\partial f}{\partial x_2}\ \cdots\ \frac{\partial f}{\partial x_n}\right]^T \tag{5.1}$$

which is a function from $\mathcal{R}^n$ to $\mathcal{R}^n$. Let $\mathbf{A}$ denote an $n \times n$ matrix, and let $\mathbf{y}$ denote an $n \times 1$ vector. Using the vector and matrix components and the chain rule, we find that

D. Torrieri, *Principles of Spread-Spectrum Communication Systems*,
https://doi.org/10.1007/978-3-319-70569-9_5

$$\nabla_x\left[\mathbf{y}^T\mathbf{x}\right]=\nabla_x\left[\mathbf{x}^T\mathbf{y}\right]=\mathbf{y} \tag{5.2}$$

$$\nabla_x\left[\mathbf{x}^T\mathbf{A}\mathbf{x}\right]=\left(\mathbf{A}+\mathbf{A}^T\right)\mathbf{x} \tag{5.3}$$

and if $\mathbf{A}$ is a symmetric matrix, then $\nabla_x\left[\mathbf{x}^T\mathbf{A}\mathbf{x}\right]=2\mathbf{A}\mathbf{x}$.

Complex Gradients

One can define a complex variable in $\mathcal{R}^2$ as a two-dimensional vector of its real and imaginary parts. A differentiable function from $\mathcal{R}^2$ into itself can then be defined in the usual manner without any allusion to the Cauchy-Riemann conditions. Instead, a complex variable can be defined as a single variable subject to complex arithmetic. Then an analytic function may be defined, provided that the Cauchy-Riemann conditions are satisfied. The benefits of analytic functions are their large number of useful properties, but they are a much more restricted set of functions than the set of differentiable functions from $\mathcal{R}^2$ into itself.

Let $\mathcal{C}$ denote the set of complex numbers. The complex variable z may be expressed in terms of its real and imaginary parts as $z=x+jy$, where $j=\sqrt{-1}$. Similarly, the complex function $f(z):\mathcal{C}\to\mathcal{C}$ may be expressed as

$$f(z)=u(x,y)+jv(x,y) \tag{5.4}$$

where $u(x,y)$ and $v(x,y)$ are real-valued functions. A complex function $f(z)$ defined in a neighborhood of the point z_0 has a *derivative* at z_0 defined by

$$f'(z_0)=\lim_{\Delta z\to 0}\frac{f(z_0+\Delta z)-f(z_0)}{\Delta z} \tag{5.5}$$

if the limit exists and is the same when z approaches z_0 along any path in the complex plane. The complex function $f(z)$ is *analytic* in a domain if $f(z)$ is differentiable at all points of the domain. If $f(z)$ is analytic in a domain, then the first partial derivatives of $u(x,y)$ and $v(x,y)$ exist and satisfy the *Cauchy-Riemann* conditions:

$$\frac{\partial u}{\partial x}=\frac{\partial v}{\partial y},\quad \frac{\partial u}{\partial y}=-\frac{\partial v}{\partial x}\,. \tag{5.6}$$

Conversely, if the real-valued functions $u(x,y)$ and $v(x,y)$ have continuous first partial derivatives that satisfy the Cauchy-Riemann equations in a domain, then $f(z)=u(x,y)+jv(x,y)$ is analytic in that domain [5].

To establish the necessity of the Cauchy-Riemann conditions, let $\Delta z=\Delta x+j\Delta y$ and $\Delta f=f(z_0+\Delta z)-f(z_0)=\Delta u+j\Delta v$. Then

$$\frac{\Delta f}{\Delta z}=\frac{\Delta u+j\Delta v}{\Delta x+j\Delta y}\,. \tag{5.7}$$

Consider $\Delta z \to 0$ using two different approaches. If we set $\Delta y = 0$ and let $\Delta x \to 0$, then

$$\lim_{\Delta z \to 0} \frac{\Delta f}{\Delta z} = \frac{\partial u}{\partial x} + j\frac{\partial v}{\partial x} \tag{5.8}$$

whereas if instead we set $\Delta x = 0$ and let $\Delta y \to 0$, then

$$\lim_{\Delta z \to 0} \frac{\Delta f}{\Delta z} = -j\frac{\partial u}{\partial y} + \frac{\partial v}{\partial y}. \tag{5.9}$$

The definition of the derivative of a complex function requires both limits to be identical. Equating the real and imaginary parts yields the Cauchy-Riemann conditions. Similar proofs establish the same differentiation rules as the standard ones in the calculus of real variables. Thus, the derivatives of sums, products, and quotients of differentiable functions are the same. The chain rule, the derivative of an exponential function, and the derivative of a variable raised to a power are the same.

Although they have many applications, analytic functions are inadequate for some scientific and engineering applications because frequently the functions of interest are functions of both z and its complex conjugate z^*. A difficulty occurs because $f(z) = z^*$ is not an analytic function of z, as verified by observing that the Cauchy-Riemann conditions are not satisfied. To remove this problem, the complex function $g(z, z^*)$ is not required to be analytic with respect to $z = x + jy$ but is considered to be a function of x and y, and a second type of derivative is defined.

A complex function $f(z^*)$ defined in a neighborhood of the point z_0^* has a *derivative* at z_0^* *with respect to* z^* defined by

$$f'\left(z_0^*\right) = \lim_{\Delta z^* \to 0} \frac{f\left(z_0^* + \Delta z^*\right) - f\left(z_0^*\right)}{\Delta z^*} \tag{5.10}$$

if the limit exists and is the same when z^* approaches z_0^* along any path in the complex plane. The complex function $f(z^*)$ is *analytic with respect to* z^* in a domain if $f(z^*)$ is differentiable at all points of the domain. A calculation of the right-hand side of (5.10) for two different paths indicates that if $f(z^*) = u(x, y) + jv(x, y)$ is analytic with respect to z^* in a domain, then the first partial derivatives of $u(x, y)$ and $v(x, y)$ exist and satisfy the conditions

$$\frac{\partial u}{\partial x} = -\frac{\partial v}{\partial y}, \quad \frac{\partial u}{\partial y} = \frac{\partial v}{\partial x}. \tag{5.11}$$

Conversely, if the real-valued functions $u(x, y)$ and $v(x, y)$ have continuous first partial derivatives that satisfy these equations in a domain, then $f(z^*) = u(x, y) + jv(x, y)$ is analytic with respect to z^* in that domain. Again, the derivatives of sums, products, and quotients of differentiable functions are the same as usual. The chain rule, the derivative of an exponential function, and the derivative of a variable raised to a power are the same. The function $f(z^*) = z$ is not an analytic function of z^*, as verified by observing that the conditions (5.11) are not satisfied.

The complex function $g(z, z^*)$ is called an *analytic function of z and z^** if $g(z, z^*)$ is an analytic function of z when z^* is held constant, and an analytic function of z^* when z is held constant. Since z and z^* are distinct functions of the x and y, the chain rule can be used to evaluate partial derivatives of $g(z, z^*)$ with respect to x and y. Since $z = x + jy$,

$$\frac{\partial z}{\partial x} = 1, \quad \frac{\partial z}{\partial y} = j, \quad \frac{\partial z^*}{\partial x} = 1, \quad \frac{\partial z^*}{\partial y} = -j\,. \tag{5.12}$$

The chain rule then implies that an analytic function of z and z^* has partial derivatives

$$\frac{\partial g}{\partial x} = \frac{\partial g}{\partial z} + \frac{\partial g}{\partial z^*} \tag{5.13}$$

$$\frac{\partial g}{\partial y} = j\frac{\partial g}{\partial z} - j\frac{\partial g}{\partial z^*} \tag{5.14}$$

from which it follows that the partial derivatives of $g(z, z^*)$ with respect to z and z^* are

$$\frac{\partial g}{\partial z} = \frac{1}{2}\left(\frac{\partial g}{\partial x} - j\,\frac{\partial g}{\partial y}\right) \tag{5.15}$$

$$\frac{\partial g}{\partial z^*} = \frac{1}{2}\left(\frac{\partial g}{\partial x} + j\,\frac{\partial g}{\partial y}\right). \tag{5.16}$$

These derivatives might not exist if $g(z, z^*)$ had been required to be analytic with respect to z or to z^*.

Let C^n denote the set of $n \times 1$ vectors with complex components. Let $g(\mathbf{z}, \mathbf{z}^*)$ denote a function of $\mathbf{z} \in C^n$ and its complex conjugate $\mathbf{z}^* \in C^n$. The function $g(\mathbf{z}, \mathbf{z}^*)$ is an analytic function of $\mathbf{z}$ and $\mathbf{z}^*$ if $g(\mathbf{z}, \mathbf{z}^*)$ is an analytic function of each component z_i when $\mathbf{z}^*$ is held constant and an analytic function of each component z_i^* when $\mathbf{z}$ is held constant. Let z_i, x_i and y_i, $i = 1, 2, \ldots, n$, denote the components of the $n \times 1$ column vectors $\mathbf{z}$, $\mathbf{x}$, and $\mathbf{y}$, respectively, where $\mathbf{z} = \mathbf{x} + j\mathbf{y}$. The *complex gradient* of $g(\mathbf{z}, \mathbf{z}^*)$ with respect to the n-dimensional complex vector $\mathbf{z}$ is defined as the column vector $\nabla_{\mathbf{z}} g$ with components $\partial g / \partial z_i$, $i = 1, 2, \ldots, n$. Similarly, $\nabla_x g$ and $\nabla_y g$ are the $n \times 1$ gradient vectors with respect to the real-valued vectors $\mathbf{x}$ and $\mathbf{y}$, respectively. The *complex gradient* of $g(\mathbf{z}, \mathbf{z}^*)$ with respect to the n-dimensional complex vector $\mathbf{z}^*$ is defined as the column vector $\nabla_{z^*} g\,(\mathbf{z}, \mathbf{z}^*)$ with components $\partial g / \partial z_i^*$, $i = 1, 2, \ldots, n$. Applying (5.15) and (5.16), we obtain

$$\nabla_z g\left(\mathbf{z},\mathbf{z}^*\right) = \frac{1}{2}\left[\nabla_x g\left(\mathbf{z},\mathbf{z}^*\right) - j\nabla_y g\left(\mathbf{z},\mathbf{z}^*\right)\right] \tag{5.17}$$

$$\nabla_{z^*} g\left(\mathbf{z},\mathbf{z}^*\right) = \frac{1}{2}\left[\nabla_x g\left(\mathbf{z},\mathbf{z}^*\right) + j\nabla_y g\left(\mathbf{z},\mathbf{z}^*\right)\right]. \tag{5.18}$$

Application of (5.17) and (5.18) indicates that if $\mathbf{b}$ is an $n \times 1$ vector and $\mathbf{A}$ is an $n \times n$ matrix, then

$$\nabla_{\mathbf{z}}\left(\mathbf{z}^H\mathbf{b}\right) = \mathbf{0},\ \nabla_{\mathbf{z}}\left(\mathbf{z}^H\mathbf{A}\mathbf{z}\right) = \mathbf{A}^T\mathbf{z}^* \tag{5.19}$$

$$\nabla_{z^*}\left(\mathbf{b}^H\mathbf{z}\right) = \mathbf{0},\ \nabla_{z^*}\left(\mathbf{z}^H\mathbf{A}\mathbf{z}\right) = \mathbf{A}\mathbf{z} \tag{5.20}$$

where the superscript H denotes the conjugate transpose. As desired, (5.19) is the result obtained if $\mathbf{z}^*$ is held constant while calculating the gradient with respect to $\mathbf{z}$, and (5.20) is the result obtained if $\mathbf{z}$ is held constant while calculating the gradient with respect to $\mathbf{z}^*$.

As a one-dimensional example, consider the real-valued function $f(z) = |z|^2 = x^2+y^2$. The Cauchy-Riemann equations are not satisfied; thus, $f(z)$ is not an analytic function of z. However, the function $g(z,z^*) = zz^*$ is an analytic function of z and z^*. Applying (5.19) and (5.20), we obtain

$$\frac{\partial g\left(z,z^*\right)}{\partial z} = z^*,\quad \frac{\partial g\left(z,z^*\right)}{\partial z^*} = z \tag{5.21}$$

which could also be calculated using (5.15) and (5.16). In contrast, if $f(z) = g(z,z^*) = z^2$, then

$$\frac{\partial g\left(z,z^*\right)}{\partial z} = 2z,\quad \frac{\partial g\left(z,z^*\right)}{\partial z^*} = 0\,. \tag{5.22}$$

5.2 Adaptive Filters

Optimal Weight Vector

Adaptive filters [22, 26, 32, 83] are linear filters with weight vectors that respond to the filter input and approximate optimal weight vectors. The input and weight vectors of an adaptive filter are

$$\mathbf{x} = \left[x_1\ x_2 \ldots x_N\right]^T,\quad \mathbf{w} = \left[w_1\ w_2 \ldots w_N\right]^T \tag{5.23}$$

where the components of the vectors may be real or complex. The output of a linear filter is the scalar

$$y = \mathbf{w}^H \mathbf{x}. \tag{5.24}$$

The derivation of the optimal filter weights that provide an estimate of the desired signal depends on the specification of a performance criterion or estimation procedure. Estimators may be derived by using the *maximum-a-posteriori* or the maximum-likelihood criteria, but the standard application of these criteria includes the restrictive assumption that any interference in $\mathbf{x}$ has a Gaussian distribution. Unconstrained estimators that depend only on the second-order moments of $\mathbf{x}$ can be derived by using other performance criteria.

The most widely used method of estimating the desired signal is based on the minimization of the *mean-square error*, which is proportional to the mean power in the *error signal*

$$\epsilon = d - y = d - \mathbf{w}^H \mathbf{x} \tag{5.25}$$

where d is the desired signal or desired response. The conditional expected value of $|\epsilon|^2$ given the value of $\mathbf{w}$ is

$$\begin{aligned} E[|\epsilon|^2|\mathbf{w}] &= E\left[\left(d - \mathbf{w}^H\mathbf{x}\right)\left(d - \mathbf{w}^H\mathbf{x}\right)^H|\mathbf{w}\right] \\ &= E[|d|^2] - \mathbf{w}^H\mathbf{R}_{xd} - \mathbf{R}_{xd}^H\mathbf{w} + \mathbf{w}^H\mathbf{R}_x\mathbf{w} \end{aligned} \tag{5.26}$$

where the *correlation matrix* of $\mathbf{x}$ is the $N \times N$ positive-semidefinite Hermitian matrix (Appendix G)

$$\mathbf{R}_x = E\left[\mathbf{x}\mathbf{x}^H\right] \tag{5.27}$$

and the $N \times 1$ *cross-correlation vector* is

$$\mathbf{R}_{xd} = E\left[\mathbf{x}d^*\right]. \tag{5.28}$$

In terms of its real part $\mathbf{w}_R$, and its imaginary part $\mathbf{w}_I$, a complex weight vector is defined as

$$\mathbf{w} = \mathbf{w}_R + j\mathbf{w}_I. \tag{5.29}$$

We define ∇_{w*}, ∇_{wr}, and ∇_{wi} as the gradients with respect to $\mathbf{w}^*, \mathbf{w}_R$, and $\mathbf{w}_I$, respectively. Equation (5.26) indicates that $E[|\epsilon|^2|\mathbf{w}]$ is an analytic function of $\mathbf{w}$ and $\mathbf{w}^*$. Since (5.18) indicates that $\nabla_{wr}g = 0$ and $\nabla_{wi}g = 0$ imply that $\nabla_{w*}g = 0$, a necessary condition for the optimal weight is obtained by setting $\nabla_{w*}E[|\epsilon|^2|\mathbf{w}] = \mathbf{0}$. Equation (5.26) yields

$$\nabla_{w*}E[|\epsilon|^2|\mathbf{w}] = \mathbf{R}_x\mathbf{w} - \mathbf{R}_{xd}. \tag{5.30}$$

Assuming that the positive-semidefinite Hermitian matrix $\mathbf{R}_x$ is positive definite and hence nonsingular (Appendix G), the necessary condition provides the *Wiener-Hopf equation* for the optimal weight vector:

$$\mathbf{w}_0 = \mathbf{R}_x^{-1}\mathbf{R}_{xd}. \tag{5.31}$$

Equations (5.26) and (5.31) imply that

$$E[|\epsilon|^2|\mathbf{w}] = \epsilon_m^2 + (\mathbf{w} - \mathbf{w}_0)^H \mathbf{R}_x (\mathbf{w} - \mathbf{w}_0) \tag{5.32}$$

where

$$\epsilon_m^2 = E[|d|^2] - \mathbf{R}_{xd}^H\mathbf{R}_x^{-1}\mathbf{R}_{xd}. \tag{5.33}$$

Since $\mathbf{R}_x$ is positive definite, the second term on the right side of (5.32) is positive if $\mathbf{w} \neq \mathbf{w}_0$. Therefore, the Wiener-Hopf equation provides a unique optimal weight vector, and ϵ_m^2 is the minimum mean-square error.

Cauchy-Schwarz Inequality

Let $\mathbf{x}$ and $\mathbf{y}$ denote $N \times 1$ vectors. Then, $||\mathbf{x}-\alpha\mathbf{y}||^2 \geq 0$ for any complex scalar α. Expanding the squared norm and substituting $\alpha = \mathbf{x}^H\mathbf{y}/\,||\mathbf{y}||^2$, where $\mathbf{x} \neq \mathbf{0}$ and $\mathbf{y} \neq \mathbf{0}$, we obtain the *Cauchy-Schwarz inequality for vectors*:

$$\left|\mathbf{x}^H\mathbf{y}\right| \leq ||\mathbf{x}|| \cdot ||\mathbf{y}|| \tag{5.34}$$

which is valid when $\mathbf{x} = \mathbf{0}$ or $\mathbf{y} = \mathbf{0}$. Equality is achieved if and only if $\mathbf{x} = k\mathbf{y}$ for some complex scalar k.

Let x_i and y_i denote the ith components of $\mathbf{x}$ and $\mathbf{y}$, respectively. Then, substitution into (5.34) gives the *Cauchy-Schwarz inequality for sequences of complex numbers:*

$$\left|\sum_{i=1}^{N} x_i^* y_i\right| \leq \left(\sum_{i=1}^{N} |x_i|^2\right)^{1/2} \left(\sum_{i=1}^{N} |y_i|^2\right)^{1/2} \tag{5.35}$$

where equality is achieved if and only if $x_i = ky_i$, $i = 1, 2, \ldots, N$, for some complex scalar k.

Method of Steepest Descent

The implementation of the Wiener-Hopf equation requires the computation of the inverse matrix $\mathbf{R}_x^{-1}$. Since time-varying signal statistics may require frequent computations of $\mathbf{R}_x^{-1}$, adaptive algorithms not entailing matrix inversion are advantageous. The *method of steepest descent* is an iterative method for solving the optimization problem by making successive estimates that improve with each iteration.

Consider the real-valued performance measure $P(\mathbf{w})$ that is to be recursively minimized by successive values of a real-valued weight vector $\mathbf{w}$. Let $\nabla_w P(\mathbf{w})$ denote the $N \times 1$ gradient vector with respect to $\mathbf{w}$. If $P(\mathbf{w})$ is a continuously differentiable function, $\mathbf{u}$ is an $N \times 1$ unit vector, and t is a scalar, then

$$P(\mathbf{w}+t\mathbf{u}) = P(\mathbf{w}) + t\mathbf{u}^T \nabla_w P(\mathbf{w}) + o\,(t) \tag{5.36}$$

where $o\,(t)\,/t \to 0$ as $t \to 0$. By the Cauchy-Schwarz inequality for vectors, the inner product $\nabla_w^T P(\mathbf{w})\mathbf{u}$ is largest for the unit vector $\mathbf{u} = \nabla_w P(\mathbf{w})/\,\|\nabla_w P(\mathbf{w})\|$, which is the unit gradient vector. Thus, $\nabla_w P(\mathbf{w})$ points locally in the direction of steepest ascent of $P(\mathbf{w})$, and $-\nabla_w P(\mathbf{w})$ points locally in the direction of steepest descent. When t is small enough, then $P(\mathbf{w}-t\nabla_w P(\mathbf{w})/\,\|\nabla_w P(\mathbf{w})\|) < P(\mathbf{w})$.

The method of steepest descent changes the weight vector at discrete-time $k+1$ along the direction of $-\nabla_w P(\mathbf{w})$ at discrete-time k, thereby tending to decrease $P(\mathbf{w})$ when $\mathbf{w} = \mathbf{w}(k)$. If the signals and weights are complex, separate steepest-descent equations can be written for the real and imaginary parts of the weight vector. Combining these equations and applying (5.18), we obtain

$$\mathbf{w}(k+1) = \mathbf{w}(k) - 2\mu(k)\nabla_{w*} P\,[\mathbf{w}(k)] \tag{5.37}$$

where the *adaptation parameter* $\mu(k)$ controls the rate of convergence and the stability, and the initial weight vector $\mathbf{w}(0)$ is arbitrary.

For complex signals and weights, a suitable performance measure is the *least-mean-square (LMS) error criterion* $P(\mathbf{w}) = E[|\epsilon|^2|\mathbf{w}]$. The application of (5.30) and (5.37) leads to the *LMS steepest-descent algorithm*:

$$\mathbf{w}(k+1) = \mathbf{w}(k) - 2\mu(k)\,[\mathbf{R}_x\mathbf{w}(k) - \mathbf{R}_{xd}] \tag{5.38}$$

where the initial weight vector $\mathbf{w}(0)$ is arbitrary, but a convenient value is $\mathbf{w}(0) = [1\ 0\ \ldots\ 0]^T$. This ideal algorithm produces a deterministic sequence of weights and does not require a matrix inversion, but it requires the knowledge of $\mathbf{R}_x$ and $\mathbf{R}_{xd}$. At each iteration, $\mu(k)$ can be determined by a line search that minimizes the performance measure. However, for computational simplicity, a constant adaptation parameter $\mu(k) = \mu$ is often preferable.

Least-Mean-Square Algorithm

The presence of interference and noise means that $\mathbf{R}_x$ and $\mathbf{R}_{xd}$ are time-varying and generally unknown; hence, stochastic-gradient algorithms are generally used instead of steepest-descent algorithms. A *stochastic-gradient algorithm* is derived from a steepest-descent algorithm by replacing the gradient of the performance measure by approximations that are more easily obtained. A stochastic-gradient algorithm includes a mechanism for tracking time variations in the desired-signal statistics.

Let $\mathbf{x}(k)$ and $d(k)$ denote the $N \times 1$ input vector and the desired response respectively, at discrete-time k, which represents a sampling time. Let

$$y(k) = \mathbf{w}^H(k)\mathbf{x}(k) \tag{5.39}$$

denote the adaptive-filter output and

$$\epsilon(k) = d(k) - y(k) = d(k) - \mathbf{w}^H(k)\mathbf{x}(k) \tag{5.40}$$

denote the estimation error. The *least-mean-square (LMS) algorithm* is the stochastic-gradient algorithm obtained from (5.38) when $\mathbf{R}_x$ is estimated by $\mathbf{x}(k)\mathbf{x}^H(k)$, $\mathbf{R}_{xd}$ is estimated by $\mathbf{x}(k)d^*(k)$, and $\mu(k) = \mu$. The LMS algorithm is

$$\mathbf{w}(k+1) = \mathbf{w}(k) + 2\mu\epsilon^*(k)\mathbf{x}(k), \quad k \geq 0 \tag{5.41}$$

where the initial weight vector $\mathbf{w}(0)$ is arbitrary, but a convenient value is $\mathbf{w}(0) = [1\ 0\ \ldots\ 0]^T$. The *adaptation constant* μ controls the rate of convergence of the algorithm. According to this algorithm, the next weight vector is obtained by adding to the present weight vector the input vector scaled by the amount of error.

Convergence of the Mean

We prove convergence of the mean weight vector of the LMS algorithm assuming that $\mathbf{x}(k)$ and $d(k)$ are jointly stationary random vectors and that $\mathbf{x}(k+1)$ is independent of $\mathbf{x}(i)$ and $d(i)$, $i \leq k$. The assumption is valid at least when the sampling times are separated by intervals that are large compared with the correlation time of the input process. This assumption and (5.41) imply that $\mathbf{w}(k)$ is independent of $\mathbf{x}(k)$ and $d(k)$. Thus, the expected value of the weight vector satisfies

$$E[\mathbf{w}(k+1)] = (1 - 2\mu\mathbf{R}_x)E[\mathbf{w}(k)] + 2\mu\mathbf{R}_{xd} \tag{5.42}$$

where

$$\mathbf{R}_x = E\left[\mathbf{x}(k)\mathbf{x}^H(k)\right], \quad \mathbf{R}_{xd} = E\left[\mathbf{x}(k)d^*(k)\right] \tag{5.43}$$

are the $N \times N$ Hermitian *correlation matrix* of $\mathbf{x}(k)$ and the $N \times 1$ *cross-correlation* vector, respectively. Let

$$\mathbf{v}(k) = \mathbf{w}(k) - \mathbf{w}_0, \ k \geq 0. \tag{5.44}$$

From (5.42) and (5.31), it follows that

$$E[\mathbf{v}(k+1)] = (\mathbf{I} - 2\mu\mathbf{R}_x)E[\mathbf{v}(k)]. \tag{5.45}$$

With an initial weight vector $\mathbf{w}(0)$, this equation implies that

$$E[\mathbf{v}(k+1)] = (\mathbf{I} - 2\mu\mathbf{R}_x)^{k+1}\mathbf{v}(0) \tag{5.46}$$

where $\mathbf{v}(0) = \mathbf{w}(0) - \mathbf{w}_0$. Assuming that the positive-semidefinite Hermitian matrix $\mathbf{R}_x$ is positive definite, it can be represented as (Appendix G)

$$\mathbf{R}_x = \mathbf{Q}\mathbf{\Lambda}\mathbf{Q}^{-1} = \mathbf{Q}\mathbf{\Lambda}\mathbf{Q}^H \tag{5.47}$$

where $\mathbf{Q}$ is the unitary modal matrix of $\mathbf{R}_x$ with eigenvectors as its columns, and $\mathbf{\Lambda}$ is the diagonal matrix of eigenvalues of $\mathbf{R}_x$. Therefore, (5.46) can be expressed as

$$\begin{aligned} E[\mathbf{v}(k+1)] &= [\mathbf{I} - 2\mu\mathbf{Q}\Lambda\mathbf{Q}^{-1}]^{k+1}\mathbf{v}(0) \\ &= \mathbf{Q}[\mathbf{I} - 2\mu\mathbf{\Lambda}]^{k+1}\mathbf{Q}^{-1}\mathbf{v}(0). \end{aligned} \tag{5.48}$$

This equation indicates that

$$\lim_{k\to\infty} [\mathbf{I} - 2\mu\mathbf{\Lambda}]^{k+1} = \mathbf{0} \tag{5.49}$$

is necessary and sufficient for $E[\mathbf{v}(k)] \to 0$ and hence for the mean weight vector to converge to the optimal weight vector of the Wiener-Hopf equation:

$$\lim_{k\to\infty} E[\mathbf{w}(k)] = \mathbf{w}_0 = \mathbf{R}_x^{-1}\mathbf{R}_{xd}. \tag{5.50}$$

A necessary and sufficient condition for (5.49) and hence (5.50) is that the diagonal elements of the diagonal matrix $[\mathbf{I}-2\mu\mathbf{\Lambda}]$ have magnitudes less than unity:

$$|1 - 2\mu\lambda_i| < 1, \ \ 1 \leq i \leq N. \tag{5.51}$$

Since $\mathbf{R}_x$ is Hermitian positive definite, its eigenvalues are positive (Appendix G); hence, (5.51) implies that $\mu > 0$ and $1 - 2\mu\lambda_{\max} > -1$, where $\lambda_{\max}$ is the largest eigenvalue. Therefore, the necessary and sufficient condition for convergence to the optimal weight vector is

$$0 < \mu < \frac{1}{\lambda_{\max}}. \tag{5.52}$$

Although stronger convergence results can be proved if the inputs are stationary processes and μ is allowed to decrease with the iteration number, making μ constant gives the adaptive system flexibility in processing nonstationary inputs.

The matrix multiplications in (5.48) indicate that during adaptation the weights undergo transients that vary as sums of terms proportional to $(1 - 2\mu\lambda_i)^k$. These transients determine the rate of convergence of the mean vector. The *time constants* $\{\tau_i\}$ of the convergence are defined so that

$$|1 - 2\mu\lambda_i|^k = \exp\left(-\frac{k}{\tau_i}\right), \quad i = 1, 2, ..., N \tag{5.53}$$

which yields

$$\tau_i = -\frac{1}{ln(|1 - 2\mu\lambda_i|)}, \quad i = 1, 2, ..., N. \tag{5.54}$$

The maximum time constant is

$$\tau_{\max} = -\frac{1}{ln(1 - 2\mu\lambda_{\min})} < \frac{1}{2\mu\lambda_{\min}}, \quad 0 < \mu < \frac{1}{\lambda_{\max}} \tag{5.55}$$

where $\lambda_{\min}$ is the smallest eigenvalue of $\mathbf{R}_x$. If μ is selected to be close to its upper bound in (5.55), then $\tau_{\max}$ increases with the *eigenvalue spread* defined as $\lambda_{\max}/\lambda_{\min}$.

Misadjustment

If the random vectors $\mathbf{w}(k)$ and $\mathbf{x}(k)$ are independent, then (5.40), (5.31), and (5.44) imply that

$$E[|\epsilon(k)|^2] = \epsilon_m^2 + E[\mathbf{v}^H(k)\mathbf{R}_x\mathbf{v}(k)] \tag{5.56}$$

where the minimum mean-square error ϵ_m^2 is given by (5.33). If $\mathbf{w}(k) = \mathbf{w}_0$, then $E[|\epsilon|^2] = \epsilon_m^2$. However, even if $E[\mathbf{v}(k)] \to \mathbf{0}$ as $k \to \infty$, it does not necessarily follow that $E[|\epsilon(k)|^2] \to \epsilon_m^2$. A measure of the extent to which the LMS algorithm fails to provide the ideal performance is the excess mean-square error, $E[|\epsilon(k)|^2] - \epsilon_m^2$. The *misadjustment* is a dimensionless measure of the performance loss defined as

$$M = \frac{\lim_{k\to\infty} E[|\epsilon(k)|^2] - \epsilon_m^2}{\epsilon_m^2}. \tag{5.57}$$

With four plausible assumptions, the misadjustment can be shown to be a function of $\mu\, tr(\mathbf{R}_x)$, where $tr(\cdot)$ denotes the trace of a matrix:

1. The jointly stationary random vectors $\mathbf{x}(k+1)$ and $d(k+1)$ are independent of $\mathbf{x}(i)$ and $d(i)$, $i \leq k$. It then follows from (5.41) that $\mathbf{w}(k)$ is independent of $\mathbf{x}(k)$ and $d(k)$.
2. The adaptation constant satisfies

$$0 < \mu < \frac{1}{tr(\mathbf{R}_x)} . \tag{5.58}$$

3. $E\left[\|\mathbf{v}(k)\|^2\right]$ converges as $k \to \infty$.
4. As $k \to \infty$, $|\epsilon(k)|^2$ and $\| \mathbf{x}(k) \|^2$ become uncorrelated so that

$$\lim_{k\to\infty} E[\|\mathbf{x}(k)\|^2 \, |\epsilon(k)|^2] = tr(\mathbf{R}_x) \left\{ \lim_{k\to\infty} E[|\epsilon(k)|^2] \right\} . \tag{5.59}$$

Assumptions 1 and *2* imply convergence of the mean weight vector, which requires (5.52), because the sum of the eigenvalues of a square matrix is equal to its trace, and hence

$$\lambda_{\max} < \sum_{i=1}^{N} \lambda_i = tr(\mathbf{R}_x). \tag{5.60}$$

The total input power is $E\left[\|\mathbf{x}(k)\|^2\right] = tr(\mathbf{R}_x)$. For *Assumption 3* to be true, a tighter restriction on μ than *Assumption 2* may be necessary. *Assumption 4* is physically plausible, but is an approximation.

Equations (5.41) and (5.44) imply that

$$\mathbf{v}(k+1) = \mathbf{v}(k) + 2\mu\epsilon^* (k)\, \mathbf{x}(k). \tag{5.61}$$

Using this equation to calculate $E\left[\|\mathbf{v}(k+1)\|^2\right]$, taking the limit as $k \to \infty$, and then using *Assumptions 3* and *4*, we obtain

$$\lim_{k\to\infty} Re\{E[\mathbf{v}^H(k)\mathbf{x}(k)\epsilon^* (k)]\} = -\mu tr(\mathbf{R}_x) \left\{ \lim_{k\to\infty} E[|\epsilon(k)|^2] \right\} . \tag{5.62}$$

Assumption 1, (5.40), (5.43), (5.44), and (5.56) yield

$$\begin{aligned} \lim_{k\to\infty} E[\mathbf{v}^H(k)\mathbf{x}(k)\epsilon^* (k)] &= \lim_{k\to\infty} E[\mathbf{v}^H(k)]\mathbf{R}_{xd} - \lim_{k\to\infty} E[\mathbf{v}^H(k)\mathbf{R}_x\mathbf{v}(k)] \\ &= \epsilon_m^2 - \lim_{k\to\infty} E[|\epsilon(k)|^2] \end{aligned} \tag{5.63}$$

which is real-valued. Substituting this equation into (5.62), we obtain

$$\lim_{k\to\infty} E[|\epsilon(k)|^2] = \frac{\epsilon_m^2}{1-\mu\, tr(\mathbf{R}_x)} \,. \tag{5.64}$$

Assumption 2 ensures that the right-hand side of this equation is positive and finite, which could not have been guaranteed if the less restrictive (5.52) had been assumed instead.

Substituting (5.64) into (5.57), we obtain

$$M = \frac{\mu\, tr(\mathbf{R}_x)}{1-\mu\, tr(\mathbf{R}_x)} \,. \tag{5.65}$$

According to this equation, increasing μ to improve the convergence rate has the side effect of increasing the misadjustment. For fixed μ, the misadjustment increases with the total input power.

Normalized Least-Mean-Square Algorithm

In some applications of the LMS algorithm, large fluctuations in $\mathbf{x}(k)$ can cause excessive fluctuations in $\mathbf{w}(k+1)$, as indicated by (5.41). This adverse behavior can be curbed by inserting an additional normalizing factor into the stochastic gradient term. The *normalized LMS algorithm* is

$$\mathbf{w}(k+1) = \mathbf{w}(k) + \frac{2\mu\epsilon^*(k)\mathbf{x}(k)}{\delta + \|\mathbf{x}(k)\|^2} \,, \quad k \geq 0 \tag{5.66}$$

where $\epsilon(k)$ is defined by (5.40), and δ is a positive constant that prevents a singularity when $\|\mathbf{x}(k)\|$ is very small.

Rayleigh Quotient

As a preliminary to the convergence analysis, we define the *Rayleigh quotient of a Hermitian matrix* $\mathbf{A}$ as

$$\rho(\mathbf{x}) = \frac{\mathbf{x}^H\mathbf{A}\mathbf{x}}{\|\mathbf{x}\|^2} \,. \tag{5.67}$$

where $\|\cdot\|$ denotes the *Euclidean norm* of a vector and $\|\mathbf{x}\|^2 = \mathbf{x}^H\mathbf{x}$. Let $\mathbf{u}_1, \ldots, \mathbf{u}_N$ denote the orthonormal eigenvectors of $\mathbf{A}$, and $\lambda_1, \ldots, \lambda_N$ the corresponding real-valued eigenvalues (Appendix G). *The vector* $\mathbf{x}$ may be expressed $\mathbf{x} = \upsilon_1\mathbf{u}_1 + \ldots + \upsilon_N\mathbf{u}_N$. Then

$$\mathbf{x}^H\mathbf{A}\mathbf{x} = \lambda_1 |v_1|^2 + \ldots + \lambda_N |v_N|^2 \leq \lambda_{\max}\left(|v_1|^2 + \ldots + |v_N|^2\right) = \lambda_{\max}\|\mathbf{x}\|^2 \tag{5.68}$$

where $\lambda_{\max}$ is the largest eigenvalue. Similarly, $\mathbf{x}^H\mathbf{A}\mathbf{x} \geq \lambda_{\min}\|\mathbf{x}\|^2$, where $\lambda_{\min}$ is the smallest eigenvalue. Thus, the Rayleigh quotient satisfies

$$\lambda_{\min} \leq \frac{\mathbf{x}^H\mathbf{A}\mathbf{x}}{\|\mathbf{x}\|^2} \leq \lambda_{\max}. \tag{5.69}$$

Since the upper bound can be attained if $\mathbf{x}$ is equal to the eigenvector associated with $\lambda_{\max}$, the maximum eigenvalue of $\mathbf{A}$ is

$$\lambda_{\max}(\mathbf{A}) = \max_{\mathbf{x}}\left[\frac{\mathbf{x}^H\mathbf{A}\mathbf{x}}{\|\mathbf{x}\|^2}\right]. \tag{5.70}$$

Convergence of the Mean

The mean weight vector of the normalized LMS algorithm converges under the same principal assumptions as convergence of the LMS algorithm. Thus, we assume that $\mathbf{x}(k)$ and $d(k)$ are jointly stationary random vectors and that $\mathbf{x}(k+1)$ is independent of $\mathbf{x}(i)$ and $d(i)$, $i \leq k$, which implies that $\mathbf{w}(k)$ is independent of $\mathbf{x}(k)$ and $d(k)$. We define the normalized correlation matrix and the normalized cross-correlation vector as

$$\mathbf{B} = E\left[\frac{\mathbf{x}(k)\mathbf{x}^H(k)}{\delta + \|\mathbf{x}(k)\|^2}\right], \quad \mathbf{A} = E\left[\frac{\mathbf{x}(k)d^*(k)}{\delta + \|\mathbf{x}(k)\|^2}\right] \tag{5.71}$$

respectively. The Hermitian positive-semidefinite matrix $\mathbf{B}$ is assumed to be positive definite and hence invertible. We define the vector

$$\mathbf{v}_1(k) = \mathbf{w}(k) - \mathbf{B}^{-1}\mathbf{A}, \ k \geq 0. \tag{5.72}$$

Equations (5.66), (5.40), and (5.72) imply that

$$\mathbf{v}_1(k+1) = \mathbf{v}_1(k) + \frac{2\mu\mathbf{x}(k)\left[d^*(k) - \mathbf{x}^H(k)\mathbf{w}(k)\right]}{\delta + \|\mathbf{x}(k)\|^2}. \tag{5.73}$$

Then (5.71) and the mutual independence of $\mathbf{w}(k)$, $\mathbf{x}(k)$, and $d(k)$ give

$$E[\mathbf{v}_1(k+1)] = (\mathbf{I} - 2\mu\mathbf{B})\, E[\mathbf{v}_1(k)]. \tag{5.74}$$

Since this equation has the same form as (5.45), a derivation similar to that leading to (5.50) implies that

$$\lim_{k\to\infty} E[\mathbf{w}(k)] = \mathbf{B}^{-1}\mathbf{A} \tag{5.75}$$

under the necessary and sufficient condition given by (5.52) with $\lambda_{\max}$ defined as the maximum eigenvalue of **B**.

Applying (5.70) and the Cauchy-Schwarz inequality, the upper bound of the maximum eigenvalue of **B** is found:

$$\begin{aligned}\lambda_{\max}(\mathbf{B}) &= \max_{\mathbf{y}} \left\{ \frac{\mathbf{y}^H\mathbf{B}\mathbf{y}}{\|\mathbf{y}\|^2} \right\} \\ &= \max_{\mathbf{y}} \left\{ E\left[\frac{\left|\mathbf{y}^H\mathbf{x}\right|^2}{\|\mathbf{y}\|^2\left(\delta + \|\mathbf{x}(k)\|^2\right)} \right] \right\} \\ &\leq \max_{\mathbf{y}} \{E[1]\} = 1. \end{aligned} \tag{5.76}$$

Therefore, a sufficient condition for the convergence of the mean weight vector of the normalized LMS algorithm is

$$0 < \mu < 1. \tag{5.77}$$

The fact that this convergence condition does not depend on the total input power is a major advantage of the normalized LMS algorithm.

The normalized mean-square error is

$$E[|\epsilon_1(k)|^2] = E\left[\frac{|\epsilon(k)|^2}{\delta + \|\mathbf{x}(k)\|^2}\right] = \epsilon_{m1}^2 + E[\mathbf{v}_1^H(k)\mathbf{B}\mathbf{v}_1(k)] \tag{5.78}$$

where the normalized minimum mean-square error is

$$\epsilon_{m1}^2 = E\left[\frac{|d|^2}{\delta + \|\mathbf{x}(k)\|^2}\right] - \mathbf{A}^H\mathbf{B}^{-1}\mathbf{A}. \tag{5.79}$$

Misadjustment

We define

$$C = E\left[\frac{\|\mathbf{x}(k)\|^2}{\delta + \|\mathbf{x}(k)\|^2}\right]. \tag{5.80}$$

With four plausible assumptions, the misadjustment can be derived:

1. The jointly stationary random vectors $\mathbf{x}(k+1)$ and $d(k+1)$ are independent of $\mathbf{x}(i)$ and $d(i)$, $i \leq k$, which implies that $\mathbf{w}(k)$ is independent of $\mathbf{x}(k)$ and $d(k)$.
2. The adaptation constant satisfies (5.77).
3. $E[\|\mathbf{v}_1(k)\|^2]$ converges as $k \to \infty$.
4. As $k \to \infty$, $|\epsilon_1(k)|^2$ and $\|\mathbf{x}(k)\|^2$ become uncorrelated so that

$$\lim_{k\to\infty} E\left[|\epsilon(k)|^2 \frac{\|\mathbf{x}(k)\|^2}{\left(\delta + \|\mathbf{x}(k)\|^2\right)^2}\right] = C\left\{\lim_{k\to\infty} E[|\epsilon_1(k)|^2]\right\}. \tag{5.81}$$

Using (5.73) to calculate $E\left[\|\mathbf{v}_1(k+1)\|^2\right]$, taking the limit as $k \to \infty$, and then using *Assumptions 3* and *4*, we obtain

$$\lim_{k\to\infty} Re\left\{E\left[\frac{\mathbf{v}_1^H(k)\epsilon^*(k)\,\mathbf{x}(k)}{\delta + \|\mathbf{x}(k)\|^2}\right]\right\} = -\mu C\left\{\lim_{k\to\infty} E[|\epsilon_1(k)|^2]\right\}. \tag{5.82}$$

Assumption 1, (5.40), (5.72), (5.75), and (5.71) yield

$$\begin{aligned}\lim_{k\to\infty} E\left[\frac{\mathbf{v}_1^H(k)\epsilon^*(k)\,\mathbf{x}(k)}{\delta + \|\mathbf{x}(k)\|^2}\right] &= -\lim_{k\to\infty} E[\mathbf{v}_1^H(k)\mathbf{B}\mathbf{v}_1(k)] \\ &= \epsilon_{m1}^2 - \lim_{k\to\infty} E[|\epsilon_1(k)|^2]\end{aligned} \tag{5.83}$$

which is real-valued. Substituting this equation into (5.82), we obtain

$$\lim_{k\to\infty} E[|\epsilon_1(k)|^2] = \frac{\epsilon_{m1}^2}{1-\mu C}. \tag{5.84}$$

Since $C \leq 1$, Assumption 2 ensures that the right-hand side of this equation is positive and finite. The misadjustment is

$$\begin{aligned}M &= \frac{\lim_{k\to\infty} E[|\epsilon_1(k)|^2] - \epsilon_{m1}^2}{\epsilon_{m1}^2} = \frac{\mu C}{1-\mu C} \\ &\leq \frac{\mu}{1-\mu}.\end{aligned} \tag{5.85}$$

This upper bound on the misadjustment is independent of the total input power. This independence provides an advantage of the normalized LMS algorithm relative to the computationally simpler LMS algorithm.

5.3 Rejection of Narrowband Interference

Narrowband interference presents a crucial problem for *spread-spectrum overlay systems*, which are systems that have been assigned a spectral band already occupied by narrowband communication systems. Jamming against tactical spread-spectrum communications is another instance of narrowband interference that may exceed the natural resistance of a practical spread-spectrum system, which has a limited spreading factor. There are a wide variety of techniques that supplement the inherent ability of a direct-sequence system to reject narrowband interference [56, 116]. All the techniques directly or indirectly exploit the spectral disparity between the narrowband interference and the wideband direct-sequence signal. The most useful methods can be classified as time-domain adaptive filtering, transform-domain processing, or nonlinear filtering techniques. The processor that implements one of the rejection methods for a direct-sequence signal follows the chip-rate sampling of the chip-matched-filter outputs in Figure 2.14 or Figure 2.18. The processor provides the input to the despreader, as shown in Figure 5.1. Since the narrowband interference is rarely known with any precision, adaptive filters are an essential part of transform-domain processing and nonlinear filtering.

Time-Domain Adaptive Filters

A *time-domain adaptive filter* [52] for interference suppression processes the baseband sample values of a received signal to adaptively estimate the interference. This estimate is subtracted from the sample values, thereby canceling the interference. The adaptive filter is primarily a predictive system that exploits the inherent predictability of a narrowband signal to form an accurate replica of it for the subtraction. Since the wideband desired signal is largely unpredictable, it does not significantly impede the prediction of a narrowband signal. When adaptive filtering is used, the processor in Figure 5.1 has the form of Figure 5.2 (a). The adaptive

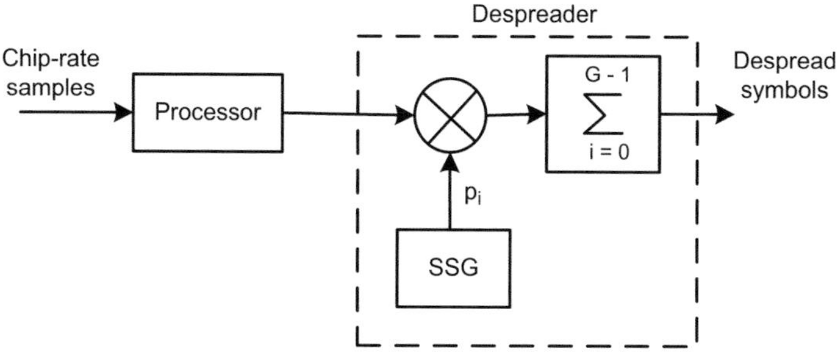

Fig. 5.1 Direct-sequence receiver with processor for rejecting narrowband interference

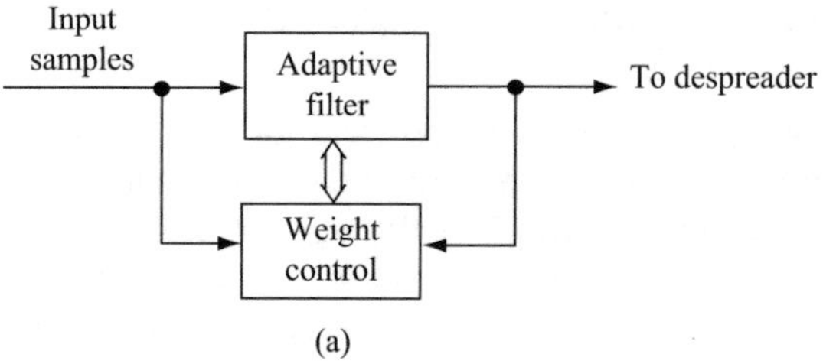

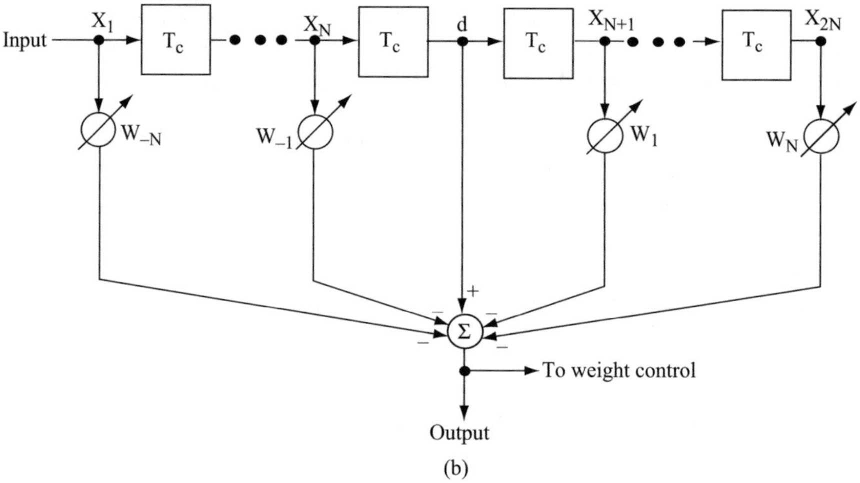

Fig. 5.2 (**a**) Processor using adaptive filter and (**b**) two-sided adaptive transversal filter

filter comprises a one-sided or two-sided digital delay line or shift register, with each stage providing a delay equal to the chip duration T_c. The outputs of the stages are multiplied by the adaptive weights.

The *two-sided adaptive filter* multiplies each tap output by a weight except for the central tap output, as diagrammed in Figure 5.2 (b). This filter is an *interpolator* in that it uses both past and future samples to estimate the value to be subtracted. The two-sided filter provides a better performance than the one-sided filter, which is a *predictor*. The adaptive algorithm of the weight-control mechanism is designed to adjust the weights so that the power in the filter output is minimized. The direct-sequence components of the tap outputs, which are delayed by integer multiples of a chip duration, are largely uncorrelated with each other, but the narrowband interference components are strongly correlated. As a result, the adaptive algorithm causes the interference cancelation in the filter output, but the direct-sequence signal is largely unaffected.

An adaptive filter with $2N + 1$ taps and $2N$ weights is shown in Figure 5.2 (b). The $2N \times 1$ input and weight vectors at iteration k are

$$\mathbf{x}(k) = [x_1(k)\, x_2(k) \ldots x_{2N}(k)]^T \tag{5.86}$$

and

$$\mathbf{w}(k) = [w_{-N}(k)\ w_{-N+1}(k) \ldots w_{-1}(k)\ w_1(k) \ldots w_N(k)]^T \tag{5.87}$$

respectively. The central tap output, which is denoted by $d(k)$ and serves as an approximation of the desired response, has been excluded from $\mathbf{x}(k)$. The LMS algorithm computes the weight vector using (5.39)-(5.41). If coherent demodulation of BPSK produces real-valued inputs to the adaptive filter, $\mathbf{x}(k)$ and $\mathbf{w}(k)$ are assumed to have real-valued components. The output of the adaptive filter is applied to the despreader.

Under certain conditions, the mean weight vector converges to the optimal weight vector $\mathbf{w}_0$ given by (5.31) and (5.43) after a number of iterations. If we assume that $\mathbf{w}(k) \rightarrow \mathbf{w}_0$, then a straightforward analysis indicates that the adaptive filter provides substantial suppression of narrowband interference [56]. Although the interference suppression increases with the number of taps, it is always incomplete if the interference has a nonzero bandwidth because a finite impulse response filter can only place a finite number of zeros in the frequency domain.

The adaptive filter is inhibited by the presence of direct-sequence components in the filter input vector $\mathbf{x}(k)$. These components can be suppressed by using decision-directed feedback, as shown in Figure 5.3. Previously detected symbols remodulate the spreading sequence delayed by G chips (long sequence) or one period of the spreading sequence (short sequence). After an amplitude compensation by a factor η, the resulting sequence provides estimates of the direct-sequence components of previous input samples. A subtraction then provides estimated sample values of the interference and noise that are largely free of direct-sequence contamination. These samples are then applied to an adaptive filter that has the form of Figure 5.2, except that it has no central tap. The adaptive filter produces refined interference estimates that are subtracted from the input samples to produce samples that have

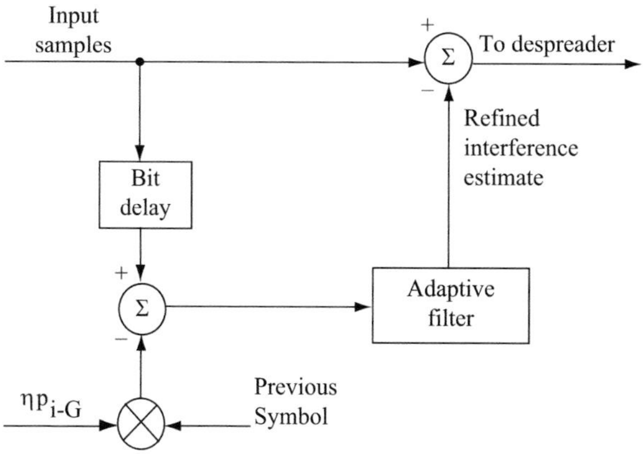

Fig. 5.3 Processor with decision-directed adaptive filter

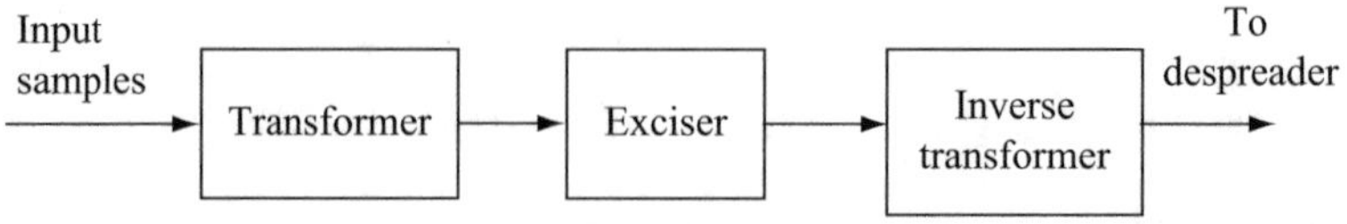

Fig. 5.4 Transform-domain filter

relatively small interference components. An erroneous symbol from the decision device causes an enhanced direct-sequence component in samples applied to the adaptive filter, and error propagation is possible. However, for moderate values of the signal-to-interference ratio at the input, the performance is not significantly degraded.

An adaptive filter is only effective after the convergence of the adaptive algorithm, which may not be able to track time-varying interference. In contrast, transform-domain processing suppresses interference almost instantaneously.

Transform-Domain Filters

The principal components of a *transform-domain filter* are depicted in Figure 5.4. The input consists of the output samples of chip-matched filters. Blocks of these samples feed a discrete-time Fourier or wavelet transformer. The transform is selected so that the transforms of the desired signal and interference are easily distinguished. The array of transform values are called *transform bins*. Ideally, the transform produces interference components that are largely confined to a few transform bins, whereas the desired signal components have nearly the same magnitude in all the transform bins. A simple exciser can then suppress the interference with little impact on the desired signal by setting the spectral weights corresponding to components in bins containing strong interference to 0, while setting all remaining spectral weights to 1. The decision as to which bins contain interference can be based on the comparison of each bin with a threshold or by selecting those transform bins with the largest average magnitudes. After the excision operation, the desired signal is largely restored by the inverse transformer.

Much better performance against stationary narrowband interference may be obtained by using a transform-domain adaptive filter as the exciser [116]. This filter adjusts a single nonbinary weight at each transform-bin output. The adaptive algorithm is designed to minimize the difference between the weighted transform and a desired signal that is the transform of the spreading sequence used by the input block of the processor. If the direct-sequence signal uses the same short spreading sequence for each data symbol and each processor input block includes the chips for a single data symbol, then the desired signal transform may be stored in a read-only memory. However, if a long spreading sequence is used, then the spreading-sequence transform must be periodically produced from outputs of the receiver's code generator. The main disadvantage of the adaptive filter is that its convergence rate may be insufficient to track rapidly time-varying interference.

A transform that operates on disjoint blocks of N input samples may be defined in terms of N orthonormal, N-component basis vectors:

$$\boldsymbol{\phi}_i = [\phi_{i1}\ \phi_{i2} \ldots \phi_{iN}]^T, \quad i = 1, 2, \ldots, N \tag{5.88}$$

which span a linear vector space of dimension N. Since the components may be complex numbers, the orthonormality implies that

$$\boldsymbol{\phi}_i^H \boldsymbol{\phi}_k = \begin{cases} 0, & i \neq k \\ 1, & i = k. \end{cases} \tag{5.89}$$

The input block $\mathbf{x} = [x_1\ x_2 \ldots x_N]^T$ may be expressed in terms of the basis as

$$\mathbf{x} = \sum_{i=1}^{N} c_i \boldsymbol{\phi}_i \tag{5.90}$$

where

$$c_i = \boldsymbol{\phi}_i^H \mathbf{x}, \ \ i = 1, 2, \ldots, N. \tag{5.91}$$

If the discrete Fourier transform is used, then $\phi_{ik} = \exp(j2\pi ik/N)$, where $j = \sqrt{-1}$.

The transformer extracts the transform vector $\mathbf{c} = [c_1\ c_2\ \ldots\ c_N]^T$ by computing

$$\mathbf{c} = \mathbf{B}^H \mathbf{x} \tag{5.92}$$

where $\mathbf{B}$ is the unitary matrix of basis vectors:

$$\mathbf{B} = [\boldsymbol{\phi}_1\ \boldsymbol{\phi}_2\ \ldots\ \boldsymbol{\phi}_N]. \tag{5.93}$$

The exciser reduces components of $\mathbf{c}$ that are excessively large and hence likely to have large interference components. The exciser computes

$$\mathbf{e} = \mathbf{W}(\mathbf{c})\,\mathbf{c} \tag{5.94}$$

where $\mathbf{W}(\mathbf{c})$ is the $N \times N$ diagonal weight matrix with diagonal elements w_i, $1 \leq i \leq N$. In general, the diagonal elements of $\mathbf{W}(\mathbf{c})$ are set to +1 or 0 by either a threshold device or an adaptive filter fed by $\mathbf{c}$.

The inverse transformer produces the excised block that is applied to the despreader:

$$\mathbf{z} = [z_1\ z_2 \ldots z_N]^T = \mathbf{B}\,\mathbf{e} = \mathbf{B}\,\mathbf{W}(\mathbf{c})\,\mathbf{c} = \mathbf{B}\,\mathbf{W}(\mathbf{c})\,\mathbf{B}^H \mathbf{x}. \tag{5.95}$$

If there were no excision, then $\mathbf{W}(\mathbf{c}) = \mathbf{I}$, $\mathbf{B}\mathbf{B}^H = \mathbf{I}$, and $\mathbf{z} = \mathbf{x}$ would result, as expected when the transformer and inverse transformer are in tandem.

Let

$$\mathbf{p} = [p_1 \, p_2 \, \ldots \, p_N]^T \tag{5.96}$$

denote a synchronous replica of N chips of the spreading sequence generated by the receiver's code generator. The despreader correlates its input block with the local replica to form the decision variable:

$$V = \mathbf{p}^T \mathbf{z}. \tag{5.97}$$

The filtering and despreading can be simultaneously performed in the transform domain because (5.95) to (5.97) give

$$V = \mathbf{p}^T \mathbf{B} \mathbf{W}(\mathbf{c})\, \mathbf{c}. \tag{5.98}$$

Extension of the Kalman Filter

By modeling the narrowband interference as part of a dynamic linear system, the linear Kalman filter can be used [83] to extract an optimal linear estimate of the interference. A subtraction of this estimate from the filter input then removes a large part of the interference from the despreader input. However, a superior nonlinear filter can be designed by approximating an extension of the Kalman filter.

Consider the estimation of an $n \times 1$ state vector $\mathbf{x}_k$ of a dynamic system based on the $r \times 1$ observation vector $\mathbf{z}_k$. Let $\boldsymbol{\phi}_k$ denote the $n \times n$ state transition matrix, $\mathbf{H}_k$ an $r \times n$ observation matrix, and $\mathbf{u}_k$ and $\mathbf{v}_k$ disturbance vectors of dimensions $n \times 1$ and $r \times 1$, respectively. According to the *linear dynamic system model*, the random state and observation vectors satisfy

$$\mathbf{x}_{k+1} = \boldsymbol{\Phi}_k \mathbf{x}_k + \mathbf{u}_k, \tag{5.99}$$

$$\mathbf{z}_k = \mathbf{H}_k \mathbf{x}_k + \mathbf{v}_k, \qquad 0 \le k < \infty \tag{5.100}$$

where all variables are real-valued, and $\mathbf{u}_k, \mathbf{v}_k, k \ge 0$, are sequences of independent, zero-mean random vectors that are also independent of each other and the initial state $\mathbf{x}_0$. The covariances of $\mathbf{u}_k$ and $\mathbf{v}_k$ are

$$E\left[\mathbf{u}_k \mathbf{u}_k^T\right] = \mathbf{Q}_k, \;\; \mathbf{R}_k = E\left[\mathbf{v}_k \mathbf{v}_k^T\right]. \tag{5.101}$$

Let $Z^k = (\mathbf{z}_1, \, \mathbf{z}_2, \, \ldots \, \mathbf{z}_k)$ denote the first k observation vectors. Let $f(\mathbf{z}_k | Z^k)$ and $f(\mathbf{x}_k | Z^k)$ denote the density functions of $\mathbf{z}_k$ and $\mathbf{x}_k$, respectively, conditioned on Z^k.

The conditional mean and covariance of $\mathbf{x}_k$ with respect to Z^k are denoted by

$$\hat{\mathbf{x}}_k = E\left[\mathbf{x}_k \left| Z^k \right.\right] \tag{5.102}$$

and

$$\mathbf{P}_k = E\left[\left(\mathbf{x}_k - \hat{\mathbf{x}}_k\right)\left(\mathbf{x}_k - \hat{\mathbf{x}}_k\right)^T \left| Z^k \right.\right] \tag{5.103}$$

respectively. The conditional mean and covariance of $\mathbf{x}_k$ with respect to Z^{k-1} are denoted by

$$\bar{\mathbf{x}}_k = E\left[\mathbf{x}_k \left| Z^{k-1} \right.\right] \tag{5.104}$$

and

$$\mathbf{M}_k = E\left[\left(\mathbf{x}_k - \bar{\mathbf{x}}_k\right)\left(\mathbf{x}_k - \bar{\mathbf{x}}_k\right)^T \left| Z^{k-1} \right.\right] \tag{5.105}$$

respectively. If $f(\mathbf{x}_k|Z^{k-1})$ is an n-dimensional Gaussian density function with mean $\bar{\mathbf{x}}_k$ and $n \times n$ covariance matrix $\mathbf{M}_k$, then (A.18) of Appendix A.1 gives

$$f(\mathbf{x}_k|Z^{k-1}) = (2\pi)^{-n/2}\left(\det \mathbf{M}_k\right)^{-1/2} \exp\left[-\frac{1}{2}(\mathbf{x}_k - \bar{\mathbf{x}}_k)^T \mathbf{M}_k^{-1}(\mathbf{x}_k - \bar{\mathbf{x}}_k)\right]. \tag{5.106}$$

From (5.100), it follows that the expectation of $\mathbf{z}_k$ conditioned on Z^{k-1} is

$$\bar{\mathbf{z}}_k = E\left[\mathbf{z}_k \left| Z^{k-1} \right.\right] = \mathbf{H}_k \bar{\mathbf{x}}_k. \tag{5.107}$$

From (5.100), (5.105), and (5.101), it follows that the covariance of $\mathbf{z}_k$ conditioned on Z^{k-1} is

$$\mathbf{L}_k = E\left[\left(\mathbf{z}_k - \bar{\mathbf{z}}_k\right)\left(\mathbf{z}_k - \bar{\mathbf{z}}_k\right)^T \left| Z^{k-1} \right.\right] = \mathbf{H}_k \mathbf{M}_k \mathbf{H}_k^T + \mathbf{R}_k. \tag{5.108}$$

The following theorem extends the Kalman filter by not assuming that $f(\mathbf{z}_k|Z^{k-1})$ is a Gaussian density function.

Theorem (Masreliez) *Consider the dynamic system described by (5.99) to (5.101). Assume that $f(\mathbf{x}_k|Z^{k-1})$ is a Gaussian density function with n-dimensional mean $\bar{\mathbf{x}}_k$ and $n \times n$ covariance matrix $\mathbf{M}_k$, and that $f(\mathbf{z}_k|Z^{k-1})$ is twice differentiable with respect to the r components of $\mathbf{z}_k$. Then, the conditional expectation $\hat{\mathbf{x}}_k$ and the conditional covariance $\mathbf{P}_k$ satisfy*

$$\hat{\mathbf{x}}_k = \bar{\mathbf{x}}_k + \mathbf{M}_k \mathbf{H}_k^T \mathbf{g}_k\left(\mathbf{z}_k\right) \tag{5.109}$$

$$\mathbf{P}_k = \mathbf{M}_k - \mathbf{M}_k \mathbf{H}_k^T \mathbf{G}_k\left(\mathbf{z}_k\right) \mathbf{H}_k \mathbf{M}_k \tag{5.110}$$

$$\mathbf{M}_{k+1} = \mathbf{\Phi}_k \mathbf{P}_k^T \mathbf{\Phi}_k^T + \mathbf{Q}_k \tag{5.111}$$

$$\bar{\mathbf{x}}_{k+1} = \mathbf{\Phi}_k \hat{\mathbf{x}}_k \tag{5.112}$$

where $\mathbf{g}_k(\mathbf{z}_k)$ *is the* $r \times 1$ *vector*

$$\mathbf{g}_k\left(\mathbf{z}_k\right) = -\frac{1}{f\left(\mathbf{z}_k \left| Z^{k-1}\right.\right)} \nabla_{\mathbf{z}_k} f\left(\mathbf{z}_k \left| Z^{k-1}\right.\right) \tag{5.113}$$

$\mathbf{G}_k(\mathbf{z}_k)$ *is an* $r \times r$ *matrix with elements*

$$\left\{\mathbf{G}_k\left(\mathbf{z}_k\right)\right\}_{i,l} = \frac{\partial \left\{\mathbf{g}_k\left(\mathbf{z}_k\right)\right\}_l}{\partial z_{k,i}} \tag{5.114}$$

and z_{kl} *is the lth component of* $\mathbf{z}_k$.

Proof When $\mathbf{x}_k$ is given, (5.100) indicates that $\mathbf{z}_k$ is independent of Z^{k-1}. Therefore, Bayes' rule gives

$$\begin{aligned} f\left(\mathbf{x}_k \left| Z^k\right.\right) &= f\left(\mathbf{x}_k \left| Z^{k-1}, \mathbf{z}_k\right.\right) \\ &= bf\left(\mathbf{x}_k \left| Z^{k-1}\right.\right) f\left(\mathbf{z}_k \left| \mathbf{x}_k\right.\right). \end{aligned} \tag{5.115}$$

where

$$b = [f(\mathbf{z}_k | Z^{k-1})]^{-1}. \tag{5.116}$$

Equations (5.102) and (5.115) yield

$$\hat{\mathbf{x}}_k - \bar{\mathbf{x}}_k = b \int_{R^n} (\mathbf{x}_k - \bar{\mathbf{x}}_k) f\left(\mathbf{z}_k \left| \mathbf{x}_k\right.\right) f\left(\mathbf{x}_k \left| Z^{k-1}\right.\right) d\mathbf{x}_k. \tag{5.117}$$

Using (5.106) in the integrand to express it in terms of $\nabla_{\mathbf{x}_k} f\left(\mathbf{x}_k \left| Z^{k-1}\right.\right)$, integrating by parts, and observing that the Gaussian density function $f\left(\mathbf{x}_k \left| Z^{k-1}\right.\right)$ is zero at its extreme points, we obtain

$$\begin{aligned} \hat{\mathbf{x}}_k - \bar{\mathbf{x}}_k &= b\mathbf{M}_k \int_{R^n} f\left(\mathbf{z}_k \left| \mathbf{x}_k\right.\right) \mathbf{M}_k^{-1} (\mathbf{x}_k - \bar{\mathbf{x}}_k) f\left(\mathbf{x}_k \left| Z^{k-1}\right.\right) d\mathbf{x}_k \\ &= -b\mathbf{M}_k \int_{R^n} f\left(\mathbf{z}_k \left| \mathbf{x}_k\right.\right) \nabla_{\mathbf{x}_k} f\left(\mathbf{x}_k \left| Z^{k-1}\right.\right) d\mathbf{x}_k \\ &= b\mathbf{M}_k \int_{R^n} f\left(\mathbf{x}_k \left| Z^{k-1}\right.\right) \nabla_{\mathbf{x}_k} f\left(\mathbf{z}_k \left| \mathbf{x}_k\right.\right) d\mathbf{x}_k \end{aligned} \tag{5.118}$$

where the $n \times 1$ gradient vector $\nabla_{\mathbf{x}_k}$ has $\partial/\partial x_{ki}$ as its ith component. Equation (5.100) implies that

$$\nabla_{\mathbf{x}_k} f(\mathbf{z}_k|\mathbf{x}_k) = \nabla_{\mathbf{x}_k} f_v(\mathbf{z}_k - \mathbf{H}_k\mathbf{x}_k) = -\mathbf{H}_k^T \nabla_{\mathbf{z}_k} f_v(\mathbf{z}_k - \mathbf{H}_k\mathbf{x}_k)$$
$$= -\mathbf{H}_k^T \nabla_{\mathbf{z}_k} f(\mathbf{z}_k|\mathbf{x}_k) \tag{5.119}$$

where $f_v(\cdot)$ is the density function of $\mathbf{v}_k$. Substitution of this equation into (5.118) gives

$$\hat{\mathbf{x}}_k - \bar{\mathbf{x}}_k = -b\mathbf{M}_k\mathbf{H}_k^T \int_{R^n} f\left(\mathbf{x}_k\left|Z^{k-1}\right.\right) \nabla_{\mathbf{z}_k} f(\mathbf{z}_k|\mathbf{x}_k)\, d\mathbf{x}_k$$
$$= -b\mathbf{M}_k\mathbf{H}_k^T \nabla_{\mathbf{z}_k} \int_{R^n} f\left(\mathbf{x}_k\left|Z^{k-1}\right.\right) f(\mathbf{z}_k|\mathbf{x}_k)\, d\mathbf{x}_k$$
$$= -b\mathbf{M}_k\mathbf{H}_k^T \nabla_{\mathbf{z}_k} \int_{R^n} f(\mathbf{z}_k|Z^{k-1}) f\left(\mathbf{x}_k\left|Z^k\right.\right) d\mathbf{x}_k$$
$$= -b\mathbf{M}_k\mathbf{H}_k^T \nabla_{\mathbf{z}_k} f(\mathbf{z}_k|Z^{k-1}) \tag{5.120}$$

where the second equality results because $f(\mathbf{x}_k|Z^{k-1})$ is not a function of $\mathbf{z}_k$, and the third equality is obtained by substituting (5.115) and (5.116). Substituting (5.113) into (5.120), we obtain (5.109).

To derive (5.110), we add and subtract $\bar{\mathbf{x}}_k$ in (5.103) and then use (5.109), which gives

$$\mathbf{P}_k = E\left[(\mathbf{x}_k - \bar{\mathbf{x}}_k)(\mathbf{x}_k - \bar{\mathbf{x}}_k)^T\left|Z^k\right.\right] - \mathbf{M}_k\mathbf{H}_k^T\mathbf{g}_k(\mathbf{z}_k)\,\mathbf{g}_k^T(\mathbf{z}_k)\,\mathbf{H}_k\mathbf{M}_k. \tag{5.121}$$

The first term, which we denote by $\mathbf{P}_{k1}$, may be evaluated in a similar manner to the derivation of (5.109). Combining several intermediate steps similar to those already explained, we obtain

$$\mathbf{P}_{k1} = b\int_{R^n} (\mathbf{x}_k - \bar{\mathbf{x}}_k) f(\mathbf{z}_k|\mathbf{x}_k) f\left(\mathbf{x}_k\left|Z^{k-1}\right.\right)(\mathbf{x}_k - \bar{\mathbf{x}}_k)^T d\mathbf{x}_k$$
$$= -b\mathbf{M}_k \int_{R^n} \nabla_{\mathbf{x}_k} f\left(\mathbf{x}_k\left|Z^{k-1}\right.\right) f(\mathbf{z}_k|\mathbf{x}_k)(\mathbf{x}_k - \bar{\mathbf{x}}_k)^T d\mathbf{x}_k$$
$$= b\mathbf{M}_k \int_{R^n} f\left(\mathbf{x}_k\left|Z^{k-1}\right.\right)\left[f(\mathbf{z}_k|\mathbf{x}_k)\mathbf{I} + \nabla_{\mathbf{x}_k} f(\mathbf{z}_k|\mathbf{x}_k)(\mathbf{x}_k - \bar{\mathbf{x}}_k)^T\right] d\mathbf{x}_k$$
$$= \mathbf{M}_k - b\mathbf{M}_k\mathbf{H}_k^T \int_{R^n} f\left(\mathbf{x}_k\left|Z^{k-1}\right.\right) \nabla_{\mathbf{z}_k} f(\mathbf{z}_k|\mathbf{x}_k)(\mathbf{x}_k - \bar{\mathbf{x}}_k)^T d\mathbf{x}_k$$
$$= \mathbf{M}_k - b\mathbf{M}_k\mathbf{H}_k^T \int_{R^n} \nabla_{\mathbf{z}_k}[f(\mathbf{z}_k|Z^{k-1}) f\left(\mathbf{x}_k\left|Z^k\right.\right)](\mathbf{x}_k - \bar{\mathbf{x}}_k)^T d\mathbf{x}_k$$
$$= \mathbf{M}_k + \mathbf{M}_k\mathbf{H}_k^T\mathbf{g}_k(\mathbf{z}_k)\,\mathbf{g}_k^T(\mathbf{z}_k)\,\mathbf{H}_k\mathbf{M}_k - \mathbf{M}_k\mathbf{H}_k^T\nabla_{\mathbf{z}_k}\mathbf{g}_k^T(\mathbf{z}_k)\,\mathbf{H}_k\mathbf{M}_k \tag{5.122}$$

where the second equality follows from a differentiation of (5.106), the third equality is obtained by an integration by parts, the fourth equality follows from (5.115) and (5.119), the fifth equality follows from (5.115), and the final equality is obtained by using the chain rule and substituting (5.113), (5.109), and (5.116). Combining (5.122) and (5.121) yields (5.110) and (5.114).

Equation (5.111) is derived by using the definition of $\mathbf{M}_{k+1}$ given by (5.105) and then substituting (5.99), (5.104), and (5.103). Equation (5.112) follows from (5.104) and (5.100). □

The filter defined by this theorem is the Kalman filter if $f(\mathbf{z}_k|Z^{k-1})$ is a Gaussian density function:

$$f(\mathbf{z}_k|Z^{k-1}) = (2\pi)^{-n/2}\left(\det \mathbf{L}_k\right)^{-1/2} \exp\left[-\frac{1}{2}(\mathbf{z}_k - \bar{\mathbf{z}}_k)^T \mathbf{L}_k^{-1}(\mathbf{z}_k - \bar{\mathbf{z}}_k)\right]. \tag{5.123}$$

Substitution of this equation, (5.107), and (5.108) into (5.113) and (5.114) implies that

$$\mathbf{g}_k(\mathbf{z}_k) = \left(\mathbf{H}_k\mathbf{M}_k\mathbf{H}_k^T + \mathbf{R}_k\right)^{-1}(\mathbf{z}_k - \mathbf{H}_k\bar{\mathbf{x}}_k) \tag{5.124}$$

$$\mathbf{G}_k(\mathbf{z}_k) = \left(\mathbf{H}_k\mathbf{M}_k\mathbf{H}_k^T + \mathbf{R}_k\right)^{-1}. \tag{5.125}$$

Substitution of these two equations into (5.109) and (5.110) yields the usual Kalman filter equations.

ACM Filter

To apply the theorem to interference suppression in a direct-sequence system with BPSK, the narrowband interference sequence $\{i_k\}$ at the filter input is modeled as an autoregressive process that satisfies

$$i_k = \sum_{l=1}^{q} \phi_l i_{k-l} + e_k \tag{5.126}$$

where e_k is an independent zero-mean random variable with variance σ_e^2, and the $\{\phi_l\}$ are known to the receiver. Since it is desired to estimate i_k, the observation noise v_k is the sum of the direct-sequence signal s_k and the independent zero-mean Gaussian noise n_k:

$$v_k = s_k + n_k. \tag{5.127}$$

The state-space representation of the system is:

$$\mathbf{x}_k = \mathbf{\Phi}\mathbf{x}_{k-1} + \mathbf{u}_k \tag{5.128}$$

$$z_k = \mathbf{H}\mathbf{x}_k + v_k$$
$$= i_k + s_k + n_k \tag{5.129}$$

where

$$\mathbf{x}_k = \begin{bmatrix} i_k & i_{k-1} & \ldots & i_{k-q+1} \end{bmatrix}^T \tag{5.130}$$

$$\mathbf{\Phi} = \begin{bmatrix} \phi_1 & \phi_2 & \ldots & \phi_{q-1} & \phi_q \\ 1 & 0 & \ldots & 0 & 0 \\ 0 & 1 & \ldots & 0 & 0 \\ \vdots & \vdots & \ldots & \vdots & \vdots \\ 0 & 0 & \ldots & 1 & 0 \end{bmatrix} \tag{5.131}$$

$$\mathbf{u}_k = [e_k \ \ 0 \ \ \ldots \ \ 0]^T \tag{5.132}$$

$$\mathbf{H} = [1 \ \ 0 \ \ \ldots \ \ 0]\,. \tag{5.133}$$

The covariance matrix $\mathbf{Q}_k$ contains a single nonzero element equal to σ_e^2. Since the first component of the state vector $\mathbf{x}_k$ is the interference i_k, the state estimate $\mathbf{H}\hat{\mathbf{x}}_k = \widehat{i_k}$ provides an interference estimate that can be subtracted from the received signal to cancel the interference.

For a random spreading sequence, $s_k = +c$ or $-c$ with equal probability. If n_k is zero-mean and Gaussian with variance σ_n^2, then v_k has the density function

$$f_v(x) = \frac{1}{2}\mathcal{N}_x\left(c,\sigma_n^2\right) + \frac{1}{2}\mathcal{N}_x\left(-c,\sigma_n^2\right) \tag{5.134}$$

where

$$\mathcal{N}_x\left(m,\sigma^2\right) = \frac{1}{\sqrt{2\pi}\sigma}\exp\left(-\frac{(x-m)^2}{2\sigma^2}\right). \tag{5.135}$$

Since i_k and n_k are independent of s_k, the expected value of z_k conditioned on Z^{k-1} and $s_k = \pm c$ is

$$E\left[z_k \left| Z^{k-1}, s_k = \pm c\right.\right] = \bar{i}_k \pm c \tag{5.136}$$

where $\bar{i}_k$ is the conditional expected value of i_k :

$$\bar{i}_k = E\left[i_k \left| Z^{k-1}\right.\right]. \tag{5.137}$$

It follows that the variance of z_k conditioned on Z^{k-1} and $s_k = \pm c$ is

$$\sigma_z^2 = E\left[\left(z_k - \bar{i}_k \mp c\right)^2 \middle| Z^{k-1}, s_k = \pm c\right]$$
$$= E\left[\left(i_k - \bar{i}_k\right)^2 \middle| Z^{k-1}\right] + \sigma_n^2. \tag{5.138}$$

Equation (5.126) indicates that i_k is not the sum of independent Gaussian random variables even if e_k has a Gaussian distribution. Therefore, i_k does not have a Gaussian distribution (Appendix A.1), and the density function $f(\mathbf{x}_k|Z^{k-1})$ is not Gaussian as required by Masreliez's theorem. However, by assuming that $f(\mathbf{x}_k|Z^{k-1})$ is approximately Gaussian, we can use the results of the theorem to derive the nonlinear *approximate conditional mean* (ACM) filter [116].

Since $f(\mathbf{x}_k|Z^{k-1})$ is approximated by a Gaussian density function and i_k is independent of s_k, $f(\mathbf{H}\mathbf{x}_k|Z^{k-1}, s_k) = f(i_k|Z^{k-1}, s_k)$ is approximated by a Gaussian density function. Since n_k has a Gaussian density function, it follows that $f\left(z_k \middle| Z^{k-1}, s_k\right) = f\left(i_k + s_k + n_k \middle| Z^{k-1}, s_k\right)$ is approximated by a Gaussian density function. Therefore, the theorem of total probability and (5.134) imply that

$$f\left(z_k \middle| Z^{k-1}\right) = \frac{1}{2}\mathcal{N}_{z_k}\left(\bar{i}_k + c, \sigma_z^2\right) + \frac{1}{2}\mathcal{N}_{z_k}\left(\bar{i}_k - c, \sigma_z^2\right)$$
$$= \frac{1}{\sqrt{2\pi}\sigma_z} \exp\left(-\frac{\epsilon_k^2 + c^2}{2\sigma_z^2}\right) \cosh\left(\frac{c\epsilon_k}{\sigma_z^2}\right) \tag{5.139}$$

where the *innovation* or prediction residual is

$$\epsilon_k = z_k - \bar{i}_k. \tag{5.140}$$

From (5.136), it follows that

$$\bar{z}_k = E\left[z_k \middle| Z^{k-1}\right] = \bar{i}_k. \tag{5.141}$$

Substitution of (5.139) into (5.113) yields

$$g_k\left(z_k\right) = \frac{1}{\sigma_z^2}\left[\epsilon_k - c \tanh\left(\frac{c\epsilon_k}{\sigma_z^2}\right)\right] \tag{5.142}$$

and the substitution of (5.142) into (5.114) yields

$$G_k\left(z_k\right) = \frac{1}{\sigma_z^2}\left[1 - \frac{c^2}{\sigma_z^2}\operatorname{sech}^2\left(\frac{c\epsilon_k}{\sigma_z^2}\right)\right]. \tag{5.143}$$

The update equations of the ACM filter are given by (5.109) to (5.112), (5.142), and (5.143). The difference between the nonlinear ACM filter and the linear Kalman filter is the presence of the nonlinear *tanh* and *sech* functions in (5.142) and (5.143).

Adaptive ACM Filter

In practical applications, the elements of the matrix $\mathbf{\Phi}$ in (5.131) are unknown and may vary with time. To cope with these problems, an adaptive algorithm that can track the interference is desirable. The *adaptive ACM filter* produces an interference estimate denoted by $\widehat{z}_k$ and an output

$$\widehat{\epsilon}_k = z_k - \widehat{z}_k \tag{5.144}$$

which ideally is $s_k + n_k$ plus a small residual of i_k. Thus, the interference, but not the noise, is suppressed.

To use the structure of the nonlinear ACM filter, we observe that the second term inside the brackets in (5.142) would be absent if s_k were absent. Therefore, $c\tanh(c\epsilon_k/\sigma_z^2)$ may be interpreted as a soft decision on the direct-sequence signal s_k. The input to the adaptive filter at time k is computed as the difference between the observation z_k and the soft decision:

$$\begin{aligned}\widetilde{z}_k &= z_k - c\tanh\left(\frac{c\widehat{\epsilon}_k}{\sigma_z^2}\right)\\ &= \widehat{z}_k + \rho\left(\widehat{\epsilon}_k\right)\end{aligned} \tag{5.145}$$

where

$$\rho\left(\widehat{\epsilon}_k\right) = \widehat{\epsilon}_k - c\tanh\left(\frac{c\widehat{\epsilon}_k}{\sigma_z^2}\right). \tag{5.146}$$

The input $\widetilde{z}_k$ is a rough estimate of the interference that is made more accurate by the adaptive filter. The architecture of the one-sided adaptive ACM filter [116] is shown in Figure 5.5. The output of the adaptive filter provides the interference estimate

$$\widehat{z}_k = \mathbf{w}^T(k)\widetilde{\mathbf{z}}(k) \tag{5.147}$$

where $\mathbf{w}(k)$ is the weight vector after iteration k and

$$\widetilde{\mathbf{z}}(k) = \left[\widetilde{z}_{k-1}\,\widetilde{z}_{k-2}\ldots\widetilde{z}_{k-N}\right]^T \tag{5.148}$$

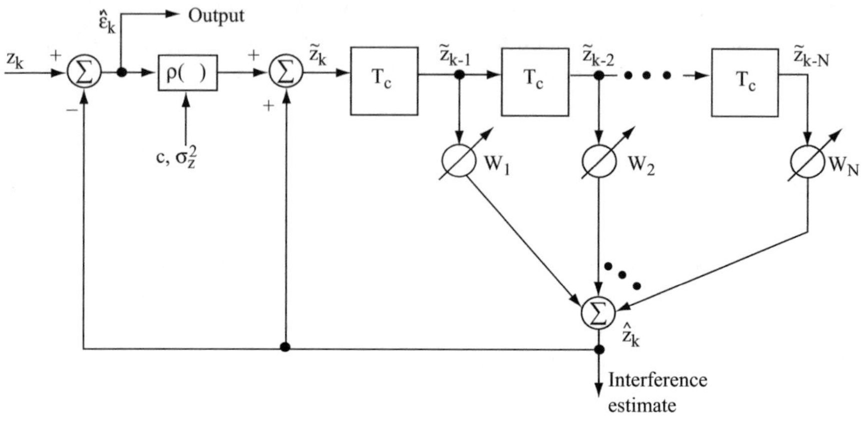

Fig. 5.5 Adaptive ACM filter

is the input vector for iteration k, which is extracted from the filter taps. When $\widetilde{\mathbf{z}}(k)$ has only a small component due to s_k, the filter can effectively track the interference, and its output $\widehat{z}_k$ provides a good estimate of this interference.

The *normalized LMS algorithm* (Section 5.2) may be used to implement the adaptive ACM filter with $\widetilde{z}_k$ serving as an approximation of the desired response i_k. The normalized LMS algorithm may be expressed as

$$\mathbf{w}(k) = \mathbf{w}(k-1) + \frac{2\mu}{\delta + ||\widetilde{\mathbf{z}}(k)||^2} (\widetilde{z}_k - \widehat{z}_k)\widetilde{\mathbf{z}}(k) \tag{5.149}$$

where μ is the adaptation constant, $0 < \mu < 1$, and δ is a positive constant.

The calculation of $\rho(\widehat{\epsilon}_k)$ requires the estimation of σ_z^2. If the $\widehat{z}_k$ produced by the adaptive filter approximates i_k, then $var(\widehat{\epsilon}_k) \approx \sigma_z^2$. Therefore, if $var(\widehat{\epsilon}_k)$ is estimated by computing the sample variance of the filter output, then the latter provides an estimate of σ_z^2.

A figure of merit for filters is the *SINR improvement,* which is the ratio of the output signal-to-interference-and-noise ratio (SINR) to the input SINR. Since the filters concerned do not change the signal power, the SINR improvement is

$$R = \frac{E\left\{|z_k - s_k|^2\right\}}{E\left\{|\widehat{\epsilon}_k - s_k|^2\right\}}. \tag{5.150}$$

In terms of this performance measure, the nonlinear adaptive ACM filter has been found to provide much better suppression of narrowband interference than the linear Kalman filter if the noise power in n_k is less than the direct-sequence signal power in s_k. If the latter condition is not satisfied, the advantage is small or absent. Disadvantages apparent from (5.146) are the requirements to estimate the parameters c and σ_z^2 and to compute or store the *tanh* function.

The preceding linear and nonlinear methods are primarily predictive methods that exploit the inherent predictability of narrowband interference. Further improvements in interference suppression are theoretically possible by using methods that were originally developed for multiuser detection (Section 6.4). Some of them can potentially be used to simultaneously suppress both narrowband interference and multiple-access interference, but they require even more computation and parameter estimation than the ACM filter, and the most powerful of the adaptive methods are practical only for short spreading sequences, if at all.

5.4 Rejection of Wideband Interference

When a direct-sequence system is part of a mobile network of similar systems, multiple-access interference (Chapter 7) by similar systems is of primary concern because this interference may be too strong for the natural resistance of a practical spread-spectrum system with a limited spreading factor. Since this type of interference has a spectrum that is similar to that of a desired direct-sequence signal, the narrowband interference-rejection filters are ineffective, and another type of adaptive filter is needed. By exploiting the known spreading sequence, an adaptive filter is potentially capable of suppressing both multiple-access interference and general wideband interference. As illustrated in Figure 5.1, chip-rate processing is performed prior to the despreading.

Lagrange Multipliers

Consider minimizing or maximizing the differentiable function $f(\mathbf{x})$ subject to the m equality constraints $g_i(\mathbf{x}) = 0$, $i = 1, ..., m$, where each $g_i(\mathbf{x})$ is differentiable. In principle, each constraint could be used to solve for a component of $\mathbf{x}$ and then proceed to minimize or maximize a function of a smaller set of independent variables. In practice, such a procedure is often difficult. The *method of Lagrange multipliers* [19, 42] provides an alternative procedure that provides local minima or maxima. We focus on minimizing $f(\mathbf{x})$, as maximizing it is equivalent to minimizing $-f(\mathbf{x})$.

Let $\mathbf{x}_0$ denote a local minimum of a continuously differentiable function $f : \mathcal{R}^n \to \mathcal{R}$ subject to $g_i(\mathbf{x}) = 0$, $i = 1, ..., m$, where each $g_i(\mathbf{x})$ is a continuously differentiable function $g_i : \mathcal{R}^n \to \mathcal{R}$, $m \leq n$, and the $\{\nabla_x g_i(\mathbf{x}_0)\}$ are linearly independent. Let

$$\mathbf{G} = \begin{bmatrix} \nabla_x^T g_1(\mathbf{x}_0) \\ \vdots \\ \nabla_x^T g_m(\mathbf{x}_0) \end{bmatrix} \tag{5.151}$$

denote an $m \times n$ matrix of gradients. Let $\mathsf{R}\left(\mathbf{G}^T\right)$ denote the range space of $\mathbf{G}^T$, which is the m-dimensional vector subspace spanned by the linearly independent $\{\nabla_x g_i(\mathbf{x}_0)\}$. Let $\mathsf{N}(\mathbf{G})$ denote the n-dimensional null space of $\mathbf{G}$, which implies that an $n \times 1$ vector $\mathbf{t} \in \mathsf{N}(\mathbf{G})$ satisfies

$$\mathbf{Gt} = \mathbf{0}. \tag{5.152}$$

Thus, any vector in $\mathsf{N}(\mathbf{G})$ is a tangent vector at the point $\mathbf{x}_0$ on the multidimensional surface defined by the constraints, and an infinitesimal movement from $\mathbf{x}_0$ along a tangent vector gives a point that remains on the surface.

Since $\mathbf{x}_0$ is a local minimum and $\nabla_x f(\mathbf{x}_0)$ has continuous components, any vector in $\mathsf{N}(\mathbf{G})$ must be perpendicular to $\nabla_x f(\mathbf{x}_0)$ so that an infinitesimal movement along the vector cannot decrease $f(\mathbf{x})$. Therefore, at a local minimum $\mathbf{x}_0$ that satisfies the constraints, it is necessary that

$$\nabla_x^T f(\mathbf{x}_0)\mathbf{t} = \mathbf{0} \tag{5.153}$$

for every vector $\mathbf{t} \in \mathsf{N}(\mathbf{G})$. Using linear algebra and (5.153), we have $\nabla_x f(\mathbf{x}_0) \in (\mathsf{N}(\mathbf{G}))^\perp = \mathsf{R}\left(\mathbf{G}^T\right)$, where $A^\perp$ denotes the subspace perpendicular to the subspace A. Therefore, $\nabla_x f(\mathbf{x}_0)$ must be a linear combination of the $\{\nabla_x g_i(\mathbf{x}_0)\}$:

$$\nabla_x f(\mathbf{x}_0) = -\sum_{i=1}^{m} \lambda_i \nabla_x g_i(\mathbf{x}_0) \tag{5.154}$$

for properly chosen *Lagrange multipliers* $\lambda_1, \ldots, \lambda_m$. This equation implies that a necessary condition for $\mathbf{x}_0$ to be a local minimum of $f(\mathbf{x})$ subject to constraints is for $\mathbf{x}_0$ to be a stationary point of the unconstrained *Lagrangian function*

$$L(\mathbf{x}, -) = f(\mathbf{x}) + \sum_{i=1}^{m} \lambda_i g_i(\mathbf{x}). \tag{5.155}$$

The *method of Lagrange multipliers* requires the stationary points of the Lagrangian function and the Lagrange multipliers that satisfy the constraints $g_i(\mathbf{x}) = 0,\ i = 1, \ldots, m$ to be found. Once the stationary points are found, it must be determined whether they locally minimize or globally minimize $f(\mathbf{x})$ or neither.

Constrained Minimum-Power Criterion

The *constrained minimum-power criterion is a* performance criterion that inherently limits the inadvertent cancelation of the desired signal while canceling interference by an adaptive filter. Let $\mathbf{p}$ denote the $G \times 1$ vector of the short spreading sequence of a desired signal. Let $\mathbf{x} = dA\mathbf{p} + \mathbf{n}$ denote the $G \times 1$ output vector of the G successive outputs of chip-matched filters that process one symbol $d = \pm 1$, where $\mathbf{n}$ is the interference and noise in the filter output and A is the amplitude. The output of a linear detector is $y = \mathbf{w}^H\mathbf{x}$, where $\mathbf{w}$ is a complex-valued $G \times 1$ weight vector. Each component of $\mathbf{p}$ has the value $+1$ or -1, and hence

$$\mathbf{p}^T\mathbf{p} = G. \tag{5.156}$$

We assume that $\mathbf{x}$ has stationary statistics and a positive-definite correlation matrix $\mathbf{R}_x$. The constrained minimum-power criterion requires the weight vector to minimize the mean output power

$$E\left[|y|^2\right] = \mathbf{w}^H\mathbf{R}_x\mathbf{w} \tag{5.157}$$

subject to the constraint

$$\mathbf{w}^H\mathbf{p} = 1. \tag{5.158}$$

By forcing this inner product to be unity, the constraint inhibits suppression of the desired signal. Thus, the power minimization tends to suppress the interference and noise.

Applying the method of Lagrange multipliers, we find the stationary points of the real-valued *Lagrangian function*

$$H\left(\mathbf{w}, \mathbf{w}^*\right) = \mathbf{w}^H\mathbf{R}_x\mathbf{w} + \gamma_1\left[\mathrm{Re}\left(\mathbf{w}^H\mathbf{p}\right) - 1\right] + \gamma_2\left[\mathrm{Im}\left(\mathbf{w}^H\mathbf{p}\right)\right] \tag{5.159}$$

where γ_1 and γ_2 are real-valued Lagrange multipliers. Defining $\gamma = \gamma_1 - j\gamma_2$, we obtain

$$\begin{aligned} H\left(\mathbf{w}, \mathbf{w}^*\right) &= \mathbf{w}^H\mathbf{R}_x\mathbf{w} + \mathrm{Re}\left(\gamma\mathbf{w}^H\mathbf{p}\right) \\ &= \mathbf{w}^H\mathbf{R}_x\mathbf{w} + \frac{1}{2}\gamma\mathbf{w}^H\mathbf{p} + \frac{1}{2}\gamma^*\mathbf{w}^T\mathbf{p} \end{aligned} \tag{5.160}$$

which is an analytic function of $\mathbf{w}$ and $\mathbf{w}^*$. Let ∇_{w*} denote the complex gradient with respect to $\mathbf{w}^*$. Then, (5.20) indicates that

$$\nabla_{w*}H\left(\mathbf{w}, \mathbf{w}^*\right) = \mathbf{R}_x\mathbf{w} + \frac{1}{2}\gamma\mathbf{p}. \tag{5.161}$$

Since the Hermitian matrix $\mathbf{R}_x$ is positive definite, its inverse exists (Appendix G). Setting $\nabla_{w*}H\left(\mathbf{w}, \mathbf{w}^*\right) = \mathbf{0}$ and applying the constraint to eliminate γ gives the necessary condition for the *optimal weight vector*:

$$\mathbf{w}_0 = \frac{\mathbf{R}_x^{-1}\mathbf{p}}{\mathbf{p}^T\mathbf{R}_x^{-1}\mathbf{p}} \tag{5.162}$$

where the denominator is a scalar.

To show that the local minimum given by (5.162) is the optimal weight vector, we combine this equation with (5.157) and (5.158), which yields

$$E\left[|y|^2\right] = (\mathbf{w} - \mathbf{w}_0)^H \mathbf{R}_x (\mathbf{w} - \mathbf{w}_0) + \left(\mathbf{p}^T \mathbf{R}_x^{-1} \mathbf{p}\right)^{-1}. \tag{5.163}$$

Since the final term is independent of $\mathbf{w}$ and $\mathbf{R}_x$ is positive definite, this equation indicates that $\mathbf{w}_0$ is the unique weight vector that minimizes $E\left[|y|^2\right]$.

Frost Algorithm

The *Frost algorithm* or *linearly constrained minimum-variance algorithm* is an adaptive algorithm that approximates the constrained optimal weight of (5.162) while avoiding the matrix inversion. Let the index k denote successive symbols of G matched-filter outputs. By the method of steepest descent and (5.161), the gradient of the Lagrangian for the kth iteration is $\mathbf{R}_x\mathbf{w}(k) + \gamma(k)\,\mathbf{p}/2$, where

$$\mathbf{R}_x = E\left[\mathbf{x}(k)\,\mathbf{x}^H(k)\right] \tag{5.164}$$

is the $N \times N$ Hermitian *correlation matrix* $\mathbf{x}(k) = d(k)\,\mathbf{p} + \mathbf{n}(k)$. The weight vector $\mathbf{w}(k)$ is updated according to

$$\mathbf{w}(k+1) = \mathbf{w}(k) - \mu\left[2\mathbf{R}_x\mathbf{w}(k) + \gamma(k)\,\mathbf{p}\right] \tag{5.165}$$

where μ is a constant that regulates the algorithm convergence rate. The Lagrange multiplier is chosen so that

$$\mathbf{p}^T\mathbf{w}(k+1) = \mathbf{w}^H(k+1)\mathbf{p} = 1. \tag{5.166}$$

Substituting (5.165) and (5.156) into (5.166), we find that

$$\gamma(k) = \frac{1}{\mu G}\left[\mathbf{p}^T\mathbf{w}(k) - 2\mu\mathbf{p}^T\mathbf{R}_x\mathbf{w}(k) - 1\right]. \tag{5.167}$$

Substituting this equation into (5.165) and exploiting that $\mathbf{p}^T\mathbf{w}(k)$ and $\mathbf{p}^T\mathbf{R}_x\mathbf{w}(k)$ are scalars, we obtain

$$\mathbf{w}(k+1) = \left(\mathbf{I} - \frac{1}{G}\mathbf{p}\mathbf{p}^T\right)\left[\mathbf{w}(k) - 2\mu\mathbf{R}_x\mathbf{w}(k)\right] + \frac{1}{G}\mathbf{p}. \tag{5.168}$$

This equation provides a deterministic *steepest-descent algorithm* that would be used if $\mathbf{R}_x$ were known.

The Frost algorithm is the *stochastic-gradient algorithm* that approximates $\mathbf{R}_x$ by $\mathbf{x}(k)\,\mathbf{x}^H(k)$ and uses the linear detector output

$$y(k) = \mathbf{w}^H(k)\,\mathbf{x}(k) \tag{5.169}$$

to determine symbol $d(k)$. A suitable choice for the initial weight vector that satisfies the constraint is $\mathbf{w}(0) = \mathbf{p}/G$. Thus, the *Frost algorithm* is

$$\mathbf{w}(0) = \frac{1}{G}\mathbf{p} \tag{5.170}$$

$$\mathbf{w}(k+1) = \left(\mathbf{I} - \frac{1}{G}\mathbf{p}\mathbf{p}^T\right)\left[\mathbf{w}(k) - 2\mu\mathbf{x}(k)\,y^*(k)\right] + \frac{1}{G}\mathbf{p}. \tag{5.171}$$

The output $\mathrm{Re}\,[y(k)]$ can be used as a symbol metric. Symbol decisions are made according to

$$\widehat{d}(k) = \mathrm{sgn}\,\{\mathrm{Re}\,[y(k)]\} \tag{5.172}$$

where the *signum function* is defined as $sgn(x) = 1, x \geq 0$, and $sgn(x) = -1, x < 0$.

Computational errors occur because of truncation, rounding, or quantization errors in the computer implementation of the algorithm. These errors may cause $\mathbf{w}^H(k)\,\mathbf{p} \neq 1$ after an iteration and could have a significant cumulative effect after a few iterations. However, the next iteration of the Frost algorithm automatically tends to correct the computational errors in the weight vector from the preceding iteration, primarily because (5.166) is used in deriving (5.167). Thus, apart from new sources of error, $\mathbf{w}^H(k+1)\mathbf{p} = 1$ even if $\mathbf{w}^H(k)\mathbf{p} \neq 1$.

Convergence of the Mean

If $\mathbf{x}(k+1)$ is independent of $\mathbf{y}(i)$, $i \leq k$, then (5.171) implies that $\mathbf{w}(k)$ and $\mathbf{x}(k)$ are independent; hence,

$$E[\mathbf{w}(k+1)] = \mathbf{A}[\mathbf{I} - 2\mu\mathbf{R}_x]E[\mathbf{w}(k)] + \frac{1}{G}\mathbf{p}, \quad k \geq 0 \tag{5.173}$$

where

$$\mathbf{A} = \left(\mathbf{I} - \frac{1}{G}\mathbf{p}\mathbf{p}^T\right). \tag{5.174}$$

Let

$$\mathbf{v}(k) = E[\mathbf{w}(k)] - \mathbf{w}_0. \tag{5.175}$$

From (5.175), (5.173), (5.162), and (5.174), we obtain

$$\mathbf{v}(k+1) = \mathbf{A}\mathbf{v}(k) - 2\mu\mathbf{A}\mathbf{R}_x\mathbf{v}(k), \quad k \geq 0. \tag{5.176}$$

Direct multiplication verifies that $\mathbf{A}^2 = \mathbf{A}$. It then follows from (5.176) that $\mathbf{A}\mathbf{v}(k) = \mathbf{v}(k)$, $k \geq 1$. It is easily verified that $\mathbf{A}\mathbf{v}(0) = \mathbf{v}(0)$. Consequently,

$$\begin{aligned}\mathbf{v}(k+1) &= [\mathbf{I} - 2\mu\mathbf{A}\mathbf{R}_x\mathbf{A}]\mathbf{v}(k) \\ &= [\mathbf{I} - 2\mu\mathbf{A}\mathbf{R}_\mathbf{x}\mathbf{A}]^{k+1}\mathbf{v}(0), \quad k \geq 0.\end{aligned} \tag{5.177}$$

Since $\mathbf{A}$ is symmetric and $\mathbf{R}_x$ is Hermitian, the matrix $\mathbf{A}\mathbf{R}_x\mathbf{A}$ is Hermitian; hence, it has a complete set of orthonormal eigenvectors. Direct calculation proves that

$$\mathbf{A}\mathbf{R}_\mathbf{x}\mathbf{A}\mathbf{p} = \mathbf{0} \tag{5.178}$$

which indicates that $\mathbf{p}$ is an eigenvector of $\mathbf{A}\mathbf{R}_x\mathbf{A}$ with an eigenvalue equal to zero. Let $\mathbf{e}_i$, $i = 1, 2, \ldots, N-1$, denote the $N-1$ remaining orthonormal eigenvectors of $\mathbf{A}\mathbf{R}_x\mathbf{A}$. The orthogonality implies that

$$\mathbf{p}^T\mathbf{e}_i = \mathbf{0}, \quad i = 1, 2, \ldots, N-1. \tag{5.179}$$

From this equation and (5.174), it follows that

$$\mathbf{A}\mathbf{e}_i = \mathbf{e}_i, \quad i = 1, 2, \ldots, N-1. \tag{5.180}$$

Let σ_i denote the eigenvalue of $\mathbf{A}\mathbf{R}_x\mathbf{A}$ associated with the unit eigenvector $\mathbf{e}_i$. Using (5.180), we obtain

$$\sigma_i = \mathbf{e}_i^H\mathbf{A}\mathbf{R}_\mathbf{x}\mathbf{A}\mathbf{e}_i = \mathbf{e}_i^H\mathbf{R}_x\mathbf{e}_i, \quad i = 1, 2, ..., N-1. \tag{5.181}$$

Since $\mathbf{e}_i$ is a unit vector, the Rayleigh quotient bounds of (5.69) imply that

$$\lambda_{\min} \leq \mathbf{e}_i^H\mathbf{R}_x\mathbf{e}_i \leq \lambda_{\max} \tag{5.182}$$

$$\sigma_{\min} \leq \mathbf{e}_i^H\mathbf{A}\mathbf{R}_\mathbf{x}\mathbf{A}\mathbf{e}_i \leq \sigma_{\max} \tag{5.183}$$

where $\lambda_{\min}$ and $\lambda_{\max}$ are the smallest and largest eigenvalues, respectively, of the positive-semidefinite Hermitian matrix $\mathbf{R}_x$, and where $\sigma_{\min}$ and $\sigma_{\max}$ are the smallest and largest eigenvalues, respectively, of $\mathbf{e}_i$, $i = 1, 2, \ldots, N-1$. If we assume that $\mathbf{R}_x$ is positive definite, then $\lambda_{\min} > 0$, and hence $\sigma_{\min} > 0$. We conclude that the $\{\mathbf{e}_i\}$ corresponds to nonzero eigenvalues.

It is easily verified that $\mathbf{p}^T\mathbf{v}(0) = \mathbf{0}$. Therefore, $\mathbf{v}(0)$ is equal to a linear combination of the $\mathbf{e}_i$, $i = 1, 2, ..., N-1$, which are the eigenvectors of $\mathbf{A}\mathbf{R}_x\mathbf{A}$

corresponding to the nonzero eigenvalues. If $\mathbf{v}(0)$ is equal to the eigenvector $\mathbf{e}_i$ with eigenvalue σ_i, then (5.177) indicates that

$$\mathbf{v}(k+1) = (1 - 2\mu\sigma_i)^{k+1}\mathbf{e}_i, \quad k \geq 0. \tag{5.184}$$

Therefore, $|1 - 2\mu\sigma_i| < 1$ for $i = 1, 2, ..., N-1$ is a necessary and sufficient condition for $\mathbf{v}(k) \to 0$; hence, the convergence of the mean weight vector to its optimal value:

$$\lim_{k\to\infty} E[\mathbf{w}(k)] = \mathbf{w}_0 = \frac{\mathbf{R}_x^{-1}\mathbf{p}}{\mathbf{p}^T\mathbf{R}_x^{-1}\mathbf{p}}. \tag{5.185}$$

Since $\sigma_{\min} > 0$, the necessary and sufficient condition for convergence is

$$0 < \mu < \frac{1}{\sigma_{\max}}. \tag{5.186}$$

Analogous to the LMS algorithm, the convergence of the mean weight vector of the Frost algorithm has transients that can be characterized by the time constants

$$\tau_i = -\frac{1}{\ln(|1 - 2\mu\sigma_i|)}, \quad i = 1, 2, ..., N-1. \tag{5.187}$$

If $0 < \mu < 1/2\sigma_{\max}$, the largest time constant is

$$\tau_{\max} = -\frac{1}{\ln(1 - 2\mu\sigma_{\min})} < \frac{1}{2\mu\sigma_{\min}}, \quad 0 < \mu < \frac{1}{2\sigma_{\max}}. \tag{5.188}$$

If μ is selected to be close to the upper bound in (5.188), then $\tau_{\max}$ increases with the *eigenvalue spread* defined as $\sigma_{\max}/\sigma_{\min}$.

An adaptive filter that uses the Frost algorithm is a type of *adaptive blind multiuser detector (Section 7.7)*. This detector is resilient in the presence of changing channel conditions and facilitates system recovery, but is practical only for known short spreading sequences.

5.5 Optimal Array

When multiple antennas are available, much more potent interference suppression is possible than can be obtained by processing the output of a single antenna. An *adaptive array* is an adaptive filter with inputs derived directly from an array of antennas. The performance criterion that has led to the most effective adaptive arrays for spread-spectrum systems is based on the maximization of the SINR; hence, that criterion is used to derive the optimal weight vector for adaptive arrays.

Consider an array with N outputs, each of which includes a different signal copy from a distinct antenna. Each array output is translated to baseband and its sampled complex envelope is extracted (Appendix D.3). Alternatively, each array output is translated to an intermediate frequency, and the sampled analytic signal is extracted. The subsequent analysis is valid for both these types of processing, but it is simplest to assume the extraction of sampled complex envelopes.

The sampled complex envelopes of the array outputs provide the inputs to a linear filter. The desired signal, interference signals, and thermal noise are modeled as independent zero-mean, wide-sense-stationary stochastic processes. Let $\mathbf{x}(i)$ denote the discrete-time vector of the N complex-valued filter inputs, where the index i denotes the sample number. This vector can be decomposed as

$$\mathbf{x}(i) = \mathbf{s}(i) + \mathbf{n}(i) \tag{5.189}$$

where $\mathbf{s}(i)$ and $\mathbf{n}(i)$ are the discrete-time vectors of the desired signal and the interference and noise, respectively. The components of both $\mathbf{s}(i)$ and $\mathbf{n}(i)$ are modeled as discrete-time jointly wide-sense-stationary processes. Let $\mathbf{w}$ denote the $N \times 1$ weight vector of a linear filter applied to the input vector. The filter output is

$$y(i) = \mathbf{w}^H \mathbf{x}(i) = y_s(i) + y_n(i) \tag{5.190}$$

where the output components due to $\mathbf{s}(i)$ and $\mathbf{n}(i)$ are

$$y_s(i) = \mathbf{w}^H \mathbf{s}(i), \;\; y_n(i) = \mathbf{w}^H \mathbf{n}(i) \tag{5.191}$$

respectively.

The $N \times N$ correlation matrices of $\mathbf{s}(i)$ and $\mathbf{n}(i)$ are defined as

$$\mathbf{R}_s = E\left[\mathbf{s}(i)\mathbf{s}^H(i)\right], \;\; \mathbf{R}_n = E\left[\mathbf{n}(i)\mathbf{n}^H(i)\right] \tag{5.192}$$

respectively. The desired-signal and interference-and-noise powers at the output are

$$p_{so} = E\left[|\, y_s(i)|^2\right] = \mathbf{w}^H \mathbf{R}_s \mathbf{w}, \;\; p_n = E\left[|\, y_n(i)|^2\right] = \mathbf{w}^H \mathbf{R}_n \mathbf{w} \tag{5.193}$$

respectively. The SINR at the filter output is

$$\rho = \frac{p_{so}}{p_n} = \frac{\mathbf{w}^H \mathbf{R}_s \mathbf{w}}{\mathbf{w}^H \mathbf{R}_n \mathbf{w}} . \tag{5.194}$$

The definitions of $\mathbf{R}_s$ and $\mathbf{R}_n$ ensure that these matrices are Hermitian nonnegative definite. Consequently, these matrices have complete sets of orthonormal eigenvectors, and their eigenvalues are real-valued and nonnegative. We assume that the positive-semidefinite Hermitian matrix $\mathbf{R}_n$ is positive definite; thus, it has positive eigenvalues. The spectral decomposition (Appendix G) of $\mathbf{R}_n$ can be expressed as

$$\mathbf{R}_n = \sum_{l=1}^{N} \lambda_l \mathbf{e}_l \mathbf{e}_l^H \tag{5.195}$$

where λ_l is an eigenvalue and $\mathbf{e}_l$ is the associated eigenvector.

To derive the weight vector that maximizes the SINR with no restriction on $\mathbf{R}_s$, we define the Hermitian matrix

$$\mathbf{A} = \sum_{l=1}^{N} \sqrt{\lambda_l}\, \mathbf{e}_l \mathbf{e}_l^H \tag{5.196}$$

where the positive square root is used. Direct calculations using the orthonormality of the $\{\mathbf{e}_l\}$ verify that

$$\mathbf{R}_n = \mathbf{A}^2 \tag{5.197}$$

and the inverse of $\mathbf{A}$ is

$$\mathbf{A}^{-1} = \sum_{l=1}^{N} \frac{1}{\sqrt{\lambda_l}} \mathbf{e}_l \mathbf{e}_l^H. \tag{5.198}$$

The matrix $\mathbf{A}$ specifies an invertible transformation of $\mathbf{w}$ into the vector

$$\mathbf{v} = \mathbf{A}\mathbf{w}. \tag{5.199}$$

We define the Hermitian matrix

$$\mathbf{C} = \mathbf{A}^{-1}\mathbf{R}_s\mathbf{A}^{-1}. \tag{5.200}$$

Then, (5.194), (5.197), (5.199), and (5.200) indicate that the SINR can be expressed as the *Rayleigh quotient*

$$\rho = \frac{\mathbf{v}^H \mathbf{C} \mathbf{v}}{\|\mathbf{v}\|^2}\,. \tag{5.201}$$

Let $\mu_{\max}$ and $\mu_{\min}$ denote the largest and smallest eigenvalues of $\mathbf{C}$, respectively. Then, (5.69) indicates that

$$\mu_{\min} \leq \rho \leq \mu_{\max}. \tag{5.202}$$

Let $\mathbf{u}$ denote the unit eigenvector of $\mathbf{C}$ associated with its largest eigenvalue $\mu_{\max}$. Thus, $\mathbf{v} = \eta \mathbf{u}$ maximizes the SINR, where η is an arbitrary constant. From (5.199) with $\mathbf{v} = \eta \mathbf{u}$, it follows that the *optimal weight vector that maximizes the SINR* is

$$\mathbf{w}_0 = \eta \mathbf{A}^{-1}\mathbf{u}. \tag{5.203}$$

The purpose of an adaptive-array algorithm is to adjust the weight vector to converge to the optimal value, which is given by (5.203) when maximization of the SINR is the performance criterion.

We assume that the desired signal is sufficiently narrowband or the antennas are close enough for the desired-signal copies in all the array outputs to be nearly aligned in time. The desired-signal input vector may be represented as

$$\mathbf{s}(i) = s(i)\mathbf{s}_0 \tag{5.204}$$

where $s(i)$ denotes the discrete-time sampled complex envelope of the desired signal, and the *steering vector* is

$$\mathbf{s}_0 = [\alpha_1 \exp(j\theta_1) \;\; \alpha_2 \exp(j\theta_2) \;\; \ldots \;\; \alpha_N \exp(j\theta_N)]^T \tag{5.205}$$

which has components that represent the relative amplitudes and phase shifts at the antenna outputs.

Example Equation (5.205) can serve as a model for a narrowband desired signal that arrives at an antenna array as a plane wave and does not experience fading. Let T_l, $l = 1, 2, \ldots, N$, denote the arrival-time delay of the desired signal at the output of antenna l relative to a fixed reference point in space. Equations (5.204) and (5.205) are valid with $\theta_l = -2\pi f_c T_l$, $l = 1, 2, \ldots, N$, where f_c is the carrier frequency of the desired signal. The α_l, $l = 1, 2, \ldots, N$, depend on the relative antenna patterns and propagation losses. If they are all equal, then the common value can be subsumed into $s(i)$. It is convenient to define the origin of a Cartesian coordinate system to coincide with the fixed reference point. Let (x_l, y_l) denote the coordinates of antenna l. If a single plane wave arrives from direction ψ relative to the normal to the array, then

$$\theta_l = \frac{2\pi f_c}{c}(x_l \sin\psi + y_l \cos\psi), \quad l = 1, 2, \ldots, N \tag{5.206}$$

where c is the speed of an electromagnetic wave. □

The substitution of (5.204) into (5.192) yields

$$\mathbf{R}_s = p_s \mathbf{s}_0 \mathbf{s}_0^H \tag{5.207}$$

where

$$p_s = E[|s(i)|^2]. \tag{5.208}$$

After substituting (5.207) into (5.200), it is observed that $\mathbf{C}$ may be factored:

$$\mathbf{C} = p_s \mathbf{A}^{-1} \mathbf{s}_0 \mathbf{s}_0^H \mathbf{A}^{-1} = \mathbf{f}\mathbf{f}^H \tag{5.209}$$

where

$$\mathbf{f} = \sqrt{p_s}\mathbf{A}^{-1}\mathbf{s}_0. \quad (5.210)$$

This factorization explicitly shows that $\mathbf{C}$ is a rank-one matrix; hence, its null space has a dimension $N - 1$. Therefore, the eigenvector of $\mathbf{C}$ associated with the only nonzero eigenvalue is $\mathbf{u} = \mathbf{f}$, and its eigenvalue is

$$\mu_{\max} = \|\mathbf{f}\|^2. \quad (5.211)$$

Substituting (5.210) into (5.203), using (5.197), and then merging $\sqrt{p_s}$ into the arbitrary constant, we obtain the optimal weight vector:

$$\mathbf{w}_0 = \eta \mathbf{R}_n^{-1}\mathbf{s}_0 \quad (5.212)$$

where η is an arbitrary constant. The maximum value of the SINR, obtained from the upper bound in (5.202), (5.211), (5.210), and (5.197), is

$$\rho_0 = p_s \mathbf{s}_0^H \mathbf{R}_n^{-1}\mathbf{s}_0. \quad (5.213)$$

5.6 Adaptive Array for Direct-Sequence Systems

If multiple antennas are available, an antenna array may be used to adaptively suppress both narrowband and wideband interference. The basic configuration of an adaptive array for spread-spectrum systems is displayed in Figure 5.6. The output of each array antenna is applied to an initial processor in a separate branch. The

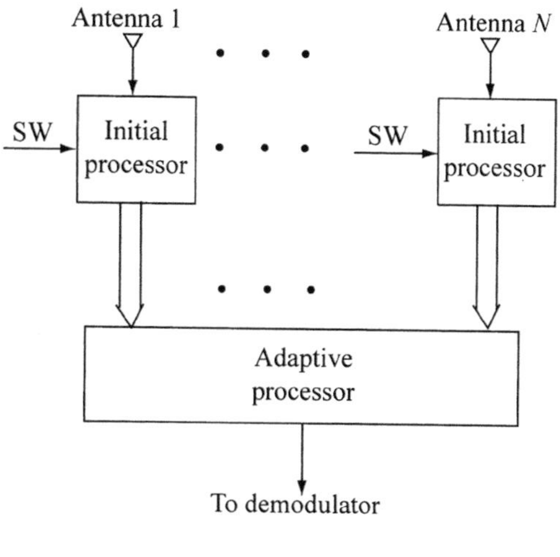

Fig. 5.6 Architecture of adaptive array for spread-spectrum system. *SW* spreading waveform

spreading waveform, which is produced by a synchronized receiver, is applied to the initial processors to enable the generation of despread discrete-time branch sequences. These sequences are applied to an adaptive processor that executes the adaptive-array algorithm.

The *maximin algorithm* is an adaptive-array algorithm that exploits the characteristics of spread-spectrum signals to provide a much larger degree of protection against strong interference than could be provided by spread spectrum alone [103]. It is a stochastic-gradient algorithm that is based on the *method of steepest ascent* and recursively increases the SINR. As indicated by its name, the maximin algorithm simultaneously maximizes the despread desired-signal components and minimizes the spectrally spread interference components. In a direct-sequence adaptive array, each despread desired-signal component has a narrowband spectrum, but each spectrally spread interference component has a wideband spectrum. This spectral difference is exploited by the maximin algorithm to estimate the interference and then cancel it.

Derivation of the Maximin Algorithm

Let $\mathbf{x}(i)$ denote the ith $N \times 1$ discrete-time vector of filtered branch outputs that provide the inputs to the maximin algorithm, where i is the sample index. The vector $\mathbf{x}(i)$ can be decomposed as $\mathbf{x}(i) = \mathbf{s}(i) + \mathbf{n}(i)$, where $\mathbf{s}(i)$ and $\mathbf{n}(i)$ are the discrete-time vectors of the desired sequence and the interference-and-noise sequence, respectively. Their $N \times N$ autocorrelation matrices are defined by (5.192). Since the interference and noise are zero-mean and statistically independent of the desired signal, and the $N \times N$ input correlation matrix is

$$\mathbf{R}_x = E[\mathbf{x}(i)\mathbf{x}^H(i)] = \mathbf{R}_s + \mathbf{R}_n. \tag{5.214}$$

The maximin processor is a linear adaptive filter that uses an $N \times 1$ weight vector $\mathbf{w}(k)$, where k is the index that denotes the weight iteration number. There are m discrete-time samples of the input vector $\mathbf{x}(i)$ for every weight iteration. The adaptive-filter output is

$$y(i) = \mathbf{w}^H(k)\,\mathbf{x}(i) = y_s(i) + y_n(i) \tag{5.215}$$

$$y_s(i) = \mathbf{w}^H(k)\,\mathbf{s}(i), \quad y_n(i) = \mathbf{w}^H(k)\,\mathbf{n}(i). \tag{5.216}$$

The adaptive-filter output power is

$$p_x(k) = E[|\,y(i)|^2] = \mathbf{w}^H(k)\mathbf{R}_x\mathbf{w}(k) = p_s(k) + p_n(k) \tag{5.217}$$

where

$$p_s(k) = \mathbf{w}^H(k)\mathbf{R}_s\mathbf{w}(k) \tag{5.218}$$

$$p_n(k) = \mathbf{w}^H(k)\mathbf{R}_n\mathbf{w}(k) \tag{5.219}$$

are the desired-sequence and interference-and-noise sequence powers, respectively. The SINR after iteration k is

$$\rho(k) = \frac{p_s(k)}{p_n(k)} = \frac{\mathbf{w}^H(k)\mathbf{R}_s\mathbf{w}(k)}{\mathbf{w}^H(k)\mathbf{R}_n\mathbf{w}(k)}. \tag{5.220}$$

The maximin algorithm changes the weight vector along the direction of the gradient of the SINR. Combining equations for the real and imaginary parts of the complex-valued weight vector, we obtain

$$\mathbf{w}(k+1) = \mathbf{w}(k) + \mu_0(k)\nabla_{w*}\rho(k) \tag{5.221}$$

where $\mu_0(k)$ is a scalar sequence that controls the rate of change of the weight vector, and $\nabla_{w*}\rho(k)$ is the complex gradient of the SINR $\rho(k)$ at iteration k. Using (5.220), we obtain

$$\nabla_{w*}\rho(k) = \rho(k)\left[\frac{\mathbf{R}_s\mathbf{w}(k)}{p_s(k)} - \frac{\mathbf{R}_n\mathbf{w}(k)}{p_n(k)}\right]. \tag{5.222}$$

Substitution of (5.214) and (5.217) into (5.222) and simplification yields

$$\nabla_{w*}\rho(k) = [\rho(k)+1]\left[\frac{\mathbf{R}_x\mathbf{w}(k)}{p_x(k)} - \frac{\mathbf{R}_n\mathbf{w}(k)}{p_n(k)}\right]. \tag{5.223}$$

Substitution of this equation into (5.221) gives the *steepest-ascent algorithm*:

$$\mathbf{w}(k+1) = \mathbf{w}(k) + \mu_0(k)[\rho(k)+1]\left[\frac{\mathbf{R}_x\mathbf{w}(k)}{p_x(k)} - \frac{\mathbf{R}_n\mathbf{w}(k)}{p_n(k)}\right]. \tag{5.224}$$

If $\mathbf{w}(k)$ is modeled as deterministic, then $\mathbf{R}_x\mathbf{w}(k) = E[\mathbf{x}(i)y^*(i)]$ and $\mathbf{R}_n\mathbf{w}(k) = E[\mathbf{n}(i)y_n^*(i)]$. Thus, we can avoid estimating the matrices $\mathbf{R}_x$ and $\mathbf{R}_n$ in (5.224) by finding estimators of $E[\mathbf{x}(i)y^*(i)]$ and $E[\mathbf{n}(i)y_n^*(i)]$. A further simplification that ultimately reduces the amount of computation by nearly a factor of two is obtained by observing that the components of $\mathbf{x}(i)$ are proportional to samples of continuous-time complex envelopes, which are modeled as zero-mean wide-sense-stationary processes. In each array branch, the thermal noise is independent of the noise in the other branches, and each desired or interference signal is a delayed version of the corresponding signal in the other branches. Therefore, the continuous-time complex envelopes are circularly symmetric (Appendix D.2), and thus

$$E[\mathbf{x}(i)\mathbf{x}^T(i)] = \mathbf{0} \tag{5.225}$$

and

$$E[\mathbf{n}(i)\mathbf{n}^T(i)] = \mathbf{0}. \tag{5.226}$$

The adaptive-filter output can be decomposed as

$$y(i) = y_r(i) + jy_i(i) \tag{5.227}$$

where $y_r(i)$ and $y_i(i)$ are the real and imaginary parts of $y(i)$, respectively. We assume that $\mathbf{w}(k)$ varies slowly relative to the symbol rate and hence approximates a deterministic vector. Then (5.190), (5.225), and the identity $\mathbf{w}^H(k)\,\mathbf{x}(i) = \mathbf{x}^T(i)\mathbf{w}^*(k)$ imply that

$$\begin{aligned} E[\mathbf{x}(i)y_r(i)] &= E\left[\mathbf{x}(i)\left\{\frac{1}{2}\mathbf{x}^T(i)\mathbf{w}^*(k) + \frac{1}{2}\mathbf{x}^H(i)\mathbf{w}(k)\right\}\right] \\ &= \frac{1}{2}E\left[\mathbf{x}(i)\mathbf{x}^H(i)\right]\mathbf{w}(k). \end{aligned} \tag{5.228}$$

This equation and (5.214) yield

$$\mathbf{R}_x\mathbf{w}(k) = 2E[\mathbf{x}(i)y_r(i)]. \tag{5.229}$$

Similarly,

$$\mathbf{R}_n\mathbf{w}(k) = 2E[\mathbf{n}(i)y_{nr}(i)] \tag{5.230}$$

where

$$y_{nr}(i) = Re[y_n(i)] = Re[\mathbf{w}^H(k)\mathbf{n}(i)] \tag{5.231}$$

is the real part of $y_n(i)$. Thus, we can avoid estimating the matrices $\mathbf{R}_x$ and $\mathbf{R}_n$ in (5.224) by finding estimators of $E[\mathbf{x}(i)y_r(i)]$ and $E[\mathbf{n}(i)y_{nr}(i)]$.

Equations (5.190) and (5.225) imply that $E[y^2(i)] = 0$, and the substitution of (5.190) yields $E[y_r^2(i)] = E[y_i^2(i)]$ and $E[y_r(i)y_i(i)] = 0$. Therefore, $p_x(k) = E[|\,y(i)|^2] = 2E[y_r^2(i)]$. These calculations and similar ones involving $p_n(k)$ give

$$p_x(k) = 2E[y_r^2(i)],\;\; p_n(k) = 2E[y_{nr}^2(i)]. \tag{5.232}$$

To derive the maximin algorithm, let $\hat{p}_x(k)$ and $\hat{p}_n(k)$ denote estimates of $E[y_r^2]$ and $E[y_{nr}^2]$, respectively, following weight iteration k. Let $\mathbf{c}_x(k)$ and $\mathbf{c}_n(k)$ denote estimates following iteration k of the input correlation vector $E[\mathbf{x}y_r]$ and the interference-and-noise correlation vector $E[\mathbf{n}y_{nr}]$, respectively. Substituting these estimates and $\mu_0(k) = \alpha(k)/[\widehat{\rho}(k) + 1]$ into (5.224), we obtain the *maximin algorithm*:

$$\mathbf{w}(k+1) = \mathbf{w}(k) + \alpha(k)\left[\frac{\mathbf{c}_x(k)}{\hat{p}_x(k)} - \frac{\mathbf{c}_n(k)}{\hat{p}_n(k)}\right], \qquad k \geq 0 \tag{5.233}$$

where $\mathbf{w}(0)$ is the deterministic initial weight vector, and $\alpha(k)$ is the *adaptation sequence*. As the adaptive weights converge, the interference components of $\mathbf{c}_x(k)$ and $\hat{p}_x(k)$ decrease. Thus, the first term within the brackets can be interpreted as a *signal term* that enables the algorithm to direct the array beam toward the desired signal. The second term within the brackets is a *noise term* that enables the algorithm to null interference signals.

The adaptation sequence $\alpha(k)$ should be chosen so that $E[\mathbf{w}(k)]$ converges to a nearly optimal steady-state value. It is also intuitively plausible that $\alpha(k)$ should decrease rapidly as $E[\mathbf{w}(k)]$ converges. A suitable candidate is

$$\alpha(k) = \alpha \frac{\hat{p}_n(k)}{\hat{t}(k)} \tag{5.234}$$

where $\hat{t}(k)$ is an estimate of the total interference-and-noise power in the passbands of the despread desired-signal copies, and α is the *adaptation constant*. The subsequent convergence analysis and simulation results confirm that this choice is effective and robust, provided that the adaptation constant is within certain numerical bounds. Simulation experiments confirm that for cyclostationary spread-spectrum signals and tone interference, the maximin algorithm suffers no performance loss due to the simplification stemming from (5.226).

The remaining issue is the choice of estimators for $\hat{t}(k), \mathbf{c}_x(k), \mathbf{c}_n(k), \hat{p}_x(k)$, and $\hat{p}_n(k)$. The specific nature of the spread-spectrum signals allows blind estimates to be made without depending on known steering vectors or reference signals.

Implementation of the Adaptive Processor

The principal components of each initial processor in Figure 5.6 are depicted in Figure 2.18. After code synchronization of the spreading sequence has been established in the receiver, the final mixing operations produce the complex-valued chip-rate branch sequence comprising the despread desired sequence, spectrally spread interference, and noise. Although frequency synchronization is necessary for downconversion of the received signal to baseband, no phase synchronization is attempted in the array branches because relative phase information must be preserved for successful beamforming and interference cancelation. The phase synchronization, which is ultimately necessary for coherent demodulation, is applied by a carrier synchronization system to the adaptive-processor output.

The chip-rate vector of branch sequences

$$\mathbf{x}_b(\ell) = [x_{b1}(\ell)\; x_{b2}(\ell) \ldots\; x_{bN}(\ell)]^T \tag{5.235}$$

where ℓ is the index of the chip-rate samples. In the maximin processor shown in Figure 5.7, each branch sequence is applied to a signal filter (SF) that estimates the desired-signal component of $\mathbf{x}_b(\ell)$, and a monitor filter (MF) that estimates the

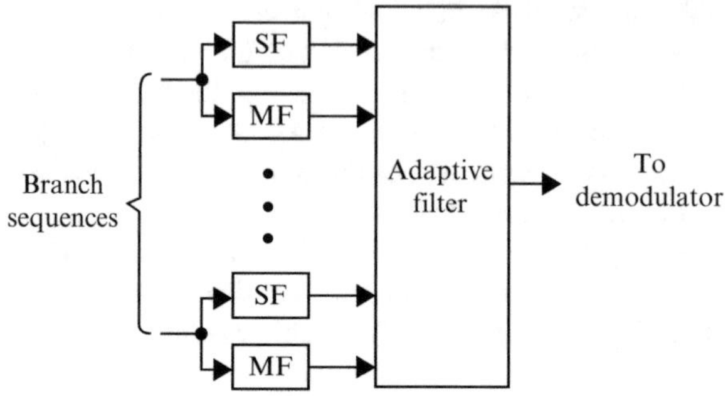

Fig. 5.7 Maximin processor for direct-sequence system. *SF* signal filter, and *MF* monitor filter

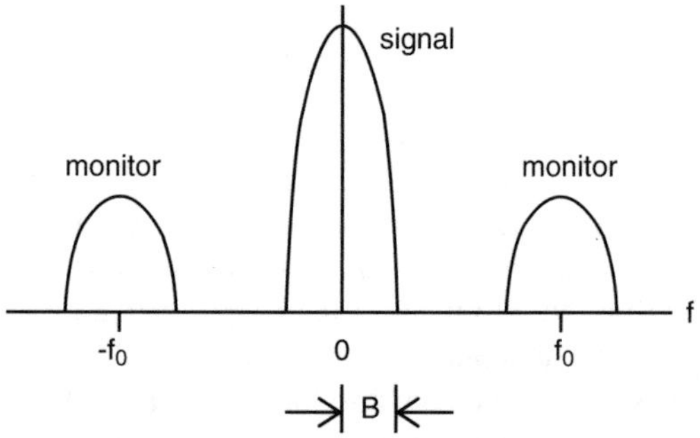

Fig. 5.8 Frequency responses of signal and monitor filters

interference component of $\mathbf{x}_b(\ell)$. For an adaptive array of N antennas, the outputs of N pairs of signal and monitor filters are applied to the adaptive filter, the output of which is applied to a digital demodulator. The maximin algorithm seeks to maximize the SINR at the input to the demodulator.

Figure 5.8 sketches the mainlobes of the frequency responses of the signal and monitor filters. Each signal filter has a frequency response $H(f)$ with one-sided bandwidth $B \approx 1/T_s$, where T_s is the data-symbol duration. The N signal-filter outputs generate $\mathbf{x}(i)$. Each monitor filter has a frequency response

$$H_1(f) = \frac{1}{2}H(f - f_o) + \frac{1}{2}H(f + f_o) \tag{5.236}$$

where the center-frequency offset is f_o, as indicated in Figure 5.8. The factor $1/2$ ensures an accurate interference-power estimate when the interference is approximately spectrally uniform over the band $|f| \leq f_o + B$. Each signal filter can have the same transfer function it would have if only a single antenna were used. Since the despread desired sequence is constant during a symbol, this signal is applied to a matched filter with a rectangular impulse response. The z-domain transfer function of the matched filter is

$$H(z) = 1 + z^{-1} + \cdots + z^{-(g-1)} = \frac{1 - z^{-g}}{1 - z^{-1}} \tag{5.237}$$

where $g = T_s/T_c$ denotes the number of chips per symbol or the spreading factor, and T_c is the chip duration. This filter, which is called an *accumulator*, has a null-to-null bandwidth equal to $2/gT_c = 2/T_s$.

Each accumulator output is sampled at the end of every symbol interval. The vector of decimated output sequences of these filters is

$$\mathbf{x}(i) = \sum_{\ell=i-g+1}^{i} \mathbf{x}_b(\ell) \tag{5.238}$$

where ℓ is the index of the chip-rate input samples and i is the index of the symbol-rate output samples. Similarly, the monitor filters, each of which has a frequency response $H_1(f)$, are *bandpass accumulators* that together produce

$$\hat{\mathbf{n}}(i) = \sum_{\ell=i-g+1}^{i} \mathbf{x}_b(\ell)\cos(2\pi f_o T_c \ell) \tag{5.239}$$

which is the vector of interference-and-noise estimates used to generate $\mathbf{c}_n(k)$ and $\hat{p}_n(k)$.

The despreading of the direct-sequence signal spreads the spectrum of the interference over the entire passband of the monitor filter if $f_o \leq (g-1)/T_s$. Any spillover or spectral splatter of the desired-signal spectrum into the monitor filter may lead to some degree of desired-signal cancelation by the adaptive algorithm. Thus, f_o must be large enough to prevent significant spectral splatter.

Adaptive Filter

The architecture of the adaptive filter is illustrated in Figure 5.9. One input vector is $\mathbf{x}(i) = \mathbf{s}(i) + \mathbf{n}(i)$, where $\mathbf{s}(i)$ and $\mathbf{n}(i)$ are the discrete-time vectors of the desired sequence and the interference-and-noise sequence, respectively. Another input vector is $\hat{\mathbf{n}}(i)$, which provides an estimate of $\mathbf{n}(i)$. There is one weight iteration after every m symbol-rate samples of the input vectors. The adaptive filter produces the output

$$y_r(i) = \text{Re}[\mathbf{w}^H(k)\mathbf{x}(i)], \quad i = km+1, \cdots, (k+1)m \tag{5.240}$$

where sample i is taken after weight iteration k. This output is applied to the demodulator and is used in the estimators

$$\mathbf{c}_x(k) = \frac{1}{m}\sum_{i=km+1}^{(k+1)m} \mathbf{x}(i)y_r(i), \qquad k \geq 0 \tag{5.241}$$

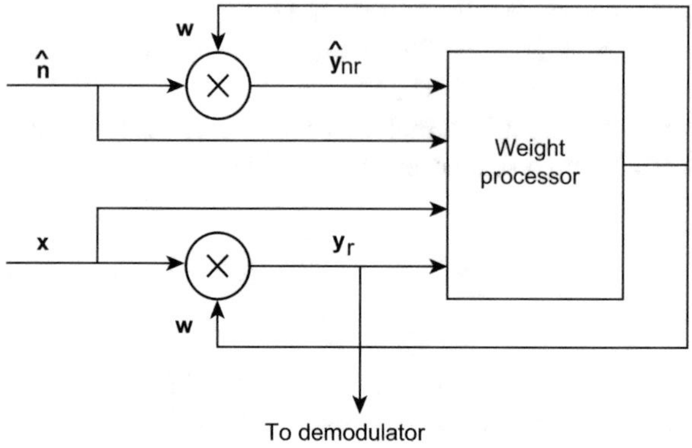

Fig. 5.9 Adaptive filter that executes maximin algorithm. Each circle with an × computes the real part of an inner product

and

$$\hat{p}_x(k) = \frac{1}{m} \sum_{i=km+1}^{(k+1)m} y_r^2(i), \qquad k \geq 0 \tag{5.242}$$

which are unbiased when $\mathbf{x}(i)$ and $y_r(i)$ are stationary processes between weight iterations.

The adaptive filter also generates

$$\hat{y}_{nr}(i) = Re[\mathbf{w}^H(k)\hat{\mathbf{n}}(i)], \quad i = km+1, \cdots, (k+1)m. \tag{5.243}$$

If the interference power is spread approximately uniformly over $|f| \leq f_o + B$, then the form of $H_1(f)$ indicates that the interference powers at the signal-filter and monitor-filter outputs are approximately equal. Therefore, using (5.226), we obtain

$$E[\hat{\mathbf{n}}(i)\hat{\mathbf{n}}^H(i)] \approx E[\mathbf{n}(i)\mathbf{n}^H(i)], \;\; E[\hat{\mathbf{n}}(i)\hat{\mathbf{n}}^T(i)] \approx E[\mathbf{n}(i)\mathbf{n}^T(i)] = 0. \tag{5.244}$$

We assume that $\mathbf{w}(k)$ varies slowly relative to the symbol rate, hence approximating a random variable that is independent of $\hat{\mathbf{n}}(i)$ and $\mathbf{n}(i)$. Then, (5.244), (5.243), and (5.231) imply that

$$E[\hat{\mathbf{n}}(i)\hat{y}_{nr}(i)] \approx E[\mathbf{n}(i)y_{nr}(i)] \tag{5.245}$$

and a suitable estimator of the interference-and-noise correlation vector at weight iteration k is

$$\mathbf{c}_n(k) = \frac{1}{m} \sum_{i=km+1}^{(k+1)m} \hat{\mathbf{n}}(i)\hat{y}_{nr}(i), \qquad k \geq 0. \tag{5.246}$$

Similarly, a suitable estimator proportional to the interference-and-noise output power is

$$\hat{p}_n(k) = \frac{1}{m} \sum_{i=km+1}^{(k+1)m} \hat{y}_{nr}^2(i), \qquad k \geq 0. \tag{5.247}$$

Both of these estimators are approximately unbiased when $\hat{\mathbf{n}}(i)$ and $\hat{y}_{nr}(i)$ are stationary processes between weight iterations.

An estimator of the total interference-and-noise power in the passbands of the despread desired-signal copies is

$$\hat{t}(k) = \frac{1}{m} \sum_{i=km+1}^{(k+1)m} \| \hat{\mathbf{n}}(i) \|^2, \quad k \geq 0 \tag{5.248}$$

which is approximately unbiased when $\hat{\mathbf{n}}(i)$ is a wide-sense-stationary process. A recursive estimator of $\hat{t}(k)$ is

$$\hat{t}(k) = \begin{cases} \mu \hat{t}(k-1) + \frac{1-\mu}{m} \sum_{i=km+1}^{(k+1)m} \| \hat{\mathbf{n}}(i) \|^2, & k \geq 1 \\ \frac{1}{m} \sum_{i=1}^{m} \| \hat{\mathbf{n}}(i) \|^2, & k = 0 \end{cases} \tag{5.249}$$

where μ is the *memory factor*, and $0 \leq \mu < 1$. As verified by simulation experiments, recursive versions of the preceding estimators merely slow the convergence of the maximin algorithm when the interference statistics are stationary. However, the recursive estimator of $\hat{t}(k)$ is useful in a nonstationary environment, such as one with pulsed interference (Section 2.6).

Code acquisition, which must be achieved before the maximin algorithm is activated, may be obtained by using an algorithm that suppresses interference until acquisition is achieved. One method is to use the estimated direction-of-arrival of the desired signal followed by beamforming to enhance the desired signal [72]. In another method, an adaptive-array algorithm exploits the high power of interference to reduce its level relative to that of a desired direct-sequence signal before code acquisition has been achieved [104]. Although the degree of interference suppression may be sufficient to enable acquisition, it is usually insufficient to enable code tracking and demodulation. Both the despreading and the maximin algorithm are needed after acquisition.

In a multipath environment, separate pairs of initial and maximin processors can potentially establish a distinct adaptive-array pattern for each resolved path. A rake demodulator (Section 6.10) can then maximize the SINR of the combined signal derived from all these patterns.

Convergence Analysis

Let $s(i)$ denote the component of $\mathbf{s}(i)$ derived from a fixed reference antenna. We assume that the array is sufficiently compact that $\mathbf{s}(i) = s(i)\mathbf{s}_0$, where $\mathbf{s}_0$ is given by (5.205). Then, (5.212) gives the optimal weight vector $\mathbf{w}_0$, and the maximum SINR ρ_0 is given by (5.213) with desired-sequence power p_s.

The highly nonlinear nature of the maximin algorithm precludes a highly rigorous convergence analysis. However, with suitable approximations and assumptions, the convergence of the mean weight vector to $\mathbf{w}_0$ can be demonstrated and bounds on the adaptation constant can be derived. We assume that the interference is wide-sense stationary and m is large enough so that (5.248) gives

$$\begin{aligned}\hat{t}(k) &\approx E[\hat{t}(k)] = E[\parallel \hat{\mathbf{n}}(i) \parallel^2] \\ &\approx E[\parallel \mathbf{n}(i) \parallel^2] = tr(\mathbf{R}_n).\end{aligned} \tag{5.250}$$

We assume that after a number of algorithm iterations k_0,

$$\begin{aligned}\frac{\hat{p}_x(k)}{\hat{p}_n(k)} &= \frac{\hat{p}_s(k)}{\hat{p}_n(k)} + 1 \\ &\approx \rho_0 + 1 \\ &= p_s \mathbf{s}_0^H \mathbf{R}_n^{-1} \mathbf{s}_0 + 1.\end{aligned} \tag{5.251}$$

Using these assumptions in (5.233) and (5.234), the maximin algorithm is approximated by

$$\mathbf{w}(k+1) = \mathbf{w}(k) + \frac{\alpha}{tr(\mathbf{R}_n)}\left[\frac{\mathbf{c}_x(k)}{\rho_0 + 1} - \mathbf{c}_n(k)\right], \quad k \geq k_0. \tag{5.252}$$

We make the approximation that $\mathbf{w}(k)$ is statistically independent of $\mathbf{x}(i)$, $\mathbf{n}(i)$, and $\widehat{\mathbf{n}}(i)$ for $i \geq km + 1$. We obtain from (5.241), (5.240), (5.225), and (5.214) that

$$E[\mathbf{c}_x(k)] = E[\mathbf{x}(i) y_r(i)] \approx \frac{1}{2}\mathbf{R}_x E[\mathbf{w}(k)]. \tag{5.253}$$

Similarly, (5.246), (5.245), (5.226), and (5.217) yield

$$E[\mathbf{c}_n(k)] \approx E[\mathbf{n}(i)) y_{nr}(i)] \approx \frac{1}{2}\mathbf{R}_n E[\mathbf{w}(k)]. \tag{5.254}$$

Taking the expected value of both sides of (5.252), substituting (5.254), (5.253), and (5.214), and simplifying algebraically, we obtain the *approximate recursive equation for the mean weight vector*:

$$E[\mathbf{w}(k+1)] = \left[\mathbf{I} - \frac{\alpha}{2tr(\mathbf{R}_n)(\rho_0+1)}\mathbf{D}\right] E[\mathbf{w}(k)], \ \ k \geq k_0 \tag{5.255}$$

where

$$\mathbf{D} = \rho_0 \mathbf{R}_n - \mathbf{R}_s = \rho_0 \mathbf{R}_n - p_s \mathbf{s}_0 \mathbf{s}_0^H . \tag{5.256}$$

Equations (5.256) and (5.213) yield

$$\mathbf{D}\mathbf{w}_0 = \mathbf{D}\mathbf{R}_n^{-1}\mathbf{s}_0 = \mathbf{0} \tag{5.257}$$

which indicates that the optimal weight vector $\mathbf{w}_0$ given by (5.212) is an eigenvector of $\mathbf{D}$, and the corresponding eigenvalue is 0. As $\mathbf{D}$ is Hermitian, it has a complete set of N orthogonal eigenvectors (Appendix G), one of which is $\mathbf{w}_0$. As

$$\mathbf{w}^H\mathbf{D}\,\mathbf{w} = \mathbf{j_0}\mathbf{w^H}\mathbf{R_n}\mathbf{w} - \mathbf{w^H}\mathbf{R_s}\mathbf{w} \geq 0 \tag{5.258}$$

for an arbitrary vector $\mathbf{w}$, $\mathbf{D}$ is positive semidefinite and hence has N nonnegative eigenvalues. We assume that only $\mathbf{w} = \mathbf{w}_0$ maximizes the SINR so that $\mathbf{w}^H\mathbf{D}\,\mathbf{w} > 0$, $\mathbf{w} \neq \mathbf{w}_0$. Because it follows that $\mathbf{D}\mathbf{w} \neq \mathbf{0}$, $\mathbf{w} \neq \mathbf{w}_0$, one of the eigenvalues of $\mathbf{D}$ is zero, and the other $N-1$ eigenvalues are positive.

To solve (5.255), we make the eigenvector decomposition

$$E[\mathbf{w}(k)] = \eta(k)\mathbf{R}_n^{-1}\mathbf{s}_0 + \sum_{l=2}^{N} a_l(k)\mathbf{e}_i \tag{5.259}$$

where each $a_l(k)$ and $\eta(k)$ are scalar functions, and $\mathbf{e}_l$ is one of the $N-1$ eigenvectors orthogonal to $\mathbf{R}_n^{-1}\mathbf{s}_0$. Substituting this equation into (5.255) and using the orthonormality of the eigenvectors, we obtain

$$\eta(k+1) = \eta(k) = \eta(k_0), \qquad k \geq k_0 \tag{5.260}$$

$$a_l(k+1) = \left[1 - \frac{\alpha\lambda_l}{2tr(\mathbf{R}_n)(\rho_0+1)}\right] a_l(k), \quad 2 \leq l \leq N, \qquad k \geq k_0 \tag{5.261}$$

where λ_l is the positive eigenvalue corresponding to $\mathbf{e}_l$. Assuming that $\eta(k_0) \neq 0$, (5.213), (5.259), and (5.260) indicate that $E[\mathbf{w}(k)] \to \mathbf{w}_0$ as $k \to \infty$ if and only if each $a_l(k) \to 0$. The solution to (5.261) is

$$a_l(k) = \left[1 - \frac{\alpha\lambda_l}{2tr(\mathbf{R}_n)(\rho_0+1)}\right]^{k-k_0} a_l(k_0), \quad 2 \leq l \leq N, \quad k \geq k_0. \tag{5.262}$$

This equation indicates that $a_l(k) \to 0$, $2 \le l \le N$, as $k \to \infty$ if and only if

$$\left| 1 - \frac{\alpha \lambda_l}{2tr(\mathbf{R}_n)(\rho_0 + 1)} \right| < 1 \,, \quad 2 \le l \le N. \tag{5.263}$$

This inequality implies that the necessary and sufficient condition for the convergence of the mean weight vector is:

$$0 < \alpha < \frac{4tr(\mathbf{R}_n)(\rho_0 + 1)}{\lambda_{max}} \tag{5.264}$$

where λ_{max} is the largest eigenvalue of $\mathbf{D}$.

As the sum of the eigenvalues of a square matrix is equal to its trace,

$$\lambda_{max} \le \sum_{i=1}^{N} \lambda_i = tr(\mathbf{D}) = \rho_0 tr(\mathbf{R}_n) - tr(\mathbf{R}_s) \le \rho_0 tr(\mathbf{R}_n). \tag{5.265}$$

Substituting this bound into (5.264) and simplifying the result, we obtain

$$0 < \alpha < 4 \tag{5.266}$$

as a sufficient (but not necessary) condition for the convergence of the mean weight vector to the optimal weight vector. Although this inequality must be regarded as an approximation because of the approximations used in its derivation, it gives at least rough guidance in the selection of the adaptation constant. The fact that the upper bound is numerical and does not depend on environmental parameters provides support for the choice of (5.234) as the adaptation sequence.

Simulation of Maximin Algorithm

In the simulation experiments, the array consists of four omnidirectional antennas located at the vertices of a square or in a uniform linear configuration. Let λ denote the wavelength corresponding to the center frequency of the desired signal, which is 3 GHz. The edge length or the separation between adjacent antennas is $d = 0.5\lambda$, $d = 1.0\lambda$, or $d = 1.5\lambda$. The direct-sequence signal with BPSK modulation and a rectangular chip waveform arrives from a direction 20° clockwise from the perpendicular to one of the edges. All signals are assumed to arrive as plane waves. The direct-sequence signal has a frequency offset equal to 1 kHz after downconversion, which models imperfect frequency synchronization. The data-symbol and spreading sequences are randomly generated for each simulation trial at the rates of 100 kbps and 10 Mbps respectively, which implies that the spreading factor is 20 dB. Perfect chip and spreading-sequence synchronization in the receiver are assumed. As indicated in Figure 5.6, the initial sampling is performed once per spreading-sequence chip. The thermal noise in each branch output is modeled as bandlimited white Gaussian noise. The signal-to-noise ratio (SNR) is 0 dB in each branch output. Each signal filter is an accumulator with a one-sided bandwidth $B = 100$ kHz. Each monitor filter is a bandpass accumulator offset by $f_o = 400$ kHz to prevent contamination by the direct-sequence signal. The maximin algorithm is implemented with $\alpha = 1$. A weight iteration occurs after each $m = 10$ data symbol. In each simulation trial, the initial weight vector of the adaptive processor is $\mathbf{w}(0) = [1\ \ 0\ \ 0\ \ 0]$, which forms an omnidirectional array pattern.

Each of 1, 2, or 3 interference signals is a tone (continuous-wave signal). After the downconversions, the tones have different initial phase shifts and residual frequency offsets equal to 10 kHz, which reflects the mismatch of the tone frequencies and the carrier frequency of the direct-sequence signal. Multiple tones do not add coherently, even if they have the same carrier frequencies, because they arrive from different directions and have different initial phase shifts. The SINR at the processor output is calculated after each sample time and then averaged over all samples in the time interval between a weight iteration and a preceding weight iteration to determine the SINR at each weight iteration. The SINR is observed to fluctuate, but tends to gradually increase until it reaches a steady-state condition with a smaller residual fluctuation.

Figure 5.10 illustrates the SINR variation versus the weight iteration number for a typical simulation trial in which one interference tone arrives at a 50° angle with an ISR equal to 30 dB. Let θ denote an arrival angle defined as the angle in the clockwise direction from the normal to one of the array edges. Let $\mathbf{s}(\theta)$ denote the *array response vector*, which is the array response to an ideal plane wave arriving

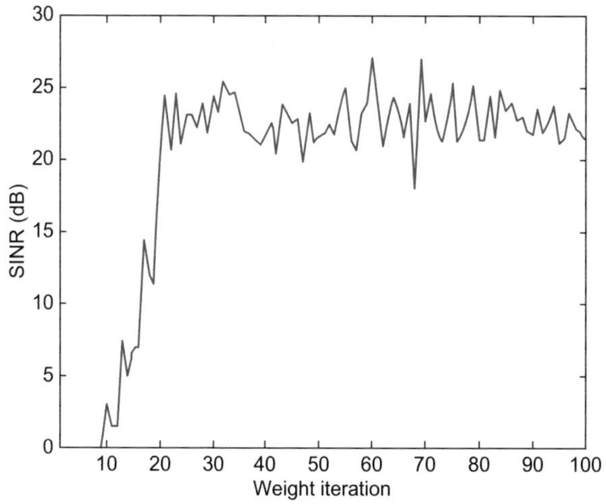

Fig. 5.10 Signal-to-interference-and-noise ratio variation in a typical simulation trial for a direct-sequence system with one interference tone at 50° and ISR = 30 dB

at angle θ. For the square array, the components of the array response vector are

$$s_{r1} = 1, \quad s_{r2} = \exp(-j2\pi \frac{d}{\lambda} \sin\theta)$$
$$s_{r3} = \exp(-j2\pi \frac{d}{\lambda} \cos\theta), \quad s_{r4} = \exp[-j2\pi \frac{d}{\lambda}(\sin\theta + \cos\theta)]. \tag{5.267}$$

The *array gain pattern* after weight iteration k is

$$G(\theta, k) = \frac{\left|\mathbf{w}^H(k)\mathbf{s}(\theta)\right|^2}{\|\mathbf{w}(k)\|^2}. \tag{5.268}$$

The array gain pattern at the end of the simulation trial of Figure 5.10 is depicted in Figure 5.11. A null deeper than −20 dB in the direction of the interference signal and a mainlobe slightly displaced from the direction of the desired signal have formed. In addition, another *grating null*, which is a low point in the array gain pattern, and *grating lobes*, which are high points in the pattern, have formed.

The results of 15 representative simulation experiments are summarized in Table 5.1. Each experiment comprises 50 trials with 100 weight iterations per trial. The first column gives the arrival angles of 1, 2, or 3 interference signals. The ISR for each interference signal is 10 dB. The second column gives the array type: square with $d = 1.0\lambda$, square with $d = 1.5\lambda$, or linear with $d = 0.5\lambda$. The SINRs expressed in decibels for the last 20 weight iterations of all the trials are used to compute the *steady-state SINR* and the *standard deviation* of the SINR. The final column gives the *crossing number*, which is the average number of weight

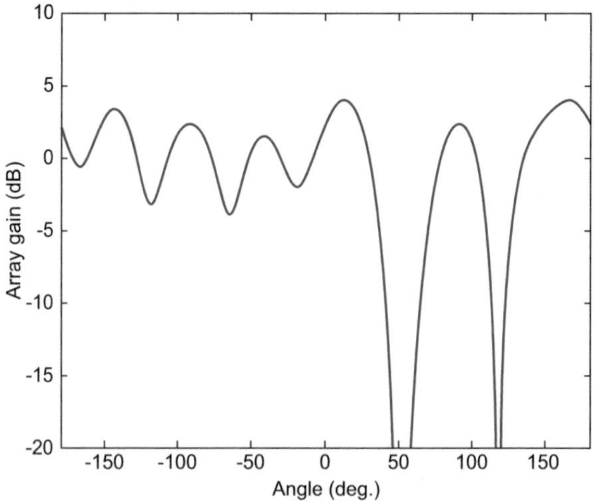

Fig. 5.11 Array gain pattern at the end of a typical simulation trial for a direct-sequence system with one interference tone at 50° and ISR = 30 dB

Table 5.1 Simulation results for interference tones with ISR = 10 dB

Arrival angles of interf. (°)	Array type	Steady-state SINR (dB)	Standard dev. (dB)	Crossing number
60	Sq, 1.0λ	25.81	1.49	3.2
60, −40	Sq, 1.0λ	22.84	2.22	8.9
60, 85	Sq, 1.0λ	24.53	1.60	5.0
60, −40, 85	Sq, 1.0λ	24.39	1.71	5.5
30	Sq, 1.0λ	20.10	1.57	16.1
60	Sq, 1.5λ	25.13	1.46	4.2
60, −40	Sq, 1.5λ	25.38	1.50	3.9
60, 85	Sq, 1.5λ	24.40	1.47	5.4
60, −40, 85	Sq, 1.5λ	24.26	1.52	5.4
30	Sq, 1.5λ	22.95	1.99	7.1
60	Linear	25.75	1.43	3.3
60, −40	Linear	24.98	1.46	4.1
60, 85	Linear	24.81	1.52	4.4
60, −40, 85	Linear	24.67	1.59	4.8
30	Linear	20.25	1.55	10.1

iterations required for the SINR to exceed 20 dB. The crossing number provides a rough measure of the relative time required for convergence to the steady state.

Table 5.1 indicates that beamforming in the direction of the desired signal and interference cancelation are achieved in a wide variety of scenarios. Additional interference signals do not necessarily slow the algorithm convergence or lower the steady-state SINR. The reason is that as the array forms one pattern lobe or null, it tends to form additional grating lobes and nulls, thereby possibly counteracting or facilitating the maximin algorithm. The results for an interference signal with a 30° arrival angle and a square array illustrate the limitations imposed by the resolution of the array. The *resolution*, which is the angular separation between the interference and desired signals that can be accommodated without great performance degradation, decreases as the array aperture increases. The square array with $d = 1.0\lambda$ and the uniform linear array with $d = 0.5\lambda$, despite having an array aperture equal to 1.5λ, have insufficient resolution for this interference. Increasing the antenna separation in a square array to 1.5λ provides a large improvement in the convergence speed and the steady-state SINR.

Many more simulation experiments [103] confirm that the maximin algorithm supplements the direct-sequence spreading factor with a large amount of additional interference suppression. The simulation results demonstrate the robust performance of the algorithm for various array configurations, numbers of interference signals, interference levels, and fading conditions.

5.7 Adaptive Array for Frequency-Hopping Systems

The *anticipative maximin algorithm* [102] is an adaptive-array algorithm that exploits both the spectral and temporal characteristics of frequency-hopping signals. The algorithm fuses multisymbol anticipative processing with the maximin algorithm to enable a high degree of cancelation of partial-band interference within a hopping band. The algorithm is executed by an adaptive maximin processor that includes a *main processor* and *anticipative filters*. The main processor observes spectral regions adjacent to the current frequency channel occupied by the frequency-hopping signal. The anticipative filters observe interference in several frequency channels that are occupied by the frequency-hopping signal after subsequent hops. Both observations are used in the interference cancelation.

Initial and Main Processors

In the adaptive array illustrated in Figure 5.6, the spreading waveform is the unmodulated frequency-hopping replica produced by a synchronized receiver, which enables the dehopping in each initial processor behind each antenna. Each of the N branch sequences is produced by the demodulator that extracts the sampled complex envelope from the dehopped signal. Let $\mathbf{x}(i)$ denote the discrete-time $N \times 1$ vector of filtered branch outputs that provides the main input to the anticipative maximin algorithm, where the index i denotes the sample number. The vector $\mathbf{x}(i)$ can be decomposed as $\mathbf{x}(i) = \mathbf{s}(i) + \mathbf{n}(i)$, where $\mathbf{s}(i)$ is the desired-signal component and $\mathbf{n}(i)$ is the interference and thermal-noise component. The maximin processor computes a weight vector $\mathbf{w}(k)$, where k is the weight iteration index. The output of the maximin processor is $y(i) = \mathbf{w}^H(k)\,\mathbf{x}(i)$.

To implement the main processor of the maximin processor, it is necessary to separate the interference $\mathbf{n}(i)$ from the total signal $\mathbf{x}(i)$. After each hop, the frequency-hopping signal has a carrier frequency f_h, and its spectrum is largely confined to a frequency channel with one-sided bandwidth B, as depicted in Figure 5.12. In each branch following an array antenna, the frequency hopping is removed, and the current frequency channel or *signal channel* is downconverted

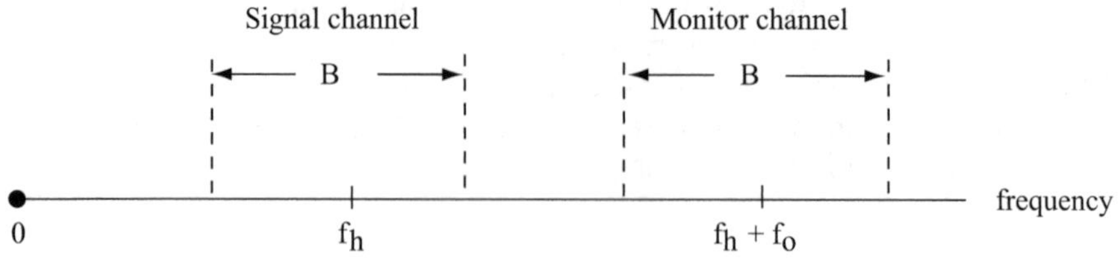

Fig. 5.12 Signal channel and monitor channel during the hop dwell time

to baseband. A *signal filter* then extracts the total signal in the signal channel. To cancel the interference embedded in $\mathbf{x}(i)$, the receiver measures the interference in a *monitor channel*, which is a nearby frequency channel or spectral region with a center frequency offset by $f_o \geq B$ from the carrier frequency. For this measurement, the processing in each branch downconverts the monitor channel to baseband. After the downconversion, a baseband *monitor filter* extracts the interference in the monitor channel. Ideally, the spectrum of an interference signal overlaps the signal and monitor channels so that the interference components in the signal-filter and monitor-filter outputs have the same second-order statistics. After a sufficient number of hops, each monitor filter processes most of the interference that originates from a particular direction and spectrally overlaps the hopping band. The outputs of all the branch signal and monitor filters are used by the anticipative maximin algorithm to enable interference cancelation and desired-signal enhancement. The frequency offset f_o must be sufficient to ensure that spectral splatter of the desired-signal spectrum into the monitor filter does not lead to significant desired-signal cancelation by the adaptive algorithm.

As shown in Section 3.5 for a frequency hopping/continuous-phase frequency-shift keying signal with modulation index h, the demodulator output for branch l is the sampled complex envelope

$$z_l(i) = \exp\left\{ j2\pi \left[\frac{f_e T_s i}{L} + \frac{h\alpha_r(i - Lr)}{2L} \right] + j\phi_{le} \right\} + n_{l1}(i),$$
$$Lr \leq i \leq Lr + L - 1, \ l = 1, 2, \ldots, N \qquad (5.269)$$

where f_e is the intermediate-frequency (IF) offset frequency, T_s is the symbol duration, L is the number of samples per symbol, α_r is the rth symbol, $n_{l1}(i)$ is the sequence of samples of the interference and noise, and the phase shift ϕ_{le} reflects the different arrival times of the desired signal at the array antennas. To capture the energy in both the signal and monitor channels, the IF filter in each branch must have a bandwidth of $2f_{\max} + f_o + B$, where $f_{\max}$ is the maximum value of $|f_e|$. It is shown in Section 3.5 that L and f_e must satisfy (3.104) and (3.105).

In the main processor illustrated in Figure 5.13, each of the N branch sequences $z_l(i)$, $l = 1, 2, \ldots, N$, is applied to a signal filter. A further downconversion of each branch sequence by f_o provides the monitor-channel sequences $m_l(i)$, $l = 1, 2, \ldots, N$, each of which is applied to a monitor-filter input. Thus, the vector of branch sequences $\mathbf{z}(i)$ is used to produce the vector of monitor-filter inputs $\mathbf{m}(i) = \mathbf{z}(i) \exp(-j2\pi f_o T_0 i)$, where $T_0 = T_s/L$ is the sampling interval. The baseband signal and monitor filters are identical, with passbands such that $|f| \leq f_{\max} + B/2$. The signal-filter and the monitor-filter outputs are components of $\mathbf{x}(i)$ and $\hat{\mathbf{n}}(i)$, respectively, the vectors applied to the adaptive filter. The vector $\hat{\mathbf{n}}(i)$ provides an estimate of the interference and noise in $\mathbf{x}(i)$.

Let $\mathbf{n}_1(i)$ denote the interference-and-noise component of $\mathbf{z}(i)$, and let $h(i)$ denote the impulse response of the signal and monitor filters. The interference-and-noise component of $\mathbf{x}(i)$ is

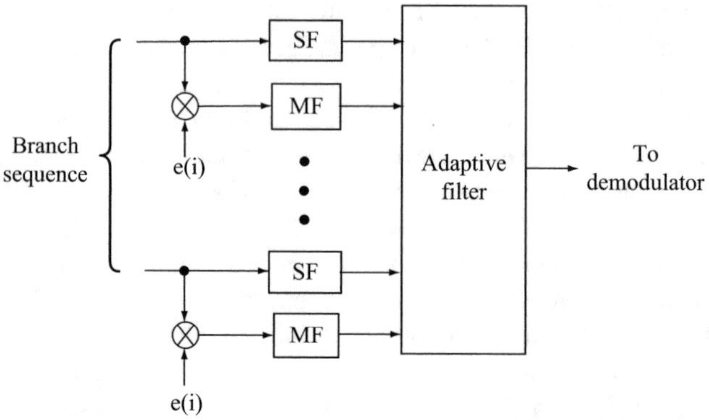

Fig. 5.13 Maximin processor for frequency-hopping system. *SF* signal filter, *MF* monitor filter, and $e(i) = \exp(-j2\pi f_o T_0 i)$

$$\mathbf{n}(i) = \sum_{\ell=0}^{i} \mathbf{n}_1(\ell)h(i-\ell) \tag{5.270}$$

and the monitor-filter outputs provide

$$\hat{\mathbf{n}}(i) = \sum_{\ell=0}^{i} \mathbf{n}_1(\ell)e^{-j2\pi f_o T_0 \ell}h(i-\ell). \tag{5.271}$$

The critical requirement of $\hat{\mathbf{n}}(i)$ is for it to have approximately the same second-order statistics as $\mathbf{n}(i)$; that is,

$$E\left[\mathbf{n}(i)\mathbf{n}^H(i)\right] \approx E\left[\hat{\mathbf{n}}(i)\hat{\mathbf{n}}^H(i)\right]. \tag{5.272}$$

This equation is satisfied for an interference signal that has similar second-order statistics in each frequency channel it occupies. As an example, suppose that $\mathbf{n}_1(i) = \mathbf{n}_t(i) + \mathbf{j}_1(i) + \mathbf{j}_2(i)$, where $\mathbf{n}_t(i)$ has white-noise components, and $\mathbf{j}_1(i)$ and $\mathbf{j}_2(i)$ are independent interference signals. We assume that $\mathbf{j}_2(i)$ resembles a frequency-shifted version of $\mathbf{j}_1(i)$; that is,

$$\mathbf{j}_2(i) \approx \mathbf{j}_1(i)\exp(j2\pi f_o T_0 i + \theta) \tag{5.273}$$

where θ is an arbitrary phase shift. Assuming that f_o is large enough, the signal filters have a negligible response to $\mathbf{j}_2(i)$, and the monitor filters have a negligible response to $\mathbf{j}_1(i)$. Therefore, $\mathbf{n}(i)$ is approximated by substituting $\mathbf{n}_1(i) = \mathbf{n}_t(i) + \mathbf{j}_1(i)$ into (5.270), and $\hat{\mathbf{n}}(i)$ is approximated by substituting $\mathbf{n}_1(i) = \mathbf{n}_t(i) + \mathbf{j}_2(i)$ into (5.271). The substitution of these approximations and (5.273) into (5.272) indicate that (5.272) is satisfied.

The architecture of the adaptive filter, which generates the estimates needed by the anticipative maximin algorithm, is the same as that illustrated in Figure 5.9. The vectors applied to the adaptive filter are $\mathbf{x}(i)$ and $\hat{\mathbf{n}}(i)$, and m samples are generated per weight iteration. The adaptive filter produces the outputs and estimates given by (5.240) to (5.247). A recursive estimator of the total interference-and-noise power in the signal-filter outputs, which is approximated by the power in the monitor-filter outputs, is given by (5.249). The recursive estimator is useful in a nonstationary environment, such as one with partial-band interference.

Anticipative Adaptive Filters

Each of ν *anticipative filters* observes a future signal channel that does not yet have the desired signal energy. Thus, these channels produce future interference estimates that potentially allow additional interference cancelation by the anticipative maximin algorithm. This capability is ultimately due to the receiver's knowledge of the frequency-hopping pattern. If $\nu \geq 1$ and $1 \leq l \leq \nu$, let f_{hl} denote the carrier frequency after the next l hops. Then, the anticipative branch l produces an output derived from the signal channel that is used after l frequency hops.

To produce the inputs to the main and anticipative filters, the output of each antenna is converted to an intermediate frequency and then applied to a *main branch* and ν parallel *anticipative branches*. The main branch produces the same outputs as previously described for the current frequency channel with carrier frequency f_h. Anticipative branch l uses similar processing to produce similar outputs, but with the different carrier frequency f_{hl} and no monitor channel ($f_o = 0$).

All the branch sequences are applied to the *anticipative maximin processor* of Figure 5.14. The main processor produces the demodulator input. The outputs of the monitor filter fed by anticipative branch l provide the $N \times 1$ vector $\hat{\mathbf{n}}_{al}(i)$, which estimates the interference-and-noise that are present in the signal channel after l frequency hops. The vector $\mathbf{x}(i)$ is provided by the signal-filter outputs of the main processor. The vectors $\hat{\mathbf{n}}_{al}(i)$ and $\mathbf{x}(i)$ are applied to anticipative adaptive filter l, which adapts its weight vector $\mathbf{w}_{al}(k)$ by using the maximin algorithm (5.233). The recursive equations executed by the anticipative filters are

$$\mathbf{w}_{al}(k+1) = \mathbf{w}_{al}(k) + \left\{ \alpha(k) \left[\frac{\mathbf{c}_x(k)}{\hat{p}_x(k)} - \frac{\mathbf{c}_n(k)}{\hat{p}_n(k)} \right] \right\}_l, \quad k \geq 0, \quad 1 \leq l \leq \nu \tag{5.274}$$

where the subscript l beneath the braces signifies computation by anticipative filter l. The adaptation constant and the memory factor used in this equation are set equal to their values in the main processor. After each hop, the weight vector associated with the new carrier frequency is transferred from anticipative filter 1 to the main processor, and the weight vector computed by filter l is transferred to filter $l-1$,

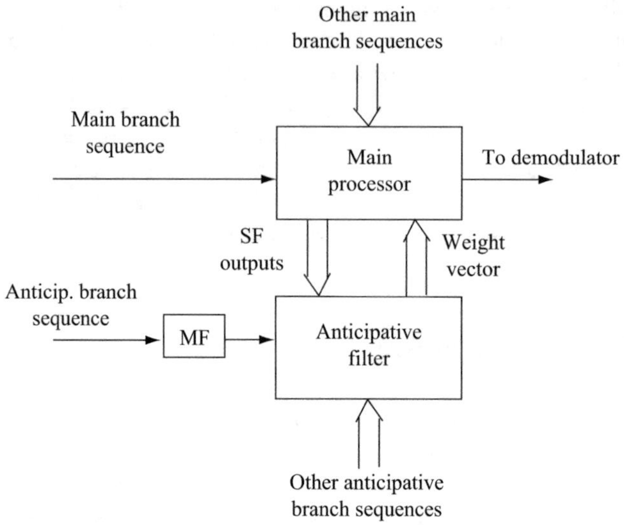

Fig. 5.14 Anticipative maximin processor for frequency-hopping system. *MF* monitor filter

where $l = 2, ..., \nu$. The weight vector of filter ν is reset to $\mathbf{w}_{a\nu}(0)$. Transfers are triggered by the clock that controls the frequency-hopping carrier transitions.

The weight vector in the main processor is updated by computing the maximin algorithm, except at sampling instants occurring at the end of a dwell interval or during the switching times of the frequency-hopping waveform. At these instants, the weight vector in the main processor is set equal to that of the anticipative filter 1. Let k_0 denote the number of iterations per hop and $n = 1, 2, \ldots$ denote the hop number. The switching times occur when $k = nk_0$ in the main processor. Thus, the algorithm in the main processor is

$$\mathbf{w}(k+1) = \mathbf{w}(k) + \alpha(k)\left[\frac{\mathbf{c}_x(k)}{\hat{p}_x(k)} - \frac{\mathbf{c}_n(k)}{\hat{p}_n(k)}\right], \quad k+1 \neq nk_0 \tag{5.275}$$

$$\mathbf{w}(nk_0) = \mathbf{w}_{a1}(nk_0), \qquad n \geq 0. \tag{5.276}$$

Equations (5.274) to (5.276) constitute the *anticipative maximin algorithm.*

The anticipative maximin algorithm expedites the convergence of the mean weight vector in the presence of partial-band interference at the cost of additional hardware and an increase in the computational requirements. The computational cost per weight iteration of the algorithm can be estimated in terms of the number of real multiplications, real additions, and real divisions. From the algorithm equations, it is found that each iteration of the anticipative maximin algorithm requires $(\nu+1)(4N+1)$ real divisions, $(\nu+1)(6Nm+6N+2m+4)$ real multiplications, and $(\nu+1)(6Nm+m-1)$ real additions if $k_0 >> 1$. The computational cost per iteration is $(\nu+1)O(mN)$ real multiplications or divisions and $(\nu+1)O(mN)$ real additions.

The computational cost per sampling interval is $(\nu + 1)O(N)$ real multiplications or divisions and $(\nu + 1)O(N)$ real additions, which is on the order of the cost of the classical LMS algorithm when $\nu = 0$.

Multipath components of the desired signal are not canceled by the anticipative maximin algorithm because their spectra occupy only the current signal channel and are insignificant in the monitor channel and subsequent signal channels. However, the beamforming generated by the algorithm often attenuates those multipath signals that arrive from directions very different from the direction of the main frequency-hopping signal.

Simulation Experiments

In the simulation experiments, the array consists of four omnidirectional antennas located at the vertices of a square. Let λ denote the wavelength corresponding to the center frequency of the desired signal, which is 3 GHz. The edge length or the separation between adjacent antennas is $d = 0.9\lambda$. All signals are assumed to arrive as plane waves. The frequency-hopping signal is modulated by binary minimum-shift keying (Section 3.3) and has a carrier frequency that is randomly chosen from a hopset. Each frequency channel, which includes a hopset frequency, has a bandwidth $B = 100$ kHz. The hopping band has a bandwidth $W_h = 30$ MHz, and there are $W_h/B = 300$ contiguous frequency channels. The hop dwell time is 1 ms. The frequency-hopping signal arrives from a direction 20° clockwise from the normal to one of the edges and has a frequency offset equal to 1 kHz after downconversion, which models imperfect frequency synchronization and the Doppler shift. Perfect timing synchronization of the local frequency-hopping replica is assumed. The sequence of data symbols is randomly generated at the rate of 100 kbps. The sample rate is 10^6 samples per second, which corresponds to 10 samples per symbol. The signal and monitor filters are digital Chebyshev filters of the second kind [68] with 3-dB bandwidths equal to B. The monitor channel is offset by $f_o = 200$ kHz, which is sufficient to prevent significant contamination by the desired signal. The anticipative maximin algorithm is implemented with $\alpha = 0.4$ and $\mu = 0.99$, values that usually provide close to the best overall performance against the modeled partial-band interference. A weight iteration occurs after each ten data symbols. The weight-iteration rate and the adaptation constant are both partly selected to ensure that the weights do not increase to excessively large values. For each simulation trial, the initial weight vector of each adaptive filter is $\mathbf{w}(0) = [1\ \ 0\ \ 0\ \ 0]$, which forms an omnidirectional array pattern. The IF filters are modeled as ideal rectangular filters. The thermal noise in each main and anticipative sequence is modeled as bandlimited complex Gaussian noise such that the SNR is 14 dB in the output of each signal filter.

Interference that occupies only a small part of the hopping band, or even frequency-hopping interference signals, can be suppressed by the adaptive array supplemented by an error-control code. If the interference is observed by a monitor filter often enough, it is rapidly suppressed by the adaptive array; if it is observed only occasionally, then the interference is primarily suppressed by the error-control code. In the simulation experiments, each of 1, 2, or 3 interference signals has its power distributed among equal-power tones (continuous-wave signals) in frequency channels that cover the fraction 0.1 of the hopping band. For each interference signal, the ISR is equal to 0 dB in each frequency channel that contains a tone. After the downconversions, the first, second, and third interference signals have different initial phase shifts and residual frequency offsets equal to 10, 13, and 16 kHz, respectively, which reflect the mismatch of the tone frequencies and the hopset carrier frequencies. Multiple interference signals do not add coherently at all antennas, even if they have the same carrier frequencies, because they arrive from different directions and have different initial phase shifts. When the interference occupies contiguous frequency channels, we assume that an interference tone in the signal filter is always accompanied by a tone in the monitor filter. The latter assumption is a good approximation for partial-band interference over a substantial fraction of the hopping band when the signal and monitor channels are close. The SINR at the processor output is calculated after each sample time and then averaged over all samples in the time interval between one weight iteration and a preceding one to determine the SINR at each weight iteration.

The results of 19 representative simulation experiments are summarized in Table 5.2. The first column gives the arrival angles of 1, 2, or 3 interference signals. The second column gives the number of anticipative adaptive filters. The third column indicates when the interference tones occupy randomly distributed rather than contiguous frequency channels. The fourth column indicates whether frequency-selective fading is present. The SINRs, expressed in decibels, for 20 simulation trials with 100 hops and 1000 weight iterations per trial are averaged over the last 20 hops of the trials to obtain the *steady-state* or *final SINR* and its standard deviation. These statistics, which are listed in the fifth and sixth columns of the table, are indicators of the degree of interference cancelation after 80 hops.

The results in Table 5.2 confirm the improved performance as the number ν of anticipative adaptive filters increases. The improvement occurs primarily because the interference in a future signal channel is processed over ν dwell intervals. When more than one interference signal is present, the adaptation becomes more difficult as the array encounters a more rapidly varying signal environment, but the array of four antennas is often able to substantially cancel three partial-band interference signals, as indicated in Table 5.2. Randomly distributed tones provide a model of sophisticated jamming. Since an interference tone in the signal channel does not necessarily imply a simultaneous interference tone in the monitor channel, the interference cancelation is impeded, and the convergence to a steady state is slowed. The anticipative maximin algorithm with $\nu = 1$ and 2 greatly improves the convergence rate, as illustrated in the table. Rows 13 to 15 illustrate the limitations imposed by the resolution of the array, which improves with increases in the array aperture or antenna separations. However, an enlarged aperture causes an increase in the number of grating lobes, which impedes the formation of nulls against two or more interference signals.

Table 5.2 Simulation results

Interference angles (°)	ν	Random	Fade	Final SINR	Final st. dev.
40	0	No	No	18.89	1.91
40	1	No	No	19.10	1.65
40	2	No	No	19.13	1.78
40, −10	0	No	No	18.53	2.16
40, −10	1	No	No	18.87	1.61
40, −10	2	No	No	19.06	1.62
40, −10	0	Yes	No	17.11	4.26
40, −10	1	Yes	No	18.61	2.04
40, −10	2	Yes	No	18.90	1.77
40, −10, 85	0	No	No	18.18	2.16
40, −10, 85	1	No	No	18.52	1.64
40, −10, 85	2	No	No	18.81	1.66
30	0	No	No	17.27	3.20
30	1	No	No	17.86	2.70
30	2	No	No	18.00	2.84
40	2	No	Yes	16.55	5.38
40, −10	2	No	Yes	16.46	5.33
40, −10, 85	2	No	Yes	16.21	5.38
30	2	No	Yes	15.41	5.85

The final four rows illustrate the impact of frequency-selective fading (Section 6.3) across the hopping band. The frequency-hopping signal is assumed to experience independent Rayleigh fading (Section 6.2) in each frequency channel. During each dwell interval, the desired-signal amplitude is multiplied by the random number A_h generated from a Rayleigh distribution. To ensure that the average power of the desired signal is unchanged by the fading, $E[A_h^2] = 1$. The interference is assumed to be unaffected by fading. Under these severe conditions, the frequency-selective fading causes significant but not extreme decreases in the SINRs and increases in their standard deviations. The sporadic low levels in the SINR can be accommodated by the error-control code in a practical communication system. Other simulation results indicate that expanding the hopping band to bandwidth $W_h = 300$ MHz (fractional bandwidth of 0.1) can be accommodated with only a slight loss.

5.8 Problems

1 Apply (5.2), (5.3), and (5.18) to derive (5.20).

2 Show that the weight vector of the LMS steepest-descent algorithm with $\mu(k) = \mu$ converges to $\mathbf{w}_0$. Are the necessary and sufficient conditions for convergence the same as those for the LMS algorithm? What is the difference between the two algorithms in terms of the weight vector?

3 Consider the performance measure $P(\mathbf{w}) = E[|\epsilon|^2] + \alpha \|\mathbf{w}\|^2$ with $\alpha > 0$, which is to be minimized. (a) What is a necessary condition for the optimal weight vector? (b) Derive the steepest-descent algorithm for this performance measure. (c) What is the corresponding stochastic-gradient algorithm? (d) Derive the conditions for the convergence of the mean weight vector. To what value does the mean weight vector converge? (e) What is the engineering justification for this choice of performance measure?

4 Consider the performance measure $P(\mathbf{w}) = E[(M - |y|)^2]$, where $y = \mathbf{w}^H\mathbf{x}$ and M is a known scalar. (a) Derive the steepest-descent algorithm for this performance measure, assuming that the gradient and expectation operations can be interchanged. (b) What is the corresponding stochastic-gradient algorithm? (c) What is the engineering justification for this choice of performance measure?

5 Assuming that $\mathbf{x}(k+1)$ is independent of $\mathbf{x}(i)$ and $d(i)$, $i \leq k$, show that $\mathbf{w}(k)$ in the LMS algorithm is independent of $\mathbf{x}(k)$.

6 Derive (5.124) and (5.125) by following the steps specified in the text.

7 Consider the soft-decision term in (5.142). What are its values as $\sigma_z \to \infty$ and as $\sigma_z \to 0$? Give an engineering interpretation of these results.

8 In the convergence analysis of the Frost algorithm, derive (5.176), and verify that $\mathbf{Av}(k) = \mathbf{v}(k)$, $k \geq 1$.

9 In the convergence analysis of the Frost algorithm, verify that $\mathbf{Av}(0) = \mathbf{v}(0)$, $\mathbf{AR}_x\mathbf{A}$ is Hermitian, and that $\mathbf{AR}_x\mathbf{Ap} = \mathbf{0}$, which indicates that $\mathbf{p}$ is an eigenvector of $\mathbf{AR}_x\mathbf{A}$ with eigenvalue equal to zero.

10 Consider the performance measure $P(\mathbf{w}) = E[|\epsilon|^2]$ subject to the constraint $\mathbf{w}^H\mathbf{p} = 1$, where $\mathbf{p}^T\mathbf{p} = G$. Use the method of Lagrange multipliers to derive the corresponding steepest-descent and stochastic-gradient algorithms. Under what condition might this algorithm be preferred to the Frost algorithm?

11 Derive (5.245), which is used in the design of an adaptive array for direct-sequence systems.

12 Verify (5.253) and (5.254), which are used in the convergence analysis of the adaptive array for direct-sequence systems.

Chapter 6
Fading and Diversity

Fading is the variation in received signal strength due to changes in the physical characteristics of the propagation medium, which alter the interaction of multipath components of the transmitted signal. The principal means of counteracting fading are *diversity methods*, which are based on the exploitation of the latent redundancy in two or more independently fading copies of the same signal. The basic concept of diversity is that even if some copies are degraded, there is a high probability that others will not be. This chapter provides a general description of the most important aspects of fading and the role of diversity methods in counteracting it. Both direct-sequence and frequency-hopping signals are shown to provide diversity. The rake demodulator, which is of central importance in most direct-sequence systems, is shown to be capable of exploiting undesired multipath signals rather than simply attempting to reject them. The multicarrier direct-sequence system and frequency-domain equalization are shown to be alternative methods of advantageously processing multipath signals.

6.1 Path Loss, Shadowing, and Fading

Free-space propagation losses of electromagnetic waves vary inversely with the square of the distance between a transmitter and a receiver. Analysis indicates that if a signal traverses a direct path and combines in the receiver with a multipath component that is perfectly reflected from a plane, then the composite received signal has a power loss proportional to the inverse of the fourth power of the distance. Thus, it is natural to seek a power-law variation for the average received power in a specified geographic area as a function of distance. For terrestrial wireless communications with frequencies between 30 MHz and 50 GHz, measurements averaged over many different positions of a transmitter and a receiver in a specified geographic area confirm that the average received power, measured in decibels and

D. Torrieri, *Principles of Spread-Spectrum Communication Systems*,
https://doi.org/10.1007/978-3-319-70569-9_6

called the *area-mean power*, does tend to vary linearly with the logarithm of the transmitter-receiver distance r. If the receiver lies in the far field of the transmitted signal, then it is found that the area-mean power, when expressed in decimal units, is approximately given by

$$p_a = p_0 \left(\frac{d}{d_0} \right)^{-\alpha}, \quad d \geq d_0 \tag{6.1}$$

where p_0 is the average received power when the distance is $d = d_0$, α is the *attenuation power law*, and d_0 is a reference distance that exceeds the minimum distance at which the receiver lies in the far field. The parameters p_0 and α are functions of the carrier frequency, antenna heights and gains, terrain characteristics, vegetation, and various characteristics of the propagation medium. Typically, the parameters vary with distance, but are constant within a range of distances. The attenuation power law increases with the carrier frequency, and typical values for microwave frequencies are in the range $3 \leq \alpha \leq 4$.

For a specific propagation path and no signal fading, the received *local-mean power* departs from the area-mean power because of *shadowing*, which is the effect of diffractions, reflections, and terrain features that are path-dependent. Only measurements can provide the local-mean power with high accuracy, but they are rarely available. Numerous path-loss models have been developed for approximate estimates of the local-mean power, and it is difficult to choose among them [65]. An alternative approach is to use a stochastic model that provides a distribution function for the local-mean power in a specified geographic area. A stochastic model greatly facilitates analysis and simulation.

Each diffraction or reflection due to obstructing terrain or an obstacle causes the signal power to be multiplied by an attenuation factor. Thus, the received signal power is often the product of many attenuation factors; hence, the logarithm of the signal power is the sum of many factors. If each factor is modeled as a uniformly bounded, independent random variable that varies from path to path, then the central limit theorem (Corollary A2, Appendix A.2) implies that the logarithm of the received signal power has an approximately normal or Gaussian distribution if the number of attenuation factors and their variances are large enough. Empirical data confirm that the ratio of the received local-mean power to the area-mean power is approximately zero-mean and has a *lognormal distribution*; that is, its logarithm has a Gaussian distribution. Thus, the local-mean power has the form

$$p_l = p_0 \left(\frac{d}{d_0} \right)^{-\alpha} 10^{\xi/10}, \quad d \geq d_0 \tag{6.2}$$

where ξ is the *shadowing factor*, and p_0 is the average received power when $\xi = 0$ and $d = d_0$. The shadowing factor is expressed in decibels and is modeled as a zero-mean random variable with a normal or Gaussian distribution and standard deviation σ_s. The *lognormal density function* of $Z = 10^{\xi/10}$ is

$$f(z) = \frac{10 \log_{10} e}{z\sqrt{2\pi\sigma_s^2}} \exp\left\{ -\frac{[10 \log_{10} z]^2}{2\sigma_s^2} \right\}. \tag{6.3}$$

The standard deviation increases with carrier frequency and terrain irregularity and sometimes exceeds 10 dB for terrestrial communications. The value of the shadowing factor for a propagation path is usually strongly correlated with its value for nearby propagation paths. For mobile communications, the typical time interval during which the shadowing factor is nearly constant corresponds to a movement of 5-10 m.

A signal experiences *fading* when the interaction of multipath components and varying channel conditions cause significant fluctuations in its received power [85], [88]. Fading may be classified as time-selective, frequency-selective, or both. *Time-selective fading* is fading caused by the movement of the transmitter or receiver or by changes in the propagation medium. *Frequency-selective fading* occurs when the delays of the multipath components significantly affect some frequencies more than others.

The *multipath components* that cause fading are generated by particle scattering, heterogeneities in the propagation medium, or reflections from small transient obstacles. These components travel along different paths before being recombined at the receiver. Because of the different time-varying delays and attenuations encountered by the multipath components, the recombined signal is a distorted version of the original transmitted signal. Fading occurs at a much faster rate than shadowing. During an observation interval in which the shadowing factor is nearly constant, the received signal power may be expressed as the product

$$p_r(t) = p_0 \left(\frac{d}{d_0}\right)^{-\alpha} 10^{\xi/10} g(t), \quad d \geq d_0 \tag{6.4}$$

where the factor $g(t)$ is due to the fading and is normalized so that $E[g(t)] = 1$.

The *Doppler shift* arises because of the relative motion between the transmitter and the receiver, which causes a change in the propagation delay. In Figure 6.1 (a), the receiver is moving at speed v during a short time interval, and the angle between the velocity vector and the propagation direction of an electromagnetic wave is ψ. Since the transmitter is moving toward the receiver, the propagation delay $d\,(t)$ at time t is shortened relative to the delay d_0 at time t_0 by $v \cos\psi(t - t_0)/c$, where c is the speed of an electromagnetic wave. Thus, the received phase increases from $2\pi f_c\,(t_0 - d_0/c)$ to

$$2\pi f_c\,[t - d\,(t)\,/c] = 2\pi[(f_c + f_d)\,(t - t_0) + f_c\,(t_0 - d_0/c)]$$

which implies that the received frequency is increased by the Doppler shift

$$f_d = f_c \frac{v}{c} \cos\psi. \tag{6.5}$$

In Figure 6.1 (b), the transmitter is moving at speed v and there is a reflecting surface that changes the arrival angle of the electromagnetic wave at the receiver. If ψ represents the angle between the velocity vector and the initial direction of the electromagnetic wave, then (6.5) again gives the Doppler shift.

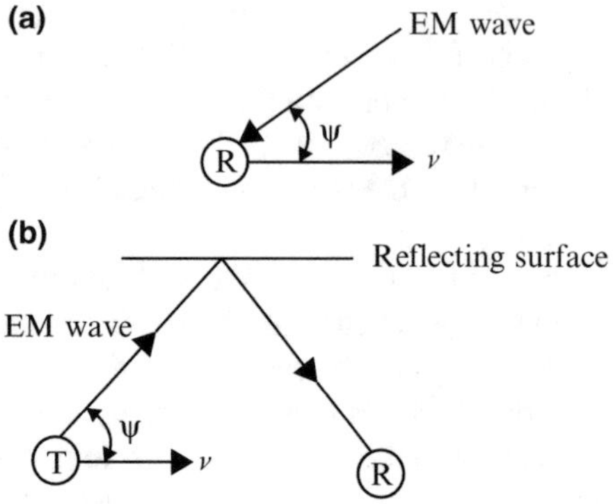

Fig. 6.1 Examples of the Doppler effect: (**a**) receiver motion and (**b**) transmitter motion and reflecting surface

The principal means of accommodating fading are provided by *diversity*, which is some form of signal redundancy. *Time diversity* is provided by channel coding or by signal copies that differ in time delay. *Frequency diversity* may be available when signal copies using different carrier frequencies experience independent or weakly correlated fading. If each signal copy is extracted from the output of a separate antenna in an antenna array, then the diversity is called *spatial diversity*. *Polarization diversity* may be obtained by using two cross-polarized antennas at the same site. Although this configuration provides compactness, it is not as potentially effective as spatial diversity because the received horizontal component of an electric field is usually much weaker than the vertical component.

A bandpass transmitted signal can be expressed as

$$s_t(t) = \text{Re}[s(t)\exp(j2\pi f_c t)] \tag{6.6}$$

where $s(t)$ denotes its complex envelope, f_c denotes its carrier frequency, and Re[·] denotes the real part. Transmission over a time-varying multipath channel of $N(t)$ paths produces a received bandpass signal that consists of the sum of $N(t)$ multipath waveforms. The ith multipath waveform is the transmitted signal delayed by time $\tau_i(t)$, multiplied by an attenuation factor $a_i(t)$ that depends on the path loss and shadowing, and shifted in frequency by the amount $f_{di}(t)$ due to the Doppler effect. Assuming that $f_{di}(t)$ is nearly constant during the path delay, the received signal may be expressed as

$$s_r(t) = \text{Re}[s_1(t)\exp(j2\pi f_c t)] \tag{6.7}$$

where the received complex envelope is

$$s_1(t) = \sum_{i=1}^{N(t)} a_i(t)\exp[j\phi_i(t)]s[t-\tau_i(t)] \tag{6.8}$$

and its phase is

$$\phi_i(t) = -2\pi f_c \tau_i(t) + 2\pi f_{di}(t)\,[t-\tau_i(t)] + \phi_{i0} \tag{6.9}$$

where ϕ_{i0} is the initial phase shift of a multipath component.

6.2 Time-Selective Fading

In this section and the next one, we consider time intervals small enough that all attenuation factors and most of the other fading parameters are approximately constants:

$$a_i(t) = a_i, \quad N(t) = N, \quad v(t) = v$$
$$f_{di}(t) = f_{di}, \quad \psi_i(t) = \psi_i, \quad \tau_i(t) = \tau_i. \tag{6.10}$$

Then, (6.5) indicates that multipath component i has Doppler shift

$$f_{di} = f_d \cos\psi_i, \quad f_d = \frac{f_0\, v}{c} \tag{6.11}$$

where f_d is the maximum Doppler shift, which occurs when $\psi_i = 0$. Equation (6.9) implies that

$$\phi_i(t+\tau) - \phi_i(t) = 2\pi f_d \tau \cos\psi_i \tag{6.12}$$

where τ is a time delay.

Time-selective fading occurs when multipath components experience different Doppler shifts, and the differences in the time delays along the various paths are small compared with the inverse of the signal bandwidth. Therefore, the received multipath components overlap in time and are called *unresolvable multipath components*. If the time origin is chosen to coincide with the average arrival time of the multipath components at a receiver and the time-delay differences are small, then the received complex envelope of (6.8) may be expressed as

$$s_1(t) \approx s(t)r(t) \tag{6.13}$$

where the *equivalent lowpass* or *equivalent baseband channel response* is

$$r(t) = \sum_{i=1}^{N} a_i \exp\left[j\phi_i(t)\right]. \tag{6.14}$$

The fluctuations in this factor cause time-selective fading at the receiver and increase the bandwidth of the received signal. If the transmitted signal is an unmodulated tone, then $s(t) = 1$, and (6.14) represents the complex envelope of the received signal.

The channel response can be decomposed as

$$r(t) = r_c(t) + jr_s(t) \tag{6.15}$$

where $j = \sqrt{-1}$ and

$$r_c(t) = \sum_{i=1}^{N} a_i \cos[\phi_i(t)], \quad r_s(t) = \sum_{i=1}^{N} a_i \sin[\phi_i(t)]. \tag{6.16}$$

If the range of the delay values is much larger than $1/f_c$, then the sensitivity of $\phi_i(t)$ to small variations in the delay τ_i makes it plausible to model the phases $\phi_i(t)$, $i = 1, 2, \ldots, N$ as random variables that are independent of each other and the attenuation factors $\{a_i\}$, and are uniformly distributed over $[0, 2\pi)$ at a specific time t. Therefore,

$$E[r_c(t)] = E[r_s(t)] = 0. \tag{6.17}$$

If the attenuation factors $\{a_i\}$ are uniformly bounded, independent random variables, then according to the central limit theorem (Corollary A2, Appendix A.2), the distributions of both $r_c(t)$ and $r_s(t)$ approach Gaussian distributions as N and the variances of both $r_c(t)$ and $r_s(t)$ increase. Thus, if N is large enough, then $r(t)$ at a specific time is modeled as a *complex Gaussian random variable*. Since the phases are independent and uniformly distributed, it follows that

$$E[r_c(t)r_s(t)] = 0 \tag{6.18}$$

$$E[r_c^2(t)] = E[r_s^2(t)] = \sigma_r^2 \tag{6.19}$$

where we define

$$\sigma_r^2 = \frac{1}{2}\sum_{i=1}^{N} E[a_i^2]. \tag{6.20}$$

Equations (6.17) to (6.19) imply that $r_c(t)$ and $r_s(t)$ are independent, identically distributed, zero-mean Gaussian random variables.

Let $\alpha = |r(t)|$ denote the fading amplitude, and $\theta(t) = \tan^{-1}[r_s(t)/r_c(t)]$ denote the phase of $r(t)$ at a specific time t. Then,

$$r(t) = \alpha e^{j\theta(t)} \tag{6.21}$$

$$\alpha^2 = r_c^2(t) + r_s^2(t) = \sum_{i=1}^{N} a_i^2. \tag{6.22}$$

From (6.15), (6.19), and (6.20), it follows that the average *fading power gain* is

$$\Omega = E[\alpha^2] = 2\sigma_r^2 = \sum_{i=1}^{N} E[a_i^2]. \tag{6.23}$$

Rayleigh, Ricean, and Nakagami Fading

As shown in Appendix E.4, since $r_c(t)$ and $r_s(t)$ are Gaussian, $\theta(t)$ has a uniform distribution over $[0, 2\pi)$, and α has the *Rayleigh* density:

$$f_\alpha(r) = \frac{2r}{\Omega} \exp\left(-\frac{r^2}{\Omega}\right) u(r) \tag{6.24}$$

where $u(r)$ is the unit step function: $u(r) = 1, r \geq 0$, and $u(r) = 0, r < 0$. The substitution of (6.21) and (6.13) into (6.7) gives

$$\begin{aligned} s_r(t) &= \text{Re}[\alpha s(t) \exp(j2\pi f_c t + j\theta(t))] \\ &= \alpha A(t) \cos[2\pi f_c t + \phi(t) + \theta(t)] \end{aligned} \tag{6.25}$$

where $A(t)$ is the amplitude and $\phi(t)$ the phase of $s(t)$, and $s_r(t)$ experiences *Rayleigh fading*. Equations (6.23) and (6.25) indicate that the instantaneous local-mean power is

$$p_l = E[s_r^2(t)] = \Omega A^2(t)/2. \tag{6.26}$$

When a line-of-sight exists between a transmitter and a receiver, a single received multipath component may be resolvable and much stronger than the other unresolvable multipath components. This strong component is called the *specular component* and the other unresolvable components are called *diffuse* or *scattered components*. As a result, the multiplicative channel response of (6.14) becomes

$$r(t) = a_0 \exp[j\phi_0(t)] + \sum_{i=1}^{N} a_i \exp[j\phi_i(t)] \tag{6.27}$$

where the first term is due to the specular component, and the summation term is due to the diffuse components. If N is large enough, then at time t, the summation term is well-approximated by a zero-mean, complex Gaussian random variable. Thus, $r(t)$ at a specific time is a complex Gaussian random variable with a nonzero mean equal to the deterministic first term, and (6.15) implies that

$$E[r_c(t)] = a_0 \cos[\phi_0(t)], \quad E[r_s(t)] = a_0 \sin[\phi_0(t)]. \tag{6.28}$$

From the independence of the $\{a_i\}$, (6.20), and (6.27), it follows that the fading amplitude $\alpha = |r(t)|$ has the average power gain given by

$$\Omega = E[\alpha^2] = a_0^2 + 2\sigma_r^2. \tag{6.29}$$

As shown in Appendix E.3, since $r_c(t)$ and $r_s(t)$ are Gaussian and $\alpha^2 = r_c^2(t) + r_s^2(t)$, α has the *Rice* density:

$$f_\alpha(r) = \frac{r}{\sigma_r^2} \exp\left\{-\frac{r^2 + a_0^2}{2\sigma_r^2}\right\} I_0\left(\frac{a_0 r}{\sigma_r^2}\right) u(r) \tag{6.30}$$

where $I_0(\cdot)$ is the modified Bessel function of the first kind and order zero (Appendix H.3). The type of fading modeled by (6.27) and (6.30) is called *Ricean fading*. The *Rice factor* is defined as

$$\kappa = \frac{a_0^2}{2\sigma_r^2} \tag{6.31}$$

which is the ratio of the specular power to the diffuse power. In terms of κ and $\Omega = 2\sigma_r^2(\kappa + 1)$, the Rice density is

$$f_\alpha(r) = \frac{2(\kappa + 1)}{\Omega} r \exp\left(-\kappa - \frac{(\kappa + 1)r^2}{\Omega}\right) I_0\left(\sqrt{\frac{\kappa(\kappa + 1)}{\Omega}} 2r\right) u(r). \tag{6.32}$$

When $\kappa = 0$, Ricean fading is the same as Rayleigh fading. When $\kappa = \infty$, there is no fading.

The *Nakagami fading* model offers more flexibility than the Rice model. The *Nakagami* density of the fading amplitude α is

$$f_\alpha(r) = \frac{2}{\Gamma(m)} \left(\frac{m}{\Omega}\right)^m r^{2m-1} \exp\left(-\frac{m}{\Omega} r^2\right) u(r), \quad m \geq \frac{1}{2} \tag{6.33}$$

where the gamma function $\Gamma(\cdot)$ is defined by (H.1) of Appendix H.1. When $m = 1$, the Nakagami density becomes the Rayleigh density, and when $m \to \infty$, there is no fading. When $m < 1$, the Nakagami density models fading that is more severe than Rayleigh fading. When $m = 1/2$, the Nakagami density becomes the one-sided Gaussian density.

Integrating over (6.33), changing the integration variable, and using (H.1), we obtain

$$E[\alpha^n] = \frac{\Gamma(m + \frac{n}{2})}{\Gamma(m)} \left(\frac{\Omega}{m}\right)^{n/2}, \quad n \geq 1. \tag{6.34}$$

A measure of the severity of the fading is $var(\alpha^2)/(E[\alpha^2])^2 = 1/m$. Equating this ratio for the Rice and Nakagami densities, it is found that the Nakagami density approximates a Rice density with Rice factor κ if

$$m = \frac{(\kappa + 1)^2}{2\kappa + 1}, \quad \kappa \geq 0. \tag{6.35}$$

Since the Nakagami model includes the Rayleigh and Rice models as special cases and provides for many other possibilities, it is not surprising that this model often fits well with empirical data.

Let $g = \alpha^2$ denote the fading power gain of a signal undergoing Nakagami fading. It follows from (6.33) that the density function of the power gain is

$$f_g(x) = \frac{1}{\Gamma(m)} \left(\frac{m}{\Omega}\right)^m x^{m-1} \exp\left(-\frac{m}{\Omega}x\right) u(x), \quad m \geq \frac{1}{2} \tag{6.36}$$

which is the *gamma density* $f(x; m/\Omega, m)$ (Appendix E.5). From (6.34), it follows that

$$E[g^n] = \frac{\Gamma(m+n)}{\Gamma(m)} \left(\frac{\Omega}{m}\right)^n, \; n \geq 1. \tag{6.37}$$

Doppler Spectrum for Isotropic Scattering

The *autocorrelation of a wide-sense-stationary complex process* $r(t)$ is defined as

$$A_r(\tau) = E[r^*(t)r(t+\tau)] \tag{6.38}$$

where the asterisk denotes the complex conjugate. The variation of the autocorrelation of the equivalent baseband channel response defined by (6.14) provides a measure of the changing channel characteristics. To interpret the meaning of (6.38), we substitute (6.15) and decompose the autocorrelation as

$$\text{Re}\{A_r(\tau)\} = E[r_c(t)r_c(t+\tau)] + E[r_s(t)r_s(t+\tau)] \tag{6.39}$$

$$\text{Im}\{A_r(\tau)\} = E[r_c(t)r_s(t+\tau)] - E[r_s(t)r_c(t+\tau)]. \tag{6.40}$$

Thus, the real part of this autocorrelation is the sum of the autocorrelations of the real and imaginary parts of $r(t)$; the imaginary part is the difference between two cross-correlations of the real and imaginary parts of $r(t)$. Substituting (6.14) into (6.38), using the independence and uniform distribution of each ϕ_i and the independence of $a_i(t) = a_i$ and ϕ_i, and then substituting (6.12), we obtain

$$A_r(\tau) = \sum_{i=1}^{N} E[a_i^2] \exp(j2\pi f_d \tau \cos\psi_i). \tag{6.41}$$

If all the received multipath components have approximately the same power and the receive antenna is omnidirectional, then (6.23) implies that $E[a_i^2] \approx \Omega/N$, $i = 1, 2, \ldots, N$, and (6.41) becomes

$$A_r(\tau) = \frac{\Omega}{N} \sum_{i=1}^{N} \exp(j2\pi f_d \tau \cos\psi_i). \tag{6.42}$$

A communication system, such as a mobile system that receives a signal from an elevated base station, may be surrounded by many scattering objects. An *isotropic scattering* model assumes that multipath components of comparable power are reflected from many different scattering objects and hence arrive from many different directions. For two-dimensional isotropic scattering, N is large, and the $\{\psi_i\}$ lie in a plane and have values that are uniformly distributed over $[0, 2\pi)$. Therefore, the summation in (6.42) can be approximated by the integral

$$A_r(\tau) \approx \frac{\Omega}{2\pi}\int_0^{2\pi} \exp(j2\pi f_d\,\tau\cos\psi)d\psi. \tag{6.43}$$

This integral has the same form as the integral representation of $J_0(\cdot)$, the Bessel function of the first kind and order zero (Appendix H.3). Thus, the *autocorrelation of the channel response for two-dimensional isotropic scattering* is

$$A_r(\tau) = \Omega J_0(2\pi f_d\,\tau). \tag{6.44}$$

The normalized autocorrelation $A_r(\tau)/A_r(0)$, which is a real-valued function of $f_d\,\tau$, is plotted in Figure 6.2. It is observed that its magnitude is less than 0.3 when $f_d\,\tau > 1$. This observation leads to definition of the *coherence time* or *correlation time* of the channel as

$$T_{coh} = \frac{1}{f_d} \tag{6.45}$$

where f_d is the maximum Doppler shift or *Doppler spread*. The coherence time is a measure of the time separation between signal samples sufficient for the samples to be largely decorrelated. If the coherence time is much longer than the duration of a channel symbol, then the fading is relatively constant over a symbol and is called *slow fading*. Conversely, if the coherence time is on the order of the duration of a channel symbol or less, then the fading is called *fast fading*.

To evaluate the Fourier transform of (6.44), we substitute the integral representation of $J_0\,(x)$ given by (H.19) of Appendix H.3, interchange the order of integration, evaluate the inner integral as a Dirac delta function, change the variables, and perform the simple remaining integration. We obtain the *Doppler power spectrum for two-dimensional isotropic scattering*:

$$S_r\,(f) = \begin{cases} \frac{\Omega}{\pi\sqrt{f_d^2-f^2}}, & |f| < f_d \\ 0, & otherwise. \end{cases} \tag{6.46}$$

The normalized Doppler spectrum $S_r(f)/S_r(0)$, which is plotted in Figure 6.3 versus f/f_d, is bandlimited by the Doppler spread f_d and tends to infinity as f approaches $\pm f_d$. The Doppler spectrum is the superposition of contributions from multipath components, each of which experiences a different Doppler shift upper bounded by f_d.

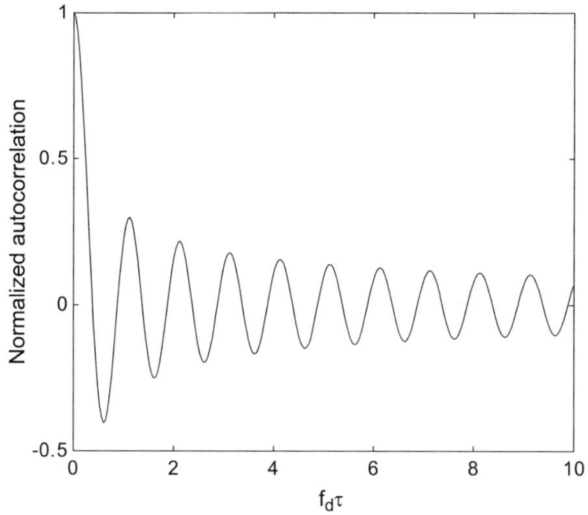

Fig. 6.2 Autocorrelation of r(t) for isotropic scattering

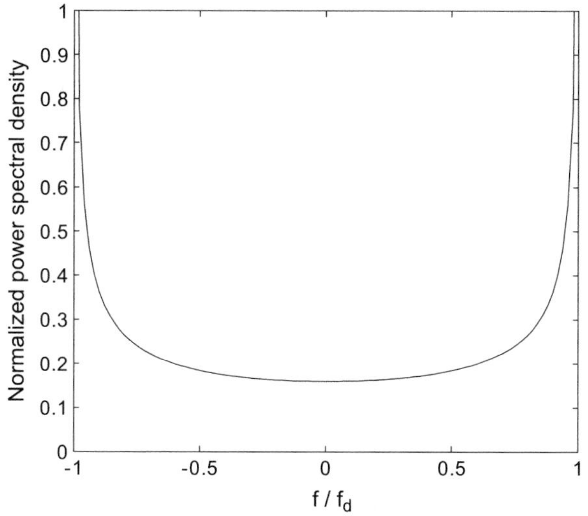

Fig. 6.3 Doppler spectrum for isotropic scattering

The power spectral density (PSD) of the received signal is calculated from (6.7), (6.13), and (6.46). For an unmodulated carrier with $s(t) = 1$, the PSD is

$$S_{\text{rec}}(f) = \frac{1}{2}S_r(f - f_c) + \frac{1}{2}S_r(f + f_c). \tag{6.47}$$

In general, when the scattering is not isotropic, the imaginary part of the autocorrelation $A_r(\tau)$ is nonzero, and the amplitude of the real part decreases much more slowly and less smoothly with increasing τ than (6.44). Both the real and imaginary parts often exhibit minor peaks for time shifts exceeding $1/f_d$. Thus, the coherence time provides only a rough characterization of the channel behavior.

Fading Rate and Fade Duration

The *fading rate* is the rate at which the fading amplitude of a received signal crosses below a specified level. Consider the signal given by (6.27) and approximated as a zero-mean, complex Gaussian process during a time interval over which the fading parameters are constant. For a level $r \geq 0$, isotropic scattering, and Ricean fading with density function given by (6.32), it can be shown that the fading rate is [88]

$$f_r = \sqrt{\frac{2\pi(\kappa + 1)}{\Omega}} f_d r \exp\left[-\kappa - \frac{(\kappa + 1)r^2}{\Omega}\right] I_0\left[\sqrt{\frac{\kappa(\kappa + 1)}{\Omega}} 2r\right] \tag{6.48}$$

where κ is the Rice factor, f_d is the Doppler spread, and Ω is the average fading power gain. For Rayleigh fading, $\kappa = 0$ and

$$f_r = \sqrt{\frac{2\pi}{\Omega}} f_d r \exp(-r^2/\Omega). \tag{6.49}$$

Equations (6.48) and (6.49) indicate that the fading rate is proportional to the Doppler spread.

Let T_f denote the average *fade duration*, which is the amount of time at which the fading amplitude remains below the specified level $r \geq 0$. Since $1/f_r$ is the average time between fades that cross below r, the product $f_r T_f$ is the fraction of the time during which a fade occurs. If the time-varying fading amplitude is assumed to be a stationary ergodic process, then this fraction is equal to $F_\alpha(r)$, the probability that the fading amplitude is below or equal to the level r. Thus,

$$T_f = \frac{F_\alpha(r)}{f_r}. \tag{6.50}$$

If the fading amplitude has a Ricean distribution, then integrating (6.32) yields

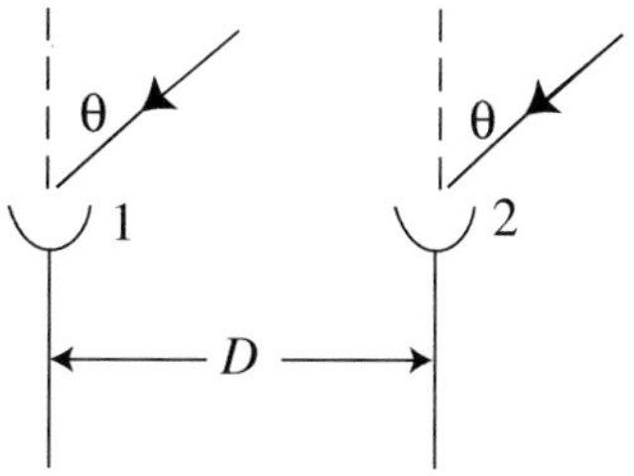

Fig. 6.4 Two antennas receiving a plane wave that results in a signal copy at each antenna

$$T_f = \frac{1 - Q_1\left(\sqrt{2\kappa}, \sqrt{\frac{2(\kappa+1)}{\Omega}}r\right)}{f_r} \tag{6.51}$$

where the first-order Marcum Q-function $Q_1(\alpha, \beta)$ is defined by (H.26) of Appendix H.3. For Rayleigh fading, $\kappa = 0$ and (6.51) becomes

$$T_f = \sqrt{\frac{\Omega}{2\pi}} \frac{\exp(r^2/\Omega) - 1}{f_d\, r}. \tag{6.52}$$

For both Ricean and Rayleigh fading, the fade duration is inversely proportional to f_d.

Spatial Diversity and Fading

To obtain spatial diversity in a fading environment, the antennas in an array at the receiver must be adequately separated so that there is little correlation between signal replicas or copies at the antennas. A few wavelengths are adequate for a mobile receiver because it tends to receive superpositions of reflected waves arriving from many random angles. Many wavelengths may be necessary for a stationary receiver located in a high position. To determine what separation is needed, consider the reception of a signal at two antennas separated by a distance D, as illustrated in Figure 6.4. If a narrowband signal arrives as an electromagnetic plane wave, then the signal copy at antenna 1 relative to antenna 2 is delayed by $D \sin\theta/c$, where θ is the arrival angle of the plane wave relative to a line perpendicular to the line joining the two antennas. Thus, if the phase at antenna i is $\phi_i(t)$, $i = 1, 2$, then

$$\phi_2(t) = \phi_1(t) + 2\pi\frac{D}{\lambda}\sin\theta \tag{6.53}$$

where $\lambda = c/f_c$ is the wavelength of the signal.

Let $\phi_{ki}(t)$ denote the phase of the complex fading amplitude of multipath component i at antenna k. Consider a time interval small enough that the fading amplitudes are constants at the two antennas, and each multipath component arrives from a fixed angle. If multipath component i of a narrowband signal arrives as a plane wave at angle ψ_i, then the phase $\phi_{2i}(t)$ of the complex fading amplitude of the component copy at antenna 2 is related to the phase $\phi_{1i}(t)$ at antenna 1 by

$$\phi_{2i}(t) = \phi_{1i}(t) + 2\pi\frac{D}{\lambda}\sin\psi_i \tag{6.54}$$

If the multipath component propagates over a distance much larger than the separation between the two antennas, then it is reasonable to assume that the attenuation a_i is identical at the two antennas.

If the range of the delay values is much larger than $1/f_c$, then the sensitivity of the phases to small delay variations makes it plausible that the phases $\phi_{1i}(t)$, $i = 1, 2, \ldots, N$, are well-modeled as independent random variables that are uniformly distributed over $[0, 2\pi)$. From (6.14), the complex fading amplitude r_k of the signal copy at antenna k when the signal is a tone is

$$r_k(t) = \sum_{i=1}^{N} a_i \exp[j\phi_{ki}(t)], \quad k = 1, 2. \tag{6.55}$$

The cross-correlation between $r_1(t)$ and $r_2(t)$ is defined as

$$C_{12}(D) = E[r_1^*(t)\, r_2(t)]. \tag{6.56}$$

Substituting (6.55) into (6.56), using the independence of each a_i and $\phi_{ki}(t)$, the independence of $\phi_{1i}(t)$ and $\phi_{2l}(t)$, $i \neq l$, and the uniform distribution of each $\phi_{1i}(t)$, and then substituting (6.54), we obtain

$$C_{12}(D) = \sum_{i=1}^{N} E[a_i^2] \exp(j2\pi D \sin\psi_i/\lambda). \tag{6.57}$$

This equation for the cross-correlation as a function of spatial separation clearly resembles (6.41) for the autocorrelation as a function of time delay. If all the multipath components have approximately the same power so that $E[a_i^2] \approx \Omega/N$, $i = 1, 2, \ldots, N$, then

$$C_{12}(D) = \frac{\Omega}{N}\sum_{i=1}^{N} \exp(j2\pi D \sin\psi_i/\lambda). \tag{6.58}$$

Applying the two-dimensional isotropic scattering model, (6.58) is approximated by an integral. As in the derivation of (6.44), the evaluation of the integral gives the real-valued cross-correlation

$$C_{12}(D) = \Omega J_0(2\pi D/\lambda). \tag{6.59}$$

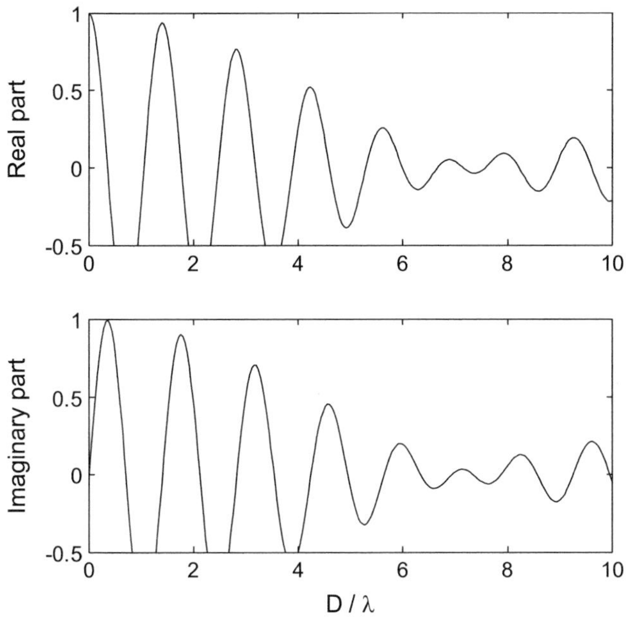

Fig. 6.5 Normalized cross-correlation for multipath components arriving between $7\pi/32$ and $9\pi/32$ radians: real and imaginary parts

This model indicates that an antenna separation of $D \geq \lambda/2$ ensures that the normalized cross-correlation $C_{12}(D)/C_{12}(0)$ is less than 0.3. A plot of the normalized cross-correlation is obtained from Figure 6.2 if the abscissa is interpreted as D/λ.

When the scattering is not isotropic or the number of scattering objects producing multipath components is small, then the real and imaginary parts of the cross-correlation decrease much more slowly with D/λ. For example, Figure 6.5 shows the real and imaginary parts of the normalized cross-correlation when the $\{\psi_i\}$ are a nearly continuous band of angles between $7\pi/32$ and $9\pi/32$ radians, so that (6.58) can be approximated by an integral over that band. Figure 6.6 depicts the real and imaginary parts of the normalized cross-correlation when $N = 9$ and the $\{\psi_i\}$ are uniformly spaced throughout the first two quadrants: $\psi_i = (i-1)\pi/8$, $i = 1, 2, \ldots, 9$. In the example shown in Figure 6.5, an antenna separation of at least 5λ is necessary to ensure approximate decorrelation of the signal copies and obtain spatial diversity. In the example of Figure 6.6, not even a separation of 10λ is adequate to ensure approximate decorrelation.

6.3 Frequency-Selective Fading

Frequency-selective fading occurs because the delays in multipath components cause them to combine destructively at some frequencies but constructively at

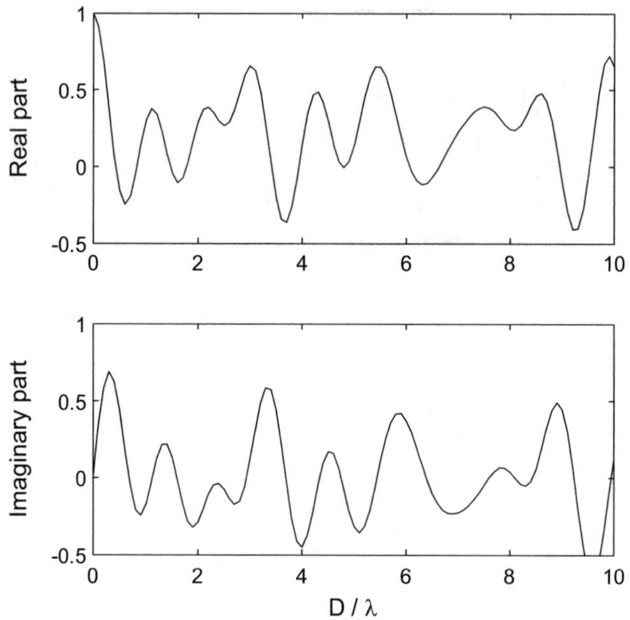

Fig. 6.6 Normalized cross-correlation for $N = 9$ multipath components arriving from uniformly spaced angles in the first two quadrants: real and imaginary parts

others. The different path delays cause *dispersion* of a received pulse in time and cause intersymbol interference between successive symbols. The multipath *delay spread* T_d is defined as the maximum delay of a significant multipath component relative to the minimum delay of a significant component; that is,

$$T_d = \max_i \tau_i - \min_i \tau_i \ , \quad i = 1, 2, \ldots, N. \tag{6.60}$$

Consider a typical delay spread small enough that (6.10) is satisfied and $f_{di}T_d << 1$. Then, (6.8) and (6.9) indicate that the received complex fading amplitude is

$$s_1(t) = \sum_{i=1}^{N} a_i \exp\left[-j2\pi(f_c\tau_i + \phi_{i0})\right] s(t - \tau_i), \quad \tau_1 \le t \le \tau_1 + T_d \tag{6.61}$$

where $\tau_1 = \min_i \tau_i$. Suppose that $BT_d << 1$, where B is the bandwidth of the complex envelope $s(t)$. Then, $s(t - \tau_i) \approx s(t - \tau_1)$, $i = 1, 2, \ldots, N$. Therefore, all multipath components fade nearly simultaneously, and

$$s_1(t) = s(t - \tau_1) \sum_{i=1}^{N} a_i \exp\left[-j2\pi(f_c\tau_i + \phi_{i0})\right], \quad \tau_1 \le t \le \tau_1 + T_d \tag{6.62}$$

which indicates that even with fading, the spectrum of $s_1(t)$ remains proportional to the spectrum of $s(t)$. Thus, this type of fading is called *frequency-nonselective* or *flat fading* and occurs if $B << B_{coh}$, where the *coherence bandwidth* is defined as

$$B_{coh} = \frac{1}{T_d} . \tag{6.63}$$

Since $B \approx 1/T_s$, where T_s is the symbol duration, flat fading occurs when the symbol duration T_s is large enough that $T_s >> T_d$.

In contrast, a signal is said to experience *frequency-selective fading* if $B > B_{coh}$, and hence $T_s < T_d$, because then the time variation or fading of the spectral components of $s(t)$ may be different. The large delay spread may cause intersymbol interference, which is accommodated by equalization in the receiver. However, if the time delays are sufficiently different among the multipath components that they are *resolvable* at the demodulator or matched-filter output, then the independently fading components provide diversity that can be exploited by a rake demodulator (Section 6.10).

To illustrate frequency-selective fading, consider the reception of two multipath components. Calculating the Fourier transform $S_1(f)$ of $s_1(t)$ using (6.61) with $L_s = 2$, we obtain

$$|S_1(f)| = \left|a_1^2 + a_2^2 + 2a_1a_2 \cos 2\pi (f + f_c) T_d\right|^{1/2} |S(f)| \tag{6.64}$$

where $T_d = \tau_1 - \tau_2$ and $S(f)$ is the Fourier transform of $s(t)$. This equation indicates that $|S_1(f)| / |S(f)|$ fluctuates over the range of f. If the range of f equals or exceeds $B_{coh} = 1/T_d$, then $|S_1(f)| / |S(f)|$ varies from $|a_1 - a_2|$ to $|a_1 + a_2|$, which is very large when $a_1 \approx a_2$.

A generalized impulse response may be used to characterize the impact of the transmission channel on the signal. The complex-valued impulse response of the channel $h(t, \tau)$ is the response at time t due to an impulse applied τ seconds earlier. The complex envelope $s_1(t)$ of the received signal is the result of the convolution of the complex envelope $s(t)$ of the transmitted signal with the baseband impulse response:

$$s_1(t) = \int_{-\infty}^{\infty} h(t, \tau)s(t - \tau)d\tau. \tag{6.65}$$

The *channel impulse response* is usually modeled as a complex-valued stochastic process:

$$h(t, \tau) = \sum_{i=1}^{N(t)} h_i(t)\delta[\tau - \tau_i(t)] \tag{6.66}$$

where $\delta[\cdot]$ is the Dirac delta function. When the fading is flat, a received signal can often be decomposed into the sum of signals reflected from several clusters of scatterers. Each cluster is the sum of a number of multipath components with nearly the same delay. In this model, $N(t)$ is the number of clusters and $\tau_i(t)$ is the distinct delay associated with the ith cluster. If the channel impulse response is time-invariant during a time interval, then $h(t,\tau) = h(0,\tau) = h(\tau)$ and

$$h(\tau) = \sum_{i=1}^{N} h_i \delta(\tau - \tau_i). \tag{6.67}$$

Therefore, if $h_i = a_i \exp[-j2\pi(f_c\tau_i + \phi_{i0})]$, then the application of (6.65) leads to (6.61).

The *wide-sense stationary, uncorrelated scattering* model is a more general channel model than (6.66) and is reasonably accurate in nearly all practical applications. The impulse response is *wide-sense stationary* if the correlation between its value at t_1 and its value at t_2 depends only on $t_1 - t_2$. Thus, the autocorrelation of the impulse response is

$$R_h(t_1, t_2, \tau_1, \tau_2) = E[h^*(t_1, \tau_1)h(t_2, \tau_2)] = R_h(t_1 - t_2, \tau_1, \tau_2). \tag{6.68}$$

Uncorrelated scattering implies that the gains and phase shifts associated with two different delays are uncorrelated so that multipath components fade independently. Extending this notion, the wide-sense stationary, uncorrelated scattering model assumes that the autocorrelation has the form

$$R_h(t_1 - t_2, \tau_1, \tau_2) = R_w(t_1 - t_2, \tau_1)\delta(\tau_1 - \tau_2) \tag{6.69}$$

which implies that

$$R_w(t_1 - t_2, \tau_1) = \int_{-\infty}^{\infty} R_h(t_1 - t_2, \tau_1, \tau_2)d\tau_2. \tag{6.70}$$

The Fourier transform of the impulse response gives the *time-varying channel frequency response*:

$$H(t,f) = \int_{-\infty}^{\infty} h(t,\tau)\exp(-j2\pi f\tau)d\tau. \tag{6.71}$$

Equation (6.68) implies that the autocorrelation of the frequency response for a wide-sense-stationary channel is

$$R_H(t_1, t_2, f_1, f_2) = E[H^*(t_1, f_1)H(t_2, f_2)] = R_H(t_1 - t_2, f_1, f_2) \tag{6.72}$$

which depends only on $t_1 - t_2$. For the *wide-sense stationary, uncorrelated scattering* model, the substitution of (6.71), (6.68), and (6.69) into (6.72) yields

$$\begin{aligned} R_H(t_1, t_2, f_1, f_2) &= \int_{-\infty}^{\infty} R_w(t_1 - t_2, \tau)\exp[-j2\pi(f_1 - f_2)\tau]d\tau \\ &= R_H(t_1 - t_2, f_1 - f_2) \end{aligned} \tag{6.73}$$

which is a function only of the differences $t_1 - t_2$ and $f_1 - f_2$.

If $t_1 = t_2$, then the autocorrelation of the frequency response is

$$R_H(0, f_1 - f_2) = \int_{-\infty}^{\infty} S_m(\tau) \exp[-j2\pi(f_1 - f_2)\tau] d\tau \tag{6.74}$$

which is the Fourier transform of the *multipath intensity profile* defined as

$$S_m(\tau) = R_w(0, \tau). \tag{6.75}$$

The form of (6.74) indicates that the coherence bandwidth B_{coh} of the channel, which is a measure of the range of $f_1 - f_2$ for which $R_H(0, f_1 - f_2)$ has a significant value, is given by the reciprocal of the range of $S_m(\tau)$. Since this range is on the order of the multipath delay spread, the multipath intensity profile is interpreted as the channel output power due to an impulse applied τ seconds earlier, and (6.63) is confirmed as a suitable definition of B_{coh} for this channel model. The multipath intensity profile has *diffuse* components if it is a piecewise continuous function and has *specular* components if it includes delta functions at specific values of the delay.

The Doppler shift is the main limitation on the channel coherence time or range of values of the difference $t_d = t_1 - t_2$ for which $R_w(t_d, 0)$ is significant. Thus, the *Doppler PSD* is defined as

$$S_D(f) = \int_{-\infty}^{\infty} R_w(t_d, 0) \exp(-j2\pi f t_d) dt_d. \tag{6.76}$$

The inverse Fourier transform of $S_D(f)$ gives the autocorrelation $R_w(t_d, 0)$. The coherence time T_{coh} of the channel, which is a measure of the range of t_d for which $R_w(t_d, 0)$ has a significant value, is given by the reciprocal of the spectral range of $S_D(f)$. Since this spectral range is on the order of the maximum Doppler shift, (6.45) is confirmed as a suitable definition of T_{coh} for this channel model.

6.4 Maximal-Ratio Combining

Diversity combiners for fading channels are designed to combine independently fading copies of the same signal in different branches. The combining is performed in such a way that the combiner output has a power level that varies much more slowly than that of a single copy. Although useless in improving communications over the AWGN channel, diversity combining improves communications over fading channels because the diversity gain is large enough to overcome any noncoherent combining loss. The three most common types of diversity combining are maximal-ratio, equal-gain, and selective combining. Since maximal-ratio and equal-gain combiners use linear combining with variable weights for each signal copy, they can be viewed as types of adaptive arrays. They differ from other adaptive antenna arrays in that they are not designed to suppress interference signals.

Consider a receiver array of L diversity branches, each of which processes a different desired-signal copy. Let $\mathbf{y} = [y_1 \ldots y_L]^T$ denote the discrete-time vector of the L complex-valued branch output samples associated with a received symbol. This vector can be decomposed as

$$\mathbf{y} = \mathbf{s} + \mathbf{n} \tag{6.77}$$

where $\mathbf{s}$ and $\mathbf{n}$ are the discrete-time vectors of the desired signal and the interference and thermal noise, respectively. Let $\mathbf{w}$ denote the $L \times 1$ weight vector of a linear combiner applied to the input vector. The combiner output is

$$z = \mathbf{w}^H \mathbf{y} = z_s + z_n \tag{6.78}$$

where the superscript H denotes the conjugate transpose, and

$$z_s = \mathbf{w}^H \mathbf{s}, \;\; z_n = \mathbf{w}^H \mathbf{n} \tag{6.79}$$

are the output components due to the desired signal and the interference and noise respectively. The components of both $\mathbf{s}$ and $\mathbf{n}$ are modeled as discrete-time, jointly wide-sense-stationary processes. The correlation matrices of the desired signal and the interference and noise are defined as the $L \times L$ matrices

$$\mathbf{R}_s = E\left[\mathbf{s}\mathbf{s}^H\right], \;\; \mathbf{R}_n = E\left[\mathbf{n}\mathbf{n}^H\right] \tag{6.80}$$

respectively.

If each desired-signal copy originates in a different receive antenna, then we assume that the desired signal is sufficiently narrowband relative to the antenna separations that the copies in all branches are nearly identical except for amplitudes and phase shifts. Let C denote the discrete-time sampled complex envelope of the desired signal in a fixed reference branch. The desired-signal input vector may be represented as

$$\mathbf{s} = C\mathbf{s}_0 \tag{6.81}$$

where the *steering vector* is

$$\mathbf{s}_0 = [\alpha_1 \exp(j\theta_1) \;\; \alpha_2 \exp(j\theta_2) \;\; \ldots \;\; \alpha_L \exp(j\theta_L)]^T. \tag{6.82}$$

and has components that represent the relative amplitudes and phase shifts in the branch outputs. The substitution of (6.81) into (6.80) yields

$$\mathbf{R}_s = \mathcal{E}_s \mathbf{s}_0 \mathbf{s}_0^H, \;\; \mathcal{E}_s = E[|C|^2] \tag{6.83}$$

where $\mathcal{E}_s$ is the energy per symbol.

The signal-to-interference-and-noise ratio (SINR) at the combiner output is

$$\rho = \frac{E\left[|z_s|^2\right]}{E\left[|z_n|^2\right]} = \frac{\mathcal{E}_s \left\|\mathbf{w}^H \mathbf{s}_0\right\|^2}{\mathbf{w}^H \mathbf{R}_n \mathbf{w}}. \tag{6.84}$$

As shown in Section 5.5, the optimal weight vector for maximizing the SINR is

$$\mathbf{w}_0 = \eta \mathbf{R}_n^{-1} \mathbf{s}_0 \tag{6.85}$$

where η is an arbitrary constant, and the maximum value of the SINR is

$$\rho_{\max} = \mathcal{E}_s \mathbf{s}_0^H \mathbf{R}_n^{-1} \mathbf{s}_0. \tag{6.86}$$

A *maximal-ratio combiner* is a linear combiner with a weight vector $\mathbf{w}_m$ that is optimal under the assumption that the components of $\mathbf{n}$ are zero-mean and uncorrelated. With this assumption, the correlation matrix $\mathbf{R}_n$ is diagonal and its ith diagonal element has the value

$$N_{0i} = E[|n_i|^2]. \tag{6.87}$$

Since $\mathbf{R}_n^{-1}$ is diagonal with diagonal elements N_{0i}^{-1}, the right-hand side of (6.85) implies that

$$\mathbf{w}_m = \eta \left[\frac{\alpha_1}{N_{01}} e^{j\theta_1} \ \frac{\alpha_2}{N_{02}} e^{j\theta_2} \ \ldots \ \frac{\alpha_L}{N_{0L}} e^{j\theta_L} \right]^T \tag{6.88}$$

which can be implemented only if the $\{\alpha_i\}$, $\{\theta_i\}$, and $\{N_{0i}\}$ can be estimated. Equations (6.79), (6.88), (6.82), and (6.83) yield the desired part of the combiner output:

$$z_s = \mathbf{w}_m^H \mathbf{s} = \eta C \sum_{i=1}^{L} \frac{\alpha_i^2}{N_{0i}}. \tag{6.89}$$

Since z_s is proportional to C, maximal-ratio combining (MRC) equalizes the phases of the signal copies in the array branches, a process called *cophasing*. The cophasing may be implemented by using a pilot signal and phase synchronization in each branch. Equation (6.86) indicates that the SINR is

$$\rho_m = \mathcal{E}_s \sum_{i=1}^{L} \frac{\alpha_i^2}{N_{0i}}. \tag{6.90}$$

In most applications, the interference-and-noise in each array branch is nearly independent of the other branches, and the powers are approximately equal so that $N_{0i} = N_0$, $i = 1, 2, \ldots, L$. If this common value is merged with the constant in (6.85) or (6.88), then the MRC weight vector is

$$\mathbf{w}_m = \eta \mathbf{s}_0 = \eta \left[\alpha_1 e^{j\theta_1} \ \alpha_2 e^{j\theta_2} \ \ldots \ \alpha_L e^{j\theta_L}\right]^T \tag{6.91}$$

the desired part of the combiner output is

$$z_s = \mathbf{w}_m^H \mathbf{s} = \eta C \sum_{i=1}^{L} \alpha_i^2 \tag{6.92}$$

and the corresponding SINR is

$$\rho_m = \frac{\mathcal{E}_s}{N_0} \sum_{i=1}^{L} \alpha_i^2. \tag{6.93}$$

Since the weight vector in (6.91) is not a function of the interference parameters, the combiner attempts no interference cancelation. The interference-and-noise signals are ignored while the combiner carries out coherent combining of the desired signal. If each α_i, $i = 1, 2, \ldots, L$, is modeled as a random variable with an identical distribution function, then (6.93) implies that

$$E[\rho_m] = L\bar{\gamma},\ \bar{\gamma} = \frac{\mathcal{E}_s}{N_0} E[\alpha_1^2]. \tag{6.94}$$

which indicates a gain in the mean SINR that is proportional to L.

Coherent BPSK and QPSK Systems

Coherent detection that exploits the availability of multiple desired-signal copies entails MRC. Consider a direct-sequence system with binary phase-shift keying (BPSK) and the reception of a single binary symbol or bit that is equally likely to be a 0 or a 1. Each received signal copy in one of L diversity branches experiences independent fading that is constant during the signal interval. We assume that the received interference and noise in each diversity branch can be modeled as independent, zero-mean, white Gaussian noise with the same two-sided PSD $N_0/2$. The unit-energy symbol waveform is the spreading waveform. The symbol-matched filter comprises the chip-matched filter, sequence generator, multiplier, and adder of Figure 2.14.

Coherent detection in each diversity branch removes the phase dependence of the received signal. Application of (1.38) and the noise analysis in Section 1.2 indicate that the sampled output of diversity branch i due to a single symbol and BPSK is

$$y_i = \alpha_i \sqrt{\mathcal{E}_b} x + n_i, \quad i = 1, 2, \ldots, L \tag{6.95}$$

where $\mathcal{E}_b$ is the desired-signal energy per bit in the absence of fading, $x = +1$ or -1 depending on the transmitted bit, each α_i is an amplitude, and n_i is independent, zero-mean, circularly symmetric, complex Gaussian noise with $E[| n_i |^2] = N_0$.

Because of its circular symmetry, n_i has independent real and imaginary components with variance $N_0/2$. Equation (6.95) and the subsequent results are the same for a BPSK system with or without direct-sequence spreading.

After using (6.95) to evaluate the log-likelihood function for all L branches and then proceeding as in Section 1.2, we find that the BPSK *symbol metric* is

$$U = \sum_{i=1}^{L} \alpha_i y_{ri} \tag{6.96}$$

where $y_{ri} = \text{Re}(y_i)$, and each term is a subsymbol metric. For hard-decision decoding, the bit decision is that $x = +1$ if $U > 0$, and $x = +1$ if $U \leq 0$. This metric has the disadvantage that it requires estimates of the $\{\alpha_i\}$.

Let $\mathbf{y} = [y_1\ y_2 \ldots y_L]^T$ denote the vector of branch outputs. The symbol metric is proportional to $\mathbf{w}_m^H[Re(\mathbf{y})]$, where $\mathbf{w}_m$ is given by (6.91) with the phase angles set to zero because of the coherent demodulation. The substitution of (6.95) into (6.96) yields

$$U = x\sqrt{\mathcal{E}_b} \sum_{i=1}^{L} \alpha_i^2 + \sum_{i=1}^{L} \alpha_i \,\text{Re}(n_i)\,. \tag{6.97}$$

Thus, the symbol metric U contains an MRC component with the same form as (6.92).

Since (6.96) is computed in either case, the implementation of the maximum-likelihood detector may use either maximal-ratio *predetection combining* before the demodulation, as illustrated in Figure 6.7 (a), or *postdetection combining* following the demodulation, as illustrated in Figure 6.7 (b). The $\{\phi_i\}$ are estimates of the phases of the input signals. Since the optimal coherent matched-filter or correlation demodulator performs a linear operation on the $\{y_i\}$, both predetection and postdetection combining provide the same symbol metric and hence the same performance.

If the $\{\alpha_i\}$ are known, the symbol metric has a Gaussian distribution with mean

$$E(U) = x\sqrt{\mathcal{E}_b} \sum_{i=1}^{L} \alpha_i^2. \tag{6.98}$$

Since the $\{n_i\}$ are independent and circularly symmetric, the $\{\text{Re}(n_i)\}$ are independent. Since $\text{Re}(n_i)$ and $\text{Im}(n_i)$ have the same variance, $E[(\text{Re}(n_i))^2] = E[|\ n_i\ |^2]/2 = N_0/2$. Therefore, the variance of U is

$$\sigma_u^2 = \frac{N_0}{2} \sum_{i=1}^{L} \alpha_i^2. \tag{6.99}$$

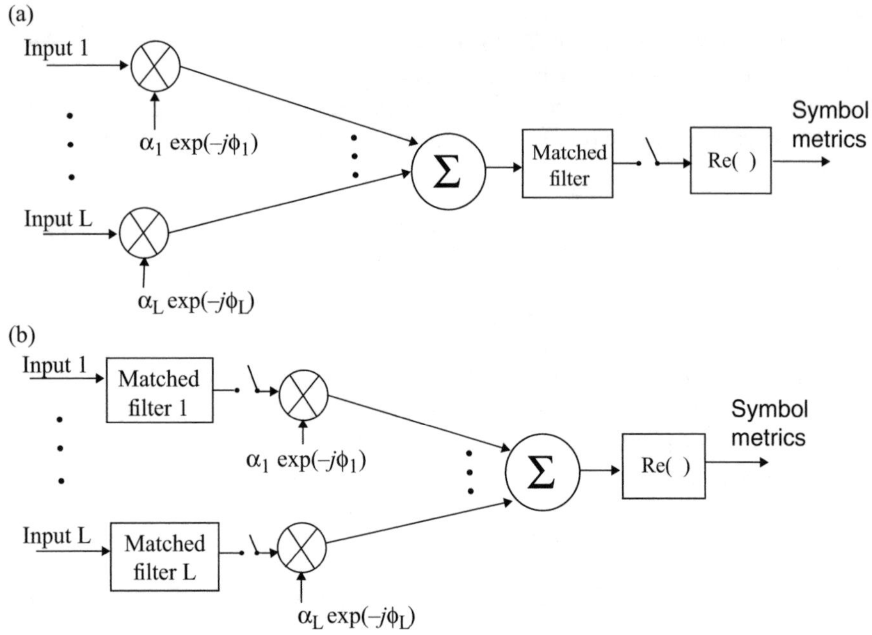

Fig. 6.7 Maximal-ratio combiners for BPSK with (**a**) predetection combining and (**b**) postdetection combining. Coherent equal-gain combiners for BPSK omit the factors $\{\alpha_i\}$

For hard-decision decoding, the bit decision is that $x = +1$ if $U > 0$. Because of the symmetry, the bit error probability is equal to the conditional bit error probability given that $x = +1$. A decision error occurs if $U < 0$. Since the symbol metric has a Gaussian conditional distribution, a standard evaluation using (6.98) and (6.99) indicates that the conditional bit error probability given the $\{\alpha_i\}$ is

$$P_{b|\boldsymbol{\alpha}}(\gamma_b) = Q(\sqrt{2\gamma_b}) \tag{6.100}$$

where $Q(x)$ is defined by (1.58) and the total bit-energy-to-noise-density ratio is

$$\gamma_b = \sum_{i=1}^{L} \gamma_i, \quad \gamma_i = \frac{\mathcal{E}_b}{N_0}\alpha_i^2. \tag{6.101}$$

The bit error probability is determined by averaging $P_{b|\boldsymbol{\alpha}}(\gamma_b)$ over the distribution of γ_b, which depends on the $\{\alpha_i\}$ and embodies the statistics of the fading channel.

Suppose that each of the $\{\alpha_i\}$ is independent with the identical Nakagami distribution. Then each α_i^2 has the gamma distribution of (6.36). As shown in Appendix E.6, since γ_b is the sum of L independent, identically distributed gamma random variables, the density function of γ_b is

$$f_{\gamma_b}(x) = \frac{m^{mL}}{\Gamma(mL)\,\bar{\gamma}^{mL}} x^{mL-1} \exp\left(-\frac{mx}{\bar{\gamma}}\right) u(x) \tag{6.102}$$

where the average bit-energy-to-noise-density ratio in each branch is

$$\bar{\gamma} = \frac{\mathcal{E}_b}{N_0} E[\alpha_1^2]. \tag{6.103}$$

The bit error probability is determined by averaging (6.100) over the density function given by (6.102). Thus,

$$P_b(L) = \int_0^\infty Q(\sqrt{2x}) \frac{m^{mL}}{\Gamma(mL)\,\bar{\gamma}^{mL}}\, x^{mL-1} \exp\left(-\frac{m\,x}{\bar{\gamma}}\right) dx. \tag{6.104}$$

We assume that m is a positive integer. Direct calculations verify that as mL is a positive integer,

$$\frac{d}{dx} Q\left(\sqrt{2x}\right) = -\frac{1}{2\sqrt{\pi}} \frac{\exp(-x)}{\sqrt{x}} \tag{6.105}$$

$$\frac{d}{dx}\left[e^{-mx/\bar{\gamma}} \sum_{i=0}^{mL-1} \frac{(mx/\bar{\gamma})^i}{i!}\right] = -\frac{m^{mL}}{(mL-1)!\bar{\gamma}^{mL}} x^{mL-1} \exp\left(-\frac{mx}{\bar{\gamma}}\right). \tag{6.106}$$

Applying integration by parts to (6.104), using (6.105), (6.106), and $Q(0) = 1/2$, we obtain

$$P_b(L) = \frac{1}{2} - \sum_{i=0}^{mL-1} \frac{m^i}{i!\bar{\gamma}^i 2\sqrt{\pi}} \int_0^\infty \exp\left[-x\left(1 + m\bar{\gamma}^{-1}\right)\right] x^{i-1/2}\, dx. \tag{6.107}$$

This integral can be evaluated in terms of the gamma function (Appendix H.1). A change of variable in (6.107) yields

$$P_b(L) = \frac{1}{2} - \frac{1}{2}\sqrt{\frac{\bar{\gamma}}{m+\bar{\gamma}}} \sum_{i=0}^{mL-1} \frac{\Gamma(i+1/2)m^i}{\sqrt{\pi}\, i!(m+\bar{\gamma})^i}. \tag{6.108}$$

Since $\Gamma(x) = (x-1)\Gamma(x-1)$ and $\Gamma(1/2) = \sqrt{\pi}$, product expansions of the gamma functions indicate that

$$\Gamma(i+1/2) = \frac{\sqrt{\pi}\,\Gamma(2i)}{2^{2i-1}\Gamma(i)} = \frac{\sqrt{\pi}\, i!}{2^{2i-1}} \binom{2i-1}{i}, \quad i \geq 1. \tag{6.109}$$

Therefore,

$$P_b(L) = \frac{1}{2} - \frac{1}{2}\sqrt{\frac{\bar{\gamma}}{m+\bar{\gamma}}} - \sqrt{\frac{\bar{\gamma}}{m+\bar{\gamma}}} \sum_{i=1}^{mL-1} \binom{2i-1}{i} \left(\frac{m}{4m+4\bar{\gamma}}\right)^i$$

$$\text{(BPSK, QPSK)} \tag{6.110}$$

which is valid for quadriphase-shift keying (QPSK) because the latter can be transmitted as two independent BPSK waveforms in phase quadrature. This expression explicitly shows the change in the bit error probability as the number of diversity branches increases. These results can be approximately related to Ricean fading by using (6.35).

Since $m = 1$ for Rayleigh fading, the preceding equations can be simplified. The bit error probability for no diversity or a single branch is

$$p = \frac{1}{2}\left(1 - \sqrt{\frac{\bar{\gamma}}{1+\bar{\gamma}}}\right) \quad \text{(Rayleigh, BPSK, QPSK).} \tag{6.111}$$

Solving this equation to determine $\bar{\gamma}$ as a function of p and then using this result in (6.110) with $m = 1$ gives

$$P_b(L) = p - (1-2p)\sum_{i=1}^{L-1}\binom{2i-1}{i}[p(1-p)]^i. \tag{6.112}$$

An alternative expression for $P_b(L)$ is

$$P_b(L) = p^L \sum_{i=0}^{L-1}\binom{L+i-1}{i}(1-p)^i \tag{6.113}$$

which can be proved to be equal to (6.112) by using mathematical induction.

To derive an upper bound on $P_b(L)$ that facilitates analysis of its asymptotic behavior, we use an identity for the sum of binomial coefficients:

$$\sum_{i=0}^{L-1}\binom{L+i-1}{i} = \binom{2L-1}{L}. \tag{6.114}$$

To prove (6.114), observe that $\binom{2L-1}{L} = \binom{2L-1}{L-1}$ is the number of ways of choosing $L-1$ distinct objects out of $2L-1$. For $0 \le k \le L-1$, a choice could also be made by selecting the first k objects, not selecting the next object, and then selecting $L-k-1$ distinct objects from the remaining $2L-1-(k+1) = 2L-k-2$ objects. Thus,

$$\binom{2L-1}{L} = \sum_{k=0}^{L-1}\binom{2L-k-2}{L-k-1} = \sum_{i=0}^{L-1}\binom{L+i-1}{i} \tag{6.115}$$

which proves the identity.

Since $1 - p \le 1$, (6.113) and (6.114) imply that

$$P_b(L) \le \binom{2L-1}{L}p^L. \tag{6.116}$$

This upper bound becomes tighter as $p \to 0$. Performing a Taylor series expansion of (6.111) in $\bar{\gamma}^{-1}$, we obtain an alternating series. Retaining only the first nonzero term in the series and then substituting it into (6.116), we obtain an upper bound:

$$P_b(L) \leq \binom{2L-1}{L}\left(\frac{1}{4^L}\right)\bar{\gamma}^{-L}. \tag{6.117}$$

This inequality motivates the following general measure of diversity. The *diversity order* is defined as

$$D_o = -\lim_{\bar{\gamma}\to\infty} \frac{\partial \ln[P_b(L)]}{\partial\, ln(\bar{\gamma})}. \tag{6.118}$$

If $\bar{\gamma} \to \infty$, (6.117) becomes an equality, and hence a BPSK or QPSK system over the Rayleigh channel has $D_o = L$.

The advantage of MRC is critically dependent on the assumption of uncorrelated fading in each diversity branch. If there is complete correlation so that the $\{\alpha_i\}$ are all equal and the fading occurs simultaneously in all the diversity branches, then (6.101) indicates that $\gamma_b = L\mathcal{E}_b\alpha_1^2/N_0$. Therefore, complete correlation is equivalent to no diversity, but an energy increase from $\mathcal{E}_b$ to $L\mathcal{E}_b$. For Rayleigh fading and complete correlation, (6.111) implies that the bit error probability is

$$P_b^{cc}(L) = \frac{1}{2}\left(1 - \sqrt{\frac{L\bar{\gamma}}{1+L\bar{\gamma}}}\right) \quad \text{(cc, BPSK, QPSK)}. \tag{6.119}$$

Graphs of the bit error probability for a single branch with no fading, L branches with independent Rayleigh fading and MRC, and L branches with completely correlated Rayleigh fading and MRC are shown in Figure 6.8. Equations (6.100), (6.111), (6.112), and (6.119) are used in generating the graphs. The independent variable is $\bar{\gamma}$ for MRC and is $\gamma_b = \mathcal{E}_b/N_0$ for the single branch with no fading. The figure demonstrates the advantages of both diversity combining and independent fading. The figure indicates that the fading is actually beneficial if the average energy-to-noise-density ratio is sufficiently low. Figure 6.9 displays the bit error probability for independent Nakagami fading with $m = 4$, BPSK, and MRC with $L = 1, 2, 3, 4$. Since the Nakagami fading with $m = 4$ is much milder than Rayleigh fading, the bit error probability is lowered significantly.

This result has an important application to receive antennas, which can be used for different purposes. Receive antennas can provide MRC diversity, but instead they can be used for beamforming. If the receive antennas are used for beamforming, they are sufficiently close that their phase-shifted outputs are highly correlated. Consequently, when $\bar{\gamma}$ is sufficiently large in an environment of fading and noise, beamforming entails a performance loss in the presence of fading relative to the potential performance with diversity combining. The main advantage of beamforming is its suppression of interference entering the sidelobes of the receive antenna pattern.

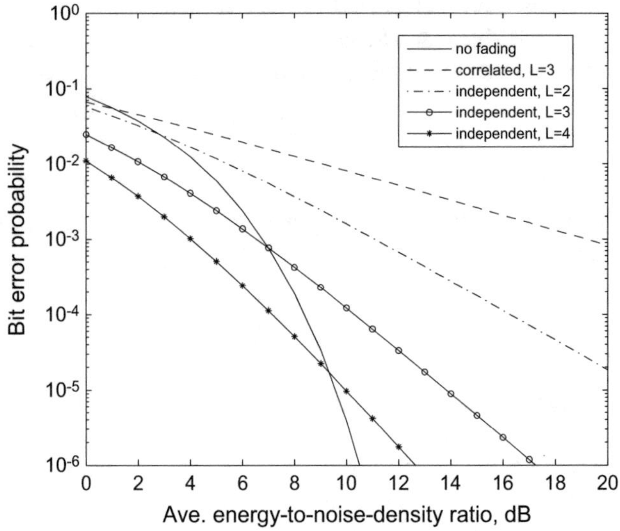

Fig. 6.8 Bit error probability of BPSK for no fading, completely correlated Rayleigh fading and MRC, and independent Rayleigh fading and MRC

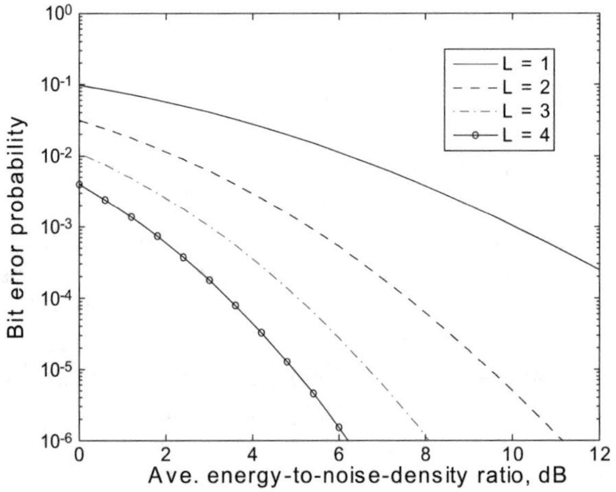

Fig. 6.9 Bit error probability of MRC with BPSK for independent Nakagami fading with $m = 4$

Coherent Orthogonal FSK Systems

Consider systems that use orthogonal signals, such as orthogonal FSK signals. One of q equal-energy orthogonal signals $s_1(t)$, $s_2(t)$, ..., $s_q(t)$, each representing $\log_2 q$ bits, is transmitted. For a direct-sequence system with q-ary code-shift keying (Section 2.7), each orthogonal signal has the form of (2.165) with $t_0 = 0$. The maximum-likelihood detector generates q symbol metrics corresponding to the q possible nonbinary symbols. The decoder decides in favor of the symbol associated with the largest of the symbol metrics. Matched filters for each of the q orthogonal signals are needed in every diversity branch. Because of the orthogonality, each filter matched to $s_r(t)$ has a zero response to $s_l(t)$, $r \neq l$, at the sampling time.

For the AWGN channel, (1.71) and the analysis in Section 1.2 indicate that when a symbol represented by $s_r(t)$ is received, matched-filter l of branch i produces the sample

$$y_{l,i} = \sqrt{\mathcal{E}_s}\alpha_i e^{j(\theta_i - \phi_i)}\delta_{l,r} + n_{l,i}, \quad l = 1, 2, \ldots, q, \quad i = 1, 2, \ldots, L \tag{6.120}$$

where $\mathcal{E}_s$ is the desired-signal energy per symbol in the absence of fading and diversity combining, $\theta_i - \phi_i$ is the difference between the received phase and the estimated phase, and each $n_{l,i}$ is an independent, zero-mean, complex Gaussian random variable. These samples provide sufficient statistics that contain all the relevant information in the received signal copies in the L diversity branches. The two-sided PSD of the interference and noise in branch i is equal to $N_0/2$, and $E[|n_{l,i}|^2] = N_0$. Because of its circular symmetry, n_{li} has independent real and imaginary components with variance $N_0/2$. For coherent detection, $\theta_i - \phi_i = 0$. The conditional density function of $y_{l,i}$ given the $\{\alpha_i\}$ and that $s_r(t)$ is received is

$$f(y_{l,i}|r, \alpha_i) = \frac{1}{\pi N_0}\exp\left[-\frac{|y_{l,i} - \sqrt{\mathcal{E}_s}\alpha_i\delta_{l,r}|^2}{N_0}\right]. \tag{6.121}$$

Since the noise in each branch is assumed to be independent, the likelihood function is the product of qL density functions given by (6.121) for $l = 1, 2, \ldots, q$ and $i = 1, 2, \ldots, L$.

Forming the log-likelihood function, observing that $\sum_r \delta_{l,r}^2 = 1$, and eliminating irrelevant terms and factors that are independent of r, we find that the *symbol metric for coherent FSK is*

$$U(r) = \sum_{i=1}^{L} \text{Re}(\alpha_i y_{ri}), \quad r = 1, 2, \ldots, q \tag{6.122}$$

which requires estimates of the $\{\alpha_i\}$. This set of q metrics, one for each of $s_1(t)$, $s_2(t)$, ..., $s_q(t)$, provides symbol metrics for soft-decision or hard-decision decoding.

For hard-decision decoding, a symbol decision is made by selecting the largest of the $\{U(r)\}$. Consider coherent binary frequency-shift keying (BFSK), for which $\mathcal{E}_s = \mathcal{E}_b$ and $q = 2$. Because of the symmetry of the model, $P_b(L)$ can be calculated by assuming that $s_1(t)$ was transmitted. With this assumption, (6.122) and (6.120) indicate that the two symbol metrics are

$$U(1) = \sqrt{\mathcal{E}_b} \sum_{i=1}^{L} \alpha_i^2 + \sum_{i=1}^{L} \alpha_i \operatorname{Re}(n_{1,i}) \tag{6.123}$$

$$U(2) = \sum_{i=1}^{L} \alpha_i \operatorname{Re}(n_{2,i}). \tag{6.124}$$

A decision error is made if $U(1) - U(2) < 0$. Since $U(1) - U(2)$ has a Gaussian conditional distribution, an evaluation indicates that the conditional bit error probability given the $\{\alpha_i\}$ is

$$P_{b|\boldsymbol{\alpha}}(\gamma_b) = Q(\sqrt{\gamma_b}) \tag{6.125}$$

where γ_b is given by (6.101).

When independent, identically distributed Nakagami fading occurs in each branch, the bit error probability is determined by averaging $P_{b|\boldsymbol{\alpha}}(\gamma_b)$ over the distribution of γ_b, which is given by (6.102). Equation (6.125) differs from (6.100) only by the replacement of $2\gamma_b$ with γ_b. Therefore, if m is a positive integer, a modification of (6.108) indicates that the bit error probability for coherent BFSK is

$$P_b(L) = \frac{1}{2} - \frac{1}{2}\sqrt{\frac{\bar{\gamma}}{2m+\bar{\gamma}}} \sum_{i=0}^{mL-1} \frac{\Gamma(i+1/2)(2m)^i}{\sqrt{\pi}\, i!(2m+\bar{\gamma})^i} \tag{6.126}$$

which explicitly shows the change in the bit error probability as the number of diversity branches increases.

For Rayleigh fading, (6.112) and (6.113) are again valid, but the bit error probability for no diversity or a single branch is

$$p = \frac{1}{2}\left(1 - \sqrt{\frac{\bar{\gamma}}{2+\bar{\gamma}}}\right) \qquad \text{(Rayleigh, coherent BFSK)} \tag{6.127}$$

where the average bit-energy-to-noise-density ratio per branch is defined by (6.103). Thus, in a fading environment, BPSK retains its usual 3 dB advantage over coherent BFSK.

Substituting (6.127) into (6.116), assuming $\bar{\gamma} > 2$, performing a Taylor series expansion in $\bar{\gamma}^{-1}$, and retaining only the first term in the alternating series, we obtain an upper bound:

$$P_b(L) \le \binom{2L-1}{L} \left(\frac{1}{2^L}\right) \bar{\gamma}^{-L}, \quad \bar{\gamma} > 2. \tag{6.128}$$

which indicates that a coherent BFSK system over the Rayleigh channel has diversity order $D_o = L$.

6.5 Equal-Gain Combining

Equal-gain combining (EGC) is the cophasing of signal copies without compensating for unequal values of the signal-to-noise-ratio (SNR) in each branch. Thus, when a narrowband desired signal experiences fading, the EGC weight vector is

$$\mathbf{w}_e = \eta[\exp(j\theta_1)\ \ \exp(j\theta_2)\ \ldots\ \exp(j\theta_L)]^T \tag{6.129}$$

where θ_i is the phase shift of the desired signal in branch i. When MRC is optimal and the values of the $\{\alpha_i/N_{0i}\}$ are unequal, EGC is suboptimal, but requires much less information about the channel. Figure 6.8 displays EGC with predetection and postdetection combining if the factors $\{\alpha_i\}$ are set equal to unity.

If the interference and noise in each array branch is zero-mean and uncorrelated with the other branches and $E[|n_i|^2] = N_0$, $i = 1, 2, \ldots, L$, then $\mathbf{R}_n$ is diagonal, and (6.84) with $\mathbf{w} = \mathbf{w}_e$ gives the output SINR:

$$\rho_e = \frac{\mathcal{E}_s}{LN_0}\left(\sum_{i=1}^{L}\alpha_i\right)^2. \tag{6.130}$$

An application of the Cauchy-Schwarz inequality for sequences of complex numbers (Section 5.2) verifies that this SINR is less than or equal to ρ_m given by (6.93).

In a Rayleigh-fading environment, each α_i, $i = 1, 2, \ldots, L$, has a Rayleigh distribution. If the desired signal in each array branch is uncorrelated with the other branches and has identical average power, then using (E.30) of Appendix E.4, we obtain

$$E[\alpha_i^2] = E[\alpha_1^2],\ \ E[\alpha_i] = \left(\frac{\pi}{4}\,E[\alpha_1^2]\right)^{1/2},\ \ i = 1, 2, \ldots, L \tag{6.131}$$

$$E[\alpha_i\alpha_k] = E[\alpha_i]E[\alpha_k] = \frac{\pi}{4}E[\alpha_1^2],\ \ i \neq k. \tag{6.132}$$

These equations and (6.130) give

$$E[\rho_e] = \left[1 + (L-1)\frac{\pi}{4}\right]\bar{\gamma},\ \bar{\gamma} = \frac{\mathcal{E}_s}{N_0}E[\alpha_1^2] \tag{6.133}$$

which indicates that the loss associated with using EGC instead of MRC is on the order of 1 dB.

In some environments, MRC is identical to EGC, but both are distinctly suboptimal because of interference correlations among the branches. Consider narrowband desired and interference signals with carrier frequency f_0 that do not experience fading and arrive as plane waves. The array antennas are sufficiently close that the steering vector $\mathbf{s}_0$ of the desired signal and the steering vector $\mathbf{J}_0$ of the interference signal can be represented by

$$\mathbf{s}_0 = \left[e^{-j2\pi f_0 \tau_1} \; e^{-j2\pi f_0 \tau_2} \; \ldots \; e^{-j2\pi f_0 \tau_L}\right]^T \tag{6.134}$$

$$\mathbf{J}_0 = \left[e^{-j2\pi f_0 \delta_1} \; e^{-j2\pi f_0 \delta_2} \; \ldots \; e^{-j2\pi f_0 \delta_L}\right]^T. \tag{6.135}$$

The noise power in each branch is equal. The correlation matrix for the interference and noise is

$$\mathbf{R}_n = N_0 \, \mathbf{I} + N_0 g \, \mathbf{J}_0 \, \mathbf{J}_0^H \tag{6.136}$$

where g is the interference-to-noise ratio in each array branch. This equation shows explicitly that the interference in one branch correlates with the interference in the other branches.

A direct matrix multiplication using $\|\mathbf{J}_0\|^2 = L$ verifies that

$$\mathbf{R}_n^{-1} = \frac{1}{N_0}\left(\mathbf{I} - \frac{g \, \mathbf{J}_0 \, \mathbf{J}_0^H}{Lg + 1}\right). \tag{6.137}$$

After merging $1/N_0$ with the constant in (6.85), it is found that the optimal weight vector is

$$\mathbf{w}_0 = \eta \left(\mathbf{s}_0 - \frac{\xi L g}{Lg + 1} \mathbf{J}_0\right) \tag{6.138}$$

where ξ is the normalized inner product

$$\xi = \frac{1}{L} \mathbf{J}_0^H \mathbf{s}_0. \tag{6.139}$$

The corresponding maximum SINR, which is calculated by substituting (6.134), (6.137), and (6.139) into (6.86), is

$$\rho_{\max} = L\gamma \left(1 - \frac{|\xi|^2 L g}{Lg + 1}\right) \tag{6.140}$$

where $\gamma = \mathcal{E}_s/N_0$ is the SNR in each branch. Equations (6.134), (6.135), and (6.139) indicate that $0 \leq |\xi| \leq 1$, and that $|\xi| = 1$ if $L = 1$. Equation (6.140) indicates that $\rho_{\max}$ decreases as $|\xi|$ increases if $L \geq 2$ and is nearly directly proportional to L if $g >> 1$.

Since the values of the SINRs in the branches are all equal, both MRC and EGC use the weight vector of (6.129), which gives $\mathbf{w}_m = \mathbf{w}_e = \eta \mathbf{s}_0$. Substituting (6.82), (6.129), and (6.136) into (6.84) gives the SINR for MRC and EGC:

$$\rho_m = \rho_e = \frac{L\gamma}{1 + |\xi|^2 L g}. \tag{6.141}$$

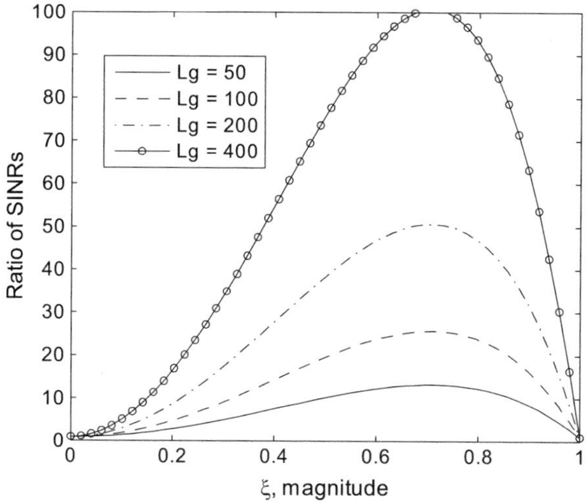

Fig. 6.10 Ratio of the maximum SINR to the maximal-ratio-combiner SINR

Both $\rho_{\max}$ and ρ_m equal $L\gamma$, the peak value, when $\xi = 0$. They both equal $L\gamma/(1+Lg)$ when $|\xi| = 1$, which occurs when both the desired and the interference signals arrive from the same direction or $L = 1$. Using calculus, it is determined that the maximum value of $\rho_{\max}/\rho_m$, which occurs when $|\xi| = 1/\sqrt{2}$, is

$$\left(\frac{\rho_{\max}}{\rho_m}\right)_{\max} = \frac{(Lg/2+1)^2}{Lg+1}, \quad L \geq 2. \tag{6.142}$$

This ratio approaches $Lg/4$ for large values of Lg. Thus, an adaptive array based on the maximization of the SINR has the potential to significantly outperform MRC or EGC if $Lg >> 1$ is assumed under the conditions of the nonfading environment. Figure 6.10 displays $\rho_{\max}/\rho_m$ as a function of $|\xi|$ for various values of Lg.

6.6 Noncoherent Combining

When accurate phase estimation is unavailable so that cophasing is not possible, then noncoherent combining potentially provides a significant performance improvement over a system with no diversity. *Noncoherent combining* is implemented as postdetection combining following noncoherent demodulation.

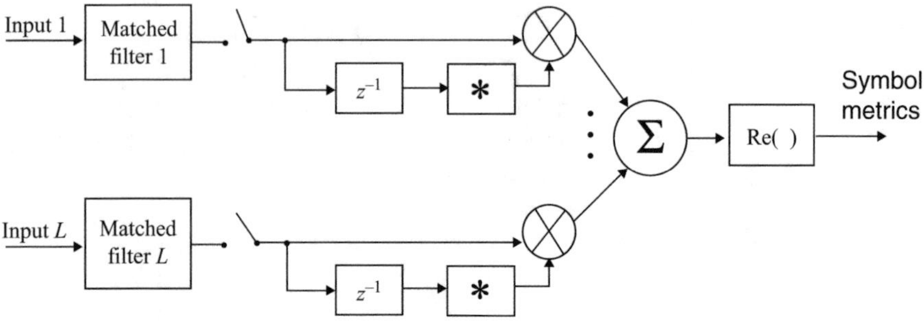

Fig. 6.11 Receiver for DPSK with postdetection noncoherent combining. The z^{-1} denotes a symbol delay; the $*$ denotes complex conjugation

Differential Phase-Shift Keying

A block diagram of a DPSK receiver with postdetection noncoherent combining is depicted in Figure 6.11. Assuming that the fading is constant over two symbols, the symbol metric is

$$U = \text{Re}\left[\sum_{i=1}^{L} y_{1i} y_{2i}^*\right] \tag{6.143}$$

where y_{1i} and y_{2i} are received symbols in branch i arising from two consecutive symbol intervals. This symbol metric has the advantage that it requires neither phase synchronization nor channel state estimation. For hard-decision decoding, a derivation [69] indicates that if the $\{\alpha_i\}$ are independent but have identical Rayleigh distributions, then $P_b(L)$ is given by (6.112), (6.113), and (6.116) with the single-branch bit error probability

$$p = \frac{1}{2(1+\bar{\gamma})} \quad \text{(DPSK)} \tag{6.144}$$

where $\bar{\gamma}$ is given by (6.103). Equation (6.144) can be directly derived by observing that the conditional bit error probability for DPSK with no diversity is $\exp(-\gamma_b)/2$ and then integrating the equation over the density function (6.102) with $L = 1$. DPSK provides the diversity order $D_o = L$. A comparison of (6.144) with (6.127) indicates that DPSK with noncoherent combining and coherent BFSK with MRC give nearly the same performance in a Rayleigh-fading environment if $\bar{\gamma} >> 1$.

Noncoherent Orthogonal FSK

Consider the transmission over the AWGN channel of the signal $s_r(t)$ representing a single symbol. To derive a receiver for noncoherent orthogonal FSK from the maximum-likelihood criterion, we assume that the $\{\alpha_i\}$ and the $\{\theta_i - \phi_i\}$ in (6.120) are random variables. Equation (1.79) and the analysis of Section 1.2 yields the conditional density function for the sampled output of matched-filter l of branch i :

$$f(y_{l,i}|r,\alpha_i) = \frac{1}{\pi N_0}\exp\left(-\frac{|y_{l,i}|^2 + \mathcal{E}_s\alpha_i^2\delta_{lr}}{N_0}\right) I_0\left(\frac{2\sqrt{\mathcal{E}_s}\alpha_i|y_{l,i}|\delta_{l,r}}{N_0}\right),$$
$$l = 1,2,\ldots,q, \quad i = 1,2,\ldots,L. \tag{6.145}$$

If the fading is statistically independent in each branch and is known to be Rayleigh, then the density function $f(y_{l,i}|r)$ may be calculated by integrating $f(y_{l,i}|r,\alpha_i)$ over the Rayleigh density given by (6.24) with $\Omega = E[\alpha_i^2]$. To evaluate this integral for $l = r$, we observe that the Rice density (E.23) of Appendix E.3 must integrate to unity, which implies that

$$\int_0^{\infty} x\exp\left(-\frac{x^2}{2b}\right) I_0\left(\frac{x\sqrt{\lambda}}{b}\right) dx = b\exp\left(\frac{\lambda}{2b}\right) \tag{6.146}$$

where λ and b are positive constants. The likelihood function of the qL-dimensional observation vector $\mathbf{y}$, which has components equal to the $\{y_{l,i}\}$, is the product of the qL density functions $\{f(y_{l,i}|r)\}$. After performing the integration and forming the product, we obtain

$$f(\mathbf{y}|r) = C\prod_{i=1}^{L}\exp\left(\frac{|y_{r,i}|^2\bar{\gamma}_i}{N_0(1+\bar{\gamma}_i)}\right) \tag{6.147}$$

where C is a constant that does not depend on r and

$$\bar{\gamma}_i = \frac{\mathcal{E}_s}{N_0}E[\alpha_i^2], \quad i = 1,2,\ldots,L. \tag{6.148}$$

To determine the transmitted symbol, we choose the value of r that maximizes the log-likelihood $\ln f(\mathbf{y}|r)$. Dropping irrelevant terms and factors, we obtain the *maximum-likelihood symbol metric for Rayleigh fading*:

$$U(r) = \sum_{i=1}^{L}|y_{r,i}|^2\left(\frac{\bar{\gamma}_i}{1+\bar{\gamma}_i}\right), \quad r = 1,2,\ldots,q \tag{6.149}$$

which requires the estimation of $\bar{\gamma}_i$ for each branch.

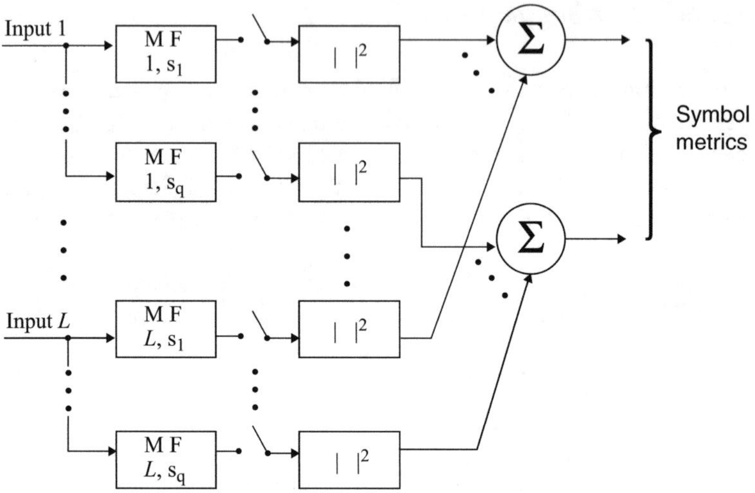

Fig. 6.12 Receiver for q-FSK with postdetection noncoherent combining. *MF* matched filter

If we assume that all the $\{\bar{\gamma}_i\}$ are equal, then (6.149) reduces to the linear *square-law metric:*

$$U(r) = \sum_{i=1}^{L} |\, y_{r,i}|^2, \quad r = 1, 2, \ldots, q. \tag{6.150}$$

This metric implies a noncoherent FSK receiver with postdetection square-law combining, which is illustrated in Figure 6.12. Each branch feeds q matched filters, and each matched filter is matched to one of the equal-energy orthogonal signals $s_1(t)$, $s_2(t)$, ..., $s_q(t)$. A major advantage of the square-law metric is that it does not require any channel-state information. If $\bar{\gamma}_i$ is large, the corresponding terms in the square-law and maximum-likelihood symbol metrics are nearly equal; if $\bar{\gamma}_i$ is small, the corresponding terms in the two metrics both tend to be insignificant compared with other terms. Thus, there is usually little penalty in using the square-law metric.

Consider hard-decision decoding of noncoherent BFSK and the square-law metric, for which $\mathcal{E}_s = \mathcal{E}_b$. Because of the symmetry of the signals, $P_b(L)$ can be calculated by assuming that $s_1(t)$ was transmitted. Given that $s_1(t)$ was transmitted, the two symbol metrics at the combiner output are

$$\begin{aligned} U(1) &= \sum_{i=1}^{L} |\sqrt{\mathcal{E}_b}\alpha_i e^{j\theta_i} + n_{1,i}|^2 \\ &= \sum_{i=1}^{L} \left(\sqrt{\mathcal{E}_b}\alpha_i \cos\theta_i + n_{1,i}^R\right)^2 + \sum_{i=1}^{L} \left(\sqrt{\mathcal{E}_b}\alpha_i \sin\theta_i + n_{1,i}^I\right)^2 \end{aligned} \tag{6.151}$$

$$U(2) = \sum_{i=1}^{L} |n_{2,i}|^2 = \sum_{i=1}^{L} \left(n_{2,i}^R\right)^2 + \sum_{i=1}^{L} \left(n_{2,i}^I\right)^2 \tag{6.152}$$

where n_{1i} and n_{2i} are the independent, complex-valued, zero-mean, Gaussian noise variables and $n_{l,i}^R$ and $n_{l,i}^I$, $l = 1, 2$, are the real and imaginary parts of $n_{l,i}$ respectively. Assuming that the noise PSD in each branch is equal to N_0, then $E[|n_{l,i}|^2] = N_0$, $l = 1, 2$. Because of its circular symmetry, $n_{l,i}$ has independent real and imaginary components, and

$$E[(n_{l,i}^R)^2] = E[(n_{l,i}^I)^2] = N_0/2\,, \quad l = 1, 2, \quad i = 1, 2, \ldots, L. \tag{6.153}$$

When independent, identically distributed, Rayleigh fading occurs in each branch, $\alpha_i \cos\theta_i$ and $\alpha_i \sin\theta_i$ are zero-mean, independent, Gaussian random variables with the same variance equal to $E[\alpha_i^2]/2 = E[\alpha_1^2]/2$, $i = 1, 2, \ldots, L$, as shown in Appendix H.4. Therefore, both $U(1)$ and $U(2)$ have central chi-squared distributions with $2L$ degrees of freedom. From (E.15), the density function of $U(l)$ is

$$f_l(x) = \frac{1}{(2\sigma_l^2)^L (L-1)!} x^{L-1} \exp\left(-\frac{x}{2\sigma_l^2}\right) u(x), \quad l = 1, 2 \tag{6.154}$$

where (6.153) and (6.103) give

$$\sigma_2^2 = E[(n_{2,i}^R)^2] = N_0/2 \tag{6.155}$$

$$\sigma_1^2 = E[(\sqrt{\mathcal{E}_b}\alpha_1 \cos\theta_i + n_{1,i}^R)^2] = N_0(1 + \bar{\gamma})/2. \tag{6.156}$$

Since an erroneous decision is made if $U(2) > U(1)$,

$$P_b(L) = \int_0^\infty \frac{x^{L-1} \exp\left(-\frac{x}{2\sigma_1^2}\right)}{(2\sigma_1^2)^L (L-1)!} \left[\int_x^\infty \frac{y^{L-1} \exp\left(-\frac{y}{2\sigma_2^2}\right)}{(2\sigma_2^2)^L (L-1)!} dy \right] dx. \tag{6.157}$$

Using (6.106) inside the brackets and integrating, we obtain

$$P_b(L) = \int_0^\infty \exp\left(-\frac{x}{2\sigma_1^2}\right) \sum_{i=0}^{L-1} \frac{(x/2\sigma_2^2)^i}{i!} \frac{x^{L-1} \exp\left(-\frac{x}{2\sigma_2^2}\right)}{(2\sigma_1^2)^L (L-1)!} dx. \tag{6.158}$$

Changing variables, applying (H.1) of Appendix H.1, and simplifying gives (6.113), where the bit error probability for $L = 1$ is

$$p = \frac{1}{2 + \bar{\gamma}} \qquad \text{(noncoherent BFSK)} \tag{6.159}$$

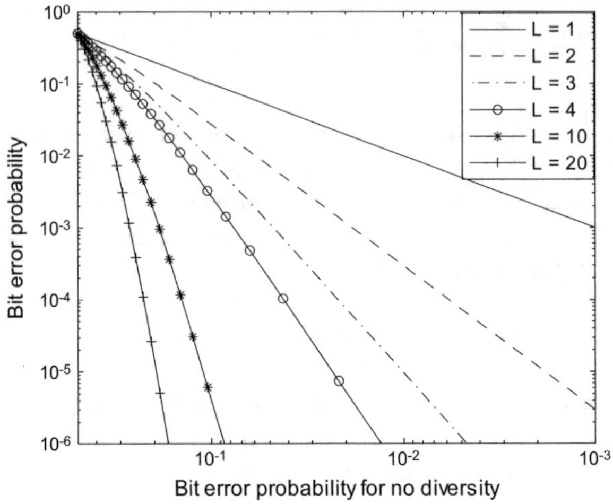

Fig. 6.13 Bit error probability over the Rayleigh channel for MRC with BPSK and coherent BFSK, and for noncoherent combining with DPSK and noncoherent BFSK

and $\bar{\gamma}$ is given by (6.103). Thus, $P_b(L)$ is once again given by (6.112), and the diversity order is $D_o = L$. Equations (6.159) and (6.144) indicate that less than 3 dB more power is needed for noncoherent BFSK to provide the same performance as DPSK in Rayleigh fading.

The analysis for Rayleigh fading has shown that (6.112) is valid for MRC with BPSK (with or without direct-sequence spreading) or coherent BFSK, and noncoherent combining with DPSK or noncoherent BFSK. Once the bit error probability p in the absence of diversity combining is determined, the bit error probability $P_b(L)$ for diversity combining in the presence of independent Rayleigh fading can be calculated from (6.112). A plot of $P_b(L)$ versus p for different values of L is displayed in Figure 6.13. This figure illustrates the diminishing returns obtained as L increases.

A plot of $P_b(L)$ versus $\bar{\gamma}$ over the Rayleigh channel is displayed in Figure 6.14 for MRC with BPSK and noncoherent combining with DPSK or noncoherent BFSK. The plot for MRC with coherent BFSK is nearly the same as that for noncoherent combining with DPSK. Since (6.116) is valid for all these modulations in the presence of independent Rayleigh fading, we find that $P_b(L)$ is asymptotically proportional to $\bar{\gamma}^{-L}$ and the diversity order is $D_o = L$ for all these modulations. Despite this asymptotic equality, the bit error probability varies substantially with modulation for the practical range $P_b(L) > 10^{-6}$.

For noncoherent q-ary orthogonal signals, such as FSK with $L \geq 2$, it can be shown that the symbol error probability $P_s(L)$ decreases slightly as q increases [69]. The price for this modest improvement is an increase in transmission bandwidth.

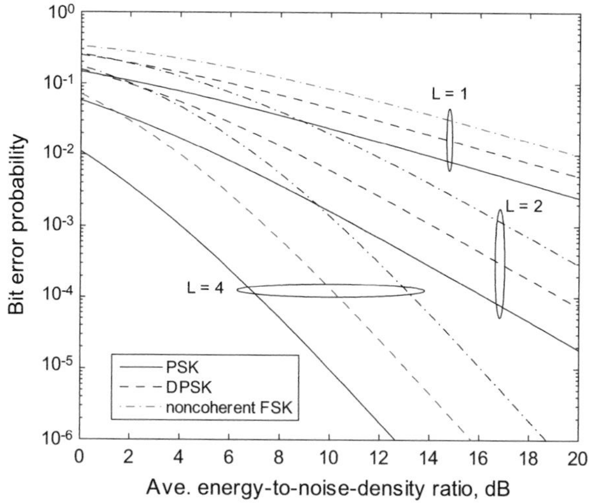

Fig. 6.14 Bit error probability over the Rayleigh channel for MRC with BPSK, and for noncoherent combining with DPSK and noncoherent BFSK

6.7 Selection Diversity

A *selection-diversity system* selects one of the diversity branches and forwards the signal in this branch for further processing. In a fading environment, selection is sensible only if the selection rate is much faster than the fading rate. A *predetection-selection system* is a type of selection-diversity system that estimates the SNR in each branch and selects the branch with the largest value. If the noise and interference levels in all the branches are nearly the same, then the total power in each branch rather than the SNR can be measured to enable the selection process, thereby allowing a major simplification. Predetection selection does not provide a performance as good as maximal-ratio combining or equal-gain combining when the interference and noise in each branch is uncorrelated with that in the other branches. However, predetection selection requires only a single demodulator, and when noises or interference signals are correlated, then predetection selection is more competitive in performance.

Consider predetection selection when the average power of the desired signal in the absence of fading is the same in each branch, and the average power of the zero-mean noise is the same in each branch. The SNR in each branch is proportional to the energy-to-noise-density ratio in branch i, which is defined as $\gamma_i = \mathcal{E}_s \alpha_i^2 / N_0$. If each of the $\{\alpha_i\}$, $i = 1, 2, \ldots, L$, has a Rayleigh distribution, then γ_i in each branch has the same expected value

$$\bar{\gamma} = \frac{\mathcal{E}_s}{N_0} E[\alpha_1^2]. \tag{6.160}$$

The results of Appendix E.4 for the square of a Rayleigh-distributed random variable indicate that each γ_i has an exponential density with the corresponding distribution function

$$F_{\gamma}(x) = \left[1 - \exp\left(-\frac{x}{\bar{\gamma}}\right)\right] u(x). \tag{6.161}$$

For predetection selection, the branch with the largest SNR is selected. Let γ_0 denote the γ_i of the selected branch:

$$\gamma_0 = \frac{\mathcal{E}_s}{N_0} \max_i \left(\alpha_i^2\right). \tag{6.162}$$

The probability that γ_0 is less than or equal to x is equal to the probability that all the $\{\gamma_i\}$ are simultaneously less than or equal to x. If the interference and noise in each branch is independent, the distribution function of γ_0 is

$$F_{\gamma_0}(x) = \left[1 - \exp\left(-\frac{x}{\bar{\gamma}}\right)\right]^L u(x). \tag{6.163}$$

The corresponding density function is

$$f_{\gamma_0}(x) = \frac{L}{\bar{\gamma}} \exp\left(-\frac{x}{\bar{\gamma}}\right) \left[1 - \exp\left(-\frac{x}{\bar{\gamma}}\right)\right]^{L-1} u(x). \tag{6.164}$$

The average γ_0 obtained by selection diversity is calculated by integrating γ_0 over the density function given by (6.164). The result is

$$\begin{aligned} E[\gamma_0] &= \int_0^\infty \frac{L}{\bar{\gamma}} x \exp\left(-\frac{x}{\bar{\gamma}}\right) \left[1 - \exp\left(-\frac{x}{\bar{\gamma}}\right)\right]^{L-1} dx \\ &= \bar{\gamma} L \int_0^\infty y e^{-y} \left[\sum_{i=0}^{L-1} \binom{L-1}{i} (-1)^i e^{-yi}\right] dy \\ &= \bar{\gamma} \sum_{i=1}^{L} \binom{L}{i} \frac{(-1)^{i+1}}{i} \\ &= \bar{\gamma} \sum_{i=1}^{L} \frac{1}{i}. \end{aligned} \tag{6.165}$$

The second equality results from a change of variable and the substitution of the binomial expansion. The third equality results from a term-by-term integration using (H.1) and an algebraic simplification. The fourth equality is obtained by applying the method of mathematical induction. Equations (6.94) and (6.133) indicate that the average SNR for predetection selection combining with $L \geq 2$ is less than that for MRC and EGC. Approximating the summation in (6.165) by an integral, it is observed that the ratio of the average SNR for MRC to that for selection diversity is approximately $L/\ln L$ for $L \geq 2$.

Consider hard-decision decoding when the modulation is BPSK and optimal coherent demodulation follows the selection process. Then the conditional bit error probability given the value of γ_0 is

$$P_b(\gamma_0) = Q(\sqrt{2\gamma_0}). \tag{6.166}$$

Therefore, using (6.164) and the binomial expansion, the bit error probability is

$$\begin{aligned} P_b(L) &= \int_0^\infty Q(\sqrt{2x})\frac{L}{\bar{\gamma}} \exp\left(-\frac{x}{\bar{\gamma}}\right)\left[1-\exp\left(-\frac{x}{\bar{\gamma}}\right)\right]^{L-1} dx \\ &= \sum_{i=0}^{L-1}\binom{L-1}{i}(-1)^i\frac{L}{\bar{\gamma}}\int_0^\infty Q(\sqrt{2x})\exp\left[-x\left(\frac{1+i}{\bar{\gamma}}\right)\right]dx. \end{aligned} \tag{6.167}$$

The last integral in this equation can be evaluated in the same manner as the one in (6.104). After changing the summation index, the result is

$$P_b(L) = \frac{1}{2}\sum_{i=1}^{L}\binom{L}{i}(-1)^{i+1}\left(1-\sqrt{\frac{\bar{\gamma}}{i+\bar{\gamma}}}\right) \quad \text{(BPSK, QPSK)}. \tag{6.168}$$

This equation is valid for QPSK since as it can be implemented as two parallel BPSK waveforms.

To obtain a simple upper bound on $P_b(L)$, we substitute the upper bound $1 - \exp(-x/\bar{\gamma}) \leq x/\bar{\gamma}$, $x \geq 0$, into the initial integral in (6.167). We find that $P_b(L)$ is upper-bounded by $L\Gamma(L)$ times an integral equal to right-hand side of (6.104) with $m = 1$. Therefore, we can upper-bound the integral by the right-hand side of (6.117) and obtain

$$P_b(L) \leq L\Gamma(L)\binom{2L-1}{L}\left(\frac{1}{4^L}\right)\bar{\gamma}^{-L} \quad \text{(BPSK, QPSK)} \tag{6.169}$$

which approaches an equality as $\bar{\gamma} \to \infty$. Thus, the diversity order is $D_o = L$, the same as it is for maximal-ratio combining with BPSK.

For coherent BFSK, the conditional bit error probability is $P_b(\gamma_0) = Q(\sqrt{\gamma_0})$. Therefore, it is found that

$$P_b(L) = \frac{1}{2}\sum_{i=1}^{L}\binom{L}{i}(-1)^{i+1}\left(1-\sqrt{\frac{\bar{\gamma}}{2i+\bar{\gamma}}}\right) \quad \text{(coherent BFSK)}. \tag{6.170}$$

Again, 3 dB more power is needed to provide to the same performance as BPSK, and the diversity order is $D_o = L$.

When DPSK is the data modulation, the conditional bit error probability is $P_b(\gamma_0) = exp(-\gamma_0)/2$. Thus, predetection selection provides the bit error probability

$$P_b(L) = \int_0^{\infty} \frac{L}{2\bar{\gamma}} \exp\left(-x\frac{1+\bar{\gamma}}{\bar{\gamma}}\right)\left[1-\exp\left(-\frac{x}{\bar{\gamma}}\right)\right]^{L-1} dx. \tag{6.171}$$

Using $t = \exp(-x/\bar{\gamma})$ to change the integration variable in (6.171) and then using (H.10) of Appendix H.2 gives

$$P_b(L) = \frac{L}{2}B(1+\bar{\gamma}, L) \qquad \text{(DPSK)} \tag{6.172}$$

where $B(\alpha, \beta)$ is the beta function.

For noncoherent orthogonal FSK, the conditional symbol error probability given the value of γ_0 is obtained from (1.93):

$$P_b(\gamma_0) = \sum_{i=1}^{q-1} \frac{(-1)^{i+1}}{i+1}\binom{q-1}{i} \exp\left(-\frac{i\gamma_0}{i+1}\right). \tag{6.173}$$

Therefore, after using $t = \exp(-x/\bar{\gamma})$ to change successive integration variables, the symbol error probability is

$$\begin{aligned} P_s(L) &= \sum_{i=1}^{q-1} \frac{(-1)^{i+1}}{i+1}\binom{q-1}{i}\frac{L}{\bar{\gamma}} \\ &\quad\times \int_0^{\infty} \exp\left(-x\frac{i+1+i\bar{\gamma}}{\bar{\gamma}\,(i+1)}\right)\left[1-\exp\left(-\frac{x}{\bar{\gamma}}\right)\right]^{L-1} dx \\ &= L\sum_{i=1}^{q-1} \frac{(-1)^{i+1}}{i+1}\binom{q-1}{i} B\left(1+\frac{i\bar{\gamma}}{i+1}, L\right) \qquad \text{(noncoherent FSK).} \end{aligned} \tag{6.174}$$

For noncoherent BFSK, the bit error probability is

$$P_b(L) = \frac{L}{2}B\left(\frac{\bar{\gamma}}{2}+1, L\right) \qquad \text{(noncoherent BFSK)} \tag{6.175}$$

which exhibits the usual 3 dB disadvantage compared with DPSK.

To obtain simple upper bounds on $P_b(L)$ for DPSK and noncoherent BFSK, we use the identity (H.11) and the inequality

$$\begin{aligned} \Gamma(x+L) &= (x+L-1)(x+L-2)\ldots\Gamma(x) \\ &\geq x^L\Gamma(x) \end{aligned} \tag{6.176}$$

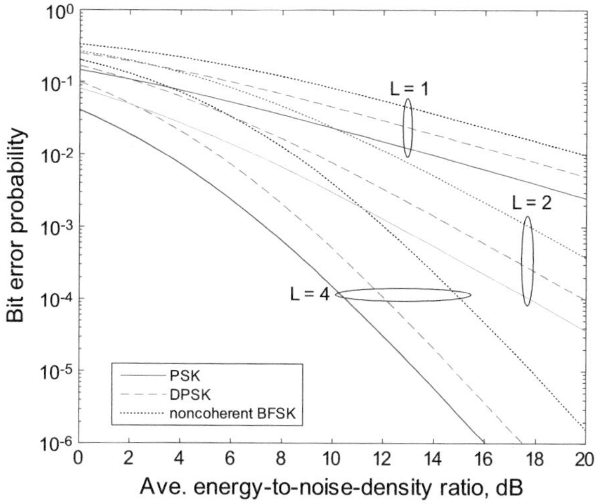

Fig. 6.15 Bit error probability over the Rayleigh channel for selection diversity with BPSK, DPSK, and noncoherent BFSK

for positive x. For DPSK, we obtain

$$P_b(L) \leq \frac{L!}{2}\bar{\gamma}^{-L} \tag{6.177}$$

and for noncoherent BFSK, we obtain

$$P_b(L) \leq 2^{L-1}L!\bar{\gamma}^{-L}. \tag{6.178}$$

These bounds are tight when $\bar{\gamma} >> L$, and the diversity order for both DPSK and noncoherent BFSK is $D_o = L$.

Figure 6.15 shows $P_b(L)$ over the Rayleigh channel as a function of $\bar{\gamma}$, assuming predetection selection with BPSK, DPSK, and noncoherent BFSK. A comparison of Figures 6.15 and 6.14 indicates the reduced gain provided by selection diversity relative to MRC and noncoherent combining.

A fundamental limitation of selection diversity is made evident by the plane-wave example in which the signal and interference steering vectors are given by (6.134) and (6.135). In this example, the SINRs are equal in all the diversity branches. Consequently, selection diversity can give no better performance than no diversity combining or the use of a single branch. In contrast, (6.141) indicates that EGC can improve the SINR significantly.

Other types of selection diversity besides predetection selection are sometimes of interest. *Postdetection selection* entails the selection of the diversity branch with the largest signal and noise power after detection. It outperforms predetection selection in general but requires as many matched filters as diversity branches. Thus, its complexity is not much less than that required for EGC. *Switch-and-stay combining (*SSC*)* or *switched combining* entails processing the output of a particular diversity branch as long as its quality measure remains above a fixed threshold. When it does not, the receiver selects another branch output and continues processing this output until the quality measure drops below the threshold. In *predetection SSC*, the quality measure is the instantaneous SNR of the connected branch. Since only one SNR is measured, predetection SSC is less complex than predetection selection but suffers a performance loss. In *postdetection SSC*, the quality measure is the same output quantity used for data detection. The optimal threshold depends on the average SNR per branch. Postdetection SSC provides a lower bit error probability than predetection SSC, and the improvement increases with both the average SNR and less severe fading [85].

6.8 Transmit Diversity

Spatial diversity may be implemented as *transmit diversity*, which uses an antenna array at the transmitter, *receive diversity*, which uses an array at the receiver, or both. Receive diversity is more effective than transmit diversity because the latter requires a power division among the transmit antennas prior to transmission. However, network requirements and practical issues may motivate the use of transmit diversity. For example, as multiple antennas are much more feasible at a base station than at a mobile, transmit diversity is usually the only type of spatial diversity in the downlink from a base station to a mobile. The configuration of a transmitter with N transmit antennas is illustrated in Figure 6.16.

Delay diversity and frequency-offset diversity are elementary forms of transmit diversity [7] that have significant practical limitations. *Delay diversity* entails the transmission of the same symbol successively from multiple antennas after appropriate delays. The received signal comprises a set of artificial multipath signals that are generated at considerable cost in power and cause multiple-access interference in other systems. *Frequency-offset diversity* transforms the transmit diversity into a type of frequency diversity by requiring each transmit antenna to use a different carrier frequency. The main practical issue is the bandwidth expansion.

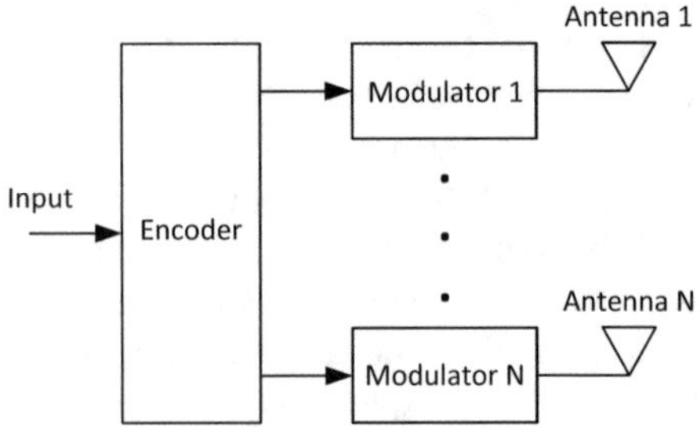

Fig. 6.16 Transmitter with N transmit antennas

Orthogonal transmit diversity, which is included in the CDMA2000 standard, transmits alternating even and odd interleaved symbols on two antennas. The deinterleaved bits generated by the different antennas provide both time diversity due to the deinterleaving and spatial diversity due to the antenna separations. The gain relative to no diversity is substantial provided that the fading is slow and the channel code is strong [87].

Space-time codes, which include *space-time block codes* (STBCs) and *space-time trellis codes* (STTCs), are transmitted by multiple antennas and improve the performance of a communication system in a fading environment without the need for either multiple receive antennas or channel-state information at the transmitter [43]. An STTC inherently combines the modulation and channel coding with transmit diversity to achieve full coding and diversity gain. However, the cost is a decoding complexity that increases with the number of antennas and exceeds that of comparable STBCs. Nonorthogonal STBCs exist that can provide full diversity at a full rate, but require more complex decoding than the separate decoding of each real-valued symbol that is possible with orthogonal STBCs. Rate-1 orthogonal STBCs for complex constellations exist only for two transmit antennas. Orthogonal STBCs for more than two transmit antennas require a code rate that is less than unity, which implies a reduced spectral efficiency.

The *Alamouti code* is by far the most widely used space-time code and is included in the CDMA2000 standard. The Alamouti code is an orthogonal STBC that provides full diversity at a full transmission rate and is decoded by maximum-likelihood decoding that entails only linear processing. Two transmit antennas and two time intervals are used to transmit two complex symbols from a PSK or quadrature amplitude modulation (QAM) constellation. The transmitted space-time codeword of length two has a code rate equal to one, the number of information symbols conveyed per time interval. A direct-sequence system multiplies each symbol by a spreading sequence prior to the modulation and transmission. Let $p_1(t)$ and $p_2(t)$ denote the spreading sequences used during successive time intervals. The 2×2 *generator matrix* representing a transmitted codeword for information symbols d_1 and d_2 is

$$\mathbf{G} = \begin{bmatrix} d_1 p_1(t) & d_2 p_1(t) \\ -d_2^* p_2(t) & d_1^* p_2(t) \end{bmatrix} \tag{6.179}$$

where each row identifies the symbols transmitted during a time interval, and each column identifies the successive symbols transmitted by one of the antennas.

Assuming a single receive antenna and AWGN, the demodulated signal during the first time interval is $r_1(t) = h_1 d_1 p_1(t) + h_2 d_2 p_1(t) + n_1(t)$, where h_i, $i = 1, 2$, is the complex channel response from the transmit antenna i to the receive antenna, and $n_1(t)$ is a complex, zero-mean, white Gaussian noise process. After despreading, sampling, and an amplitude normalization, the observation at the end of the first time interval is $r_1 = h_1 d_1 + h_2 d_2 + n_1$, where n_1 is complex zero-mean Gaussian noise. Similarly, assuming that the channel does not change during two time intervals, the observation at the end of the second time interval is $r_2 = -h_1 d_2^* + h_2 d_1^* + n_2$, where n_2 is complex zero-mean Gaussian noise that is independent of n_1. These two observations are combined in the vector $\mathbf{y}_r = \left[r_1 \; r_2^*\right]^T$. Then

$$\mathbf{y}_r = \mathbf{Hd} + \mathbf{n} \tag{6.180}$$

where $\mathbf{d} = [d_1 \; d_2]^T$, $\mathbf{n} = \left[n_1 \; n_2^*\right]^T$, and the channel matrix is

$$\mathbf{H} = \begin{bmatrix} h_1 & h_2 \\ -h_2^* & -h_1^* \end{bmatrix}. \tag{6.181}$$

Let $\mathcal{E}_s$ denote the average energy per symbol received from both transmit antennas. The power splitting between the two transmit antennas implies that $E\left[|d_k|^2\right] = \mathcal{E}_s/2$, $k = 1, 2$. In the presence of AWGN with power spectral density $N_0/2$, the analysis of Section 1.2 indicates that $\mathbf{n}$ is the zero-mean Gaussian random vector with independent components and covariance matrix $E\left[\mathbf{nn}^H\right] = N_0\mathbf{I}$.

The matrix $\mathbf{H}$ satisfies the *orthogonality condition*:

$$\mathbf{H}^H\mathbf{H} = ||\mathbf{h}||^2\mathbf{I} \tag{6.182}$$

where $||\mathbf{h}||$ denotes the Euclidean norm of $\mathbf{h} = [h_1 \; h_2]$, and $\mathbf{I}$ is the 2×2 identity matrix. Therefore, the receiver computes the 2×1 vector

$$\mathbf{y} = \mathbf{H}^H\mathbf{y}_r = \mathbf{d}||\mathbf{h}||^2 + \mathbf{n}_1 \tag{6.183}$$

where $\mathbf{n}_1 = \mathbf{H}^H\mathbf{n}$ is a zero-mean Gaussian random vector, as the components of $\mathbf{n}$ are independent (Appendix A.1). Its covariance matrix is

$$E\left[\mathbf{n}_1\mathbf{n}_1^H\right] = N_0||\mathbf{h}||^2\mathbf{I}. \tag{6.184}$$

Equation (6.183) indicates that the maximum-likelihood decision for d_k is separately obtained by finding the value of d_k that minimizes $\left|y_k - d_k||\mathbf{h}||^2\right|$, $k = 1, 2$. Since each noise component is independent, each symbol decision is decoupled from the other one, and there is no intersymbol interference.

The components of (6.183) may be expressed as

$$y_k = d_k \sum_{i=1}^{2} \alpha_i^2 + n_{1k}, \quad k = 1, 2 \tag{6.185}$$

where $\alpha_i = |h_i|$ and n_{1k} is the kth component of $\mathbf{n}_1$. The desired part of y_k is similar to that in (6.92) obtained by maximal-ratio combining with two signals and indicative of diversity order 2. Thus, the bit error probabilities for BPSK, QPSK, and coherent BFSK derived for MRC in the presence of Rayleigh fading are applicable with one important change. Since $E\left[|d_k|^2\right] = \mathcal{E}_s/2$ because of the power splitting between the two transmit antennas, $\bar{\gamma}$ must be replaced by $\bar{\gamma}/2$ in the equations.

The Alamouti STBC with a generator matrix given by (6.179) provides diversity order $2L$ when there are L receive antennas. Let $\mathbf{h}_i$, $i = 1, 2$ denote an $L \times 1$ vector, each component of which is the complex channel response from transmit antenna i to a receive antenna. After despreading, sampling, and an amplitude normalization of each receive-antenna output, the observation at the end of the first time interval is the $L \times 1$ vector $\mathbf{r}_1 = \mathbf{h}_1 d_1 + \mathbf{h}_2 d_2 + \mathbf{n}_{a1}$, where each component of the $L \times 1$ vector $\mathbf{n}_{a1}$ is complex zero-mean Gaussian noise. Similarly, assuming that the channel does not change during the two time intervals, the observation at the end of the second time interval is $\mathbf{r}_2 = -\mathbf{h}_1 d_2^* + \mathbf{h}_2 d_1^* + \mathbf{n}_{a2}$, where each component of $\mathbf{n}_{a2}$ is complex zero-mean Gaussian noise, and all components of $\mathbf{n}_{a1}$ and $\mathbf{n}_{a2}$ are independent of each other. The combined observation vector is given by (6.180) with the $2L \times 1$ vectors

$$\mathbf{y}_r = \left[\mathbf{r}_1^T \; \mathbf{r}_2^T\right]^T, \; \mathbf{n} = \left[\mathbf{n}_{a1}^T \; \mathbf{n}_{a2}^T\right]^T \tag{6.186}$$

and the $2L \times 2$ channel matrix

$$\mathbf{H} = \begin{bmatrix} \mathbf{h}_1 & \mathbf{h}_2 \\ -\mathbf{h}_2^* & -\mathbf{h}_1^* \end{bmatrix}. \tag{6.187}$$

The Gaussian noise vector $\mathbf{n}$ is zero-mean with the $2L \times 2L$ covariance matrix $E\left[\mathbf{n}\mathbf{n}^H\right] = N_0\mathbf{I}$, where is $\mathbf{I}$ the $2L \times 2L$ identity matrix. The orthogonality condition (6.182) is satisfied if

$$\mathbf{h} = \left[\mathbf{h}_1^T \; \mathbf{h}_2^T\right]^T, \; ||\mathbf{h}||^2 = ||\mathbf{h}_1||^2 + ||\mathbf{h}_2||^2. \tag{6.188}$$

The receiver computes the 2×1 vector given by (6.183), and the maximum-likelihood decision for d_k is separately obtained by finding the value of d_k that minimizes $\left| y_k - d_k ||\mathbf{h}||^2 \right|$, $k = 1, 2$. The components of (6.183) may be expressed as

$$y_k = d_k \sum_{i=1}^{2L} \alpha_i^2 + n_{1k}, \quad k = 1, 2 \tag{6.189}$$

which has the same form as maximal-ratio combining and indicates diversity order $2L$ and no intersymbol interference. Again, the bit error probabilities for BPSK, QPSK, and coherent BFSK derived for MRC in the presence of fading are applicable if $\bar{\gamma}$ is replaced by $\bar{\gamma}/2$ in the equations.

Transmit antenna selection (TAS) is a form of transmit diversity in which a subset of the transmit antennas that produce the largest output SNR at the receiver are selected for transmission [93]. Since fewer transmit antennas are activated, TAS is able to reduce the number of radio-frequency devices that are needed in the transmitter. The cost is that, in contrast to the space-time codes, TAS requires channel-state information at the transmitter. This information comprises the indices of the transmit antennas that the receiver has selected for maximizing the SNR. Since a single-antenna TAS is able to concentrate the transmit power in one antenna and does not need to distribute the power among all the available antennas, it can outperform the Alamouti code and other STBCs of the same diversity order over a flat-fading channel.

6.9 Channel Codes and Fading

Coherent Systems

Consider a BPSK system with or without direct-sequence spreading, as in Section 6.4. An (n, k) linear block code is used with soft-decision decoding, where n is the number of code symbols and k is the number of information symbols. If the channel symbols are interleaved to a depth beyond the coherence time of the channel, then the symbols fade independently. As a result, a channel code provides a form of time diversity.

Assuming that the fading is constant over a symbol interval, let α_i denote the fading amplitude of symbol i. We assume that the received interference and noise in each diversity branch can be modeled as independent, zero-mean, white Gaussian noise with the same two-sided power spectral density $N_0/2$. Equation (1.52) and the noise analysis of Section 1.2 indicate that the *codeword metric* for codeword c in a BPSK system is

$$U(c) = \sum_{i=1}^{n} \alpha_i x_{ci} y_{ri}, \quad c = 1, 2, \ldots, 2^k. \tag{6.190}$$

where $y_{ri} = \mathrm{Re}\,(y_i)$, $x_{ci} = +1$ or -1, the conditional density function of y_{ri} is

$$f\left(y_{ri} \mid x_{ci}\right) = \frac{1}{\sqrt{\pi N_0}} \exp\left[-\frac{\left(y_{ri} - \alpha_i \sqrt{\mathcal{E}_s} x_{ci}\right)^2}{N_0}\right], \quad i = 1, 2, \ldots, n \tag{6.191}$$

and $\mathcal{E}_s$ is the energy per symbol in the absence of fading. If each of the $\{\alpha_i\}$ is independent with the identical distribution, then the average energy-to-noise-density ratio per binary code symbol is

$$\bar{\gamma}_s = \frac{\mathcal{E}_s}{N_0} E[\alpha_1^2] = \frac{r\mathcal{E}_b}{N_0} E[\alpha_1^2] = r\bar{\gamma} \quad \text{(binary symbols)} \tag{6.192}$$

where $\mathcal{E}_b$ is the information-bit energy, r is the code rate, and $\bar{\gamma}$ is the average bit-energy-to-noise-density ratio.

For a linear block code, the error probabilities may be calculated by assuming that the all-zero codeword denoted by $c = 1$ was transmitted. The two-codeword error probability is equal to the probability that $U(c) - U(1) > 0, c \neq 1$. This probability depends only on the d terms that differ, where d is the weight of codeword c. Thus, d has the same role as L in uncoded MRC, and a derivation closely following that of (6.108) implies that the two-codeword error probability in the presence of Nakagami fading with positive integer m is

$$P_2(d) = \frac{1}{2} - \frac{1}{2}\sqrt{\frac{\bar{\gamma}}{m+\bar{\gamma}}}\sum_{i=0}^{md-1}\frac{\Gamma(i+1/2)m^i}{\sqrt{\pi}\,i!(m+\bar{\gamma})^i} \quad \text{(binary symbols)}. \tag{6.193}$$

For Rayleigh fading,

$$P_2(d) = P_s - (1-2P_s)\sum_{i=1}^{d-1}\binom{2i-1}{i}[P_s(1-P_s)]^i \tag{6.194}$$

where the symbol error probability is

$$P_s = \frac{1}{2}\left(1 - \sqrt{\frac{\bar{\gamma}_s}{1+\bar{\gamma}_s}}\right) \quad \text{(BPSK, QPSK)}. \tag{6.195}$$

The same equations are valid for both BPSK and QPSK because the latter can be transmitted as two independent BPSK waveforms in phase quadrature.

From (6.194), the equality of the right-hand sides of (6.112) and (6.113), and (6.114), we obtain

$$P_2(d) \leq \binom{2d-1}{d}P_s^d. \tag{6.196}$$

As indicated in (1.31), an upper bound on the information-symbol error probability for soft-decision decoding is given by

$$P_{is} \leq \sum_{d=d_m}^{n}\frac{d}{n}A_d P_2(d) \tag{6.197}$$

where A_d denotes the number of codewords with weight d. Substituting (6.196) yields

$$\begin{aligned} P_{is} &\leq \sum_{d=d_m}^{n}\frac{d}{n}A_d\binom{2d-1}{d}P_s^d \\ &\lesssim \binom{2d_m-1}{d_m}\frac{d_m A_{d_m}}{n}P_s^{d_m}, \ \bar{\gamma}_s >> 1 \end{aligned} \tag{6.198}$$

where the final inequality reflects the domination of the first term in the series as $\bar{\gamma}_s \to \infty$. A Taylor series expansion of (6.195) in $\bar{\gamma}_s^{-1}$ and a truncation of the alternating series indicates that

$$P_s \leq \frac{1}{4\bar{\gamma}_s}, \quad \bar{\gamma}_s \geq 1. \tag{6.199}$$

Therefore,

$$P_{is} \lesssim \binom{2d_m - 1}{d_m} \frac{d_m A_{d_m}}{n 4^{d_m}} \bar{\gamma}_s^{-d_m} \tag{6.200}$$

which indicates that a binary block code, coherent BPSK, and maximum-likelihood decoding provide the diversity order $D_o = d_m$.

For coherent BFSK, a derivation closely following that of (6.126) implies that the two-codeword error probability in the presence of independent, identically distributed Nakagami fading of each codeword symbol is

$$P_2(d) = \frac{1}{2} - \frac{1}{2}\sqrt{\frac{\bar{\gamma}_s}{2m + \bar{\gamma}_s}} \sum_{i=0}^{md-1} \frac{\Gamma(i + 1/2)(2m)^i}{\sqrt{\pi}\, i!(2m + \bar{\gamma}_s)^i} \tag{6.201}$$

where $\bar{\gamma}_s$ is given by (6.192), and P_{is} is upper bounded by (6.197). For Rayleigh fading, $P_2(d)$ is again given by (6.194) provided that

$$P_s = \frac{1}{2}\left(1 - \sqrt{\frac{\bar{\gamma}_s}{2 + \bar{\gamma}_s}}\right) \quad \text{(coherent BFSK)} \tag{6.202}$$

which indicates a 3-dB disadvantage relative to BPSK. A Taylor series expansion of this equation in $\bar{\gamma}_s^{-1}$, a truncation of the alternating series, and substitution into (6.198) indicates that the diversity order of BFSK is $D_o = d_m$. At the cost of bandwidth, the performance of coherent q-FSK is similar to that of coherent BPSK if $q = 4$ and superior to coherent BPSK if $q \geq 8$.

Noncoherent Orthogonal FSK Systems

When fast fading makes it impossible to obtain accurate estimates of the $\{\alpha_i\}$ and $\{\theta_i\}$, noncoherent orthogonal FSK is a suitable modulation. The square-law metric has the major advantage that it does not require any channel-state information. For noncoherent BFSK, a derivation closely following that of (6.158) implies that the two-codeword error probability in the presence of independent, identically distributed Rayleigh fading of each codeword symbol is again given by (6.194), provided that

$$P_s = \frac{1}{2 + \bar{\gamma}_s} \quad \text{(noncoherent BFSK)} \tag{6.203}$$

where $\bar{\gamma}_s$ is given by (6.192), and P_{is} is upper bounded by (6.197). The diversity order is $D_o = d_m$.

A comparison of (6.195) and (6.203) indicates that for large values of $r\bar{\gamma}$ and the same block code, BPSK and QPSK have an advantage of approximately 6 dB over noncoherent BFSK in a fading environment. Thus, the fading accentuates the advantage that exists for the AWGN channel. However, BPSK and noncoherent 16-FSK provide approximately the same performance, and noncoherent FSK provides superior performance if $q \geq 32$ at the cost of bandwidth.

For hard-decision decoding, the symbol error probability P_s is given by (6.195) for coherent BPSK, (6.202) for coherent BFSK, (6.203) for noncoherent BFSK, or (6.144) for DPSK. For loosely packed codes, P_{is} is approximated by (1.23), whereas it is approximated by (1.22) for tightly packed codes.

Figure 6.17 illustrates $P_b = P_{is}$ for an extended Golay (24,12) code without diversity combining and P_b for MRC with $L = 1, 4, 5$, and 6 diversity branches and no coding. A Rayleigh-fading channel and BPSK are assumed. The extended Golay (24,12) code is tightly packed with 12 information bits, $r = 1/2$, $d_m = 8$, and $t = 3$. The values of A_d in (6.197) are listed in Table 1.3. The figure indicates the benefits of coding, particularly when the desired P_b is low. At $P_b = 10^{-3}$, the Golay (24,12) code with hard decisions provides an 11-dB advantage in $\bar{\gamma}$ over uncoded BPSK with no diversity (MRC, $L = 1$); with soft decisions, the advantage becomes 16 dB. The advantage of soft-decision decoding relative to hard-decision decoding increases to more than 10 dB at $P_b = 10^{-7}$, a vast gain over the approximately 2-dB advantage of soft-decision decoding for the AWGN channel. At $P_b = 10^{-9}$, the Golay (24,12) code with soft decisions outperforms uncoded MRC with $L = 5$ and nears the performance of uncoded MRC with $L = 6$. However, since $A_{d_m} = A_8 = 759$, the diversity order of the Golay (24,12) code does not reach the theoretical limit $D_o = d_m = 8$ even for $P_b = 10^{-9}$. For noncoherent BFSK, all the graphs in the figure are shifted approximately 6 dB to the right when $P_b \leq 10^{-3}$.

Since the soft-decision decoding of long block codes is usually impractical, convolutional codes are more likely to give a good performance over a fading channel. The metrics are basically the same as they are for block codes with the same modulation, but they are evaluated over path segments that diverge from the correct path through the trellis and then merge with it subsequently (Section 1.3). The linearity of binary convolutional codes ensures that the all-zero path can be assumed to be the correct one when calculating the decoding error probability. Let d denote the Hamming distance of an incorrect path from the correct all-zero path. If perfect symbol interleaving is assumed and BPSK is used, then the probability

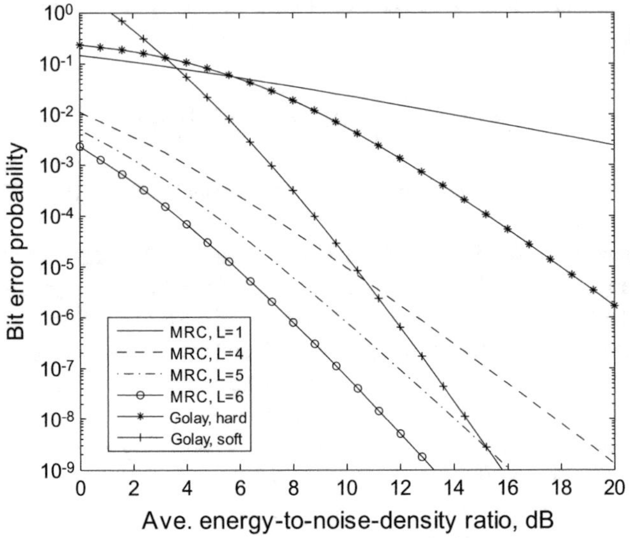

Fig. 6.17 Information-bit error probability for extended Golay (24,12) code with soft and hard decisions, coherent BPSK modulation, and Rayleigh fading, and for maximal-ratio combining with L = 1, 4, 5, and 6

of error in the comparison of two paths with an unmerged segment is $P_2(d)$ given by (6.194). As shown in Section 1.3, the probability of an information-bit error in soft-decision decoding is upper bounded by

$$P_b \leq \frac{1}{k} \sum_{d=d_f}^{\infty} B(d)P_2(d) \tag{6.204}$$

where $B(d)$ is the number of information-bit errors over all paths with unmerged segments at Hamming distance d, k is the number of information bits per trellis branch, and d_f is the minimum free distance, which is the minimum Hamming distance between any two convolutional codewords. This upper bound approaches $B_{d_f}P_2(d_f)/k$ as $P_b \to 0$; thus, the diversity order is $D_o = d_f$. In general, d_f increases with the constraint length of the convolutional code.

If each encoder output bit is repeated n_r times, then the minimum distance of the convolutional code increases to $n_r d_f$ without a change in the constraint length, but at the cost of a bandwidth expansion by the factor n_r. From (6.204), we infer that for the code with repeated bits,

$$P_b \leq \frac{1}{k} \sum_{d=d_f}^{\infty} B(d)P_2(n_r d) \tag{6.205}$$

where $B(d)$ refers to the original code. The diversity order is $D_o = n_r d_f$ if P_b and $B(d_f)/k$ are small.

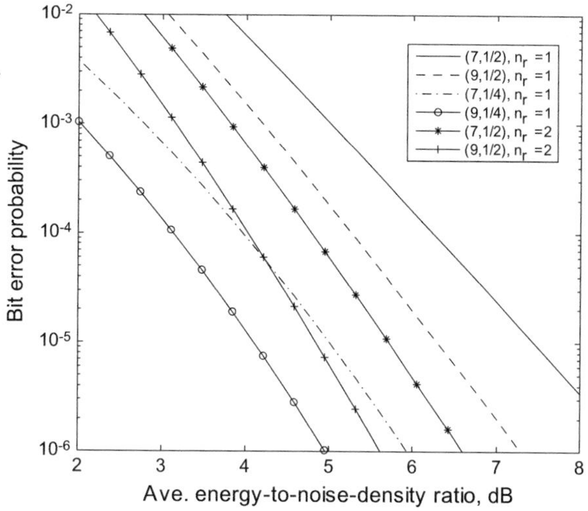

Fig. 6.18 Information-bit error probability for Rayleigh fading, coherent BPSK, and binary convolutional codes with various values of (K, r) and n_r

Figure 6.18 illustrates P_b as a function of $\bar{\gamma}$ for the Rayleigh-fading channel and binary convolutional codes with different values of the constraint length K, the code rate r, and the number of repetitions n_r. Relations (6.205) and (6.194) with $k = 1$ are used, and the $\{B(d)\}$ are taken from the listings for seven terms in Tables 1.4 and 1.5. The figure indicates that an increase in the constraint length provides a much greater performance improvement for the Rayleigh-fading channel than the increase does for the AWGN channel. For a fixed constraint length, the rate-1/4 codes give a better performance than the rate-1/2 codes with $n_r = 2$, which require the same bandwidth but are less complex to implement. The latter two codes require twice the bandwidth of the rate-1/2 code with no repetitions.

Other issues exist for trellis-coded modulation (Section 1.3), which provides a coding gain without a bandwidth expansion. If parallel state transitions occur in the trellis, then $d_f = 1$, which implies that the code provides no diversity protection against fading. Thus, for fading communications, a conventional trellis code with distinct transitions from each state to all other states must be selected. Since fading causes large amplitude variations, multiphase PSK is usually a better choice than multilevel QAM for the symbol modulation. The optimal trellis decoder uses coherent detection and requires an estimate of the channel attenuation.

Turbo, low-density parity-check and serially concatenated codes with iterative decoding based on the *maximum a posteriori* criterion can provide excellent performance in the presence of fading if the system can accommodate the decoding delay and computational complexity. Even without iterative decoding, a serially concatenated code with an outer Reed-Solomon code and an inner binary convolutional code (Section 1.5) can be effective against Rayleigh fading. In the worst case, each output bit error of the inner decoder causes a separate symbol error at the input to the Reed-Solomon decoder. Therefore, an upper bound on P_b is given by (1.148) and (1.147). For coherent BPSK with soft-decision decoding, $P_2(d)$ is

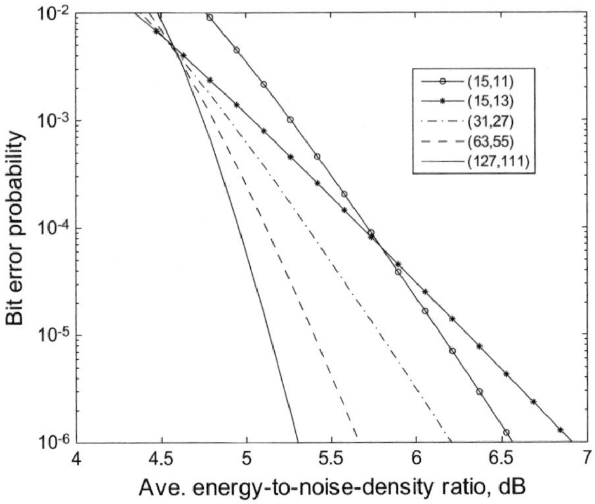

Fig. 6.19 Information-bit error probability for Rayleigh fading, coherent BPSK, soft decisions, and concatenated codes comprising an inner binary convolutional code with K = 7 and $r_1 = 1/2$, and various Reed-Solomon (n, k) outer codes

given by (6.194) and (6.195), and $\bar{\gamma}_s$ is given by (6.192). The concatenated code has a code rate $r = r_1 r_0$, where r_1 is the inner-code rate and r_0 is the outer-code rate.

Figure 6.19 depicts examples of the upper bound on P_b as a function $\bar{\gamma}$ for Rayleigh fading, coherent BPSK, soft decisions, an inner binary convolutional code with $K = 7$, $r_1 = 1/2$, and $k = 1$, and various Reed-Solomon (n, k) outer codes. The required bandwidth is B_u/r, where B_u is the uncoded BPSK bandwidth. Thus, the codes of the figure require a bandwidth less than $3B_u$.

Bit-Interleaved Coded Modulation

The performance of a channel code over a fading channel depends on the minimum Hamming distance, whereas the performance over the AWGN channel depends on the minimum Euclidean distance. For binary modulations, such as BPSK and BFSK, the Euclidean distance increases monotonically with the Hamming distance. For nonbinary modulations, the increase in one of these distances often decreases the other one. BICM, which is described in Section 1.7, increases the minimum Hamming distance, and hence the diversity order, of a code because two trellis paths or codewords tend to have more distinct bits than distinct nonbinary symbols. To compensate for the decrease in the minimum Euclidean distance, BICM with iterative decoding and demodulation (BICM-ID) may be used, as explained in Section 1.7. BICM-ID introduces flexibility into communication systems using nonbinary alphabets over an AWGN channel with a variable level of fading. Since small alphabets are used in BPSK and QPSK modulations, BICM and BICM-ID add little to direct-sequence systems. In contrast, frequency-hopping systems can exploit large alphabets and noncoherent CPFSK modulation; hence, BICM and BICM-ID are often effective. A detailed description, analysis, and simulation of a frequency-hopping system with BICM-ID is presented in Section 9.4.

6.10 Rake Demodulator

To compensate for the effects of fading, spread-spectrum systems exploit the different types of diversity that are available. If the multipath components accompanying a direct-sequence signal are delayed by more than one chip, then the approximate independence of the chips ensures that the multipath interference is suppressed by at least the spreading factor (Section 7.2). However, since multipath signals carry information, they are a potential resource to be exploited rather than merely rejected. A *rake demodulator* provides *path diversity* by coherently combining the resolvable multipath components present during frequency-selective fading, which occurs when the chip rate of the spreading sequence exceeds the coherence bandwidth.

An idealized sketch of the output of a baseband matched filter that is matched to a symbol of duration T_s and receives three multipath components of the signal to which it is matched is shown in Figure 6.20. The duration of the response of a matched filter to a multipath component is on the order of the duration of the mainlobe of the autocorrelation function, which is on the order of the chip duration T_c. Multipath components that produce distinguishable matched-filter output pulses are said to be *resolvable*. Thus, three multipath components are resolvable if their relative delays are greater than T_c, as depicted in the figure. A necessary condition for at least two resolvable multipath components is that T_c is less than the multipath delay spread T_d. Since the signal bandwidth is $W \approx 1/T_c$, (6.63) implies that $W > B_{coh}$ is required. Thus, both frequency-selective fading and resolvable multipath components are associated with wideband signals. There are at most $\lfloor T_d W \rfloor + 1$ resolvable components, where $\lfloor x \rfloor$ denotes the largest integer less than or equal to x.

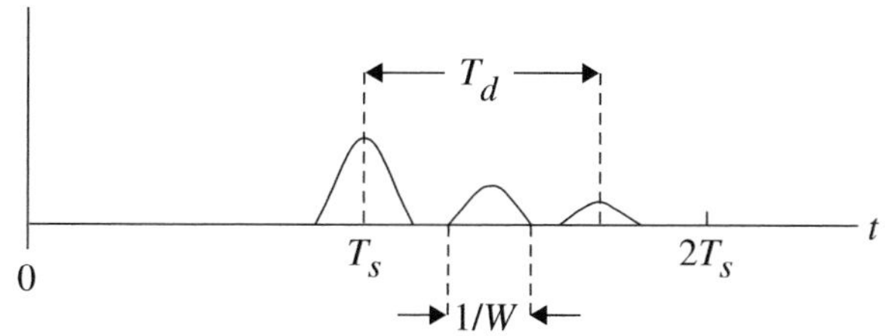

Fig. 6.20 Response of the matched filter to input with three resolvable multipath components

As observed in the figure, the required condition for insignificant intersymbol interference at the sampling times is

$$T_d + T_c < T_s. \tag{6.206}$$

If this condition is not satisfied, then rake demodulation becomes impractical, and some type of equalization is needed in the receiver (Section 6.12).

Consider a multipath channel with frequency-selective fading slow enough that its time variations are negligible over an observation interval. To harness the energy in all the multipath components, a receiver should decide which signal was transmitted among M baseband candidates, $s_1(t)$, $s_2(t)$, …, $s_M(t)$, only after processing all the received multipath components of the signal. If the channel impulse response is time-invariant over the time interval of interest, then for each symbol the receiver selects among the M baseband signals or complex envelopes, which have the form

$$v_k(t) = \sum_{i=1}^{L} C_i s_k(t-\tau_i), \quad k = 1, 2, \ldots, M, \quad 0 \le t \le T_s + T_d \tag{6.207}$$

where L is the number of multipath components, τ_i is the delay of component i, and the channel parameter C_i is the *complex fading amplitude* or *fading coefficient* that represents the attenuation and phase shift of component i.

In the following analysis for the AWGN channel, we assume that (6.206) is satisfied and that the M possible signals are orthogonal to each other. The receiver uses a separate baseband matched filter or correlator for each possible desired signal including its multipath components. Thus, if $s_k(t)$ is the kth symbol waveform, $k = 1, 2, \ldots, M$, then the kth matched filter is matched to the signal $v_k(t)$ in (6.207). Each matched-filter output sampled at $t = T_s + T_d$ provides a symbol metric for soft-decision or hard-decision decoding. The received signal for a symbol can be expressed as

$$r_{1,k}(t) = \text{Re}\left[\sqrt{2\mathcal{E}_s}\, v_k(t) e^{j2\pi f_c t}\right] + n(t), \quad 0 \le t \le T_s + T_d \tag{6.208}$$

where $n(t)$ is the zero-mean, white Gaussian noise with PSD equal to $N_0/2$, $\mathcal{E}_s$ is the signal energy when $v_k(t) = s_k(t)$, and the symbol energy for all the waveforms is

$$\int_0^{T_s} |s_k(t)|^2 dt = 1, \quad k = 1, 2, \ldots, M. \tag{6.209}$$

The orthogonality of symbol waveforms implies that

$$\int_0^{T_s} s_r(t) s_l^*(t) dt = 0, \quad r \ne l. \tag{6.210}$$

We assume that each of the $\{s_l(t)\}$ has a spectrum confined to $|f| < f_c$.

A frequency translation or *downconversion* of $r_{1,k}(t)$ to baseband provides

$$r(t) = r_{1,k}(t)e^{-j2\pi f_c t}. \tag{6.211}$$

Using complex conjugates to eliminate undesired phase shifts, matched-filter k produces the symbol metric

$$U(k) = \text{Re}\left[\sum_{i=1}^{L} C_i^* \int_0^{T_s+T_d} r(t)s_k^*(t-\tau_i)dt\right] \tag{6.212}$$

where the real part is taken to eliminate noise that is orthogonal to the desired signal. A receiver implementation based on this equation would require M delay lines and M matched filters. A practical receiver implementation that requires only a single delay line and M matched filters is derived by changing variables in (6.212) and using the fact that $s_k(t)$ is zero outside the interval $[0, T_s)$. The result is the kth symbol metric

$$U(k) = \text{Re}\left[\sum_{i=1}^{L} C_i^* \int_0^{T_s} r(t+\tau_i)s_k^*(t)dt\right]. \tag{6.213}$$

For frequency-selective fading and resolvable multipath components, a simplifying assumption is that each delay is an integer multiple of $1/W$. Accordingly, L is increased to equal the maximum number of resolvable components, and we set $\tau_i = (i-1)/W$, $i = 1, 2, \ldots, L$, and $(L-1)/W \approx \tau_m$, where τ_m is the maximum delay. As a result, some of the $\{C_i\}$ may be equal to zero. The kth symbol metric becomes

$$U(k) = \text{Re}\left[\sum_{i=1}^{L} C_i^* \int_0^{T_s} r(t+(i-1)/W)s_k^*(\tau)dt\right]. \tag{6.214}$$

A receiver based on these symbol metrics, which is called a *rake demodulator or rake receiver*, is diagrammed in Figure 6.21. The received signal enters a delay line, which may be implemented as a shift register with memory stages (Section 2.2) or as a surface acoustic wave (SAW) delay line (Section 2.8). Since $r(t)$ is designated as the output of the final stage or tap, the sampling occurs at $t = T_s$. Equation (6.214) indicates that the receiver must provide accurate estimates of the complex conjugate of the complex fading amplitude C_i^* and the associated delay $\tau_i \approx (i-1)/W$ for $i = 1, 2, \ldots, L$.

An alternative configuration uses a separate rake demodulator for each of the M symbol metrics. As shown in Figure 6.22, sampled matched-filter outputs of the kth rake demodulator are applied to $L_s \leq L$ parallel *fingers* that separately process each resolvable multipath component with significant power. The finger outputs are recombined to produce one of the M symbol metrics.

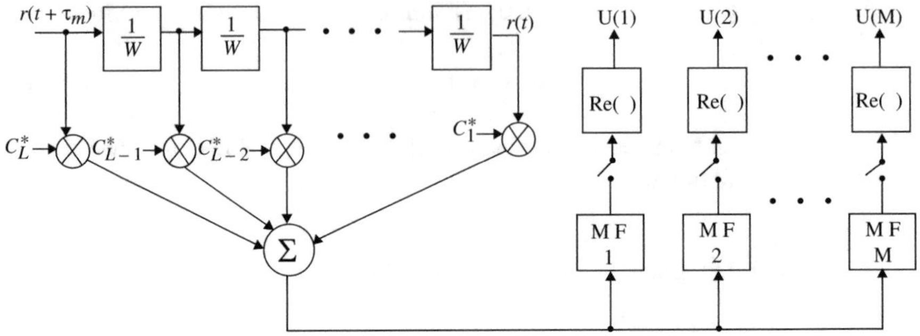

Fig. 6.21 Rake demodulator for M orthogonal pulses

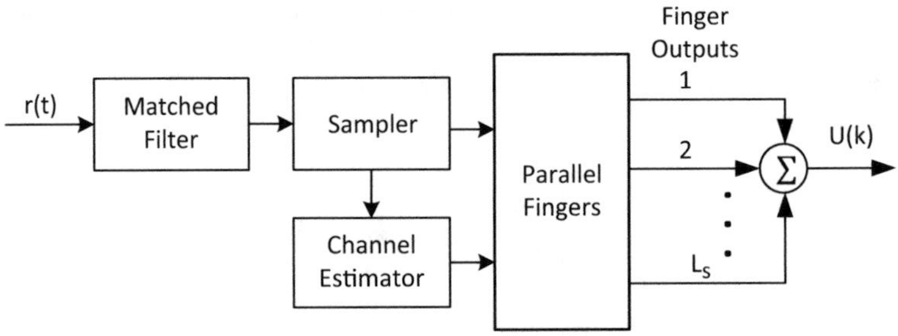

Fig. 6.22 Rake demodulator for generating each of the *M* decision variables or symbol metrics

When direct-sequence signals with chip duration T_c are transmitted, as assumed henceforth, each matched filter is a chip-matched filter. When BPSK is used, $M = 1$, whereas $M \geq 2$ if code-shift keying (Section 2.7) with $t_0 = 0$ is used. The output of the chip-matched filter is sampled at the rate m/T_c, where $m > 1$. A high sampling rate allows different sample subsets to be used in processing different multipath components. Each subset has a timing that most closely approximates the delay of its associated multipath component.

The *channel estimator* must estimate the complex fading amplitudes $\{C_i^*\}$ and delays $\{\tau_i\}$ of the multipath components that are required by the rake demodulator. The joint maximum-likelihood estimation of these amplitudes and delays is computationally prohibitive. Therefore, each significant multipath component must be separately acquired and tracked to provide estimated amplitudes $\left\{\widehat{C}_i^*\right\}$ and delays $\{\widehat{\tau}_i\}$ of L_s multipath components to the corresponding fingers. An unmodulated direct-sequence pilot signal with a distinct spreading sequence is transmitted to facilitate the channel estimation. The channel estimator uses an acquisition correlator similar to that of Section 4.4 to estimate the positions and magnitudes of correlation peaks that indicate the delays and strengths of the multipath components. The parameters of the strongest multipath component are found, and regenerated matched-filter samples due to this component are subtracted

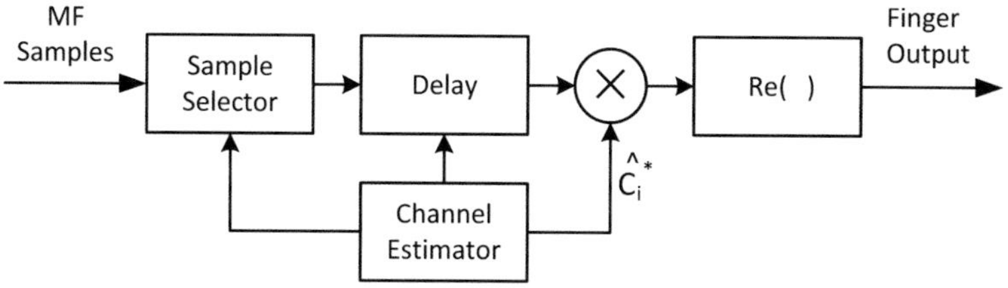

Fig. 6.23 Finger of a rake demodulator

from the input matched-filter samples. Then the parameters of the second strongest component are found, the previous procedure is repeated, and the next strongest components are processed in the order of their relative strengths. This successive estimation of the parameters of each multipath component and the subtractions limit the impact of the interference among the multipath components. Channel estimates must be updated at a rate exceeding the fading rate of (6.48) or (6.49).

The processing within the ith finger is illustrated in Figure 6.23. The matched filter produces a number of output pulses in response to the multipath components, as illustrated in Figure 6.20, and these pulses generally do not overlap. Since the ith pulse is delayed by τ_i relative to the main pulse, the ith finger delays its processing by $T_d - \widehat{\tau}_i$. As a result, all the finger output samples can be aligned in time and constructively combined after weighting the ith set of samples by $\widehat{C}_i^*$. If the delay spread T_d is close to or exceeds the symbol duration, then the rake demodulator must be followed by an equalizer.

The transmission of a pilot signal diverts energy from the modulated direct-sequence signals that carry information. Thus, a pilot signal is particularly useful when it is shared by many users so that the energy allocation to the pilot signal is a minor loss. The downlink from a base station to the mobiles of a cellular network (Section 8.5) is an example of efficient sharing.

Path crosstalk is interference in a rake finger associated with one multipath component caused by a multipath component that is associated with another rake finger. For the path crosstalk to be negligible when $s_k(t)$ is a direct-sequence signal with chip duration $T_c = 1/W$, it is necessary that

$$\int_0^{T_s} s_k(t + i/W) s_k(t) dt << \mathcal{E}_s, \quad i = 1, 2, \ldots, L-1. \tag{6.215}$$

When the data modulation is binary antipodal or BPSK, only a single symbol waveform $s(t)$ and its associated symbol metric U are needed. Let x represent the transmitted bit. Then, the received signal for a single symbol is

$$v(t) = \sum_{i=1}^{L} C_i x s(t - \tau_i), \quad 0 \leq t \leq T_s + T_d. \tag{6.216}$$

We assume negligible path crosstalk and the AWGN channel. Substituting (6.207), (6.208), and (6.211) into (6.214), and then applying (6.209) and (6.215), we obtain the symbol or bit metric:

$$U = \frac{x\mathcal{E}_b}{\sqrt{2}} \sum_{i=1}^{L} \alpha_i^2 + n_r \tag{6.217}$$

where $\mathcal{E}_b = \mathcal{E}_s$ is the average energy per bit, $\alpha_i = |C_i|$, and n_r is a zero-mean Gaussian random variable with variance

$$var\left(n_r^2\right) = \frac{\mathcal{E}_b N_0}{4} \sum_{i=1}^{L} \alpha_i^2 \tag{6.218}$$

which is calculated by the methods described in Section 1.2. Equation (6.217) indicates that the ideal rake demodulator with no path crosstalk produces a symbol metric that is essentially a version of MRC.

If hard decisions are made on the received bits, then the bit decision is that $x = +1$ if $U > 0$. Because of the symmetry, the bit error probability is equal to the conditional bit error probability given that $x = +1$, which implies that a decision error is made if $U < 0$. Since the bit metric has a Gaussian conditional distribution, a standard evaluation yields the conditional bit error probability given the $\{\alpha_i\}$:

$$P_{b|\boldsymbol{\alpha}}(\mathsf{g}) = Q(\sqrt{2\mathsf{g}}) \tag{6.219}$$

$$\mathsf{g} = \sum_{i=1}^{L} \gamma_i \,, \quad \gamma_i = \frac{\mathcal{E}_b}{N_0} \alpha_i^2 . \tag{6.220}$$

For a rake demodulator, each of the $\{\alpha_i\}$ is associated with a different multipath component that fades independently. If each α_i has a Rayleigh distribution, then each γ_i has the exponential density (Appendix E.4)

$$f_{\gamma_i}(x) = \frac{1}{\bar{\gamma}_i} \exp\left(-\frac{x}{\bar{\gamma}_i}\right) u(x), \quad i = 1, 2, \ldots, L \tag{6.221}$$

where the average energy-to-noise-density ratio for a symbol in branch i is

$$\bar{\gamma}_i = \frac{\mathcal{E}_b}{N_0} E[\alpha_i^2], \quad i = 1, 2, \ldots, L. \tag{6.222}$$

If each multipath component fades independently so that each of the $\{\gamma_i\}$ is statistically independent, then g is the sum of independent, exponentially distributed random variables. However, (6.102) cannot be used because the multipath components have distinct amplitudes, and hence $\bar{\gamma}_r \neq \bar{\gamma}_s$ when $r \neq s$.

The Laplace transform of (6.221) is

$$\mathcal{L}_i(s) = \frac{1}{1 + s\bar{\gamma}_i}. \tag{6.223}$$

Since g is the sum of independent random variables, Theorem B3 of Appendix B.2 implies that its Laplace transform is

$$\mathcal{L}_{\mathsf{g}}(s) = \prod_{i=1}^{L} \frac{1}{1 + s\bar{\gamma}_i}. \tag{6.224}$$

Expanding the right-hand side of this equation in a partial-fraction expansion and identifying the inverse Laplace transform of each term, we obtain the density function of g:

$$f_{\mathsf{g}}(x) = \sum_{i=1}^{L} \frac{A_i}{\bar{\gamma}_i} \exp\left(-\frac{x}{\bar{\gamma}_i}\right) u(x) \tag{6.225}$$

where $\bar{\gamma}_r \neq \bar{\gamma}_s$ when $r \neq s$, and

$$A_i = \begin{cases} \prod\limits_{\substack{k=1 \\ k \neq i}}^{L} \frac{\bar{\gamma}_i}{\bar{\gamma}_i - \bar{\gamma}_k}, & L \geq 2 \\ 1, & L = 1. \end{cases} \tag{6.226}$$

Direct integrations using (H.1) of Appendix H.1 and algebra yield the distribution function

$$F_{\mathsf{g}}(x) = 1 - \sum_{i=1}^{L} A_i \exp\left(-\frac{x}{\bar{\gamma}_i}\right), \quad x \geq 0 \tag{6.227}$$

and the moments

$$E[\mathsf{g}^n] = \Gamma(n+1) \sum_{i=1}^{N} A_i \bar{\gamma}_i^n, \quad n \geq 0. \tag{6.228}$$

The bit error probability for Rayleigh fading is determined by averaging the conditional bit error probability $P_{b|\alpha}(\mathsf{g})$ given by (6.219) over the density function given by (6.225). An integration similar to that leading to (6.108) yields

$$P_b(L) = \frac{1}{2} \sum_{i=1}^{L} A_i \left(1 - \sqrt{\frac{\bar{\gamma}_i}{1 + \bar{\gamma}_i}}\right) \quad \text{(Rayleigh, BPSK)}. \tag{6.229}$$

If $\bar{\gamma}_L \approx 0$, $L \geq 2$, then $P_b(L) \approx P_b(L-1)$, which indicates that no great harm is done when a rake demodulator based on MRC includes in its combining an input without a desired-signal component.

The estimation of the channel parameters needed in a rake demodulator becomes more difficult as the fading rate increases. When the estimation errors are large, an option is to use a rake demodulator that avoids channel-parameter estimation by abandoning MRC and using noncoherent combining. The form of this rake demodulator for DPSK or FSK is again depicted in Figures 6.22 and 6.23, except that only the timing information is generated by the channel estimator and the final weighting is eliminated.

Consider the rake demodulator for the noncoherent detection of $M = 2$ orthogonal spreading waveforms with negligible path crosstalk. The two symbol metrics are defined by (6.214). For the AWGN channel and Rayleigh fading, the symbol metrics $U(1)$ and $U(2)$ are given by (6.151) and (6.152). The density function for $U(2)$ is given by (6.154) with $L = M = 2$. However, the density function $f_1(x)$ for $U(1)$ must account for the differing energy levels of the multipath components. A derivation similar to that of (6.225) indicates that

$$f_1(x) = \sum_{i=1}^{L} C_i \exp\left[-\frac{x}{N_0(1+\bar{\gamma}_i)}\right] u(x) \tag{6.230}$$

where

$$C_i = \begin{cases} \prod\limits_{\substack{k=1 \\ k\neq i}}^{L} \frac{1}{\bar{\gamma}_i - \bar{\gamma}_k}, & L \geq 2 \\ \frac{1}{N_0(1+\bar{\gamma}_1)}, & L = 1 \end{cases} \tag{6.231}$$

$\bar{\gamma}_r \neq \bar{\gamma}_s$ when $r \neq s$, and $\bar{\gamma}_i$ is defined by (6.222). If hard decisions are made on the received binary symbols, an erroneous decision is made if $U(2) > U(1)$, and hence the bit error probability is

$$P_b(L) = \sum_{i=1}^{L} C_i \int_0^{\infty} \exp\left[-\frac{x}{N_0(1+\bar{\gamma}_i)}\right] \int_x^{\infty} \frac{y^{L-1}\exp\left(-\frac{y}{N_0}\right)}{(N_0)^L (L-1)!} dydx. \tag{6.232}$$

Using (H.5) and (H.8) of Appendix H.1 to evaluate the inner integral, changing the remaining integration variable, applying (H.1), and simplifying yields the symbol error probability for two orthogonal signals and a rake demodulator with postdetection noncoherent combining:

$$P_b(L) = \sum_{i=1}^{L} B_i \left[1 - \left(\frac{1+\bar{\gamma}_i}{2+\bar{\gamma}_i}\right)^L\right] \quad \text{(Rayleigh, orthogonal)} \tag{6.233}$$

where

$$B_i = \begin{cases} \prod_{\substack{k=1 \\ k \neq i}}^{L} \frac{1+\bar{\gamma}_i}{\bar{\gamma}_i - \bar{\gamma}_k}, & L \geq 2 \\ 1, & L = 1. \end{cases} \tag{6.234}$$

Evaluations of (6.233) and (6.229) indicate that noncoherent combining with two orthogonal signals has a power disadvantage on the order of 6 dB compared with MRC and coherent BPSK.

For dual rake combining with $L = 2$ and two orthogonal signals, (6.233) reduces to

$$P_b(2) = \frac{8 + 5\bar{\gamma}_1 + 5\bar{\gamma}_2 + 3\bar{\gamma}_1\bar{\gamma}_2}{(2+\bar{\gamma}_1)^2(2+\bar{\gamma}_2)^2}. \tag{6.235}$$

If $\bar{\gamma}_2 = 0$, then

$$P_b(2) = \frac{2 + \frac{5}{4}\bar{\gamma}_1}{(2+\bar{\gamma}_1)^2} \geq \frac{1}{2+\bar{\gamma}_1} = P_b(1). \tag{6.236}$$

This result illustrates the performance degradation that results when a rake combiner has an input that provides no desired-signal component. In the absence of a desired-signal component, this input contributes only noise to the combiner. For large values of $\bar{\gamma}_1$, the extraneous noise causes a loss of almost 1 dB. As previously observed, this loss does not occur when MRC and coherent BPSK are used.

The processing of a multipath component requires channel estimation. When a practical channel estimator is used, only a few components are likely to have a signal-to-interference ratio (SIR) that is high enough to be useful in the rake combining. Typically, three significant multipath components are available in mobile networks. To assess the potential performance of the rake demodulator, we assume that the principal multipath component has average energy-to-noise-density ratio $\bar{\gamma}_1 = \mathcal{E}_b\overline{\alpha_1^2}/N_0$ and that $L = 4$ components are received and processed. The three minor multipath components have relative energy-to-noise-density ratios specified by the *multipath intensity vector*

$$\mathbf{M} = \left(\frac{\bar{\gamma}_2}{\bar{\gamma}_1}, \frac{\bar{\gamma}_3}{\bar{\gamma}_1}, \frac{\bar{\gamma}_4}{\bar{\gamma}_1}\right) = \left(\frac{\overline{\alpha_2^2}}{\overline{\alpha_1^2}}, \frac{\overline{\alpha_3^2}}{\overline{\alpha_1^2}}, \frac{\overline{\alpha_4^2}}{\overline{\alpha_1^2}}\right) \tag{6.237}$$

where $\overline{\alpha_i^2} = E\left[\alpha_i^2\right]$.

Figure 6.24 plots the bit error probability $P_b(4)$ for coherent BPSK, an ideal rake demodulator, and the AWGN channel as a function of $\bar{\gamma}_1$, which is given by (6.229). The vector $\mathbf{M} = (1, 0, 0)$ represents the hypothetical environment in which a single additional multipath component has the same power as the main component. Expressing the components in decibels, $\mathbf{M} = (-4, -8, -12)$ dB

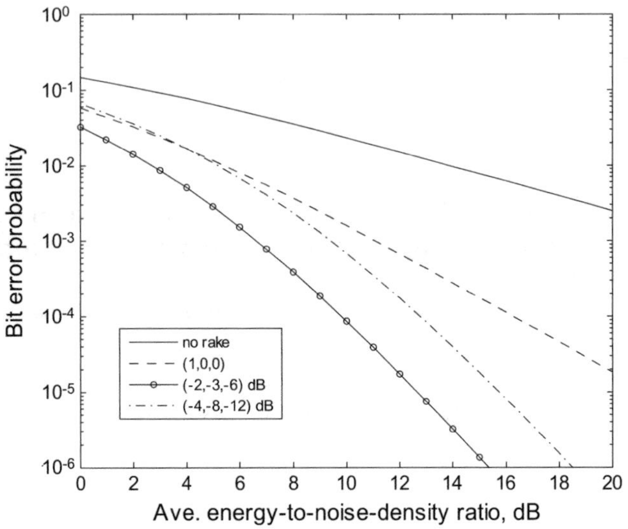

Fig. 6.24 Bit error probability for coherent BPSK and rake demodulators with $L = 4$ multipath components and different multipath intensity vectors

represents the minor multipath intensities typical of a rural environment, and $\mathbf{M} = (-2, -3, -6)$ dB represents a typical urban environment. The figure indicates that despite 2.1 dB less power in the minor components, the rural environment generally provides a lower symbol error probability than the hypothetical one. The superior performance in the urban environment relative to the rural environment is primarily due to its 3.5 dB of additional power in the minor multipath components.

This figure and other numerical data establish two basic features of single-carrier direct-sequence systems with ideal rake demodulators that experience negligible path crosstalk.

1. System performance improves as the total energy in the minor multipath components increases. The underlying reason is that the rake demodulator of the single-carrier system harnesses energy that would otherwise be unavailable.
2. When the total energy in the minor multipath components is fixed, the system performance improves as the number of resolved multipath components L increases and as the energy becomes uniformly distributed among these components.

An increase in the number of resolved components L is potentially beneficial if it is caused by natural changes in the physical environment that generate additional multipath components. However, an increase in L due to an increase in the bandwidth W is not always beneficial [34]. Although new components provide additional diversity and may exhibit the more favorable Ricean fading rather than Rayleigh fading, the average power per multipath component decreases because some composite components fragment into more numerous but weaker components.

Hence, the estimation of the channel parameters becomes more difficult, and the fading of some multipath components may be highly correlated rather than independent.

The number of fingers in an ideal rake demodulator equals the number of significant resolvable multipath components, which is constantly changing in a mobile communications receiver. Rather than attempting to implement all the fingers that may sometimes be desirable, a more practical alternative is to implement a fixed number of fingers independent of the number of multipath components. *Generalized selection diversity* entails selecting the L_c strongest resolvable components among the L resolvable ones and then applying MRC or noncoherent combining of these L_c components, thereby discarding the $L - L_c$ components with the lowest SNRs. Analysis [85] indicates that diminishing returns are obtained as L_c increases, but for a fixed value of L_c, the performance improves as L increases, provided that the strongest components can be isolated.

If an adaptive array produces a directional beam to reject interference or enhance the desired signal (Section 7.10), it also reduces the delay spread of the significant multipath components of the desired signal, because components arriving from angles outside the beam are greatly attenuated. As a result, the potential benefit of a rake demodulator diminishes. Another procedure is to assign a separate set of adaptive weights to each significant multipath component. Consequently, the adaptive array can form separate array patterns, each of which enhances a particular multipath component while nulling other components. The set of enhanced components is then applied to the rake demodulator [95].

6.11 Frequency Hopping and Diversity

Rake demodulators are not useful in frequency-hopping systems because of the relatively narrow bandwidth of the frequency channels and the required readjustment to a new channel impulse response each time the carrier frequency hops. Frequency-hopping systems exploit diversity through the inherent frequency-selective fading ensured by the periodic frequency changes. To demonstrate the effectiveness of frequency hopping in compensating for fading, consider a transmitted signal that has two frequency hops per code symbol and undergoes independent Nakagami fading with the same parameter values during the two consecutive dwell intervals. Let g_1 and g_2 denote random variables equal to the power gains at the receiver during the first and second hop dwell intervals, respectively. From (6.36), it follows that these power gains have gamma densities given by

$$f_{g_i}(x) = \frac{1}{\Gamma(m)} \left(\frac{m}{\Omega}\right)^m x^{m-1} \exp\left(-\frac{m}{\Omega}x\right) u(x), \; m \geq \frac{1}{2}, \; i = 1, 2 \tag{6.238}$$

where the average power gain during each dwell interval is $E[g_i] = \Omega$. The average received power gain of the code symbol over the two dwell intervals is

$g_s = (g_1 + g_2)/2$. If the fading is independent during each of the dwell intervals, then the results of Appendix E.5 indicate that the density function of g_s is

$$f_{g_s}(x) = \frac{1}{\Gamma(2m)}\left(\frac{2m}{\Omega}\right)^{2m} x^{2m-1} \exp\left(-\frac{2m}{\Omega}x\right) u(x), \ m \geq \frac{1}{2} \tag{6.239}$$

which is a gamma density with $E[g_s] = \Omega$ but a different variance. Equation (6.37) implies that

$$\frac{var(g_s)}{var(g_i)} = \frac{1}{2}, \ m \geq \frac{1}{2} \tag{6.240}$$

which indicates that the power variation due to the fading is reduced by the frequency hopping.

Suppose that the coherence bandwidth for a received interference signal is large enough that it undergoes flat Nakagami fading with average power gain Ω_1 and parameter m_1 during the two dwell intervals. Let h denote the average interference power gain over the two intervals. Its density function is

$$f_h(x) = \frac{1}{\Gamma(2m_1)}\left(\frac{2m_1}{\Omega_1}\right)^{2m_1} x^{2m_1-1} \exp\left(-\frac{2m_1}{\Omega_1}x\right) u(x), \ m_1 \geq \frac{1}{2} \tag{6.241}$$

and $E[h] = \Omega_1$. If the desired signal and interference are independent, then a straightforward integration indicates that the average SIR is

$$E\left[\frac{g_s}{h}\right] = \Omega E\left[h^{-1}\right] = \frac{2m_1\Omega}{(2m_1 - 1)\Omega_1}, \ m_1 > \frac{1}{2}. \tag{6.242}$$

Since the average SIR with no fading is Ω/Ω_1, (6.242) indicates that in this example the frequency hopping increases the average SIR when the desired signal experiences independent frequency-selective fading, but the interference signal does not.

Interleaving of the code symbols over many dwell intervals provides a high level of diversity to frequency-hopping systems operating over a frequency-selective fading channel. Let F_s denote the minimum separation between adjacent carrier frequencies in a hopset. A necessary condition for nearly independent symbol errors is

$$F_s \gtrsim \max(B, B_{coh}) \tag{6.243}$$

where B_{coh} is the coherence bandwidth of the fading channel and B is the bandwidth of a frequency channel. If (6.243) is not satisfied, there is a performance loss due to the correlated symbol errors. If $B < B_{coh}$, equalization is not necessary because the fading is flat over each frequency channel. If $B > B_{coh}$, either equalization may be used to prevent intersymbol interference or a multicarrier modulation may be combined with the frequency hopping. Frequency-hopping systems usually do not exploit the Doppler spread of the channel because any additional diversity due to time-selective fading is insignificant.

Let n denote the number of code symbols that are interleaved, and let M denote the number of frequency channels in the hopset. For each of these symbols to fade independently with a high probability, n distinct frequency channels and $n \leq M$ are necessary. Let $T_{\mathrm{del}} \geq (n-1)T_h + T_s$ denote the maximum tolerable processing delay. Combining these inequalities, we find that

$$n \leq \min\left(M,\ 1 + \frac{T_{\mathrm{del}} - T_s}{T_h}\right) \tag{6.244}$$

is required. If this inequality is not satisfied, then some performance degradation results.

6.12 Multicarrier Direct-Sequence Systems

When the data rate is sufficiently high that (6.206) is not satisfied, then the rake demodulator is impractical and a single-carrier system requires channel equalization to cope with the intersymbol interference. To minimize the intersymbol interference, information can be transmitted over multiple subchannels of the allocated channel bandwidth. A *multicarrier direct-sequence system* partitions the bandwidth into subchannels, each of which is used by a separate direct-sequence signal with a distinct subcarrier frequency. Whereas a single-carrier system provides diversity by using a rake demodulator that combines several multipath signals, a multicarrier system provides diversity or multiplexing by processing parallel correlator outputs, each of which is associated with a different subcarrier. A multicarrier system does not need a rake demodulator and has the potential ability to avoid transmissions in subchannels with strong interference or where the multicarrier signal might interfere with other signals. This feature has a counterpart in frequency-hopping systems.

A typical multicarrier system divides the signal power among L *subcarriers*, each with a frequency $f_k = k/T_c$, where k is an integer and $k \geq 1$. The subcarrier signals have a sufficiently small bandwidth that they do not experience any significant frequency-selective fading if T_c exceeds the delay spread. If the chip waveforms are rectangular, then these subcarrier frequencies ensure L orthogonal subcarrier signals, which can be verified by a calculation similar to that leading to (3.67). Although the orthogonality prevents *self-interference* among the subcarrier signals, its effectiveness is reduced by multipath components and Doppler shifts. An alternative is to use bandlimited subcarrier signals to minimize the self-interference without requiring orthogonality.

One type of multicarrier direct-sequence system partitions the data symbols among the subcarriers, as illustrated in Figure 6.25. The transmitter uses the serial-to-parallel (S/P) converter to convert a stream of code or data symbols $d(t)$ into

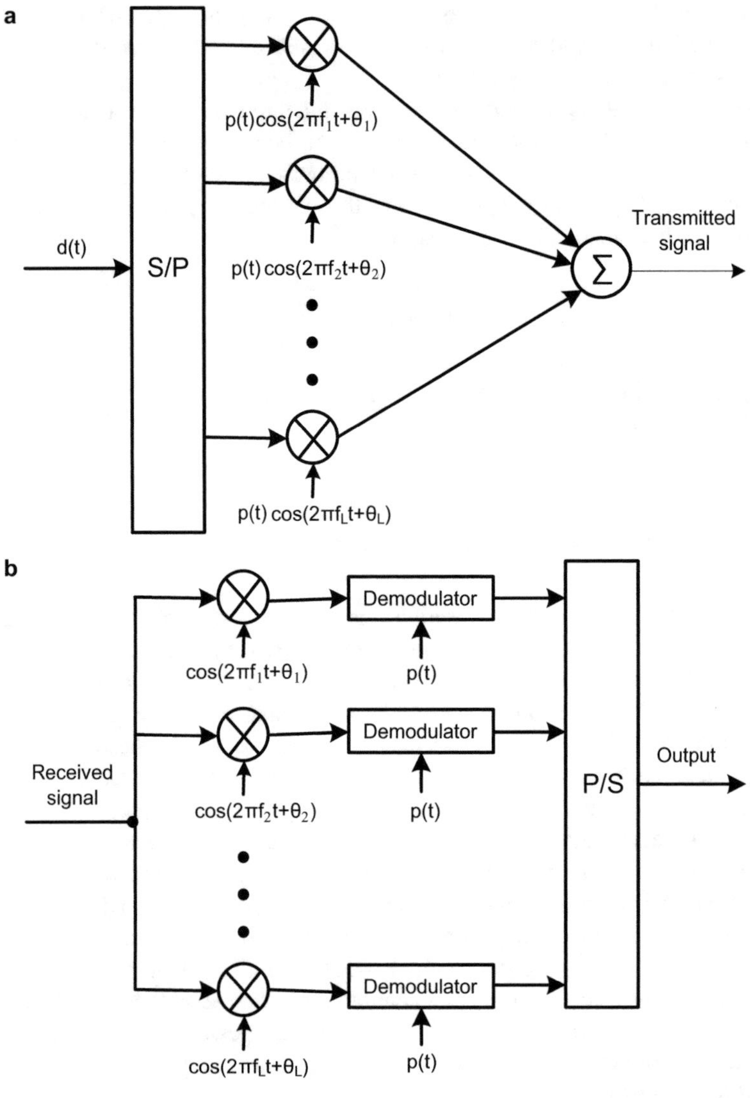

Fig. 6.25 Multicarrier direct-sequence system for data multiplexing: (**a**) transmitter and (**b**) receiver

L parallel substreams of different data symbols, each of which is multiplied by the spreading waveform $p(t)$ and one of the subcarriers. Both the chip rate and the data rate for each subchannel are reduced by the same factor L; hence, the spreading factor provided by each subcarrier remains unchanged relative to the single-carrier system with the same total bandwidth. The reduced data-symbol rate in each subchannel reduces the intersymbol interference in the subchannel. Each demodulator in the receiver uses the despreading to suppress interference in the

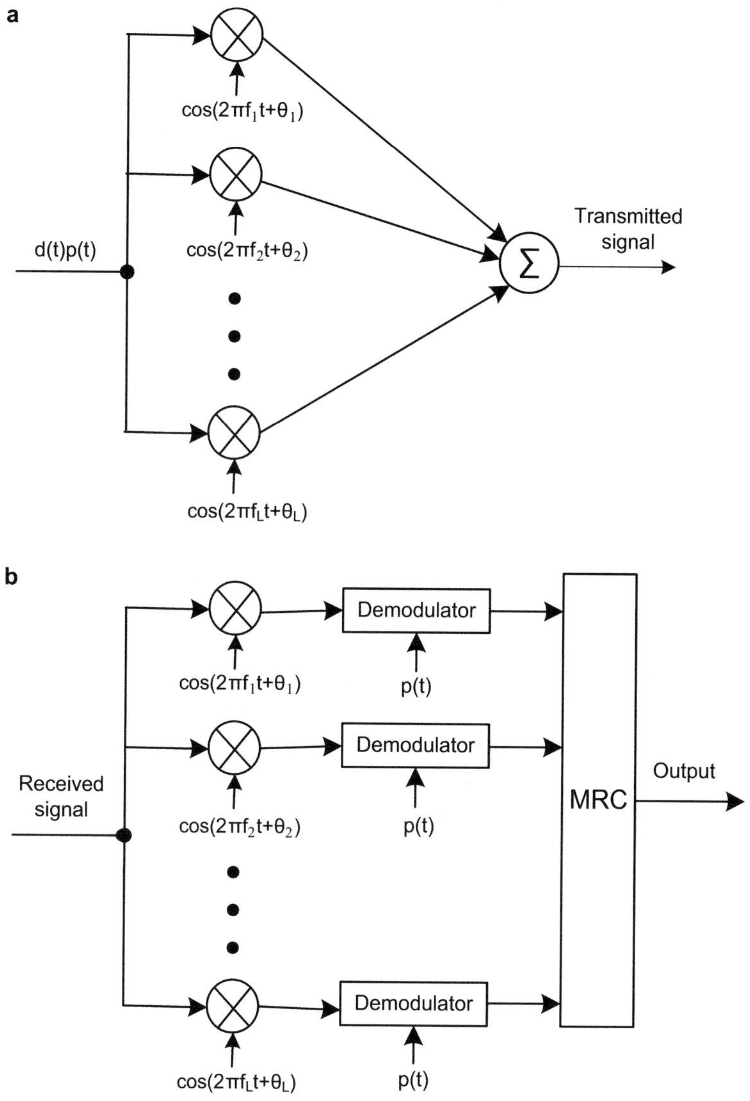

Fig. 6.26 Multicarrier direct-sequence system for frequency diversity: (**a**) transmitter and (**b**) receiver

spectral vicinity of its subcarrier. The parallel-to-serial (P/S) converter restores the stream of data symbols. The cost of this efficient multiplexing with low intersymbol interference is the large amount of hardware and the high peak-to-average power ratio for the transmitted signal.

Another type of multicarrier direct-sequence system provides frequency diversity instead of high throughput. The transmitter has the form of Figure 6.26 (a), and the

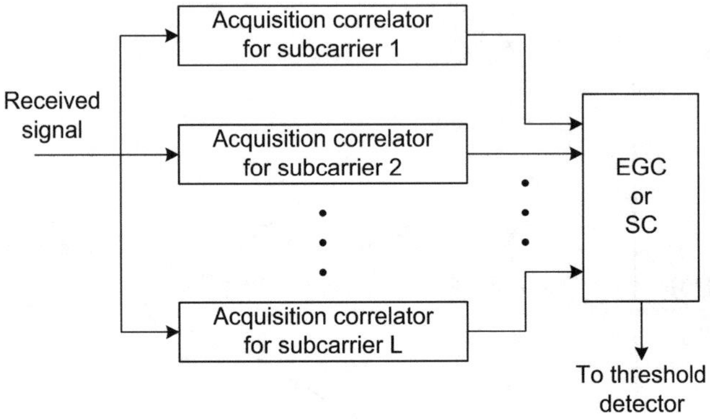

Fig. 6.27 Code acquisition system for multicarrier direct-sequence system. *EGC* equal-gain combining, *SC* selection combining

product $d(t)p(t)$ simultaneously modulates L subcarriers. The chip rate and hence the spreading factor for each subcarrier of this system are reduced by the factor L relative to a single-carrier direct-sequence system. The receiver has L parallel demodulators, one for each subcarrier, and has the form of Figure 6.26 (b). Each demodulator provides despreading to suppress interference, and the demodulator outputs provide the inputs to a maximal-ratio combiner (MRC). With appropriate feedback, the transmitter can omit a subcarrier associated with an interfered subchannel and redistribute the saved power among the remaining subcarriers.

Figure 6.27 diagrams the code acquisition system for a multicarrier direct-sequence system. During acquisition, each subcarrier signal carries the same acquisition sequence. In each branch, the received subcarrier signal is downconverted and then applied to an acquisition correlator (Section 4.4). The outputs of all the correlators are jointly processed by either an equal-gain combiner (EGC) or a selection combiner (SC) to produce the decision variable applied to a threshold detector. The output of the threshold detector indicates when acquisition of the spreading sequence has been achieved. An analysis [121] indicates that the acquisition performance of this system is superior to that of a single-carrier direct-sequence system with an identical bandwidth.

6.13 Multicarrier CDMA Systems

Multicarrier systems that transmit multiple continuous-time waveforms using subcarriers have extravagant hardware requirements. A much more practical multicarrier system, which is called a *multicarrier code-division multiple-access CDMA* (MC-CDMA) system, adapts the efficient digital implementation of orthogonal frequency-division multiplexing (OFDM), but differs from OFDM [7, 29] in that substantial frequency diversity is provided. In an MC-CDMA system, each data

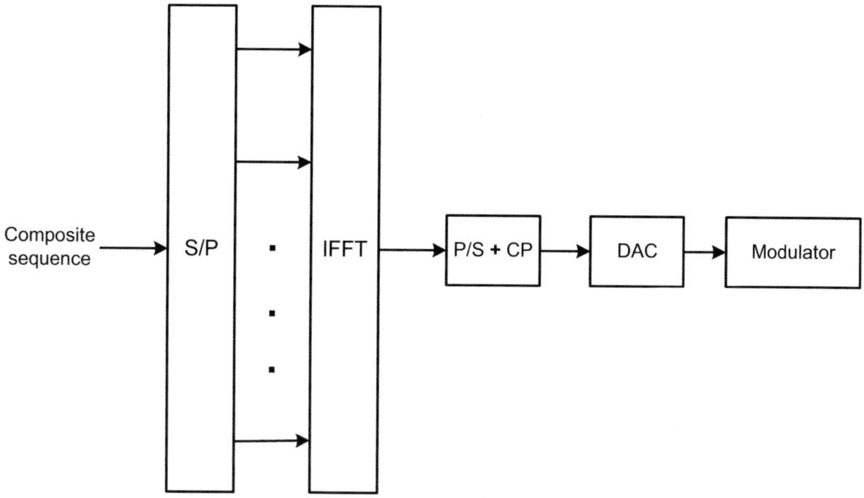

Fig. 6.28 Principal components of the transmitter of an MC-CDMA system. *CP* cyclic prefix, *DAC* digital-to-analog converter

symbol is modulated by a spreading sequence of G chips that is converted into G parallel data-modulated chip samples. Each of these G samples modulates a different sampled subcarrier so that the spreading occurs in the frequency domain. This system is considered a CDMA system because it can use multiple spreading sequences to mitigate multiple-access interference among multiple users of the same transmitter and receiver or among multiple symbols of a single user. More general *CDMA* systems that accommodate multiple-access interference from disparate sources are presented in Chapter 7.

The principal components of the transmitter of an MC-CDMA system are depicted in Figure 6.28. Each of the N data symbols are modulated by a separate orthogonal spreading sequence with spreading factor G. Users requiring higher data rates may use more than one data symbol and spreading sequence. We consider a synchronous MC-CDMA system for downlink communications over the AWGN channel. The data symbols and the spreading-sequence chips of all users are all synchronized in time.

Consider a *block* of N simultaneously transmitted data symbols. The input to the S/P converter is the composite sequence

$$s_i = \sum_{n=0}^{N-1} d_n p_{n,i}, \quad i = 0, 1, \ldots, G-1 \tag{6.245}$$

where d_n is proportional to the nth binary data symbol, and $p_{n,i}$ is sample i of the G samples of the spreading-sequence of symbol n. The data symbols and spreading sequences take the values

$$d_n = \pm 1,\ p_{n,i} = \pm 1, \quad i = 0, 1, \ldots, G-1,\ n = 0, 1, \ldots, N-1. \tag{6.246}$$

The vector $\mathbf{p}_n$, which represents the spreading sequence of symbol n, and the vector $\mathbf{d}$, which represents the data symbols, are

$$\mathbf{p}_n = [p_{n,0}\ p_{n,1}\ \ldots\ p_{n,G-1}\ ]^T, \quad \mathbf{d} = [d_0\ d_1\ \ldots\ d_{N-1}]^T. \tag{6.247}$$

After the composite sequence is received by the S/P converter, its ith output is given by (6.245), and its duration is equal to the symbol duration T_s. The G parallel outputs of the S/P converter may be represented by the vector

$$\mathbf{Pd} = \sum_{n=0}^{N-1} \mathbf{p}_n d_n \tag{6.248}$$

where column n of the $G \times G$ matrix $\mathbf{P}$ is $\mathbf{p}_n$. Orthogonality of the spreading sequences and (6.246) imply that

$$\mathbf{P}^T\mathbf{P} = G\mathbf{I} \tag{6.249}$$

where $\mathbf{I}$ is the $G \times G$ identity matrix.

A G-point *discrete Fourier transform* of a $G\times 1$ vector $\mathbf{x}$ of discrete-time samples is $\mathbf{Fx}$, where $\mathbf{F}$ is the $G \times G$ matrix

$$\mathbf{F} = \frac{1}{\sqrt{G}} \begin{bmatrix} 1 & 1 & 1 & \cdots & 1 \\ 1 & V & V^2 & \cdots & V^{G-1} \\ \vdots & \vdots & \vdots & \vdots & \vdots \\ 1 & V^{G-1} & V^{2(G-1)} & \cdots & V^{(G-1)^2} \end{bmatrix}, \quad V = \exp(-j2\pi/G) \tag{6.250}$$

and $j = \sqrt{-1}$. If T is the time between samples, then component i of $\mathbf{Fx}$ approximates the continuous-time Fourier transform at frequency i/T of the continuous-time signal from which $\mathbf{x}$ is obtained. An evaluation using the sums of finite geometric series and the periodicity of the complex exponentials verifies that

$$\mathbf{F}^H\mathbf{F} = \mathbf{F}\mathbf{F}^H = \mathbf{I} \tag{6.251}$$

which indicates that $\mathbf{F}$ is a unitary matrix, and hence $\mathbf{F}^{-1} = \mathbf{F}^H$. Thus, $\mathbf{F}^H$ represents the G-point *inverse discrete Fourier transform.*

In the MC-CDMA transmitter, the G S/P converter outputs are applied to an *inverse fast Fourier transformer* (IFFT), which implements an inverse discrete Fourier transform. The G parallel outputs of the IFFT are represented as components of the vector $\mathbf{x} = [x_{G-1}x_{G-2}\ \ldots x_0]^T$, which transforms $\mathbf{Pd}$ according to

$$\mathbf{x} = \mathbf{F}^H\mathbf{Pd} = \sum_{n=0}^{N-1} \left(\mathbf{F}^H\mathbf{P}\right)_{:,n} d_n \tag{6.252}$$

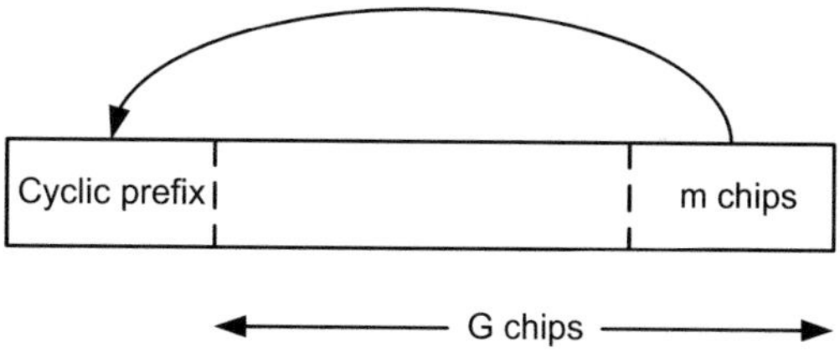

Fig. 6.29 Appending the cyclic prefix prior to the data stream

where $\mathbf{B}_{:,n}$ denotes the nth column of $\mathbf{B}$, and the sum is over the columns of $\mathbf{F}^H\mathbf{P}$. The IFFT transforms each component of $\mathbf{Pd}$ as if it were associated with a distinct subcarrier frequency. The P/S converter converts the components of $\mathbf{x}$ into a serial stream at the original chip rate $1/T_c$, where $T_c = T_s/G$ is the chip duration of the composite sequence. The vector $\mathbf{x}$ represents one block of data, and successive blocks are transmitted.

A *guard interval* of duration mT_c is inserted between blocks to prevent *intersymbol interference* between symbols in adjacent blocks if the multipath delay spread is less than mT_c. The guard interval is implemented by appending an m-sample *cyclic prefix* with $m \leq G-1$ prior to the samples of each block, as illustrated in Figure 6.29. After the insertion of the cyclic prefix, the resultant sequence with $m+G$ chips associated with one set of N aligned symbols is $\overline{x}_{-m}, \overline{x}_{-m+1}, \ldots, \overline{x}_{G-1}$, where

$$\overline{x}_i = x_k, \;\; k = i \; modulo\text{-}G, \;\; -m \leq i \leq -1$$
$$\overline{x}_i = x_i, \;\; 0 \leq i \leq G-1. \tag{6.253}$$

This sequence is applied to a digital-to-analog converter (DAC) and then an upconverter for transmission, as shown in Figure 6.28. The transmitted signal uses a single carrier frequency and in-phase and quadrature components.

Since appending the cyclic prefix causes each symbol to be associated with $m+G$ transmitted chips, the transmitted energy per chip is reduced by the *prefix factor*

$$c = \frac{G}{m+G} \tag{6.254}$$

which indicates that $G >> m$ is needed in a practical system.

The channel impulse response is assumed to have the form of (6.67). For symbol d_k, the impulse response due to the transmitter, channel, and receiver is assumed to be

$$h(\tau) = A_k \sum_{i=0}^{m} h_i \delta(\tau - iT_c - \tau_{\min}) \tag{6.255}$$

where $\tau_{\min}$ is the minimum signal delay, and mT_c is the *multipath delay spread* or duration of the channel impulse response. The real-valued factor $A_k \geq 0$ is not necessarily the same for all symbols because they may be transmitted with unequal power levels. Some of the coefficients $\{h_i\}$ may be zero, depending on how many significant multipath components exist. Let

$$\mathbf{h} = [h_0 \dots h_m \dots 0]^T \tag{6.256}$$

denote the $G \times 1$ vector of impulse-response coefficients, at least $G - m - 1$ of which are zero. The coefficients are normalized so that

$$\| \mathbf{h} \|^2 = 1. \tag{6.257}$$

Because of this normalization and the cyclic prefix, the received energy in all the multipath components due to symbol d_k is $\mathcal{E}_{sk} = (m + G)\, A_k^2$; hence

$$A_k^2 = \frac{c\mathcal{E}_{sk}}{G} \tag{6.258}$$

where c is the cyclic factor defined by (6.254).

The principal components of the MC-CDMA receiver are diagrammed in Figure 6.30. We define the $G \times 1$ vector

$$\begin{aligned}\widetilde{\mathbf{x}} &= \sum_{n=0}^{N-1} \left(\mathbf{F}^H\mathbf{P}\right)_{:,n} A_n d_n \\ &= \mathbf{F}^H\mathbf{P}\mathbf{A}\mathbf{d}\end{aligned} \tag{6.259}$$

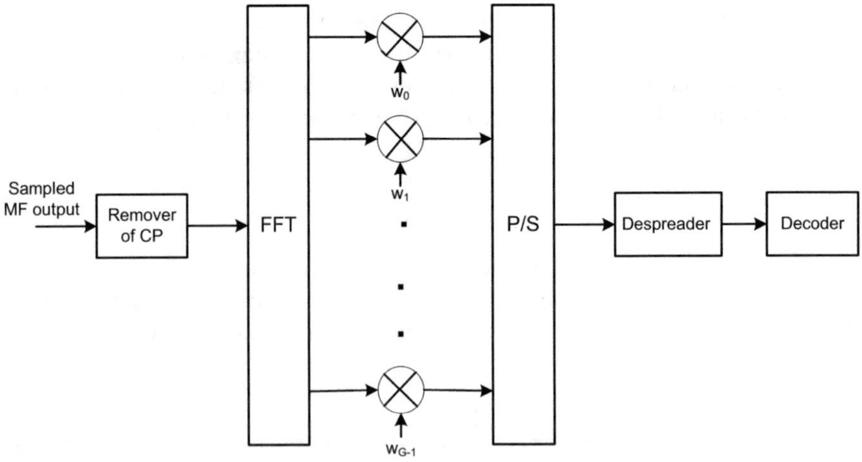

Fig. 6.30 Principal components of the MC-CDMA receiver. *CP* cyclic prefix, *MF* matched filter

where the sum is over the columns of $\mathbf{F}^H\mathbf{P}$, and $\mathbf{A}$ is the diagonal matrix with A_n as its nth diagonal element. We define the $(G+m)\times 1$ vector $\widehat{\mathbf{x}} = [\widehat{x}_{G-1}\ \widehat{x}_{G-2}\ \ldots\ \widehat{x}_{-m}]^T$ with components

$$\widehat{x}_i = \widetilde{x}_k,\ \ k = i\ modulo\text{-}G,\ -m \le i \le -1.$$
$$\widehat{x}_i = \widetilde{x}_i,\ \ 0 \le i \le G-1. \tag{6.260}$$

We assume the ideal coherent demodulation of a pulse amplitude modulation signal described in Section 1.2. The channel impulse response (6.255) implies that the sampled complex-valued matched-filter output is

$$y_i = \sum_{n=0}^{N-1} d_n \sum_{l=0}^{m} h_l \widehat{x}_{i-l,n} + \overline{n}_i, \quad -m \le i \le G-1 \tag{6.261}$$

where $\overline{n}_i$ is sampled noise. The m-sample cyclic prefix of the matched-filter output samples is discarded because these samples are corrupted by the previous block. The remaining G samples of a block are represented by the G-dimensional received vector

$$\overline{\mathbf{y}} = [y_{G-1}\ y_{G-2}\ \ldots\ y_0]^T = \mathbf{H}_1 + \overline{\mathbf{n}} \tag{6.262}$$

where $\mathbf{H}_1$ is the $G\times(G+m)$ matrix

$$\mathbf{H}_1 = \begin{bmatrix} h_0 & h_1 & \cdots & h_m & 0 & \cdots & 0 \\ 0 & h_0 & \cdots & h_{m-1} & h_m & \ddots & 0 \\ \vdots & \vdots & \vdots & \vdots & \vdots & \vdots & \vdots \\ 0 & \cdots & 0 & h_0 & \cdots & h_{m-1} & h_m \end{bmatrix} \tag{6.263}$$

and $\overline{\mathbf{n}}$ is the G-dimensional vector of noise samples.

Since the final m components of $\widehat{\mathbf{x}}$ constitute the cyclic prefix, which is derived from $\widetilde{\mathbf{x}}$ as indicated by (6.260), we find that the received vector may be represented by

$$\overline{\mathbf{y}} = \mathbf{H}\widetilde{\mathbf{x}} + \overline{\mathbf{n}} \tag{6.264}$$

where $\mathbf{H}$ is the $G\times G$ matrix

$$\mathbf{H} = \begin{bmatrix} h_0 & h_1 & \cdots & h_m & 0 & \cdots & 0 \\ 0 & h_0 & \cdots & h_{m-1} & h_m & \cdots & 0 \\ \vdots & \vdots & \vdots & \vdots & \vdots & \vdots & \vdots \\ 0 & \cdots & 0 & h_0 & \cdots & h_{m-1} & h_m \\ \vdots & \vdots & \vdots & \vdots & \vdots & \vdots & \vdots \\ h_2 & h_3 & \cdots & h_{m-2} & \cdots & h_0 & h_1 \\ h_1 & h_2 & \cdots & h_{m-1} & \cdots & 0 & h_0 \end{bmatrix}. \tag{6.265}$$

This matrix is a *circulant* matrix, which is a matrix in which each row is obtained from the previous one by circularly shifting the latter to the right by one element. The form of $\mathbf{H}$ indicates that although the cyclic prefix has been removed, it affects $\mathbf{H}$, and hence influences the received vector $\overline{\mathbf{y}}$. As shown in Section 1.2 for coherent demodulation of the received signal and AWGN, $\overline{\mathbf{n}}$ is a zero-mean, circularly symmetric, Gaussian random vector with

$$E\left[\overline{\mathbf{n}}\overline{\mathbf{n}}^H\right] = N_0\mathbf{I},\ E\left[\overline{\mathbf{n}}\overline{\mathbf{n}}^T\right] = \mathbf{0} \tag{6.266}$$

where $N_0/2$ is the two-sided noise PSD.

Each column of $\mathbf{F}^H$ has the form

$$\mathbf{f}_i = \frac{1}{\sqrt{G}}[1\ \ V^{-i}\ \ V^{-2i}\ \ \cdots\ \ V^{-(G-1)i}]^T,\ \ i = 0, 1, \cdots, G-1. \tag{6.267}$$

The evaluation of $\mathbf{Hf}_i$ using the fact that $V^G = 1$ proves that

$$\mathbf{Hf}_i = \lambda_i\mathbf{f}_i,\ \ i = 0, 1, \cdots, G-1 \tag{6.268}$$

$$\lambda_i = \sum_{k=0}^{m} h_k V^{-ki},\ \ i = 0, 1, \cdots, G-1 \tag{6.269}$$

which indicate that $\mathbf{f}_i$ is an eigenvector of $\mathbf{H}$ with the associated eigenvalue λ_i. Let $\boldsymbol{\lambda}$ denote the $G \times 1$ *eigenvalue vector*

$$\boldsymbol{\lambda} = [\lambda_0\ \lambda_1\ \ldots\ \lambda_{G-1}]^T. \tag{6.270}$$

Then (6.269) implies that

$$\boldsymbol{\lambda} = \sqrt{G}\mathbf{F}^H\mathbf{h} \tag{6.271}$$

which indicates that *the energy in the time-domain components of the impulse response has been distributed among the eigenvalues of* $\mathbf{H}$, and

$$\mathbf{h} = \frac{1}{\sqrt{G}}\mathbf{F}\boldsymbol{\lambda}. \tag{6.272}$$

Equations (6.257) and (6.272) indicate that

$$\|\boldsymbol{\lambda}\|^2 = G. \tag{6.273}$$

Since $\mathbf{F}$ is nonsingular, the $\{\mathbf{f}_i\}$ are linearly independent. Therefore, $\mathbf{H}$ is diagonalizable, and (6.268) implies that $\mathbf{H}\mathbf{F}^H = \mathbf{F}^H\boldsymbol{\Lambda}$ or

$$\mathbf{H} = \mathbf{F}^H\boldsymbol{\Lambda}\mathbf{F} \tag{6.274}$$

where

$$\boldsymbol{\Lambda} = diag\,(\boldsymbol{\lambda}) \tag{6.275}$$

is the diagonal matrix with λ_i as its ith diagonal element, $i = 0, 1, \cdots, G-1$. This diagonalization is possible because of the way the cyclic prefix is defined, and hence provides the motivation for the definition of the cyclic prefix.

As indicated in Figure 6.30, after an S/P conversion, the received vector is applied to a *fast Fourier transformer* (FFT), which implements the *discrete Fourier transform*. The G parallel FFT outputs constitute the vector

$$\mathbf{y} = \mathbf{F}\bar{\mathbf{y}}. \tag{6.276}$$

The substitution of (6.264), (6.274), (6.259), and (6.251) into (6.276) yields

$$\mathbf{y} = \boldsymbol{\Lambda}\mathbf{P}\mathbf{A}\mathbf{d}+\mathbf{n} \tag{6.277}$$

where $\mathbf{n} = \mathbf{F}\bar{\mathbf{n}}$ is a zero-mean, independent, and circularly symmetric Gaussian random vector with

$$E\left[\mathbf{n}\mathbf{n}^H\right] = N_0\mathbf{I},\, E\left[\mathbf{n}\mathbf{n}^T\right] = \mathbf{0}. \tag{6.278}$$

Equalization

Equalization is the process by which the effect of the communication channel on $\mathbf{b} = \mathbf{PAd}$ is compensated. The *linear equalizer* computes the estimator

$$\widehat{\mathbf{b}} = \mathrm{Re}\,(\mathbf{W}\mathbf{y}) \tag{6.279}$$

where $\mathbf{W}$ is a $G \times G$ diagonal matrix with diagonal elements $w_i = W_{ii}$, $i = 0, 1, \cdots, G-1$, and $\widehat{\mathbf{b}} = [\widehat{b}_0 \;\; \widehat{b}_1 \; \ldots \; \widehat{b}_{N-1}]^T$. The diagonal elements are called the *weights* of the equalizer. The imaginary part of $\mathbf{Wy}$ is discarded because $\widehat{\mathbf{b}}$ is an estimator of the real-valued $\mathbf{b}$. As shown in Figure 6.30, the equalized FFT outputs $\{\hat{b}_i\}$ are applied to a P/S converter that feeds its output to the despreader. The despreader output is the $N \times 1$ vector

$$\mathbf{s} = \mathbf{P}^T\widehat{\mathbf{b}} = \mathbf{P}^T\,\mathrm{Re}\,(\mathbf{W}\mathbf{y}) \tag{6.280}$$

and the estimator of the data symbols is

$$\widehat{\mathbf{d}} = sgn\,(\mathbf{s})\,. \tag{6.281}$$

The substitution of (6.279), (6.277), and (6.248) into (6.280) indicates that

$$\mathbf{s} = \mathbf{P}^T\,\text{Re}(\mathbf{W}\boldsymbol{\Lambda})\mathbf{P}\mathbf{A}\mathbf{d} + \mathbf{P}^T\,\text{Re}\,(\mathbf{W}\mathbf{n})\,. \tag{6.282}$$

The kth component of $\mathbf{s}$ is

$$s_k = A_k d_k \sum_{i=0}^{G-1} \text{Re}(w_i\lambda_i) + \sum_{n=0,n\neq k}^{N-1} A_n d_n \mathbf{p}_k^T\,\text{Re}\,(\mathbf{W}\boldsymbol{\Lambda})\,\mathbf{p}_n + \mathbf{p}_k^T\,\text{Re}\,(\mathbf{W}\mathbf{n})\,. \tag{6.283}$$

Zero-Forcing Equalizer

A *zero-forcing* (ZF) equalizer uses

$$\mathbf{W} = \boldsymbol{\Lambda}^{-1} \quad \text{(ZF)}. \tag{6.284}$$

After substituting (6.249) and (6.284) into (6.282), we obtain

$$\mathbf{s} = G\mathbf{A}\mathbf{d} + \mathbf{P}^T\text{Re}\left(\boldsymbol{\Lambda}^{-1}\mathbf{n}\right) \tag{6.285}$$

which indicates that the zero-forcing equalizer allows the unbiased estimation of each data symbol without interference from the other symbols. The kth component of $\mathbf{s}$ is

$$s_k = GA_k d_k + n_{sk},\; n_{sk} = \mathbf{p}_k^T\text{Re}\left(\boldsymbol{\Lambda}^{-1}\mathbf{n}\right). \tag{6.286}$$

The problem with the zero-forcing equalizer is that if $|\lambda_i|$ is low for some i, then s_k is degraded by a large amount of noise.

The *trace* of matrix $\mathbf{A}$, which is the sum of its diagonal elements, is denoted by $tr(\mathbf{A})$. By direct substitution of the definitions of matrix multiplication and the trace of a matrix, we obtain the *trace* identities:

$$tr\,(\mathbf{AB}) = tr\,(\mathbf{BA}) \tag{6.287}$$

$$tr\left\{E\left[\mathbf{z}\mathbf{z}^H\right]\right\} = E\left[\mathbf{z}^H\mathbf{z}\right] = E\left[\|\mathbf{z}\|^2\right] \tag{6.288}$$

for compatible matrices $\mathbf{A}$ and $\mathbf{B}$, and for any vector $\mathbf{z}$.

Since the components of $\mathbf{n}$ are independent, zero-mean, and Gaussian, n_{sk} is a zero-mean Gaussian random variable. Let $|\boldsymbol{\Lambda}|^{-2}$ denote the diagonal matrix with $|\lambda_i|^{-2}$ as its ith diagonal element. Equations (6.288) and (6.278) yield

$$var\,(n_{sk}) = \frac{N_0}{2} tr\left(\mathbf{p}_k^T\,|\boldsymbol{\Lambda}|^{-2}\,\mathbf{p}_k\right)$$

$$= \frac{N_0}{2} \sum_{i=0}^{G-1} |\lambda_i|^{-2}. \quad (6.289)$$

From (6.286), (6.289), and (6.258), we obtain the SINR for data symbol k provided by the ZF equalizer:

$$\gamma_{sk} = \frac{\{E[s_k]\}^2}{var(n_{sk})}$$
$$= \frac{2c\mathcal{E}_{sk}}{N_0} \frac{G}{\sum_{i=0}^{G-1} |\lambda_i|^{-2}} \quad \text{(ZF)}. \quad (6.290)$$

The noise is Gaussian, and hence when a hard decision is made, the symbol error probability is

$$P_s(k) = Q\left(\sqrt{\frac{2c\mathcal{E}_{sk}}{N_0} \frac{G}{\sum_{i=0}^{G-1} |\lambda_i|^{-2}}}\right) \quad \text{(ZF)}. \quad (6.291)$$

MRC Equalizer

The noise amplification of the zero-forcing equalizer is avoided by using *MRC* equalization. An *MRC equalizer* maximizes the SNR of each data symbol. By considering only the first and third terms in (6.283), we find that the SNR for data symbol k is

$$\gamma_k = \frac{2c\mathcal{E}_{sk}}{GN_0} \frac{\left|\sum_{i=0}^{G-1} \text{Re}(w_i\lambda_i)\right|^2}{\sum_{i=0}^{G-1} |w_i|^2} \le \frac{2c\mathcal{E}_{sk}}{GN_0} \frac{\left|\sum_{i=0}^{G-1} w_i\lambda_i\right|^2}{\sum_{i=0}^{G-1} |w_i|^2}. \quad (6.292)$$

Application of the Cauchy-Schwarz inequality for sequences of complex numbers (Section 5.2) indicates that γ_k is maximized if

$$w_i = \eta\lambda_i^* \quad \text{(MRC)} \quad (6.293)$$

where η is an arbitrary constant. Therefore, $\mathbf{W}$ is the diagonal matrix

$$\mathbf{W} = \eta\mathbf{\Lambda}^* \quad \text{(MRC)} \quad (6.294)$$

which is independent of the particular symbol. Substituting (6.293) and (6.273) into (6.292), we find that the *SNR for* data symbol k *provided by the MRC equalizer* is

$$\gamma_k = \frac{2c\mathcal{E}_{sk}}{N_0} \quad \text{(MRC)} \quad (6.295)$$

which indicates that *all the energy in the multipath components of signal k is recovered.*

Equation (6.295) gives the SINR when a single signal is received. For a *single signal* with symbol energy $\mathcal{E}_s$ and an MRC equalizer, the noise is Gaussian, and hence when a hard decision is made, (6.295) indicates that the symbol error probability is

$$P_s = Q\left(\sqrt{\frac{2c\mathcal{E}_s}{N_0}}\right). \tag{6.296}$$

As indicated by comparing (6.296) with (6.219) and (6.220), the single-carrier direct-sequence system with rake combining and the MC-CDMA system with an MRC equalizer give approximately the same performance for $N = 1$ except for two principal factors. One is the MC-CDMA system loss due to the prefix factor $c \leq 1$, which accounts for the energy allocated to the cyclic prefix. The other, which is generally much more significant, is the loss in the single-carrier direct-sequence system because of path crosstalk, which has been neglected in the derivation of the performance of the rake demodulator. Aside from these factors, we observe the remarkable fact that *the MC-CDMA receiver recovers the same energy captured from the multipath components by the rake demodulator.*

MMSE Equalizer

The *minimum mean-square error* (MMSE) *equalizer or linear detector* equalizes the effect of the communication channel on $\mathbf{b} = \mathbf{PAd}$ with a diagonal matrix $\mathbf{W}$ such that the mean-square error

$$MSE = E[\| \mathbf{b} - \mathbf{Wy} \|^2] \tag{6.297}$$

is minimized under the assumption that the spreading sequences are approximated as independent random binary sequences. The trace identities indicate that

$$MSE = tr\left\{E[(\mathbf{b} - \mathbf{Wy})(\mathbf{b} - \mathbf{Wy})^H]\right\}. \tag{6.298}$$

We define the $G \times G$ positive-semidefinite Hermitian matrix

$$\mathbf{R}_b = E[\mathbf{bb}^{\mathrm{T}}] = E[\mathbf{PAdd}^T\mathbf{AP}^T]. \tag{6.299}$$

Using (6.277) and (6.278), we find that the $G \times G$ Hermitian *correlation matrix* is

$$\mathbf{R}_y = E[\mathbf{yy}^H] = N_0\mathbf{I} + \boldsymbol{\Lambda}\ \mathbf{R}_b\ \boldsymbol{\Lambda}^* \tag{6.300}$$

which indicates that $\mathbf{R}_y$ is positive definite and hence invertible (Appendix G). An expansion of (6.298) and substitution of (6.277) yields

$$\begin{aligned} MSE &= tr\left[\mathbf{W}\,\mathbf{R}_y\,\mathbf{W}^H - \mathbf{W}\,\mathbf{\Lambda}\,\mathbf{R}_b - \mathbf{R}_b\,\mathbf{\Lambda}^*\,\mathbf{W}^* \; + \mathbf{R}_b\right] \\ &= tr\left[\mathbf{B}\mathbf{R}_y\,\mathbf{B}^H\right] + \mathbf{C} \\ &= tr\left[\mathbf{R}_y\,\mathbf{B}^H\mathbf{B}\right] + \mathbf{C} \end{aligned} \tag{6.301}$$

where

$$\mathbf{B} = \mathbf{W} - \mathbf{R}_b\,\mathbf{\Lambda}^*\,\mathbf{R}_y^{-1} \tag{6.302}$$

$$\mathbf{C} = tr\left[\mathbf{R}_b - \mathbf{R}_b\,\mathbf{\Lambda}^*\,\mathbf{R}_y^{-1}\mathbf{\Lambda}\,\mathbf{R}_b\right]. \tag{6.303}$$

Since the Hermitian matrix $\mathbf{R}_y$ is positive definite, $\mathbf{B}\mathbf{R}_y\,\mathbf{B}^H$ is Hermitian positive semidefinite, which implies that $tr\left[\mathbf{B}\mathbf{R}_y\,\mathbf{B}^H\right] = tr\left[\mathbf{R}_y\,\mathbf{B}^H\mathbf{B}\right] \geq 0$.

If the spreading sequences are modeled as independent random binary sequences, then

$$E\left[\mathbf{p}_n\mathbf{p}_m^T\right] = \mathbf{0},\; n \neq m,\; E\left[\mathbf{p}_n\mathbf{p}_n^T\right] = \mathbf{I}. \tag{6.304}$$

Equations (6.299), (6.248), (6.258), and (6.304) imply that

$$\mathbf{R}_b = \frac{c\mathcal{E}_t}{G}\mathbf{I} \tag{6.305}$$

where

$$\mathcal{E}_t = \sum_{n=0}^{N-1} \mathcal{E}_{sn} \tag{6.306}$$

is the total energy of all N symbols. Substituting (6.305) into (6.300), we obtain the diagonal matrix

$$\mathbf{R}_y = E[\mathbf{y}\mathbf{y}^H] = N_0\mathbf{I} + \frac{c\mathcal{E}_t}{G}\,|\mathbf{\Lambda}|^2. \tag{6.307}$$

Therefore, $\mathbf{B}$ is the diagonal matrix

$$\mathbf{B} = \mathbf{W} - \frac{c\mathcal{E}_t}{GN_0}\mathbf{\Lambda}^*\left[\mathbf{I} + \frac{c\mathcal{E}_t}{GN_0}\,|\mathbf{\Lambda}|^2\right]^{-1} \tag{6.308}$$

and hence $\mathbf{B}^H\mathbf{B}$ is a diagonal matrix with nonnegative diagonal elements. Since $\mathbf{R}_y$ has positive diagonal elements, $tr\left[\mathbf{R}_y\,\mathbf{B}^H\mathbf{B}\right] = 0$ if and only if $\mathbf{B} = \mathbf{0}$. Thus, the unique MMSE equalizer is

$$\mathbf{W} = \frac{c\mathcal{E}_t}{GN_0}\mathbf{\Lambda}^*\left[\mathbf{I} + \frac{c\mathcal{E}_t}{GN_0}\,|\mathbf{\Lambda}|^2\right]^{-1} \quad \text{(MMSE)}. \tag{6.309}$$

The MRC and MMSE equalizers produce data-symbol estimators that are degraded by the presence of the other data symbols. However, the noise is usually not amplified by the processing; hence, these equalizers are usually preferred over the zero-forcing equalizer. If N is sufficiently large that

$$\frac{c\mathcal{E}_t}{N_0} \gg \frac{G}{\min_i |\lambda_i|^2} \tag{6.310}$$

then (6.309) indicates that the MMSE equalizer approximates the zero-forcing equalizer. If we assume that

$$\frac{c\mathcal{E}_t}{N_0} \ll \frac{G}{\max_i |\lambda_i|^2} \tag{6.311}$$

then (6.309) indicates that the MMSE equalizer is proportional to the MRC equalizer.

Performance Analysis

The SINR for MRC and MMSE equalizers obtained from (6.283) depends on the N particular spreading sequences selected and does not provide much insight into how the equalizers compare. An approximate but more useful general equation for the SINR can be obtained by modeling the N spreading sequences as independent random binary sequences. The noise is independent of the spreading sequences, and the diagonal matrix $\mathbf{W}\mathbf{\Lambda}$ has real-valued elements for both equalizers. Therefore, (6.283), (6.287), and (6.304) imply that

$$\begin{aligned} var(s_k) &= \sum_{n=0,n\neq k}^{N-1} \sum_{m=0,m\neq k}^{N-1} A_n A_m d_n d_m E[\mathbf{p}_k^T \mathbf{W}\mathbf{\Lambda} \mathbf{p}_n \mathbf{p}_m^T \mathbf{W}\mathbf{\Lambda} \mathbf{p}_k] + \frac{N_0}{2} \sum_{i=0}^{G-1} |w_i|^2 \\ &= \frac{c\mathcal{E}_{t/k}}{G} tr[(\mathbf{W}\mathbf{\Lambda})^2] + \frac{N_0}{2} \sum_{i=0}^{G-1} |w_i|^2 \end{aligned} \tag{6.312}$$

where $\mathcal{E}_{t/k}$ is the *total symbol energy of the multiple-access interference*:

$$\mathcal{E}_{t/k} = \sum_{n=0,n\neq k}^{N-1} \mathcal{E}_{sn} = \mathcal{E}_t - \mathcal{E}_{sk}. \tag{6.313}$$

The SINR for data symbol k is

$$\gamma_{sk} = \frac{\frac{2c\mathcal{E}_{sk}}{GN_0} [tr(\mathbf{W}\mathbf{\Lambda})]^2}{\frac{2c\mathcal{E}_{t/k}}{GN_0} tr[(\mathbf{W}\mathbf{\Lambda})^2] + \sum_{i=0}^{G-1} |w_i|^2}. \tag{6.314}$$

For an MRC equalizer, the substitution of (6.294) and (6.273) into (6.314) yields

$$\gamma_{sk} = \frac{\frac{2c\mathcal{E}_{sk}}{N_0}G}{\frac{2c\mathcal{E}_{t/k}}{GN_0}\sum_{i=0}^{G-1}|\lambda_i|^4 + G} \quad \text{(MRC)} \tag{6.315}$$

which reduces to (6.295) when $\mathcal{E}_{t/k} = 0$. For an MMSE equalizer, the substitution of (6.309) into (6.314) yields

$$\gamma_{sk} = \frac{\frac{2c\mathcal{E}_{sk}}{GN_0}\left[\sum_{i=0}^{G-1}|\lambda_i|^2\left(1+\frac{c\mathcal{E}_t}{GN_0}|\lambda_i|^2\right)^{-1}\right]^2}{\frac{2c\mathcal{E}_{t/k}}{GN_0}\sum_{i=0}^{G-1}|\lambda_i|^4\left(1+\frac{c\mathcal{E}_t}{GN_0}|\lambda_i|^2\right)^{-2} + \sum_{i=0}^{G-1}|\lambda_i|^2\left(1+\frac{c\mathcal{E}_t}{GN_0}|\lambda_i|^2\right)^{-2}} \quad \text{(MMSE)}. \tag{6.316}$$

If the spreading sequences are modeled as independent random binary sequences, then the middle term of (6.283) is the sum of $N-1$ independent, identically distributed random variables, each of which has a finite mean and variance. Therefore, the central limit theorem (Corollary A1, Appendix A.2) implies that the distribution of s_k is approximately the Gaussian distribution when N is large; hence

$$P_s(k) \approx Q\left(\sqrt{\gamma_{sk}}\right). \tag{6.317}$$

A frequency-selective fading channel is characterized by several significant multipath components with a relatively large delay spread and small coherence time. Since $h_k > 0$ for several values of k, (6.271) and (6.250) indicate that it is likely that $\max_i |\lambda_i| >> \min_i |\lambda_i|$. In contrast, if a flat fading channel has $h_k > 0$ for a single value of k, (6.271) and (6.250) indicate that $|\lambda_i| = 1, i = 0, 1, \cdots, G-1$.

As an example of the performance of MC-CDMA systems, we evaluate P_s for the frequency-selective fading channel with

$$G = 64,\ \mathbf{h} = \frac{4}{\sqrt{21}}[1\ \ 0.5\ \ \text{-}0.25\ 0\ \ldots\ 0]^T,\ \ \|\,\mathbf{h}\,\|^2 = 1.$$

All N data symbols have the same energy so that $\mathcal{E}_{t/k} = (N-1)\mathcal{E}_{sk}$. Calculations give $\max_i |\lambda_i| = 5.59 \min_i |\lambda_i|$, which indicates that the channel is strongly frequency selective. Figure 6.31 illustrates $P_s(k)$ as a function of N for ZF, MRC, and MMSE equalizers, and $c\mathcal{E}_{sk}/N_0 = 10$ dB and 13 dB. It is observed that the MMSE equalizer slightly outperforms the MRC equalizer in this example. The MMSE equalizer provides a better performance than the ZF equalizer when $N \leq 8$ if $c\mathcal{E}_{sk}/N_0 = 10$ dB, and when $N \leq 5$ if $c\mathcal{E}_{sk}/N_0 = 13$ dB.

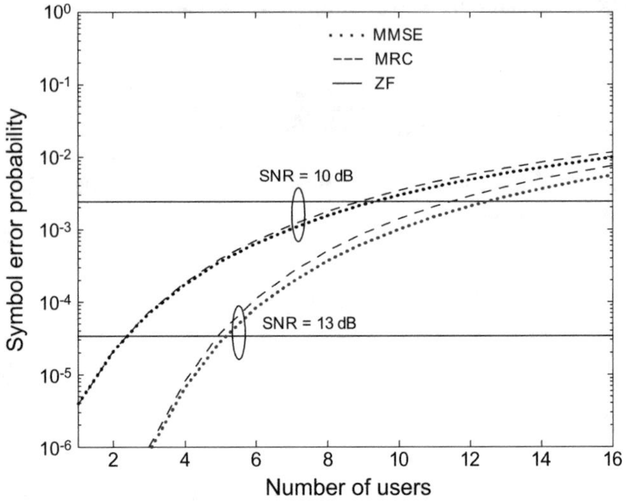

Fig. 6.31 Symbol error probability for multiuser MC-CDMA system as a function of N for $G = 64$, SNR = $c\mathcal{E}_{sk}/N_0 = 10$ dB, and SNR = $c\mathcal{E}_{sk}/N_0 = 13$ dB

Channel Estimation

The implementation of equalizers requires *channel estimates* of the $\{\lambda_i\}$, which may be determined by transmitting known pilot sequences. Accordingly, let $\mathbf{b}_a = [b_{a0}\ b_{a1} \ldots b_{a,G-1}]^T$ denote a known $G \times 1$ vector of a pilot sequence with $b_{ai} = \pm 1$, and let $\mathbf{B}$ denote a $G \times G$ diagonal matrix with diagonal elements

$$B_{ii} = b_{ai}, \quad i = 0, 1, \ldots, G-1. \tag{6.318}$$

When $\mathbf{b}_a$ is the received vector, (6.277) indicates that the FFT output vector at the input to the equalizer is

$$\mathbf{y} = \boldsymbol{\Lambda}\mathbf{b}_a + \mathbf{n}. \tag{6.319}$$

The vector $\boldsymbol{\lambda}$ can be directly estimated by computing $\mathbf{By}$. However, by first estimating the vector $\mathbf{h}$ and using the fact that it has at most m nonzero components, we can eliminate some noise and then produce a better estimate of $\boldsymbol{\lambda}$. Equation (6.272) indicates that a rough estimator of $\mathbf{h}$ is the $G \times 1$ vector

$$\widehat{\mathbf{h}}_r = G^{-1/2}\mathbf{FBy}. \tag{6.320}$$

Substituting (6.318) and (6.319) into (6.320) and using (6.272), we find that $\widehat{\mathbf{h}}_r = \mathbf{h} + G^{-1/2}\mathbf{FBn}$. The final $G-m-1$ components of $\widehat{\mathbf{h}}_r$ would be zero, like those of $\mathbf{h}$, in the absence of noise but are nonzero in the presence of noise. The final $G-m-1$

components are set to zero by the refined estimator

$$\widehat{\mathbf{h}}= \mathbf{I}_{m+1}\widehat{\mathbf{h}}_r \tag{6.321}$$

where $\mathbf{I}_{m+1}$ is the $G \times G$ diagonal matrix with its first $m + 1$ diagonal values equal to 1 and its remaining diagonal values set equal to 0. Since $\mathbf{I}_{m+1}$ has no effect on $\mathbf{h}$,

$$\widehat{\mathbf{h}} = \mathbf{h}+G^{-1/2}\mathbf{I}_{\mathbf{m+1}}\mathbf{FBn}. \tag{6.322}$$

Equation (6.271) suggests that a refined estimator of $\boldsymbol{\lambda}$ is

$$\widehat{\boldsymbol{\lambda}} = G^{1/2}\mathbf{F}^H\widehat{\mathbf{h}}. \tag{6.323}$$

Substituting (6.320) and (6.321) into (6.323), we obtain the *channel estimator*:

$$\widehat{\boldsymbol{\lambda}} = \mathbf{F}^H\mathbf{I}_{m+1}\mathbf{FBy} \tag{6.324}$$

where the $G \times G$ product matrix $\mathbf{F}^H\mathbf{I}_{m+1}\mathbf{FB}$ can be stored in the receiver. Substituting (6.322), (6.271), and (6.272) into (6.323), we find that

$$\widehat{\boldsymbol{\lambda}} = \boldsymbol{\lambda}+\mathbf{n}_e \tag{6.325}$$

where $\mathbf{n}_e = \mathbf{F}^H\mathbf{I}_{m+1}\mathbf{FBn}$. Since $\mathbf{n}_e$ is zero-mean, (6.325) indicates that $\widehat{\boldsymbol{\lambda}}$ defined by (6.324) provides an unbiased estimate of $\boldsymbol{\lambda}$ that can be used to calculate the weights in Figure 6.30. The covariance matrix of $\mathbf{n}_e$ is

$$\mathbf{R}_{ne} = E[\mathbf{n}_e\mathbf{n}_e^H] = N_0\mathbf{F}^H\mathbf{I}_{m+1}\mathbf{F}. \tag{6.326}$$

Applying (6.287), (6.288), and (6.251), we obtain the total noise power

$$\begin{aligned} E\left[\|\mathbf{n}_e\|^2\right] &= tr\left(\mathbf{R}_{ne}\right) \\ &= (m+1)\,N_0 \end{aligned} \tag{6.327}$$

which indicates that the channel estimator of (6.324) results in a reduction in the total noise power by the factor $(m + 1)\,/G$ if knowledge of the multipath delay spread is available in the receiver.

Peak-to-Average-Power Ratio

The chips $\overline{x}_{-m}, \overline{x}_{-m+1}, \ldots, \overline{x}_{G-1}$ applied to the DAC in the MC-CDMA transmitter have potentially large amplitude variations because each chip is the sum of complex numbers with different phases; hence, these numbers may combine either constructively or destructively. The DAC output is applied to the transmitter's power amplifier. The amplifier produces an output power that is approximately a linear function of the input power when the input power is relatively low, but the function becomes highly nonlinear as the input power increases. If the input power level is nearly constant, then operation in the nonlinear region allows the highest transmitted power level, and hence potentially the best receiver performance. However, if the input power has large variations, then the nonlinear function causes signal distortion, excessive radiation into other spectral regions, and intersymbol interference. If the power amplifier operates in its linear region then these problems are largely absent even if the amplifier input has considerable power variations. Therefore, it is necessary to reduce the power variations at the input to the power amplifier enough that the power amplifier nearly always operates in its linear region, but near the onset of its nonlinear region.

The *peak-to-average-power ratio* (PAPR) of a transmitted signal over a time interval is defined as the ratio of the maximum instantaneous power of a signal to its average value during the interval. In an MC-CDMA system, the complex envelope $x(t)$ of a signal transmitted over the time interval $\mathcal{I}$ with duration T has a PAPR defined as

$$PAPR\left[x(t)\right] = \frac{\max\limits_{\mathcal{I}} |x(t)|^2}{\frac{1}{T}\int_{\mathcal{I}} |x(t)|^2\, dt}. \tag{6.328}$$

Assuming that all symbols are transmitted with the same power, we define the *discrete-time PAPR* of a transmitted block of $m + G$ chips as

$$PAPR\left[\{\overline{x}_i\}\right] = \frac{\max\limits_{-m\leq i\leq G-1} |\overline{x}_i|^2}{\frac{1}{G+m}\sum_{i=-m}^{G-1} |\overline{x}_i|^2} \tag{6.329}$$

where the $\{\overline{x}_i\}$ are defined by (6.252) and (6.253). Because the pulse shaping of each transmitted chip generates nonrectangular transmitted pulses or waveforms that increase the PAPR relative to its value without pulse shaping, $PAPR\left[\{\overline{x}_i\}\right] < PAPR\left[x(t)\right]$. However, oversampling at a rate higher than the chip rate can reduce the disparity.

To derive an approximate distribution function for $PAPR\left[\{\overline{x}_i\}\right]$, we replace the denominator in (6.329) with $E\left[|\overline{x}_i|^2\right]$ so that

$$PAPR\left[\{\overline{x}_i\}\right] \approx \max_{-m\leq i\leq G-1} \left\{\frac{|\overline{x}_i|^2}{E\left[|\overline{x}_i|^2\right]}\right\}. \tag{6.330}$$

We assume that each $\overline{x}_i$ is an independent, identically distributed, zero-mean, complex random variable. Equations (6.252) and (6.253) imply that both the real and imaginary parts of $\overline{x}_i$ are sums of uniformly bounded, independent random variables and have variances that $\to \infty$ as $G+m \to \infty$. Therefore, if $G+m$ is large, the central limit theorem (Corollary A2, Appendix A.2) indicates that the real and imaginary parts of $\bar{x}_i$ have distributions that are approximately Gaussian with variances equal to $E\left[|\bar{x}_i|^2\right]/2$. As shown in Appendix E.4, $|\bar{x}_i|$ then has a Rayleigh distribution with a variance equal to $E\left[|\bar{x}_i|^2\right]$, and $|\bar{x}_i|^2 / E\left[|\bar{x}_i|^2\right]$ has an exponential distribution $F(z) = 1 - \exp(-z)$ and a mean equal to 1. Equation (6.330) then implies that $PAPR[\{\bar{x}_i\}]$ has a distribution function approximated by $[1 - \exp(-z)]^{G+m}$. The probability that $PAPR[\{\bar{x}_i\}]$ exceeds z is

$$P[PAPR > z] \approx 1 - [1 - \exp(-z)]^{G+m}. \tag{6.331}$$

This equation indicates that an excessive PAPR is probable even for small values of the spreading factor. For example, $P[PAPR > 4] > 0.168$ if $G + m > 10$. Since a large PAPR drives the power amplifier into its nonlinear region or saturation, some method of PAPR reduction is needed to maintain a relatively high average input power while reducing the peak input power.

Clipping entails the distortion of the transmitted signal so that high magnitude peaks occur rarely at the input of the power amplifier. Effective clipping reduces PAPR at the cost of an acceptable loss in receiver performance. A clipper limits the magnitude of an input signal to a clipping level ζ if this magnitude exceeds ζ and does not otherwise change the signal. The nonlinear clipping causes the generation of increased out-of-band radiation, which can be suppressed by filtering the clipper output. The signal distortion due to the clipping potentially causes increased bit errors in the receiver, but most of these errors can be corrected by the channel code.

Numerous PAPR reduction techniques for OFDM systems have been proposed [76], and some of these techniques might be adapted to MC-CDMA systems. However, relative to clipping, other techniques require much more computational complexity and sometimes a bandwidth expansion.

6.14 DS-CDMA-FDE Systems

The *direct-sequence code-division multiple-access system with frequency-domain equalization* (DS-CDMA-FDE system) preserves a favorable PAPR by eliminating the IFFT in the transmitter and using a single carrier for transmission. Both the FFT and IFFT are performed in the receiver as part of the equalization. Although subsequently we set the spreading factor equal to the FFT window size, this equality is not required [1, 2].

The transmitter of the DS-CDMA-FDE system has the form of Figure 6.28 without the IFFT. The vector of G chips associated with a set of N aligned symbols is

$$\mathbf{x} = [x_{G-1} x_{G-2} \ldots x_0]^T$$

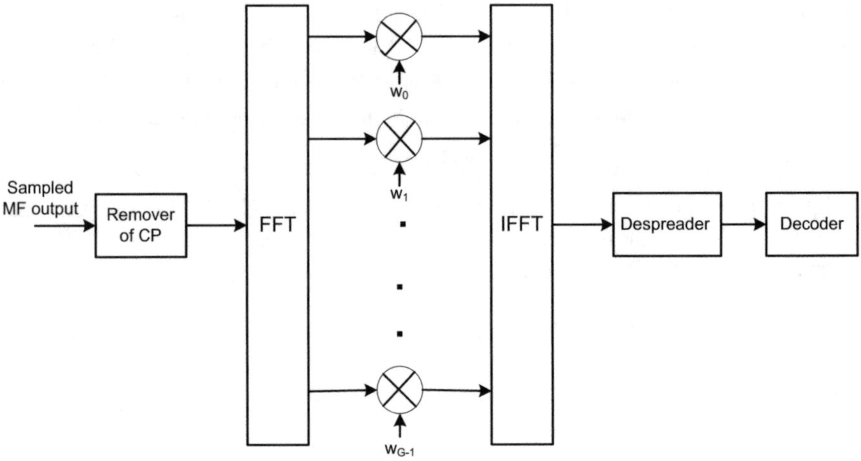

Fig. 6.32 Receiver of DS-CDMA-FDE system. *CP* cyclic prefix, *MF* matched filter

$$= \mathbf{Pd} = \sum_{n=0}^{N-1} \mathbf{p}_n d_n. \tag{6.332}$$

After insertion of the cyclic prefix to prevent intersymbol interference, the sequence $\overline{\mathbf{x}}$ applied to the DAC has components

$$\overline{x}_i = x_k, \quad k = i \; modulo\text{-}G, \quad -m \leq i \leq G-1 \tag{6.333}$$

where the vector $\mathbf{b} = \mathbf{Pd}$, as defined by (6.248). The cyclic prefix ensures that there is negligible intersymbol interference in the receiver. A BPSK signal is transmitted.

The principal components of the receiver are diagrammed in Figure 6.32. After coherent demodulation, the m-sample cyclic prefix of the matched-filter output samples is discarded because these samples are corrupted by the previous data block. The remaining samples constitute the components of the received vector $\overline{\mathbf{y}}$. Assuming the same channel model as in Section 6.13 and substituting (6.274), we obtain

$$\begin{aligned} \overline{\mathbf{y}} &= \mathbf{HPAd} + \overline{\mathbf{n}} \\ &= \mathbf{F}^H \mathbf{\Lambda} \mathbf{FPAd} + \overline{\mathbf{n}} \end{aligned} \tag{6.334}$$

where $\overline{\mathbf{n}}$ is the Gaussian noise vector with covariance matrix $E[\overline{\mathbf{n}}\overline{\mathbf{n}}^H] = N_0\mathbf{I}$. As indicated in Figure 6.32, the samples are applied to an S/P conversion and FFT. The G parallel FFT outputs constitute the vector

$$\mathbf{y} = \mathbf{F}\overline{\mathbf{y}}$$

$$= \mathbf{\Lambda FPAd} + \mathbf{n} \tag{6.335}$$

where $\mathbf{n} = \mathbf{F\overline{n}}$ is a zero-mean, independent, and circularly symmetric Gaussian random vector with statistics given by (6.278).

Equalization

The *equalizer* computes the vector $\mathbf{Wy}$, where $\mathbf{W}$ is a diagonal matrix with diagonal elements $w_i = W_{ii}$. The IFFT produces the vector $\mathbf{F}^H \mathbf{Wy}$, which is applied to a P/S converter that feeds its output stream to the despreader. The despreader output is

$$\begin{aligned} \mathbf{s} &= \mathbf{P}^T \operatorname{Re}(\mathbf{F}^H \mathbf{Wy}) \\ &= \mathbf{P}^T \operatorname{Re}(\mathbf{F}^H \mathbf{W\Lambda FPAd}) + \mathbf{P}^{\mathrm{T}} \operatorname{Re}(\mathbf{F}^H \mathbf{Wn}) \end{aligned} \tag{6.336}$$

and the estimator of the data symbols is given by (6.281). The kth component of $\mathbf{s}$ is

$$s_k = A_k d_k \operatorname{Re}(\mathbf{u}^H \mathbf{m}) + \sum_{n=0, n\neq k}^{N-1} A_n d_n \operatorname{Re}(\mathbf{u}^H \mathbf{\Lambda F p}_n) + \operatorname{Re}(\mathbf{u}^H \mathbf{n}) \tag{6.337}$$

where $\mathbf{u}$ and $\mathbf{m}$ are the $G \times 1$ vectors defined as

$$\mathbf{u} = \mathbf{W}^* \mathbf{F p}_k, \quad \mathbf{m} = \mathbf{\Lambda F p}_k. \tag{6.338}$$

Using (6.278) and (6.288), we find that the noise term $n_{sk} = \operatorname{Re}(\mathbf{u}^H \mathbf{n})$ is a zero-mean Gaussian random variable with variance

$$var\,(n_{sk}) = \frac{N_0}{2} \|\mathbf{u}\|^2 = \frac{N_0}{2} \left(\mathbf{p}_k^T \mathbf{F}^H \, |\mathbf{W}|^2 \, \mathbf{F p}_k \right). \tag{6.339}$$

Zero-Forcing Equalizer

A *ZF* equalizer uses $\mathbf{W} = \Lambda^{-1}$. Since the spreading sequences are orthogonal, this equalizer provides

$$\mathbf{s} = G\mathbf{Ad} + \mathbf{P}^{\mathrm{T}} \operatorname{Re}(\mathbf{F}^H \, \mathbf{\Lambda}^{-1} \mathbf{n}). \tag{6.340}$$

Thus, the zero-forcing equalizer allows the recovery of the data symbols without intersymbol interference at the cost of noise enhancement when one of the $\{\lambda_i\}$ is small. The kth component of $\mathbf{s}$ is

$$s_k = GA_k d_k + n_{sk}, \; n_{sk} = \mathbf{p}_k^T \operatorname{Re}(\mathbf{F}^H \, \mathbf{\Lambda}^{-1} \mathbf{n}). \tag{6.341}$$

From (6.341), (6.339), and (6.258), we obtain the SINR for data symbol k provided by the zero-forcing equalizer:

$$\gamma_{sk} = \frac{2c\mathcal{E}_{sk}}{N_0} \frac{G}{\mathbf{p}_k^T \mathbf{F}^H \, |\mathbf{\Lambda}|^{-2} \, \mathbf{F}\mathbf{p}_k} \quad \text{(ZF)} \tag{6.342}$$

where $\mathcal{E}_{sk}$ is the symbol energy of d_k. The noise is Gaussian, and hence when a hard decision is made, the symbol error probability is

$$P_s(k) = Q\left(\sqrt{\frac{2c\mathcal{E}_{sk}}{N_0} \frac{G}{\mathbf{p}_k^T \mathbf{F}^H \, |\mathbf{\Lambda}|^{-2} \, \mathbf{F}\mathbf{p}_k}}\right) \quad \text{(ZF).} \tag{6.343}$$

MRC and MMSE Equalizers

An *MRC equalizer* maximizes the signal-to-noise ratio (SNR) of each data symbol. From the first and third terms of (6.337) and (6.339), we find that the SNR for symbol k is

$$\gamma_k = \frac{2c\mathcal{E}_{sk}}{N_0} \frac{[\mathrm{Re}(\mathbf{u}^H \mathbf{m})]^2}{G \, \|\mathbf{u}\|^2} . \tag{6.344}$$

Application of the Cauchy-Schwarz inequality for vectors (Section 5.2) indicates that

$$\mathrm{Re}(\mathbf{u}^H \mathbf{m}) \le \left|\mathbf{u}^H \mathbf{m}\right| \le \|\mathbf{u}\| \, \|\mathbf{m}\| \tag{6.345}$$

with equality only if $\mathbf{u} = \eta \mathbf{m}$, where η is an arbitrary constant. Thus, the MRC equalizer uses $\mathbf{W} = \mathbf{\Lambda}^*$, and *the SNR provided by the MRC equalizer* is

$$\gamma_k = \frac{2c\mathcal{E}_{sk}}{N_0} \frac{\mathbf{p}_k^T \mathbf{F}^H \, |\mathbf{\Lambda}|^2 \, \mathbf{F}\mathbf{p}_k}{G} \quad \text{(MRC).} \tag{6.346}$$

For a *single symbol* with total energy $\mathcal{E}_s$, spreading sequence $\mathbf{p}$, and an MRC equalizer, the noise is Gaussian; hence, when a hard decision is made, the symbol error probability is

$$P_s = Q\left(\sqrt{\frac{2c\mathcal{E}_s}{N_0} \frac{\mathbf{p}^T \mathbf{F}^H \, |\mathbf{\Lambda}|^2 \, \mathbf{F}\mathbf{p}}{G}}\right) . \tag{6.347}$$

Since the equalizer output is applied to an IFFT, the *MMSE equalizer or linear detector* equalizes with diagonal matrix $\mathbf{W}$ such that the MSE = $E[\| \, \mathbf{FPAd} - \mathbf{Wy} \, \|^2]$ is minimized. If each chip in a spreading sequence is modeled as an independent, zero-mean binary random variable, then a derivation similar to the previous one for the MC-CDMA system indicates that the MMSE equalizer uses the weights given by (6.309). Thus, the ZF, MRC, and MMSE equalizers are the same for the MC-CDMA and DS-CDMA-FDE systems.

Performance Analysis

To compare the MRC and MMSE equalizers, an approximate but useful general equation for the SINR is obtained by modeling the N spreading sequences as independent, random binary sequences satisfying (6.304). Under this model, only the first term in (6.337) has a nonzero mean, and the first term is uncorrelated with the other terms. We define the real-valued matrix

$$\mathbf{D} = \mathrm{Re}(\mathbf{F^H W \Lambda F}) \tag{6.348}$$

which is symmetric because the diagonal matrix $\mathbf{W\Lambda}$ is real-valued for all three equalizers. It follows from (6.348) and (6.250) that

$$tr\,(\mathbf{D}) = tr\,(\mathbf{W\Lambda}) \tag{6.349}$$

and

$$\sum_i D_{i,i}^2 = G^{-1}\,[tr\,(\mathbf{W\Lambda})]^2\,. \tag{6.350}$$

Using (6.287), (6.304), and (6.349), we obtain

$$E\,[s_k] = A_k d_k tr\,(\mathbf{W\Lambda})\,. \tag{6.351}$$

The variance of s_k is

$$var\,(s_k) = var\,(si) + var\,(mai) + \frac{N_0}{2}\sum_{i=0}^{G-1}|w_i|^2 \tag{6.352}$$

where $var\,(si)$ is the variance of the first term and is due to the self-interference, $var\,(mai)$ is the variance of the second term and is due to the multiple-access interference, and the final term is due to the noise.

Under the binary sequence model, $E[p_k\,(i)\,p_k\,(l)\,p_k\,(m)\,p_k\,(n)] = 0$ unless the indices are the set $I_1 = (n = m, l = i)$ or $I_2 = (n = l, m = i)$ or $I_3 = (n = i, m = l)$. Since

$$\begin{aligned} I_1 \cup I_2 \cup I_3 &= I_1 + I_2 + I_3 - I_1 \cap I_2 - I_1 \cap I_3 - I_2 \cap I_3 + I_1 \cap I_2 \cap I_3 \\ &= I_1 + I_2 + I_3 - 2\,(I_1 \cap I_2) \end{aligned} \tag{6.353}$$

we obtain

$$
\begin{aligned}
var\,(si) &= \frac{c\mathcal{E}_{sk}}{G}\{E[(\mathbf{p}_k^T\mathbf{D}\mathbf{p}_k)^2] - [tr\,(\mathbf{W}\boldsymbol{\Lambda})]^2\} \\
&= \frac{c\mathcal{E}_{sk}}{G}\left\{E\left[\sum_{i,l,m,n} p_k(i)\,D_{i,l}p_k(l)\,p_k(m)\,D_{m,n}p_k(n)\right] - [tr\,(\mathbf{W}\boldsymbol{\Lambda})]^2\right\} \\
&= \frac{c\mathcal{E}_{sk}}{G}\left\{\sum_{i,m} D_{i,i}D_{m,m} + \sum_{i,l} D_{i,l}D_{i,l} + \sum_{i,l} D_{i,l}D_{l,i} - 2\sum_{i} D_{i,i}^2 - [tr\,(\mathbf{W}\boldsymbol{\Lambda})]^2\right\} \\
&= \frac{2c\mathcal{E}_{sk}}{G}\left\{tr\left(\mathbf{D}^2\right) - G^{-1}\,[tr\,(\mathbf{W}\boldsymbol{\Lambda})]^2\right\}
\end{aligned}
\tag{6.354}
$$

where the final equality follows from the symmetry of $\mathbf{D}$ and (6.350). The independence of the $\{\mathbf{p}_n\}$ implies that $var\,(mai)$ is equal to the sum of variances, and we obtain

$$
\begin{aligned}
var\,(mai) &= \sum_{n=0,n\neq k}^{N-1} \frac{c\mathcal{E}_{sn}}{G} E[\{\mathbf{p}_k^T\,\mathrm{Re}[\mathbf{F}^H\mathbf{W}\boldsymbol{\Lambda}\mathbf{F}]\mathbf{p}_n\}^2] \\
&= \sum_{n=0,n\neq k}^{N-1} \frac{c\mathcal{E}_{sn}}{G} E[\mathbf{p}_k^T\mathbf{D}\mathbf{p}_n\mathbf{p}_n^T\mathbf{D}\mathbf{p}_k] \\
&= \frac{c\mathcal{E}_{t/k}}{G} E[\mathbf{p}_k^T\mathbf{D}^2\mathbf{p}_k] \\
&= \frac{c\mathcal{E}_{t/k}}{G} tr\left(\mathbf{D}^2\right)
\end{aligned}
\tag{6.355}
$$

where the third and fourth equalities use (6.304), the fourth equality uses (6.288), and $\mathcal{E}_{t/k}$ is defined by (6.313).

The SINR of symbol k is equal to the ratio of the square of $E\,[s_k]$ to $var\,(s_k)$. Thus, the SINR for MRC and MMSE equalizers is

$$
\gamma_{sk} = \frac{\frac{2c\mathcal{E}_{sk}}{GN_0}\,[tr\,(\mathbf{W}\boldsymbol{\Lambda})]^2}{\frac{2c\mathcal{E}_{t/k}}{GN_0} tr\,(\mathbf{D}^2) + \frac{2}{N_0} var\,(si) + \sum_{i=0}^{G-1} |w_i|^2}\,. \tag{6.356}
$$

The contribution of $var\,(si)$ in the denominator is due to self-interference, which occurs in the DS-CDMA-FDE system but has no counterpart in the MC-CDMA system. Consequently, a comparison with (6.314) indicates that the MC-CDMA is advantageous when $\mathcal{E}_{t/k}$ is small. However, when $\mathcal{E}_{t/k} >> \mathcal{E}_{sk}$, the performances of the two systems are similar.

If the spreading sequences are modeled as independent random binary sequences, then the middle term of (6.283) is the sum of $N-1$ independent, identically distributed random variables each of which has a finite mean and variance.

Therefore, the central limit theorem (Corollary A1, Appendix A.2) implies that the distribution of s_k is approximately given by (6.317) when N is large.

Channel Estimation

The implementation of the equalization requires the estimation of the vector $\boldsymbol{\lambda}$, which can be accomplished in the DS-CDMA-FDE system by a method similar to that used in the MC-CDMA system. Let $\mathbf{b}_a = [b_{a0}\ b_{a1} \ldots b_{a,G-1}]^T$ denote a known $G \times 1$ vector of binary pilot chips transmitted during some block, and let $\mathbf{x}_a = \mathbf{F}\mathbf{b}_a = [x_{a0}\ x_{a1} \ldots x_{a,G-1}]^T$ denote the corresponding $G \times 1$ discrete Fourier transform vector. Let $\mathbf{X}$ denote a $G \times G$ diagonal matrix with diagonal elements

$$X_{ii} = x_{ai}^* / |x_{ai}|^2, \quad i = 0, 1, \ldots, G-1. \tag{6.357}$$

When $\mathbf{b}_a$ is the transmitted vector, the FFT output vector at the input of the equalizer is

$$\begin{aligned}\mathbf{y} &= \mathbf{F}\mathbf{H}\mathbf{b}_\mathbf{a}+\mathbf{n}\\ &= \boldsymbol{\Lambda}\mathbf{F}\mathbf{b}_\mathbf{a}+\mathbf{n}\\ &=\boldsymbol{\Lambda}\mathbf{x}_a+\mathbf{n}\end{aligned} \tag{6.358}$$

where $E[\mathbf{n}\mathbf{n}^H] = N_0\mathbf{I}$.

The vector $\boldsymbol{\lambda}$ could be estimated as $\mathbf{X}\mathbf{y}$, but noise can be eliminated by first estimating $\mathbf{h}$, which has at most m nonzero components. A rough estimator of $\mathbf{h}$ is the $G \times 1$ vector

$$\widehat{\mathbf{h}}_r = G^{-1/2}\mathbf{F}\mathbf{X}\mathbf{y}. \tag{6.359}$$

The final $G - m - 1$ components are set to zero by the refined estimator

$$\widehat{\mathbf{h}} = \mathbf{I}_{m+1}\widehat{\mathbf{h}}_r \tag{6.360}$$

where $\mathbf{I}_{m+1}$ is the $G \times G$ diagonal matrix with its first $m + 1$ diagonal values equal to 1 and its remaining diagonal values set equal to 0. A refined estimator of $\boldsymbol{\lambda}$ is

$$\begin{aligned}\widehat{\boldsymbol{\lambda}} &= G^{1/2}\mathbf{F}^H\widehat{\mathbf{h}}.\\ &= \mathbf{F}^H\mathbf{I}_{m+1}\mathbf{F}\mathbf{X}\mathbf{y}\end{aligned} \tag{6.361}$$

where the $G\times G$ product matrix $\mathbf{F}^H\mathbf{I}_{m+1}\mathbf{F}\mathbf{X}$ can be stored in the receiver. Substituting (6.358), (6.271), and (6.272) into (6.362), we obtain

$$\widehat{\boldsymbol{\lambda}} = \boldsymbol{\lambda}+\mathbf{n}_e \tag{6.362}$$

where $\mathbf{n}_e = \mathbf{F}^H \mathbf{I}_{m+1} \mathbf{F} \mathbf{X} \mathbf{n}$. The covariance matrix of $\mathbf{n}_e$ is given by (6.326), and the total noise power is given by (6.327). Thus, knowledge of the multipath delay spread enables the channel estimator of (6.361) to reduce the total noise power by the factor $(m+1)/G$.

Comparisons

Simulation and numerical results indicate that when the same equalizers are used, the DS-CDMA-FDE and MC-CDMA systems with the same equalizers provide nearly the same performance [1, 2]. Both systems benefit from the use of joint antenna diversity and equalization, but the performance improvement hinges on accurate calculations of the discrete Fourier transforms.

Although FDE using MRC is essentially rake combining in the spectral domain, there are practical differences. As the frequency selectivity increases, the number of paths with significant power increases, thereby increasing the required number of rake fingers. In contrast, the complexity of FDE implementation is independent of the frequency selectivity. When (6.311) is not satisfied, simulation results indicate that FDE with MMSE usually provides better performance than FDE with MRC.

The OFDM system does not provide the diversity gain of the DS-CDMA-FDE and MC-CDMA systems. However, when channel coding and interleaving are used, an OFDM system provides not only a higher throughput but also time diversity and more coding gain than the DS-CDMA-FDE and MC-CDMA systems.

6.15 Problems

1 Give an alternative derivation of (6.46). We observe that the total received Doppler power $S_r(f)\,|df|$ in the spectral band $[f, f+df]$ corresponds to arrival angles determined by $f_d \cos\theta = f$. For $|\theta| \leq \pi$, $S_r(f)\,|df| = P(\theta)\,|d\theta| + P(-\theta)\,|d\theta|$, where $P(\theta)$ is the power density arriving from angle θ. Assume that the received power arrives uniformly spread over all angles $|\theta| \leq \pi$.

2 Use mathematical induction to prove that the right-hand side of (6.113) is equal to the right-hand side of (6.112).

3 Use Taylor series expansions to calculate the ratio of $P_b(L)$ for independent Rayleigh fading to $P_b^{cc}(L)$ for completely correlated Rayleigh fading when $\bar{\gamma} >> 1$. Observe that the ratio is proportional to $\bar{\gamma}^{-L+1}$, which clearly shows the large disparity in performance between a system with completely correlated fading and one with independent fading when $\bar{\gamma}$ is sufficiently large.

4 Derive an explicit equation for the constant C in (6.147). This constant does not depend on which orthogonal signal was transmitted.

5 Use mathematical induction to prove the fourth equality of (6.165).

6 Three multipath components arrive at a direct-sequence receiver moving at 30 m/s relative to the transmitter. The second and third multipath components travel over paths 200 and 250 m longer than the first component. (a) If the chip rate is equal to the bandwidth of the received signal, what is the minimum chip rate required to resolve all components? (b) Let t_e denote the time required to estimate the relative delay of a multipath component, and let v denote the relative radial velocity of a receiver relative to a transmitter. Then vt_e/c is the change in delay that occurs during the estimation procedure, where c is the speed of an electromagnetic wave. How much time can the receiver allocate to the estimation of the component delays?

7 Consider dual rake combining and Rayleigh fading. Compare BPSK and MRC with two noncoherent orthogonal signals and EGC by deriving approximate equations for $P_b(2)$ when $\gamma_1 >> 1 >> \gamma_2$. Show that BPSK and MRC provide a power advantage of more than 6 dB.

8 Consider dual rake combining and Rayleigh fading. (a) For two noncoherent orthogonal signals and EGC, find the lower bound on $\overline{\gamma}_2$ such that $P_s(2) \leq P_s(1)$. (b) What is the physical reason why $P_s(2) = P_s(1)$ when $\gamma_2 = 0$ for MRC but not for noncoherent combining?

9 Verify that (6.269) gives the eigenvalues of **H**.

10 An MC-CDMA system uses $G = 8$ chips per data symbol and transmits over a communication channel with $\mathbf{h} = \sqrt{4/5}\,[1\ \ 0\ \ 0.5]^T$. (a) Compute the eigenvalues of **F**, and assess the frequency selectivity of the channel by examining their magnitudes. (b) Evaluate γ_{sk} for MRC and ZF time-domain equalizers in terms of $x = c\mathcal{E}_{sk}/N_0$ and $y = c\mathcal{E}_{t/k}/N_0$. What is the maximum y as a function of x for which the MRC equalizer outperforms the ZF equalizer?

11 Compare MC-CDMA systems with time-domain equalization and a single signal. Use the Cauchy-Schwarz inequality to show that the SINR for ZF is less than or equal to the SINR for MRC.

12 An MC-CDMA system receives $m = G$ equal multipath components so that $h_i = 1/\sqrt{G}$, $0 \leq i \leq G-1$. Which of the three equalizers provides the largest SINR?

13 Compare the SINRs of the MC-CDMA and DS-CDMA-FDE systems for all three equalizers when each receives a single multipath component. Thus, $h_0 = 1/\sqrt{G}$, and $h_i = 0$, $1 \leq i \leq G-1$. Which equalizer provides the largest SINR?

14 Prove that the DS-CDMA-FDE system with the MMSE equalizer uses the weights given by (6.309).

Chapter 7
Code-Division Multiple Access

Multiple access is the ability of many users to communicate with each other while sharing a common transmission medium. Wireless multiple-access communications are facilitated if the transmitted signals are orthogonal or separable in some sense. Signals may be separated in time (*time-division multiple access* or TDMA), frequency (*frequency-division multiple access* or FDMA), or code (*code-division multiple access* or CDMA). This chapter presents the general characteristics of *direct-sequence* CDMA (DS-CDMA) and *frequency-hopping* CDMA (FH-CDMA) systems. The use of spread-spectrum modulation with CDMA allows the simultaneous transmission of signals from multiple users in the same frequency band. All signals use the entire allocated spectrum, but the spreading sequences or frequency-hopping patterns differ. Information theory indicates that in an isolated cell, CDMA systems achieve the same spectral efficiency as TDMA or FDMA systems only if optimal multiuser detection is used. However, even with single-user detection, CDMA has advantages for mobile communication networks because it eliminates the need for frequency and time-slot coordination, allows carrier-frequency reuse in adjacent cells, imposes no sharp upper bound on the number of users, and provides resistance to interference and interception. In this chapter, the vast potential and practical difficulties of spread-spectrum multiuser detectors, such as optimal, decorrelating, minimum mean-square error, or adaptive detectors, are described and assessed. The tradeoffs and design issues of direct-sequence multiple-input multiple-output with spatial multiplexing or beamforming are determined.

7.1 Implications of Information Theory

Information theory provides a means of ascertaining the potential benefits and trade-offs in using the various multiple-access methods. The main results stem from evaluations of the channel capacity. The *channel capacity* of an additive white

D. Torrieri, *Principles of Spread-Spectrum Communication Systems*,
https://doi.org/10.1007/978-3-319-70569-9_7

Gaussian noise (AWGN) channel with noise variance $\mathcal{N}$ and power constraint $\mathcal{P}$ is defined as the maximum of the average mutual information $I(X;Y)$ over all possible source-code or input-symbol distribution functions. For the one-dimensional AWGN channel with continuously distributed real-valued input and output symbols, fundamental results of information theory [20] are that the optimal input-symbol distribution is Gaussian and that the *one-dimensional channel capacity* is

$$\mathcal{C} = \frac{1}{2}\log_2\left(1 + \frac{\mathcal{P}}{\mathcal{N}}\right) \tag{7.1}$$

in *bits per channel use*. For the AWGN channel with a power constraint and a code-rate $\mathcal{R}$, there exists a sequence of codes such that the maximal probability of error tends to be zero if $\mathcal{R} \leq \mathcal{C}$.

For the two-dimensional AWGN channel with continuously distributed complex-valued input and output symbols, the real and imaginary components of the symbols are affected by independent Gaussian noises with the same power $\mathcal{N}$. The channel capacity is the sum of the capacities of the two components. The channel capacity is maximized if the total symbol power $\mathcal{P}$ is allocated equally to the two components. Therefore, the *two-dimensional channel capacity* is

$$\mathcal{C} = \log_2\left(1 + \frac{\mathcal{P}}{2\mathcal{N}}\right). \tag{7.2}$$

Since the total noise power is $2\mathcal{N}$, the signal-to-noise ratio (SNR) for the two-dimensional channel is $\mathcal{P}/2\mathcal{N}$. Thus, the maximum rate $\mathcal{R}$ supported by the two-dimensional AWGN channel is $\log_2(1 + SNR)$.

Consider a bandlimited AWGN channel for which signals are bandlimited by the one-sided bandwidth W Hz. If the two-sided power spectral density (PSD) of the noise is $N_0/2$, then the noise power is $\mathcal{N} = N_0 W$. The sampling theorem (Appendix D.4) indicates that bandlimited signals are completely determined by samples spaced $1/2W$ seconds apart. The autocorrelation function of the noise, calculated from its PSD, indicates that each noise sample is an independent, identically distributed Gaussian random variable. Since the channel can be used independently $2W$ times per second, (7.1) implies that the one-dimensional channel capacity is

$$\mathrm{C} = W\log_2\left(1 + \frac{\mathcal{P}}{N_0 W}\right) \tag{7.3}$$

in *bits per second*, whereas (7.2) implies that the two-dimensional channel capacity is

$$\mathrm{C} = 2W\log_2\left(1 + \frac{\mathcal{P}}{2N_0 W}\right) \tag{7.4}$$

in *bits per second*.

Consider the two-dimensional AWGN multiple-access channel that has m users with powers $\mathcal{P}_1, \mathcal{P}_2, \ldots, \mathcal{P}_m$ and noise with power $\mathcal{N} = N_0 W$ at the receiver. The analysis and results are similar for the analogous one-dimensional AWGN multiple-access channel. Define

$$\mathrm{C}(x) = 2\log_2(1+x)\,. \tag{7.5}$$

Information theory indicates that for a low probability of error in the receiver, the rates $\mathcal{R}_1, \mathcal{R}_2, \ldots, \mathcal{R}_m$ of the source codes, in bits per second, are bounded by the inequalities [20]

$$\mathcal{R}_i \leq W\mathrm{C}\left(\frac{\mathcal{P}_i}{2N_0 W}\right) \tag{7.6}$$

for $i = 1, 2, \ldots, m$ and

$$\sum_{i\in S}\mathcal{R}_i \leq W\mathrm{C}\left(\frac{\sum_{i\in S}\mathcal{P}_i}{2N_0 W}\right) \tag{7.7}$$

where S is any subset of the m source codes of the m users. Inequality (7.6) follows from the upper bound on an individual rate that applies even when the other users cause no interference. Inequality (7.7) indicates that the sum of the rates cannot exceed the rate achieved by a single code with a received power equal to the sum of the m powers. Inequalities (7.6) and (7.7) restrict the rates $\mathcal{R}_1, \mathcal{R}_2, \ldots, \mathcal{R}_m$ to a bounded region within an m-dimensional rate hyperspace. Because of (7.7), not all of the individual rates can attain the upper bounds of (7.6). The underlying cause is the unavoidable mutual interference among the users.

Assume that all m users cooperate to simultaneously transmit their signals to a single receiver, S comprises all m source codes, and an interfering signal can be modeled as AWGN. A *multiuser detector* jointly demodulates and decodes all the received signals. Points on the boundary of the rate hyperspace can be attained by implementing a multistage decoding process that executes *successive interference cancelation*. In the initial stage, code 1 is decoded with each code $j > 1$ regarded as AWGN. Then, (7.6) implies that there is a low probability of error for code 1 if

$$\mathcal{R}_1 = W\mathrm{C}\left(\frac{\mathcal{P}_1}{\sum_{i\in S, i\neq 1}\mathcal{P}_i + 2N_0 W}\right). \tag{7.8}$$

After code 1 is subtracted from the received signal, the remaining codes can be extracted with a low probability of error if (7.6) is satisfied for $i = 2, 3, \ldots, m$ and

$$\begin{aligned}\sum_{i\in S, i\neq 1}\mathcal{R}_i &= \sum_{i\in S}\mathcal{R}_i - \mathcal{R}_1 \\ &\leq W\mathrm{C}\left(\frac{\sum_{i\in S}\mathcal{P}_i}{2N_0 W}\right) - W\mathrm{C}\left(\frac{\mathcal{P}_1}{\sum_{i\in S, i\neq 1}\mathcal{P}_i + 2N_0 W}\right).\end{aligned} \tag{7.9}$$

Using (7.5) in (7.9), we find that

$$\sum_{i\in S, i\neq 1} \mathcal{R}_i \leq W\mathrm{C}\left(\frac{\sum_{i\in S, i\neq 1} \mathcal{P}_i}{2N_0 W}\right). \tag{7.10}$$

Further stages of the successive interference cancelation proceed in a similar manner. Thus, a multiuser detector can provide a low probability of error if the code rates are suitably constrained.

A *single-user detector* or *conventional detector* ignores the presence of the interfering signals and does not execute successive interference cancelation. Interfering signals are modeled as additional Gaussian noise. Therefore, if S includes all m source codes, the rates of the source codes are bounded by

$$\mathcal{R}_i \leq W\mathrm{C}\left(\frac{\mathcal{P}_i}{\sum_{j\in S, j\neq i} \mathcal{P}_j + 2N_0 W}\right) \tag{7.11}$$

for $i = 1, 2, \ldots, m$. In general, a conventional detector limits the code rates much more than a multiuser detector. For example, if all m sources have the same power $\mathcal{P}$ and rate $\mathcal{R}$, then (7.11) implies that a conventional detector requires that

$$\mathcal{R} \leq W\mathrm{C}\left(\frac{\mathcal{P}}{(m-1)\mathcal{P} + 2N_0 W}\right) \tag{7.12}$$

whereas (7.7) implies that a multiuser detector requires that

$$\mathcal{R} \leq \frac{W}{m}\mathrm{C}\left(\frac{m\mathcal{P}}{2N_0 W}\right). \tag{7.13}$$

Inequality (7.13) is directly applicable to a DS-CDMA network in which all m users have the same power $\mathcal{P}$ and rate $\mathcal{R}$. In a TDMA network with equal time slots and common rate $\mathcal{R}$ allocated to all users, each user transmits $1/m$ of the time with power $m\mathcal{P}$ during an assigned slot and zero otherwise. Therefore, $\mathcal{R}$ must satisfy (7.13).

In an FDMA network with an equal bandwidth W/m for each spectral band and equal rate $\mathcal{R}$ allocated to all users, each user can transmit simultaneously in distinct spectral bands with power $\mathcal{P}$. Since the bandwidth of each spectral band is W/m, the noise power received by each user is $2N_0 W/m$. Thus, application of (7.6) indicates that $\mathcal{R}$ must satisfy (7.13). Consider an FH-CDMA network with frequency channels that have the same bandwidth W/m. We assume that all users transmit with the same power $\mathcal{P}$, but their frequency-hopping patterns are synchronous so that no collisions occur in the same frequency channel. With these assumptions, the FH-CDMA network is equivalent to an FDMA network with periodic variations in the spectral allocations; hence, $\mathcal{R}$ must satisfy (7.13).

These results indicate that FDMA, TDMA, DS-CDMA, and FH-CDMA networks impose the same upper bound on the achievable code rates, but the latter three networks have important practical limitations. A TDMA network cannot increase mP beyond the peak transmitter power that can be sustained. A DS-CDMA network must use multiuser detection for (7.13) to be applicable. If it does not, the achievable rate for a low probability of error is constrained by the more restrictive (7.12). FH-CDMA networks require synchronous operation with orthogonal frequency-hopping patterns to ensure that no collisions occur.

A competitive and frequently adopted alternative to DS-CDMA for CDMA networks with frequency-selective fading is orthogonal frequency-division multiplexing (OFDM), which uses orthogonal subcarriers with less frequency separation than that achieved by classical FDMA. OFDM transmits code symbols over parallel narrowband channels with flat fading that can be easily equalized. By selecting the symbol duration in an OFDM system to be significantly larger than the channel dispersion and using a cyclic prefix, intersymbol interference can be avoided. As the code-symbol duration decreases, OFDM can add more subcarriers and preserve the duration of a transmitted OFDM symbol. Among the disadvantages of OFDM are its need for the cyclic prefix, which reduces the spectral efficiency, and its sensitivity to the peak-to-average-power ratio.

7.2 Spreading Sequences for DS-CDMA

The *periodic autocorrelation* of a periodic complex-valued or polyphase sequence with period N is defined as

$$\theta_p(l) = \frac{1}{N}\sum_{n=0}^{N-1} p_n p_{n+l}^* \tag{7.14}$$

where p_n is the nth component of the sequence, and the asterisk denotes the complex conjugate. The *periodic cross-correlation* of periodic complex-valued or polyphase sequences $\mathbf{p}$ and $\mathbf{q}$ with the same period N is defined as

$$\theta_{pq}(l) = \frac{1}{N}\sum_{n=0}^{N-1} p_n q_{n+l}^* = \frac{1}{N}\sum_{n=0}^{N-1} p_{n-l} q_n^* \,. \tag{7.15}$$

Two sequences are *orthogonal* if $\theta_{pq}(0) = 0$.

Let $\mathbf{a} = (\ldots, a_0, a_1, \ldots)$ and $\mathbf{b} = (\ldots, b_0, b_1, \ldots)$ denote binary sequences with components in $GF(2)$. The sequences $\mathbf{a}$ and $\mathbf{b}$ are mapped into antipodal sequences $\mathbf{p}$ and $\mathbf{q}$, respectively, with components in $\{-1, +1\}$ by means of the transformation

$$p_i = (-1)^{a_i+1}, \quad q_i = (-1)^{b_i+1}. \tag{7.16}$$

After this mapping, the periodic autocorrelation of a periodic binary sequence **p** with period N is defined by (7.14). The *periodic cross-correlation* of periodic binary sequences **a** and **b** with the same period N is defined as the periodic cross-correlation of the antipodal sequences **p** and **q**, which is defined by (7.15). Substitution of (7.16) into (7.15) indicates that the periodic cross-correlation of **a** and **b** is given by

$$\theta_{pq}(l) = \frac{A_l - D_l}{N} \tag{7.17}$$

where A_l denotes the number of agreements in the corresponding components of **b** and the shifted sequence $\mathbf{a}(l)$, and D_l denotes the number of disagreements.

Orthonormal Sequences

In constructing N orthonormal sequences of length N with components in $\{-1, +1\}$, the first issue is to establish the values of N for which the construction is possible. Equation (7.17) indicates that N must be an even number. To find further necessary conditions, suppose that N orthonormal sequences are represented as rows in an $N \times N$ matrix $\mathbf{H}$ with elements equal to +1 or -1. An interchange of the rows or columns of $\mathbf{H}$ or a multiplication of a row or column by -1 does not change the orthonormality. By multiplying by -1 those columns with a -1 in the first row, we obtain an $\mathbf{H}$ with a first row that has all elements equal to +1. The orthonormality condition then requires each of the $N-1$ remaining rows to have $N/2$ elements equal to +1 and $N/2$ elements equal to -1. An appropriate interchange of columns establishes a second row with its first $N/2$ elements equal to +1 and its remaining $N/2$ elements equal to -1. If $N \geq 4$ and there are α elements equal to +1 in the first half of the third row and $N/2 - \alpha$ in the second half, then the orthogonality of the second and third rows requires that $\alpha = N/2 - \alpha$, which implies that $\alpha = N/4$. Since α must be an integer, N must be divisible by 4. Thus, *a necessary condition for the existence of N orthonormal sequences of length N is that $N = 2$ or N is a multiple of 4.*

This necessary condition is not a sufficient condition. However, a specific construction procedure establishes the existence of $2^n \times 2^n$ matrices with orthonormal rows for $n \geq 1$. Two binary sequences, each of length 2, are orthogonal if each sequence is described by one of the rows of the 2×2 matrix

$$\mathbf{H}_1 = \begin{bmatrix} +1 & +1 \\ +1 & -1 \end{bmatrix}. \tag{7.18}$$

A set of 2^n mutually orthogonal sequences, each of length 2^n, is obtained by using the rows of the matrix

$$\mathbf{H}_n = \begin{bmatrix} \mathbf{H}_{n-1} & \mathbf{H}_{n-1} \\ \mathbf{H}_{n-1} & \bar{\mathbf{H}}_{n-1} \end{bmatrix}, \quad n \geq 2 \tag{7.19}$$

where $\bar{\mathbf{H}}_{n-1}$ is the *complement* of $\mathbf{H}_{n-1}$, obtained by replacing each +1 and −1 by −1 and +1, respectively, and $\mathbf{H}_1$ is defined by (7.18). Any pair of rows in $\mathbf{H}_n$ differ in exactly 2^{n-1} columns, thereby ensuring orthogonality of the corresponding sequences. The $2^n \times 2^n$ matrix $\mathbf{H}_n$, which is called a *Hadamard matrix*, can be used to generate 2^n *orthogonal spreading sequences* for synchronous direct-sequence communications. The orthogonal spreading sequences generated from a Hadamard matrix are called *Walsh sequences*.

Multirate CDMA systems, which are used for multimedia applications of CDMA2000 and wideband CDMA (WCDMA), have a variety of code-symbol rates for various services and users. The code-symbol rates might be accommodated using repetition codes, code puncturing, or the multicode method described in Section 7.9. Another way of accommodating different code-symbol rates is to maintain the chip rate, and hence the bandwidth, for all users while varying the spreading factors in accordance with the code-symbol rates. The spreading sequences are selected to be orthogonal to each other despite differences in the spreading factors. A tree-structured set of orthogonal sequences called the *orthogonal variable-spreading-factor (OVSF) codes* can be generated recursively and enable the receiver to completely avoid multiple-access interference among the users [3].

Let $\mathbf{C}_N(n)$ denote the row vector representing the *nth* OVSF sequence with spreading factor N, where $n = 1, 2, \ldots, N$, and $N = 2^k$ for some nonnegative integer k. The set of N sequences with N chips is derived by concatenating sequences from the set of $N/2$ sequences with $N/2$ chips:

$$\begin{aligned}
\mathbf{C}_N(1) &= [\mathbf{C}_{N/2}(1)\ \mathbf{C}_{N/2}(1)] \\
\mathbf{C}_N(2) &= [\mathbf{C}_{N/2}(1)\ \overline{\mathbf{C}}_{N/2}(1)] \\
&\ \ \vdots \\
\mathbf{C}_N(N-1) &= [\mathbf{C}_{N/2}(N/2)\ \mathbf{C}_{N/2}(N/2)] \\
\mathbf{C}_N(N) &= [\mathbf{C}_{N/2}(N/2)\ \overline{\mathbf{C}}_{N/2}(N/2)].
\end{aligned} \tag{7.20}$$

For example, $\mathbf{C}_{16}(4)$ is produced by concatenating $\mathbf{C}_8(2)$ and $\overline{\mathbf{C}}_8(2)$, thereby doubling the number of chips per code symbol to 16. A sequence used in the recursive generation of a longer sequence is called a *mother code* of the longer sequence. Equation (7.20) indicates that all the sequences with N chips are orthogonal to each other, and these sequences constitute a set of orthogonal Walsh sequences.

Let R denote the code-symbol rate supported by an OVSF sequence of length N. Since the chip rate is maintained while the spreading factor decreases from N to 1, the corresponding code-symbol rate increases from R to NR. A tree diagram illustrating the hierarchy of sequences is shown in Figure 7.1. Each $\mathbf{C}_N(n)$ is orthogonal to concatenations of all sequences $\mathbf{C}_{N/2}(n'), \mathbf{C}_{N/4}(n''), \ldots$ and their complements except for its mother codes. For example, $\mathbf{C}_{16}(3)$ is not orthogonal to its mother codes $\mathbf{C}_8(2)$, $\mathbf{C}_4(1)$, or $\mathbf{C}_2(1)$. If $\mathbf{C}_8(3)$ is assigned to a user requesting

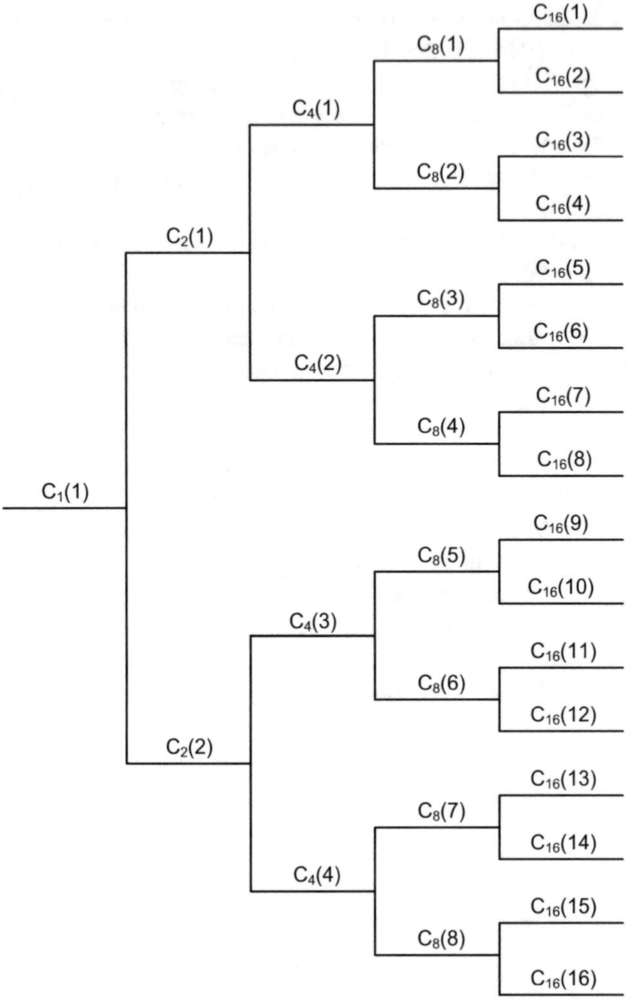

Fig. 7.1 Tree diagram of orthogonal variable-spreading-factor code

a code-symbol rate twice that of a user assigned a sequence of 16 chips, then the sequences $\mathbf{C}_{16}(5)$ and $\mathbf{C}_{16}(6)$ descended from $\mathbf{C}_8(3)$ cannot be assigned to other users requesting lower code-symbol rates, and the mother codes of $\mathbf{C}_8(3)$ cannot be assigned to other users requesting higher code-symbol rates.

The multirate capacity of a multirate DS-CDMA system using an OVSF code is the maximum code-symbol rate that the system can accommodate. Achieving this multirate capacity may require high code-rate users to transmit with high powers to compensate for the low spreading factors. The unavailability or *blocking* of ancestors and descendants may cause some new code-rate requests to be rejected even though the system has sufficient capacity to accept them. Thus, potential multirate capacity is wasted. Another source of wasted capacity and an inflexibility for rate-matching is due to the quantization of code-symbol rates and spreading factors that must be powers of 2. A number of code assignment schemes have been proposed to reduce or even eliminate the wasted capacity [81]. A basic limitation of multirate DS-CDMA systems is the implementation complexity due to the need for multiple rake demodulators, each of which responds to a different spreading sequence.

Polyphase Sequences

Quaternary direct-sequence systems may use complex binary spreading sequences (Section 2.5), which are pairs of short binary sequences, in a DS-CDMA network. An alternative for direct-sequence systems is to use complex-valued *polyphase sequences* that are not derived from pairs of binary sequences but have better periodic correlation functions.

Symbols of polyphase sequences are powers of the complex qth root of unity, which is

$$\Omega = \exp\left(j\frac{2\pi}{q}\right) \tag{7.21}$$

where $j = \sqrt{-1}$. The complex spreading or signature sequence $\mathbf{p}$ of period N has symbols given by

$$p_i = \Omega^{a_i} e^{j\phi}, \qquad a_i \in Z_q = \{0, 1, 2, \ldots, q-1\}, \qquad i = 1, 2, \ldots, N \tag{7.22}$$

where Z_q is a set of real-valued exponents, and ϕ is an arbitrary phase chosen for convenience. If p_i is specified by the exponent a_i and q_i is specified by the exponent b_i, then the periodic cross-correlation between sequences $\mathbf{p}$ and $\mathbf{q}$ is

$$\theta_{pq}(k) = \frac{1}{N}\sum_{i=0}^{N-1} \Omega^{a_i - b_i}. \tag{7.23}$$

Polyphase sequences with $q = 2$ are real-valued binary antipodal sequences, and polyphase sequences with $q = 4$ are complex-valued quaternary sequences.

The sequences in a family of polyphase sequences with period $N = 2^m - 1$ can be generated by a shift register with nonlinear feedback. The feedback coefficients $\{c_i\}$ are determined by the *characteristic polynomial*, which is defined as

$$f(x) = 1 + \sum_{i=1}^{m} c_i x^i, \; c_i \in Z_q, \; c_m = 1. \tag{7.24}$$

The shift-register output sequence $\{a_i\}$ satisfies a linear recurrence relation with the same form as (2.20). Each output symbol $a_i \in Z_q$ is converted to p_i according to (7.22).

For example, a family of quaternary sequences with $m = 3$ and period $N = 7$ has the characteristic polynomial $f(x) = 1+2x+3x^2+x^3$. A feedback shift register that implements the sequences of the family is depicted in Figure 7.2 (a), where all operations are modulo-4. The generation of a particular sequence is illustrated in Figure 7.2 (b).

Different quaternary sequences may be generated by loading the shift register with any nonzero initial contents and then cycling the shift register through its full period $N = 2^m - 1$. Since the shift register has $4^m - 1$ nonzero states, there are $M = (4^m - 1)/(2^m - 1) = 2^m + 1$ *cyclically distinct* members of the family. Each cyclically distinct family member may be generated by loading the shift register with any nonzero triple that is not a state occurring during the generation of another family member.

A polyphase spreading sequence multiplies a complex-valued code sequence to produce the transmitted sequence. The generation of one chip of a transmitted quaternary sequence is represented in Figure 2.22 of Section 2.5, and a representation of the receiver in terms of complex variables is illustrated in Figure 2.23. Quaternary sequences ensure balanced power in the in-phase and quadrature branches of the transmitter, which limits the peak-to-average power fluctuations.

Since polyphase sequences have more favorable periodic autocorrelations and cross-correlations than pairs of binary sequences, they provide a potential advantage in code synchronization systems (Chapter 4). In contrast, polyphase sequences do not provide significantly smaller error probabilities in asynchronous DS-CDMA networks because system performance is determined by the aperiodic autocorrelations and cross-correlations [50, 120].

Welch Bound

For a set S of M periodic polyphase sequences of length N, let $\theta_{\max}$ denote the peak magnitude of the periodic cross-correlations or autocorrelations:

$$\theta_{\max} = \max\left\{\left|\theta_{pq}(k)\right| : 0 \le k \le N-1;\ \mathbf{p},\mathbf{q} \in S;\ \mathbf{p} \neq \mathbf{q} \text{ or } k \neq 0\right\}. \tag{7.25}$$

Theorem 1 *A set S of M periodic polyphase sequences of length N has*

$$\theta_{\max} \ge \sqrt{\frac{M-1}{MN-1}}\,. \tag{7.26}$$

Proof Consider an extended set S_e of MN sequences $\mathbf{p}_i$, $i = 1, 2, \ldots, MN$ comprising the N distinct shifted sequences derived from each of the sequences

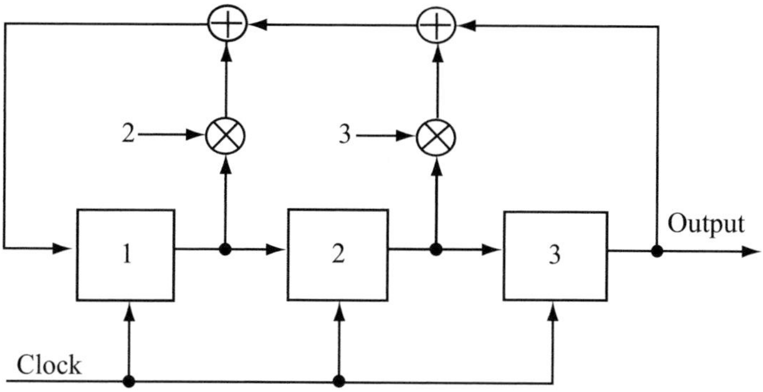

(a)

Shift	Contents		
	Stage 1	Stage 2	Stage 3
Initial	0	0	1
1	1	0	0
2	2	1	0
3	3	2	1
4	1	3	2
5	1	1	3
6	0	1	1
7	0	0	1

(b)

Fig. 7.2 (**a**) Feedback shift register for quaternary sequence and (**b**) contents after successive shifts

in S. The periodic cross-correlation of sequences $\mathbf{p}_i$ and $\mathbf{p}_l$ in S_e is

$$\psi_{i,l} = \frac{1}{N} \sum_{n=1}^{N} p_{i,n} p_{l,n}^* \tag{7.27}$$

and

$$\theta_{\max} = \max \{ |\psi_{i,l}| : \mathbf{p}_i \in S_e, \ \mathbf{p}_l \in S_e, \ i \neq l \} .$$

Define the double summation

$$Z = \sum_{i=1}^{MN} \sum_{l=1}^{MN} \psi_{i,l}^2 . \tag{7.28}$$

Separating the MN terms for which $\psi_{i,i} = 1$ and then bounding the remaining $MN(MN-1)$ terms yields

$$Z \leq MN + MN(MN-1)\theta_{\max}^2 \, . \tag{7.29}$$

Substituting (7.27) into (7.28), interchanging summations, and omitting the terms for which $m \neq n$, we obtain

$$\begin{aligned} Z &= \frac{1}{N^2} \sum_{n=1}^{N} \sum_{m=1}^{N} \sum_{i=1}^{MN} p_{i,n} p_{i,m}^* \sum_{l=1}^{MN} p_{l,n} p_{l,m}^* \\ &= \frac{1}{N^2} \sum_{n=1}^{N} \sum_{m=1}^{N} \left(\sum_{i=1}^{MN} p_{i,n} p_{i,m}^* \right)^2 \\ &\geq \frac{1}{N^2} \sum_{n=1}^{N} \left(\sum_{i=1}^{MN} |p_{i,n}|^2 \right)^2 \\ &= M^2 N. \end{aligned}$$

Combining this inequality with (7.29) gives (7.26). □

The lower bound in (7.26) is known as the *Welch bound*. It approaches $1/\sqrt{N}$ for large values of M and N.

Gold and Kasami Sequences

Only small subsets of maximal sequences can be found with $\theta_{\max}$ close to the Welch bound. Large sets of binary sequences with $\theta_{\max}$ approaching the Welch bound can be obtained by combining maximal sequences with subsequences of these sequences. If q is a positive integer, the new binary sequence $\mathbf{b}$ formed by taking every qth bit of binary sequence $\mathbf{a}$ is known as a *decimation* of $\mathbf{a}$ by q, and the components of the two sequences are related by $b_i = a_{qi}$. Let $gcd(x, y)$ denote the greatest common divisor of x and y. If the original sequence $\mathbf{a}$ has a period N and the new sequence $\mathbf{b}$ is not identically zero, then $\mathbf{b}$ has period $N/gcd(N, q)$. If $gcd(N, q) = 1$, then the decimation is called a *proper decimation*. Following a proper decimation, the bits of $\mathbf{b}$ do not repeat themselves until every bit of $\mathbf{a}$ has been sampled. Therefore, $\mathbf{b}$ and $\mathbf{a}$ have the same period N.

If $\mathbf{a}$ is a maximal sequence, then if each bit of $\mathbf{a}$ is sampled, $\mathbf{b}$ is a maximal sequence. The sequences $\mathbf{a}$ and $\mathbf{b}$ are mapped into antipodal sequences $\mathbf{p}$ and $\mathbf{q}$, respectively, with components in $\{-1, +1\}$ by means of the transformation of (7.16). A *preferred pair* of antipodal maximal sequences with period $2^m - 1$ are a pair with a periodic cross-correlation that takes only the three values $-t(m)/N, -1/N$, and $[t(m) - 2]/N$, where

$$t(m) = 2^{\lfloor (m+2)/2 \rfloor} + 1 \tag{7.30}$$

and $\lfloor x \rfloor$ denotes the integer part of the real number x.

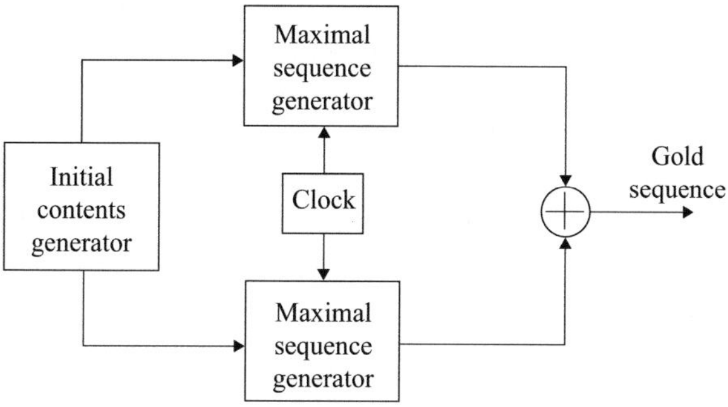

Fig. 7.3 Gold sequence generator

The *Gold sequences* [28] are a large set of sequences with period $N = 2^m - 1$ that may be generated by the modulo-2 addition of preferred pairs when m is odd or $m = 2$ modulo-4. One sequence of the preferred pair is a decimation by q of the other sequence. The positive integer q is either $q = 2^k + 1$ or $q = 2^{2k} - 2^k + 1$, where k is a positive integer such that $gcd(m, k) = 1$ when m is odd and $gcd(m, k) = 2$ when $m = 2$ modulo-4. Since the periodic cross-correlation between any two Gold sequences in a set can take only three values, the peak magnitude of the periodic cross-correlation between any two Gold sequences of period $N = 2^m - 1$ is

$$\theta_{max} = \frac{t(m)}{2^m - 1} . \tag{7.31}$$

For large values of m, θ_{max} for Gold sequences exceeds the Welch bound by a factor of $\sqrt{2}$ for m odd and a factor of 2 for m even.

One form of a Gold sequence generator is shown in Figure 7.3. If each maximal sequence generator has m stages, different Gold sequences in a set are generated by selecting the initial state of one maximal sequence generator and then shifting the initial state of the other generator. Since any shift from 0 to $2^m - 2$ results in a different Gold sequence, $2^m - 1$ different Gold sequences can be produced by the system shown in Figure 7.3. Gold sequences identical to maximal sequences are produced by setting the state of one of the maximal sequence generators to zero. Altogether, there are $2^m + 1$ different Gold sequences, each with a period of $2^m - 1$, in the set.

An example of a set of Gold sequences is the set generated by the primitive characteristic polynomials

$$f_1(x) = 1 + x^3 + x^7, \quad f_2(x) = 1 + x + x^2 + x^3 + x^7 \tag{7.32}$$

which specify a preferred pair of maximal sequences. Since $m = 7$, there are 129 Gold sequences with period 127 in this set, and (7.31) gives $\theta_{max} = 0.134$. Equation (2.43) indicates that there are only 18 maximal sequences with $m = 7$. For this set of 18 sequences, calculations indicate that $\theta_{max} = 0.323$. If $\theta_{max} = 0.134$ is desired for a set of maximal sequences with $m = 7$, then the set has only six sequences. This result illustrates the much greater utility of Gold sequences in CDMA networks with many subscribers.

As shown in Section 2.5, the generating function of the output sequence generated by a linear feedback shift register with characteristic polynomial $f(x)$ may be expressed in the form

$$G(x) = \frac{\phi(x)}{f(x)} \tag{7.33}$$

where the degree of $\phi(x)$ is less than the degree of $f(x)$, and

$$\phi(x) = \sum_{i=0}^{m-1} x^i \left(a_i + \sum_{k=1}^{i} c_k a_{i-k} \right) \tag{7.34}$$

where the $\{c_k\}$ are the coefficients of $f(x)$, and the $\{a_k\}$ are the initial contents. If the sequence generators shown in Figure 7.2 have the primitive characteristic polynomials $f_1(x)$ and $f_2(x)$ of degree m, then the *generating function for the Gold sequence* is

$$\begin{aligned} G(x) &= \frac{\phi_1(x)}{f_1(x)} + \frac{\phi_2(x)}{f_2(x)} \\ &= \frac{\phi_1(x)f_2(x) + \phi_2(x)f_1(x)}{f_1(x)f_2(x)} . \end{aligned} \tag{7.35}$$

Since the degrees of both $\phi_1(x)$ and $\phi_2(x)$ are less than m, the degree of the numerator of $G(x)$ must be less than $2m$. Since the product $f_1(x)f_2(x)$ has the form of a characteristic polynomial of degree $2m$, this product defines the feedback coefficients of a single linear feedback shift register with $2m$ stages that can generate the Gold sequences. The initial state of the register for any particular sequence can be determined by equating each coefficient in the numerator of (7.35) with the corresponding coefficient in (7.34) and then solving $2m$ linear equations. Thus, *a Gold sequence of period* $2^m - 1$ *can be generated by a single linear feedback shift register with* $2m$ stages.

A *small set of Kasami sequences* [28] comprises $2^{m/2}$ sequences with period $2^m - 1$ if m is even. To generate a set, a maximal sequence **a** with period $N = 2^m - 1$ is decimated by $q = 2^{m/2} + 1$ to form a binary sequence **b** with period $N/gcd(N, q) = 2^{m/2} - 1$. The modulo-2 addition of **a** and any cyclic shift of **b** from 0 to $2^{m/2} - 2$ provides a Kasami sequence. By including sequence **a**, we obtain a set of $2^{m/2}$ Kasami sequences with period $2^m - 1$. The periodic cross-correlation between any two Kasami sequences in a set can only take the values $-s(m)/N, -1/N$, or $[s(m) - 2]/N$, where

$$s(m) = 2^{m/2} + 1. \tag{7.36}$$

The peak magnitude of the periodic cross-correlation between any two Kasami sequences is

$$\theta_{max} = \frac{s(m)}{N} = \frac{1}{2^{m/2} - 1} . \tag{7.37}$$

The Kasami sequences are optimal in the sense that θ_{max} has the minimum value for any set of sequences of the same size and period. For proof, we observe that if $M = 2^{m/2}$ and $N = 2^m - 1 = M^2 - 1$, $m \geq 2$, then the Welch bound implies that

$$\begin{aligned} N\theta_{max} &\geq N\sqrt{\frac{M-1}{MN-1}} = \sqrt{\frac{N(M-1)}{M-N^{-1}}} \\ &= \sqrt{\frac{(M+1)(M-1)^2}{M-N^{-1}}} \\ &> M - 1 = 2^{m/2} - 1. \end{aligned} \tag{7.38}$$

Since N is an odd integer, A_l–D_l in (7.17) must be an odd integer; hence, $N\theta_{max}$ must be an odd integer. Since $2^{m/2} + 1$ is the smallest odd integer greater than $2^{m/2} - 1$, it follows that $M = 2^{m/2}$ periodic antipodal sequences of length $N = 2^m - 1$, $m \geq 2$, *require*

$$N\theta_{max} \geq 2^{m/2} + 1. \tag{7.39}$$

Since Kasami sequences have $N\theta_{max} = 2^{m/2} + 1$, they are optimal given their size and period.

As an example, let $m = 10$. There are 60 maximal sequences, 1025 Gold sequences, and 32 Kasami sequences with period 1023. The peak periodic cross-correlations are 0.37, 0.06, and 0.03, respectively.

A *large set of Kasami sequences* [28] comprises $2^{m/2}(2^m+1)$ sequences if $m = 2$ modulo-4 and $2^{m/2}(2^m + 1) - 1$ sequences if $m = 0$ modulo-4. The sequences have period $2^m - 1$. To generate a set, a maximal sequence **a** with period $N = 2^m - 1$ is decimated by $q = 2^{m/2} + 1$ to form a binary sequence **b** with period $N/gcd(N, q) = 2^{m/2} - 1$ and then decimated by $q_1 = 2^{(m+2)/2} + 1$ to form another binary sequence **c** with period $N/gcd(N, q_1)$. The modulo-2 addition of **a**, a cyclic shift of **b**, and a cyclic shift of **c** provide a Kasami sequence with period N. The periodic cross-correlations between any two Kasami sequences in a set can only take the values $-1/N$, $-t(m)/N$, $[t(m) - 2]/N$, $-s(m)/N$, or $[s(m) - 2]/N$. A large set of Kasami sequences includes both a small set of Kasami sequences and a set of Gold sequences as subsets. Since $t(m) \geq s(m)$, the value of θ_{max} for a large set is

$$\theta_{max} = \frac{t(m)}{2^m - 1} = \frac{2^{\lfloor (m+2)/2 \rfloor} + 1}{2^m - 1} . \tag{7.40}$$

This value is suboptimal, but the large size of these sets makes them an attractive option for asynchronous CDMA networks.

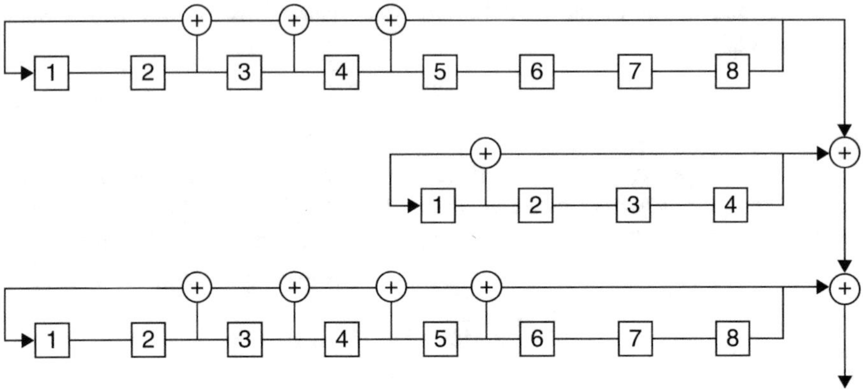

Fig. 7.4 Generator of Kasami sequences with period 255

A generator of a large set of 4111 Kasami sequences with $m = 8$ and period 255 is illustrated in Figure 7.4. The two shift registers at the top of the figure by themselves generate a small set of 16 Kasami sequences with $m = 8$ and period 15. The top 8-stage shift register generates a maximal sequence with period 255, and the 4-stage shift register below it generates a maximal sequence with period 15. The bottom shift register generates a nonmaximal sequence with period 85.

In a network of similar systems, interfering sequences are substantially suppressed during acquisition when code modulation is absent if the periodic cross-correlations among sequences are small, as they are if all the sequences are Gold or Kasami sequences. Some large families of polyphase sequences have the potential to provide better acquisition performance than the Gold or Kasami sequences. For a positive integer m, a family $\mathcal{A}$ of $M = N + 2$ quaternary or Z_4 sequences, each of period $N = 2^m - 1$, with θ_{max} that asymptotically approaches the Welch bound, has been identified [31]. In contrast, a small set of binary Kasami sequences has only $\sqrt{N+1}$ sequences.

7.3 Synchronous and Asynchronous Communications

Consider a DS-CDMA network with K users in which BPSK is coherently detected and every receiver has the form of Figure 2.14. For a particular user of interest, the binary code-symbol modulation is $d(t)$, the binary spreading sequence is $\{p_n\}$ with $p_n \in \{-1, +1\}$, and $\psi(t)$ is the chip waveform. The desired signal that arrives at a receiver is

$$s(t) = \sqrt{2\mathcal{E}_s}d(t)p(t)\cos 2\pi f_c t \tag{7.41}$$

where $\mathcal{E}_s$ is the energy per binary channel symbol of duration T_s, and the spreading waveform is

$$p(t) = \sum_{n=-\infty}^{\infty} p_n \psi(t - nT_c). \tag{7.42}$$

The chip waveforms are assumed to be identical throughout the network, largely confined to an interval of duration T_c, and normalized so that

$$\int_0^{T_c} \psi^2(t)dt = \frac{1}{G} \tag{7.43}$$

where $G = T_s/T_c$ is the spreading factor.

The multiple-access interference that enters a receiver synchronized to a desired signal is modeled as

$$i(t) = \sum_{k=1}^{K-1} \sqrt{2\mathcal{E}_k} d_k(t - \tau_k) q_k(t - \tau_k) \cos(2\pi f_c t + \phi_k) \tag{7.44}$$

where $K - 1$ is the number of interfering direct-sequence signals, $\mathcal{E}_k$ is the received energy per symbol in interference signal k, $d_k(t)$ is its code-symbol modulation, $q_k(t)$ is its spreading waveform, τ_k is its relative delay, and ϕ_k is the phase shift of interference signal k including the effect of carrier time delay. Each spreading waveform of an interference signal has the form

$$q_k(t) = \sum_{n=-\infty}^{\infty} q_{k,n} \psi(t - nT_c), \quad i = 1, 2, \ldots, K - 1 \tag{7.45}$$

where $q_{k,n} \in \{-1, +1\}$, and the $\{q_{k,n}\}$ are the spreading sequences. The K spreading sequences in the network are often called *signature sequences*.

In the receiver, carrier removal is followed by chip-matched filtering. If d_0 is the desired symbol over $[0, T_s]$, (2.78) and (7.44) indicate that the decision metric corresponding to d_0 is

$$V = d_0 \sqrt{\mathcal{E}_s} + V_1 + V_2 \tag{7.46}$$

where the component due to the multiple-access interference is

$$V_1 = \sum_{i=0}^{G-1} p_i J_i \tag{7.47}$$

the component due to the noise is

$$V_2 = \sum_{i=0}^{G-1} p_i N_{s,i} \tag{7.48}$$

and

$$J_i = \sum_{k=1}^{K-1} \sqrt{\mathcal{E}_k} \cos\phi_k \int_{iT_c}^{(i+1)T_c} d_k(t-\tau_k) q_k(t-\tau_k)\,\psi(t-iT_c)\,dt \tag{7.49}$$

$$N_{s,i} = \sqrt{2} \int_{iT_c}^{(i+1)T_c} n(t)\psi(t-iT_c) \cos 2\pi f_c t\, dt. \tag{7.50}$$

Synchronous Communications

Synchronous communication signals are generated typically when a single station transmits to multiple mobiles, as in the downlinks of cellular networks (Section 8.4). We consider synchronous communication signals such that all code symbols have duration T_s, symbol and chip transitions are aligned at the receiver input, and short spreading sequences with period $N = G$ extend over each code symbol. Then, $\tau_k = 0$, $k = 1, 2, \ldots, K-1$, and $d_k(t) = d_k$ is constant over the integration interval $[0, T_s]$. Thus, for synchronous communications, the substitution of (7.49), (7.45), and (7.43) into (7.47) with $N = G$ yields

$$V_1 = \frac{1}{G} \sum_{k=1}^{K-1} \lambda_k \sum_{n=0}^{G-1} p_n q_{k,n} = \sum_{k=1}^{K-1} \lambda_k \theta_{p,k}(0) \tag{7.51}$$

where

$$\lambda_k = \sqrt{\mathcal{E}_k} d_k \cos\phi_k \tag{7.52}$$

and $\theta_{p,k}(l)$ is the periodic cross-correlation of the desired spreading sequence and interfering sequence k. If the K synchronous spreading sequences are mutually orthogonal, then $V_1 = 0$ and the multiple-access interference is suppressed at the receiver. A large number of synchronous multiple-access interference signals can be suppressed in a network if *orthonormal spreading sequences* are used.

Asynchronous Communications

Asynchronous communication signals are typically generated in a receiver when mobiles independently transmit to a receiver, as in the uplinks of cellular networks (Section 8.4). The symbol transitions of *asynchronous* multiple-access signals at a receiver are not simultaneous, usually because of changing path-length differences among the various communication links. Since the spreading sequences are shifted relative to each other, sets of periodic sequences with small periodic cross-correlations for any relative shifts are necessary, but not sufficient, to limit the effect of multiple-access interference. Walsh sequences, orthogonal variable-spreading-factor codes, and maximal sequences usually do not provide cross-correlations small enough for practical applications.

The *aperiodic autocorrelation of a polyphase sequence* $\{p_n\}_{n=0}^{G-1}$ of length G is defined as

$$A(p,\nu) = \begin{cases} \frac{1}{G}\sum\limits_{n=0}^{G-1-\nu} p_{n+\nu}p_n^*, & 0 \le \nu \le G-1 \\ \frac{1}{G}\sum\limits_{n=0}^{G-1+\nu} p_n p_{n-\nu}^*, & -G+1 \le \nu < 0 \\ 0, & |\nu| \ge G. \end{cases} \tag{7.53}$$

The *aperiodic cross-correlation of two polyphase sequences* $\{p_n\}_{n=0}^{G-1}$ and $\{q_{k,n}\}_{n=0}^{G-1}$ of length G is defined as

$$A(p,q_k,\nu) = \begin{cases} \frac{1}{G}\sum\limits_{n=0}^{G-1-\nu} p_{n+\nu}q_{k,n}^*, & 0 \le \nu \le G-1 \\ \frac{1}{G}\sum\limits_{n=0}^{G-1+\nu} p_n q_{k,n-\nu}^*, & -G+1 \le \nu < 0 \\ 0, & |\nu| \ge G \end{cases} \tag{7.54}$$

and $A(p,p,\nu) = A(p,\nu)$.

Consider a DS-CDMA network with K users in which BPSK is coherently detected and every receiver has the form of Figure 2.14. Let $\mathbf{d}_k = (d_{-1}^{(k)}, d_0^{(k)})$ denote the vector of the two binary code symbols of asynchronous multiple-access interference signals k that are received during the detection of a symbol of the desired signal. Equations (7.47), (7.42), and (7.49) imply that

$$V_1 = \sum_{k=1}^{K-1}\sqrt{\mathcal{E}_k}\cos\phi_k\left[d_{-1}^{(k)}R_{pk}(\tau_k) + d_0^{(k)}\hat{R}_{pk}(\tau_k)\right],\ 0 \le \tau_k \le T_s \tag{7.55}$$

where the *continuous-time partial cross-correlation functions* are

$$R_{pk}(\tau_k) = \int_0^{\tau_k} p(t)q_k(t-\tau_k)dt \tag{7.56}$$

$$\hat{R}_{pk}(\tau_k) = \int_{\tau_k}^{T_s} p(t)q_k(t-\tau_k)dt. \tag{7.57}$$

Let $\tau_k = \nu_k T_c + \epsilon_k$, where ν_k is an integer such that $0 \le \nu_k \le G-1$, and $0 \le \epsilon_k < T_c$. Assuming binary spreading sequences of period $N = G$ and the normalized rectangular chip waveform,

$$\psi(t) = \begin{cases} \frac{1}{\sqrt{T_s}}, & 0 \leq t < T_c \\ 0, & \text{otherwise} \end{cases} \tag{7.58}$$

a derivation analogous to that leading to (2.51) gives

$$R_{pk}(\tau_k) = A(p, q_k, \nu_k - G) + [A(p, q_k, \nu_k + 1 - G) - A(p, q_k, \nu_k - G)] \frac{\epsilon_k}{T_c} \tag{7.59}$$

$$\hat{R}_{pk}(\tau_k) = A(p, q_k, \nu_k) + [A(p, q_k, \nu_k + 1) - A(p, q_k, \nu_k)] \frac{\epsilon_k}{T_c} \tag{7.60}$$

which shows the dependence of V_1 on the aperiodic cross-correlations. If interference signal k is a multipath component of the desired signal, then $q_k(t) = p(t)$ and (7.56) and (7.57) are the *continuous-time autocorrelation functions* of $p(t)$.

The nonzero sidelobes of the aperiodic autocorrelation degrade the performance against multipath interference, whereas the nonzero cross-correlations degrade the performance against multiple-access interference. For most sets of sequences, the aperiodic cross-correlations are larger than the periodic cross-correlations. By the proper selection of the sequences and their relative phases, a system performance can be obtained that is slightly better than that attainable with sequences with good periodic cross-correlations or random sequences. However, the number of suitable sequences is too small for most applications.

If all the spreading sequences are short, and the power levels of all received signals are equal, then the symbol error probability can be approximated and bounded by functions of the aperiodic cross-correlations [70, 71], but the process is complicated. An alternative analytical approach is to model the spreading sequences as random binary sequences, as is done for long sequences.

In a network with multiple-access and/or multipath interference, code synchronization is impaired by the aperiodic cross-correlations and autocorrelation sidelobes. Code acquisition is impaired because V_c and V_s in (4.87) and (4.88) of Section 4.4 have additional terms, each of which is proportional to the aperiodic cross-correlation between the desired signal and an interference signal.

7.4 Alternative Spreading Sequences and Systems

Complementary Codes

The Welch bound (7.26) indicates that it is not possible for a set of conventional spreading sequences to be mutually orthogonal in the sense of having zero periodic cross-correlations. To overcome this limitation to DS-CDMA network performance, spreading sequences may be generated as composite sequences called *complemen-*

tary codes [89]. Each complementary code comprises a set or *flock* of elementary sequences that are separately transmitted, received, and applied to separate correlators. The combined correlator outputs provide zero aperiodic autocorrelation sidelobes and zero aperiodic cross-correlations with other complementary codes. Consequently, a complementary-coded CDMA (CC-CDMA) system is a DS-CDMA system that is capable of largely suppressing asynchronous multipath and multiple-access interference.

Consider a CC-CDMA network of K users, each of which uses a complementary code instead of a conventional spreading sequence. Let $\mathbf{C}(k) = \{\mathbf{c}_l(k)\}_{l=1}^{L}$ denote the kth complementary code, $k = 1, 2, \ldots, K$, which comprises a flock of L elementary sequences, each of which has the form $\mathbf{c}_l(k) = [c_{l,1}(k), c_{l,2}(k), \ldots, c_{l,N}(k)]$, $l = 1, 2, \ldots, L$, where N is the length of an elementary sequence. The spreading factor, which determines the amount of spectral spreading due to the complementary code, is LN. The L elementary sequences are transmitted to the receiver in L-independent subchannels. The receiver performs separate chip-matched filtering of each elementary sequence and separate correlations with receiver-generated elementary sequences. The results of the separate correlations are combined to generate the symbol metrics for synchronization and decoding. The capability of the receiver to synchronize with the desired complementary code and to reject multiple-access interference from other complementary codes depends on the *complementary aperiodic correlation function*. This function is defined as

$$\Theta\left[\mathbf{C}(k_1), \mathbf{C}(k_2), \nu\right] = \sum_{l=1}^{L} A\left[\mathbf{c}_l(k_1), \mathbf{c}_l(k_2), \nu\right] \tag{7.61}$$

where $A\left[\mathbf{c}_l(k_1), \mathbf{c}_l(k_2), \nu\right]$, which is defined by (7.53) and (7.54), is the aperiodic autocorrelation function of $\mathbf{c}_l(k_1)$ if $k_1 = k_2$ and the aperiodic cross-correlation function of $\mathbf{c}_l(k_1)$ and $\mathbf{c}_l(k_2)$ if $k_1 \neq k_2$. Ideal complementary codes provide

$$\Theta\left[\mathbf{C}(k_1), \mathbf{C}(k_2), \nu\right] = \begin{cases} LN, & k_1 = k_2,\ \nu = 0 \\ 0, & otherwise \end{cases} \tag{7.62}$$

which overcomes the limitations of the Welch bound on a set of conventional spreading sequences.

There are two different ways to transmit elementary codes via independent subchannels: time-division multiplexing and frequency-division multiplexing. For time-division multiplexing, the L elementary sequences are serially transmitted in different time slots, and guard intervals prevent the overlap of adjacent sequences because of multipath propagation, propagation delay, or timing inaccuracies. For frequency-division multiplexing, the L elementary sequences are transmitted simultaneously by different carrier frequencies in nonoverlapping spectral regions. Both time-division and frequency-division multiplexing lower the spectral efficiency of a direct-sequence system because of the guard intervals and the carrier separations, respectively.

Although the complementary codes are highly desirable in theory, there are formidable practical obstacles to their implementation. A primary limitation is the small number of users that can be supported because of the relatively small number of complementary codes for a specified flock size. Time-selective or frequency-selective fading is likely to undermine the accurate calculation of the complementary aperiodic correlation function. Timing synchronization among the subchannel outputs is a significant problem, and there are other practical issues [89].

Chaotic Spread-Spectrum Systems

Chaotic signals are deterministic wideband signals that resemble noise waveforms and are derived from nonlinear dynamic systems. *Chaotic sequences* are generated by a discrete-time sampling and nonlinear mapping of chaotic signals. These aperiodic sequences are potentially desirable as spreading sequences in *chaotic spread-spectrum systems* because of their favorable aperiodic cross-correlation properties. After reproducing the chaotic sequences in the receiver, a demodulator based on correlation can be used [92].

The main problem impeding the practical implementation of chaotic spread-spectrum systems is the development of synchronization systems that operate efficiently in the presence of noise. Chaos-shift keying is a coherent chaotic modulation that is potentially effective but relies on the recovery of a noise-like chaotic carrier, which is very difficult in noisy environments. A much more practical modulation is noncoherent differential chaos-shift keying (DCSK), which sacrifices 6 dB in performance over the AWGN channel but eliminates part of the synchronization problem. A DCSK receiver requires only a differentially coherent demodulator that processes successive reference and information signals during each symbol interval and does not need to reproduce the chaotic carrier. Limitations of DCSK include a relatively low code-rate and the requirement of a radio-frequency delay line. Although these limitations can be overcome by more elaborate variants of DCSK, the complexity and the 6 dB performance loss remain significant disadvantages [25].

Ultra-Wideband Systems

An ultra-wideband (UWB) system is one with a 10-dB bandwidth exceeding 500 MHz or a fractional bandwidth, which is the ratio of the 10-dB bandwidth to the center frequency, exceeding 0.2. Direct-sequence systems are not practical at such bandwidths because of the hardware requirements associated with the carrier-frequency modulation. The UWB systems achieve their wide bandwidths by transmitting short pulses that directly generate a wide-bandwidth signal [23]. Consequently, UWB systems do not need the mixers and oscillators required for upconversion and downconversion in direct-sequence systems.

Some methods of direct-sequence technology are applicable to a major class of UWB systems called *direct-sequence UWB systems*. In these systems, a spreading sequence combined with a data symbol controls the polarities of a series of pulses that represent a data symbol.

The large bandwidth of UWB signals means that a potentially large number of distinct multipath components arrive at the receiver. The result is that UWB systems benefit from a potentially high diversity order. However, the realization of the diversity benefit is impeded by two major factors. First, the power in each transmitted pulse is severely constrained to prevent the disruption of other communication systems. Second, each multipath component arriving at the receiver usually has a small fraction of the energy in the transmitted pulse. Therefore, a rake demodulator requires a large number of fingers and is difficult to implement because of the synchronization requirements. Primarily because of the power constraints, the main application of UWB communication systems is for high-throughput indoor communications.

7.5 Systems with Random Spreading Sequences

If all the spreading sequences in a network of asynchronous CDMA systems have a common period equal to the code-symbol duration, then by the proper selection of the sequences and their relative phases, a system performance can be obtained that is better than that theoretically attainable with random sequences. However, the performance advantage is small, the number of suitable sequences is too small for many applications, and long sequences that extend over many code symbols provide more system security. Furthermore, long sequences ensure that successive code symbols are covered by different sequences, thereby limiting the time duration of an unfavorable cross-correlation due to multiple-access interference. Modeling long sequences as random spreading sequences is clearly desirable, but even if short sequences are used, the random-sequence model gives fairly accurate performance predictions.

The analysis and comparisons of CDMA systems are greatly facilitated by applying Jensen's inequality.

Jensen's Inequality

A function $g(\cdot)$ defined on an open interval I is *convex* if

$$g(px + (1-p)y) \leq pg(x) + (1-p)g(y) \tag{7.63}$$

for x, y in I and $0 \leq p \leq 1$. Suppose that $g(x)$ has a continuous, nondecreasing derivative $g'(x)$ on I. The inequality is valid if $p = 0$ or 1. If $x \geq y$ and $0 \leq p < 1$,

$$g(px+(1-p)y)-g(y)=\int_{y}^{px+(1-p)y} g'(z)dz \le p(x-y)g'(px+(1-p)y)$$
$$\le \frac{p}{1-p}\int_{px+(1-p)y}^{x} g'(z)dz$$
$$=\frac{p}{1-p}[g(x)-g(px+(1-p)y)]. \tag{7.64}$$

Simplifying this result, we obtain (7.63). If $y \ge x$, a similar analysis again yields (7.63). Thus, *if* $g(x)$ *has a continuous, nondecreasing derivative on* I*, it is convex. If* $g(x)$ *has a nonnegative second derivative on* I*, it is convex.*

If the inequalities in (7.63) are replaced by equalities, then $g(x)$ is said to be *strictly convex*. If $g(x)$ *has a continuous, increasing derivative or a positive second derivative on* I*, it is strictly convex.*

Lemma 1 *If* $g(x)$ *is a convex function on the open interval* I*, then*

$$g(y) \ge g(x)+g^{-}(x)(y-x) \tag{7.65}$$

for all y, x *in* I*, where* $g^{-}(x)$ *is the left derivative of* $g(x)$.

Proof If $y-x \ge z > 0$, then substituting $p = 1-z/(y-x)$ into (7.63) gives

$$g(x+z) \le \left(1-\frac{z}{y-x}\right)g(x)+\frac{z}{y-x}g(y)$$

which yields

$$\frac{g(x+z)-g(x)}{z} \le \frac{g(y)-g(x)}{y-x}, \quad y-x \ge z > 0. \tag{7.66}$$

If $v > 0$ and $z > 0$, then (7.63) with $p = z/v+z$ implies that

$$g(x) \le \frac{z}{v+z}g(x-v)+\frac{v}{v+z}g(x+z)$$

and hence

$$\frac{g(x)-g(x-v)}{v} \le \frac{g(x+z)-g(x)}{z}, \quad v, z > 0. \tag{7.67}$$

Inequality (7.66) indicates that the ratio $[g(y)-g(x)]/(y-x)$ decreases monotonically as $y \to x$ from above, and (7.67) implies that this ratio has a lower bound. Therefore, the right derivative $g^{+}(x)$ exists on I.

If $x-y \ge v > 0$, then (7.63) with $p = 1-v/(x-y)$ implies that

$$g(x-v) \le \left(1-\frac{v}{x-y}\right)g(x)+\frac{v}{x-y}g(y)$$

which yields

$$\frac{g(x)-g(y)}{x-y} \leq \frac{g(x)-g(x-v)}{v}, \quad x-y \geq v > 0. \tag{7.68}$$

This inequality indicates that the ratio $[g(x)-g(y)]/(x-y)$ increases monotonically as $y \to x$ from below, and (7.67) implies that this ratio has an upper bound. Therefore, the left derivative $g^{-}(x)$ exists on I. Taking the limits as $z \to 0$ and $v \to 0$ in (7.67) yields

$$g^{-}(x) \leq g^{+}(x). \tag{7.69}$$

Taking the limits as $z \to 0$ and $v \to 0$ in (7.66) and (7.68), respectively, and then using (7.69), we find that (7.65) is valid for all y, x in I. □

Jensen's Inequality If X is a random variable with a finite expected value $E[X]$, and $g(\cdot)$ is a convex function on an open interval containing the range of X, then

$$E[g(X)] \geq g(E[X]). \tag{7.70}$$

Proof Set $y = X$ and $x = E[X]$ in (7.65), which gives $g(X) \geq g(E[X]) + g^{-}(E[X])(X - E[X])$. Taking the expected values of the random variables on both sides of this inequality gives Jensen's inequality. □

Direct-Sequence Systems with BPSK

Consider the direct-sequence system with coherent BPSK. The spreading sequences of the desired signal and interference signals are modeled as independent random binary sequences with chip duration T_c. The code-symbol sequences of the interference signals are modeled as independent random binary sequences with code-symbol duration $T_s = GT_c$. Therefore, the code-symbol sequences can be subsumed into the random binary spreading sequences with no loss of generality. Since $q_k(t)$ is determined by an independent, random spreading sequence, only time delays modulo-T_c are significant; thus, we can assume that $0 \leq \tau_i < T_c$ in (7.49) without loss of generality. Therefore, the substitution of (7.45) into (7.49) yields

$$J_i = \sum_{k=1}^{K-1} \sqrt{\mathcal{E}_k} \cos\phi_k \begin{bmatrix} q_{k,i-1}\int_{iT_c}^{iT_c+\tau_i} \psi(t-iT_c)\psi[t-(i-1)T_c-\tau_k]dt \\ +q_{k,i}\int_{iT_c+\tau_i}^{(i+1)T_c} \psi(t-iT_c)\psi(t-iT_c-\tau_k)dt \end{bmatrix}. \tag{7.71}$$

The *partial autocorrelation* for the normalized chip waveform is defined as

$$R_{\psi}(s) = T_s \int_0^s \psi(t)\psi(t+T_c-s)dt, \quad 0 \leq s < T_c. \tag{7.72}$$

Substitution into (7.71) and appropriate changes of variables in the integrals yield

$$J_i = \sum_{k=1}^{K-1} \frac{\sqrt{\mathcal{E}_i}}{T_s} \cos\phi_k \left[q_{k,i-1} R_\psi(\tau_k) + q_{k,i} R_\psi(T_c - \tau_k) \right]. \tag{7.73}$$

For rectangular chips in the spreading waveform, the substitution of (7.58) into (7.72) gives

$$R_\psi(s) = s \quad \text{(rectangular)}. \tag{7.74}$$

For *sinusoidal chips* in the spreading waveform,

$$\psi(t) = \begin{cases} \sqrt{\frac{2}{T_s}} \sin\left(\frac{\pi}{T_c} t\right), & 0 \le t \le T_c \\ 0, & \text{otherwise.} \end{cases} \tag{7.75}$$

Substituting this equation into (7.72), using a trigonometric identity, and performing the integrations, we obtain

$$R_\psi(s) = \frac{T_c}{\pi} \sin\left(\frac{\pi}{T_c} s\right) - s \cos\left(\frac{\pi}{T_c} s\right) \quad \text{(sinusoidal)}. \tag{7.76}$$

When $\boldsymbol{\phi} = (\phi_1, \phi_2, \ldots, \phi_{K-1})$ and $\boldsymbol{\tau} = (\tau_1, \tau_2, \ldots, \tau_{K-1})$ are given, (7.73) indicates that J_i and J_{i+l}, $|l| > 1$, are statistically independent. However, since adjacent terms J_i and J_{i+1} contain the same random variable $q_{k,i}$, it does not appear at first that adjacent terms in (7.47) are statistically independent, even when $\boldsymbol{\phi}$ and $\boldsymbol{\tau}$ are given. The following lemma [100] resolves this issue.

Lemma 2 *Suppose that $\{\alpha_i\}$ and $\{\beta_i\}$ are statistically independent, random binary sequences. Let x and y denote arbitrary constants. Then, $\alpha_i \beta_j x$ and $\alpha_i \beta_k y$ are statistically independent random variables when $j \neq k$.*

Proof Let $P(\alpha_i\beta_j x = a, \alpha_i\beta_k y = b)$ denote the joint probability that $\alpha_i\beta_j x = a$ and $\alpha_i\beta_k y = b$. From the theorem of total probability, it follows that

$$\begin{aligned} &P(\alpha_i\beta_j x = a, \alpha_i\beta_k y = b) \\ &\quad = P(\alpha_i\beta_j x = a, \alpha_i\beta_k y = b, \alpha_i = 1) + P(\alpha_i\beta_j x = a, \alpha_i\beta_k y = b, \alpha_i = -1) \\ &\quad = P(\beta_j x = a, \beta_k y = b, \alpha_i = 1) + P(\beta_j x = -a, \beta_k y = -b, \alpha_i = -1). \end{aligned}$$

From the independence of $\{\alpha_i\}$ and $\{\beta_j\}$ and the fact that they are random binary sequences, we obtain a simplification for $j \neq k$, $x \neq 0$, and $y \neq 0$:

$$\begin{aligned} &P(\alpha_i\beta_j x = a, \alpha_i\beta_k y = b) \\ &\quad = P(\beta_j x = a) P(\beta_k y = b) P(\alpha_i = 1) + P(\beta_j x = -a) P(\beta_k y = -b) P(\alpha_i = -1) \\ &\quad = \frac{1}{2} P\left(\beta_j = \frac{a}{x}\right) P\left(\beta_k = \frac{b}{y}\right) + \frac{1}{2} P\left(\beta_j = -\frac{a}{x}\right) P\left(\beta_k = -\frac{b}{y}\right). \end{aligned}$$

Since β_j equals +1 or −1 with equal probability, $P(\beta_j = a/x) = P(\beta_j = -a/x)$ and thus

$$P(\alpha_i\beta_j x = a, \alpha_i\beta_k y = b) = P\left(\beta_j = \frac{a}{x}\right) P\left(\beta_k = \frac{b}{y}\right)$$
$$= P(\beta_j x = a)P(\beta_k y = b).$$

A similar calculation gives

$$P(\alpha_i\beta_j x = a)P(\alpha_i\beta_k y = b) = P(\beta_j x = a)P(\beta_k y = b).$$

Therefore,

$$P(\alpha_i\beta_j x = a, \alpha_i\beta_k y = b) = P(\alpha_i\beta_j x = a)P(\alpha_i\beta_k y = b)$$

which satisfies the definition of statistical independence of $\alpha_i\beta_j x$ and $\alpha_i\beta_k y$. The same relation is trivial to establish for $x = 0$ or $y = 0$. □

The lemma indicates that when $\boldsymbol{\phi}$ and $\boldsymbol{\tau}$ are given, the terms in (7.47) are statistically independent. Since $q_k(t)$ is determined by an independent, random spreading sequence, the $\{q_{k,i}\}$ are identically distributed; hence, the $\{J_i\}$ are identically distributed. Since the $\{p_i\}$ are identically distributed, each term of V_1 is identically distributed. Since $p_i^2 = 1$, (7.47) and (7.71) imply that the conditional variance of V_1 is

$$\text{var}(V_1) = \sum_{i=0}^{G-1} \text{var}(J_i) = \sum_{k=1}^{K-1} \frac{\mathcal{E}_k}{G} h(\tau_k) \cos^2 \phi_k \tag{7.77}$$

where

$$h(\tau_k) = \frac{1}{T_c^2}[R_\psi^2(\tau_k) + R_\psi^2(T_c - \tau_k)] \tag{7.78}$$

is the *chip function*. Using (7.74) and (7.76), we find that for rectangular chip waveforms,

$$h(\tau_k) = \frac{2\tau_k^2 - 2\tau_k T_c + T_c^2}{T_c^2} \tag{7.79}$$

and for sinusoidal chip waveforms,

$$h(\tau_k) = \frac{2\left(\frac{T_c}{\pi}\right)^2 \sin^2(\frac{\pi}{T_c}\tau_k) + \left(2\tau_k^2 - 2\tau_k T_c + T_c^2\right)\cos^2(\frac{\pi}{T_c}\tau_k) + \frac{T_c^2 - \tau_k T_c}{\pi}\sin(\frac{2\pi}{T_c}\tau_k)}{T_c^2}. \tag{7.80}$$

Since the terms of V_1 in (7.47) are independent, identically distributed, zero-mean random variables when $\boldsymbol{\phi}$ and $\boldsymbol{\tau}$ are given, the central limit theorem (Corollary A1, Appendix A.2) implies that $V_1/\sqrt{\text{var}(V_1)}$ converges in distribution as $G \to \infty$ to a Gaussian random variable with mean 0 and variance 1. Thus, when $\boldsymbol{\phi}$ and $\boldsymbol{\tau}$ are given, the conditional distribution of V_1 is approximately Gaussian when G is large. Since the noise component has a Gaussian distribution and is independent of V_1, V has an approximate Gaussian distribution with

$$E[V] = d_0\sqrt{\mathcal{E}_s}, \; var\,(V) = \frac{N_0}{2} + var\,(V_1)\,. \tag{7.81}$$

We derive the symbol error probability for a system in which hard decisions are made on successive symbol metrics to produce a symbol sequence that is applied to a hard-decision decoder. The Gaussian distribution of the symbol metric V implies that the conditional symbol error probability given $\boldsymbol{\phi}$ and $\boldsymbol{\tau}$ is

$$P_s(\phi, \tau) = Q\left[\sqrt{\frac{2\mathcal{E}_s}{N_{0e}(\boldsymbol{\phi}, \boldsymbol{\tau})}}\right] \tag{7.82}$$

where $Q(x)$ is defined by (1.58), and the *equivalent noise PSD* is defined as

$$N_{0e}(\boldsymbol{\phi}, \boldsymbol{\tau}) = N_0 + \frac{2}{G}\sum_{k=1}^{K-1} \mathcal{E}_k h\,(\tau_k)\cos^2\phi_k. \tag{7.83}$$

For an asynchronous network, we assume that the time delays are independent and uniformly distributed over $[0, T_c)$ and that the phase angles θ_k, $k = 1, 2, \ldots, K-1$, are uniformly distributed over $[0, 2\pi)$. Therefore, the symbol error probability is

$$P_s = \left(\frac{2}{\pi T_c}\right)^{K-1} \int_0^{\pi/2} \cdots \int_0^{\pi/2} \int_0^{T_c} \cdots \int_0^{T_c} P_s(\boldsymbol{\phi}, \boldsymbol{\tau}) d\boldsymbol{\phi}\, d\boldsymbol{\tau} \tag{7.84}$$

where the fact that $\cos^2\phi_k$ takes all its possible values over $[0,\pi/2)$ has been used to shorten the integration intervals. The absence of sequence parameters ensures that the amount of computation required for (7.84) is much less than that required for P_s for a short deterministic spreading sequence. Nevertheless, the computational requirements are large enough that it is highly desirable to find an accurate approximation that entails less computation. The conditional symbol error probability given $\boldsymbol{\phi}$ is defined as

$$P_s(\boldsymbol{\phi}) = \left(\frac{1}{T_c}\right)^{K-1} \int_0^{T_c} \cdots \int_0^{T_c} P_s(\boldsymbol{\phi}, \boldsymbol{\tau}) d\boldsymbol{\tau}\,. \tag{7.85}$$

A closed-form approximation to $P_s(\boldsymbol{\phi})$ greatly simplifies the computation of P_s, which reduces to

$$P_s = \left(\frac{2}{\pi}\right)^{K-1} \int_0^{\pi/2} \cdots \int_0^{\pi/2} P_s(\boldsymbol{\phi}) d\boldsymbol{\phi}\,. \tag{7.86}$$

To approximate $P_s(\boldsymbol{\phi})$, we first obtain upper and lower bounds.

For either rectangular or sinusoidal chip waveforms, (7.79), (7.80), and elementary calculus establish that

$$h(\tau_k) \leq 1. \tag{7.87}$$

Using this upper bound successively in (7.83), (7.82), and (7.85), and performing the trivial integrations that result, we obtain

$$P_s(\boldsymbol{\phi}) \leq Q\left[\sqrt{\frac{2\mathcal{E}_s}{N_{0u}(\boldsymbol{\phi})}}\right] \tag{7.88}$$

where

$$N_{0u}(\boldsymbol{\phi}) = N_0 + \frac{2}{G}\sum_{k=1}^{K-1} \mathcal{E}_k \cos^2 \phi_k. \tag{7.89}$$

To obtain a lower bound on $P_s(\boldsymbol{\phi})$, the successive integrals in (7.85) are interpreted as successive expected values over the $\{\tau_k\}$, where each τ_k is uniformly distributed over $[0, T_c)$. With this interpretation, integrations using (7.79) and (7.80) give

$$E[h(\tau_k)] = \frac{1}{T_c}\int_0^{T_c} h(\tau_k)\, d\tau_k = h \tag{7.90}$$

where the *chip factor* is

$$h = \begin{cases} \frac{2}{3}, & \textit{rectangular chip} \\ \frac{1}{3} + \frac{5}{2\pi^2}, & \textit{sinusoidal chip.} \end{cases} \tag{7.91}$$

The function $P_s(\phi, \tau)$ of (7.82) has the form

$$g(x) = Q\left(ax^{-1/2}\right), \; a \geq 0. \tag{7.92}$$

Since the second derivative of $g(x)$ is nonnegative over the interval such that $0 < x \leq 1/3$, $g(x)$ is a convex function over that interval. Relations (7.83), (7.87), and $\cos^2 \phi_k \leq 1$ yield a sufficient condition for the convexity of $P_s(\phi, \tau)$ for all $\boldsymbol{\tau}$ and $\boldsymbol{\phi}$, which is

$$\mathcal{E}_s \geq \frac{3}{2}\left(N_0 + \frac{2\mathcal{E}_t}{G}\right) \tag{7.93}$$

where the total received interference energy per symbol is

$$\mathcal{E}_t = \sum_{k=1}^{K-1} \mathcal{E}_k. \tag{7.94}$$

Application of Jensen's inequality successively to each component of $\boldsymbol{\tau}$ in (7.85) and the substitution of (7.90) yield

$$P_s(\boldsymbol{\phi}) \geq Q\left[\sqrt{\frac{2\mathcal{E}_s}{N_{0l}(\boldsymbol{\phi})}}\right] \tag{7.95}$$

where

$$N_{0l}(\boldsymbol{\phi}) = N_0 + \frac{2h}{G}\sum_{k=1}^{K-1}\mathcal{E}_k\cos^2\phi_k. \tag{7.96}$$

If N_0 is negligible, then (7.96) and (7.89) give $N_{0l}/N_{0u} = h$. Thus, a good approximation is provided by

$$P_s(\boldsymbol{\phi}) \approx Q\left[\sqrt{\frac{2\mathcal{E}_s}{N_{0a}(\boldsymbol{\phi})}}\right] \tag{7.97}$$

where

$$N_{0a}(\boldsymbol{\phi}) = N_0 + \frac{2\sqrt{h}}{G}\sum_{k=1}^{K-1}\mathcal{E}_k\cos^2\phi_k. \tag{7.98}$$

If N_0 is negligible, then $N_{0u}/N_{0a} = N_{0a}/N_{0l} = 1/\sqrt{h}$. Therefore, in terms of the value of $\mathcal{E}_s$ needed to ensure a given $P_s(\phi)$, the error in using approximation (7.97) instead of (7.85) is bounded by $10\log_{10}(1/\sqrt{h})$ in decibels, which equals 0.88 dB for rectangular chip waveforms and 1.16 dB for sinusoidal chip waveforms. In practice, the error is expected to be only a few tenths of a decibel because $N_0 \neq 0$ and P_s coincides with neither the upper nor the lower bound.

As an example, suppose that rectangular chip waveforms with $h = 2/3$ are used, $\mathcal{E}_s/N_0 = 15$ dB, and $K = 2$. Figure 7.5 illustrates four different evaluations of P_s as a function of $G\mathcal{E}_s/\mathcal{E}_1$, the *despread signal-to-interference ratio*, which is the signal-to-interference ratio after taking into account the beneficial results from the despreading in the receiver. The accurate approximation is computed from (7.82) and (7.84), the upper bound from (7.88) and (7.86), the lower bound from (7.95) and (7.86), and the simple approximation from (7.97) and (7.86). The figure shows that the accurate approximation moves from the lower bound toward the simple approximation as the symbol error probability decreases. For $P_s = 10^{-5}$, the simple approximation is less than 0.3 dB in error relative to the accurate approximation.

Figure 7.6 compares the symbol error probabilities for $K = 2$ to $K = 4$, rectangular chip waveforms and $\mathcal{E}_s/N_0 = 15$ dB. The simple approximation is used for P_s, and the abscissa shows $G\mathcal{E}_s/\mathcal{E}_1$, where $\mathcal{E}_1$ is the symbol energy of each equal-power interfering signal. The figure shows that P_s increases with K, but the shift in P_s is mitigated somewhat because the interference signals tend to partially cancel each other.

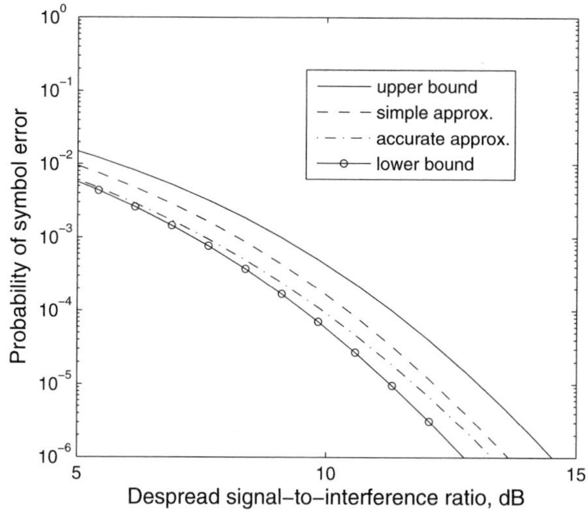

Fig. 7.5 Symbol error probability of a direct-sequence system with BPSK in the presence of a single multiple-access interference signal and $\mathcal{E}_s/N_0 = 15\,\text{dB}$

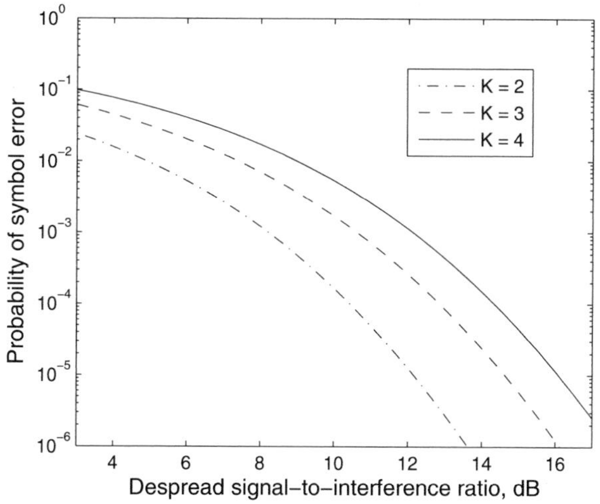

Fig. 7.6 Symbol error probability of a direct-sequence system with BPSK in the presence of $K-1$ equal-power multiple-access interference signals and $\mathcal{E}_s/N_0 = 15\,\text{dB}$

The preceding bounding methods can be extended to the bounds on $P_s(\boldsymbol{\phi})$ by observing that $\cos^2 \phi_k \leq 1$ and $h \leq 1$ for the upper bound and interpreting successive integrals as successive expected values over the $\{\phi_k\}$, where each ϕ_k is uniformly distributed over $[0, \pi/2)$. We obtain

$$Q\left(\sqrt{\frac{2\mathcal{E}_s}{N_0 + h\mathcal{E}_t/G}}\right) \leq P_s \leq Q\left[\sqrt{\frac{2\mathcal{E}_s}{N_0 + 2\mathcal{E}_t/G}}\right]. \tag{7.99}$$

A simple approximation is provided by

$$P_s \approx Q\left(\sqrt{\frac{2\mathcal{E}_s}{N_0 + \sqrt{2h}\,\mathcal{E}_t/G}}\right). \tag{7.100}$$

If P_s is specified, then the error in the required $G\mathcal{E}_s/\mathcal{E}_t$ caused by using (7.100) instead of (7.84) is bounded by $10\log_{10}\sqrt{2/h}$ in decibels. Thus, the error is bounded by 2.39 dB for rectangular chip waveforms and 2.66 dB for sinusoidal ones.

The lower bound in (7.99) gives the same result as that often called the *standard Gaussian approximation*, in which V_1 in (7.47) is assumed to be approximately Gaussian, each ϕ_k in (7.73) is assumed to be uniformly distributed over $[0, 2\pi)$, and each τ_k is assumed to be uniformly distributed over $[0, T_c)$. This approximation gives an optimistic result for P_s that can be as much as 4.77 dB in error for rectangular chip waveforms according to (7.99). The substantial improvement in accuracy provided by (7.97) or (7.82) is due to the application of the Gaussian approximation only after conditioning V_1 on given values of $\boldsymbol{\phi}$ and $\boldsymbol{\tau}$. The accurate approximation given by (7.82) is a version of what is often called the *improved Gaussian approximation*.

Figure 7.7 illustrates the symbol error probability for three interferers, each with equal received symbol energy, rectangular chip waveforms, and $\mathcal{E}_s/N_0 = 15$ dB as a function of $G\mathcal{E}_s/\mathcal{E}_t$. The graphs show the standard Gaussian approximation of (7.99), the simple approximation of (7.100), and the upper and lower bounds given by (7.88), (7.95), and (7.86). The large error in the standard Gaussian approximation is evident. The simple approximation is reasonably accurate if $10^{-6} \leq P_s \leq 10^{-2}$.

For *synchronous* networks, (7.82) and (7.83) can be simplified because the $\{\tau_k\}$ are all zero. For either rectangular or sinusoidal chip waveforms, we obtain

$$P_s(\boldsymbol{\phi}) = Q\left[\sqrt{\frac{2\mathcal{E}_s}{N_{0e}(\boldsymbol{\phi})}}\right] \tag{7.101}$$

where

$$N_{0e}(\boldsymbol{\phi}) = N_0 + \frac{2}{G}\sum_{k=1}^{K-1} \mathcal{E}_k \cos^2 \phi_k. \tag{7.102}$$

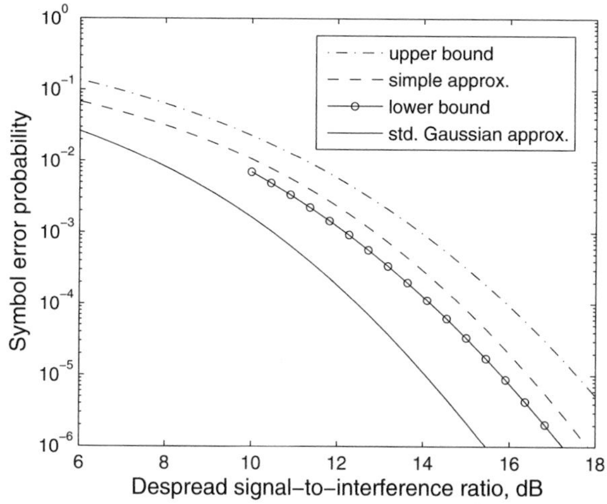

Fig. 7.7 Symbol error probability of a direct-sequence system with BPSK in the presence of three equal-power multiple-access interference signals and $\mathcal{E}_s/N_0 = 15\,\text{dB}$

Since $P_s(\boldsymbol{\phi})$ given by (7.101) equals the upper bound in (7.88), we observe that the *symbol error probability for a synchronous network is equal to or exceeds the symbol error probability for a similar asynchronous network when random spreading sequences are used*. This phenomenon is due to the increased bandwidth of a despread asynchronous interference signal relative to the desired signal, which allows increased filtering of the interference.

Quadriphase Direct-Sequence Systems

Consider a DS-CDMA network of quadriphase direct-sequence systems, each of which uses dual quadriphase-shift keying (QPSK) and random spreading sequences. As described in Section 2.5, each direct-sequence signal is

$$s(t) = \sqrt{\mathcal{E}_s}d_1(t)p_1(t)\cos 2\pi f_c t + \sqrt{\mathcal{E}_s}d_2(t)p_2(t)\sin 2\pi f_c t \tag{7.103}$$

where $\mathcal{E}_s$ is the received energy per binary channel-symbol component and per code bit. The multiple-access interference is

$$i(t) = \sum_{k=1}^{K-1}[\sqrt{\mathcal{E}_k}\, q_{1k}(t-\tau_k)\cos(2\pi f_c t + \phi_k) + \sqrt{\mathcal{E}_k}\, q_{2k}(t-\tau_k)\sin(2\pi f_c t + \phi_k)] \tag{7.104}$$

where $q_{1k}(t)$ and $q_{2k}(t)$ both have the form of (7.45) and incorporate the code modulation, and $\mathcal{E}_k$ is the received energy per binary channel-symbol component and per code bit of interference signal k.

The symbol metrics are given by

$$V = d_{10}\sqrt{2\mathcal{E}_s} + \sum_{i=0}^{G_1-1} p_{1,i}J_i + \sum_{\nu=0}^{G_1-1} p_{1,i}N_{si} \tag{7.105}$$

$$U = d_{20}\sqrt{2\mathcal{E}_s} + \sum_{i=0}^{G_1-1} p_{2,i}J_i' + \sum_{i=0}^{G_1-1} p_{2,i}N_i' \tag{7.106}$$

where J_i, N_i, J_i', and N_i' are defined by (2.76), (2.77), (2.124), and (2.125), respectively. The substitution of (7.104), (7.45), and (7.72) into (2.76) yields

$$J_i = \sum_{k=1}^{K-1} \sqrt{\frac{\mathcal{E}_k}{2T_s^2}} \{\cos\phi_k[q_{1k,i-1}R_\psi(\tau_k) + q_{1k,i}R_\psi(T_c - \tau_k)]$$
$$- \sin\phi_k[q_{2k,i-1}R_\psi(\tau_k) + q_{2k,i}R_\psi(T_c - \tau_k)]\}. \tag{7.107}$$

Let V_1 and U_1 denote the interference terms in (7.105) and (7.106), respectively. Lemma 2, (7.107), (7.78), and analogous results for the $\{J_i'\}$ yield the variances of the interference terms of the symbol metrics:

$$\text{var}(V_1) = \text{var}(U_1) = \sum_{k=1}^{K-1} \frac{2\mathcal{E}_k}{G_1} h(\tau_k) \tag{7.108}$$

where $h(\tau_k)$ is given by (7.78). The noise variances and the means are given by (2.127) and (2.126).

We derive the symbol error probability for a system in which hard decisions are made on successive symbol metrics to produce a symbol sequence that is applied to a hard-decision decoder. Since all variances and means are independent of $\boldsymbol{\phi}$, the Gaussian approximation yields a $P_s(\boldsymbol{\phi}, \boldsymbol{\tau})$ that is independent of $\boldsymbol{\phi}$:

$$P_s = \left(\frac{1}{T_c}\right)^{K-1} \int_0^{T_c} \cdots \int_0^{T_c} Q\left[\sqrt{\frac{2\mathcal{E}_s}{N_{0e}(\boldsymbol{\tau})}}\right] d\boldsymbol{\tau} \tag{7.109}$$

where

$$N_{0e}(\boldsymbol{\tau}) = N_0 + \frac{2}{G_1} \sum_{k=1}^{K-1} \mathcal{E}_k h(\tau_k). \tag{7.110}$$

This equation indicates that each interference signal has its power reduced by the factor $G/h(\tau_k)$. Since $h(\tau_k) < 1$ in general, $G/h(\tau_k)$ reflects the increased interference suppression due to the chip waveform and the random timing offsets of the interference signals. Since a similar analysis for direct-sequence systems with balanced QPSK yields (7.110) again, *both quadriphase systems perform equally well against multiple-access interference.*

Application of the previous bounding and approximation methods to (7.109) yields

$$Q\left(\sqrt{\frac{2\mathcal{E}_s}{N_0 + 2h\mathcal{E}_t/G_1}}\right) \le P_s \le Q\left(\sqrt{\frac{2\mathcal{E}_s}{N_0 + 2\mathcal{E}_t/G_1}}\right) \tag{7.111}$$

where the total interference energy $\mathcal{E}_t$ is defined by (7.94). A sufficient condition for the validity of the lower bound is

$$\mathcal{E}_s \ge \frac{3}{2}\left(N_0 + 2\mathcal{E}_t/G_1\right). \tag{7.112}$$

A simple approximation that limits the error in the required $G_1\mathcal{E}_s/2\mathcal{E}_t$ for a specified P_s to $10\log_{10}(1/\sqrt{h})$ is

$$P_s \approx Q\left(\sqrt{\frac{2\mathcal{E}_s}{N_0 + \sqrt{h}2\mathcal{E}_t/G_1}}\right). \tag{7.113}$$

This approximation introduces errors bounded by 0.88 dB and 1.16 dB for rectangular and sinusoidal chip waveforms, respectively. In (7.111) and (7.113), only the total interference power is relevant, not how it is distributed among the individual interference signals.

To compare asynchronous quadriphase direct-sequence systems with asynchronous systems using BPSK, we find a lower bound on P_s for direct-sequence systems with BPSK. Substituting (7.82) into (7.84) and applying Jensen's inequality successively to the integrations over ϕ_k, $k = 1, 2, \ldots, K-1$, we find that a lower bound on P_s is given by the right-hand side of (7.109) if (7.112) is satisfied. This result implies that *asynchronous quadriphase direct-sequence systems are more resistant to multiple-access interference than asynchronous direct-sequence systems with BPSK.*

Figure 7.8 illustrates P_s for an asynchronous quadriphase direct-sequence system in the presence of three interferers, each with equal received symbol energy, rectangular chip waveforms with $h = 2/3$, and $\mathcal{E}_s/N_0 = 15$ dB. The graphs represent the accurate approximation of (7.109), the simple approximation of (7.113), and the bounds of (7.111) as functions of $G_1\mathcal{E}_s/2\mathcal{E}_t$. A comparison of Figures 7.8 and 7.7 indicates the advantage of a quadriphase system.

For *synchronous networks* with either rectangular or sinusoidal chip waveforms, we set the $\{\tau_k)$ equal to zero in (7.109) and obtain

$$P_s = Q\left(\sqrt{\frac{2\mathcal{E}_s}{N_0 + 2\mathcal{E}_t/G_1}}\right). \tag{7.114}$$

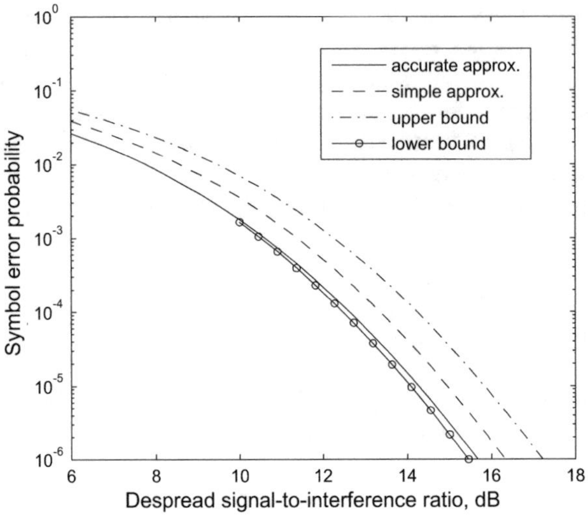

Fig. 7.8 Symbol error probability of a quadriphase direct-sequence system in the presence of three equal-power multiple-access interference signals and $\mathcal{E}_s/N_0 = 15\,\mathrm{dB}$

As this equation coincides with the upper bound in (7.111), we conclude that *asynchronous networks accommodate more multiple-access interference than similar synchronous networks using quadriphase direct-sequence signals with random spreading sequences.*

The equations for P_s allow the evaluation of the average information-bit error probability P_b or its upper bound for channel codes with hard-decision decoders and some soft-decision decoders. For example, consider a quadriphase direct-sequence system with a binary convolutional code and soft-decision decoding. The upper bound on P_b is given by (1.117) of Section 1.3, where $P_2(l)$ is calculated by averaging

$$P_2(l\ |\boldsymbol{\phi}, \boldsymbol{\tau}) = Q\left(\sqrt{\frac{2\mathcal{E}_s l}{N_{0e}(\boldsymbol{\phi}, \boldsymbol{\tau})}}\right) \tag{7.115}$$

over the distributions of $\boldsymbol{\phi}$ and $\boldsymbol{\tau}$. This equation is also valid for direct-sequence systems with BPSK.

When turbo or LDPC codes are used, P_b is a function of $\mathcal{E}_s/N_{0e}(\boldsymbol{\phi}, \boldsymbol{\tau})$. Since x^{-1} is a convex function, Jensen's inequality implies that the average signal-to-interference-and-noise ratio (SINR) γ has the lower bound given by

$$\gamma = E\left[\frac{\mathcal{E}_s}{N_{0e}(\boldsymbol{\phi}, \boldsymbol{\tau})}\right] \geq \frac{\mathcal{E}_s}{E\left[N_{0e}(\boldsymbol{\phi}, \boldsymbol{\tau})\right]}. \tag{7.116}$$

For uniform distributions of $\boldsymbol{\phi}$ and $\boldsymbol{\tau}$, (7.110) implies that for asynchronous communications and quadriphase direct-sequence systems,

$$\gamma \geq \frac{\mathcal{E}_s}{N_0 + 2h\mathcal{E}_t/G_1} \tag{7.117}$$

which indicates that the direct-sequence receiver reduces the power of each interference signal by at least the factor G_1/h on average. Similarly, this inequality is valid for direct-sequence systems with BPSK if we replace G_1 with G, and is valid for synchronous communications if we set $h = 1$.

7.6 Frequency-Hopping Patterns for FH-CDMA

When two or more frequency-hopping signals using the same frequency channel are received simultaneously, they are said to *collide*. To ensure synchronization and to maximize the throughput of an FH-CDMA network, we have to minimize the number of collisions due to all sources. An elementary method of limiting collisions due to signals of other network users is to partition the hopping band among the network users so that each user transmits over a smaller band accessed by no other users and hence experiences no collisions. However, this approach ignores the primary purpose of frequency hopping, which is to avoid all interference. Any reduction in the hopping band of a frequency-hopping system weakens its resistance to interference due to non-network users.

Assuming a common hopset and no partitioning of the hopping band, the primary method of collision minimization is to attempt to make all hop dwell times largely coincide and then optimize the choice of the set of frequency-hopping patterns that are used by the network systems. A mathematically tractable criterion for optimization of the patterns is the minimization of the partial cross-correlations among the patterns, under the assumption that hop dwell intervals coincide.

Consider the hopset of M frequencies: $F = \{f_1, f_2, \ldots, f_M\}$. Let L denote the *correlation window length*, which is the number of hops processed for acquisition or dehopping after acquisition. For two periodic frequency-hopping patterns $X = \{x_i\}$, $Y = \{y_i\}$ of length n, their *partial Hamming cross-correlation* is defined as

$$H_{X,Y}(j, k\,|L) = \sum_{i=0}^{L-1} h\left[x_{i+j} y_{i+j+k}\right], \quad 0 \leq j, k \leq L-1, \; 1 \leq L \leq n \tag{7.118}$$

where $h[x_n y_m] = 1$ if $x_n = y_m$, and 0 otherwise. The partial Hamming cross-correlation is a measure of the number of collisions between two patterns, and it is desirable to use a set of long frequency-hopping patterns with favorable partial Hamming correlations. The partial Hamming autocorrelation of X is $H_{X,X}(j, k\,|L)$.

The maximum partial Hamming cross-correlation of patterns X and Y over all correlation windows of length L is

$$H(X,Y;L) = \max_{0 \le j \le L-1} \max_{0 \le k \le L-1} H_{X,Y}(j,k\,|L) \tag{7.119}$$

and the maximum partial Hamming autocorrelation of pattern X over all correlation windows of length L is

$$H(X;L) = \max_{0 \le j \le L-1} \max_{0 \le k \le L-1} H_{X,X}(j,k\,|L) \ . \tag{7.120}$$

Let P_N denote a set of N distinct frequency-hopping patterns, each using hopset F and having length n. For any L, the *maximum nontrivial partial Hamming correlation* of the pattern set is defined as

$$\mathcal{M}(P_N;L) = \max\left\{\max_{X \in P_N} H(X;L)\,,\ \max_{X,Y \in P_N, X \neq Y} H(X,Y;L)\right\} . \tag{7.121}$$

The selection of a P_N that minimizes $\mathcal{M}(P_N;L)$ tends to minimize both collisions among asynchronous frequency-hopping patterns and undesirable sidelobe peaks in the autocorrelation of a pattern that might hinder frequency-hopping acquisition in a receiver.

Periodic Hamming Cross-Correlation

Consider frequency-hopping acquisition with a matched filter in each receiver of an FH-CDMA network (Section 4.7). The correlation window length L, which is the number of hops stored in each matched filter, is equal to pattern length n. Since $L = n$ and $H_{X,Y}(k,j\,|n)$ is independent of j, the measure of the number of collisions between two patterns is the *periodic Hamming cross-correlation:*

$$H_{X,Y}(k) = \sum_{i=0}^{n-1} h\left[x_i y_{i+k}\right], \ \ 0 \le k \le n-1 \ . \tag{7.122}$$

The periodic Hamming autocorrelation of X is $H_{X,X}(k)$. The *maximum nontrivial periodic Hamming correlation* of the pattern set is defined as

$$\mathcal{M}(P_N) = \max\left\{\max_{X \in P_N} H(X)\,,\ \max_{X,Y \in P_N, X \neq Y} H(X,Y)\right\} \tag{7.123}$$

where

$$H(X) = \max_{1 \le k \le n-1} H_{X,X}(k) \tag{7.124}$$

$$H(X,Y) = \max_{0 \le k \le n-1} H_{X,Y}(k) \ . \tag{7.125}$$

Theorem 2 *For positive integers* N, M, n, *and* $I = \lfloor nN/M \rfloor$, *the* Peng-Fan bounds *[63] are*

$$(n-1)\, MH\,(X) + nM\,(N-1)\, H\,(X,Y) \geq (nN - M)\, n \tag{7.126}$$

$$(n-1)\, NH\,(X) + nN\,(N-1)\, H\,(X,Y) \geq 2InN - (I+1)\, IM \tag{7.127}$$

and

$$\mathcal{M}\,(P_N) \geq \left\lceil \frac{(nN-M)\, n}{(nN-1)\, M} \right\rceil \tag{7.128}$$

$$\mathcal{M}\,(P_N) \geq \left\lceil \frac{2InN - (I+1)\, IM}{(nN-1)\, N} \right\rceil . \tag{7.129}$$

Proof For any two patterns in P_N, let

$$P_{X,Y}\,(N) = \sum_{k=0}^{n-1} H_{X,Y}\,(k) \ . \tag{7.130}$$

Summing over the patterns $X, Y \epsilon P_N$, we obtain

$$\sum_{X,Y\epsilon P_N} P_{X,Y}\,(N) = \sum_{X\epsilon P} H_{X,X}\,(0) + \sum_{X\epsilon P}\sum_{k=1}^{n-1} H_{X,X}\,(k) + \sum_{X,Y\epsilon P_N, X\neq Y}\ \sum_{k=0}^{n-1} H_{X,Y}\,(k) \ . \tag{7.131}$$

The substitution of (7.124), (7.125), and $H_{X,X}\,(0) = n$ into (7.131) yields

$$\sum_{X,Y\epsilon P_N} P_{X,Y}\,(N) \leq nN + (n-1)\, NH\,(X) + nN\,(N-1)\, H\,(X,Y) \ . \tag{7.132}$$

Let $m_X\,(f)$ denote the number of times that hopset frequency f appears in the pattern X. In terms of $m_X\,(f)$, $P_{X,Y}\,(N)$ may be expressed as

$$P_{X,Y}\,(N) = \sum_{f=f_1}^{f_M} m_X\,(f)\, m_Y\,(f) \ .$$

Therefore,

$$\sum_{X,Y\epsilon P_N} P_{X,Y}\,(N) = \sum_{i=1}^{M} g_i^2 \tag{7.133}$$

where

$$g_i = \sum_{X \epsilon P} m_X(f_i) \ .$$

A lower bound on $\sum_{i=1}^{M} g_i^2$ may be established by observing that

$$\sum_{i=1}^{M} g_i = \sum_{X \epsilon P_N} \sum_{i=1}^{M} m_X(f_i)$$
$$= \sum_{X \epsilon P_N} n = nN \ .$$

Applying the method of Lagrange multipliers (Section 5.4) to the minimization of $\sum_{i=1}^{M} g_i^2$ subject to the constraint $\sum_{i=1}^{M} g_i = nN$, we find that

$$\sum_{X,Y \epsilon P_N} P_{X,Y}(N) \geq \frac{n^2 N^2}{M} \ . \tag{7.134}$$

To obtain a tighter alternative lower bound, we use the constraint that the $\{g_i\}$ must be nonnegative integers. Order these integers $\{g_i\}$ so that $0 \leq g_1 \leq g_2 \cdots \leq g_M$. If $\{g_i\}$ is a sequence that minimizes $\sum_{i=1}^{M} g_i^2$ and $g_M - g_1 > 1$, then construct the sequence $\{p_i\}$ of nonnegative integers such that $p_i = g_i$, $i = 2, 3, \ldots, M-1$; $p_1 = g_1 + 1$; $p_M = g_M - 1$. Then,

$$\sum_{i=1}^{M} g_i^2 - \sum_{i=1}^{M} p_i^2 = 2(g_M - g_1 - 1) > 0$$

which is a contradiction. Thus, if $\{g_i\}$ is a sequence that minimizes $\sum_{i=1}^{M} g_i^2$, then $g_M = g_1 + 1$ or $g_M = g_1$. Therefore, the minimizing sequence has the form

$$g_1 = g_2 = \cdots = g_{M-r} = I, \quad g_{M-r+1} = g_{M-r+2} = \cdots = g_M = I + 1 \tag{7.135}$$

where I is a nonnegative integer and $0 \leq r < M$. The constraint $\sum_{i=1}^{M} g_i = nN$ requires that $nN = IM + r$, which implies that $I = \lfloor nN/M \rfloor$. Thus, (7.133), (7.135), and $r = nN - IM$ imply that

$$\sum_{X,Y \epsilon P_N} P_{X,Y}(N) \geq (2I + 1)\, nN - (I + 1)\, IM \ . \tag{7.136}$$

Combining (7.132), (7.134), and (7.136), we obtain (7.126) and (7.127). Substituting $H(X) \leq \mathcal{M}(P_N)$ and $H(X,Y) \leq \mathcal{M}(P_N)$ and recognizing that $\mathcal{M}(P_N)$ must be an integer, we obtain (7.128) and (7.129). □

A set of patterns P_N is considered *optimal with respect to the periodic Hamming cross-correlation* if $\mathcal{M}(P_N)$ is equal to the larger or common value of the two lower bounds of the theorem, which are equal when nN/M is an integer. Many of these optimal sets of patterns are known. To ensure that all frequencies in the hopset are used in each frequency-hopping pattern, it is necessary that $n \geq M$.

Example 1 Consider the selection of frequency-hopping patterns when $N = M = 4$ and $n = 7$, which imply that $\mathcal{M}(P_4) \geq 2$. The hopset is $F = \{f_1, f_2, f_3, f_4\}$, and the set of patterns is $P_4 = \{X_1, X_2, X_3, X_4\}$. If the set of patterns is

$$X_1 = f_1, f_2, f_3, f_2, f_4, f_4, f_3 \quad X_2 = f_2, f_1, f_4, f_1, f_3, f_3, f_4$$
$$X_3 = f_3, f_4, f_1, f_4, f_2, f_2, f_1 \quad X_4 = f_4, f_3, f_2, f_3, f_1, f_1, f_2$$

we find that

$$\max_{X \in P_N} H(X) = 1, \qquad \max_{X,Y \in P_N, X \neq Y} H(X,Y) = 2 \tag{7.137}$$

and hence $\mathcal{M}(P_4) = 2$, which verifies optimality. □

Partial Hamming Cross-Correlation

When serial-search acquisition is used (Section 4.2), $L \leq n$, and an *optimal set of frequency-hopping patterns* is defined to be a set that achieves one of the lower bounds in the following theorem [122] for any correlation window length such that $1 \leq L \leq n$.

Theorem 3 *For positive integers* N, M, n, *and* $I = \lfloor nN/M \rfloor$, *and any window length such that* $1 \leq L \leq n$,

$$\mathcal{M}(P_N; L) \geq \left\lceil \frac{L}{n} \cdot \frac{(nN - M)\, n}{(nN - 1)\, M} \right\rceil \tag{7.138}$$

$$\mathcal{M}(P_N; L) \geq \left\lceil \frac{L}{n} \cdot \frac{2InN - (I+1)\, IM}{(nN - 1)\, N} \right\rceil . \tag{7.139}$$

Proof Define

$$\mathcal{S}(L; j) = \sum_{X,Y \in P_N} \sum_{k=0}^{n-1} H_{X,Y}(j, k \,|L) . \tag{7.140}$$

Substitution of (7.118), an interchange of summations, evaluation of the outer summation, and then substitution of (7.122) and (7.130) yield

$$\begin{aligned}\sum_{j=0}^{n-1} \mathcal{S}(L;j) &= \sum_{i=0}^{L-1} \sum_{X,Y \epsilon P_N} \sum_{k=0}^{n-1} \sum_{j=0}^{n-1} h\left[x_{i+j} y_{i+j+k}\right] \\ &= L \sum_{X,Y \epsilon P_N} \sum_{k=0}^{n-1} \sum_{j=0}^{n-1} h\left[x_j y_{j+k}\right] \\ &= L \sum_{X,Y \epsilon P_N} \sum_{k=0}^{n-1} H_{X,Y}(k) \\ &= L \sum_{X,Y \epsilon P_N} P_{X,Y}(N) \ . \end{aligned} \tag{7.141}$$

Expanding the right-hand side of (7.140), we obtain

$$\begin{aligned}\mathcal{S}(L;j) &= \sum_{X \epsilon P_N} H_{X,X}(j,0\,|L) + \sum_{X \epsilon P_N} \sum_{k=1}^{n-1} H_{X,X}(j,k\,|L) \\ &\quad + \sum_{X,Y \epsilon P_N X \neq Y} \sum_{k=0}^{n-1} H_{X,Y}(j,k\,|L) \\ &\leq ML + M(n-1)\,H(X;L) + M(M-1)\,nH(X,Y;L) \\ &\leq ML + M(Mn-1)\,\mathcal{M}(P_N;L) \ . \end{aligned}$$

Therefore,

$$\sum_{j=0}^{n-1} \mathcal{S}(L;j) \leq nML + nM(Mn-1)\,\mathcal{M}(P_N;L) \ . \tag{7.142}$$

Combining (7.141) and (7.142), we obtain

$$L \sum_{X,Y \epsilon P_N} P_{X,Y}(N) \leq nML + nM(Mn-1)\,\mathcal{M}(P_N;L) \ . \tag{7.143}$$

The successive substitution of (7.134) and (7.136) into (7.143) and the recognition that $\mathcal{M}(P_N;L)$ must be an integer yields (7.138) and (7.139). □

Sets of optimal frequency-hopping patterns that achieve the lower bounds of Theorem 3 for any correlation window length such that $1 \leq L \leq n$ have been found ([12] and the references therein). These pattern sets also achieve the Peng-Fan bounds. However, pattern sets that achieve the Peng-Fan bounds do not necessarily achieve the bounds of Theorem 3 when $L \neq n$.

Uniform Patterns

The primary purpose of frequency hopping is to avoid interference, but the spectral distribution of the non-network interference is generally unknown. Therefore, it is prudent to choose $n = mM$, where m is a positive integer, and to ensure that a period of a hopping pattern includes every frequency in the hopset the same number of times. If $n = mM$ and all hopset frequencies are used equally, the frequency-hopping pattern is called a *uniform frequency-hopping pattern*. To protect against pattern reproduction by an opponent, a large linear span (Section 3.1) of the sequence generating the uniform pattern is required, which implies that $n = mM >> 1$. Theorem 3 implies that an *optimal set* of uniform frequency-hopping patterns has

$$\mathcal{M}\left(P_N;L\right)=\left\lceil\frac{L\left(mN-1\right)}{mMN-1}\right\rceil,\quad n=mM\,. \tag{7.144}$$

The *maximum collision rate* between two distinct patterns in P_N during L hops is

$$C\left(M,N,L\right)=\frac{\max_{X,Y\in P_N,X\neq Y}H\left(X,Y\right)}{L}\,. \tag{7.145}$$

For an *optimal set* of uniform frequency-hopping patterns,

$$\begin{aligned}C\left(M,N,L\right)&\leq\frac{\mathcal{M}\left(P_N;L\right)}{L}\\&\leq\frac{1}{L}\left\lceil\frac{L}{M}\right\rceil\\&\leq\frac{1}{L}+\frac{1}{M},\quad n=mM\,.\end{aligned} \tag{7.146}$$

If $L = n = mM$, we find that

$$C\left(M,N,n\right)\leq\frac{1}{M},\quad n=mM\,. \tag{7.147}$$

The probability of a collision at any instant between two patterns when random frequency-hopping patterns are used is $1/M$. Let Z denote the number of collisions between two patterns during $n = mM$ hops. The expected value of Z is $E[Z] = n/M = m$, and the average collision rate is $1/M$. A straightforward calculation of the variance of Z for independent, random frequency-hopping patterns proves that it is equal to $(M-1)m/M$, and hence the standard deviation of the collision rate is $s = \sqrt{(M-1)/mM^3} \leq 1/\sqrt{m}M$. This result indicates that the collision rate for $L = nM$ hops of two random patterns seldom exceeds the upper bound on $C(M,N,n)$, the collision rate when the optimal uniform patterns are used. Since this bound is generally tight, modeling the frequency-hopping patterns as random is a good approximation when calculating collision rates or probabilities, and this model is used in the subsequent analyses.

7.7 Multiuser Detectors for DS-CDMA Systems

The conventional direct-sequence receiver for a single user, which only requires knowledge of the spreading sequence of the desired signal, is suboptimal against multiple-access interference. The primary susceptibility of the conventional direct-sequence receiver is the *near-far problem,* which refers to the possibility that the transmitter of a desired signal may be much farther from the receiving antenna than interfering transmitters. The received power of the desired signal may be much lower than that of the interfering signals, and the spreading factor may be insufficient to overcome this disadvantage. If a potential near-far problem exists, power control (Section 8.4) may be used to limit its impact. However, power control is imperfect, entails a substantial overhead cost, and is not feasible for ad hoc communication networks. Even if the power control is perfect, the remaining interference causes a nonzero *error floor*, which is a minimum bit error probability that exists when the thermal noise is zero. Thus, an alternative to the conventional receiver is desirable.

A *multiuser detector* [40, 69, 84, 115] is a receiver that exploits the deterministic structure of multiple-access interference or uses joint processing of a set of multiple-access signals. An optimal multiuser detector almost completely eliminates the multiple-access interference and hence the near-far problem, thereby rendering power control unnecessary, but implementation of such a detector is prohibitively complex, especially when long spreading sequences are used. Since all received signals from mobiles within a cell are demodulated at its base station, a suboptimal multiuser detector is more feasible for an uplink than it is for a downlink. Multipath components can be accommodated as separate interference signals, and rake combining may precede the multiuser detection. The combination of a multiuser detector with iterative channel estimation, demodulation, and decoding (Chapter 9) is potentially very effective against multiple-access interference if the practical impediments can be resolved.

Optimal Detectors

A multiuser detector is *jointly optimal* if it makes collective symbol decisions for K received signals based on the *maximum a posteriori* (MAP) criterion. A multiuser detector is *individually optimal* if it selects the most probable set of symbols of a single desired signal based on the MAP criterion, thereby providing the minimum symbol error probability. In nearly all applications, jointly optimal decisions are preferable because of their lower complexity and because both types of decisions agree with very high probability unless the symbol error probability is very high. Assuming that equally likely symbols are transmitted, the jointly optimal MAP detector is the same as the jointly optimal maximum-likelihood detector, which is henceforth called the *optimal detector*.

Consider a DS-CDMA terminal that receives K BPSK signals, each of which has a common carrier frequency f_c, chip duration T_c, chip waveform $\psi(t)$, and number of spreading-sequence chips per symbol G. Information bits are mapped into code symbols. The chip waveform is assumed to have unit energy in a symbol interval, as in (7.43), and cause negligible intersymbol interference. Let $d_k(t)$ denote the unit-magnitude code symbols of signal k transmitted by user k. For signal k, let $\mathcal{E}_k$ denote the received energy per symbol, $p_{k,i} = \pm 1$ the chip i of the spreading sequence, τ_k the timing offset relative to some reference signal, and θ_k the phase at a reference time. Consider the single symbol interval $0 \leq t \leq T_s$ of the reference signal at the receiver. For *asynchronous communications* over the AWGN channel, the composite received signal for BPSK signals over a symbol interval is

$$y(t) = \sum_{k=0}^{K-1} \sqrt{2\mathcal{E}_k} d_k(t-\tau_k) \sum_{i=0}^{G-1} p_{k,i}\psi(t - iT_c - \tau_k)\cos(2\pi f_c t + \theta_k) + n(t),$$
$$0 \leq t \leq T_s \tag{7.148}$$

where $n(t)$ is the white Gaussian noise. The channel effects are included in $\mathcal{E}_k$ and θ_k. A frequency translation or downconversion to baseband is followed by matched filtering.

For *synchronous communications* over the AWGN channel, the symbols are aligned in time, and the detection of each symbol of the desired signal is independent of the other symbols. If accurate symbol or timing synchronization is implemented, the optimal detector can be determined by considering a single symbol interval $0 \leq t \leq T_s$. For *synchronous communications,* the composite received signal is

$$y(t) = \sum_{k=0}^{K-1} \sqrt{2\mathcal{E}_k} d_k \sum_{i=0}^{G-1} p_{k,i}\psi(t - \nu T_c)\cos(2\pi f_c t + \theta_k) + n(t), \quad 0 \leq t \leq T_s \tag{7.149}$$

where $d_k = \pm 1$. The downconversion, which is implemented by a quadrature downconverter (Section 2.5), is represented by the multiplication of the received signal by $\sqrt{2}\exp(-j2\pi f_c t)$, where the factor $\sqrt{2}$ has been inserted for mathematical convenience and $j = \sqrt{-1}$. The in-phase and quadrature decomposition is necessary because it is not necessarily assumed that any of the K signals is synchronized with the receiver-generated carrier signal. The downconverter outputs are applied to parallel chip-matched filters, the outputs of which are sampled at the chip rate.

As shown in Section 2.4, after chip-matched filtering and discarding of a negligible integral, the demodulated chip-rate sequence associated with the synchronous received symbols is

$$y_i = \sum_{k=0}^{K-1} A_k d_k p_{k,i} + n_i, \quad i = 0, 1, \ldots, G-1 \tag{7.150}$$

where the complex symbol amplitude for user k is

$$A_k = G^{-1}\sqrt{\mathcal{E}_k}\exp(j\theta_k) \tag{7.151}$$

and the noise sample is

$$n_i = \sqrt{2}\int_{iT_c}^{(i+1)T_c} n(t)\psi(t-iT_c)\exp(-j2\pi f_c t)\,dt,\ \ i = 0, 1, \ldots, G-1. \tag{7.152}$$

Let $\mathbf{y}$ and $\mathbf{n}$ denote the $G \times 1$ demodulated-sequence and noise vectors with components defined by (7.150) and (7.152), respectively. Let $\mathbf{d} = [d_0 \ldots d_{K-1}]^T$ denote the $K \times 1$ vector of code symbols. The demodulated vector may be represented as

$$\mathbf{y} = \mathbf{PAd} + \mathbf{n} \tag{7.153}$$

where column k of the $G \times K$ matrix $\mathbf{P}$ is the vector $\mathbf{p}_k = [p_{k,0}\ p_{k,1}\ \ldots\ p_{k,G-1}]^T$ that represents the spreading sequence of user k, and $\mathbf{A}$ is the $K \times K$ diagonal matrix with A_k as its kth diagonal element.

Since the $\{n_\upsilon\}$ are limits of Riemann sums that are linear combinations of the zero-mean, white Gaussian noise $n(t)$, Theorem A1 of Appendix A.1 implies that $\mathbf{n}$ is a zero-mean, complex Gaussian random vector. If the two-sided PSD of $n(t)$ is $N_0/2$,then the $G \times G$ correlation matrix of $\mathbf{n}$ is

$$E\left[\mathbf{nn}^H\right] = G^{-1}N_0\mathbf{I} \tag{7.154}$$

where the superscript H denotes the conjugate transpose. If the bandwidth of the chip waveform is less than f_c, $\mathbf{n}$ is a circularly symmetric, complex Gaussian random vector with

$$E\left[\mathbf{nn}^T\right] = \mathbf{0}\,. \tag{7.155}$$

Equations (7.154) and (7.155) indicate that the $2G$ real and imaginary Gaussian components of $\mathbf{n}$ are uncorrelated and hence independent.

Assuming that all possible values of the symbol vector $\mathbf{d}$ are equally likely, the *optimal multiuser detector* is defined as the maximum-likelihood detector, which selects the value of $\mathbf{d}$ that maximizes the log-likelihood function. Since the noise vector is a complex Gaussian random vector, calculations in Section 1.2 show that maximum-likelihood detection for synchronous communications is the same as selecting the value of $\mathbf{d}$ that minimizes the Euclidean norm

$$\Lambda(\mathbf{d}) = ||\mathbf{y} - \mathbf{PAd}||^2 \tag{7.156}$$

subject to the constraint that $\mathbf{d} \in \mathbf{D}$, where $\mathbf{D}$ is the set of $K \times 1$ vectors such that each element is ± 1. Thus, the optimal detector uses G successive samples of the output of a complex chip-matched filter to compute the decision vector

$$\widehat{\mathbf{d}} = \arg \min_{\mathbf{d} \in \mathbf{D}} ||\mathbf{y} - \mathbf{PAd}||^2 . \tag{7.157}$$

Expanding (7.157), dropping the term $||\mathbf{y}||^2$, which is irrelevant to the selection of $\mathbf{d}$, and using the fact that $\mathbf{A}$ is diagonal and hence $\mathbf{A}^{\mathbf{H}} = \mathbf{A}^*$, we find that the maximum-likelihood detector for synchronous communications selects

$$\widehat{\mathbf{d}} = \arg \min_{\mathbf{d} \in \mathbf{D}} [C(\mathbf{d})] \tag{7.158}$$

where the *correlation metric* is the real-valued function

$$C(\mathbf{d}) = \mathbf{d}^T \mathbf{A}^* \mathbf{P}^{\mathbf{T}} \mathbf{PAd} - 2\mathbf{d}^T \operatorname{Re}(\mathbf{A}^* \mathbf{P}^{\mathbf{T}} \mathbf{y}). \tag{7.159}$$

To evaluate $C(\mathbf{d})$, the K spreading sequences must be known so that $\mathbf{P}$ can be calculated, and the K complex signal amplitudes must be estimated. Short spreading sequences are necessary or $\mathbf{P}$ must change with each symbol, which greatly increases the implementation complexity. In principle, (7.158) can be evaluated by an exhaustive search of all values of $\mathbf{d} \in \mathbf{D}$, but this search requires a computational complexity that increases exponentially with K, and hence is only feasible for small values of K and G.

For *asynchronous communications*, the design of the maximum-likelihood detector is immensely complicated by the timing offsets of the K signals in (7.148). A timing offset implies that a desired symbol overlaps two consecutive symbols from each interference signal. Consequently, an entire message or codeword of N correlated code symbols from each of the K users must be processed, and decisions must be made about NK binary symbols. The vector $\mathbf{d}$ is $NK \times 1$ with the first N elements representing the symbols of signal 1, the second N elements representing the symbols of signal 2, and so forth. The detector must estimate the transmission delays of all K multiple-access signals and estimate the partial cross-correlations among the signals. The detector must select K symbol sequences, each of length N, corresponding to the maximum-likelihood criterion. Recursive algorithms similar to the Viterbi algorithm simplify computations by exploiting the fact that each received symbol overlaps at most $2(K - 1)$ other symbols. Nevertheless, the computational complexity increases exponentially with K.

In view of both the computational requirements and the parameters that must be estimated, it is highly unlikely that the optimal asynchronous multiuser detector will have practical applications. Subsequently, alternative suboptimal multiuser detectors for DS-CDMA systems with complexities that increase linearly with K are considered. All of them use a despreader bank.

Conventional Single-User Detector

A *conventional single-user detector* is designed under the assumption that multiuser interference is equivalent to additional zero-mean, white Gaussian noise. Assuming synchronous BPSK signals, the real-valued received vector for K signals is given by (7.153). Since the multiuser interference is modeled as white Gaussian noise, the kth conventional detector makes a maximum-likelihood bit decision by computing

$$\widehat{d}_k = \arg \min_{\mathbf{d_k} \in (+1,-1)} \|\mathbf{y} - \mathbf{p_k} A_k d_k\|^2 . \tag{7.160}$$

Expanding the right-hand side of this equation, using $\mathbf{p}_k^T \mathbf{p}_k = G$, and dropping an irrelevant term and an irrelevant factor that do not affect the decision, we find that

$$\begin{aligned} \widehat{d}_k &= \arg \max_{\mathbf{d_k} \in (+1,-1)} \left[d_k \operatorname{Re}\left(A_k^* \mathbf{p}_k^T \mathbf{y} \right) \right] \\ &= \operatorname{sgn}\left[\operatorname{Re}\left(A_k^* \mathbf{p}_k^T \mathbf{y} \right) \right] \end{aligned} \tag{7.161}$$

where the *signum function* is defined as $sgn(x) = 1, x \geq 0$, and $sgn(x) = -1, x < 0$. If we assume that the detector does phase synchronization of the kth signal, then A_k^* may be replaced by $A_k > 0$, which then becomes irrelevant. Therefore,

$$\widehat{d}_k = \operatorname{sgn}[\mathbf{p}_k^T \mathbf{y}_r] \tag{7.162}$$

where $\mathbf{y}_r = \operatorname{Re}(\mathbf{y})$.

Let $\mathbf{d}_{ke}$ denote the vector of $K - 1$ components of $\mathbf{d}$ excluding d_k. Substituting (7.153) into (7.162), we obtain

$$\widehat{d}_k = \operatorname{sgn}[G A_k d_k + G A_k B\left(\mathbf{d}_{ke}\right) + Re(\mathbf{p}_k^T \mathbf{n})] \tag{7.163}$$

where $Re(\mathbf{p}_k^T \mathbf{n})$ is a zero-mean Gaussian random variable,

$$B\left(\mathbf{d}_{ke}\right) = \sum_{\substack{i=0 \\ i \neq k}}^{K-1} d_i R_{ki} \frac{\operatorname{Re}(A_i)}{A_k} . \tag{7.164}$$

and the *correlation matrix* is defined as the real-valued matrix

$$\mathbf{R} = G^{-1} \mathbf{P}^{\mathbf{T}} \mathbf{P}. \tag{7.165}$$

Assuming equally likely code symbols, we derive the symbol error probability for user k as a function of the correlation matrix. Let S_k denote the set of the 2^{K-1} distinct vectors that $\mathbf{d}_{ke}$ can equal. Let $\mathbf{d}_{kn} \in \mathsf{S}_k$ denote the nth one of those distinct vectors. Conditioning on $\mathbf{d}_{ke} = \mathbf{d}_{kn}$, $1 \leq n \leq 2^{K-1}$, we find that a symbol error occurs if the Gaussian noise term $Re(\mathbf{p}_k^T \mathbf{n})$ exceeds the first two terms in (7.163). By symmetry, we can assume that $d_k = 1$ in the evaluation of the symbol error probability. Using (7.155) and (7.154), we obtain

$$\begin{aligned} var[Re(\mathbf{p}_k^T \mathbf{n})] &= \frac{1}{4} E[\mathbf{p}_k^T (\mathbf{n} + \mathbf{n}^*)(\mathbf{n} + \mathbf{n}^*)^T \mathbf{p}_k] \\ &= \frac{N_0}{2} \,. \end{aligned} \tag{7.166}$$

Therefore, the conditional symbol error probability for user k is

$$P_s\left(k|\mathbf{d}_{kn}\right) = Q\left(\sqrt{\frac{2\mathcal{E}_k}{N_0}}[1 + B\left(\mathbf{d}_{kn}\right)]\right) \,. \tag{7.167}$$

If all $\mathbf{d}_{kn} \in \mathsf{S}_k$ are equally likely, then the symbol error probability for user k is

$$P_s(k) = 2^{-(K-1)} \sum_{n=1}^{2^{K-1}} P_s\left(k|\mathbf{d}_{kn}\right) \,. \tag{7.168}$$

Decorrelating Detector

The complexity of the optimal multiuser detector of (7.158) is reduced by ignoring the finite signal alphabet of the code symbols and instead allowing them to be differentiable functions. The *decorrelating detector* for synchronous BPSK signals is derived by initially maximizing the correlation metric $C(\mathbf{d})$ of (7.159) without any constraint on $\mathbf{d}$.

Applying (5.2) and (5.3), we find that the gradient of $C(\mathbf{d})$ with respect to the K-dimensional vector $\mathbf{d}$ is

$$\nabla_{\mathbf{d}} C(\mathbf{d}) = \mathbf{F}\mathbf{d} - 2\,\mathrm{Re}(\mathbf{A}^*\mathbf{P}^{\mathbf{T}}\mathbf{y}) \tag{7.169}$$

where

$$\mathbf{F} = \mathbf{A}^*\mathbf{P}^{\mathbf{T}}\mathbf{P}\mathbf{A} + \mathbf{A}\mathbf{P}^{\mathbf{T}}\mathbf{P}\mathbf{A}^* = 2\,\mathrm{Re}\left(\mathbf{A}^*\mathbf{P}^{\mathbf{T}}\mathbf{P}\mathbf{A}\right). \tag{7.170}$$

If $G \geq K$, then we assume that the $G \times K$ matrix $\mathbf{P}$ has full rank K, which implies that $\mathbf{P}^{\mathbf{T}}\mathbf{P}$ is full rank and hence invertible. Assuming that the diagonal elements of the diagonal matrix $\mathbf{A}$ are nonzero, the $K \times K$ matrix $\mathbf{PA}$ has full rank K. Therefore, the Hermitian matrices $\mathbf{A}^*\mathbf{P}^{\mathbf{T}}\mathbf{P}\mathbf{A}$ and $\mathbf{A}\mathbf{P}^{\mathbf{T}}\mathbf{P}\mathbf{A}^*$ are positive definite, and hence $\mathbf{F}$ is Hermitian, positive definite, and invertible (Appendix G). Applying the necessary condition that $\nabla_{\mathbf{d}} C(\mathbf{d}) = \mathbf{0}$ to find the stationary point $\mathbf{d}_s$, we obtain

$$\mathbf{d}_s = 2\mathbf{F}^{-1}\,\mathrm{Re}(\mathbf{A}^*\mathbf{P}^{\mathbf{T}}\mathbf{y}). \tag{7.171}$$

The solution $\mathbf{d}_s$ corresponds to the minimum of $C(\mathbf{d})$ because (7.159) may be expressed in terms of $\mathbf{d}_s$ as

$$C(\mathbf{d}) = \frac{1}{2}(\mathbf{d} - \mathbf{d}_s)^T \mathbf{F}(\mathbf{d} - \mathbf{d}_s) - \frac{1}{2}\mathbf{d}_s^T \mathbf{F}\mathbf{d}_s . \tag{7.172}$$

Since $\mathbf{F}$ is a Hermitian positive-definite matrix, $C(\mathbf{d})$ has a unique minimum at $\mathbf{d} = \mathbf{d}_s$.

Substituting (7.153) into (7.171), we obtain

$$\mathbf{d}_s = \mathbf{d} + \mathbf{n}_s \tag{7.173}$$

where the noise vector is

$$\mathbf{n}_s = 2\mathbf{F}^{-1}\,\text{Re}(\mathbf{A}^*\mathbf{P}^\mathbf{T}\mathbf{n}). \tag{7.174}$$

Equation (7.173) shows that the components of $\mathbf{d}_s$ are decorrelated from each other because the kth component of $\mathbf{d}_s$ depends only on the kth component of $\mathbf{d}$, and not on the other components of $\mathbf{d}$. Thus, the decorrelating detector, which uses an estimator based on $\mathbf{d}_s$, is a type of *zero-forcing estimator* that eliminates multiple-access interference from each of its components. Since the multiple-access interference is completely removed, the near-far problem does not exist, and the *decorrelating detector* is *near-far resistant.*

Since $d_k = \pm 1$, a decorrelating detector that makes hard decisions computes

$$\widehat{\mathbf{d}} = \text{sgn}(\mathbf{d}_s) = \text{sgn}\left[2\mathbf{F}^{-1}\,\text{Re}(\mathbf{A}^*\mathbf{P}^\mathbf{T}\mathbf{y})\right] \tag{7.175}$$

where $sgn(\mathbf{u})$ denotes the vector of signum functions of the components of $\mathbf{u}$. If a direct-sequence system applies the output of a decorrelating detector to a channel decoder, then it is desirable for the detector to make soft rather than hard decisions. Soft symbol metrics can be produced by eliminating the signum function in (7.175) or replacing it with a nonlinear function $\mathcal{N}(\cdot)$. Thus, the input to the decoder becomes

$$\widehat{\mathbf{d}} = \mathcal{N}\left[2\mathbf{F}^{-1}\,\text{Re}(\mathbf{A}^*\mathbf{P}^\mathbf{T}\mathbf{y})\right]. \tag{7.176}$$

Since $d_k = \pm 1$, plausible choices for $\mathcal{N}(\cdot)$ are the *clipping function* defined as

$$\mathsf{c}(x) = \begin{cases} +1, & x \geq 1 \\ x, & -1 < x < 1 \\ -1, & x \leq -1 \end{cases} \tag{7.177}$$

and the hyperbolic tangent function $\tanh(\alpha x)$. If $\alpha = 1$, then $\tanh(x) \approx \mathsf{c}(x)$; if $a >> 1$, then $\tanh(\alpha x) \to sgn(x)$.

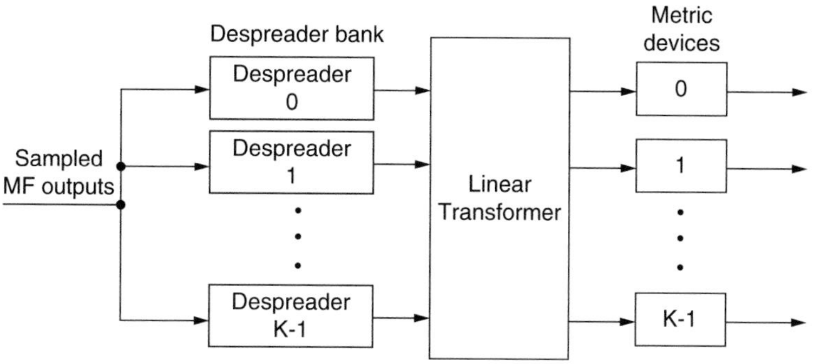

Fig. 7.9 Architecture of complete decorrelating or MMSE detector for synchronous DS-CDMA communications. Despreader bank comprises K parallel correlators

The complete decorrelating detector that implements (7.176) for K synchronous users has the form diagrammed in Figure 7.9. The G sampled outputs of the complex-valued chip-matched filter are applied to a despreader bank of K parallel despreaders, which provide the K-dimensional vector $\mathbf{P}^{\mathbf{T}}\mathbf{y}$. The linear transformer performs a linear transformation on $\mathbf{P}^{\mathbf{T}}\mathbf{y}$ to compute the K-dimensional vector $\mathbf{d}_s$. Each of the K metric devices either makes a hard decision or applies a nonlinear function to provide one of the K symbol metrics.

Consider the hard-decision detection of symbol d_k, $0 \leq k \leq K-1$. Substituting (7.153) into (7.176) implies that the kth component of $\widehat{\mathbf{d}}$ is

$$\widehat{d}_k = \operatorname{sgn}(d_k + n_{sk}) \tag{7.178}$$

where n_{sk} is the kth component of $\mathbf{n}_s$. Equation (7.174) indicates that $\mathbf{n}_s$ is a Gaussian random vector (see Appendix A.1) because the vector $\mathbf{n}$ has components that are independent zero-mean Gaussian random variables. From (7.174) and (7.154), it follows that the real-valued correlation matrix of $\mathbf{n}_s$ is

$$E\left[\mathbf{n}_s\mathbf{n}_s^T\right] = \frac{N_0}{G}\mathbf{F}^{-1} \tag{7.179}$$

and the substitution of (7.165) gives

$$var\,(n_{sk}) = \frac{N_0}{2\mathcal{E}_k}R_{kk}^{-1} \tag{7.180}$$

where R_{kk}^{-1} denotes element k, k of $\mathbf{R}^{-1}$, and $\mathcal{E}_k = |A_k|^2\,G^2$ is the symbol energy of the kth user. The noise causes an error if it causes the argument of the signum function in (7.178) to have a different sign than d_k. Since this noise has a Gaussian density function, the symbol error probability for symbol k of user k is

$$P_s(k) = Q\left(\sqrt{\frac{2\mathcal{E}_k}{N_0 R_{kk}^{-1}}}\right), \quad k = 0, 1, \ldots, K-1. \tag{7.181}$$

In the absence of multiple-access interference, $R_{kk}^{-1} = 1$. Thus, the presence of multiple-access interference requires an increase in energy by the factor R_{kk}^{-1} if a specified symbol error probability is to be maintained.

Element i, k of $\mathbf{R}$ is

$$R_{ik} = \frac{\mathbf{p}_i^T \mathbf{p}_k}{G} = \frac{\mathcal{A}_{ik} - \mathcal{D}_{ik}}{G}, \; i \neq k \tag{7.182}$$

where $\mathcal{A}_{ik}$ denotes the number of agreements in the corresponding bits of $\mathbf{p}_i$ and $\mathbf{p}_k$, and $\mathcal{D}_{ik}$ denotes the number of disagreements. As the spreading factor G increases, it becomes increasingly likely that R_{ik} is small for $i \neq k$; hence, $\mathbf{R}$ approximates a unit diagonal matrix, even if the spreading sequences are not selected to be orthogonal. If the spreading sequences are modeled as random binary sequences, then $E\left[R_{ik}\right] = 0$ and $var\left[R_{ik}\right] = G^{-1}$, $i \neq k$, which shows the advantage of a large spreading factor.

Example 2 Consider synchronous communications with $K = 2$, $R_{01} = R_{10} = \rho$, and $|\rho| < 1$. The correlation matrix and its inverse are

$$\mathbf{R} = \begin{bmatrix} 1 & \rho \\ \rho & 1 \end{bmatrix}, \quad \mathbf{R}^{-1} = \frac{1}{1-\rho^2}\begin{bmatrix} 1 & -\rho \\ -\rho & 1 \end{bmatrix}. \tag{7.183}$$

Since $R_{00}^{-1} = R_{11}^{-1} = (1-\rho^2)^{-1}$, (7.181) yields the symbol error rate for user k:

$$P_s(k) = Q\left(\sqrt{\frac{2\mathcal{E}_k\left(1-\rho^2\right)}{N_0}}\right), \quad k = 0, 1. \tag{7.184}$$

If $|\rho| \leq 1/2$, the required increase in $\mathcal{E}_k/N_0$ for each $P_s(k)$ to accommodate the multiple-access interference is less than 1.25 dB.

To compare the symbol error probabilities of the decorrelating and conventional detectors for this example, we observe that for equally likely values of d_k, (7.167) and (7.168) yield the symbol error probability for the conventional detector:

$$\begin{aligned} P_s(k) &= \frac{1}{2}Q\left(\sqrt{\frac{2\mathcal{E}_k}{N_0}}\left(1-|\rho C|\right)\right) + \frac{1}{2}Q\left(\sqrt{\frac{2\mathcal{E}_k}{N_0}}\left(1+|\rho C|\right)\right) \\ &\leq Q\left(\sqrt{\frac{2\mathcal{E}_k}{N_0}}\left(1-|\rho C|\right)\right), \quad k = 0, 1 \end{aligned} \tag{7.185}$$

where $C = \text{Re}(A_{1-k})/A_k$. Comparing this upper bound with (7.184), we observe that the conventional detector has a lower $P_s(k)$ than the decorrelating detector if

$$|C| < \frac{1-\sqrt{1-\rho^2}}{|\rho|}, \quad \rho \neq 0. \tag{7.186}$$

Since $0 < |\rho| < 1$, the right-hand side of (7.186) is upper bounded by unity; hence, $|C| > 1$ is necessary for the decorrelating detector to be advantageous relative to the conventional detector in this example.

If we assume that the amplitudes are real-valued, then by applying (7.176) and discarding irrelevant factors, we find that the symbol decisions of the decorrelating detector are $\widehat{d}_0 = sgn(r_0 - \rho r_1)$ and $\widehat{d}_1 = sgn(r_1 - \rho r_0)$, where $[r_0\ r_1]^T = \mathbf{P}^T \operatorname{Re}(\mathbf{y})$. □

For asynchronous communications, an entire message or codeword of N correlated code symbols from each of the K users must be processed because of the timing offsets. The symbol vector $\mathbf{d}$ and the correlation matrix $\mathbf{R}$ of the decorrelating detector are $NK \times 1$ and $NK \times NK$, respectively. Compared with the optimal detector, the decorrelating detector offers greatly reduced, but still formidable, computational requirements. Application of the decorrelating detector in practical asynchronous communications is doubtful.

Minimum-Mean-Square-Error Detector

The *minimum-mean-square-error* (MMSE) detector is the receiver that uses a linear transformation of $\boldsymbol{y}$ by the $K \times K$ matrix $\mathbf{L}_0$ such that the metric

$$M = E\left[\|\mathbf{d} - \mathbf{L}\mathbf{y}\|^2\right] \tag{7.187}$$

is minimized when $\mathbf{L} = \mathbf{L}_0$. We assume synchronous BPSK signals. Therefore, $\mathbf{d}$ is real-valued and hence the decision statistic or symbol metric is based on

$$\mathbf{d}_m = \operatorname{Re}(\mathbf{L}_0 \boldsymbol{y}). \tag{7.188}$$

We define the matrices

$$\mathbf{R}_{\mathbf{dy}} = E\left[\mathbf{d}\mathbf{y}^H\right], \quad \mathbf{R}_y = \left\{E\left[\boldsymbol{y}\boldsymbol{y}^H\right]\right\}. \tag{7.189}$$

If the code symbols are independent and equally likely to be $+1$ or -1, then $E[\mathbf{d}\mathbf{d}^T] = \mathbf{I}$, where $\mathbf{I}$ is the identity matrix. Using this result, (7.153), (7.154), $E[\mathbf{n}] = \mathbf{0}$, and the independence of $\mathbf{d}$ and $\mathbf{n}$, we obtain

$$\mathbf{R}_{\mathbf{dy}} = \mathbf{A}^*\mathbf{P}^T, \quad \mathbf{R}_y = \mathbf{P}\,|\mathbf{A}|^2\,\mathbf{P}^T + G^{-1}N_0\mathbf{I} \tag{7.190}$$

where $|\mathbf{A}|^2$ is the diagonal matrix with $|A_k|^2$ as its kth component. From identity (6.287), (7.187), and (7.190), it follows that

$$M = tr\left\{E[\mathbf{d}\mathbf{d}^{\mathbf{T}}]+\mathbf{L}\mathbf{R}_y\mathbf{L}^H-\mathbf{A}^*\mathbf{P}^T\mathbf{L}^H - \mathbf{L}\mathbf{P}\mathbf{A}\right\}. \tag{7.191}$$

Equation (7.190) indicates that the $G \times G$ matrix $\mathbf{R}_y$ is symmetric positive definite. Therefore, $\mathbf{R}_y^{-1}$ exists, and after completing the square, we find that

$$M = tr\left\{E[\mathbf{d}\mathbf{d}^T] - \mathbf{A}^*\mathbf{P}^T\mathbf{R}_{\mathbf{y}}^{-1}\mathbf{P}\mathbf{A}\right\} + tr\left\{(\mathbf{L}-\mathbf{L}_0)\,\mathbf{R}_y\,(\mathbf{L}-\mathbf{L}_0)^H\right\} \tag{7.192}$$

where

$$\begin{aligned}\mathbf{L}_0 &= \mathbf{A}^*\mathbf{P}^T\mathbf{R}_{\mathbf{y}}^{-1} \\ &= \mathbf{A}^*\mathbf{P}^T\left(\mathbf{P}\,|\mathbf{A}|^2\,\mathbf{P}^T+G^{-1}N_0\mathbf{I}\right)^{-1}.\end{aligned} \tag{7.193}$$

Since $\mathbf{R}_y$ is symmetric positive definite, $(\mathbf{L}-\mathbf{L}_0)\,\mathbf{R}_y\,(\mathbf{L}-\mathbf{L}_0)^H$ is Hermitian positive semidefinite, which implies that it has nonnegative eigenvalues and hence a nonnegative trace (Appendix G). Let $\mathbf{r}_i$ denote the ith row of $\mathbf{L}-\mathbf{L}_0$. Then

$$tr\left\{(\mathbf{L}-\mathbf{L}_0)\,\mathbf{R}_y\,(\mathbf{L}-\mathbf{L}_0)^H\right\} = \sum_{i=1}^{K}\mathbf{r}_i\mathbf{R}_y\mathbf{r}_i^H \geq 0. \tag{7.194}$$

Since $\mathbf{R}_y$ is symmetric positive definite, each term in the sum equals zero if and only if, $\mathbf{r}_i = \mathbf{0}$, $1 \leq i \leq K$. Therefore, the trace has its minimum value of zero if and only if $\mathbf{L} = \mathbf{L}_0$; hence, the minimum value of M is attained if and only if $\mathbf{L} = \mathbf{L}_0$.

Direct matrix multiplication proves the identity

$$\left(\mathbf{P}^T\mathbf{P}+G^{-1}N_0|\mathbf{A}|^{-2}\right)|\mathbf{A}|^2\mathbf{P}^T = \mathbf{P}^T\left(\mathbf{P}\,|\mathbf{A}|^2\mathbf{P}^T+G^{-1}N_0\mathbf{I}\right). \tag{7.195}$$

Combining this identity with (7.193), we obtain

$$\mathbf{L}_0 = G^{-1}\mathbf{A}^{-1}\mathbf{Q}\mathbf{P}^T \tag{7.196}$$

where

$$\mathbf{Q} = (\mathbf{R}+\frac{N_0}{G^2}|\mathbf{A}|^{-2})^{-1}. \tag{7.197}$$

Substitution of (7.196) into (7.188) gives

$$\mathbf{d}_m = \text{Re}\left(G^{-1}\mathbf{A}^{-1}\mathbf{Q}\mathbf{P}^T\mathbf{y}\right). \tag{7.198}$$

Since $d_k = \pm 1$, the output of the MMSE detector and the input to the decoder is

$$\widehat{\mathbf{d}} = \mathcal{N}[\,\text{Re}\left(G^{-1}\mathbf{A}^{-1}\mathbf{Q}\mathbf{P}^T\mathbf{y}\right)] \tag{7.199}$$

where $\mathcal{N}(\mathbf{x}) = sgn(\mathbf{x})$ for hard-decision decoding, and $\mathcal{N}(\mathbf{x})$ is a nonlinear function, such as c$(\beta\mathbf{x})$ or $\tanh(\alpha\mathbf{x})$, for soft-decision decoding. Thus, the MMSE detector has the structure of Figure 7.9 but requires a different linear transformation than the decorrelating detector.

Substituting (7.153) into (7.199), we obtain

$$\widehat{\mathbf{d}} = \mathcal{N}\left[\mathrm{Re}\left(\mathbf{A}^{-1}\mathbf{QRAd}\right) + \mathbf{n}_m\right] \tag{7.200}$$

where

$$\mathbf{n}_m = \mathrm{Re}\left(G^{-1}\mathbf{A}^{-1}\mathbf{QP}^{\mathrm{T}}\mathbf{n}\right). \tag{7.201}$$

Equation (7.201) indicates that $\mathbf{n}_m$ is a Gaussian random vector (see Appendix A.1), because the vector $\mathbf{n}$ has components that are independent zero-mean Gaussian random variables. From (7.201), (7.155), and (7.154), it follows that the correlation matrix of $\mathbf{n}_m$ is

$$E\left[\mathbf{n}_m\mathbf{n}_m^H\right] = \frac{N_0}{2G^2}\mathbf{A}^{-1}\mathbf{QRQ}\left(\mathbf{A}^*\right)^{-1}. \tag{7.202}$$

Consider the hard-decision detection of symbol d_k, $0 \le k \le K-1$. If we assume that the detector does phase synchronization of the kth signal, then A_k^* may be replaced by $A_k > 0$, which then becomes irrelevant. Equation (7.200) implies that the kth component of $\widehat{\mathbf{d}}$ is

$$\widehat{d}_k = \mathrm{sgn}\left(u_k\right) \tag{7.203}$$

where

$$u_k = (\mathbf{QR})_{kk}d_k + B\left(\mathbf{d}_{ke}\right) + n_{mk} \tag{7.204}$$

$$B\left(\mathbf{d}_{ke}\right) = \sum_{\substack{i=0\\ i\neq k}}^{K-1}(\mathbf{QR})_{ki}\frac{\mathrm{Re}\left(A_i\right)}{A_k}d_i \tag{7.205}$$

and $\mathbf{d}_{ke}$ denotes the vector of $K-1$ components of $\mathbf{d}$ excluding d_k. The variance of n_{mk} is

$$var\left(n_{mk}\right) = \frac{N_0}{2\mathcal{E}_k}(\mathbf{QRQ})_{kk} \tag{7.206}$$

where $\mathcal{E}_k = |A_k|^2 G^2$ is the symbol energy.

By symmetry, we can assume that $d_k = 1$ in the evaluation of the symbol error probability. Let S_k denote the set of the 2^{K-1} distinct vectors that $\mathbf{d}_{ke}$ can equal. Let $\mathbf{d}_{kn} \in \mathsf{S}_k$ denote the nth of those distinct vectors. Conditioning on $\mathbf{d}_{ke} = \mathbf{d}_{kn}$, $1 \le n \le 2^{K-1}$, we find that a symbol error occurs if n_{mk} causes u_k to have a different sign than d_k. Since the noise has a Gaussian density function, the conditional symbol error probability at the output of decision-device k is

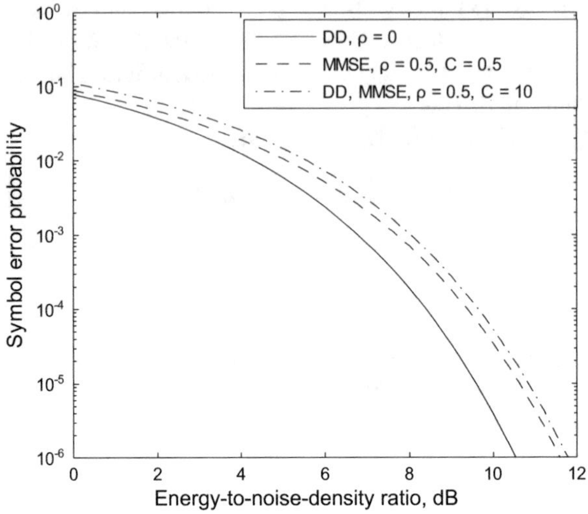

Fig. 7.10 Symbol error probability for DS-CDMA systems with decorrelating detector (DD) and MMSE detector when there is one multiple-access interference signal (K = 2)

$$P_s\left(k|\mathbf{d}_{kn}\right) = Q\left(\sqrt{\frac{2\mathcal{E}_k}{N_0(\mathbf{QRQ})_{kk}}}\left[(\mathbf{QR})_{kk} + B\left(\mathbf{d}_{kn}\right)\right]\right). \tag{7.207}$$

If all symbol sets are equally likely, then the symbol error probability for user k is given by (7.168). In contrast to the decorrelating detector, the MMSE detector has a symbol error probability that depends on the symbols and spreading sequences of the other users.

Example 3 Consider multiuser detection for $K = 2$, $C = \text{Re}\left(A_{1-k}\right)/A_k$, and correlation matrix given by (7.183). Calculations of the symbol error probability using (7.184), (7.207), and (7.168) indicate that the performance of the MMSE detector degrades with increasing $|\rho|$ but is close to that of the decorrelating detector, which does not depend on C, when $C \geq 1$. In contrast, if the conventional detector is used, (7.185) indicates that the symbol error probability rapidly degrades as C increases. Representative plots are illustrated in Figure 7.10. The decorrelating detector with $\rho = 0$ has the same symbol error probability as a system with no multiuser detection and no multiple-access interference. □

The MMSE and decorrelating detectors have almost the same computational requirements, and they both have equalizer counterparts, but they differ in several ways. The MMSE detector does not obliterate the multiple-access interference; hence, it does not completely eliminate the near-far problem, but does not accentuate the noise to the degree to which the decorrelating detector does. Although the MMSE detector tends to suppress strong interference signals, it also suppresses the desired signal to the degree that its spreading sequence is correlated with the spreading sequences of strong interference signals. For practical scenarios, the symbol error probability of the MMSE detector generally tends to be lower than that provided by the decorrelating detector. As $N_0 \to 0$, the MMSE estimate approaches the decorrelating detector estimate. Therefore, the MMSE detector is *asymptotically near-far resistant*. As N_0 increases, the MMSE estimate approaches that of the conventional detector, and hence provides diminished near-far resistance.

For either the MMSE or decorrelating linear detectors to be practical for synchronous communications, it is important for the spreading sequences to be short. Short sequences ensure that the correlation matrix $\mathbf{R}$ is constant for multiple symbols. The price of short sequences is a security loss and the occasional but sometimes persistent performance loss due to a particular set of relative signal delays. Even with short spreading sequences, the estimation of $\mathbf{A}$ presents an obstacle to the implementation of the MMSE detector.

Like the decorrelating detector, the MMSE detector is impractical for asynchronous multiuser detection because the computational requirements increase rapidly as N and K increase. Even more significant is the requirement that the receiver must know or estimate the delays of the spreading sequences at the receiver input.

Synchronous versions of multiuser detectors require the synchronized timing of the multiple-access signals. The synchronous model certainly is applicable to a downlink in a cellular network. In that case, however, the use of short orthogonal spreading sequences renders a multiuser detector unnecessary, as a correlator in each receiver is sufficient to eliminate the multiple-access interference (Section 7.8).

Adaptive Multiuser Detector

An *adaptive multiuser detector* is an adaptive system that does not require explicit receiver knowledge of either the spreading sequences or the timing of the multiple-access interference signals. At the cost of a reduction in spectral efficiency, the adaptive multiuser detector learns by processing a known *training sequence* of L_t pilot symbols during a *training phase*. Each pilot symbol is represented by a spreading sequence of length G. The use of short spreading sequences affords the opportunity for an adaptive multiuser detector to essentially learn the sequence cross-correlations and thereby to suppress the interference. However, the requirement for short spreading sequences limits the applications of adaptive multiuser detection; e.g., the WCDMA (Wideband CDMA) and CDMA2000 standards do not support it.

The LMS algorithm (Section 5.2) may be used as the adaptive algorithm in the adaptive multiuser detector. The nth symbol of the known training sequence for user k is denoted by $d_k(n)$, $n = 0, 1, \ldots, L_t - 1$. The nth vector of G chip-matched-filter outputs, which is produced during the reception of symbol n, is denoted by $\mathbf{p}(n)$, $n \geq 0$. Assuming that chip synchronization has been established, the LMS algorithm iteratively updates the G-dimensional weight vector

$$\mathbf{w}(n+1) = \mathbf{w}(n) + 2\mu\epsilon^*(n)\mathbf{p}(n), \quad n \geq 0 \tag{7.208}$$

where μ is a constant that regulates the algorithm convergence rate,

$$\epsilon(n) = \overline{d}_k(n) - \mathbf{w}^H(n)\mathbf{p}(n) \tag{7.209}$$

and $\overline{d}_k(n) = d_k(n)$, $n = 0, 1, \ldots, L_t - 1$. This training phase is followed by a *decision-directed phase* that continues the adaptation by feeding back symbol decisions

$$\widehat{d}_k(n) = \text{sgn}[\text{Re}(\mathbf{w}^H(n)\mathbf{p}(n))] \tag{7.210}$$

and using $\overline{d}_k(n) = \widehat{d}_k(n)$, $n \geq L_t$. Adaptive detectors can potentially achieve much better performance than the conventional detector, at least if the transmission channel is time-invariant, but coping with fast fading and fluctuating interference sometimes requires elaborate modifications.

A *blind multiuser detector* does not require pilot symbols or training sequences. Instead of training, blind multiuser detectors only require knowledge of the spreading sequence of the desired signal and its timing, which is no more information than is required by the simpler, conventional single-user system. Short spreading sequences are necessary, as long spreading sequences do not possess the cyclostationarity that makes possible the advanced signal processing techniques used by blind multiuser detectors. An *adaptive blind multiuser detector* is necessary to accommodate changing channel conditions and system recovery, but entails some performance loss and complexity increase relative to adaptive multiuser detectors with a training phase. Several adaptive algorithms, including the Frost algorithm (Section 5.4), can be used in adaptive blind multiuser detectors. The Frost algorithm uses the known spreading sequence to constrain desired-signal cancelation while canceling the interference.

Multiuser Detector for Frequency Hopping

In an FH-CDMA network, the rough coordination of the transmit times of the users and sufficiently long switching times limits or eliminates the collisions among users and makes multiuser processing unnecessary. In an asynchronous network, multiuser detection may be desirable, but it is much more challenging for frequency-hopping systems than for direct-sequence systems. An optimal multiuser detector requires the receiver to know the hopping patterns and hop transition times of all users to be detected and can simultaneously demodulate the signals at all carrier frequencies. These requirements are completely unrealistic for any practical network of frequency-hopping systems. A much more practical multiuser detector exploits differences in the hop transition times of the users in an asynchronous frequency-hopping network [94]. These differences expose portions of the desired and interfering signals in a way that can be exploited by an iterative demodulator, channel estimator, and decoder. The receiver synchronizes with the frequency-hopping pattern of the desired signal and estimates the timing information of the interfering signals. An important part of the processing is the use of the expectation-maximization algorithm (Section 9.1) for channel estimation. For both nonfading and Rayleigh fading channels, this multiuser detector accommodates more strong interference signals than a conventional frequency-hopping receiver, but alternative and more robust frequency-hopping systems are possible (Section 9.4).

7.8 Interference Cancelers

An *interference canceler* is a multiuser detector that explicitly estimates the interference signals and then subtracts them from the received signal to produce a desired signal. Although suboptimal compared with ideal multiuser detection, multiuser interference cancelers bear much less of an implementation burden and still provide considerable interference suppression and alleviation of the near-far problem. An interference canceler is by far the most practical multiuser detector for an asynchronous DS-CDMA network, especially if long spreading sequences are planned.

Implementation of an interference canceler entails having stored spreading sequences of all desired and potentially interfering signals and a means of synchronizing with received interference signals. Accurate power control is still needed at least during initial synchronization and to avoid overloading the front end of the receiver.

Interference cancelers [7, 40] may be classified as *successive interference cancelers* in which the subtractions are performed sequentially, *parallel interference cancelers* in which the subtractions are performed simultaneously, or hybrids of these types. Only the basic structures and features of the successive and parallel cancelers are presented subsequently, but a large number of alternative versions, some of them hybrids, adaptive, or blind, have been proposed.

Successive Interference Canceler

Figure 7.11 is a functional block diagram of a successive interference canceler, which uses successive replica generations and subtractions to produce estimates of the symbol streams transmitted by the K users. The K parallel outputs of a despreader bank are applied to K parallel detectors that order the K signals according to their estimated power levels. This ordering determines the placement of the detector-generators in the figure, which are ordered according to descending power levels. Detector-generator k corresponds to the kth strongest signal.

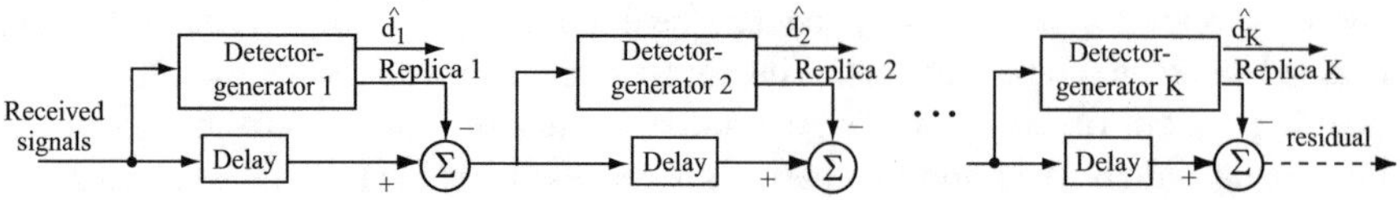

Fig. 7.11 Successive interference canceler with K detector-generators to produce signal estimates for subtraction

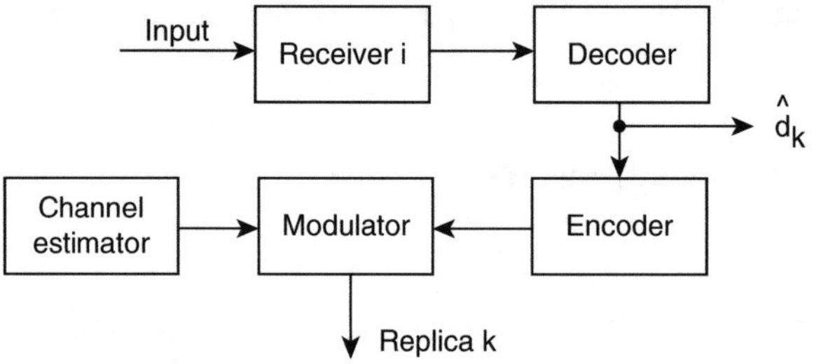

Fig. 7.12 Structure of detector-generator for signal k

Detector-generator k, which has the structure depicted in Figure 7.12, produces a replica of each received symbol of signal k. Receiver k may be a conventional detector with the form of Figure 2.14 for BPSK modulation or take the form of Figure 2.20 for quaternary modulation. The decoder provides a hard estimate $\widehat{d}_k$ of each symbol. These symbols are encoded and applied to a modulator, which generates a replica of signal k by adding the spreading sequence, modulating the chip waveforms, and adjusting the signal timing. The channel estimator, which may use known pilot or training symbols to determine the channel response, provides the estimated channel amplitude, phase, and relative timing that are applied to the modulator to compensate for the effects of the propagation channel. Over a symbol interval, replica k has the form

$$R_k(t) = \sqrt{2\widehat{\mathcal{E}}_k}\widehat{d}_k \sum_{i=0}^{G-1} p_{k,i}\psi(t - iT_c - \widehat{\tau}_k)\cos(2\pi f_c t + \widehat{\theta}_k),\ 1 \leq k \leq K \tag{7.211}$$

where $\widehat{\mathcal{E}}_k$, $\widehat{d}_k$, $\widehat{\tau}_k$, and $\widehat{\theta}_k$ are estimates of $\mathcal{E}_k$, d_k, τ_k, and θ_k, respectively. Replica k is sent to a subtractor that produces the difference-signal

$$D_k(t) = D_{k-1}(t - T_s) - R_k(t)\,. \tag{7.212}$$

Because of the previous successive subtractions, $D_k(t)$ has most of the interference due to signals $k, k-1, \ldots, 1$ eliminated. If all the preceding replicas are nearly exact, then $D_k(t)$, which is applied to detector-generator $k+1$, is

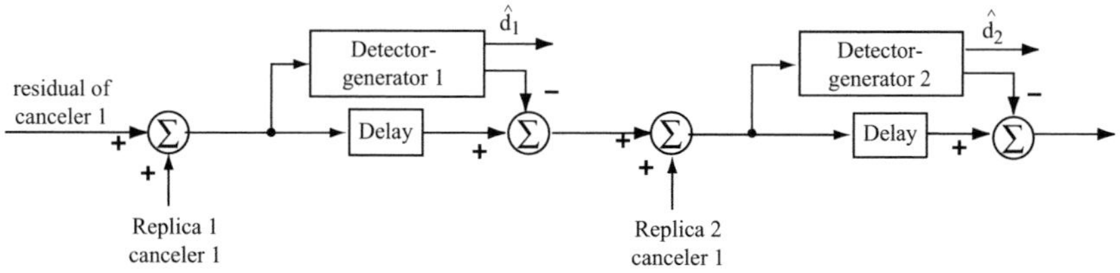

Fig. 7.13 Second canceler of multistage canceler using successive interference cancelers

$$D_k(t) = \sum_{l=k+1}^{K} \sqrt{2\mathcal{E}_l} d_l \sum_{i=0}^{G-1} p_{l,i} \psi(t - iT_c - \tau_l) \cos(2\pi f_c t + \theta_l). \tag{7.213}$$

The first canceler stage eliminates the strongest signal, thereby immediately alleviating the near-far problem for weaker signals while exploiting the superior detectability of the strongest signal. The amount of interference removal prior to the detection of a signal increases from the strongest received signal to the weakest one.

Any error in the replica generation adversely affects subsequent symbol estimates and replicas. If the decoder of a detector-generator makes a symbol error, then the amplitude of the interference that enters the next stage of the canceler of Figure 7.11 is doubled. Since each delay in Figure 7.11 exceeds one symbol in duration, the overall processing delay of the successive interference canceler is one of its disadvantages. The delay introduced, the impact of cancelation errors, and the implementation complexity may limit the number of useful canceler stages to fewer than K, but usually only a few interference signals need to be canceled to obtain the bulk of the available performance gain [117]. At low SINRs, inaccurate cancelations may cause the canceler to lose its advantage over the conventional detector.

A *multistage interference canceler* comprising more than one successive interference canceler potentially improves performance by repeated cancelations if the delay and complexity can be accommodated. The second canceler or stage of a multistage canceler is illustrated in Figure 7.13. The input is the residual of canceler 1, which is shown in Figure 7.11. As shown in Figure 7.13, replica 1 of canceler 1 is added to the residual to produce a sum signal that is applied to detector-generator 1. Since most of the interference has been removed from the residual, detector-generator 1 can produce an improved estimate $\widehat{d}_1$ and an improved replica of signal 1, which is then subtracted from the sum signal. The resulting difference signal contains less interference than the corresponding difference signal in Figure 7.11. Subsequently, other replicas from canceler 1 are added and corresponding improved replicas are subtracted. The final estimated symbol streams are produced by the final canceler. Rake combining of multipath components may be incorporated into a multistage or single-stage interference canceler to improve performance in a fading environment [82].

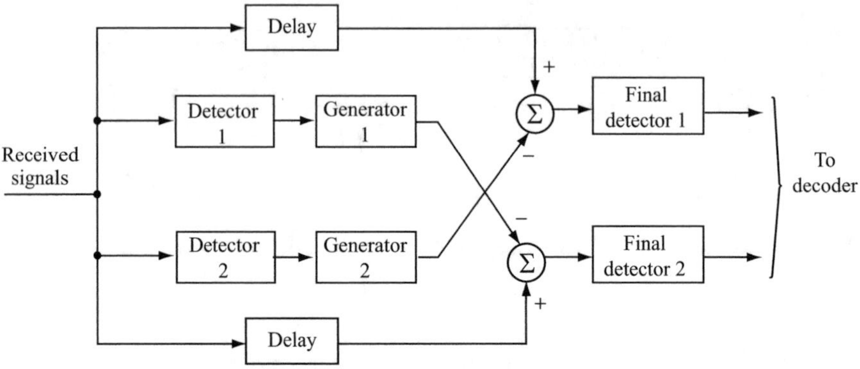

Fig. 7.14 Parallel interference canceler for two signals

Parallel Interference Canceler

A parallel interference canceler detects, generates, and then subtracts all multiple-access interference signals from each desired signal simultaneously. A parallel interference canceler for two signals is diagrammed in Figure 7.14. Each detector-generator pair may be implemented as shown in Figure 7.12. Each of the final detectors includes a digital matched filter and a decision device that produces soft or hard decisions, which are applied to the decoder.

Since each strong signal enters each detector-generator simultaneously and the initial detections influence the final ones, the parallel interference canceler is not as effective at suppressing the near-far problem as the successive interference canceler. However, if a DS-CDMA network uses power control, the first several stages of a successive interference canceler may provide less accurate detected symbols than those provided by a one-stage parallel interference canceler. The primary reason is the presence of uncanceled multiple-access interference in each replica generated by a successive interference canceler.

Consider synchronous BPSK signals and the model of Section 7.7. The $G \times 1$ received vector $\mathbf{y}$ is given by (7.153), and the output of the despreader bank is $\mathbf{P}^T\mathbf{y}$. We define

$$\mathbf{B} = \mathbf{P}^T\mathbf{P}\mathbf{A}. \tag{7.214}$$

Using (7.153), we observe that $\mathbf{P}^T\mathbf{y} = \mathbf{B}\mathbf{d} + \mathbf{P}^T\mathbf{n}$. This equation motivates the realization of parallel interference cancelation as the estimator $\widehat{\mathbf{d}}_1$ obtained by solving the equation

$$\mathbf{B}\widehat{\mathbf{d}}_1 = \mathbf{P}^T\mathbf{y} \tag{7.215}$$

which provides

$$\widehat{\mathbf{d}}_1 = \mathbf{B}^{-1}\mathbf{P}^T\mathbf{y} \tag{7.216}$$

if **B** is invertible. Since it is known that **d** is real-valued, $\text{Re}\left(\widehat{\mathbf{d}}_1\right)$ provides an improved estimator, and a suitable final estimator is

$$\widehat{\mathbf{d}} = \mathcal{N}\left[\text{Re}\left(\mathbf{B}^{-1}\mathbf{P}^T\mathbf{y}\right)\right] \tag{7.217}$$

where $\mathcal{N}(x)$ equals $\text{c}(x)$ or $\tanh(\alpha x)$ for symbol metrics, or $\mathcal{N}(x) = sgn(x)$ for hard decisions. However, the complexity or number of multiplications required to compute $\mathbf{B}^{-1}$ is $O\left(K^3\right)$, which might be prohibitively large.

To reduce the complexity, we rewrite (7.215) in the form

$$\mathbf{B}_d\widehat{\mathbf{d}}_1 = \mathbf{P}^T\mathbf{y} - (\mathbf{B} - \mathbf{B}_d)\,\widehat{\mathbf{d}}_1 \tag{7.218}$$

where $\mathbf{B}_d$ is the diagonal matrix with diagonal elements equal to those of **B**. We solve (7.218) iteratively by computing

$$\widehat{\mathbf{d}}_1(1) = \mathbf{B}_d^{-1}\mathbf{P}^T\mathbf{y}$$
$$\widehat{\mathbf{d}}_1(i+1) = \mathbf{B}_d^{-1}\left[\mathbf{P}^T\mathbf{y} - (\mathbf{B} - \mathbf{B}_d)\,\widehat{\mathbf{d}}_1(i)\right], \; i \geq 1. \tag{7.219}$$

Since the matrix inversion is now trivial, we need $O\left(K^2\right)$ multiplications per iteration. Thus, if the number of iterations is many fewer than K, we have a reduction in computation relative to (7.216).

A *multistage parallel interference canceler* provides successively improved parallel inputs to successive stages that generate successively improved parallel outputs. Figure 7.15 shows the multistage canceler for two signals. The initial stage may consist of a parallel interference canceler, successive interference canceler, decorrelating detector, or MMSE detector. Each subsequent stage has the form of Figure 7.14. Increasing the number of stages is not always useful because decision errors increase, and at some point cause an overall performance degradation.

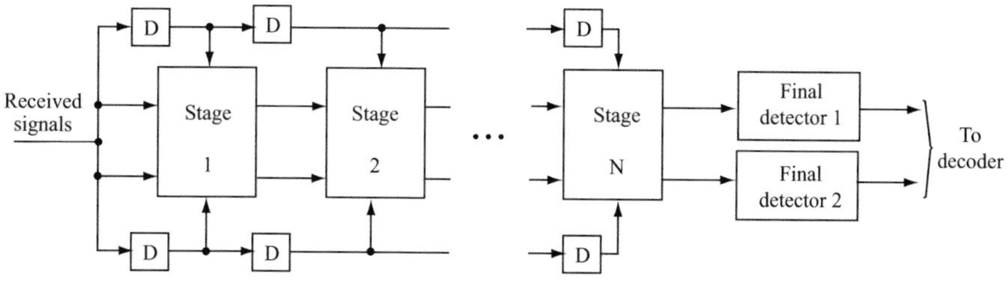

Fig. 7.15 Multistage parallel interference canceler for two signals. *D* delay

7.9 Multicode DS-CDMA Systems

A *multicode DS-CDMA system* or *multirate DS-CDMA system* provides multiple data rates at a fixed chip rate and spreading factor. The transmitter provides multiple orthogonal spreading sequences to a subscriber who requires a code-symbol transmission rate higher than the standard rate defined by the fixed chip rate and spreading factor. High-rate code-symbol sequences are divided into several distinct standard-rate symbol sequences, each of which is modulated by a distinct spreading sequence to create a distinct composite chip-rate sequence. The composite sequences are synchronously combined and modulate the same sinusoidal carrier. By coherently demodulating the carrier and using orthogonal spreading sequences, the receiver can suppress interference among the distinct composite sequences. In contrast, a conventional single-rate CDMA system operating at the selected high rate would require a bandwidth expansion or a reduced spreading factor. The primary disadvantage of the multicode method is the high peak-to-average-power ratio that may require the use of a reduction method (see Section 6.13).

Consider a multicode DS-CDMA system that provides a high-rate transmission with K synchronous streams, each of which represents a component of the $K \times 1$ data vector $\mathbf{d}$. The maximum-likelihood multicode detector, which has the same form as the maximum-likelihood multiuser detector, computes (7.158) and (7.159). Since the spreading sequences are orthonormal and the received composite signal is coherently demodulated, $\mathbf{P^T P} = G\mathbf{I}$ and the amplitudes are positive. Therefore, an equivalent decision vector is

$$\widehat{\mathbf{d}} = \arg\max_{\mathbf{d}\in\mathbf{D}} \left(\mathbf{d}^T\mathbf{A}^2\mathbf{d} - 2\mathbf{d}^T\mathbf{A}\mathbf{P}^{\mathbf{T}}\mathbf{y}_r\right) \tag{7.220}$$

where $\mathbf{y}_r = \mathrm{Re}\,(\mathbf{y})$. Since $\mathbf{A}^2$ is diagonal, $\mathbf{d}^T\mathbf{A}^2\mathbf{d}$ is irrelevant to the decision, and we find that an equivalent decision vector is

$$\widehat{\mathbf{d}} = \arg\max_{\mathbf{d}\in\mathbf{D}} \left(\mathbf{d}^T\mathbf{A}\mathbf{P}^{\mathbf{T}}\mathbf{y}_r\right). \tag{7.221}$$

If $\mathbf{d} \in \mathbf{D}$, then $\mathbf{d}^T\mathbf{A}\mathbf{P}^{\mathbf{T}}\mathbf{y}_r$ is maximized when every term in the inner product of $\mathbf{d}$ and $\mathbf{A}\mathbf{P}^{\mathbf{T}}\mathbf{y}_r$ is positive, which implies that an equivalent maximum-likelihood decision vector is $\widehat{\mathbf{d}} = \mathrm{sgn}[\mathbf{A}\mathbf{P}^{\mathbf{T}}\mathbf{y_r}]$. Since the amplitudes are positive, the maximum-likelihood decision vector is

$$\widehat{\mathbf{d}} = \mathrm{sgn}\left(\mathbf{P}^{\mathbf{T}}\mathbf{y}_r\right). \tag{7.222}$$

Substituting (7.153), we obtain

$$\widehat{\mathbf{d}} = \mathrm{sgn}\left(G\mathbf{A}\mathbf{d} + \mathbf{P}^{\mathbf{T}}\mathbf{n}_r\right) \tag{7.223}$$

which indicates that a symbol estimate $\widehat{d}_k$ is unaffected by interference from the other symbols. The maximum-likelihood decision for symbol k is

$$\widehat{d}_k = \text{sgn}\left(\mathbf{p}_k^T \mathbf{y}_r\right) \tag{7.224}$$

and the symbol error probability for coded messages is given by (1.60).

For soft-decision decoding, we can use $\widehat{\mathbf{d}} = \mathbf{P}^{\mathbf{T}}\mathbf{y}_r$ as the vector of symbol metrics, or we can use

$$\widehat{\mathbf{d}} = \mathcal{N}\left(\mathbf{P}^{\mathbf{T}}\mathbf{y}_r\right) \tag{7.225}$$

where $\mathcal{N}(\cdot)$ is a nonlinear function. Suitable nonlinear functions include *tanh* $(\alpha\mathbf{x})$ or $\text{c}(\mathbf{x})$.

7.10 Multiple-Input Multiple-Output Systems

A *multiple-input multiple-output* (MIMO) system comprises one or more transmitters with multiple antennas and one or more receivers with multiple antennas. The multiple antennas enable increased throughput, increased spatial diversity, or stronger received signals, but the degree to which any of these features is implemented entails limitations on the other features. Diversity techniques, such as space-time coding, exploit the independent fading in the multiple paths from transmit antennas to receive antennas. These techniques do not require channel-state information (CSI), but cannot provide a high throughput or a significant array gain. MIMO systems that provide high throughput use spatial multiplexing, and those that provide gain use beamforming, but hybrid combinations are also possible.

If a MIMO system has only multiple receive antennas, it is called a *single-input multiple-output* (SIMO) system. The adaptive arrays of Sections 5.6 and 5.7, which can be used in either SIMO or MIMO systems, use multiple receive antennas for beamforming that suppresses interference entering the sidelobes of the receive array pattern. Diversity combining methods for receive antennas in SIMO or MIMO systems, which are described in Sections 6.4–6.7, compensate for fading and antenna imperfections. If a MIMO system has only multiple transmit antennas, it is called a *multiple-input single-output* (MISO) system. The use of multiple transmit antennas in providing spatial diversity or a space-time code, which can be used in either MISO or MIMO systems, is described in Section 6.8. The receiver of a MISO system can serve as a multiuser detector for direct-sequence communications, as described in Section 7.7.

Spatial Multiplexing

Multiple-input multiple-output systems can increase the throughput by using *spatial multiplexing*, which is the transmission of different symbols or signals simultaneously as separate signals through different transmit antennas and the reception of the transmitted signals at each receive antenna. In each of these paths between transmit and receive antenna pairs, a signal encounters a different channel. Therefore, another potential advantage of spatial multiplexing stems from the diverse propagation paths, which can provide a diversity gain. The receiver reconstructs the transmitted symbols or signals by using CSI.

Consider a MIMO system that has N_t transmit antennas and N_r receive antennas and uses spatial multiplexing for direct-sequence communications. The transmit antennas and receive antennas are closely spaced enough that all N_t signals arriving at each receive antenna can be considered synchronous. Consider a symbol interval during which each of the N_t symbols is received by each of the N_r receive antennas. The MIMO receiver prior to the despreaders has the form illustrated in Figure 7.16. Each receive antenna is followed by a chip-matched filter. Chip-rate sampling of its output by an analog-to-digital converter produces a demodulated sequence of length G. All N_r demodulated sequences are applied to each despreader, as illustrated in Figure 7.17. Despreader k and maximal-ratio combiner k produce a symbol metric or decision for symbol k. We assume that the transmitted signals have bandwidths less than the channel coherence bandwidth, and thus the fading is flat. To accommodate frequency-selective fading, each chip-matched filter must be embedded in an equalizer that compensates for the frequency selectivity.

Assuming that accurate timing synchronization provides a synchronous system, we can analyze the MIMO receiver by extending the analysis for the optimal multiuser detector in Section 7.7. Let

$$\begin{aligned} \mathbf{y}_m &= [y_{m,0}\ y_{m,1}\ \ldots\ y_{m,G-1}]^T \\ \mathbf{n}_m &= [n_{m,0}\ n_{m,1}\ \ldots\ n_{m,G-1}]^T \\ \mathbf{d} &= [d_0\ d_1\ \ldots\ d_{N_t-1}]^T \end{aligned} \tag{7.226}$$

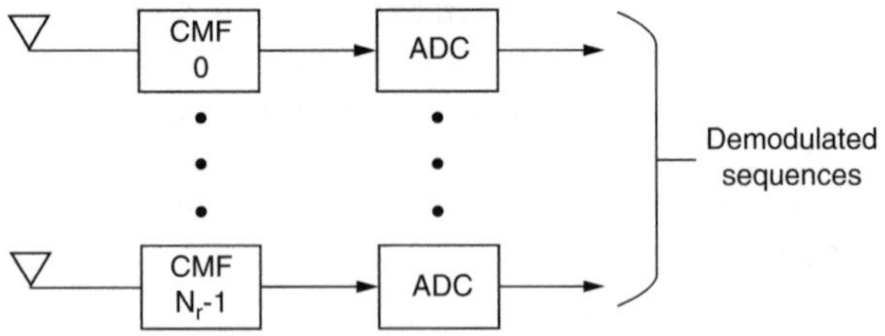

Fig. 7.16 MIMO receiver preceding despreaders. Its inputs are derived from N_r receive antennas. *CMF* chip-matched filter. *ADC* analog-to-digital converter

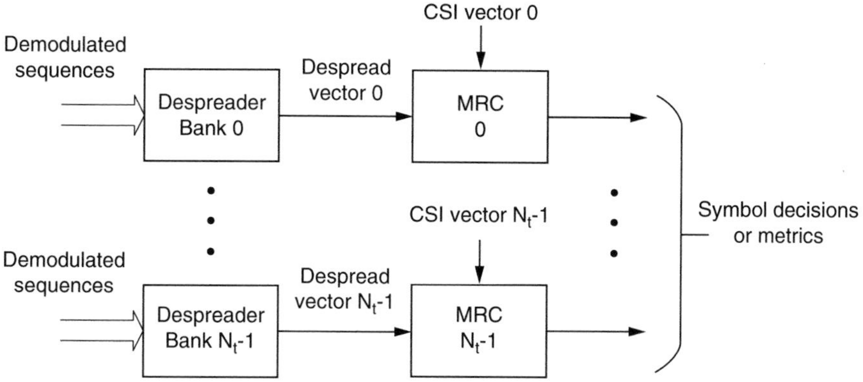

Fig. 7.17 Despreaders and maximal-ratio combiners of a MIMO receiver. *MRC* maximal-ratio combiner

denote the demodulated sequence of antenna m, the accompanying noise sequence of antenna m, and N_t simultaneously transmitted code symbols with values $d_i = \pm 1$ respectively. As in (7.151), the complex amplitude at receive antenna m due to the signal transmitted by transmit antenna k is

$$A_{k,m} = G^{-1}\sqrt{\mathcal{E}_{k,m}}\exp\left(j\theta_{k,m}\right) \tag{7.227}$$

where $\mathcal{E}_{k,m}$ is the energy per symbol of signal k at antenna m, and $\theta_{k,m}$ is the phase of signal k at antenna m. Equations (7.153) and (7.154) may be applied to the demodulated sequence at receive antenna m corresponding to a set of N_t synchronous transmitted symbols. The demodulated sequence is the $G \times 1$ vector

$$\mathbf{y}_m = \mathbf{PA}_m\mathbf{d} + \mathbf{n}_m,\ 0 \le m \le N_r - 1 \tag{7.228}$$

where column k of the $G \times N_t$ matrix $\mathbf{P}$ is the vector $\mathbf{p}_k = [p_{k,0}\ p_{k,1}\ \cdots\ p_{k,G-1}]^T$ that represents the spreading sequence modulating symbol k, $p_{k,i} = \pm 1$, and $\mathbf{A}_m$ is the $N_t \times N_t$ diagonal matrix with $A_{k,m}$ as its kth diagonal element. Assuming that the noise in each receive antenna is independent zero-mean, white Gaussian noise with two-sided noise PSD $N_0/2$, $G \times 1$ vector $\mathbf{n}_m$ is a zero-mean, circularly symmetric, Gaussian random vector with

$$E\left[\mathbf{n}_m\mathbf{n}_m{}^H\right] = G^{-1}N_0\mathbf{I},\ E\left[\mathbf{n}_m\mathbf{n}_m{}^T\right] = \mathbf{0} \tag{7.229}$$

and

$$E\left[\mathbf{n}_m\mathbf{n}_p{}^H\right] = E\left[\mathbf{n}_m\mathbf{n}_p{}^T\right] = \mathbf{0},\ m \neq p. \tag{7.230}$$

Each of the N_t despreader banks of Figure 7.17 receives the N_r sequences in the columns of the $G \times N_r$ matrix

$$\mathbf{Y} = \left[\mathbf{y}_1 \; \mathbf{y}_2 \; \ldots \; \mathbf{y}_{N_r}\right] \tag{7.231}$$

and has N_r identical despreaders. The $N_r \times 1$ despread vector produced by the despreader bank k is

$$\boldsymbol{\zeta}_k = \mathbf{Y}^T \mathbf{p}_k \, . \tag{7.232}$$

If the spreading sequences are orthonormal, then the substitution of (7.228) and (7.231) into (7.232) and a rearrangement yields

$$\boldsymbol{\zeta}_k = d_k G \mathbf{B}_k + \mathbf{n}_{ak} \tag{7.233}$$

where

$$\mathbf{B}_k = \left[A_{k,0} \; A_{k,1} \; \ldots \; A_{k,N_r-1}\right]^T \tag{7.234}$$

$$E\left[\mathbf{n}_{ak}\mathbf{n}_{ak}{}^H\right] = N_0 \mathbf{I}, \; E\left[\mathbf{n}_{ak}\mathbf{n}_{ak}{}^T\right] = \mathbf{0}. \tag{7.235}$$

The maximum-likelihood decision for symbol k is

$$\widehat{d}_k = \arg \min_{\mathbf{d_k} \in (+1,-1)} ||\boldsymbol{\zeta}_k - d_k G \mathbf{B}_k||^2 \, . \tag{7.236}$$

Expanding this equation and dropping terms that are irrelevant to the decision, we find that the maximum-likelihood detector selects

$$\begin{aligned} \widehat{d}_k &= \arg \max_{\mathbf{d_k} \in (+1,-1)} \left[d_k \, \text{Re}(\mathbf{B}_k^H \boldsymbol{\zeta}_k)\right] \\ &= \text{sgn}[\text{Re}(\mathbf{B}_k^H \boldsymbol{\zeta}_k)] \end{aligned} \tag{7.237}$$

for hard-decision decoding of symbol k. This equation implies that each despread vector is applied to a maximal-ratio combiner (Section 6.4), as illustrated in Figure 7.17. The maximal-ratio combiner computes $\mathbf{B}_k^H \boldsymbol{\zeta}_k$, which optimally combines the components of the vector $\boldsymbol{\zeta}_k$.

For soft-decision decoding, the symbol metric for symbol k is computed as

$$\widehat{d}_k = \mathcal{N}[\text{Re}(\mathbf{B}_k^H \boldsymbol{\zeta}_k)] \tag{7.238}$$

where $\mathcal{N}(\cdot)$ is a nonlinear function. Suitable nonlinear functions include $\tanh(\alpha x)$ or $\text{c}\left[\left(G\mathbf{B}_k^H \mathbf{B}_k\right)^{-1} x\right]$, where $\text{c}(x)$ is the clipping function of (7.177), and $\left(G\mathbf{B}_k^H \mathbf{B}_k\right)^{-1}$ has been inserted so that $\widehat{d}_k = d_k$ in the absence of noise.

Since $\mathbf{B}_k$ is unknown, CSI is used to produce the estimate $\widehat{\mathbf{B}}_k$. Assuming that $\widehat{\mathbf{B}}_k \approx \mathbf{B}_k$, substituting (7.233) into (7.237), and using (7.235), we find that

$$\widehat{d}_k = d_k G \sum_{m=0}^{N_r-1} |A_{k,m}|^2 + n_k, \; n_k = \text{Re}(\mathbf{B}_k^H \mathbf{n}_{ak}) \tag{7.239}$$

where n_k is a zero-mean Gaussian random variable. Applying (7.235), we obtain

$$E\left[n_k^2\right] = N_0 \sum_{m=0}^{N_r-1} |A_{k,m}|^2 . \tag{7.240}$$

For the detected symbol $\widehat{d}_k$, a straightforward calculation using (7.239), (7.240), and (7.227) gives the symbol error probability:

$$P_s(k) = Q\left(\sqrt{\frac{2}{N_0}\sum_{m=0}^{N_r-1} \mathcal{E}_{k,m}}\right) \tag{7.241}$$

which explicitly shows the gain due to the maximal ratio combining. If the $\{\mathcal{E}_{k,m}\}$ are all equal, then the MIMO system provides an array gain equal to N_r and a throughput gain equal to N_t. If the N_r paths from the kth transmit antenna to the receive antennas fade independently, then the sum in (7.241) provides a diversity gain of order N_r.

The multicode method may be used in a spatial multiplexing MIMO system by transmitting K multicode signals through each transmit antenna. Each of the N_t sets of K multicode signals can use the same K orthogonal spreading sequences. To enable the receiver to distinguish among the N_t sets, an orthogonal *scrambling sequence* is assigned to each transmit antenna. Each scrambling sequence is added to each of the K spreading sequences.

In a spatial multiplexing system, the separation of the received signals is enabled by CSI. In a closed-loop system, a transmitter sends pilot signals through all its transmit antennas to potential receivers, each of which extracts the CSI for all paths to its receive antennas and then sends quantized estimates of the CSI to the transmitter. The reciprocity of signal paths when time-division duplexing is used enables a transmitter to use an open-loop system with direct measurements of the CSI. If a transmitter has accurate CSI, it can use *precoding* of a signal and divide its transmitted power efficiently among its transmit antennas.

The degradation of a received pilot signal because of the presence of other pilot signals is called *pilot contamination*. When highly correlated transmission channels or high-speed mobiles cause severe pilot contamination, the absence of accurate CSI may make the scheduling of transmissions necessary.

Spatial multiplexing with maximum-likelihood detection has a complexity that increases rapidly with the number of transmit antennas and their required CSI, and sphere decoding may have to be used to alleviate this problem. The partitioning of the transmitter power among the multiple transmit antennas lowers the SNR for each path to each receive antenna. Low SINRs, insufficient decorrelation among closely spaced antennas, and insufficient diversity provided by the communication channel impede spatial multiplexing and may render beamforming a better option for MIMO communications.

Beamforming

Beamforming by multiple transmit or multiple receive antennas is the formation of one or more spatially narrow beams. Each beam is produced by a separate set of weights behind the antennas and provides enhanced power gain, spatial discrimination, and interference suppression for one or more desired signals. The power gain of each beam is proportional to the number of antennas in its subarray. Narrow beams block multipath components and thereby lessen the severity of fading. Beamforming requires much less CSI than spatial multiplexing and is thus much less susceptible to pilot contamination and multiuser or multiple-access interference. Although a beam provides no diversity gain, beamforming may be preferable to spatial multiplexing when the paths from the transmit to the receive antennas are primarily line-of-sight and hence few multipath signals are significant, or when the paths experience highly correlated fading, which inhibits spatial multiplexing.

Beamforming can generate fixed or adaptive beams. An adaptive beam for receiving is formed by either estimating the arrival angle of a desired signal or indirectly adjusting to it by using an adaptive algorithm, as described in Sections 5.6 and 5.7. If an arrival angle is estimated or a transmission direction is selected, a beam can be formed or steered in the desired direction by setting the values of weights behind the antennas. In *codebook-based beamforming*, the weights are found as precomputed entries, each of which corresponds to a specific direction. The number of entries in the codebook determines the accuracy of the beam alignment with the desired direction.

The principal difficulty in adaptive beamforming for a MIMO system is the delay resulting from the need for the spatial alignment of both the transmit and the receive beams. The initial search for alignment entails the exchange of pilot signals between the transmit and receive antennas and the steering of both arrays over a range of angles in their codebooks. This delay increases as the beams become narrower to avoid interference. Another difficulty is that the mutual contamination of pilot signals and their multipath components may cause large beam-steering errors.

The beamforming of a set of disjoint fixed beams that cover all feasible directions is called *sectorization*. The advantages of sectorization at one end of a communication link are the reduction of the beam-alignment delay, pilot-signal contamination, and beam-steering errors. The primary disadvantage of sectorization is the loss of flexibility inherent in adaptive arrays. The role of sectorization in cellular CDMA networks is described in Chapter 8.

A *hybrid array* combines the outputs of multiple beams. This combining of beam outputs adds some spatial multiplexing capability to the benefits of beamforming.

When a large number of antennas and wide bandwidths are used, the hybrid array allows a reduction in the required number of power-consuming analog-to-digital converters. They can be relegated to processing the beam outputs while the beamforming is performed with analog devices. The cost is some loss in beamforming capability.

7.11 Problems

1 Apply information theory to multiple-access communications over an AWGN channel. (a) Show that if the bandwidth is infinite, then all users can send messages at their individual capacities, which implies that the interference among users can be avoided. (b) Verify (7.9) by proving that

$$\mathrm{C}\left(\frac{\sum_{i\in S}\mathcal{P}_i}{2N_0W}\right)=\mathrm{C}\left(\frac{\mathcal{P}_1}{\sum_{i\in S,i\neq 1}\mathcal{P}_i+2N_0W}\right)+\mathrm{C}\left(\frac{\sum_{i\in S,i\neq 1}\mathcal{P}_i}{2N_0W}\right).$$

2 A Gold sequence is constructed from a maximal sequence with characteristic polynomial $1 + x^2 + x^3$. The second sequence is obtained by decimation of the maximal sequence by $q = 3$. (a) Find one period of each of the two sequences, and show that the second sequence is maximal. (b) List the seven cross-correlation values of this pair of sequences. Show that they are a preferred pair.

3 The characteristic polynomials for generating Gold sequences of length 7 are: $f_1(x) = 1 + x + x^3$ and $f_2(x) = 1 + x^2 + x^3$. (a) What is a general expression for the generating function of an arbitrary Gold sequence? Choose the fixed nonzero initial state for the first maximal sequence generator so that $a_0 = a_1 = 0$, and $a_2 = 1$. (b) What is the generating function for the Gold sequence generated by adding the sequences generated by $f_1(x)$ and $f_2(x)$ when both maximal sequence generators have the same arbitrary initial state?

4 A small set of Kasami sequences is formed by starting with the maximal sequence generated by the characteristic polynomial $1 + x^2 + x^3 + x^4 + x^8$. After decimation by q, a second sequence with a characteristic polynomial $1 + x + x^4$ is found. (a) What is the value of q? How many sequences are in the set? What is the period of each sequence? What is the peak magnitude of the periodic cross-correlation? Draw a block diagram of the generator of the small Kasami set. (b) Prove whether the second sequence is maximal.

5 The small set of the preceding problem is extended to a large set of Kasami sequences by a decimation of the original maximal sequence by q_1. A third sequence with a characteristic polynomial $1 + x^2 + x^3 + x^4 + x^5 + x^7 + x^8$ is found. (a) What is the value of q_1? How many sequences are in the large set? What is the period of each sequence? What is the peak magnitude of the periodic cross-correlation?

Draw a block diagram of the generator of the large Kasami set. (b) Prove whether the third sequence is maximal.

6 Use the periodicity of the spreading sequences to derive (7.59).

7 Apply Jensen's inequality to (2.133) for rectangular chip waveforms. Let $X = \cos 2\phi$ and use the fact that $E[\cos 2\phi] = 0$ to obtain a lower bound identical to the right-hand side of (2.137). Thus, the *balanced QPSK* system, for which $d_1(t) = d_2(t)$, provides a lower symbol error probability against tone interference than the dual QPSK system, for which $d_1(t) \neq d_2(t)$. What is the lower bound on $\mathcal{E}_s$ that provides a sufficient convexity condition for all ϕ and f_d?

8 Prove (7.87) and (7.91) for a rectangular chip waveform.

9 Use bounding and approximation methods to establish (7.111).

10 (a) Verify (7.134). (b) How many Hamming correlations need to be checked to verify (7.137)? Check that a few of them are consistent with (7.137).

11 Let Z denote the number of collisions between two random frequency-hopping patterns during L hops. Show that the variance of Z for independent, random frequency-hopping patterns that use M frequency channels is equal to $(M-1)L/M^2$.

12 Consider the optimal multiuser detector for two synchronous direct-sequence signals with BPSK over the AWGN channel. Assume that phase synchronization is performed and that the two spreading sequences are identical. For what values of the vector $\mathbf{P}^T\mathbf{y}$ does the detector decide that $\mathbf{d} = [+1, +1]$ was transmitted?

13 Derive the optimal multiuser detector for synchronous direct-sequence signals with BPSK over the AWGN channel when all the spreading sequences are mutually orthogonal. (a) Show that the detector has the form $\widehat{\mathbf{d}} = sgn\,[f\,(\mathbf{x})]$. (b) Show that the detector decouples the code symbols in the sense that the code symbol decision of one user is not influenced by the other code symbols.

14 Consider the conventional detector for two synchronous users. Evaluate $P_s(0)$ as $N_0 \to 0$ for the three cases: $|\rho C| < 1$, $|\rho C| > 1$, and $|\rho C| = 1$.

15 Under certain circumstances, the noise in a conventional detector can be beneficial against the multiple-access interference. Consider the conventional detector for two synchronous users with $|\rho C| > 1$. Find the noise level that minimizes $P_s(0)$.

16 Assume that the amplitudes in Example 2 are positive. Show that the symbol decisions of the decorrelating detector are $\widehat{d}_0 = sgn(r_0 - \rho r_1)$ and $\widehat{d}_1 = sgn(r_1 - \rho r_0)$, where $[r_0\ r_1]^T = \mathbf{P}^T\,\mathrm{Re}\,(\mathbf{y})$.

17 Consider the MMSE and decorrelating detectors for synchronous users with orthonormal spreading sequences. Show that the symbol error probability is identical for both detectors.

Chapter 8
Mobile Ad Hoc and Cellular Networks

The impact of multiple-access interference in mobile ad hoc and cellular networks with direct-sequence code-division multiple access (DS-CDMA) and frequency-hopping code-division multiple access (FH-CDMA) systems is analyzed in this chapter. Phenomena and issues that become prominent in mobile networks using spread spectrum include exclusion zones, guard zones, power control, rate control, network policies, sectorization, and the selection of various spread-spectrum parameters. The outage probability, which is the fundamental network performance metric, is derived for both ad hoc and cellular networks and both DS-CDMA and FH-CDMA systems. Acquisition and synchronization methods that are needed within a cellular DS-CDMA network are addressed.

8.1 Conditional Outage Probability

In a mobile network, the most useful link performance metric is the *outage probability*, which is the probability that the link does not currently support communications with a specified reliability or quality. Since link performance can generally be related to the signal-to-interference-and-noise ratio (SINR) at the receiver, an *outage* is said to occur if the instantaneous SINR of a system is less than a specified threshold, which may be adjusted to account for any diversity, rake combining, or channel code. The outage criterion has the advantage that it simplifies the analysis and does not require explicit specification of the code-symbol modulation or channel coding.

In this section, a model of wireless networks based on *deterministic geometry* is used to derive a closed-form equation for the *conditional outage probability* at the receiver of a reference node [107]. The conditioning is with respect to the location of the interfering nodes and their shadowing. The expression averages over the fading, which has timescales much faster than that of the shadowing or node location.

D. Torrieri, *Principles of Spread-Spectrum Communication Systems*,
https://doi.org/10.1007/978-3-319-70569-9_8

The channel from each node to the reference receiver may have its own distinct Nakagami fading parameter (Section 6.2), and the ability to vary the Nakagami parameters can be used to model differing line-of-sight conditions between the reference receiver and each node. The closed-form equation is then applied to the assessment of the effects of various network parameters on outage probabilities.

The network comprises $M+2$ nodes that include a reference receiver X_{M+1}, a desired or reference transmitter X_0, and M interfering nodes $X_1, \ldots, X_M$. The scalar X_i represents the ith node, the vector $\mathbf{X}_i$ represents its location, and $||\mathbf{X}_i - \mathbf{X}_{M+1}||$ is the distance from the ith node to the reference receiver. The nodes can be located in any arbitrary two- or three-dimensional regions. Two-dimensional coordinates are conveniently represented by allowing $\mathbf{X}_i$ to assume a complex value, where the real component is the East-West coordinate and the imaginary component is the North-South coordinate. Each mobile uses a single omnidirectional antenna.

When direct-sequence spreading is used, long spreading sequences are assumed and modeled as random binary sequences with chip duration T_c. The spreading factor G directly reduces the interference power. The multiple-access interference is assumed to be asynchronous, and the power from each interfering X_i is further reduced by the chip function $h(\tau_i)$, which is a function of the chip waveform and the timing offset τ_i of X_i's spreading sequence relative to that of the desired signal. Since only timing offsets modulo-T_c are relevant, $0 \leq \tau_i < T_c$. In a network of quadriphase direct-sequence systems, a multiple-access interference signal with received power $\mathcal{I}_i$ before despreading is reduced after despreading to the level $\mathcal{I}_i h(\tau_i)/G$, where $h(\tau_i)$ is given by (7.78). Thus, the interference power is reduced by the *effective spreading factor* $G_i = G/h(\tau_i)$ while the despreading does not significantly affect the desired-signal power. Assuming that τ_i has a uniform distribution over $[0, T_c]$, the lower bound of (7.117) indicates that the direct-sequence receiver reduces the power of each interference signal by at least the factor G/h on average, where h is the chip factor defined by (7.91).

The power of the desired signal from the reference transmitter at the reference receiver is

$$\rho_0 = P_0 g_0 10^{\xi_0/10} f\left(||\mathbf{X}_0 - \mathbf{X}_{M+1}||\right) \tag{8.1}$$

and the power of the interference from X_i at the reference receiver is (Section 6.1)

$$\rho_i = \frac{P_i}{G_i} g_i 10^{\xi_i/10} f\left(||\mathbf{X}_i - \mathbf{X}_{M+1}||\right), \ 1 \leq i \leq M \tag{8.2}$$

where $P_i, 0 \leq i \leq M$, is the received power at the reference distance d_0 (assumed to be sufficiently far that the signals are in the far field) before despreading when fading and shadowing are absent, g_i is the power gain due to fading, ξ_i is a shadowing factor, and $f(\cdot)$ is a path-loss function. The path-loss function of distance d is expressed as the power law

$$f(d) = \left(\frac{d}{d_0}\right)^{-\alpha}, \quad d \geq d_0 \tag{8.3}$$

where $\alpha \geq 2$ is the path-loss exponent. The $\{g_i\}$ are independent with unit-mean, but are not necessarily identically distributed because the channels from the different $\{X_i\}$ to the reference receiver may undergo fading with different distributions. For analytical tractability and close agreement with measured fading statistics, Nakagami fading is assumed, and $g_i = a_i^2$, where a_i is Nakagami with parameter m_i. When the channel between X_i and the reference receiver undergoes Rayleigh fading, $m_i = 1$ and the corresponding g_i is exponentially distributed. The shadowing on the link from one node to another is determined by the local terrain. If the shadowing is modeled as having a lognormal distribution, the $\{\xi_i\}$ are independent zero-mean Gaussian random variables (Section 6.1). In the absence of shadowing, $\xi_i = 0$.

We assume that the $\{g_i\}$ remain fixed for the duration of a time interval but vary independently from interval to interval (block fading). The *activity probability* p_i is the probability that the ith node transmits in the same time interval as the desired signal. The $\{p_i\}$ can be used to model frequency hopping, voice-activity factors, controlled silences, or failed link transmissions and the resulting retransmission attempts. The $\{p_i\}$ need not be the same; for instance, carrier-sense multiple access (CSMA) protocols can be modeled by setting $p_i = 0$ only when a mobile lies within the CSMA guard zone of another active mobile.

The instantaneous SINR at the reference receiver is given by:

$$\gamma = \frac{\rho_0}{\mathcal{N} + \sum_{i=1}^{M} I_i \rho_i} \tag{8.4}$$

where ρ_0 is the received power of the desired signal, $\mathcal{N}$ is the noise power, and the indicator I_i is a Bernoulli random variable with probability $P[I_i = 1] = p_i$ and $P[I_i = 0] = 1 - p_i$. The substitution of (8.1)-(8.3) into (8.4) yields

$$\gamma = \frac{g_0 \Omega_0}{\Gamma^{-1} + \sum_{i=1}^{M} I_i g_i \Omega_i}, \quad \Gamma = \frac{d_0^{\alpha} P_0}{\mathcal{N}} \tag{8.5}$$

where

$$\Omega_i = \begin{cases} 10^{\xi_0/10} ||\mathbf{X}_0 - \mathbf{X}_{M+1}||^{-\alpha}, & i = 0 \\ \frac{P_i}{G_i P_0} 10^{\xi_i/10} ||\mathbf{X}_i - \mathbf{X}_{M+1}||^{-\alpha}, & i > 0 \end{cases} \tag{8.6}$$

is the normalized power of X_i, and Γ is the signal-to-noise ratio (SNR) when the reference transmitter X_0 is at a unit distance from the reference receiver X_{M+1} and fading and shadowing are absent. When direct-sequence spreading is not used, $G_i = 1$.

Two preliminary results are needed in the analysis. Consider the expansion of $(x_1 + x_2 + \ldots + x_k)^n$, where n is a positive integer. A typical term in the expansion is $x_1^{n_1} x_2^{n_2} \ldots x_k^{n_k}$, where $\{n_i\}$ are nonnegative integers such that $n_1 + n_2 + \ldots + n_k = n$. The number of times this term appears is

$$\binom{n}{n_1}\binom{n-n_1}{n_2}\cdots\binom{n-\sum_{i=1}^{k-2} n_i}{n_{k-1}}\binom{n_k}{n_k} = \frac{n!}{n_1! n_2! \ldots n_k!} \tag{8.7}$$

because we may count the appearances by selecting x_1 from n_1 of the n factors, selecting x_2 from n_2 of the remaining factors, and so forth. Thus, we obtain the *multinomial expansion*:

$$(x_1 + x_2 + \ldots + x_k)^n = \sum_{n_i : \sum_{i=1}^{k} n_i = n} \frac{n!}{n_1! n_2! \ldots n_k!} x_1^{n_1} x_2^{n_2} \ldots x_k^{n_k} \tag{8.8}$$

where the principal sum is over the nonnegative integers that have a sum equal to n.

For Nakagami fading m_i and $E[g_i] = 1$, it follows from (6.33) and elementary probability that the density function of each random variable $g_i = a_i^2$ is given by the *gamma density* (Appendix E.5):

$$f_{g_i}(x) = \frac{m_i^{m_i}}{\Gamma(m_i)} x^{m_i - 1} \exp(-m_i x)\, u(x) \tag{8.9}$$

where $u(x)$ is the unit step function: $u(x) = 1, x \geq 0$, and $u(x) = 0$, otherwise.

In the subsequent analysis [107], the spatial extent of the network and number of nodes are finite. Each node has an arbitrary location distribution with an allowance for the node's duty factor, shadowing, exclusion zones, and possible guard zones. Let $\boldsymbol{\Omega} = [\Omega_0, \ldots, \Omega_M]$ represent the set of normalized powers given by (8.6). An *outage* occurs when the SINR γ falls below an SINR threshold β required for reliable reception of a signal, and γ is given for the particular $\boldsymbol{\Omega}$ by (8.5). It follows that the outage probability for the given $\boldsymbol{\Omega}$ is

$$\epsilon(\boldsymbol{\Omega}) = P\left[\gamma \leq \beta | \boldsymbol{\Omega}\right]. \tag{8.10}$$

Because it is conditioned on $\boldsymbol{\Omega}$, the outage probability is conditioned on the network geometry and shadowing factors, which have dynamics over timescales that are much slower than the fading.

Substituting (8.5) into (8.10), and rearranging yields

$$\epsilon(\boldsymbol{\Omega}) = P\left[\beta^{-1} g_0 \Omega_0 - \sum_{i=1}^{M} I_i g_i \Omega_i \leq \Gamma^{-1} \middle| \boldsymbol{\Omega}\right]. \tag{8.11}$$

By defining

$$\mathsf{S} = \beta^{-1} g_0 \Omega_0, \ \mathsf{Y}_i = I_i g_i \Omega_i \tag{8.12}$$

$$\mathsf{Z} = \mathsf{S} - \sum_{i=1}^{M} \mathsf{Y}_i \tag{8.13}$$

the outage probability may be expressed as

$$\epsilon(\boldsymbol{\Omega}) = P\left[\mathsf{Z} \leq \Gamma^{-1} | \boldsymbol{\Omega}\right] = F_{\mathsf{Z}}\left(\Gamma^{-1} | \boldsymbol{\Omega}\right) \tag{8.14}$$

which is the distribution function of Z conditioned on $\mathbf{\Omega}$ and evaluated at Γ^{-1}. Let $f_{\mathsf{S},\mathbf{Y}}(s,\mathbf{y}|\mathbf{\Omega})$ denote the joint density function of S and the vector $\mathbf{Y} = (\mathsf{Y}_1,\ldots,\mathsf{Y}_M)$ conditioned on $\mathbf{\Omega}$. Equation (8.14) implies that

$$\begin{aligned}
1-\epsilon(\mathbf{\Omega}) &= P[\mathsf{Z} > z|\mathbf{\Omega}] = P\left[\mathsf{S} > z + \sum_{i=1}^{M}\mathsf{Y}_i|\mathbf{\Omega}\right] \\
&= \underset{\mathbb{R}^M}{\int\cdots\int}\int_{z+\sum_{i=1}^{M}y_i}^{\infty} f_{\mathsf{S},\mathbf{Y}}(s,\mathbf{y}|\mathbf{\Omega})ds d\mathbf{y} \\
&= \underset{\mathbb{R}^M}{\int\cdots\int}\int_{z+\sum_{i=1}^{M}y_i}^{\infty} f_{\mathsf{S}}(s|\mathbf{\Omega},\mathbf{y})f_{\mathbf{Y}}(\mathbf{y}|\mathbf{\Omega})ds d\mathbf{y} \qquad (8.15)
\end{aligned}$$

where $z = \Gamma^{-1}$, $f_{\mathbf{Y}}(\mathbf{y}|\mathbf{\Omega})$ is the joint density function of $\mathbf{Y}$ conditioned on $\mathbf{\Omega}$,$f_{\mathsf{S}}(s|\mathbf{\Omega},\mathbf{y})$ is the density function of S conditioned on $(\mathbf{\Omega},\mathbf{y})$, and the outer integral is over M-dimensional space.

All channels are assumed to fade independently. Since S is independent of $\mathbf{y}$ and $\Omega_i, i \neq 0, f_{\mathsf{S}}(s|\mathbf{\Omega},\mathbf{y}) = f_{\mathsf{S}}(s|\Omega_0)$, where $f_{\mathsf{S}}(s|\Omega_0)$ is the density function of S conditioned on Ω_0. Since the $\{\mathsf{Y}_i\}$ are independent of each other and each Y_i is independent of $\Omega_k, k \neq i$, we have $f_{\mathbf{Y}}(\mathbf{y}|\mathbf{\Omega})=\prod_{i=1}^{M} f_i(y_i)$, where $f_i(y_i) = f_{\mathsf{Y}_i}(y_i|\Omega_i)$ is the density function of Y_i conditioned on Ω_i. Since density functions are nonnegative and integrable, according to Fubini's theorem (Appendix C.1) the order of integration is interchangeable; hence,

$$\epsilon(\mathbf{\Omega}) = 1 - \underset{\mathbb{R}^M}{\int\cdots\int}\left[\int_{z+\sum_{i=1}^{M}y_i}^{\infty} f_{\mathsf{S}}(s|\Omega_0)ds\right]\prod_{i=1}^{M} f_i(y_i)dy_i. \qquad (8.16)$$

The density function of the gamma-distributed S with Nakagami parameter m_0 is

$$f_{\mathsf{S}}(s|\Omega_0) = \frac{\left(\frac{\beta m_0}{\Omega_0}\right)^{m_0}}{(m_0-1)!} s^{m_0-1}\exp(-\beta m_0 s)u(s). \qquad (8.17)$$

Successive integrations by parts and the assumption that m_0 is a positive integer provide the evaluation of the inner integral:

$$\int_{z+\sum_{i=1}^{M}y_i}^{\infty} f_{\mathsf{S}}(s|\Omega_0)ds = \exp\left\{-\frac{\beta m_0}{\Omega_0}(z+\sum_{i=1}^{M}y_i)\right\}\sum_{s=0}^{m_0-1}\frac{1}{s!}\left[\frac{\beta m_0}{\Omega_0}(z+\sum_{i=1}^{M}y_i)\right]^s. \qquad (8.18)$$

Defining $\beta_0 = \beta m_0/\Omega_0$ and substituting (8.18) into (8.16) yields

$$\epsilon(\mathbf{\Omega}) = 1 - e^{-\beta_0 z}\sum_{s=0}^{m_0-1}\frac{(\beta_0 z)^s}{s!}\underset{\mathbb{R}^M}{\int\cdots\int} e^{-\beta_0\sum_{i=1}^{M}y_i}\left(1+z^{-1}\sum_{i=1}^{M}y_i\right)^s\prod_{i=1}^{M} f_i(y_i)dy_i. \qquad (8.19)$$

Since s is a positive integer, the binomial theorem indicates that

$$\left(1+z^{-1}\sum_{i=1}^{M}y_i\right)^s=\sum_{t=0}^{s}\binom{s}{t}z^{-t}\left(\sum_{i=1}^{M}y_i\right)^t. \tag{8.20}$$

The multinomial expansion (8.8) yields

$$\left(\sum_{i=1}^{M}y_i\right)^t=\sum_{\substack{\ell_i\geq 0\\ \sum_{i=1}^{M}\ell_i=t}}t!\left(\prod_{i=1}^{M}\frac{y_i^{\ell_i}}{\ell_i!}\right) \tag{8.21}$$

where the summation on the right-hand side is over all sets of nonnegative indices that sum to t. Substituting (8.20) and (8.21) into (8.19) and bringing the exponential into the product, we obtain

$$\begin{aligned}\epsilon(\boldsymbol{\Omega})=1-e^{-\beta_0 z}&\sum_{s=0}^{m_0-1}\frac{(\beta_0 z)^s}{s!}\sum_{t=0}^{s}\binom{s}{t}z^{-t}t!\\ &\times\sum_{\substack{\ell_i\geq 0\\ \sum_{i=1}^{M}\ell_i=t}}\int\cdots\int_{\mathbb{R}^M}\left(\prod_{i=1}^{M}e^{-\beta_0 y_i}\frac{y_i^{\ell_i}}{\ell_i!}\right)\prod_{i=1}^{M}f_i(y_i)dy_i.\end{aligned} \tag{8.22}$$

Using the fact that the $\{\mathsf{Y}_i\}$ are nonnegative, we obtain

$$\epsilon(\boldsymbol{\Omega})=1-e^{-\beta_0 z}\sum_{s=0}^{m_0-1}\frac{(\beta_0 z)^s}{s!}\sum_{t=0}^{s}\binom{s}{t}z^{-t}t!\sum_{\substack{\ell_i\geq 0\\ \sum_{i=1}^{M}\ell_i=t}}\prod_{i=1}^{M}\int_0^{\infty}\frac{y^{\ell_i}}{\ell_i!}e^{-\beta_0 y}f_i(y)dy. \tag{8.23}$$

Using (8.9) and the activity probability p_i, we find that the conditional density function $f_i(y)$ is

$$f_i(y)=f_{\mathsf{Y}_i}(y|\Omega_i)=(1-p_i)\delta(y)+p_i\left(\frac{m_i}{\Omega_i}\right)^{m_i}\frac{1}{\Gamma(m_i)}y^{m_i-1}e^{-ym_i/\Omega_i}u(y) \tag{8.24}$$

where $\delta(y)$ is the Dirac delta function. Substituting this equation, the integral in (8.23) is

$$\begin{aligned}\int_0^{\infty}\frac{y^{\ell_i}}{\ell_i!}e^{-\beta_0 y}f_i(y)dy&=(1-p_i)\delta_{\ell_i}+\left(\frac{p_i\Gamma(\ell_i+m_i)}{\ell_i!\Gamma(m_i)}\right)\left(\frac{\Omega_i}{m_i}\right)^{\ell_i}\\ &\times\left(\beta_0\frac{\Omega_i}{m_i}+1\right)^{-(m_i+\ell_i)}\end{aligned} \tag{8.25}$$

where δ_ℓ is the Kronecker delta function, equal to 1 when $\ell = 0$, and equal to 0 otherwise. Substituting (8.25) into (8.23) and using

$$\binom{s}{t}\left(\frac{t!}{s!}\right) = \left(\frac{s!}{t!(s-t)!}\right)\left(\frac{t!}{s!}\right) = \frac{1}{(s-t)!} \tag{8.26}$$

gives

$$\begin{aligned}\epsilon(\boldsymbol{\Omega}) = 1 - e^{-\beta_0 z} &\sum_{s=0}^{m_0-1} (\beta_0 z)^s \sum_{t=0}^{s} \frac{z^{-t}}{(s-t)!} \\ \times &\sum_{\substack{\ell_i \geq 0 \\ \sum_{i=1}^{M} \ell_i = t}} \prod_{i=1}^{M} \left[(1-p_i)\delta_{\ell_i} + \frac{p_i \Gamma(\ell_i + m_i) \left(\frac{\Omega_i}{m_i}\right)^{\ell_i}}{\ell_i! \Gamma(m_i) \left(\beta_0 \frac{\Omega_i}{m_i} + 1\right)^{(m_i+\ell_i)}} \right].\end{aligned} \tag{8.27}$$

This equation may be written as

$$\begin{aligned}\epsilon(\boldsymbol{\Omega}) &= 1 - e^{-\beta_0 z} \sum_{s=0}^{m_0-1} (\beta_0 z)^s \sum_{t=0}^{s} \frac{z^{-t}}{(s-t)!} H_t(\boldsymbol{\Omega}), \\ \beta_0 &= \frac{\beta m_0}{\Omega_0}, \quad z = \Gamma^{-1}\end{aligned} \tag{8.28}$$

where m_0 is a positive integer,

$$H_t(\boldsymbol{\Omega}) = \sum_{\substack{\ell_i \geq 0 \\ \sum_{i=1}^{M} \ell_i = t}} \prod_{i=1}^{M} G_{\ell_i}(i) \tag{8.29}$$

the summation in (8.29) is over all sets of indices that sum to t,

$$G_\ell(i) = \begin{cases} 1 - p_i(1 - \Psi_i^{m_i}), & \ell = 0 \\ \frac{p_i \Gamma(\ell + m_i)}{\ell! \Gamma(m_i)} \left(\frac{\Omega_i}{m_i}\right)^{\ell} \Psi_i^{m_i+\ell}, & \ell > 0 \end{cases} \tag{8.30}$$

and

$$\Psi_i = \left(\beta_0 \frac{\Omega_i}{m_i} + 1\right)^{-1} \qquad i = 1, 2, \ldots, M. \tag{8.31}$$

Equation (8.29) may be efficiently computed as follows. For each possible $t = \{0, .., m_0 - 1\}$, precompute a matrix $\mathcal{I}_t$ that has rows containing all sets of non-negative indices $\{\ell_1, \ldots, \ell_M\}$ that sum to t. There are

$$\binom{t+M-1}{t} \tag{8.32}$$

rows and M columns in $\mathcal{I}_t$. The $\mathcal{I}_t$s may be reused whenever the same M is considered. Compute a row vector $\boldsymbol{\Psi}$ containing the Ψ_i. For each possible $\ell = \{0, .., m_0 - 1\}$, compute (8.30) using $\boldsymbol{\Psi}$ and place the resulting row into an $m_0 \times M$ matrix $\mathbf{G}$. Each term of (8.29) can be found by using the corresponding row from $\mathcal{I}_t$ as an index into $\mathbf{G}$. Taking the product along the length of the resulting row vector gives the corresponding term of the summation. More generally, the entire $\mathcal{I}_t$ matrix can be used to index $\mathbf{G}$. To be consistent with matrix-based languages, such as MATLAB, denote the result of the operation as $\mathbf{G}(\mathcal{I}_t)$. Taking the product along the rows of $\mathbf{G}(\mathcal{I}_t)$ and then the sum down the resulting column gives (8.29).

In the subsequent examples based on deterministic geometry, we assume that the nodes are mobiles that lie within a circular region of radius r_{net}. A circular *exclusion zone* of radius $r_{ex} \geq d_0$ surrounds the reference receiver, and no mobiles are permitted within the exclusion zone. The exclusion zone is based on the spacing that occurs in actual mobile networks. For instance, when the radios are mounted on separate vehicles, there is a need for crash avoidance by maintaining a minimum vehicle separation. A small exclusion zone maintained by visual sightings exists in practical networks, but a more reliable and extensive one can be established by equipping each mobile with a global positioning system (GPS) and periodically broadcasting each mobile's GPS coordinates. Mobiles that receive those messages could compare their locations with those in the messages and alter their movements accordingly.

Consider the network topology shown in the upper right corner of Figure 8.1. The reference receiver is at the center of the network, the corresponding reference transmitter is located to its right, and $M = 28$ interfering mobiles are within an annular region with inner radius $r_{ex} = 0.05$ and outer radius $r_{net} = 1$.

Mobiles are placed successively according to a *uniform clustering* model as follows. Let $X_i = r_i e^{j\theta_i}$ represent the location of the ith mobile. A pair of independent random variables (y_i, z_i) is selected from the uniform distribution over $[0, 1]$. From these variables, the location is initially selected according to a uniform spatial distribution over a disk with radius r_{net} by setting $r_i = \sqrt{y_i} r_{net}$ and $\theta_i = 2\pi z_i$. If the corresponding X_i falls within an exclusion zone of one of the $i-1$ previous mobile locations, then a new random location is assigned to the ith mobile as many times as necessary until it falls outside any exclusion zone.

In the following examples, the $\{\Omega_i\}$ are determined by assuming an attenuation power-law exponent $\alpha = 3.5$, a common transmit power $P_i = P_0$ for all i, no shadowing, and that the reference transmitter is at distance $||\mathbf{X}_0 - \mathbf{X}_{M+1}|| = 0.1$ from the reference receiver. We assume that $p_i = 0.5$ for all i, and that the SINR threshold is $\beta = 0\,\text{dB}$. This threshold corresponds to a maximum code-symbol rate equal to one information bit per channel symbol, as indicated by (7.2).

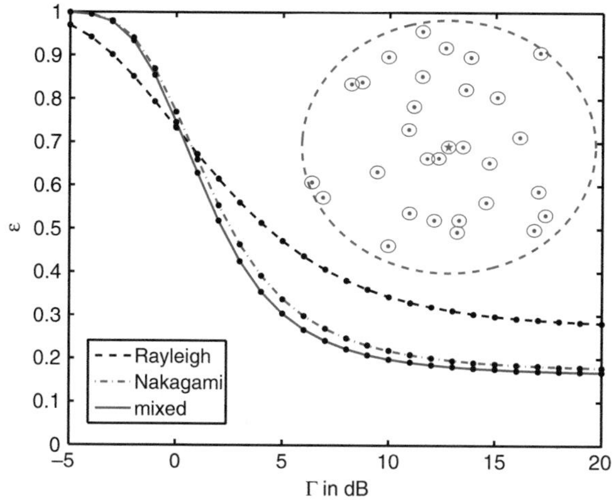

Fig. 8.1 Example network, which is drawn according to the uniform clustering model, and the outage probability as a function of the SNR Γ, conditioned on the pictured network topology. Performance is shown for three fading models without spreading or shadowing. Analytical expressions are plotted by lines while dots represent simulation results (with one million trials per point) [107]

Example 1 Suppose that spread-spectrum modulation is not used ($G = h = 1$) and that all signals undergo Rayleigh fading. Then, $m_i = 1$ for all i, $\beta_0 = \beta/\Omega_0 = \beta||\mathbf{X}_0 - \mathbf{X}_{M+1}||^{\alpha}$, $\Omega_i = ||\mathbf{X}_i - \mathbf{X}_{M+1}||^{-\alpha}$, and (8.28) specializes to

$$\epsilon(\mathbf{\Omega}) = 1 - e^{-\beta_0 \Gamma^{-1}} \prod_{i=1}^{M} \frac{1 + \beta_0(1 - p_i)\Omega_i}{1 + \beta_0 \Omega_i} \tag{8.33}$$

which can be easily evaluated for any given realization of $\mathbf{\Omega}$. The outage probability is shown along with the spatial locations of the mobiles in Figure 8.1. Also shown is the outage probability generated by simulation, which involves randomly generating the mobile locations and the exponentially distributed power gains $g_0, \ldots, g_M$. As can be seen in the figure, the analytical and simulation results coincide, which is to be expected because (8.33) is exact. Any discrepancy between the curves could be attributed to the finite number of Monte Carlo trials (one million trials were executed per SNR point). □

Example 2 Now suppose that the link between the source and receiver undergoes Nakagami fading with parameter $m_0 = 4$, which is much milder than Rayleigh fading. The outage probability, found using (8.28) through (8.30), is also plotted in Figure 8.1. The figure shows two choices for the Nakagami parameter of the interfering mobiles: $m_i = 1$ and $m_i = 4, i = 1, 2, \ldots, M$. The $m_i = 4$ case, denoted by "Nakagami" in the figure legend, represents the situation where the reference transmitter and interfering mobiles are equally visible to the receiver. The $m_i = 1$

case, denoted by "mixed" in the figure legend, represents a more typical situation where the interfering mobiles are not in the line-of-sight. As with the previous example, the analytical curves are verified by simulations involving one million Monte Carlo trials per SNR point. □

8.2 DS-CDMA Mobile Ad Hoc Networks

A *mobile ad hoc network* or *peer-to-peer network* comprises autonomous mobiles that communicate without a centralized control or assistance. Communications between two mobiles are either direct or relayed by other mobiles. Mobile ad hoc networks, which have both commercial and military applications, possess no supporting infrastructure, fixed or mobile. In addition to being essential when a cellular infrastructure is not possible, ad hoc networks provide more robustness and flexibility than cellular networks.

In DS-CDMA mobile ad hoc networks, the mobiles of multiple users simultaneously transmit direct-sequence signals in the same frequency band. All signals use the entire allocated spectrum, but the spreading sequences differ. DS-CDMA is advantageous for ad hoc networks because it eliminates the need for any frequency or time-slot coordination, imposes no sharp upper bound on the number of mobiles, directly benefits from inactive terminals in the network, and is capable of efficiently implementing sporadic data traffic, intermittent voice signals, multibeam arrays, and reassignments to accommodate variable data rates. Furthermore, DS-CDMA systems are inherently resistant to interference, interception, and frequency-selective fading.

Example 3 The high outage probabilities of Examples 1 and 2 can be reduced by using a spread spectrum. Suppose that direct-sequence signals are used with a spreading factor G and common chip function $h(\tau_i) = h$ so that $G_i = G/h$. The other parameter values Examples 1 and 2 remain the same. In Figure 8.2, the outage probability is shown for DS-CDMA networks using three different spreading factors and $h = 2/3$, and for an unspread network ($G_i = 1$). A mixed fading model ($m_0 = 4$ and $m_i = 1$ for $i \geq 1$) is used. From this plot, a dramatic reduction in outage probability when using direct-sequence spreading can be observed. □

Example 4 Because it is conditioned on $\boldsymbol{\Omega}$, the outage probability varies from one network realization to the next. The variability in outage probability is illustrated in Figure 8.3, which shows the outage probability for ten different network realizations and no spreading. One of the networks is that shown in Figure 8.1, whereas the other nine networks were each realized in the same manner, i.e., with $M = 28$ interfering mobiles drawn from a uniform clustering process with $r_{ex} = 0.05$ and $r_{net} = 1$ and the reference transmitter placed at distance $||\mathbf{X}_0 - \mathbf{X}_{M+1}|| = 0.1$ from the reference receiver. The same set of parameters (α, β, p_i, P_i, and m_i) used to generate the mixed-fading results of Example 2 were used again. From the figure, it can be seen that the outage probabilities of different network realizations can vary dramatically. □

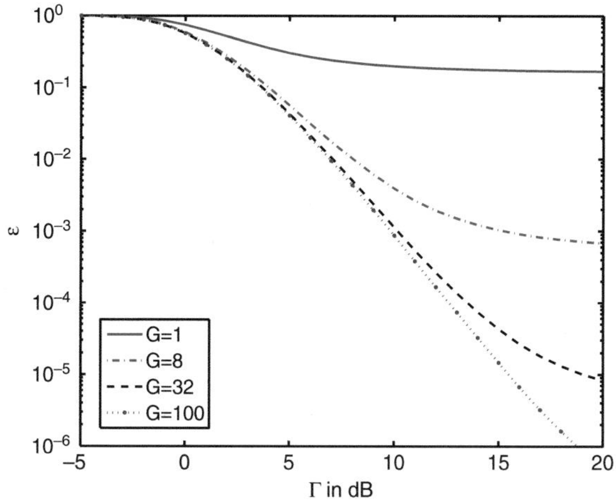

Fig. 8.2 Outage probability as a function of SNR Γ, conditioned on the network shown in Figure 8.1 with the mixed-fading model. Performance is shown for several values of the spreading factor G [107]

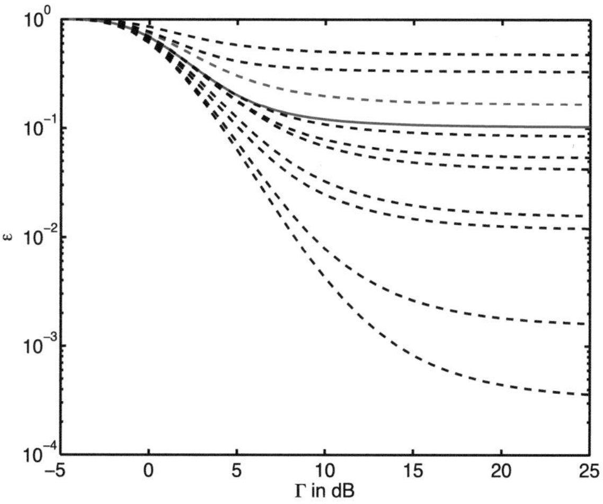

Fig. 8.3 Outage probability conditioned on ten different network realizations. A uniform clustering model is assumed with $r_{net} = 1.0$, $r_{ex} = 0.05$, $M = 28$, mixed fading, $\alpha = 3.5$, and no spreading. The conditional outage probabilities are indicated by dashed lines. The average outage probability over 10,000 network realizations $\bar{\epsilon}$ is shown by the solid line [107]

In addition to the locations of the interfering mobiles, $\mathbf{\Omega}$ depends on the realization of the shadowing. The shadowing factors $\{\xi_i\}$ can be modeled as random variables with any arbitrary distributions and need not be the same for all i. In subsequent examples, lognormal shadowing is assumed, and the shadow factors are independent, identically distributed, and zero-mean Gaussian random variables with a common standard deviation σ_s.

The conditioning on $\boldsymbol{\Omega}$ can be removed by averaging the conditional outage probability ϵ with respect to many network geometries, thereby producing the *spatially averaged outage probability or average outage probability*. This probability is useful for assessing the average effects of parameter variations. The averaging can be performed analytically only under certain limitations [107]. For more general cases of interest, the average outage probability can be estimated through Monte Carlo simulation by generating many different networks and hence many different $\boldsymbol{\Omega}$ vectors, computing the outage probability for each network, and taking the numerical average. Suppose that N networks are generated, and let ϵ_i denote the outage probability of the ith network, which has normalized powers expressed as the vector $\boldsymbol{\Omega}_i$. The average outage probability is

$$\bar{\epsilon} = \frac{1}{N}\sum_{i=1}^{N}\epsilon_i. \tag{8.34}$$

As an example, the solid line in Figure 8.3 shows the corresponding average outage probability for $N = 10{,}000$ network realizations.

Example 5 In a finite network, the average outage probability depends on the location of the reference receiver. Table 8.1 explores the change in performance when the reference receiver moves from the center of the radius-r_{net} circular network to the perimeter of the network. The SNR was set to $\Gamma = 10$ dB, a mixed-fading

Table 8.1 Average outage probability when the receiver is at the center ($\bar{\epsilon}_c$) and on the perimeter ($\bar{\epsilon}_p$) of the network for various M, α, G, and σ_s [107]

Parameters				Outage probabilities	
M	α	G	σ_s	$\bar{\epsilon}_c$	$\bar{\epsilon}_p$
30	3	1	0	0.1528	0.0608
			8	0.2102	0.0940
		32	0	0.0017	0.0012
			8	0.0112	0.0085
	4	1	0	0.1113	0.0459
			8	0.1410	0.0636
		32	0	0.0028	0.0017
			8	0.0123	0.0089
60	3	1	0	0.3395	0.1328
			8	0.4102	0.1892
		32	0	0.0030	0.0017
			8	0.0163	0.0107
	4	1	0	0.2333	0.0954
			8	0.2769	0.1247
		32	0	0.0052	0.0027
			8	0.0184	0.0117

channel model was assumed, and other parameter values were $r_{ex} = 0.05$, $r_{net} = 1$, $\beta = 0\,\text{dB}$, and $p_i = 0.5$. The interfering mobiles were placed according to the uniform clustering model and the reference transmitter was placed at distance $||\mathbf{X}_0 - \mathbf{X}_{M+1}|| = 0.1$ from the reference receiver. For each set of values of the parameters G, α, σ_s, and M, the outage probability at the network center $\bar{\epsilon}_c$ and at the network perimeter $\bar{\epsilon}_p$ were computed by averaging over $N = 10{,}000$ realizations of mobile placement and shadowing. Two values of each parameter were considered: $G = \{1, 32\}, \alpha = \{3, 4\}, \sigma_s = \{0, 8\}, M = \{30, 60\}$. The table indicates that $\bar{\epsilon}_p$ is considerably less than $\bar{\epsilon}_c$ in the finite network. This reduction in outage probability is more significant for the unspread network, and is less pronounced with increasing G. Both $\bar{\epsilon}_p$ and $\bar{\epsilon}_c$ increase as M and σ_s increase and G decreases.

As α increases, both $\bar{\epsilon}_p$ and $\bar{\epsilon}_c$ increase when $G = 32$, but decrease when $G = 1$. This difference occurs because spread-spectrum systems are less susceptible to the near-far problem than unspread ones. The increase in α is not enough to cause a significant increase in the already high outage probability for unspread systems in those realizations with interfering mobiles close enough to the reference receiver to cause a near-far problem. In the same realizations, the less susceptible spread-spectrum systems do experience a significantly increased outage probability. □

A useful metric for quantifying the spatial variability is the probability that the conditional outage probability ϵ is either above or below a threshold ϵ_T. In particular, $P[\epsilon > \epsilon_T]$ represents the fraction of network realizations that fail to meet a minimum required outage probability at the reference receiver and can be construed as a *network* outage probability. The complement of the network outage probability $P[\epsilon \leq \epsilon_T]$ is the distribution function of ϵ.

Example 6 The distribution function $P[\epsilon \leq \epsilon_T]$ is shown in Figure 8.4 for the three fading models without spreading and for the mixed-fading model with spreading. Each curve was computed by generating $N = 10{,}000$ networks with $M = 28$ interfering mobiles drawn from a uniform clustering process with $r_{ex} = 0.05$, $r_{net} = 1$, $||\mathbf{X}_0 - \mathbf{X}_{M+1}|| = 0.1$, $\Gamma = 5\,\text{dB}$, and no shadowing. For each network, the outage probability for the link from a source to a receiver was computed and compared with the threshold ϵ_T. The curves show the fraction of networks with an ϵ that does not exceed the threshold. The curves become steeper with increasing G, which shows that spreading has the effect of making performance less sensitive to the particular network topology. □

Shadowing can be incorporated into the model by simply drawing an appropriate set of independent shadowing factors $\{\xi_i\}$ for each network realization and using them to compute the $\{\Omega_i\}$ according to (8.6).

Example 7 In Figure 8.5, shadowing was applied to the same set of $N = 10{,}000$ networks used to generate Figure 8.4. Two standard deviations were considered for the lognormal shadowing: $\sigma_s = 2\,\text{dB}$ and $\sigma_s = 8$ dB, and again $\alpha = 3.5$. For each shadowed network realization, the outage probability was computed for the mixed-fading model without spreading ($G = 1$). All other parameter values are the same as those used to produce Figure 8.4. The figure shows $P[\epsilon \leq \epsilon_T]$ for both the shadowed and unshadowed realizations. The presence of shadowing and increases

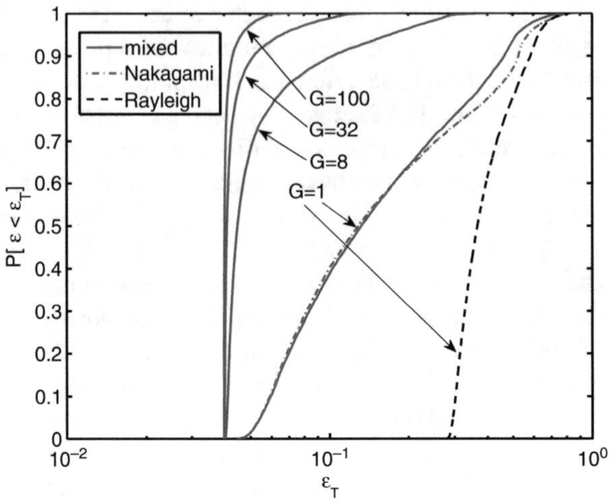

Fig. 8.4 Probability that the conditional outage probability ϵ is below the threshold outage probability ϵ_T in a network with $r_{net} = 1.0$, $r_{ex} = 0.05$, $M = 28$, and $\Gamma = 5\,\text{dB}$. $N = 10{,}000$ network realizations were drawn to produce the figure. Results are shown for the three fading models without spreading ($G = 1$) and for the mixed-fading model with spreading ($G = \{8, 32, 100\}$) [107]

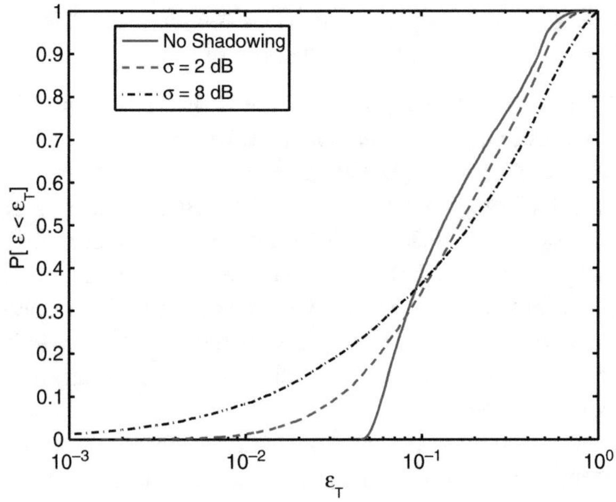

Fig. 8.5 Probability that the conditional outage probability ϵ is below the threshold outage probability ϵ_T in a network with $r_{ex} = 0.05$, $r_{net} = 1.0$, $M = 28$, $\Gamma = 5\,\text{dB}$, mixed fading, and no spreading ($G = 1$). $N = 10{,}000$ network realizations were drawn to produce the figure. Curves are shown for no shadowing and for shadowing with two values of σ_s [107]

in σ_s increase the variability of the conditional outage probability, as indicated by the reduced slope of the distribution functions. Shadowing does not significantly alter the average outage probability in this example. For low thresholds, such as $\epsilon_T < 0.1$, performance is actually better with shadowing than without. This behavior is due to the fact that the shadowing sometimes may cause the reference signal power to be much higher than it would be without shadowing. □

The central issue in DS-CDMA mobile ad hoc networks is the prevention of a near-far problem. If all mobiles transmit at the same power level, then the received power at a receiver is higher for transmitters near the receiving antenna. There is a near-far problem because transmitters that are far from the receiving antenna may be at a substantial power disadvantage, and the spreading factor may not be large enough to allow satisfactory reception of their signals. The solution to the near-far problem in cellular networks (Section 8.3) is *power control*, which is the control or regulation of the power levels received from signal sources. However, the absence of a centralized control of an ad hoc network renders any attempted power control local rather than pervasive and generally infeasible. Multiuser detection in DS-CDMA networks, such as interference cancelation (Section 7.5), reduces but does not eliminate the near-far problem. Even if an interference canceler can suppress a large amount of interference, the residual interference due to imperfect channel estimation may prevent acquisition.

The IEEE 802.11 standard uses CSMA with collision avoidance in its medium-access control protocol for ad hoc networks. The implementation entails the exchange of request-to-send (RTS) and clear-to-send (CTS) handshake packets between a transmitter and receiver during their initial phase of communication that precedes the subsequent data and acknowledgment packets. The receipt of the RTS/CTS packets with sufficient power levels by nearby mobiles causes them to inhibit their own transmissions, which would produce interference in the receiver of interest. The transmission of separate CTS packets in addition to the RTS packets decreases the possibility of subsequent signal collisions at the receiver because of nearby hidden terminals that do not sense the RTS packets. Thus, the RTS/CTS packets essentially establish *CSMA guard zones* surrounding a transmitter and receiver, hence preventing a near-far problem except during the initial reception of an RTS packet. The interference at the receiver is restricted to concurrent transmissions generated by mobiles outside the guard zones.

The major advantage of the exclusion zone compared with a CSMA guard zone is that the exclusion zone prevents near-far problems at receivers while not inhibiting any potential concurrent transmissions. Another advantage of an exclusion zone is enhanced network connectivity because of the inherent constraint on the clustering of mobiles. The CSMA guard zone offers additional near-far protection beyond that offered by the exclusion zone, but at the cost of reduced network transmission capacity, as shown subsequently. However, CSMA guard zones are useful for operating environments that do not permit large exclusion zones, such as networks with low mobility or a high density of mobile terminals.

When CSMA is used in a network, the CSMA guard zone usually encompasses the exclusion zone. Although both zones may cover arbitrary regions, they are modeled as circular regions in the subsequent examples for computational convenience, and the region of the CSMA guard zone that lies outside the exclusion zone is an annular ring. The existence of an annular ring enhances the near-far protection at the cost of inhibiting potential concurrent transmissions within the annular ring. The radii of the exclusion zone and the CSMA guard zone are denoted by r_{ex} and r_{g}, respectively.

An analysis of the impact of guard zones that encompass exclusion zones [108] begins with an initial placement of the mobiles according to the uniform clustering model. In generating a network realization, potentially interfering mobiles within guard zones are deactivated according to the following procedure. First, the reference transmitter X_0 is activated. Next, each potentially interfering mobile is considered in the order in which it was placed. For each mobile, a check is made to see if it is in the guard zone of a previously placed active mobile. Since mobiles are indexed according to the order of placement, X_1 is first considered for possible deactivation; if it lies in the guard zone of X_0, it is deactivated, and otherwise it is activated. The process repeats for each subsequent X_i, deactivating it if it falls within the guard zone of any active $X_j, j < i$, or otherwise activating it.

Figure 8.6 displays an example of a network realization. The reference receiver is placed at the origin, the reference transmitter is at $X_0 = 1/6$ to its right, and $M = 30$ mobiles are placed according to the uniform clustering model, each with an exclusion zone (not shown) of radius $r_{\mathsf{ex}} = 1/12$. Active mobiles are indicated by filled circles, and deactivated mobiles are indicated by unfilled circles. A guard zone of radius $r_{\mathsf{g}} = 1/4$ surrounds each active mobile, as depicted by dashed circles. When CSMA guard zones are used, the other mobiles within the guard zone of an active mobile are deactivated. The reference receiver has not been assigned a CSMA guard zone, which reflects the fact that it has none while it is receiving the initial RTS. In Figure 8.6, 15 mobiles have been deactivated, and the remaining 15 mobiles remain active.

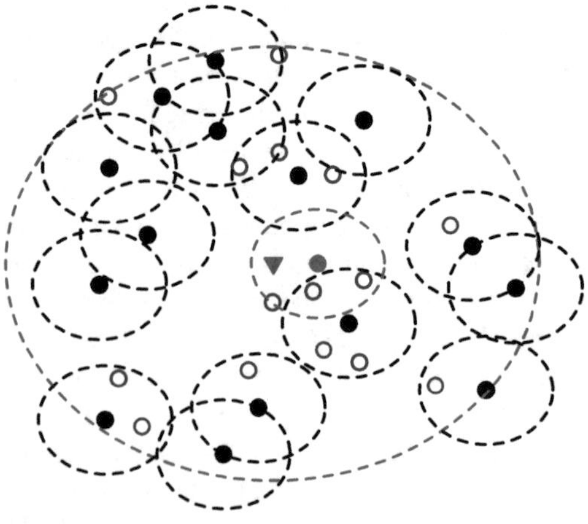

Fig. 8.6 Example network realization [108]

Although the outage probability is improved with CSMA because of the deactivation of potentially interfering mobiles, the overall network becomes less efficient because of the suppression of transmissions. The network efficiency can be quantified by the *area spectral efficiency*:

$$\mathcal{A} = \lambda(1 - \bar{\epsilon})R \tag{8.35}$$

where λ is the mobile density, which is the number of active mobiles per unit area, $\bar{\epsilon}$ is defined by (8.34), and R is the code rate in units of information bits per channel use. Generating $\boldsymbol{\Omega}_i$ for each network realization involves not only placing the mobiles according to the uniform clustering model, but also realizing the shadowing and deactivating mobiles that lie within the CSMA guard zones of the active mobiles. The area spectral efficiency represents the spatially averaged maximum network throughput per unit area. Increasing the size of the guard zone generally reduces area spectral efficiency because of fewer simultaneous transmissions.

For a given value of M, the mobile density without a CSMA guard zone remains fixed as all mobiles remain active. However, with a CSMA guard zone, the number of potentially interfering mobiles is random with a value that depends on the value of r_g, the locations of the mobiles, and their order of placement, which affects how they are deactivated. As with the average outage probability, Monte Carlo simulation is used to estimate the area spectral efficiency.

In the following examples, we assume that the SNR is $\Gamma = 10$ dB, all channels undergo mixed fading ($m_0 = 3$ and $m_i = 1, i \geq 1$) with lognormal shadowing ($\sigma_s = 8$ dB), and the SINR threshold is $\beta = 0$ dB. Once the mobile locations $\{X_i\}$ are realized, the $\{\Omega_i\}$ are determined by assuming a path-loss exponent $\alpha = 3.5$ and a common transmit power ($P_i = P_0$ for all i). The reference receiver is at the center of the network, and the spatially averaged outage probability is computed by averaging over $N = 10{,}000$ network realizations. Although the model permits nonidentical spreading factors, we assume that each *effective spreading factor* $G_i = G_\mathsf{e} = G/h$ is constant for all interference signals. Both spread and unspread systems are considered, with $G_\mathsf{e} = 1$ for the unspread system and $G_\mathsf{e} = 48$ for the spread system, corresponding to a typical direct-sequence waveform with $G = 32$ and $h(\tau_i) = h = 2/3$. Although the model permits nonidentical p_i in the range $[0, 1]$, the value $p_i = 0.5$ for all active X_i is chosen, corresponding to a half-duplex mobile terminal with a full input buffer and a symmetric data transmission rate to a peer terminal.

Example 8 To investigate the influence of the exclusion-zone radius r_ex and guard-zone radius r_g on the network performance, the outage probability and area spectral efficiency were determined over a range of r_ex and r_g. To remove the dependence on the transmitter-receiver separation, the guard and exclusion zones were normalized with respect to $||\mathbf{X}_0 - \mathbf{X}_{M+1}|| = ||\mathbf{X}_0||$. The normalized guard-zone radius was varied over $1/2 \leq r_\mathsf{g}/||\mathbf{X}_0|| \leq 3$, and three representative values of the normalized exclusion-zone radius are selected: $r_\mathsf{ex}/||\mathbf{X}_0|| = \{1/4, 1/2, 3/4\}$. The number of potentially interfering mobiles was set to $M = 30$ and $r_\mathsf{net}/||\mathbf{X}_0|| = 6$.

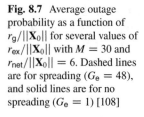

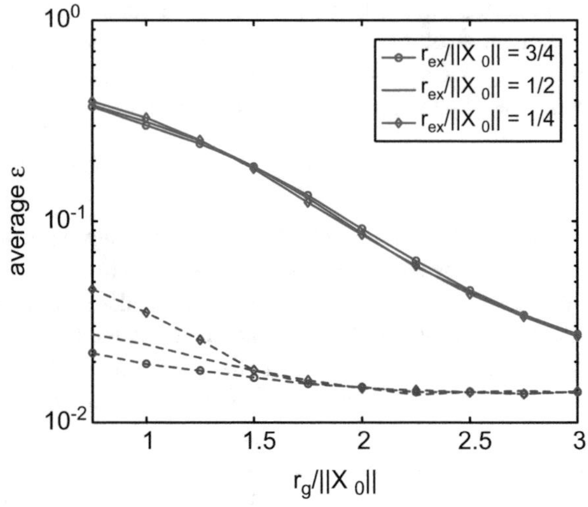

Fig. 8.7 Average outage probability as a function of $r_g/||\mathbf{X}_0||$ for several values of $r_{ex}/||\mathbf{X}_0||$ with $M = 30$ and $r_{net}/||\mathbf{X}_0|| = 6$. Dashed lines are for spreading ($G_e = 48$), and solid lines are for no spreading ($G_e = 1$) [108]

Figure 8.7 shows the spatially averaged outage probability $\bar{\epsilon}$. For the unspread system, the outage probability is insensitive to the exclusion-zone radius, but very sensitive to the guard-zone radius. This sensitivity underscores the importance of a guard zone for an unspread network. Except when the guard-zone radius is relatively small, the outage probability of the spread system is insensitive to both the exclusion-zone and the guard-zone radii.

Figure 8.8 shows the normalized area spectral efficiency $\mathcal{A}/R$ as a function of the guard-zone radius. Although the outage probability at the reference receiver in an unspread network is insensitive to r_{ex}, Figure 8.8 shows that the area spectral efficiency is sensitive to r_{ex}, especially at low r_g. As r_{ex} increases, there are fewer nearby interfering mobiles that get deactivated by the guard zone. Thus, more mobiles remain active, and the area spectral efficiency increases even though the outage probability remains fixed. A similar behavior is seen for the spread network, which for small r_g has a significantly higher area spectral efficiency than the unspread network because of the lower outage probability. The main limitation to increasing r_{ex} is that it must be small enough for there to be no significant impediment to the movements of mobiles. For both the spread and unspread network, the area spectral efficiency diminishes quickly with increasing r_g because of the increased number of silenced mobiles. At high r_g, the area spectral efficiency is insensitive to the spreading factor and exclusion-zone radius. □

Example 9 The previous example assumes that the distance between the reference transmitter and receiver is fixed with respect to the network radius. However, performance depends on this distance. Figure 8.9 shows the normalized area spectral efficiency as a function of the distance $||\mathbf{X}_0||$ between the reference transmitter and receiver. All distances are normalized to the network radius so that $r_{net} = 1$. The exclusion-zone radius is set to $r_{ex} = 1/12$, and both unspread and spread ($G_e = 48$)

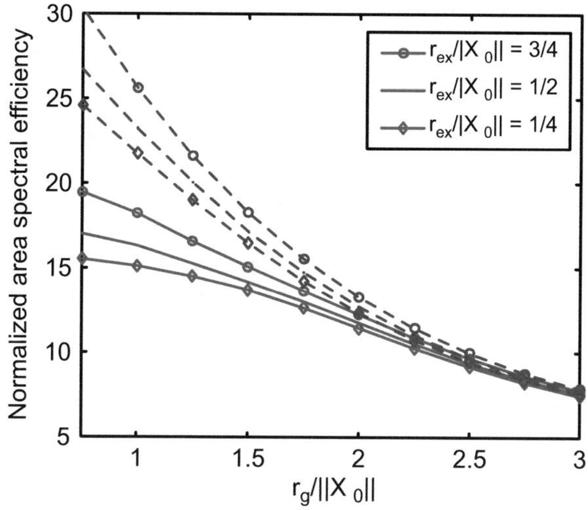

Fig. 8.8 Normalized area spectral efficiency as a function of $r_g/||\mathbf{X}_0||$ for several values of $r_{ex}/||\mathbf{X}_0||$ with $M = 30$ and $r_{net}/||\mathbf{X}_0|| = 6$. Dashed lines are for spreading ($G_e = 48$), and solid lines are for no spreading ($G_e = 1$) [108]

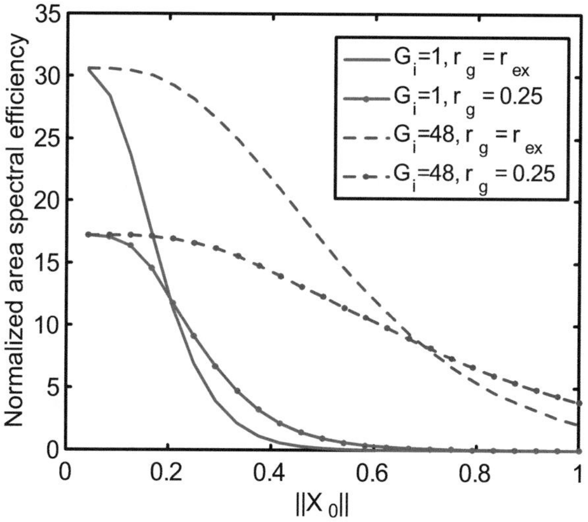

Fig. 8.9 Normalized area spectral efficiency for $r_{ex} = 1/12$ and different transmit distances $||X_0||$. All distances are normalized to the network radius [108]

networks are considered. Results are shown for a CSMA guard zone with radius $r_g = 1/4$ and for a system that uses no additional CSMA guard zone.

It is observed that increasing the transmission distance reduces the area spectral efficiency because of the increase in the number of interfering mobiles that are closer to the receiver than the reference transmitter, but this reduction is more gradual with the spread system than the unspread one. As $||\mathbf{X}_0||$ increases, an increased guard zone alleviates the potential near-far problems. Consequently, the area spectral efficiency degrades at a more gradual rate when CSMA is used, and at sufficiently large transmitter distances, a system with CSMA outperforms a system without it.

The CSMA guard zone decreases area spectral efficiency at short transmission distances, but increases it at long distances. This variation occurs because at a long transmission distance, the received signal power from the far reference transmitter is weak, whereas the received powers from the nearby interfering mobiles are relatively high. To overcome this near-far problem, nearby interfering mobiles need to be deactivated so that the SINR threshold can be be met. However, at short distances, the signal power from the nearby reference transmitter is already strong enough for deactivation of the interfering mobiles to be unnecessary and harmful to the area spectral efficiency because of a reduction of simultaneous transmissions. □

8.3 DS-CDMA Cellular Networks

In a *cellular network*, a geographic region is partitioned into cells. A base station that includes a transmitter and receiver is located within each cell. Figure 8.10 depicts an ideal cellular network in which the cells have equal hexagonal areas and the base stations are at the centers. Each *mobile* in the network is *associated with* or *connected to* a specific base station that coordinates the radio communications of the mobile. That base station is the one from which the mobile receives the strongest signal. The base stations collectively act as a switching center for the mobiles and communicate among themselves by wirelines in most applications. For the cellular configuration of Figure 8.10, most of the mobiles in a cell would be associated with the base station at the center of the cell. Cellular networks with DS-CDMA allow universal frequency reuse in that the same carrier frequency and spectral band are shared by all the cells. Distinctions among the direct-sequence signals are possible because each signal is assigned a unique spreading sequence.

In networks using CDMA2000 and WCDMA, each base station transmits an unmodulated spreading sequence as a pilot signal to enable the association of mobiles with base stations. By comparing the pilot signals from several base stations in a process known as *soft handoff*, a mobile decides which signal is strongest relative to the interference at any instant. A *soft handoff* uses a form of selection diversity (Section 6.7) to ensure that a mobile is served by the most suitable base station most of the time. In the mobile's receiver, an upper threshold determines which pilot signals are strong enough for further processing, and hence which base stations qualify for possible association. A lower threshold determines when a base

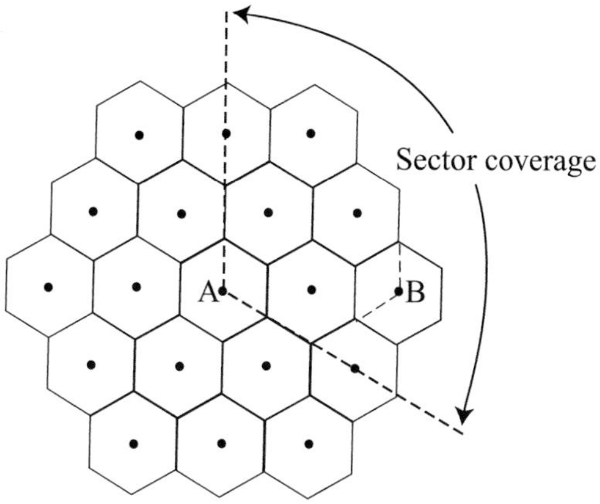

Fig. 8.10 Geometry of a cellular network with a base station at the center of each hexagon. Two concentric tiers of cells surrounding a central cell are shown

station ceases to qualify or a transfer of association is warranted. The cost of soft handoffs is the need to detect and process several simultaneously received pilot signals.

Sectorization is the provision of a set of disjoint fixed beams that cover all feasible directions. A base-station *sector* is defined as the range of angles from which a directional sector beam can receive signals. A mobile within a sector is said to be *covered* by the sector beam. Cells may be divided into sectors by using several directional sector beams, each covering disjoint angles, at the base stations. Typically, there are three sectors with $2\pi/3$ radians in each angular sector, but more sectors are viable as more antennas in the base-station array become feasible at higher frequencies. Sectorization enables coordination of spectral assignments and scheduling among the mobiles associated with each sector. Other advantages of base-station sectorization are the reduction of the beam-alignment delay, pilot-signal contamination, and beam-steering errors, and the avoidance of interference signals.

An ideal *sector antenna* has a uniform gain over the covered sector and negligible sidelobes. Figure 8.10 depicts the directions covered by sector antenna A at the base station of the central cell. The mobiles in the covered portion of the central cell are associated with sector antenna A. Only mobiles in the directions covered by the sector antenna can cause intracell or intercell multiple-access interference on the *reverse link* or *uplink* from a mobile to its associated sector antenna. Only a sector antenna serving a cell sector oriented toward a mobile, such as sector antenna B in the figure, can cause multiple-access interference on the *forward link* or *downlink* to a mobile from its associated sector antenna, such as sector antenna A in the figure. Thus, the numbers of interfering signals on both the uplink and the downlink are reduced approximately by a factor equal to the number of sectors. The mobile antennas are generally omnidirectional.

Cell Search for Downlinks

The process of code acquisition and timing synchronization in the downlinks of cellular DS-CDMA networks, which is called *cell search*, is more elaborate than the methods used for ad hoc networks or point-to-point communications (Chapter 4). To facilitate the identification of a base station controlling communications with a mobile, each spreading sequence for a downlink is formed as the product or concatenation of two sequences often called the scrambling and channelization codes. A *scrambling sequence or scrambling code* is a spreading sequence that identifies a particular base station or cell when the code is acquired by mobiles associated with the cell. A *channelization code* is the spreading sequence of a mobile that allows the mobile's receiver to extract messages to the mobile while suppressing messages intended for other mobiles within the same cell.

If the set of base stations use the GPS or some other common timing source, then each scrambling code may be generated from a distinct starting point within a common long spreading sequence. The common timing source prevents a timing ambiguity that may lead to code ambiguity. After a mobile receiver identifies the scrambling sequence and synchronizes with it, the receiver despreads the scrambling code. Since the timing of the scrambling code determines the timing of the channelization code, the receiver then despreads the channelization code to extract the message. Walsh or other orthogonal sequences (Section 7.2) are suitable as channelization codes for downlinks.

The WCDMA system provides an example of the cell-search process by which the scrambling code is acquired or tracked, and its timing is identified. Cell search comprises the three stages of *slot synchronization, frame synchronization with code-group identification, and scrambling-code determination* [49]. Each stage processes one of three sequence types simultaneously transmitted using a single carrier frequency. The three sequence types are the primary synchronization code (PSC), the secondary synchronization codes (SSCs), and the scrambling code. The first two slots of the basic frame structure for cell search are illustrated in Figure 8.11. Each frame comprises 15 slots, each of which has 2560 chips. The PSC and each of the SSCs are 256 chips long and are only transmitted after a slot boundary; hence these codes have 10% duty factors. The chip rate is 3.84 megachips per second.

All cells within the network use a common PSC that is always transmitted in the same position within each slot, as shown in Figure 8.11. During the first stage of each cell search, the PSC provides each mobile with a means of detecting the slot boundaries of the base-station signal with the largest power at the mobile. Let $c_{psc}(i)$ denote the ith bit of the PSC generated by the receiver with assumed slot and frame boundaries. The received signal with the embedded PSC is applied to a quadrature downconverter and chip-matched filters, the outputs of which are sampled at the chip rate to produce the complex-valued received sequence. Let $r(q)$ denote the qth bit of the received sequence. Let the integer $s \in [0, 14]$ denote the slot number of the received PSC relative to the receiver-generated PSC. Let the integer $h \in [0, 2559]$ denote a hypothetical offset of the received slot boundary relative to the

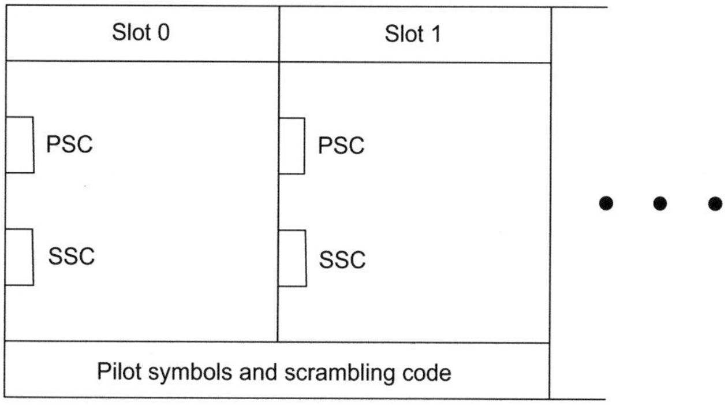

Fig. 8.11 First two slots of wideband code-division multiple access (W-CDMA) frame

presumed slot boundary used by the receiver while generating $c_{psc}(i)$. To estimate the true offset and thus synchronize with the slot boundaries, the receiver correlates the received sequence with each possible offset h and computes

$$y_{psc}(h) = \sum_{s=0}^{N-1} \left| \sum_{i=0}^{255} r(h + 2560s + i)\, c_{psc}(i) \right|, \quad h = 0, 1, \ldots, 2559 \tag{8.36}$$

where the noncoherent combining over $N > 1$ receiver-generated slots is performed to accommodate a low signal-to-interference ratio and obtain a time-diversity gain. The hypothesis $\widehat{h}$ that maximizes $y_{psc}(h)$ is selected as the identifier of the slot boundaries of the received signal.

During the second stage of the cell search, frame synchronization is achieved by estimating s, and the degree of code uncertainty is reduced by identifying the *code group* of the scrambling code. To enable simultaneous frame synchronization and code-group identification, a $(15, 3)$ *comma-free* Reed-Solomon (CFRS) code is used to determine the 15 SSCs assigned to each frame. The comma-free property implies that any cyclic shift of a codeword is not another codeword, and the minimum Hamming distance between codewords is 13. Each CFRS codeword of 15 symbols is drawn from a codebook of 64 codewords. Each codeword symbol is one of 16 possible symbols, which are represented by 16 orthogonal SSCs. Each CFRS codeword defines a code group g, which comprises eight scrambling codes.

Starting at the slot boundary of each slot, as determined by $\widehat{h}$, the received sequence correlates with each of the 16 possible SSCs. For each value of s, the PSC output $= y_{psc}\left(\widehat{h}, s\right)$ is used to provide a phase reference to correct the phase of the SSC correlations. The output of the nth coherent SSC correlator is

$$y_{ssc}\left(\widehat{h}, s, n\right) = \mathrm{Re}\left[y_{psc}^{*}\left(\widehat{h}, s\right) \sum_{i=0}^{255} r\left(2560s + \widehat{h} + i\right) c_{ssc}(n, i) \right]$$

$$n \in [0, 15], \; s \in [0, 14] \tag{8.37}$$

where $c_{ssc}(n, i)$ is the nth SSC. A hard decision on the SSC and corresponding CFRS symbol is made for each s by selecting the n that maximizes $y_{ssc}\left(\widehat{h}, s, n\right)$. After a 15-slot duration, hard-decision decoding of the CFRS codeword is used to determine an estimate $\widehat{g}$ of the code group and an estimate $\widehat{s}$ of the slot number.

As indicated in Figure 8.11, all slots of a frame are fully occupied by downlink pilot symbols representing the scrambling code that identifies the cell. There are 10 quadriphase-shift keying (QPSK)-modulated pilot symbols in each slot, and each symbol is spread by 256 chips. The spreading sequence is one of the orthogonal variable-spreading-factor codes (Section 7.2). After the slot and frame synchronization have been established, the received sequence correlates with the eight scrambling codes of group $\widehat{g}$. For each pilot symbol $m \in [0, 149]$ in a frame and each scrambling code $k \in [0, 7]$, the receiver computes

$$y_{scr}\left(\widehat{h}, \widehat{s}, m, k\right) = \left|\sum_{i=0}^{255} r\left(2560\widehat{s} + \widehat{h} + 256m + i\right) c_{scr}^{*}(k, 256m + i)\right|$$
$$m \in [0, 149],\ k \in [0, 7] \tag{8.38}$$

where $c_{scr}(k, l)$ is chip l of the kth scrambling code. For each pilot symbol m, the scrambling code k with the maximum $y_{scr}\left(\widehat{h}, \widehat{s}, m, k\right)$ receives one vote, and a majority vote among the 150 pilot symbols determines a candidate scrambling code. Since the acceptance of the incorrect scrambling code is costly and disruptive for a mobile, the candidate scrambling code is accepted only if its number of votes exceeds a predetermined threshold designed to maintain an acceptable probability of a false alarm. If the threshold is exceeded, the cell search is complete, and the despreading of messages with the scrambling and channelization codes commences.

Inaccurate frequency estimation by a receiver causes large errors in its clock, which may cause the failure of the initial cell search. To protect against large clock errors, more elaborate methods of cell search can be used [49].

Adaptive Rate Control

It is a fundamental result of information theory that a lowering of the code rate lowers the SINR required for the successful reception of the transmitted signal [20], but the cost is a reduced throughput or spectral efficiency. This fact may be exploited to adapt the code rate to the channel state, which determines the SINR at the receiver. *Adaptive rate control* lowers the code rate when the channel state is unfavorable and raises the code rate and hence the throughput when the channel state is favorable. If the number of code symbols per packet is fixed, the code rate is changed by decreasing or increasing the number of information bits per packet. Adaptive rate control may be supplemented by *adaptive modulation*, which entails a change in the signal constellation or alphabet in response to the channel state, but changes in the modulation are generally more difficult to implement. Adapting the spreading factor is possible in principle, but presents the practical problem of a changing bandwidth.

To adapt the code rate to channel conditions, either the SINR is estimated or some function of the channel state is measured at the receiver, and then the measurement or a selected code rate is sent to the transmitter. The ratio $\widehat{P}_v/\widehat{V}_v$ calculated from (8.44) to (8.47) below provides an estimator of the SINR after the despreading. If a rake demodulator (Section 6.10) is used, then the effective SINR is estimated by adding the SINR estimates generated by each finger of the rake demodulator. There are several other methods of SINR or SNR estimation [62]. Measured functions of the channel state used in adaptive rate control include bit, packet, and frame successes or failures. In the IEEE 802.11 standard, the channel measurement comprises a certain number of consecutive transmission successes or a frame loss.

8.4 DS-CDMA Cellular Uplinks

Uplink Power Control

In cellular DS-CDMA networks, the near-far problem is critical only on the uplink because on the downlink, the base station transmits orthogonal signals synchronously to each mobile associated with it. For cellular networks, the usual solution to the near-far problem of uplinks is *power control*, whereby all mobiles in a cell or sector regulate their power levels so that the powers arriving at a base station are all equal. Synchronous, orthogonal uplink signals in a sector ensure the absence of near-far problems and intrasector interference in principle. However, power control is needed to limit the potential intrasector interference caused by asynchronous uplink signals, synchronization errors in synchronous uplink signals, strong multipath signals, and hardware imperfections. Since solving the near-far problem is essential to the viability of a cellular DS-CDMA network, the accuracy of the power control is a crucial issue.

An *open-loop method of power control* in a cellular network causes a mobile to adjust its transmitted power in accordance with changes in the average received power of a *pilot signal* transmitted by the base station. Open-loop power control is effective if the propagation losses on the uplinks and downlinks are nearly the same. Whether they are or not is influenced by the duplexing method used to allow transmissions on both links. *Time-division duplexing* is a half-duplex method that assigns closely spaced but distinct time slots to the uplinks and downlinks. When time-division duplexing is used, the propagation losses on the uplinks and downlinks are nearly the same if the duplexing is sufficiently fast compared with changes in the network topology and fading. Thus, time-division duplexing is potentially compatible with an open-loop method of power control.

A full-duplex method allows the simultaneous transmission and reception of signals, but must isolate the received signals from the transmitted signals because of the vast disparities in power levels. Most cellular systems use *frequency-division duplexing*, which is a full-duplex method that assigns different frequencies to the uplinks and downlinks, thereby spectrally separating signals that could not be accommodated simultaneously in the same frequency band. When frequency-division duplexing is used, the frequency separation is generally wide enough that the channel transfer functions of the uplink and downlink are different. This lack of *link reciprocity* implies that average power measurements over the downlink do not provide reliable information for regulating the average power of subsequent uplink transmissions. Thus, a closed-loop method of power control is required when frequency-division duplexing is used.

A *closed-loop method of power control*, which is used by the WCDMA and CDMA2000 systems, attempts to compensate for both the propagation losses and the fading by a feedback mechanism. Closed-loop power control requires a base station to dynamically track the received power of a desired signal from a mobile and then transmit appropriate power-control information to that mobile. As the fading rate increases, the tracking ability and hence the power-control accuracy decline. This imperfect power control in the presence of fast fading is partially compensated for by the increased time diversity provided by the interleaving and channel coding, but some degree of power control must be maintained.

To implement closed-loop power control [16], the base station receives N known test bits from a mobile. These bits are processed to provide a power estimate, which is subtracted from a desired received power. The difference determines one or more power-control bits that are transmitted to the mobile every N received bits. The mobile then adjusts its transmitted power in accordance with the power-control bits.

When the received instantaneous power of the desired signal from a mobile is tracked, there are four principal error components. They are the quantization error due to the stepping of the transmitted power level at the mobile, the error introduced in the decoding of the power-control information at the mobile, the error in the power estimation at the base station, and the error caused by the processing and propagation delay. The processing and propagation delay is a source of error because the multipath propagation conditions change during the execution of the closed-loop power-control algorithm. The propagation delay is generally negligible compared with the processing delay. The processing-delay and the power-estimation errors are generally much larger than other errors.

Let s denote the maximum speed of a mobile in the network, f_c the carrier frequency of its direct-sequence transmitted signal, and c the speed of an electromagnetic wave. We assume that this signal has a bandwidth that is only a few percent of f_c, so that the effect of the bandwidth is negligible. The maximum Doppler shift or Doppler spread is $f_d = f_c s/c$, which is proportional to the fading rate. To obtain a small processing-delay error requires nearly constant values of the channel attenuation during the processing and propagation delay T_p. Thus, this delay must be much less than the coherence time (Section 6.2), which implies that

$$T_p << 1/f_d. \tag{8.39}$$

For example, if $s = 30$ m/s and $f_c = 1$ GHz, then $T_p << 10$ ms is required. Inequality (8.39) indicates that accurate power control becomes more difficult as the carrier frequency, and hence the Doppler spread, increases. To compensate for an increased Doppler spread, one might increase the spreading factor to decrease the impact of interference or increase the density of cells so that the network encounters more benign fading.

Let $\widehat{P}$ denote an estimate of p_0, the average received signal power from a mobile. Let σ_p^2 denote the variance of $\widehat{P}$. A lower bound on σ_p^2 can be determined by modeling the multiple-access interference as a white Gaussian process that increases the one-sided power spectral density (PSD) of the noise from N_0 to N_{0e}. The received signal from a mobile that is to be power-controlled has the form $\sqrt{p_0 T_m} s(t)$, where the observation interval is $[0, T_m]$, and

$$\int_0^{T_m} s^2(t)dt = 1. \tag{8.40}$$

Application of the Cramer-Rao inequality (F.48) of Appendix F.4 provides a lower bound on the variance σ_p^2 of an unbiased estimator of the power:

$$\sigma_p^2 \geq \left\{ \frac{2}{N_{0e}} \int_0^{T_m} \left[\frac{\partial}{\partial p_0} (\sqrt{p_0 T_m} s(t)) \right]^2 dt \right\}^{-1}. \tag{8.41}$$

Evaluating the lower bound and using (8.40), we obtain $\sigma_p^2 \geq N_{0e} p_0 / T_m$. Thus, if σ_p / p_0 is a specified measure of estimation accuracy, then

$$T_p > T_m \geq \frac{N_{0e}}{p_0} \left(\frac{\sigma_p}{p_0} \right)^{-2} \tag{8.42}$$

is required. Together, (8.39) and (8.42) not only constrain T_p but also indicate the Doppler spread that can be accommodated.

An iterative method of power estimation can be derived by applying the expectation-maximization algorithm (Section 9.1), which provides approximate maximum-likelihood estimates. Consider a direct-sequence system with QPSK. A binary sequence $x_i = \pm 1, i = 1, \ldots, N$, of test bits is transmitted over the additive white Gaussian noise (AWGN) channel. As shown in Section 2.5, the despreading produces the received vector $\mathbf{y} = [y_1 \ldots y_N]^T$ given by

$$\mathbf{y} = \sqrt{2\mathcal{E}_b}\mathbf{x} + \mathbf{n} \tag{8.43}$$

where $\mathcal{E}_b = p_0 T_b$ is the energy per bit, p_0 is the average power, T_b is the bit duration, N is the total number of inphase and quadrature bits in the two receiver branches, $\mathbf{x} = [x_1 \ldots x_N]^T$ is the vector of these bits, and $\mathbf{n} = [n_1 \ldots n_N]^T$ is the interference-and-noise vector. The components of $\mathbf{n}$ are assumed to approximate independent, identically distributed, zero-mean, Gaussian random variables with variance $V = N_{0e}$.

Let $\left\{\widehat{P}_i\right\}$ and $\left\{\widehat{V}_i\right\}$ denote sequences of estimators of p_0 and V, respectively, that are obtained from the received vector by applying the iterative expectation-maximization (EM) algorithm. As described in Section 9.1, we obtain

$$\widehat{P}_i = \frac{1}{2T_b}\left(\hat{A}_i\right)^2 \tag{8.44}$$

$$\hat{A}_{i+1} = \frac{1}{N}\sum_{k=1}^{N} y_k \tanh\left(\frac{\hat{A}_i y_k}{\widehat{V}_i}\right) \tag{8.45}$$

$$\widehat{V}_{i+1} = \frac{1}{N}\sum_{k=1}^{N} y_k^2 - \hat{A}_{i+1}^2. \tag{8.46}$$

A suitable set of initial values are

$$\hat{A}_0 = \frac{1}{N}\sum_{k=1}^{N} |y_k|, \quad \widehat{V}_0 = \frac{1}{N}\sum_{k=1}^{N} y_k^2 - \hat{A}_0^2. \tag{8.47}$$

After ν iterations, the final EM estimators $\widehat{P}_\nu$ and $\widehat{V}_\nu$ based on the received vector are computed. Then, the old received vector is discarded, and a new received vector is used to compute new EM estimators.

Network Topology

Although the subsequent analytical methods are applicable for arbitrary topologies, the cellular network model in this section enforces a minimum separation among the base stations for each *network realization*, which comprises a base-station placement, a mobile placement, and a shadowing realization. The model for both mobile and base-station placement is the *uniform clustering* model (Section 8.2), which entails a uniformly distributed placement within the network after certain regions have been excluded. The base stations and mobiles are confined to a finite area, which is assumed to be the interior of a circle of radius r_{net} and area πr_{net}^2. An *exclusion zone* of radius r_{bs} surrounds each base station, and no other base stations are allowed within this zone. Similarly, an exclusion zone of radius r_{m}, where no other mobiles are allowed, surrounds each mobile. The minimum separation between mobiles is generally much smaller than the minimum separation between the base stations. Base-station exclusion zones are primarily determined by economic considerations and the terrain, whereas mobile exclusion zones are determined by the need to avoid physical collisions. One way that the value of r_{bs} can be selected is by examining the locations of base stations in real networks and identifying the value of r_{bs} that provides the best statistical fit.

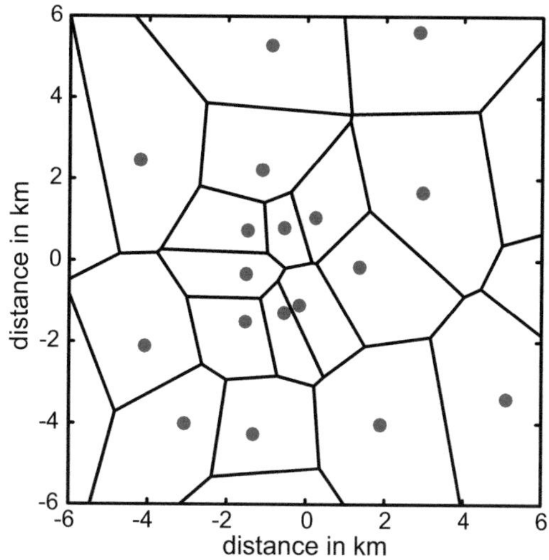

Fig. 8.12 Actual base station locations from a current cellular deployment. Base stations are represented by large circles, and cell boundaries are represented by thick lines [109]

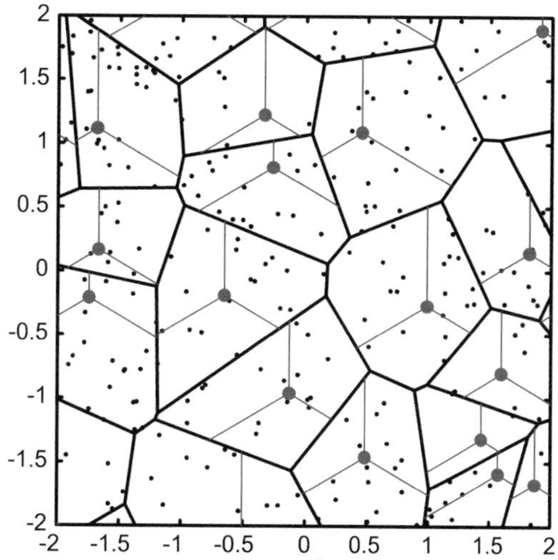

Fig. 8.13 Simulated base-station locations using a base-station exclusion zone $r_{bs} = 0.25$. The simulated positions of the mobiles are represented by small dots, the sector boundaries are represented by light lines, and the average cell load is 16 mobiles per cell. The mobile exclusion radius is $r_m = 0.01$, and no mobile is within distance $d_0 < r_m$ of any base station [109]

Figure 8.12 depicts the locations of actual base stations in a small city with hilly terrain. The base-station locations are given by the large filled circles, and the Voronoi cells are indicated in the figure. The minimum base-station separation is observed to be about 0.43 km. Figure 8.13 depicts a portion of a randomly generated network with average number of mobiles per cell equal to 16, a base-station exclusion radius $r_{bs} = 0.25$, and a mobile exclusion radius $r_m = 0.01$. We assume that r_{bs} and r_m exceed reference distance d_0, and that there are no mobiles within distance d_0 of any base station. The locations of the mobiles are represented by small dots, and dashed lines indicate the angular coverage of sector antennas. The general similarity of the two figures lends credibility to the uniform clustering model.

The downlinks of a cellular DS-CDMA network use orthogonal spreading sequences. Because the number of orthogonal spreading sequences available to a cell or cell sector is limited to G, the number of mobiles M_j with either uplink or downlink service using the base-station or sector antenna S_j is also limited to $M_j \leq G$. If there are $M_j > G$ mobiles, then some of these mobiles will either be refused service by antenna S_j or given service at a lower rate (through the use of an additional time-multiplexing). Two policies for handling this situation are the *denial policy* and the *reselection policy*. With the denial policy, the $M_j - G$ mobiles with the greatest path losses to the base station are denied service, in which case they are not associated with any sector antenna. With the reselection policy, each of the $M_j - G$ mobiles in an overloaded cell sector attempts to connect to the sector antenna with the next-lowest path loss out to a maximum reassociation distance d_{max}. If no suitably oriented sector antenna is available within distance d_{max}, the mobile is denied service. The tradeoff entailed in keeping these mobiles connected under the reselection policy is that the downlink area spectral efficiency decreases slightly for rate control and significantly for power control (since the base station must direct much of its transmit power toward the distant reassociated mobiles).

Outage Probability

For the subsequent analysis [109] and examples of DS-CDMA uplinks, three ideal sector antennas and sectors per base station, each covering $2\pi/3$ radians, are assumed. The mobile antennas are assumed to be omnidirectional. The network comprises C base stations and cells, $3C$ sectors $\{S_1, \ldots, S_{3C}\}$, and M mobiles $\{X_1, \ldots, X_M\}$. The scalar S_k represents the kth sector antenna, and the vector $\mathbf{S}_k$ represents its location. The scalar X_i represents the ith mobile and the vector $\mathbf{X}_i$ represents its location. Let $\mathcal{A}_k$ denote the set of mobiles *covered* by base-station or sector antenna S_k, and let $\mathcal{X}_k \subset \mathcal{A}_k$ denote the set of mobiles associated with antenna S_k. Each mobile is served by a single base-station or sector antenna. Let $\mathsf{g}(i)$ denote a function that returns the index of the antenna serving X_i so that $X_i \in \mathcal{X}_k$ if $\mathsf{g}(i) = k$. Usually, the antenna $S_{\mathsf{g}(i)}$ that serves mobile X_i is selected from among those that cover X_i to be the one with a minimum path loss in the absence of fading from X_i to the antenna. Thus, the antenna index is

$$\mathsf{g}(i) = \underset{k}{\operatorname{argmax}} \left\{ 10^{\xi_{i,k}/10} f\left(||\mathbf{S}_k - \mathbf{X}_i||\right), \; X_i \in \mathcal{A}_k \right\} \tag{8.48}$$

where $\xi_{i,k}$ is the shadowing factor for the link from X_i to S_k, and $f(\cdot)$ is given by (8.3). In the absence of shadowing, it is the antenna that is closest to X_i. In the presence of shadowing, a mobile may actually be associated with an antenna that is more distant than the closest one if the shadowing conditions are sufficiently better.

Consider a reference receiver of a sector antenna that receives a desired signal from a reference mobile within its cell and sector. Both intracell and intercell interference are received from other mobiles within the covered angle of the sector, but interference from mobiles outside the covered angle is negligible. Duplexing is assumed to prevent interference from other sector antennas. The varying propagation delays cause interference signals to be asynchronous with respect to the desired signal. We assume that a DS-CDMA network of asynchronous quadriphase direct-sequence systems has a constant effective spreading factor (Section 8.4) equal to G/h at each base-station, sector, or mobile receiver in the network.

Let $X_r \in \mathcal{X}_{g(r)}$ denote a reference mobile that transmits a desired signal to a reference receiver at sector antenna $S_{g(r)}$. The power of X_r at the reference receiver of $S_{g(r)}$ is not significantly affected by the spreading factor. The power of X_i, $i \neq r$, at the reference receiver, which is nonzero only if $X_i \in \mathcal{A}_{g(r)}$, is reduced by the factor $G_i = G/h$. We assume that path loss has a power-law dependence on distance and is perturbed by shadowing. When accounting for fading and path loss, the despread instantaneous power of X_i at the reference receiver of $S_{g(r)}$ is

$$\rho_i = \begin{cases} P_r g_r 10^{\xi_r/10} f\left(||\mathbf{S}_{g(r)} - \mathbf{X}_r||\right), & i = r \\ \left(\frac{h}{G}\right) P_i g_i 10^{\xi_i/10} f\left(||\mathbf{S}_{g(r)} - \mathbf{X}_i||\right), & i : X_i \in \mathcal{A}_{g(r)} \backslash X_r \\ 0, & i : X_i \notin \mathcal{A}_{g(r)} \end{cases} \tag{8.49}$$

where g_i is the power gain at the reference receiver due to Nakagami fading with parameter m_i, ξ_i is the *shadowing factor for the link from* X_i to $S_{g(r)}$, P_i is the power transmitted by X_i, and $\mathcal{A}_{g(r)} \backslash X_r$ is set $\mathcal{A}_{g(r)}$ with element X_r removed (required since X_r does not interfere with itself). We assume that the $\{g_i\}$ remain fixed for the duration of a time interval but vary independently from interval to interval (block fading).

The *activity probability* p_i is the probability that the ith mobile transmits during the same time interval as the reference signal. The instantaneous SINR at the reference receiver of sector antenna $S_{g(r)}$ when the desired signal is from $X_r \in \mathcal{X}_{g(r)}$ is

$$\gamma = \frac{\rho_r}{\mathcal{N} + \sum_{i=1, i \neq r}^{M} I_i \rho_i} \tag{8.50}$$

where $\mathcal{N}$ is the noise power, and I_i is a Bernoulli variable with probability $P[I_i = 1] = p_i$ and $P[I_i = 0] = 1 - p_i$. Substituting (8.49) and (8.3) into (8.50) yields

$$\gamma = \frac{g_r \Omega_r}{\Gamma^{-1} + \sum_{i=1, i \neq r}^{M} I_i g_i \Omega_i}, \quad \Gamma = \frac{d_0^{\alpha} P_r}{\mathcal{N}} \tag{8.51}$$

where Γ is the SNR due to a mobile located at unit distance when fading and shadowing are absent, and

$$\Omega_i = \begin{cases} 10^{\xi_r/10}||\mathbf{S}_{g(r)} - \mathbf{X}_r||^{-\alpha}, & i = r \\ \frac{hP_i}{GP_r}10^{\xi_i/10}||\mathbf{S}_{g(r)} - \mathbf{X}_i||^{-\alpha}, & i : X_i \in \mathcal{A}_{g(r)} \backslash X_r \\ 0, & i : X_i \notin \mathcal{A}_{g(r)} \end{cases} \tag{8.52}$$

is the normalized mean despread power of X_i received at $S_{g(r)}$, where the normalization is with respect to P_r. The set of $\{\Omega_i\}$ for reference receiver $S_{g(r)}$ is represented by the vector $\mathbf{\Omega} = \{\Omega_1, \ldots, \Omega_M\}$.

Let β denote the minimum instantaneous SINR required for reliable reception of a signal from X_r at the receiver of its serving sector antenna $S_{g(r)}$. An *outage* occurs when the SINR of a signal from X_r falls below β. The value of β is a function of the rate R of the uplink, which is expressed in units of information bits per channel use. The relationship between β and R depends on the modulation and coding schemes used, and typically only a discrete set of R can be selected.

Let m_0 denote the positive-integer Nakagami parameter for the link from X_r to $S_{g(r)}$, and let m_i denote the Nakagami parameter for the link from X_i to $S_{g(r)}, i \neq r$. Conditioning on $\mathbf{\Omega}$, the outage probability $\epsilon(\mathbf{\Omega})$ of a desired signal that arrives at the reference receiver $S_{g(r)}$ is given by (8.28), (8.30), and (8.31) with $\Omega_0 = \Omega_r$, and

$$H_t(\mathbf{\Omega}) = \sum_{\substack{\ell_i \geq 0 \\ \sum_{i=1}^{M} \ell_i = t}} \prod_{i=1, i \neq r}^{M} G_{\ell_i}(i). \tag{8.53}$$

A typical power-control policy for DS-CDMA networks is to select the transmit power $\{P_i\}$ for all mobiles in the set $\mathcal{X}_j$ such that, after compensation for shadowing and power-law attenuation, each mobile's transmission is received at sector antenna S_j with the same average power P_a. For such a power-control policy, each mobile in $\mathcal{X}_j$ transmits with an average power P_i that satisfies

$$P_i 10^{\xi_{i,j}/10} f\left(||\mathbf{S}_j - \mathbf{X}_i||\right) = P_a, \;\; X_i \in \mathcal{X}_j \tag{8.54}$$

where $\xi_{i,j}$ is the shadowing factor for the link from X_i to S_j, and $f(\cdot)$ is given by (8.3). To accomplish the power-control policy, sector-antenna receivers estimate the average received powers of their associated mobiles. Feedback of these estimates enables the associated mobiles to change their transmitted powers so that all received powers are approximately equal on average.

To provide mobiles with some flexibility in exploiting favorable channel conditions, *fractional power control* of the constrained local-mean power implies that

$$P_i \left[10^{\xi_{i,j}/10} f\left(||\mathbf{S}_j - \mathbf{X}_i||\right)\right]^{\delta} = P_a, \;\; X_i \in \mathcal{X}_j, \;\; 0 < \delta < 1 \tag{8.55}$$

where δ is the *power-control parameter*. If $\delta = 0$, there is no power control, and transmitter powers are all equal. If $\delta = 1$, full power control forces the received local-mean powers from all mobiles to be equal. If $0 < \delta < 1$, then decreasing

δ improves the performance of some mobiles in a sector while increasing the interference in neighboring sectors.

For a reference mobile X_r, the interference at the reference receiver of sector antenna $S_{g(r)}$ is from the mobiles in the set $\mathcal{A}_{g(r)} \backslash X_r$. The mobiles in this set can be partitioned into two subsets. The first subset $\mathcal{X}_{g(r)} \backslash X_r$ comprises the *intracell interferers*, which are the other mobiles in the same cell and sector as the reference mobile. The second subset $\mathcal{A}_{g(r)} \backslash \mathcal{X}_{g(r)}$ comprises the *intercell interferers*, which are the mobiles covered by sector antenna $S_{g(r)}$ but associated with a cell sector other than $\mathcal{X}_{g(r)}$. Since both the intracell and the intercell interference signals arrive asynchronously, they cannot be suppressed by using orthogonal spreading sequences.

Consider intracell interference. Since P_r and P_i are obtained from (8.55) for all mobiles in $\mathcal{X}_{g(r)}$,

$$\frac{P_i}{P_r} = \left[\frac{10^{\xi_r/10} ||\mathbf{S}_{g(r)} - \mathbf{X}_r||^{-\alpha}}{10^{\xi_i/10} ||S_{g(r)} - X_i||^{-\alpha}} \right]^{\delta}, \quad X_i \in \mathcal{X}_{g(r)} \backslash X_r \tag{8.56}$$

where (8.3) has been used. Therefore, (8.52) implies that the normalized received power of the intracell interferers is

$$\Omega_i = \frac{h}{G} \Omega_r \left[\frac{10^{\xi_{i,}/10} ||\mathbf{S}_{g(r)} - \mathbf{X}_i||^{-\alpha}}{10^{\xi_r/10} ||\mathbf{S}_{g(r)} - \mathbf{X}_i||^{-\alpha}} \right]^{1-\delta}, \quad X_i \in \mathcal{X}_{g(r)} \backslash X_r. \tag{8.57}$$

Although the number of mobiles $M_{g(r)}$ in the cell sector must be known to compute the outage probability, the locations of these mobiles in the cell are irrelevant to the computation of the $\{\Omega_i\}$ of the intracell interferers.

Considering intercell interference, the set $\mathcal{A}_{g(r)} \backslash \mathcal{X}_{g(r)}$ can be further partitioned into sets $\mathcal{A}_{g(r)} \cap \mathcal{X}_k$, $k \neq g(r)$, containing the mobiles covered by sector antenna $S_{g(r)}$ and associated with some other sector antenna S_k. For those mobiles in $\mathcal{A}_{g(r)} \cap \mathcal{X}_k$, fractional power control implies that

$$P_i \left[10^{\xi_{i,\mathsf{k}}/10} f\left(||\mathbf{S}_{\mathsf{k}} - \mathbf{X}_i|| \right) \right]^{\delta} = P_a, \;\; X_i \in \mathcal{X}_k \cap A_{g(r)}, \;\; k \neq g(r) \tag{8.58}$$

where $\xi_{i,k}$ is the shadowing factor for the link from X_i to S_k. Substituting (8.58), (8.54) with $i = r$, and (8.3) into (8.52) yields

$$\Omega_i = \frac{h}{G} 10^{\xi_i'/10} \left(\frac{||\mathbf{S}_{g(r)} - \mathbf{X}_i|| ||\mathbf{S}_{g(r)} - \mathbf{X}_r||}{||\mathbf{S}_k - \mathbf{X}_i||} \right)^{-\alpha}$$

$$\xi_i' = \xi_i + \xi_r - \xi_{i,k}, \;\; X_i \in \mathcal{X}_k \cap A_{g(r)}, \;\; k \neq g(r), \; i \in \mathcal{A}_{g(r)} \backslash \mathcal{X}_{g(r)} \tag{8.59}$$

which is the normalized intercell interference power at the reference sector antenna due to interference from mobile i.

Uplink Rate Control

In addition to controlling the transmitted power, the rate R_k of each uplink k needs to be selected. Due to the irregular network geometry, which results in cell sectors of variable areas and numbers of intracell mobiles, the amount of interference received by a sector antenna can vary dramatically from one sector to another. With a fixed rate, or equivalently, a fixed SINR threshold β for each sector k, the result is a highly variable outage probability ϵ_k. An alternative to using a fixed rate for the entire network is to adapt the rate of each uplink to satisfy an outage constraint or maximize the throughput of each uplink. Assuming the use of a capacity-approaching code, two-dimensional signals transmitted over an AWGN channel, and Gaussian interference, (7.2) indicates that the SINR threshold corresponding to rate R nearly equal to the channel capacity is

$$\beta = 2^R - 1. \tag{8.60}$$

Several performance measures are of interest. The *average outage probability* over all M uplinks of the network is

$$\mathbb{E}[\epsilon] = \frac{1}{M}\sum_{k=1}^{M} \epsilon_k. \tag{8.61}$$

The *throughput* of the kth uplink is

$$T_k = R_k(1 - \epsilon_k) \tag{8.62}$$

and represents the bits per channel use of successful transmissions. The *average throughput* over all M uplinks of the network is

$$\mathbb{E}[T] = \frac{1}{M}\sum_{k=1}^{M} R_k(1 - \epsilon_k). \tag{8.63}$$

Example 10 To illustrate the influence of rate on performance, consider the following example. The network has $C = 50$ base stations and $M = 400$ mobiles placed in a circular network of radius $r_{\text{net}} = 2$. The base-station exclusion zones have radius $r_{\text{bs}} = 0.25$, whereas the mobile exclusion zones have radius $r_{\text{m}} = 0.01$. The spreading factor is $G = 16$, and the chip factor is $h = 2/3$. Since $M/C = G/2$, the network is characterized as being *half loaded*. The SNR is $\Gamma = 10\,\text{dB}$, the activity factor is $p_i = 1$, the path-loss exponent is $\alpha = 3$, the power-control parameter is $\delta = 1$, and the shadowing has a lognormal distribution with standard deviation $\sigma_s = 8\,\text{dB}$. A *distance-dependent fading* model is assumed, where the Nakagami parameter m_i of the link from X_i to S_j is

$$m_i = \begin{cases} 3, & ||\mathbf{S}_j - \mathbf{X}_i|| \le r_{\text{bs}}/2 \\ 2, & r_{\text{bs}}/2 < ||\mathbf{S}_j - \mathbf{X}_i|| \le r_{\text{bs}} \\ 1, & ||\mathbf{S}_j - \mathbf{X}_i|| > r_{\text{bs}}. \end{cases} \tag{8.64}$$

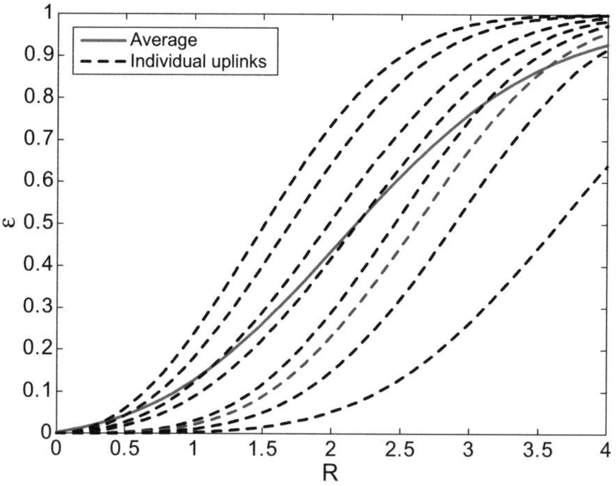

Fig. 8.14 Outage probability of eight randomly selected uplinks (dashed lines) along with the average outage probability for the entire network (solid line). The results are for a half-loaded network ($M/C = G/2$), with distance-dependent fading and shadowing ($\sigma_s = 8$ dB) and are shown as a function of the rate R [109]

The distance-dependent-fading model characterizes the situations where mobiles close to the base station are in the line-of-sight, but mobiles farther away are not.

Figure 8.14 shows the outage probability as a function of rate R. The dashed lines in Figure 8.14 were generated by selecting eight random uplinks and computing the outage probability for each using the threshold given by (8.60). Despite the use of power control, there is considerable variability in the outage probability. The outage probabilities $\{\epsilon_k\}$ were computed for all M uplinks in the network, and the average outage probability $\mathbb{E}[\epsilon]$ is displayed as a solid line in the figure. Figure 8.15 shows the throughputs as functions of the rate for the same eight uplinks with outages shown in Figure 8.14 and shows the average throughput $\mathbb{E}[T]$ computed for all M uplinks in the network. □

A *fixed-rate policy* requires all uplinks in the system to use the same rate: $R_k = R$ for all uplinks. A *maximal-throughput fixed-rate* (MTFR) policy selects the fixed rate that maximizes the average throughput. With respect to the example shown in Figure 8.15, this choice corresponds to selecting the R that maximizes the solid curve, which occurs at $R = 1.81$. However, at the rate that maximizes throughput, the corresponding outage probability could be unacceptably high. When $R = 1.81$ in the example, the corresponding average outage probability is $\mathbb{E}[\epsilon] = 0.37$, which is too high for many applications. As an alternative to maximizing throughput, the *outage-constrained fixed-rate* (OCFR) policy selects the rate R that satisfies an outage constraint ζ so that $\mathbb{E}[\epsilon] \leq \zeta$. For instance, setting $R = 0.84$ in the example satisfies an average outage constraint $\zeta = 0.1$ with equality.

If R is selected to satisfy an average outage constraint, the outage probabilities of the individual uplinks vary. Furthermore, selecting R to maximize the average

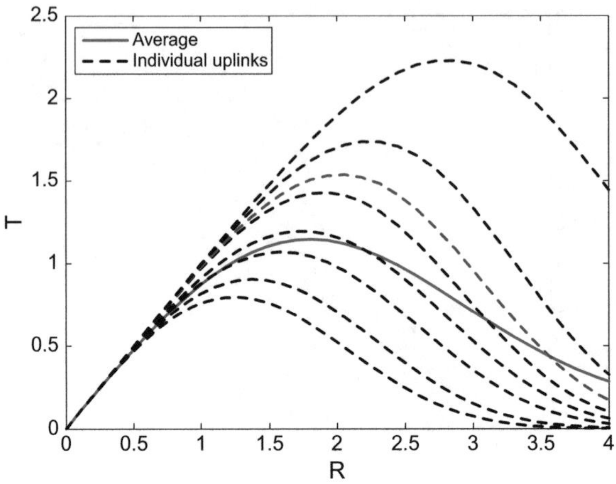

Fig. 8.15 Throughput of eight randomly selected uplinks (dashed lines) along with the average throughput for the entire network (solid line). System parameters are the same as those used to generate Figure 8.14 [109]

throughput does not generally maximize the throughput of the individual uplinks. These issues can be alleviated by selecting each rate R_k independently for the different uplinks. The *outage-constrained variable-rate* (OCVR) policy selects each uplink rate to satisfy the outage constraint that $\epsilon_k \leq \zeta$ for all k. Alternatively, the *maximal-throughput variable-rate* (MTVR) policy selects each uplink rate to maximize the throughput of the uplink, i.e., $R_k = \arg\max T_k$ for each uplink, where the maximization is over all possible rates. Both policies can be implemented by having the base station track the outage probabilities or throughputs of each uplink and feeding back rate-control commands to ensure that the target performance is achieved. The outage probability can easily be found by encoding the data with an error-detection code and declaring an outage whenever a frame fails a check.

For both the OCVR and MTVR rate-control policies, we assume that the code rate is adapted by maintaining the duration of channel symbols while varying the number of information bits per channel symbol. The spreading factor G and symbol rate are held constant; thus, there is no change in bandwidth. However, a major drawback with rate control is that the rates required to maintain a specified outage probability vary significantly among the mobiles in the network. This variation results in low throughput for some mobiles, particularly those located at the edges of the cells, whereas other mobiles have a high throughput. Unequal throughputs may not be acceptable when a mobile may be stuck or parked near a cell edge for a long time, and an interior mobile may have more allocated throughput than it needs.

Although the outage probability, throughput, and rate characterize the performance of a single uplink, they do not quantify the total data flow in the network because they do not account for the number of uplink users that are served. By taking into account the number of mobiles per unit area, the total data flow in a given area can be characterized by the *average area spectral efficiency*, defined as

$$\overline{\mathcal{A}} = \lambda \mathbb{E}[T] = \frac{\lambda}{M} \sum_{k=1}^{M} R_k(1 - \epsilon_k) \tag{8.65}$$

where $\lambda = M/\pi r_{\mathsf{net}}^2$ is the maximum density of transmissions in the network, and the units are bits per channel use per unit area. Average area spectral efficiency can be interpreted as the maximum spatial efficiency of transmissions, i.e., the maximum rate of successful data transmission per unit area.

In the following example, performance metrics are calculated by using a Monte Carlo approach with 1000 simulation trials as follows. In each simulation trial, a realization of the network is obtained by placing C base stations and M mobiles within the disk of radius r_{net} according to the uniform clustering model with minimum base-station separation r_{bs} and minimum mobile separation r_{m}. The path loss from each base station to each mobile is computed by applying randomly generated shadowing factors. The set of mobiles associated with each cell sector is determined. Assuming that the number of mobiles served in a cell sector cannot exceed G, which is the number of orthogonal sequences available for the downlink, any excess in the number of mobiles in the cell sector results in service denials. At each sector antenna, the power-control policy is applied to determine the power the antenna receives from each mobile that it serves. In each cell sector, the rate-control policy is applied to determine the rate and threshold.

For each uplink, the outage probability $\epsilon(\boldsymbol{\Omega})$ is computed for the rate-control policies and using (8.57) and (8.59). Equation (8.61) is applied to calculate the average outage probability for the network realization. Equation (8.62) is applied to calculate the throughputs according to the MTFR, OCFR, MTVR, or OCVR network policies, and then (8.63) is applied to compute the average throughput for the network realization. Finally, multiplying by the mobile density λ and averaging over all simulated network realizations with different topologies yield the average area spectral efficiency $\overline{\mathcal{A}}$.

Example 11 The network has $C = 50$ base stations placed in a circular network of radius $r_{\mathsf{net}} = 2$, and the base-station exclusion zones are set to have radius $r_{\mathsf{bs}} = 0.25$. A variable number M of mobiles are placed within the network using exclusion zones of radius $r_{\mathsf{m}} = 0.01$. The SNR is $\Gamma = 10\,\text{dB}$, the activity factor is $p_i = 1$, and the path-loss exponent is $\alpha = 3$. Two fading models are considered: *Rayleigh fading*, where $m_i = 1$ for all i, and *distance-dependent fading*, which is described by (8.64). Both unshadowed and shadowed ($\sigma_s = 8\,\text{dB}$) environments are considered. The chip factor is $h = 2/3$, and the spreading factor is $G = 16$. Figure 8.16 illustrates the variability of the $\{\epsilon_k\}$ with respect to all uplinks and 1000 simulation trials under a OCFR policy by plotting its complementary distribution function with $R = 2$, three network loads, distance-dependent fading, and shadowing.

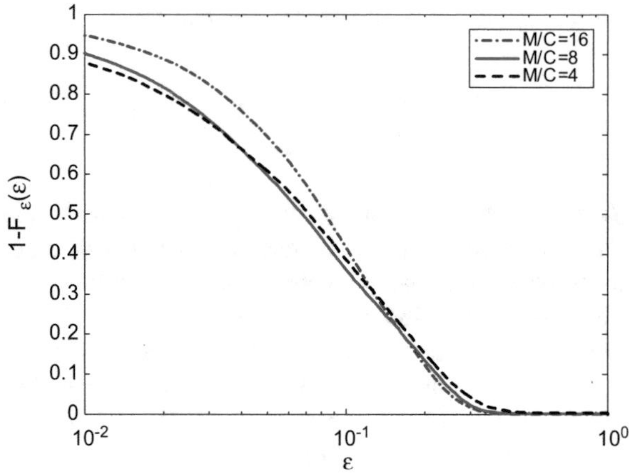

Fig. 8.16 Complementary distribution function of the outage probability using an OCFR policy, $R = 2$, three network loads, distance-dependent fading, and shadowing ($\sigma_s = 8$ dB) [109]

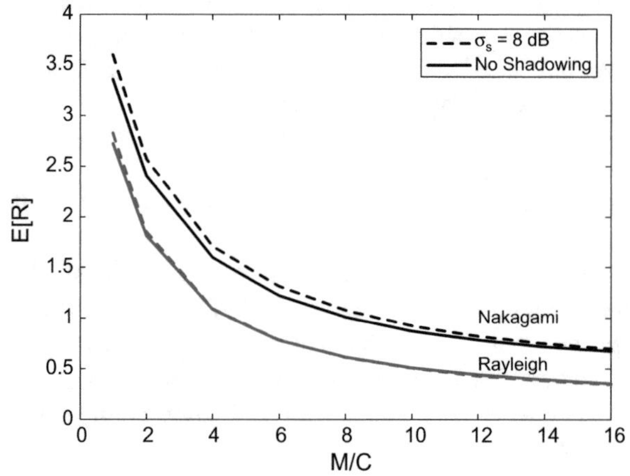

Fig. 8.17 Average rate of the OCVR policy as a function of the load M/C for both Rayleigh and distance-dependent Nakagami fading, and both shadowed ($\sigma_s = 8$ dB) and unshadowed cases [109]

With the OCVR policy, the rate R_k (equivalently, β_k) of each uplink is selected such that the outage probability does not exceed $\zeta = 0.1$. Although the outage probability is fixed, the rates of the uplinks are variable. Let $E[R]$ denote the average rate over all uplinks and 1000 simulation trials. In Figure 8.17, $E[R]$ is shown as a function of the load M/C. In Figure 8.18, the variability of R_k is illustrated by showing the complementary distribution function of the rate for a fully loaded system ($M/C = G = 16$) in both Rayleigh fading and distance-dependent Nakagami fading, and both with and without shadowing. The fairness of the system

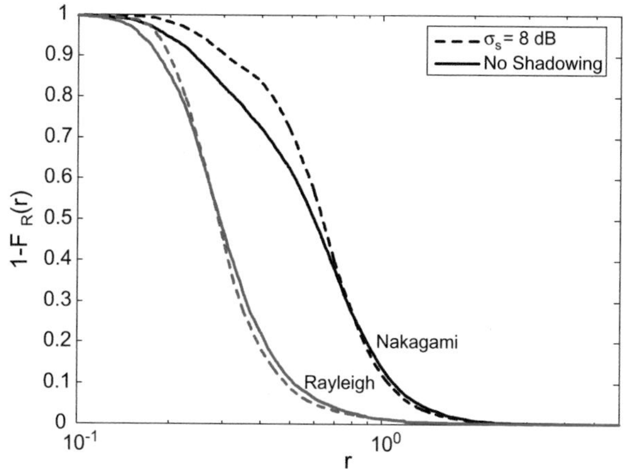

Fig. 8.18 Complementary distribution function of the rate for a fully loaded system ($M/C = G$) under the OCVR policy in Rayleigh and distance-dependent Nakagami fading, and both shadowed ($\sigma_s = 8$ dB) and unshadowed cases [109]

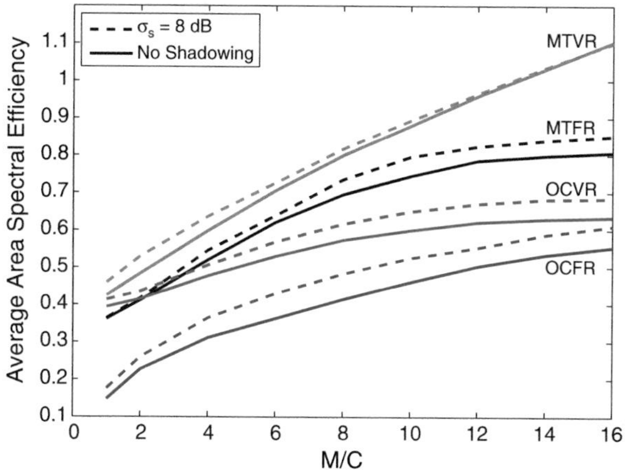

Fig. 8.19 Average area spectral efficiency for the four network policies as a function of the load M/C for distance-dependent fading and both shadowed ($\sigma_s = 8$ dB) and unshadowed cases [109]

can be determined from this figure, which shows the percentage of uplinks that meet a particular rate requirement. In both figures, the major impact of the much more severe Rayleigh fading is apparent.

Figure 8.19 shows the average area spectral efficiency $\overline{\mathcal{A}}$ of the four network policies in distance-dependent fading, both with and without shadowing, as a function of the load M/C. For the OCFR and MTFR policies, the optimal rates are determined for each simulation trial, and then $\bar{\mathcal{A}}$ is computed by averaging over 1000 simulation trials. For the OCVR and MTVR policies, the uplink rate

R_k of each uplink is maximized subject to an outage constraint or maximum-throughput requirement, respectively, and then $\bar{A}$ is computed by averaging over 1000 simulation trials. Although the average area spectral efficiencies of the MTFR and MTVR policies are potentially superior to those of the OCFR and OCVR policies, this advantage comes at the cost of variable and high values of the $\{\epsilon_k\}$, which are generally too large for most applications. The bottom pair of curves in Figure 8.19 indicate that the OCVR policy has a higher average area spectral efficiency than the OCFR policy.

Calculations indicate that increases in $\alpha, G/h$, and/or r_{bs} cause increases in $\overline{A}$ for all four network policies in distance-dependent fading, both with and without shadowing. □

8.5 DS-CDMA Cellular Downlinks

A DS-CDMA downlink differs from an uplink in at least four significant ways. First, the sources of interference are many mobiles for an uplink, whereas the sources are a few base stations for a downlink. Second, the orthogonality and the synchronous timing of all transmitted signals prevents significant intracell interference on the downlinks. When the uplink signals arriving at a base station are asynchronous or there are synchronization errors, orthogonal signals do not prevent intracell interference. Third, sectorization is a critical factor in uplink performance, whereas it is expendable or at most of minor importance in downlink performance. Fourth, base stations are equipped with better transmit high-power amplifiers and receive low-noise amplifiers than the mobiles. The net effect is that the operational SNR is typically more than 5 dB lower for an uplink than for a downlink. The overall impact of the differences between uplinks and downlinks is superior performance for the downlinks.

Along with all the signals transmitted to mobiles associated with it, a base station transmits a pilot signal over the downlinks. A mobile, which is usually associated with the base station from which it receives the largest pilot signal, uses the pilot to identify a base station or sector, to initiate uplink power control, to estimate the attenuation, phase shift, and delay of each significant multipath component, and to assess the downlink power-allocation requirement of the mobile.

A base station synchronously combines and transmits the pilot and all the signals destined for mobiles associated with it. Consequently, all the signals fade together at each mobile location, and the use of orthogonal spreading sequences (Section 7.2) prevents intracell interference and hence a near-far problem on a downlink. There is interference caused by asynchronously arriving multipath components. However, since these components are weaker than the main signal, they are suppressed by the despreading process, and hence have a negligible effect. Intercell interference, which is the dominant source of performance degradation, is caused by interference signals from other base stations that arrive at a mobile asynchronously and fade independently.

Outage Probability

In the analysis of downlink performance [113], we assume that there is no sectorization. The network comprises C cellular base stations $\{S_1, \ldots, S_C\}$ and M mobiles $\{X_1, \ldots, X_M\}$ placed in the interior of a circle of radius r_{net} and area $A_{net} = \pi r_{net}^2$. The signal transmitted by base station S_i to mobile $X_k \in \mathcal{X}_i$ has average power $P_{i,k}$. We assume that all base stations transmit with a common total power P_0 such that

$$\frac{1}{1-f_p} \sum_{k:X_k \in \mathcal{X}_i} P_{i,k} = P_0, \quad i = 1, 2, \ldots, C \tag{8.66}$$

where f_p is the fraction of the base-station power reserved for pilot signals needed for synchronization and channel estimation. The signal transmitted by base station S_i to mobile $X_k \notin \mathcal{X}_i$ has average power P_0.

The despreading process in the receiver of the reference mobile reduces the interference from base stations other than that associated with the mobile. After accounting for fading, path loss, and the effective spreading factor, the despread instantaneous powers transmitted by $S_{g(r)}$ and S_i at reference mobile X_r are

$$\rho_r = P_{j,r} g_j 10^{\xi_{j,r}/10} f\left(||\mathbf{S}_j - \mathbf{X}_r||\right), \; j = \mathsf{g}(r) \tag{8.67}$$

$$\rho_i = \frac{h}{G} P_0 g_i 10^{\xi_{i,r}/10} f\left(||\mathbf{S}_i - \mathbf{X}_r||\right) \; i \neq j \tag{8.68}$$

respectively, where g_i is the power gain due to fading, and $\xi_{i,r}$ is the shadowing factor for the link from S_i to the reference mobile, $f(\cdot)$ is a path-loss function given by (8.3), and $j = \mathsf{g}(r)$ is defined by (8.48).

The *activity probability* p_i is the probability that a base-station is actively transmitting. The instantaneous SINR at the reference mobile X_r associated with base station S_{j} is

$$\gamma = \frac{\rho_{\mathrm{r}}}{\mathcal{N} + \sum_{i=1, i\neq j}^{M} I_i \rho_i} \tag{8.69}$$

where $\mathcal{N}$ is the noise power, and I_i is a Bernoulli variable with probability $P[I_i = 1] = p_i$ and $P[I_i = 0] = 1 - p_i$. Substituting (8.67), (8.68), and (8.3) into (8.69) yields

$$\gamma = \frac{g_{\mathrm{r}} \Omega_r}{\Gamma^{-1} + \sum_{i=1, i\neq j}^{M} I_i g_i \Omega_i}, \quad \Gamma = \frac{d_0^{\alpha} P_0}{\mathcal{N}} \tag{8.70}$$

where Γ is the SNR at a mobile located at unit distance from its sector antenna when fading and shadowing are absent, and

$$\Omega_r = \frac{P_{\mathrm{j},r}}{P_0} 10^{\xi_{j,r}/10} ||\mathbf{S}_\mathrm{j} - \mathbf{X}_r||^{-\alpha},\ j = \mathsf{g}(r) \tag{8.71}$$

$$\Omega_i = \frac{h}{G} 10^{\xi_{i,r}/10} ||\mathbf{S}_i - \mathbf{X}_r||^{-\alpha},\ i \neq j. \tag{8.72}$$

Let β denote the minimum SINR required by the reference mobile X_r for reliable reception and $\mathbf{\Omega} = \{\Omega_1, \ldots, \Omega_C\}$ represent the set of normalized despread base-station powers received. An *outage* occurs when the SINR falls below β. Let m_0 denote the positive-integer Nakagami parameter for the link from S_j to X_r, and let m_i denote the Nakagami parameter for the link from S_i to X_r, $i \neq \mathrm{j}$. Conditioning on $\mathbf{\Omega}$, the outage probability $\epsilon(\mathbf{\Omega})$ of a desired signal from S_j that arrives at the reference receiver of X_r is given by (8.28), (8.30), and (8.31) with $\Omega_0 = \Omega_\mathrm{r}$, $M = C$, and

$$H_t(\mathbf{\Omega}) = \sum_{\substack{\ell_i \geq 0 \\ \sum_{i=1}^{M} \ell_i = t}} \prod_{i=1, i\neq \mathrm{j}}^{C} G_{\ell_i}(i). \tag{8.73}$$

Downlink Rate Control

A key consideration in the operation of the network is the manner in which the total power P_0 transmitted by a base station is shared by the mobiles it serves. A simple and efficient way of allocating P_0 is with an *equal-share* policy, which involves base station S_i transmitting to mobile $X_k \in \mathcal{X}_i$ with power

$$P_{i,k} = \frac{(1 - f_p) P_0}{K_i},\quad X_k \in \mathcal{X}_i \tag{8.74}$$

where K_i is the number of mobiles associated with S_i. Under this policy, the downlink SINR varies dramatically from close mobiles to more distant ones. If a common SINR threshold is used by all mobiles, then the outage probability is likewise highly variable. Instead of using a common threshold, the threshold β_k of mobile X_k can be selected individually such that the outage probability of mobile X_k satisfies the constraint $\epsilon_k = \hat{\epsilon}$.

For a given threshold β_k, there is a corresponding transmission rate R_k that can be supported. Assuming the use of a capacity-approaching code, two-dimensional signals transmitted over an AWGN channel, and Gaussian interference, (8.60) indicates that the maximum rate supported by a threshold β_k is $R_k = \log_2(1 + \beta_k)$. With an equal power share, the number of mobiles K_i in the cell is first determined, and then the power share given to each mobile is found from (8.74). For each mobile in the cell, the corresponding β_k that achieves the outage constraint $\epsilon_k = \hat{\epsilon}$ is found by inverting the equation for $\epsilon(\mathbf{\Omega})$. Once β_k is found, the supported R_k is found.

The rate R_k is adapted by changing the number of information bits per channel symbol. The spreading factor G and symbol rate are held constant; thus, there is no change in bandwidth. Because this policy involves fixing the transmit power and then determining the rate for each mobile that satisfies the outage constraint, it is called *downlink rate control*.

Under outage constraint $\epsilon_k = \hat{\epsilon}$, the performance of a given network realization is largely determined by the set of achieved rates $\{R_k\}$ of the M mobiles in the network. Because the network realization is random, it follows that the set of rates is also random. Let the random variable R represent the rate of an arbitrary mobile. The statistics of R can be found for a given class of networks using a Monte Carlo approach as follows. First, a realization of the network is produced by placing C base stations and M mobiles within the network according to the uniform clustering model with minimum base-station separation r_{bs} and minimum mobile separation r_m. The path loss from each base station to each mobile is computed with randomly generated shadowing factors if shadowing is present. The set of mobiles associated with each base station is determined with a reselection policy if one is used. At each base station, the power-allocation policy is applied to determine the power it transmits to each mobile that it serves. For downlink rate control, after setting the outage probability equal to the outage constraint, an inversion provides the SINR threshold for each mobile in the cell. For each SINR threshold, the corresponding code rate is computed.

Downlink Power Control

Downlink power control entails power allocation by the base station in a manner that meets the requirements of the individual mobiles associated with it. Although there is no near-far problem on the downlinks, power control is still desirable to enhance the received power of a mobile during severe fading or when a mobile is near a cell edge. A major drawback with rate control is that the rates provided to the different mobiles in the network vary significantly. This variation results in unfairness to some mobiles, particularly those located at the edges of the cells. To ensure fairness, R_k could be constrained to be the same for all $X_k \in \mathcal{X}_i$. The common rate of a given cell is found by determining the value of R_k that allows the outage constraint $\epsilon_k = \hat{\epsilon}$ for all $X_k \in \mathcal{X}_i$ and the power constraint (8.66) to be simultaneously met. Because the power transmitted to each mobile is varied while holding the rate constant for all mobiles in the cell, this policy is called *downlink power control*. Although the rate is the same for all mobiles within a given cell, it may vary from cell to cell.

Let $E[R]$ represent the mean value of the variable R, which can be found by numerically averaging the values of R over the mobiles and the network realizations. Under this definition of $E[R]$ and an outage constraint, the *average area spectral efficiency* $\overline{\mathcal{A}}$ *for downlinks* is defined as

$$\overline{\mathcal{A}} = \lambda\,(1 - \hat{\epsilon})\,E[R] \tag{8.75}$$

where $\lambda = C/A_{net}$ is the maximum density of base-station transmissions in the network.

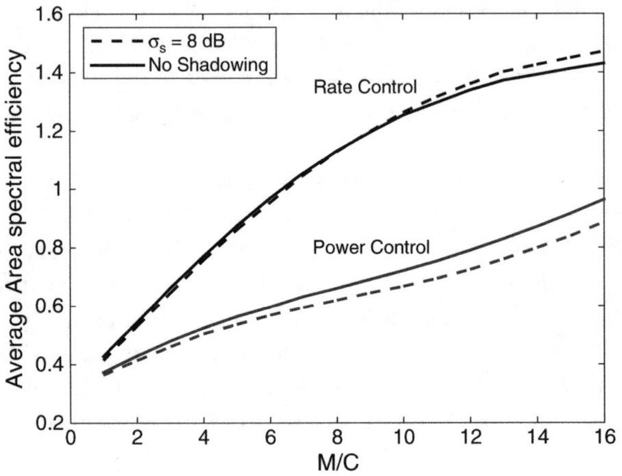

Fig. 8.20 Average area spectral efficiency as a function of M/C with rate control and power control [113]

Example 12 Consider a network with $C = 50$ base stations placed in a network of radius $r_{net} = 2$ with base-station exclusion zones of radius $r_{bs} = 0.25$. The *activity probability* is $p_i = 1$ for all base stations. A variable number M of mobiles are placed within the network using exclusion zones of radius $r_m = 0.01$. The outage constraint is set to $\hat{\epsilon} = 0.1$, and both power control and rate control are considered. A mobile in an overloaded cell is denied service. The SNR is set to $\Gamma = 10\,\text{dB}$, the fraction of power devoted to pilots is $f_p = 0.1$, and the spreading factor is set to $G = 16$ with chip factor $h = 2/3$. The propagation environment is characterized by a path-loss exponent $\alpha = 3$, and the Nakagami parameters are $m_{i,k} = 3$ for $i = \mathsf{g}(k)$ and $m_{i,k} = 1$ for $i \neq \mathsf{g}(k)$; that is, the signal from the serving base station experiences milder fading than the signals from the interfering base stations. This model is realistic because the signal from the serving base station is likely to be in the line of sight, whereas typically the interfering base stations are not.

Figure 8.20 shows the average area spectral efficiency, as a function of the ratio M/C, for rate control and power control in an unshadowed environment and in the presence of lognormal shadowing with $\sigma_s = 8$ dB. The figure shows that the average area spectral efficiency under rate control is higher than it is under power control. This disparity occurs because mobiles that are close to the base station are allocated extremely high rates under rate control, whereas with power control, mobiles close to the base station must be allocated the same rate as mobiles at the edge of the cell. As the network becomes denser (M/C increases), shadowing actually improves the performance with rate control, whereas it degrades the performance with power control. This improvement occurs because shadowing can sometimes cause the signal power of the base station serving a mobile to increase, while the powers of the interfering base stations are reduced. The effect of favorable shadowing is equivalent to the mobile being located closer to its serving base station. When this

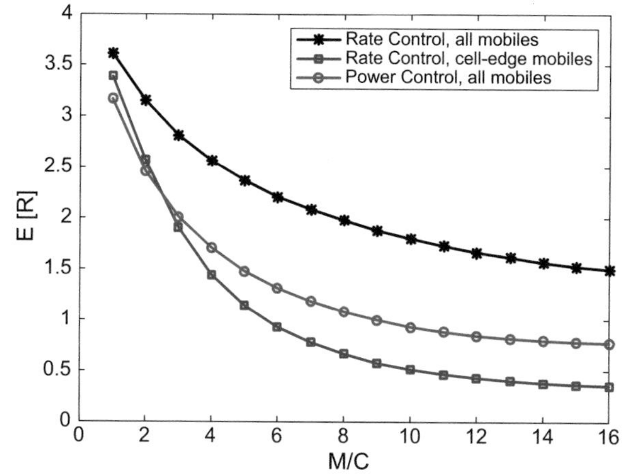

Fig. 8.21 Average rate as a function of M/C in the presence of shadowing with rate control and power control. For rate control, averaging is performed over all mobiles and over just the cell-edge mobiles. With power control, all mobiles in a cell are given the same rate [113]

occurs with rate control, the rate is increased, sometimes by a very large amount. Although shadowing does not induce extremely favorable conditions very often, when it does the improvement in rate is significant enough to cause the average to increase. In contrast, a single mobile with favorable shadowing conditions operating under power control continues to receive at the same code rate.

Although rate control offers a higher average rate than power control, the rates it offers are much more variable. This behavior can be seen in Figure 8.21, which compares the rates of all mobiles with those located at the cell edges. In particular, the figure shows the rate averaged across all mobiles for both power control and rate control, in addition to the rate averaged across just the cell-edge mobiles for rate control, where the cell-edge mobiles are defined to be the 5% of the mobiles that are furthest from their serving base station. The average rate of cell-edge mobiles is not shown for power control because each cell-edge mobile has the same rate as all the mobiles in the same cell. As seen in Figure 8.21, the performance of cell-edge mobiles is worse with rate control than it is with power control. □

Increasing the spreading factor G allows more interference to be suppressed, resulting in a reduced outage probability for a particular SINR threshold. This improvement can be used to increase the threshold, or equivalently, increase the rate.

8.6 FH-CDMA Mobile Ad Hoc Networks

Three major advantages of frequency hopping are that it can be implemented over a much larger frequency band than is possible with direct-sequence spreading, that the band can be divided into noncontiguous segments, and that frequency hopping provides inherent resistance to the near-far problem. These advantages of frequency hopping, particularly the near-far advantage, are decisive in many applications. For example, the Bluetooth system and combat net radios use frequency hopping primarily to avoid the near-far problem. Frequency hopping may be added to almost any communication system to strengthen it against interference or fading. Thus, frequency hopping may be applied to the set of carriers used in a multicarrier CDMA system or the subcarriers of an orthogonal frequency-division multiplexing (OFDM) system.

Since the probability of a collision in an asynchronous network can be decreased by increasing the number of frequency channels, it is highly desirable to choose a data modulation that has a compact spectrum and produces little spectral splatter. Continuous-phase frequency-shift keying (CPFSK) has these characteristics and provides a constant-envelope signal (Section 3.3). CPFSK is characterized by its modulation order, which is the number of possible tones, and by its modulation index h, which is the normalized tone spacing. For a fixed modulation order, the selection of h involves a tradeoff between bandwidth and performance. When CPSK is the data modulation in an FH-CDMA network, the number of frequency channels and hence resistance to frequency-hopping multiple-access interference generally increases with decreasing h, whereas the bit error probability for the AWGN channel generally decreases with increasing h.

Only a portion of the power in a frequency-hopping signal lies within its selected frequency channel. Let ψ represent the *fractional in-band power*, which is the fraction of power in the occupied frequency channel (Section 3.3). We assume that a fraction $K_s = (1-\psi)/2$, called the *adjacent-channel splatter ratio*, spills into each of the frequency channels that are adjacent to the one selected by a mobile. For most practical systems, this is a reasonable assumption, as the fraction of power that spills into frequency channels beyond the adjacent ones is negligible. A typical choice that limits the spectral splatter into an adjacent frequency channel is $\psi = 0.99$. However, the bit rate may be increased while fixing the channel bandwidth, thereby decreasing ψ and increasing K_s. Although the resulting increased adjacent-channel interference negatively affects performance, the increased bit rate can be used to support a lower-rate error-correcting code, thereby improving performance. Thus, there is a fundamental tradeoff involved in determining the fractional in-band power that should be contained within each frequency channel.

In the subsequent analysis and optimization for FH-CDMA mobile ad hoc networks [91], the basic network model is the same as in Section 8.1. The reference receiver is X_{M+1}, the reference transmitter is X_0, the network lies within a circle of radius r_{net}, and a nonzero radius r_{ex} defines an exclusion zone or guard zone. The fading gains $\{g_i\}$ remain fixed for the duration of a hop, but vary independently from hop to hop (block fading). Although the $\{g_i\}$ are independent, the channel from each transmitting mobile to the reference receiver can have a distinct Nakagami parameter m_i. A hopping band of W Hz is divided into L contiguous frequency channels, each of bandwidth W/L Hz. The mobiles independently select their transmit frequencies with equal probability. The source X_0 selects a frequency channel at the edge of the band with probability $2/L$ and a frequency channel in the interior of the band with probability $(L-2)/L$. Let $D_i \leq 1$ be the *duty factor* of X_i, which is the probability that the mobile transmits any signal.

We make the worst-case assumption that the hop dwell times of all network mobiles coincide. Two types of collisions are possible, *co-channel* collisions, which involve the source and interfering mobile selecting the same frequency channel, and *adjacent-channel* collisions, which involve the source and interfering mobile selecting adjacent channels. Let p_{c} and p_{a} denote the probabilities of a co-channel collision and an adjacent-channel collision respectively. Assume that $D_i = D$ is constant for every mobile. If X_i, $1 \leq i \leq M$, transmits a signal, then it uses the same frequency as X_0 with probability $1/L$. Since X_i transmits with probability D, the probability that it induces a co-channel collision is $p_{\mathrm{c}} = D/L$. The probability that X_i uses a frequency channel adjacent to that selected by X_0 is $1/L$ if X_0 selected a frequency channel at the edge of the band (in which case, there is only one adjacent channel); otherwise, the probability is $2/L$ (since there are two adjacent channels). It follows that, for a randomly chosen channel, the probability that X_i, $1 \leq i \leq M$, induces an adjacent-channel collision is

$$p_{\mathrm{a}} = D\left[\left(\frac{2}{L}\right)\left(\frac{L-2}{L}\right) + \left(\frac{1}{L}\right)\left(\frac{2}{L}\right)\right] = \frac{2D(L-1)}{L^2}. \tag{8.76}$$

Under the model described above, the instantaneous SINR at the reference receiver is

$$\gamma = \frac{\psi \rho_0}{\mathcal{N} + \sum_{i=1}^{M} I_i \rho_i} \tag{8.77}$$

where $\mathcal{N}$ is the noise power, ρ_i is given by (8.2), and I_i is a discrete random variable that may take three values:

$$I_i = \begin{cases} \psi & \text{with probability } p_{\mathrm{c}} \\ K_s & \text{with probability } p_{\mathrm{a}} \\ 0 & \text{with probability } p_{\mathrm{n}} \end{cases} \tag{8.78}$$

where the probability of no collision is

$$p_{\mathrm{n}} = 1 - p_{\mathrm{c}} - p_{\mathrm{a}} = 1 - \frac{D(3L-2)}{L^2}. \tag{8.79}$$

Adjacent-channel interference can be neglected by setting $\psi = 1$ and $K_s = 0$.

Substituting (8.2), (8.3), and $\widetilde{P}_i = P_i$ (since $G_i = 1$ for frequency hopping) into (8.77) and dividing the numerator and denominator by $d_0^{\alpha} P_0$, the SINR is

$$\gamma = \frac{\psi g_0 \Omega_0}{\Gamma^{-1} + \sum_{i=1}^{M} g_i I_i \Omega_i}, \quad \Gamma = d_0^{\alpha} P_0 / \mathcal{N} \tag{8.80}$$

where Γ is the SNR when the transmitter is at unit distance and the fading and shadowing are absent, and

$$\Omega_i = \begin{cases} 10^{\xi_0/10} \left\| \mathbf{X}_0 - \mathbf{X}_{M+1} \right\|^{-\alpha} & i = 0 \\ \frac{P_i}{P_0} 10^{\xi_i/10} \left\| \mathbf{X}_i - \mathbf{X}_{M+1} \right\|^{-\alpha} & i \geq 1 \end{cases} \tag{8.81}$$

is the normalized power of X_i at the reference receiver.

Conditional Outage Probability

Let β denote the minimum SINR required for reliable reception and $\boldsymbol{\Omega} = \{\Omega_0, \ldots, \Omega_M\}$ represent the set of normalized received powers. An *outage* occurs when the SINR falls below β. Accounting for the statistics of I_i in using (8.9) to derive the conditional density function of $Y_i = g_i I_i \Omega_i$, we obtain

$$\begin{aligned} f_i(y) &= f_{Y_i}(y|\Omega_i) \\ &= p_n \delta(y) + \frac{y^{m_i - 1}}{\Gamma(m_i)} \left[p_c \left(\frac{m_i}{\psi \Omega_i} \right)^{m_i} e^{-\frac{y m_i}{\psi \Omega_i}} + p_a \left(\frac{m_i}{K_s \Omega_i} \right)^{m_i} e^{-\frac{y m_i}{K_s \Omega_i}} \right] u(y). \end{aligned} \tag{8.82}$$

As a result, if m_0 is a positive integer, a slight modification of the derivation of (8.19) indicates that the outage probability is

$$\epsilon(\boldsymbol{\Omega}) = 1 - e^{-\beta_0 z} \sum_{s=0}^{m_0 - 1} (\beta_0 z)^s \sum_{t=0}^{s} \frac{z^{-t}}{(s-t)!} H_t(\boldsymbol{\Omega}), \; \beta_0 = \frac{\beta m_0}{\Omega_0}, \quad z = \Gamma^{-1} \tag{8.83}$$

$$H_t(\boldsymbol{\Omega}) = \sum_{\substack{\ell_i \geq 0 \\ \sum_{i=1}^{M} \ell_i = t}} \prod_{i=1}^{M} G_{\ell_i}(i) \tag{8.84}$$

$$G_{\ell_i}(\Omega_i) = p_n \delta_{\ell_i} + \frac{\Gamma(\ell_i + m_i)}{\ell_i! \Gamma(m_i)} \left[p_c \phi_i(\psi) + p_a \phi_i(K_s) \right] \tag{8.85}$$

$$\phi_i(x) = \left(\frac{x \Omega_i}{m_i} \right)^{\ell_i} \left(\frac{x \beta_0 \Omega_i}{m_i} + 1 \right)^{-(m_i + \ell_i)} \tag{8.86}$$

where δ_ℓ is the Kronecker delta function.

Rate Adaptation

Let $C(h, \gamma)$ denote the channel capacity when the modulation is CPFSK with modulation index h. The channel is in an outage when $C(h, \gamma) \leq R$. Assuming Gaussian interference, the channel is conditionally Gaussian during a particular hop, and the modulation-constrained AWGN capacity can be used for $C(h, \gamma)$. For any value of h and SINR threshold β, the achievable code rate is $R = C(h, \beta)$. For any h and R, the SINR threshold required to support that rate is obtained by inverting $R = C(h, \beta)$. We assume that a mobile is able to estimate the channel and appropriately adapt its rate so that $P[\gamma \leq \beta] < \hat{\epsilon}$, where $\hat{\epsilon}$ is a constraint on the channel outage.

Modulation-Constrained Area Spectral Density

The maximum data transmission rate is determined by the bandwidth W/L of a frequency channel, the fractional in-band power ψ, the spectral efficiency of the modulation, and the code rate. Let η denote the *spectral efficiency of the modulation*, given in symbols per second per Hertz, and defined by the symbol rate divided by the $100\psi\%$-power bandwidth of the modulation. Since we assume many symbols per hop, the spectral efficiency of CPFSK can be found by numerically integrating (3.92) and then inverting the result. To emphasize the dependence of η on h and ψ, we denote the spectral efficiency of CPFSK as $\eta(h, \psi)$ subsequently. When combined with a rate-R code, the spectral efficiency becomes $R\eta(h, \psi)$ bits per second per Hertz (bps/Hz), where R is the ratio of information bits to code symbols. Since the signal occupies a frequency channel with $100\psi\%$-power bandwidth W/L Hz, the throughput supported by a single link operating with a duty factor D and outage probability $\hat{\epsilon}$ is

$$T = \frac{WRD\eta(h, \psi)(1 - \hat{\epsilon})}{L} \tag{8.87}$$

bps. Averaging the throughput over N network topologies, we obtain

$$\widetilde{T} = \frac{WD\eta(h, \psi)(1 - \hat{\epsilon})}{L}\widetilde{R} \tag{8.88}$$

where the average code rate is

$$\widetilde{R} = \frac{1}{N}\sum_{n=1}^{N} R_n. \tag{8.89}$$

Multiplying $\mathbb{E}[T]$ by the mobile density $\lambda = M/A_{net}$ and dividing by the system bandwidth W gives the *normalized modulation-constrained area spectral efficiency* (MASE):

$$\tau(\lambda) = \frac{\lambda D\eta(h, \psi)(1-\hat{\epsilon})\widetilde{R}}{L} \tag{8.90}$$

which has units of bps/Hz per unit area. The normalized MASE is the normalized area spectral efficiency resulting from a *conditional* outage probability $\hat{\epsilon}$, where the conditioning is with respect to the network topology. As a performance measure, the normalized MASE explicitly takes into account the code rate, the spectral efficiency of the modulation, and the number of frequency channels.

Optimization Algorithm

For an FH-CDMA ad hoc network that uses CPSK modulation and adaptive-rate coding, there is an optimal set of (L, h, ψ) that maximizes the normalized MASE. For the optimization, the density of interference per unit area λ and the duty factor D have fixed values. The normalized MASE is computed for positive integer values of L fewer than 1000, values of h quantized to a spacing of 0.01 over the range $0 \leq h \leq 1$, and values of Ψ quantized to a spacing of 0.005 over the range $0.90 \leq \Psi \leq 0.99$. For each set of (L, h, ψ) considered, the average code rate $\mathbb{E}[R]$ and the normalized MASE can be found by using a Monte Carlo approach as follows. A realization of the network is produced by placing M mobiles within an area A_{net} according to the specified spatial distribution. The path loss from each mobile to the reference receiver is computed with randomly generated shadowing factors. After setting $\epsilon(\boldsymbol{\Omega})$ given by (8.83) equal to the outage constraint $\hat{\epsilon}$, an inversion of the equation determines the adequate SINR threshold β. The rate $R_n = C(h, \beta)$ that corresponds to the SINR threshold for the nth topology is computed, and its value is stored. The entire process is repeated for a large number of networks N. The average code rate, which is defined as (8.89), is used to find the spectral efficiency $\eta(h, \Psi)$ corresponding to the current h and Ψ, and then (8.90) provides the normalized MASE. Finally, the optimization is completed by performing an exhaustive search over the sets of (L, h, ψ) to determine the maximum value of the normalized MASE and the corresponding values of L, h, and Ψ.

Example 13 Consider a network with the reference receiver at the origin and interfering mobiles drawn from a uniform distribution within an annular region. The region has inner radius $r_{\mathsf{ex}} = 0.25$ and two outer radii are considered: $r_{\mathsf{net}} = 2$ and $r_{\mathsf{net}} = 4$. The optimization is performed by considering $N = $10,000 different network topologies. The path-loss exponent is $\alpha = 3$, the duty factor is $D = 1$, the source is located at a unit distance away from the receiver at the origin, the SNR is $\Gamma = 10\,\text{dB}$, and when a shadowed scenario is examined, the standard deviation

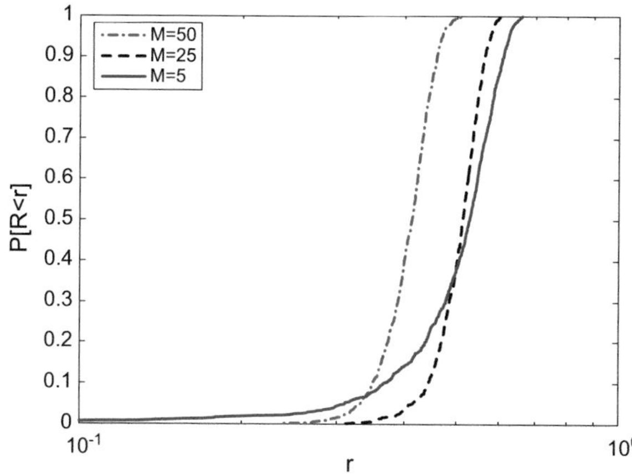

Fig. 8.22 Distribution function of the code rate when the network is optimized [91]

of the lognormal distribution is $\sigma_s = 8$ dB. Binary CPFSK is used, and the rate is adapted to satisfy a typical outage constraint of $\hat{\epsilon} = 0.1$. Three fading models are considered: Rayleigh fading ($m_i = 1$ for all i), Nakagami fading ($m_i = 4$ for all i), and mixed fading ($m_0 = 4$ and $m_i = 1$ for $i \geq 1$).

Figure 8.22 illustrates the variability of the code rate through its distribution function in the presence of mixed fading and shadowing. The figure shows three curves, corresponding to a dense network ($M = 50$), a moderately dense network ($M = 25$), and a sparse network ($M = 5$). The outer radius of the annular network is $r_{\text{net}} = 2$. The network parameters are independently optimized for each mobile density. The figure shows that the distribution function of the code rate is quite steep for all three scenarios. This characteristic is indicative of the fairness in the rate selection because most nodes have similar rates. However, the curves are steeper for dense networks than for sparse networks. The reason is the scarcity of interfering mobiles in a sparse network and hence the increased variability in the SINR and code rate from one realization to the next.

Figures 8.23, 8.24, and 8.25 show the optimal normalized MASE τ'_{opt} for a dense network ($M = 50$ and $r_{\text{net}} = 2$) as functions of the number of frequency channels L, the modulation index h, and the fractional in-band power Ψ respectively. These three figures demonstrate the effects of each of the three parameters of the optimization. In Figures 8.23 and 8.24, $\Psi = 0.96$ is assumed, and for each value of the parameter L or h, respectively, the other parameter is chosen to maximize the normalized MASE. In Figure 8.25, the other two parameters are jointly optimized for each value of Ψ.

The results of the optimization are shown in Table 8.2. For each set of network conditions, the (L, h, Ψ) that maximize the normalized MASE are listed along with the corresponding τ'_{opt}. For comparison, the normalized MASE τ_{sub} is shown for the parameters $(L, h, \Psi) = (200, 0.5, 0.99)$. The comparison illustrates the importance of parameter optimization and indicates that the selection of optimal parameters may improve the normalized MASE by a factor greater than two relative to an arbitrary parameter selection.

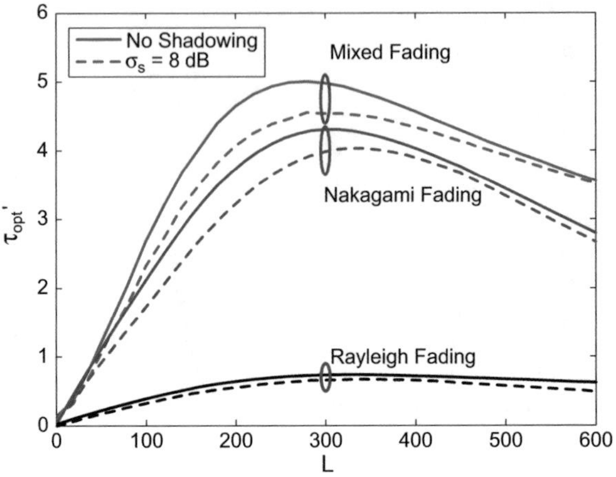

Fig. 8.23 Optimal normalized MASE as a function of the number of frequency channels L when $\Psi = 0.96$. A dense network is considered ($M = 50, r_{\text{net}} = 2$) [91]

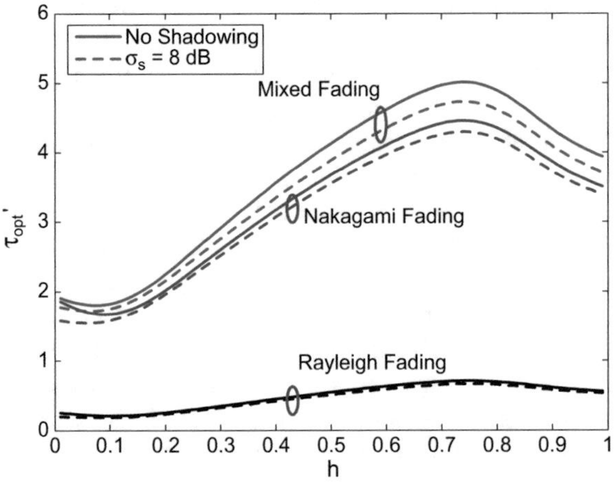

Fig. 8.24 Optimal normalized MASE as a function of the modulation index h when $\Psi = 0.96$. A dense network is considered ($M = 50, r_{\text{net}} = 2$) [91]

Optimization results are given for different fading channel models and for both a shadowed and an unshadowed scenario. Performance is very poor for the pessimistic assumption of Rayleigh fading, but it improves for a more realistic mixed fading, which implies a line of sight between the source and destination but not for the other links. Shadowing is always detrimental, and even though it leads to a higher code rate, it requires a larger number of frequency channels. An increase in the density of mobiles per unit area leads to a lower modulation index, but an increase in the normalized MASE, code rate, number of frequency channels, and fractional in-band power. □

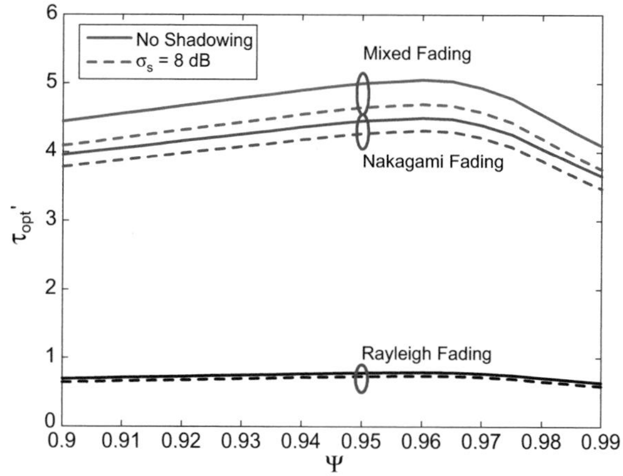

Fig. 8.25 Optimal normalized MASE as a function of the fractional in-band power Ψ. For each Ψ, both the number of frequency channels and the modulation index are varied to maximize the normalized MASE. A dense network is considered ($M = 50, r_{\text{net}} = 2$) [91]

Table 8.2 Results of the optimization for $M = 50$ interferers

r_{net}	Channel	L	$\mathbb{E}[R]$	h	Ψ	τ'_{opt}	τ_{sub}
2	R/U	315	0.07	0.80	0.96	0.79	0.50
	N/U	280	0.36	0.80	0.96	4.42	2.99
	M/U	279	0.41	0.80	0.96	5.00	3.56
	R/S	320	0.07	0.80	0.96	0.76	0.40
	N/S	290	0.42	0.80	0.96	4.24	2.27
	M/S	290	0.38	0.80	0.96	4.70	3.02
4	R/U	73	0.05	0.84	0.95	0.63	0.32
	N/U	80	0.35	0.84	0.95	3.74	1.87
	M/U	75	0.36	0.84	0.95	4.09	1.91
	R/S	95	0.05	0.84	0.95	0.49	0.28
	N/S	130	0.40	0.84	0.95	2.65	1.75
	M/S	100	0.35	0.84	0.95	3.03	1.80

The normalized MASE τ is in units of bps/kHz-m^2. Channel abbreviations: R indicates Rayleigh fading, N indicates Nakagami fading ($m_i = 4$, for all i), M indicates mixed fading ($m_0 = 4$ and $m_i = 1$ for $i \geq 1$), U indicates an unshadowed environment, and S indicates shadowing ($\sigma_s = 8$ dB) [91]

8.7 FH-CDMA Cellular Networks

In an FH-CDMA cellular network, synchronous and orthogonal frequency-hopping patterns can be simultaneously transmitted by a base station so that there is no intracell interference on the downlinks within a cell sector. By choosing frequency-hopping patterns with M hopset frequencies that are circularly rotated versions of each other, as many as M frequency-hopping signals can be simultaneously transmitted; hence, as many as M active mobiles can be accommodated.

If the cells are small enough and the switching times between hops are large enough, it may be practical to synchronize the transmit times of the mobiles in a cell sufficiently to avoid collisions at the base station of the uplink signals and thereby prevent intracell interference. The synchronization may be implemented by using the measured arrival times of the downlink signals at the mobiles. If there are L frequency channels, but fewer than $\lfloor L/2 \rfloor$ active mobiles are present in a cell, then separate frequency-hopping patterns can be selected so that there is no intracell adjacent-channel interference on either the uplinks or the downlinks.

Intercell interference cannot be entirely prevented because the patterns associated with different cells or sectors are asynchronous signals. Consider a sector B with N_b mobiles covered by sector antenna A. Since the mobiles in sector B independently and asynchronously use a set of any N_b carrier frequencies in the network hopset with equal probability, the probability that a mobile in the covered sector B produces interference in the transmission channel of an uplink in sector A is

$$p_c = \frac{DN_b}{L}. \tag{8.91}$$

This equation also gives the probability that sector antenna B oriented toward a mobile in sector A produces interference in the transmission channel of the downlink of the mobile. Because of orthogonality within each sector, no more than one signal from a sector produces interference in the transmission channel of either link in another sector.

The FH-CDMA networks largely avoid the near-far problem by continually changing the carrier frequencies so that frequency collisions become brief, unusual events. Thus, power control in an FH-CDMA network is unnecessary, and all mobiles may transmit at the same power level. When power control is used, it tends to benefit signals from mobiles far from an associated sector antenna, while degrading signals from mobiles close to it so that even perfect power control typically increases system capacity by only a small amount. There are good reasons to forego this slight potential advantage and not use power control. The required overhead may be excessive. If the geolocation of mobiles is performed by using measurements at two or more base stations, then the power control may result in significantly less signal power arriving at one or more base stations and the consequent loss of geolocation accuracy.

The most common FH-CDMA network has features of both ad hoc and cellular networks and uses *Bluetooth* technology. A *piconet* comprises a master Bluetooth device that communicates with seven or fewer slave devices. Any Bluetooth device can assume the role of master or slave in a particular piconet and may even be the master of one piconet while serving as a slave in another. The master of a piconet chooses the slave with which it communicates in a time slot. The slave synchronizes with the timing and frequency-hopping pattern of the master, which makes regular transmissions. Intracell interference is avoided on the uplinks and downlinks of a piconet because the master communicates with one slave at a time.

The DS-CDMA cellular networks are preferable to FH-CDMA cellular networks because coherent demodulation is available. The performance of similar FH-CDMA networks is potentially inferior because a noncoherent demodulation is usually necessary. Thus, FH-CDMA cellular networks are seldom implemented except as hybrid Bluetooth networks.

Long-term evolution (LTE) is a fourth-generation standard for the wireless communication of high-speed data. It uses OFDM for multiple access instead of CDMA, and different frequency subbands may be assigned to different users. During a two-slot subframe, frequency hopping is sometimes used on the uplink to obtain diversity by changing a user's subband allocation during the first slot to another one during the second slot.

8.8 Problems

1

(a) For CDMA ad hoc networks, show that if $m_0 = 1$, then

$$\epsilon(\mathbf{\Omega}) = 1 - \exp\left(-\frac{\beta z}{\Omega_0}\right) \prod_{i=1}^{M} \left[1 - p_i + p_i \left(\frac{m_i}{m_i + \beta\Omega_i/\Omega_0}\right)^{m_i}\right].$$

(b) If $m_0 = 1$ and $m_i \to \infty, 1 \le i \le M$, what is the equation for $\epsilon(\mathbf{\Omega})$?
(c) Verify (8.33) for $m_0 = m_i = 1, 1 \le i \le M$.

2 (a) For CDMA ad hoc networks with $\Omega_0 \to \infty$, for which Nakagami parameters does $\epsilon(\mathbf{\Omega}) \to 0$? (b) For CDMA ad hoc networks with $\Omega_0 \to 0$, for what Nakagami parameters does $\epsilon(\mathbf{\Omega}) \to 1$?

3 For CDMA ad hoc networks, derive an equation for $\epsilon(\mathbf{\Omega})$ when $\Omega_0 \neq 0$ and $\Omega_i \to \infty, i \ge 1$, and show that $\epsilon(\mathbf{\Omega}) \to 1$ regardless of the Nakagami parameters if and only if $p_i = 1$ for some i.

4 For CDMA ad hoc networks with $z = \Gamma^{-1} = 0$, derive an equation for $\epsilon(\mathbf{\Omega})$.

5 Consider a cellular network in which there are M active mobiles that are independently located throughout the network. The probability that a mobile is located in a particular cell sector is λ/M. Therefore, the distribution of K active mobiles in the cell sector is given by the binomial distribution

$$P(K = k) = \binom{M}{k} \left(\frac{\lambda}{M}\right)^k \left(1 - \frac{\lambda}{M}\right)^{M-k}$$

and $E[K] = \lambda$. (a) Show that as $M \to \infty$, this distribution approaches the Poisson distribution:

$$P(K = k) = \frac{\exp(-\lambda)\lambda^k}{k!}, \quad k = 0, 1, 2, \ldots, M.$$

(b) What is the conditional probability that $K = k$ given that $K \geq 1$?

6 Consider the acquisition of the scrambling code associated with a cell. Suppose that there are K hypothesized scrambling codes in a group, the number of symbols in a codeword is C, and the acceptance threshold is T votes. Assume that the correct scrambling code of a cell is not transmitted and that votes for each of the K codes are equally likely. Use the union bound to derive an upper bound on the probability of a false alarm.

7 Consider an uplink of a DS-CDMA cellular network that has full power control with $\delta = 1$, $m_0 = 1$, and $m_i = p_i = 1, 1 \leq i \leq M, i \neq r$. Show that as $\beta \to 0$,

$$\epsilon(\boldsymbol{\Omega}) \to \beta \left[\frac{1}{\Omega_r \Gamma} + \frac{M-1}{G/h} \right].$$

8 Consider a downlink of cell j in a DS-CDMA cellular network that has $m_0 = 1$ and $m_i = p_i = 1, 1 \leq i \leq C, i \neq j$. Show that as $\beta \to 0$,

$$\epsilon(\boldsymbol{\Omega}) \to \beta \left[\frac{1}{\Omega_r \Gamma} + \frac{\sum_{i=1, i \neq j}^{C} \Omega_i}{\Omega_r} \right].$$

9 Consider the downlinks in a cell j of a DS-CDMA network. There is no fading, no noise, and the required SINR γ_k of the downlink to X_k in the cell is γ. Show that the requirement is met only if

$$\gamma \leq \frac{(1 - f_p)G}{h} \left[\sum_{k: X_k \in \mathcal{X}_j} \frac{\sum_{i=1, i \neq j}^{M} I_i P_{i,k} 10^{\xi_{i,k}/10} ||S_i - X_k||^{-\alpha}}{10^{\xi_{j,k}/10} ||S_j - X_k||^{-\alpha}} \right]^{-1}.$$

10 Consider an FH-CDMA ad hoc network that has $m_0 = m_i = 1, 1 \leq i \leq M$. Show that as $\beta \to 0$,

$$\epsilon(\boldsymbol{\Omega}) \to \beta \left[\frac{1}{\Omega_0 \Gamma} + \frac{(p_c \psi + p_a K_s) \sum_{i=1}^{M} \Omega_i}{\Omega_0} \right].$$

11 Consider an FH-CDMA network with two mobiles that communicate with a target mobile or base station. The modulation is ideal DPSK, the target mobile uses noncoherent combining with M diversity branches, the number of frequency channels is $L = 10$, and both the receiver noise and the spectral splatter are negligible. Both signals arriving at the target mobile encounter Rayleigh fading, but one signal has SINR $\bar{\gamma} = 10\,\text{dB}$, and the other signal has SINR $\bar{\gamma} = -10\,\text{dB}$. From (6.112) and (6.144), the channel-symbol error probability is

$$P_s(M) = p - (1-2p)\sum_{i=1}^{M-1} \binom{2i-1}{i} [p(1-p)]^i,\; p = \frac{1}{2(1+\bar{\gamma})}.$$

If power control is used, then $\bar{\gamma} = 0\,\text{dB}$ for both signals. For $M = 1$ and $M = 2$, assess the relative merits for each mobile of introducing power control. Assume that an acceptable average channel-bit error probability is 0.02.

12 Consider the transmission channel T of an uplink in sector A of an FH-CDMA cellular network that uses a hopset with $L \geq 3$ frequencies. Sector B is covered by sector antenna A. The $N_b \leq L$ mobiles in sector B use distinct carrier frequencies in the network hopset with equal probability, and $D = 1$. Let $N_1 = 1$ if a signal from sector B uses T; let $N_1 = 0$ if no signal does. (a) Show that the probability that some mobile in the covered sector B produces interference in exactly one of the adjacent channels of T is

$$P_a = \frac{2\,(N_b - N_1)\,(L - N_b + N_1)}{L\,(L-1)},\quad L - 1 \geq N_b - N_1 \geq 0.$$

(b) Show that the probability that some mobile in the covered sector B produces interference in both adjacent channels of T is

$$P_b = \frac{(N_b - N_1)\,(N_b - N_1 - 1)}{L\,(L-1)},\quad L - 1 \geq N_b - N_1 \geq 0.$$

Chapter 9
Iterative Channel Estimation, Demodulation, and Decoding

The estimation of channel parameters, such as the fading amplitude and the power spectral density (PSD) of the interference and noise, is essential to the effective use of soft-decision decoding. Channel estimation may be implemented by the transmission of pilot signals that are processed by the receiver, but pilot signals entail overhead costs, such as the loss of data throughput. Deriving maximum-likelihood channel estimates directly from the received data symbols is often prohibitively difficult. There is an effective alternative when turbo or low-density parity-check codes are used. The expectation-maximization algorithm, which is derived and explained in this chapter, provides an iterative approximate solution to the maximum-likelihood equations and is inherently compatible with iterative demodulation and decoding. Two examples of advanced spread-spectrum systems that apply iterative channel estimation, demodulation, and decoding are described and analyzed. These systems provide good illustrations of the calculations required in the design of advanced systems.

9.1 Expectation-Maximization Algorithm

The expectation-maximization (EM) algorithm offers a low-complexity iterative approach to maximum-likelihood estimation [42, 47]. A substantial body of literature exists on EM-based techniques for channel estimation, data detection, and multiuser detection [17]. In this chapter, random vectors are represented by upper-case letters while realizations of the random vectors are represented by lower-case letters.

Maximum-likelihood estimation of a nonrandom parameter vector $\boldsymbol{\theta}$ of m parameters is obtained by maximizing the conditional density function $g(\mathbf{y}|\boldsymbol{\theta})$ of a random vector $\mathbf{Y} = [Y_1 \cdots Y_n]^T$ of the observations. Since the logarithm is a

D. Torrieri, *Principles of Spread-Spectrum Communication Systems*,
https://doi.org/10.1007/978-3-319-70569-9_9

monotonic function of its argument, the maximum-likelihood estimate $\hat{\boldsymbol{\theta}}_{ml}$ of $\boldsymbol{\theta}$ may be expressed as

$$\hat{\boldsymbol{\theta}}_{ml} = \arg\max_{\theta} \ln g(\mathbf{y}|\boldsymbol{\theta}) \tag{9.1}$$

where $\ln g(\mathbf{y}|\boldsymbol{\theta})$ is the *log-likelihood function*. When the likelihood function is differentiable, the maximum-likelihood estimate is the solution of the *likelihood equation*:

$$\nabla_{\boldsymbol{\theta}} \ln g(\mathbf{y}|\boldsymbol{\theta})\Big|_{\theta=\boldsymbol{\theta}_{ml}} = \mathbf{0} \tag{9.2}$$

where $\nabla_{\boldsymbol{\theta}}$ is the gradient vector with respect to $\boldsymbol{\theta}$:

$$\nabla_{\boldsymbol{\theta}} = \left[\frac{\partial}{\partial\theta_1} \cdots \frac{\partial}{\partial\theta_n}\right]^T. \tag{9.3}$$

When (9.2) cannot be solved in closed form, it can sometimes be solved iteratively by applying Newton's method or fixed-point methods. When an iterative maximum-likelihood solution is intractable, an alternative procedure is the *EM* algorithm, which has the major advantage that it requires no calculations of gradients or Hessians.

The EM algorithm is based on considering the set of observations $\{Y_i\}$ forming the random data vector $\mathbf{Y}$ as a subset of or derived from a larger data set $\{Z_i\}$ forming a random data vector $\mathbf{Z}$ such that the maximization of its conditional density function $f(\mathbf{z}|\boldsymbol{\theta})$ is mathematically tractable. The data vector $\mathbf{Y}$ is called the *incomplete* data vector, and the data vector $\mathbf{Z}$ is called the *complete* data vector. The complete data vector $\mathbf{Z}$ is related to each observation Y_k by a many-to-one transformation. Since this transformation is not invertible, there is no unique mapping from $\mathbf{Y}$ to $\mathbf{Z}$. The function $\ln f(\mathbf{z}|\boldsymbol{\theta})$ does not directly provide a useful estimate of $\boldsymbol{\theta}$ because $\mathbf{Z}$ is not observable, so the expectation of $\ln f(\mathbf{Z}|\boldsymbol{\theta})$ given both $\mathbf{Y} = \mathbf{y}$ and an estimate of $\boldsymbol{\theta}$ is iteratively maximized by the EM algorithm.

Since $\mathbf{Y}$ is determined by $\mathbf{Z}$, the joint conditional density function is $f(\mathbf{z}, \mathbf{y}|\boldsymbol{\theta}) = f(\mathbf{z}|\boldsymbol{\theta})$, and the definition of a conditional density function implies that

$$f(\mathbf{z}|\mathbf{y}, \boldsymbol{\theta}) = \frac{f(\mathbf{z}|\boldsymbol{\theta})}{g(\mathbf{y}|\boldsymbol{\theta})} \tag{9.4}$$

where $g(\mathbf{y}|\boldsymbol{\theta})$ is the conditional density function of $\mathbf{Y}$ given $\boldsymbol{\theta}$. Therefore,

$$\ln g(\mathbf{y}|\boldsymbol{\theta}) = \ln f(\mathbf{z}|\boldsymbol{\theta}) - \ln f(\mathbf{z}|\mathbf{y}, \boldsymbol{\theta}). \tag{9.5}$$

Beginning with an initial estimate $\hat{\boldsymbol{\theta}}_0$, the EM algorithm computes successive estimates $\hat{\boldsymbol{\theta}}_i$ that increase the value of $\ln f(\mathbf{y}|\hat{\boldsymbol{\theta}}_i)$. Let

$$E_{\mathbf{z}|\mathbf{y},\widehat{\boldsymbol{\theta}}_i}[h(\mathbf{Z})] = \int h(\mathbf{z}) f(\mathbf{z}|\mathbf{y}, \widehat{\boldsymbol{\theta}}_i) d\mathbf{z} \tag{9.6}$$

denote the expectation of $h(\mathbf{Z})$ with respect to the conditional density function $f(\mathbf{z}|\mathbf{y}, \hat{\boldsymbol{\theta}}_i)$, where $h(\mathbf{Z})$ is some function of the random vector $\mathbf{Z}$. Integrating both sides of (9.5) over $f(\mathbf{z}|\mathbf{y}, \hat{\boldsymbol{\theta}}_i)$ yields

$$\ln g(\mathbf{y}|\boldsymbol{\theta}) = \chi(\boldsymbol{\theta}, \widehat{\boldsymbol{\theta}}) - E_{\mathbf{z}|\mathbf{y},\widehat{\boldsymbol{\theta}}_i}[\ln f(\mathbf{Z}|\mathbf{Y} = \mathbf{y}, \boldsymbol{\theta})] \tag{9.7}$$

where

$$\chi(\boldsymbol{\theta}, \widehat{\boldsymbol{\theta}}) = E_{\mathbf{z}|\mathbf{y},\hat{\boldsymbol{\theta}}}[\ln f(\mathbf{Z}|\theta)]. \tag{9.8}$$

Lemma *For any $\hat{\boldsymbol{\theta}}_{i+1}$,*

$$E_{\mathbf{z}|\mathbf{y},\hat{\boldsymbol{\theta}}_i}[\ln f(\mathbf{Z}|\mathbf{Y} = \mathbf{y}, \widehat{\boldsymbol{\theta}}_{i+1})] \leq E_{\mathbf{z}|\mathbf{y},\hat{\boldsymbol{\theta}}_i}[\ln f(\mathbf{Z}|\mathbf{Y} = \mathbf{y}, \widehat{\boldsymbol{\theta}}_i)] \tag{9.9}$$

with equality if and only if $f(\mathbf{z}|\mathbf{y}, \widehat{\boldsymbol{\theta}}_{i+1}) = f(\mathbf{z}|\mathbf{y}, \widehat{\boldsymbol{\theta}}_i)$.

Proof As $\ln a - \ln b = \ln a/b$,

$$\begin{aligned} &E_{\mathbf{z}|\mathbf{y},\hat{\boldsymbol{\theta}}_i}[\ln f(\mathbf{Z}|\mathbf{Y} = \mathbf{y}, \hat{\boldsymbol{\theta}}_{i+1})] - E_{\mathbf{z}|\mathbf{y},\hat{\boldsymbol{\theta}}_i}[\ln f(\mathbf{Z}|\mathbf{Y} = \mathbf{y}, \widehat{\boldsymbol{\theta}}_i)] \\ &\quad = E_{\mathbf{z}|\mathbf{y},\hat{\boldsymbol{\theta}}_i}\left[\ln \frac{f(\mathbf{Z}|\mathbf{Y} = \mathbf{y}, \hat{\boldsymbol{\theta}}_{i+1})}{f(\mathbf{Z}|\mathbf{Y} = \mathbf{y}, \widehat{\boldsymbol{\theta}}_i)}\right]. \end{aligned}$$

As $\ln x = x - 1$ when $x = 1$ and a double differentiation proves that $\ln x$ is a concave function of x for $x > 0$, $\ln x \leq x - 1$ for $x > 0$ with equality if and only if $x = 1$. Therefore,

$$\begin{aligned} E_{\mathbf{z}|\mathbf{y},\hat{\boldsymbol{\theta}}_i}\left[\ln \frac{f(\mathbf{Z}|\mathbf{Y} = \mathbf{y}, \hat{\boldsymbol{\theta}}_{i+1})}{f(\mathbf{Z}|\mathbf{Y} = \mathbf{y}, \hat{\boldsymbol{\theta}}_i)}\right] &\leq \int \left[\frac{f(\mathbf{z}|\mathbf{y}, \hat{\boldsymbol{\theta}}_{i+1})}{f(\mathbf{z}|\mathbf{y}, \hat{\boldsymbol{\theta}}_i)} - 1\right] f(\mathbf{z}|\mathbf{y}, \hat{\boldsymbol{\theta}}_i) d\mathbf{z} \\ &= 0 \end{aligned}$$

which proves the inequality. The condition for equality follows directly. □

Theorem 1 *If successive estimates are computed as*

$$\hat{\theta}_{i+1} = \arg\max_{\boldsymbol{\theta}} \chi(\boldsymbol{\theta}, \hat{\boldsymbol{\theta}}_i) \tag{9.10}$$

and $\ln g(\mathbf{y}|\boldsymbol{\theta})$ has an upper bound, then the sequence $\{\ln g(\mathbf{y}|\hat{\boldsymbol{\theta}}_i)\}$ converges to a limit as $i \to \infty$.

Proof The hypothesis of the theorem implies that

$$\chi(\hat{\boldsymbol{\theta}}_{i+1}, \hat{\boldsymbol{\theta}}_i) \geq \chi(\hat{\boldsymbol{\theta}}_i, \hat{\boldsymbol{\theta}}_i).$$

This inequality, the lemma, (9.7), and (9.8) imply that

$$\ln g(\mathbf{y}|\hat{\theta}_{i+1}) \geq \ln g(\mathbf{y}|\hat{\boldsymbol{\theta}}_i)$$

with equality if and only if

$$\chi(\hat{\boldsymbol{\theta}}_{i+1}, \hat{\boldsymbol{\theta}}_i) = \chi(\hat{\boldsymbol{\theta}}_i, \hat{\boldsymbol{\theta}}_i), \ \ f(\mathbf{z}|\mathbf{y},\hat{\boldsymbol{\theta}}_{i+1}) = f(\mathbf{z}|\mathbf{y}, \hat{\boldsymbol{\theta}}_i).$$

A bounded monotonically increasing sequence converges to a limit. Therefore, if $\ln g(\mathbf{y}|\hat{\boldsymbol{\theta}}_i)$ has an upper bound, then it converges to a limit as $i \to \infty$. □

This theorem leads directly to the EM algorithm, which has two primary computational steps: an expectation (*E-step*) and a maximization (*M-step*).

EM Algorithm

1. Set $i = 0$ and select the initial estimate $\hat{\boldsymbol{\theta}}_0$.
2. E-step: Compute $\chi(\boldsymbol{\theta}, \hat{\boldsymbol{\theta}}_i) = E_{\mathbf{z}|\mathbf{y},\hat{\boldsymbol{\theta}}_i}[\ln f(\mathbf{Z}|\boldsymbol{\theta})]$.
3. M-step: Compute $\hat{\boldsymbol{\theta}}_{i+1} = \arg\max_{\boldsymbol{\theta}} \chi(\boldsymbol{\theta}, \hat{\boldsymbol{\theta}}_i)$.
4. If i does not exceed some preset maximum and $\|\hat{\boldsymbol{\theta}}_{i+1} - \hat{\boldsymbol{\theta}}_i\| > \epsilon$, where ϵ is preset positive number, then return to the E-step. Otherwise, terminate the iterations. □

If $\ln g(\mathbf{y}|\boldsymbol{\theta})$ is not concave, the sequence $\{\ln g(\mathbf{y}|\widehat{\boldsymbol{\theta}}_i)\}$ may converge to a local maximum rather than a global maximum of $\ln g(\mathbf{y}|\boldsymbol{\theta})$, and $\hat{\boldsymbol{\theta}}$ may not converge to $\hat{\boldsymbol{\theta}}_{ml}$. Furthermore, an iteration of the EM algorithm might produce a jump of $\hat{\boldsymbol{\theta}}_i$ from the vicinity of one local maximum to the vicinity of another of $\ln g(\mathbf{y}|\hat{\boldsymbol{\theta}}_i)$. Thus, it may be necessary to execute the EM algorithm with several different values of the initial vector $\boldsymbol{\theta}_0$ to ensure that $\hat{\boldsymbol{\theta}}_i \to \hat{\boldsymbol{\theta}}_{ml}$ or a close approximation of it.

In many applications, $\mathbf{Y}$ is part of the complete data vector $\mathbf{Z} = [\mathbf{Y}\ \mathbf{X}]^T$, where $\mathbf{X}$ is called the *missing* data vector. In this case, $f(\mathbf{z}|\mathbf{y}, \boldsymbol{\theta}) = h(\mathbf{x}|\mathbf{y}, \boldsymbol{\theta})$, where $h(\mathbf{x}|\mathbf{y}, \boldsymbol{\theta})$ is the conditional density function of $\mathbf{X}$ given $\mathbf{Y} = \mathbf{y}$ and $\boldsymbol{\theta}$. Therefore, (9.8) becomes

$$\chi(\boldsymbol{\theta}, \hat{\boldsymbol{\theta}}_i) = E_{\mathbf{x}|\mathbf{y},\hat{\boldsymbol{\theta}}_i}[\ln f(\mathbf{Z}|\boldsymbol{\theta})]. \tag{9.11}$$

Bayes' rule gives

$$h(\mathbf{x}|\mathbf{y}, \hat{\boldsymbol{\theta}}_i) = \frac{g(\mathbf{y}|\mathbf{x}, \hat{\boldsymbol{\theta}}_i)h(\mathbf{x}|\hat{\boldsymbol{\theta}}_i)}{g(\mathbf{y}|\hat{\boldsymbol{\theta}}_i)} \tag{9.12}$$

which is used in the evaluation of the expected value in (9.11).

In applications with discrete random variables, probability functions replace some of the density functions.

Example 1 As a practical example, consider a random binary sequence X_i, $i = 1, \ldots, n$, transmitted over the additive white Gaussian noise (AWGN) channel. The received random vector $\mathbf{Y} = [Y_1 \ldots Y_n]^T$ is

$$\mathbf{Y} = A\mathbf{X} + \mathbf{N} \tag{9.13}$$

where A is a positive constant amplitude, $\mathbf{X} = [X_1 \ldots X_n]^T$ is the vector of bits, and $\mathbf{N} = [N_1 \ldots N_n]^T$ is the vector of noise samples, which are assumed to be independent, identically distributed, zero-mean, Gaussian random variables with variance v. The bits are assumed to be independent of each other and the parameters, and $X_i = 1$ with probability $1/2$ and $X_i = -1$ with probability $1/2$. Thus, density functions involving $\mathbf{X}$ are replaced by probability functions. It is desired to estimate the parameter vector $\boldsymbol{\theta} = [A \;\; v]^T$.

Equation (9.13), the independent bits, and the independent Gaussian noise imply that

$$g(y_k|x_k, \boldsymbol{\theta}) = (2\pi v)^{-1/2} \exp\left[-\frac{(y_k - Ax_k)^2}{2v}\right] \tag{9.14}$$

and

$$\begin{aligned} g(\mathbf{y}|\mathbf{x}, \boldsymbol{\theta}) &= \prod_{k=1}^{n} g(y_k|x_k, \boldsymbol{\theta}) \\ &= (2\pi v)^{-n/2} \prod_{k=1}^{n} \exp\left[-\frac{(y_k - Ax_k)^2}{2v}\right]. \end{aligned} \tag{9.15}$$

Since $\mathbf{X}$ is independent of $\boldsymbol{\theta}$, the probability that $\mathbf{X} = \mathbf{x}$ is

$$p(\mathbf{x}) = 2^{-n}. \tag{9.16}$$

Since the components of $\mathbf{Y}$ are independent given $\boldsymbol{\theta}$,

$$g(\mathbf{y}|\boldsymbol{\theta}) = 2^{-n} \prod_{k=1}^{n} [g(y_k|x_k = 1, \boldsymbol{\theta}) + g(y_k|x_k = -1, \boldsymbol{\theta})]. \tag{9.17}$$

Substitution of this equation into the likelihood equation (9.2) results in a mathematically intractable set of equations. Thus, we apply the EM algorithm to estimate $\boldsymbol{\theta}$.

We define the complete data vector as $\mathbf{Z} = [\mathbf{Y} \; \mathbf{X}]^T$. From (9.15) and (9.16), we obtain

$$\begin{aligned} \ln f(\mathbf{z}|\boldsymbol{\theta}) &= \ln g(\mathbf{y}|\mathbf{x}, \boldsymbol{\theta}) - n \ln 2 \\ &= C - \frac{n}{2} \ln v - \frac{1}{2v} \sum_{k=1}^{n} (y_k - Ax_k)^2 \end{aligned} \tag{9.18}$$

where C is a constant that does not depend on the parameters A and v. The conditional probability that $\mathbf{X} = \mathbf{x}$ given $\mathbf{Y} = \mathbf{y}$ and $\hat{\boldsymbol{\theta}}_i = [\hat{A}_i \; \hat{v}_i]^T$ is

$$p(\mathbf{x}|\mathbf{y}, \hat{\boldsymbol{\theta}}_i) = \frac{2^{-n} g(\mathbf{y}|\mathbf{x}, \hat{\boldsymbol{\theta}}_i)}{g(\mathbf{y}|\hat{\boldsymbol{\theta}}_i)}. \tag{9.19}$$

The substitution of (9.18) into (9.11) yields

$$\chi(\boldsymbol{\theta},\hat{\boldsymbol{\theta}}_i) = C - \frac{n}{2}\ln v - \frac{1}{2v}\sum_{k=1}^{n}\sum_{\mathbf{x}:\{x_l=\pm 1\}}(y_k - Ax_k)^2 p(\mathbf{x}|\mathbf{y},\hat{\boldsymbol{\theta}}_i) \tag{9.20}$$

where the summation is over 2^n possible vectors. Substituting (9.14), (9.15), and (9.17) into (9.19), we obtain

$$\begin{aligned} p(\mathbf{x}|\mathbf{y},\hat{\boldsymbol{\theta}}_i) &= p(\mathbf{x}/k|\mathbf{y}/k,\hat{\boldsymbol{\theta}}_i)\frac{g(y_k|x_k,\boldsymbol{\theta})}{g(y_k|x_k=1,\hat{\boldsymbol{\theta}}_i)+g(y_k|x_k=-1,\hat{\boldsymbol{\theta}}_i)} \\ &= p(\mathbf{x}/k|\mathbf{y}/k,\hat{\boldsymbol{\theta}}_i)\frac{\exp\left(\frac{\hat{A}_i y_k x_k}{\hat{v}_i}\right)}{2\cosh\left(\frac{\hat{A}_i y_k}{\hat{v}_i}\right)} \end{aligned} \tag{9.21}$$

where $\mathbf{x}/k$ denotes the vector $\mathbf{x}$ excluding the component x_k, and $\mathbf{y}/k$ denotes the vector $\mathbf{y}$ excluding the component y_k. Therefore,

$$\begin{aligned} &\sum_{\mathbf{x}:\{x_l=\pm 1\}}(y_k - Ax_k)^2 p(\mathbf{x}|\mathbf{y},\hat{\boldsymbol{\theta}}_i) \\ &= \sum_{x_k=\pm 1}(y_k - Ax_k)^2\frac{\exp\left(\frac{\hat{A}_i y_k x_k}{\hat{v}_i}\right)}{2\cosh\left(\frac{\hat{A}_i y_k}{\hat{v}_i}\right)} \\ &= \frac{(y_k - A)^2\exp\left(\frac{\hat{A}_i y_k}{\hat{v}_i}\right) + (y_k + A)^2\exp\left(-\frac{\hat{A}_i y_k}{\hat{v}_i}\right)}{2\cosh\left(\frac{\hat{A}_i y_k}{\hat{v}_i}\right)} \\ &= A^2 + y_k^2 - 2Ay_k\tanh\left(\frac{\hat{A}_i y_k}{\hat{v}_i}\right). \end{aligned} \tag{9.22}$$

The substitution of this equation into (9.20) gives

$$\chi(\boldsymbol{\theta},\hat{\boldsymbol{\theta}}_i) = C - \frac{n}{2}\ln v - \frac{nA^2}{2v} - \frac{1}{2v}\sum_{k=1}^{n}\left[y_k^2 - 2Ay_k\tanh\left(\frac{\hat{A}_i y_k}{\hat{v}_i}\right)\right] \tag{9.23}$$

which completes the E-step.

The estimates that maximize $\chi(\boldsymbol{\theta},\hat{\boldsymbol{\theta}}_i)$ are obtained by solving

$$\frac{\partial\chi(\boldsymbol{\theta},\hat{\boldsymbol{\theta}}_i)}{\partial A}\Big|_{\boldsymbol{\theta}=\hat{\boldsymbol{\theta}}_{i+1}} = 0, \qquad \frac{\partial\chi(\boldsymbol{\theta},\hat{\boldsymbol{\theta}}_i)}{\partial v}\Big|_{\boldsymbol{\theta}=\hat{\boldsymbol{\theta}}_{i+1}} = 0. \tag{9.24}$$

Combining the solutions, we obtain the M-step:

$$\hat{A}_{i+1} = \frac{1}{n}\sum_{k=1}^{n} y_k \tanh\left(\frac{\hat{A}_i y_k}{\hat{v}_i}\right) \tag{9.25}$$

$$\hat{v}_{i+1} = \frac{1}{n}\sum_{k=1}^{n} y_k^2 - \hat{A}_{i+1}^2. \tag{9.26}$$

A suitable set of initial values are

$$\hat{A}_0 = \frac{1}{n}\sum_{k=1}^{n} |y_k|, \quad \hat{v}_0 = \frac{1}{n}\sum_{k=1}^{n} y_k^2 - \hat{A}_0^2 \tag{9.27}$$

which completes the algorithm specification. □

Fixed-Point Iteration

In the maximization step of the expectation-maximization method, derivatives are calculated and set equal to zero. After algebraic transformations, we can often obtain one or more equations of the form $f(x) = 0$ that must be solved. The solution is the value x_s, such that $f(x_s) = 0$ when $x = x_s$. If $f(x)$ is a polynomial of degree 3 or higher or if $f(x)$ includes transcendental functions, then a closed-form solution or formula for x_s may not exist, and an approximate calculation of x_s may be necessary.

The *fixed-point iteration method* [39] is a method that does not require the calculation of the derivative of $f(x)$, which may be difficult. To use the method, $f(x) = 0$ is algebraically transformed into an equation of the form

$$x = g(x). \tag{9.28}$$

Then the solution x_s such that

$$x_s = g(x_s) \tag{9.29}$$

which implies that $f(x_s) = 0$, is computed iteratively. After an initial estimate x_0 of the solution, the *fixed-point iteration* is

$$x_{n+1} = g(x_n), \quad n \geq 0 \tag{9.30}$$

which converges to the solution x_s under certain conditions. This solution is called the *fixed point* of $g(x)$ because $g(x_s) = x_s$.

Sufficient convergence conditions for the fixed-point iteration are established by the following theorem. Let $g'(x)$ denote the derivative of $g(x)$ with respect to x.

Theorem 2 *Suppose that $g(x)$ has a continuous derivative such that $|g'(x)| \leq K < 1$ in an interval I_0. If $x_s \in I_0$ and $x_s = g(x_s)$, then the fixed-point iteration converges so that $x_n \to x_s$ as $n \to \infty$ for any $x_0 \in I_0$.*

Proof According to the mean-value theorem of calculus, there is a number $u \in (x_s, x_n)$ such that

$$g(x_n) - g(x_s) = g'(u)(x_n - x_s), \quad n \geq 1. \tag{9.31}$$

Since by hypothesis $|g'(x)| \leq K$ for $(x_s, x_n) \in I_0$,

$$|g(x_n) - g(x_s)| \leq K|x_n - x_s|. \tag{9.32}$$

Equations (9.29), (9.30), and (9.32) yield

$$|x_n - x_s| = |g(x_{n-1}) - g(x_s)| \leq K|x_{n-1} - x_s|. \tag{9.33}$$

Repeated application of this inequality implies that

$$|x_n - x_s| \leq K^n|x_0 - x_s|. \tag{9.34}$$

Since $K < 1$, $K^n \to 0$ and $|x_n - x_s| \to 0$ as $n \to \infty$. Therefore, $x_n \to x_s$ as $n \to \infty$. □

The algebraic transformation from the equation $f(x) = 0$ to the equation $x = g(x)$ can usually be performed in more than one way. The value of K such that $|g'(x)| < K$ for $x \in I_0$ is generally different for each transformation. Among those transformations for which $K < 1$, thereby ensuring convergence of the fixed-point iteration, choosing the specific transformation with the smallest value of K maximizes the speed of convergence.

Example 2 Suppose that $f(x) = x^2 + ax + b = 0$ where $x \neq 0$ and $a \neq 0$. Then, one algebraic transformation gives $x = -(x^2 + b)/a$. For this transformation, $g'(x) = -2x/a$ and convergence occurs if $|x| < |a|/2$. A second algebraic transformation gives $x = -a - bx^{-1}$. Then, $g'(x) = b/x^2$ and convergence occurs if $|x| > \sqrt{|b|}$. The two intervals in which convergence of the fixed-point iteration is ensured do not intersect if $a^2 < 4|b|$. □

9.2 Direct-Sequence Systems

The accuracy of channel-state information (CSI) at the receiver is critical for coherent demodulation and efficient soft-decision decoding (Chapter 1). Cellular protocols, such as wideband code division multiple access (WCDMA) and LTE, specify the use of pilot-assisted channel estimation (PACE). *Pilot symbols* are known symbols either multiplexed with or superimposed onto the transmitted data in the time or frequency domain. Pilot symbols have the associated disadvantages of a loss in spectral or power efficiency and an unsuitability for fast-fading channels with a coherence time shorter than the duration of the pilot symbols. Although their primary role in cellular standards is channel estimation, pilot symbols often play a secondary role in cell, frame, or symbol synchronization, but alternative methods of synchronization may be used [7, 69]. *Blind channel-estimation methods,* which typically use second-order statistics of the received symbols for channel estimation, avoid the implementation cost of pilot symbols, but entail performance losses. Improved performance is achieved by using channel estimation based on the EM algorithm. In this section, a direct-sequence system featuring both iterative EM channel estimation and iterative detection and decoding without *any* pilot symbols is described [106]. The channel estimation includes an estimation of the received *interference PSD*, which is due to both the thermal noise and the time-varying interference. An accurate estimate enables improved interference suppression by the decoder.

Encoding, Modulation, and Channel Estimation

Each $1 \times K$ message vector $\mathbf{m} = [m(1) \ldots m(K)]$ is encoded into a $1 \times N$ codeword $\mathbf{c} = [c(1) \ldots c(N)]$ using a systematic, extended, irregular repeat-accumulate (IRA) code (Section 1.8). IRA codes offer a combination of the linear complexity of turbo encoding and the lower complexity of LDPC decoding without compromising performance. A rate-$1/2$ IRA code is constructed using *density evolution* [80] with maximum node degrees $d_v = 8$ and $d_c = 7$ in (1.230) and (1.231). We obtain a strong IRA code with degree distributions

$$
\begin{aligned}
v(x) &= 0.00008 + 0.31522x + 0.34085x^2 + 0.0.06126x^6 + 0.28258x^7 \\
\chi(x) &= 0.62302x^5 + 0.37698x^6. \qquad (9.35)
\end{aligned}
$$

The IRA systematic parity-check matrix and the generator matrix are defined by (1.232) and (1.234), respectively.

Figure 9.1 shows the block diagram of a dual quaternary DS-CDMA transmitter (Section 2.5) comprising a QPSK modulator and a direct-sequence spreading generator that multiplies orthogonal chip sequences with the in-phase and quadrature modulator inputs. Gray-labeled QPSK (Section 1.7) is used with two encoded bits mapped into a modulation symbol $d_i \in \{\pm 1, \pm j\}$, $i = 1, \ldots, N/2$, $j = \sqrt{-1}$. Although QPSK is assumed, the subsequent analysis and simulation is easily extended to q-ary QAM. Parallel streams of code bits, which represent the real and imaginary components of d_i, are each spread using Gold sequences

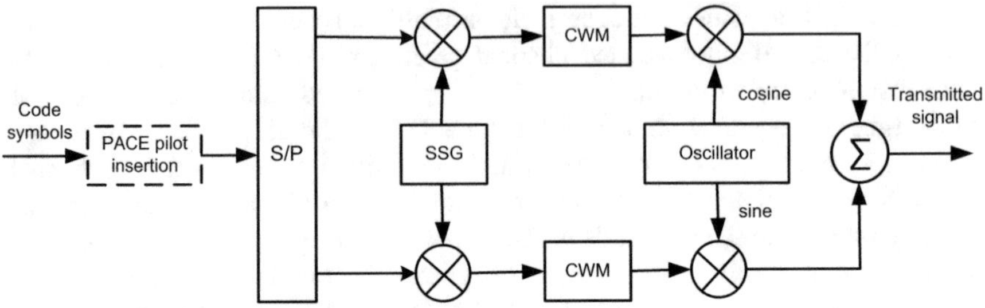

Fig. 9.1 DS-CDMA transmitter with QPSK. *S/P* serial-to-parallel converter, *SSG* spreading-sequence generator, *CWM* chip-waveform modulator

before rectangular pulse-shaping by the chip-waveform modulator. In practice, an intermediate frequency is used prior to the carrier-frequency upconversion, but the upconversion from baseband to the intermediate frequency is omitted for clarity in Figure 9.1.

No channel interleaving is applied to the IRA code because of the inherent interleaving characteristics of the IRA encoder, which can be represented as a repetition code concatenated with an interleaver and an accumulator. The interleaver is essentially embedded within the matrix $\mathbf{H}_1$ that is part of the generator matrix defined by (1.234).

Each codeword or frame comprises $N/2$ QPSK code symbols. There are G spreading-sequence chips for the in-phase and quadrature components of each QPSK symbol, where G is the *component spreading factor*. Each of these frames has two different types of subframes or blocks. A *fading block* comprises N_b code bits or $N_b/2$ QPSK symbols over which the fading amplitude is assumed to be constant. An *interference block* comprises N_{ib} code bits or $N_{ib}/2$ QPSK symbols over which the interference level is assumed to be constant.

Iterative Receiver Structure

Figure 9.2 shows a block diagram of the dual-quaternary iterative receiver. The received signal is downconverted, passed through chip-matched filters, and despread by a synchronized spreading sequence in each branch. The synchronization system is omitted in the figure for clarity. Accurate synchronization in the receiver is assumed to prevent self-interference between the two spreading sequences of the desired signal.

Consider a scenario with flat fading and multiple-access interference. The complex fading amplitude associated with both spreading sequences during a fading block is

$$B = \sqrt{\mathcal{E}_s}\alpha e^{j\theta} \tag{9.36}$$

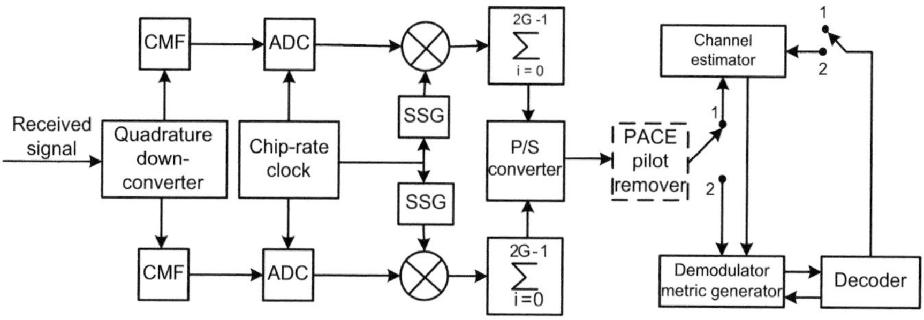

Fig. 9.2 Iterative DS-CDMA receiver with channel estimator. *CMF* chip-matched filter, *ADC* analog-to-digital converter, *SSG* spreading-sequence generator, *P/S* parallel-to-serial

where $\mathcal{E}_s$ is the average energy per QPSK symbol, α is the magnitude of the fading amplitude with $E\left[\alpha^2\right] = 1$, and θ is the unknown fading-induced channel phase. Let $N_0/2$ denote the two-sided PSD of the white Gaussian noise. During a fading block, the complex envelope of the received signal at the ith symbol time, which appears as two successive outputs of the parallel-to-serial converter, may be expressed as

$$y_i = Bd_i + J_i + n_i, \; 1 \leq i \leq \frac{N_b}{2} \tag{9.37}$$

where d_i is the complex transmitted code symbol of the desired signal, n_i is a complex zero-mean, circularly symmetric, Gaussian noise sample with $E\left[|n_i|^2\right] = N_0$ (Section 1.2), and J_i is the interference after the demodulation. If pilot symbols are received, they are removed and applied to the channel estimator. The time-varying multiple-access interference is assumed to be generated by interfering signals with a structure identical to the desired signal, although the spreading sequences differ and the complex fading amplitudes are independent.

A major benefit of the direct-sequence spread spectrum is that the despreading in the receiver tends to whiten the interference PSD over the code-symbol passband, and the subsequent filtering tends to produce a residual interference with an approximate Gaussian distribution (Section 2.4). Thus, the combined interference and thermal noise is assumed to have a two-sided PSD $A/2$ that is constant over each block of N_{ib} code bits but varies from block to block. This model leads to the derivation of an EM estimator for A that is used in the demodulator metrics and leads to suppression of the interference.

A *receiver iteration* is defined as a decoder iteration followed by internal EM iterations in the channel estimator of Figure 9.2, and then a single demodulator metric generation. Let r denote the index for the internal EM iteration, $r = 1, \ldots, r_{max}$; let l denote the index for the closed-loop receiver iteration, $l = 1, \ldots, l_{max}$.

Let $\boldsymbol{\theta}_{(r)}^{(l)} = \left(\widehat{B}_{(r)}^{(l)}, \widehat{A}_{(r)}^{(l)}\right)$ represent the estimates of the complex fading amplitude and interference PSD parameters at the rth EM iteration during the lth overall receiver iteration. EM iterations commence after the initial channel estimation and decoding, which is obtained while the switch in Figure 9.2 is set to position 1. The subsequent receiver iterations are performed while the switch is set to position 2 to refine the initial channel estimate with the aid of soft feedback from the channel decoder.

Application of EM Algorithm

The direct calculation of the maximum-likelihood channel-vector estimator $\boldsymbol{\theta}$ from a received data vector $\mathbf{Y} = [Y(1)\ldots Y(N_d)]$ of N_d code symbols is not feasible because the computational complexity increases exponentially with N_d. Instead, the EM algorithm is used with the *complete received data vector* defined as $\mathbf{Z} = (\mathbf{Y}, \mathbf{D})$, where the missing data vector $\mathbf{D}$ is the transmitted signal vector.

Since $\mathbf{D}$ is independent of the parameter vector $\boldsymbol{\theta}$,

$$\ln f(\mathbf{z}|\boldsymbol{\theta}) = \ln f(\mathbf{y}|\mathbf{d}, \boldsymbol{\theta}) + \ln f(\mathbf{d})\,. \tag{9.38}$$

Assuming independent symbols and zero-mean, circularly symmetric, white Gaussian interference and noise, we obtain

$$f(\mathbf{y}|\mathbf{d}, \boldsymbol{\theta}) = \frac{1}{(\pi A)^{N_d}} \exp\left(-\sum_{i=1}^{N_d} \frac{(|y_i - Bd_i|^2)}{A}\right). \tag{9.39}$$

Therefore, since $|d_i|^2 = 1$,

$$\ln f(\mathbf{y}|\mathbf{d}, \boldsymbol{\theta}) = -N_d \cdot \ln(A) - \frac{1}{A}\sum_{i=1}^{N_d}\left[|y_i|^2 + |B|^2 - 2\,\mathrm{Re}\left(y_i{}^* B d_i\right)\right] \tag{9.40}$$

where an irrelevant constant has been dropped.

E-Step The E-step requires the calculation of the conditional expectation of the conditional log-likelihood of $\mathbf{Z} = (\mathbf{Y}, \mathbf{D})$:

$$\chi\left(\boldsymbol{\theta}, \boldsymbol{\theta}_{(r)}^{(l)}\right) = E_{\mathbf{z}|\mathbf{y},\boldsymbol{\theta}_{(r)}^{(l)}}\left[\ln f(\mathbf{Z}|\boldsymbol{\theta})\right] \tag{9.41}$$

where $\boldsymbol{\theta}_{(r)}^{(l)}$ is the previous estimate. Using (9.38) and (9.40) and observing that $\ln f(\mathbf{d})$ in (9.38) is independent of $\boldsymbol{\theta}$, and hence irrelevant to the subsequent maximization, we obtain

$$\chi\left(\boldsymbol{\theta}, \boldsymbol{\theta}_{(r)}^{(l)}\right) = -N_d \cdot \ln(A) - \frac{1}{A}\sum_{i=1}^{N_d}\left[|y_i|^2 + |B|^2 - 2\,\mathrm{Re}\left(y_i{}^* B \bar{d}_{(r)}^{(l)}(i)\right)\right] \tag{9.42}$$

where

$$\overline{d}_{(r)}^{(l)}(i) = E_{\mathbf{z}|\mathbf{y},\boldsymbol{\theta}_{(r)}^{(l)}}[D(i)] = E_{\mathbf{d}|\mathbf{y},\boldsymbol{\theta}_{(r)}^{(l)}}[D(i)]. \tag{9.43}$$

We assume the independence of each transmitted symbol $D(i)$ and the independence of $D(i)$ and $\boldsymbol{\theta}_{(r)}^{(l)}$. Using Bayes' rule and the fact that (9.39) can be expressed as a product of N_d factors, we obtain

$$\overline{d}_{(r)}^{(l)}(i) = E_{d_i|y_i,\boldsymbol{\theta}_{(r)}^{(l)}}[D(i)] \tag{9.44}$$

$$h\left(d_i|y_i,\boldsymbol{\theta}_{(r)}^{(l)}\right) = \frac{g\left(y_i|\, d_i,\boldsymbol{\theta}_{(r)}^{(l)}\right)}{g\left(y_i|\boldsymbol{\theta}_{(r)}^{(l)}\right)} P\left(D_i = d_i\right) \tag{9.45}$$

$$g\left(y_i|d_i,\boldsymbol{\theta}_{(r)}^{(l)}\right) = \frac{1}{\pi\widehat{A}_{(r)}^{(l)}} \exp\left(-\frac{|y_i - \widehat{B}_{(r)}^{(l)} d_i|^2}{\widehat{A}_{(r)}^{(l)}}\right) \tag{9.46}$$

where $h\,(d_i|\cdot)$ is a conditional probability of $D(i)$, and $g\,(y_i|\cdot)$ is a conditional density function of $Y(i)$.

M-Step Taking the derivative of (9.42) with respect to the real and imaginary parts of the complex-valued B, and then setting the results equal to zero, we obtain the estimate of the complex fading amplitude at iteration $r+1$ as

$$\mathrm{Re}\left(\widehat{B}_{(r+1)}^{(l)}\right) = \frac{1}{N_d}\sum_{i=1}^{N_d} \mathrm{Re}\left(y_i{}^*\overline{d}_{(r)}^{(l)}(i)\right) \tag{9.47}$$

$$\mathrm{Im}\left(\widehat{B}_{(r+1)}^{(l)}\right) = -\frac{1}{N_d}\sum_{i=1}^{N_d} \mathrm{Im}\left(y_i{}^*\overline{d}_{(r)}^{(l)}(i)\right). \tag{9.48}$$

Similarly, maximizing (9.42) with respect to A leads to

$$\widehat{A}_{(r+1)}^{(l)} = \frac{1}{N_d}\sum_{i=1}^{N_d} \left|y_i - \widehat{B}_{(r+1)}^{(l)}\overline{d}_{(r)}^{(l)}(i)\right|^2. \tag{9.49}$$

These equations indicate that the unknown parameters can be estimated once $\overline{d}_{(r)}^{(l)}(i)$ has been estimated.

Since the probabilities $s_1 = P[D_i = +1]$, $s_2 = P[D_i = +j]$, $s_3 = P[D_i = -1]$, and $s_4 = P[D_i = -j]$ are unknown, the decoder estimates them by using the code-symbol probabilities $s_\beta^{(l)}$, $\beta = 1, 2, 3, 4$, obtained from the soft outputs of the channel decoder after receiver iteration l. Using these estimated probabilities, (9.44), and (9.45), we obtain

$$\overline{d}_{(r)}^{(l)}(i) \simeq \left[g\left(y_i|\boldsymbol{\theta}_{(r)}^{(l)}\right)\right]^{-1} [s_1^{(l)} g\left(y_i|\ 1,\boldsymbol{\theta}_{(r)}^{(l)}\right) + js_2^{(l)} g\left(y_i|\ j,\boldsymbol{\theta}_{(r)}^{(l)}\right)$$
$$-s_3^{(l)} g\left(y_i|\ -1,\boldsymbol{\theta}_{(r)}^{(l)}\right) - js_4^{(l)} g\left(y_i|\ -j,\boldsymbol{\theta}_{(r)}^{(l)}\right)] \tag{9.50}$$

where

$$g\left(y_i|\boldsymbol{\theta}_{(r)}^{(l)}\right) \simeq s_1^{(l)} g\left(y_i|\ 1,\boldsymbol{\theta}_{(r)}^{(l)}\right) + s_2^{(l)} g\left(y_i|\ j,\boldsymbol{\theta}_{(r)}^{(l)}\right)$$
$$+s_3^{(l)} g\left(y_i|\ -1,\boldsymbol{\theta}_{(r)}^{(l)}\right) + s_4^{(l)} g\left(y_i|\ -j,\boldsymbol{\theta}_{(r)}^{(l)}\right). \tag{9.51}$$

Substituting (9.46) into (9.50) and (9.51), we find that the expectation of $D(i)$ at the rth EM and lth receiver iteration is calculated as

$$\overline{d}_{(r)}^{(l)}(i) \simeq \frac{s_1^{(l)} R_{1,(r)}^{(l)} + js_2^{(l)} R_{2,(r)}^{(l)} - s_3^{(l)} R_{3,(r)}^{(l)} - js_4^{(l)} R_{4,(r)}^{(l)}}{\sum\limits_{\beta=1}^{4} s_\beta^{(l)} R_{\beta,(r)}^{(l)}} \tag{9.52}$$

where likelihood-ratio $R_{\beta,(r)}^{(l)}$ depends on the current channel estimates as

$$R_{1,(r)}^{(l)} = \exp\left[\frac{2}{\widehat{A}_{(r)}^{(l)}} \operatorname{Re}(\widehat{B}_{(r)}^{(l)} y_i)\right], \quad R_{2,(r)}^{(l)} = \exp\left[\frac{2}{\widehat{A}_{(r)}^{(l)}} \operatorname{Im}(\widehat{B}_{(r)}^{(l)} y_i)\right]$$

$$R_{3,(r)}^{(l)} = \exp\left[-\frac{2}{\widehat{A}_{(r)}^{(l)}} \operatorname{Re}(\widehat{B}_{(r)}^{(l)} y_i)\right], \quad R_{4,(r)}^{(l)} = \exp\left[-\frac{2}{\widehat{A}_{(r)}^{(l)}} \operatorname{Im}(\widehat{B}_{(r)}^{(l)} y_i)\right] \tag{9.53}$$

and

$$\widehat{B}_{(0)}^{(l)} = \widehat{B}_{(r_{\max})}^{(l-1)}, \quad 1 \le l \le l_{\max}. \tag{9.54}$$

Methods for obtaining the initial estimates $\left(\widehat{B}_{(r_{max})}^{(0)}, \widehat{A}_{(r_{\max})}^{(0)}\right)$ of each fading block are described subsequently.

Therefore, for a given receiver iteration, $\overline{d}_{(r)}^{(l)}(i)$ and $R_{\beta,(r)}^{(l)}$ are updated r_{max} times using decoder feedback $s_\beta^{(l)}$. In the next receiver iteration, after channel-code re-estimation, the fading-amplitude and interference PSD estimates are updated, and then used by the demodulator and channel decoder to recompute $\overline{d}_{(r)}^{(l+1)}(i)$ and $R_{\beta,(r)}^{(l+1)}$. This process is repeated again for r_{max} EM iterations, and the aforementioned cycles continue similarly for subsequent receiver iterations.

In estimating the fading parameters, we set $N_d = N_b/2$; in estimating A, we choose $N_{ib} \leq N_b$ and set $N_d = N_{ib}/2$. The EM estimator first finds the value of $\widehat{B}_{(r)}^{(l)}$ for a fading block of size N_b by using (9.47), (9.48), and (9.52)-(9.54), none of which require $\widehat{A}_{(r)}^{(l)}$. Then, it finds the value of $\widehat{A}_{(r)}^{(l)}$ for each smaller or equal interference block of size N_{ib} using (9.49) with the value of $\widehat{B}_{(r)}^{(l)}$ found for the larger or equal fading block.

When pilot symbols are used, $\overline{d}_{(r)}^{(l)}(i) = d_i$ for each known pilot symbol, and there are no EM iterations if only known pilot symbols are processed in calculating the channel estimates. The number of EM iterations and the receiver latency are reduced by applying a *stopping criterion*. Iterations stop once $\widehat{B}_{(r)}^{(l)}$ is within a specified fraction of its value at the end of the previous iteration or a specified maximum number of iterations is reached.

The estimates $\widehat{A}_{(r_{\max})}^{(l)}$ and $\widehat{B}_{(r_{\max})}^{(l)}$ and decoder log-likelihood ratios for each of the QPSK bits are fed back to the demodulator as part of the iterative demodulation and decoding (Section 1.7). The demodulator then computes extrinsic log-likelihood ratios given by (1.215) that are applied to the channel decoder. The constellation labeling of a QPSK symbol $d_i \in \{\pm 1, \pm j\}$ by bits $b_1(i)$ and $b_2(i)$ is the following. The symbol $d_i = +1$ is labeled 00; the symbol $d_i = +j$ is labeled 01; the symbol $d_i = -1$ is labeled 11; the symbol $d_i = -j$ is labeled 10. Let $b_1(i)$ and $b_2(i)$ denote the bits of symbol d_i, and v_1, v_2 the corresponding log-likelihood ratios that are fed back by the channel decoder after receiver iteration l. Partition the set of possible symbols of d_i into two disjoint sets $\mathcal{D}_i^{(1)}$ and $\mathcal{D}_i^{(0)}$, where $\mathcal{D}_i^{(b)}$ contains all symbols labeled with $b_1(i) = b$. Substituting (9.46) into (1.215), accounting for the different notation in (1.215) by setting $k \to i$, $s \to d_i$, $l = 1$, and $m = 2$ in (1.215), and canceling common factors, we obtain the *extrinsic log-likelihood ratio* for $b_1(i)$:

$$z_1^{(l)}(i) = \ln \left[\frac{\sum\limits_{d_i \in \mathcal{D}_i^{(1)}} \exp\left\{ \frac{2}{\widehat{A}_{(r_{\max})}^{(l)}} \operatorname{Re}\left[\widehat{B}_{(r_{\max})}^{(l)} y_i^* d_i \right] + b_2(i)\, v_2 \right\}}{\sum\limits_{d_i \in \mathcal{D}_i^{(0)}} \exp\left\{ \frac{2}{\widehat{A}_{(r_{\max})}^{(l)}} \operatorname{Re}\left[\widehat{B}_{(r_{\max})}^{(l)} y_i^* d_i \right] + b_2(i)\, v_2 \right\}} \right] \tag{9.55}$$

where both sums are over two symbols.

Let

$$F^{(l)}(i) = \frac{2}{\widehat{A}_{(r_{\max})}^{(l)}} \operatorname{Re}\left[\widehat{B}_{(r_{\max})}^{(l)} y_i^* \right] \tag{9.56}$$

$$G^{(l)}(i) = \frac{2}{\widehat{A}_{(r_{\max})}^{(l)}} \operatorname{Im}\left[\widehat{B}_{(r_{\max})}^{(l)} y_i^* \right]. \tag{9.57}$$

In the sum in the numerator of (9.55), $b_1(i) = 1$ implies that $d_i = -j$ or -1. If $d_i = -j$, then the argument of the exponential function is equal to $G^{(l)}(i)$; if $d_i = -1$, then the argument is equal to $-F^{(l)}(i)+v_2$. In the sum in the denominator, $b_1(i) = 0$ implies that $d_i = +j$ or $+1$. If $d_i = +j$, then the argument of the exponential function is equal to $-G^{(l)}(i)+v_2$; if $d_i = +1$, then the argument is equal to $F^{(l)}(i)$. Similar calculations provide the extrinsic log-likelihood ratio for $b_2(i)$. Therefore, the demodulation metrics (extrinsic log-likelihood ratios) $z_\nu^{(l)}(i)$, $\nu = 1, 2$, for bits $1, 2$ of symbol i that are applied to the channel decoder are

$$z_1^{(l)}(i) = \ln\left\{\frac{\exp\left[G^{(l)}(i)\right] + \exp\left[-F^{(l)}(i)+v_2\right]}{\exp\left[F^{(l)}(i)\right] + \exp\left[-G^{(l)}(i)+v_2\right]}\right\} \tag{9.58}$$

$$z_2^{(l)}(i) = \ln\left\{\frac{\exp\left[-G^{(l)}(i)\right] + \exp\left[-F^{(l)}(i)+v_1\right]}{\exp\left[F^{(l)}(i)\right] + \exp\left[G^{(l)}(i)+v_1\right]}\right\}. \tag{9.59}$$

Perfect Phase Information at Receiver

The carrier synchronization provided by a phase-locked loop in several cellular standards, such as CDMA2000, can be exploited to obviate the need to estimate the channel phase. Assuming perfect phase information at the receiver, the fading amplitude is real-valued and nonnegative, and (9.48) does not have to be computed. The EM algorithm generates updated channel estimates according to (9.47) and (9.49) *after* the initial coherent demodulation and decoding that precedes the first receiver iteration. Blind initial estimates for each fading block can be obtained from the received symbols as

$$\widehat{B}^{(0)}_{(r_{max})} = \frac{2}{N_b}\sum_{i=1}^{N_b/2} |y_i| \tag{9.60}$$

$$\widehat{A}^{(0)}_{(r_{\max})} = \max\left[P_s - \left(\widehat{B}^{(0)}_{(r_{max})}\right)^2, c\left(\widehat{B}^{(0)}_{(r_{max})}\right)^2\right] \tag{9.61}$$

where

$$P_s = \frac{2}{N_b}\sum_{i=1}^{N_b/2} |y_i|^2 \tag{9.62}$$

represents the average power of the received symbols, and $P_s - \left(\widehat{B}^{(0)}_{(r_{max})}\right)^2$ is the difference between that power and the average power of a desired symbol. Equation (9.60) would provide a perfect estimate in the absence of interference and noise. The parameter $c > 0$ is chosen such that $\left(\widehat{B}^{(0)}_{(r_{max})}\right)^2 / \widehat{A}^{(0)}_{(r_{\max})}$ does not exceed some maximum value, and here a constant $c = 0.1$ is always used for simplicity. This approach to the initial channel estimates is called *blind method I* in the sequel.

Although the EM estimation is a relatively low-complexity iterative approach to maximum-likelihood estimation, it consumes a much larger number of floating-point operations than pilot-assisted schemes do. To evaluate the complexity of the EM estimator in terms of required real additions and multiplications per block of N_d code symbols, each complex addition is equated to 2 real additions, each complex multiplication is equated to 4 real multiplications, and divisions are equated to multiplications. Equations (9.47)-(9.49) require $6N_d+4$ real additions and $12N_d+4$ real multiplications per EM iteration. Equations (9.58) and (9.59) require 6 real additions, 30 real multiplications, and the computation of 4 exponentials per EM iteration. Each of these calculations is repeated for each of $l_{\max}r_{\max}$ total EM iterations. The initial estimates calculated using (9.60)-(9.62), which only need to be computed once prior to the first EM iterations, require $2N_d$ real additions, $8N_d+7$ real multiplications, and the computation of the maximum of two real numbers. A PACE receiver that uses only pilot symbols for channel estimation requires $6N_d+4$ real multiplications and $12N_d+4$ real multiplications to compute (9.47)-(9.49) once and does not need to compute the other equations. Thus, EM estimation increases the amount of computation for channel estimation by a factor of more than $l_{\max}r_{\max}$ relative to PACE.

No Phase Information at Receiver

The initial channel estimates in (9.60) and (9.61) for blind method I are expected to be degraded significantly when the phase information is unknown, because an arbitrary initial phase value (e.g., 0 radians) must be assumed. To circumvent this problem, the initial receiver iteration consists of hard-decision demodulation and channel decoding, after which each decoded bit is used as $\overline{d}^{(0)}_{(r_{max})}(i)$ in (9.47)-(9.49). This step is followed by the regular EM estimation process in subsequent receiver iterations. This approach to the initial channel estimates, which is referred to as *blind method II* in the sequel, results in increased receiver latency relative to the previous method when phase information is available.

Options for Blind Methods

When the frame duration of a system with PACE is fixed but the pilot symbols are not transmitted, the following options are available for blind methods I and II:

- (case A) an increase in the number of transmitted information symbols
- (case B) an increase in the duration of transmitted symbols
- (case C) an increase in the number of transmitted parity bits (lowered IRA code rate).

These options offset the loss in system performance due to the degraded channel estimation obtained from blind methods I and II with respect to PACE. Assuming that the no-pilot cases A, B, and C have the same transmitted frame duration as the frame with pilot symbols, cases A, B, and C provide the most favorable throughput, spectral efficiency, and bit error probability, respectively.

To compare the options, simulations were conducted. In all the simulations, the codeword blocks have 2200 bits, and the bit rate is 100 kbs. The iterative PACE system considered for comparison uses a rate-$1/2$ IRA code with $K = 1000$ information bits, $N = 2000$ code bits, and 200 pilot symbols, which implies a 9.1% pilot-symbol overhead. In most of the simulations, except where stated, the fading blocks have $N_b = 40$ bits. Increasing the fading-block sizes increases the accuracy of the EM estimators, but decreasing the fading-block sizes allows closer tracking of the channel parameters and includes more diversity in the receiver computations.

The number of closed-loop receiver iterations is set to $l_{max} = 9$, as there is insignificant performance improvement for $l_{max} > 9$. The number of internal EM iterations is $r_{\max} = 10$. The IRA code does not use channel interleaving and is decoded by the sum-product algorithm (Section 1.8). The component spreading factor is $G = 31$, and the mobile velocity is 120 km/h unless otherwise stated. For each of the representative scenarios tested, 5000 Monte Carlo simulation trials were conducted.

Flat fading is assumed in most of the simulations, whereas a frequency-selective channel is examined in the final simulation. The fading in a block correlates with the fading in the other blocks. The correlated fading model uses the autocorrelation of the channel response for two-dimensional isotropic scattering given by (6.44). The complex fading amplitude during block n is computed as

$$B_n = \sqrt{J_0(2\pi f_d T_f)}B_{n-1} + \sqrt{1 - J_0(2\pi f_d T_f)}B_{dn}, \quad B_1 = B_{d1} \tag{9.63}$$

where f_d is the Doppler shift defined by (6.5), T_f is the duration of a fading block, and B_{dn} is a complex fading amplitude selected for block n from the complex zero-mean, circularly symmetric, Gaussian distribution. For this distribution, the magnitude of the amplitude has a Rayleigh distribution, and the phase has a uniform distribution.

The bit error rate (BER), which is equal to the information-bit error probability (Section 1.1), is calculated as a function of the energy-to-noise-density ratio $\mathcal{E}_b/N_0$, where $\mathcal{E}_b$ is the energy per information bit in the PACE system. The *information throughput* is a vital performance criterion in addition to the BER. One of the primary motivations in removing pilot symbols is the expectation of achieving greater information throughput, even though the BER may be degraded marginally. The information throughput is defined as

$$\mathcal{T} = \frac{\text{information bits in a codeword}}{\text{codeword duration}} \times (1 - BER) \quad \text{bps}. \tag{9.64}$$

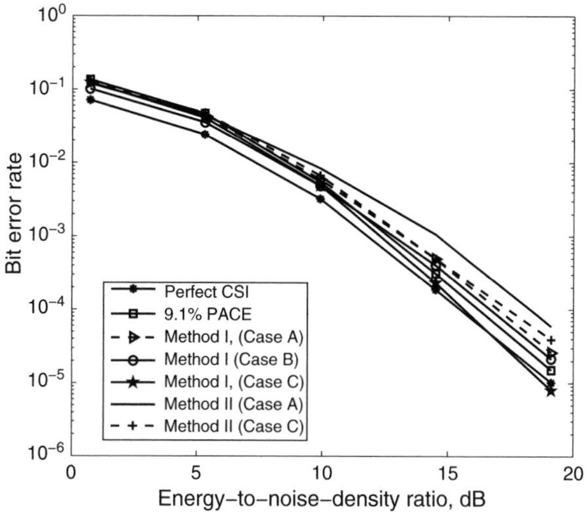

Fig. 9.3 BER versus $\mathcal{E}_b/N_0$ for IRA-coded iterative receiver in a single-user environment with a perfectly estimated phase [106]

Single-User Environment, Perfect Phase Knowledge

Figures 9.3 and 9.4 illustrate the performance when there is a single signal received with perfect phase knowledge. Figure 9.3 displays the BER versus $\mathcal{E}_b/N_0$ for an IRA-coded iterative receiver operating with perfect CSI, PACE, blind method I with cases *A*, *B*, and *C*, and blind method II with cases *A* and *C*. The key observation is that blind method II is worse than method I by 2 dB at $BER = 10^{-3}$ for both case *A* and case *C*, which illustrates the sensitivity of the EM algorithm to the accuracy of the initial estimates.

The addition of extra parity bits to blind method I (case *C*, rate-1000/2200) offers the greatest improvement in BER, surpassing even the rate-1/2 code with perfect CSI at high $\mathcal{E}_b/N_0$. The increase in the number of information symbols (case *A*) results in the worst BER performance with a separation of 1 and 0.5 dB from PACE and case *B* at $BER = 10^{-3}$ respectively. The various scenarios featured in the figure were also tested under a slow-fading channel with a mobile velocity of 10 km/h, which implies a reduction in the maximum Doppler shift by a factor of 12. It was observed that all the BER curves were shifted toward the right by as much as 7 dB at $BER = 10^{-3}$ because of the loss of diversity among the fading blocks, but the overall trends among the different cases remained the same.

Figure 9.4 exhibits information throughput $\mathcal{T}$ versus $\mathcal{E}_b/N_0$ for the IRA-coded iterative receiver with the scenarios of Figure 9.3. The throughput advantage of case *A* is achieved even though no pilot symbols are used at all; that is, the initial estimation is blind. It is evident that increasing the symbol duration or adding additional parity information does not give the blind methods any significant advantage in throughput over PACE. Both blind methods with cases *B*, *C* and PACE provide about 20% less throughput than the receiver with perfect CSI.

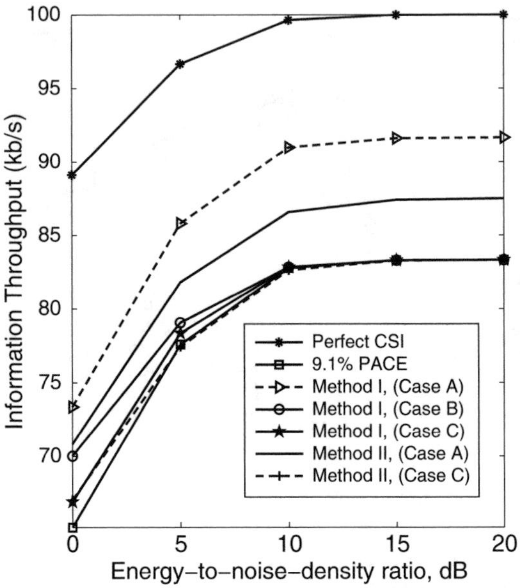

Fig. 9.4 Information throughput versus $\mathcal{E}_b/N_0$ for an IRA-coded iterative receiver in single-user environment with a perfectly estimated phase [106]

Multiuser Environment, Unknown Phase

A four-signal interference environment with equal mean bit energies for all signals at the receiver, $\mathcal{E}_b/N_0 = 20$ dB, and no phase information at the receiver is examined next. We assume that both the interference levels and the unknown phase are constant during each subframe. Each interference signal experiences independent correlated fading and uses independent data and Gold sequences with respect to the desired signal. The simulation uses chip-synchronous interference signals, which is a worst-case assumption (Section 7.3). Two variations of channel estimation are examined here: *partially adaptive* with only complex fading amplitude $\widehat{B}_{(i)}^{(l)}$ estimated using (9.47) and (9.48), and $\widehat{A}_{(r)}^{(l)}$ set equal to N_0 for all subframes; and *fully adaptive* estimation of both $\widehat{B}_{(r)}^{(l)}$ and $\widehat{A}_{(r)}^{(l)}$ using (9.47), (9.48), and (9.49).

Figure 9.5 displays IRA-coded BER versus $\mathcal{E}_b/N_0$ for partially and fully adaptive channel estimation per fading block and case C for both blind methods. The mismatch of $\widehat{A}$ and the true value of A at the demodulator and decoder results in a high error floor for the partially adaptive cases. The intuition behind the error floor is that the partially adaptive estimator overestimates the true signal-to-interference-and-noise ratio (SINR) by disregarding the multiple-access interference, with the degree of overestimation increasing with SINR. The fully adaptive estimation offers a more accurate SINR estimate and hence suppresses interference and reduces the error floor significantly. This interference suppression is achieved without using the far more elaborate multiuser and interference-cancelation methods (Sections 7.7 and 7.8) that could be implemented in a DS-CDMA receiver. For both partially and fully adaptive estimation, it is observed that blind method II now outperforms blind method I because of better phase estimation, whereas both blind methods outperform PACE at $BER = 10^{-3}$ because of the added parity information.

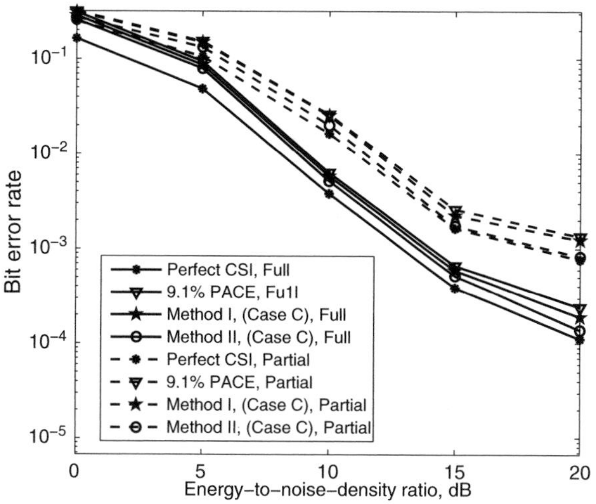

Fig. 9.5 BER versus $\mathcal{E}_b/N_0$ for IRA-coded iterative receiver affected by multiple-access interference from four mobiles, fully and partially adaptive estimation, and unknown phase [106]

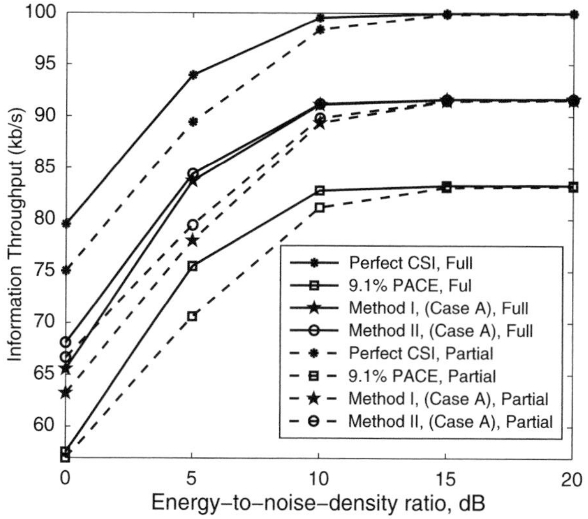

Fig. 9.6 Information throughput versus $\mathcal{E}_b/N_0$ for an IRA-coded iterative receiver affected by multiple-access interference from four mobiles, fully and partially adaptive estimation, and unknown phase [106]

Figure 9.6 demonstrates the IRA-coded receiver throughput offered by the blind methods with case A compared with PACE under multiple-access interference. The blind methods always provide a better throughput compared with PACE; for example, blind method I with case A is superior by 9% to both PACE scenarios when $\mathcal{E}_b/N_0 > 5$ dB. It is observed that both partial and fully adaptive estimation methods offer a similar asymptotic throughput, which indicates that partial channel estimation may be sufficient for applications with a non-stringent BER criterion. On the other hand, error-critical applications requiring less than $BER = 10^{-3}$ must use the fully adaptive channel estimation, as seen from Figure 9.5.

Varying Fading-Block Size, Unknown Phase

In urban mobile environments, the phase can be expected to change significantly after approximately $0.01/f_d$ - $0.04/f_d$ seconds, where f_d is the maximum Doppler shift. For the assumed mobile velocity of 120 km/h, this time range corresponds to roughly 10-40 code bits at 100 kbs. The fading and interference block sizes $N_b = N_{ib}$ are therefore varied accordingly, and *no* phase information is assumed to be available at the receiver for the next set of results.

Figure 9.7 displays fully adaptive IRA-coded BER versus $\mathcal{E}_b/N_0$ for blind methods I and II with case C, 9.1% PACE, and perfect CSI decoding for $N_b = 10$ and 40 in a single-user environment. An improvement of 1 to 2 dB is observed for all methods for the smaller fading-block size of $N_b = 10$ because of the increased fading diversity. The throughput with case A is shown in Figure 9.8. It is observed that the throughput gains of the blind methods over PACE (roughly 9% at medium to high $\mathcal{E}_b/N_0$) are preserved, even when the phase is initially unknown at the receiver.

Varying Multiple-Access Interference, Unknown Phase

Figure 9.9 displays IRA-coded iterative receiver performance for blind method II, case C with three and six multiple-access interference signals and equal mean

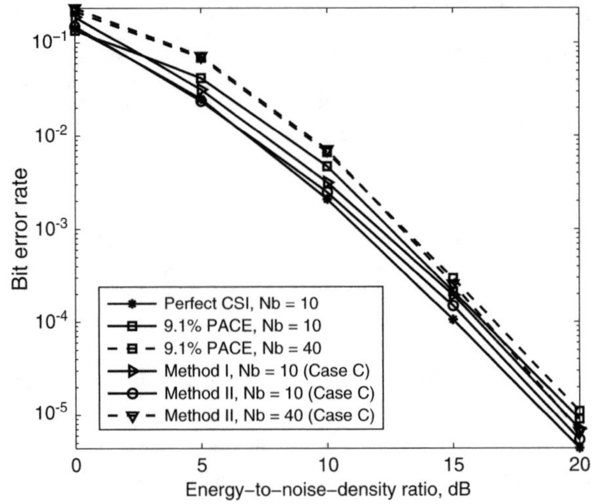

Fig. 9.7 BER versus $\mathcal{E}_b/N_0$ for an IRA-coded iterative receiver in a single-user environment, varying N_b, unknown phase [106]

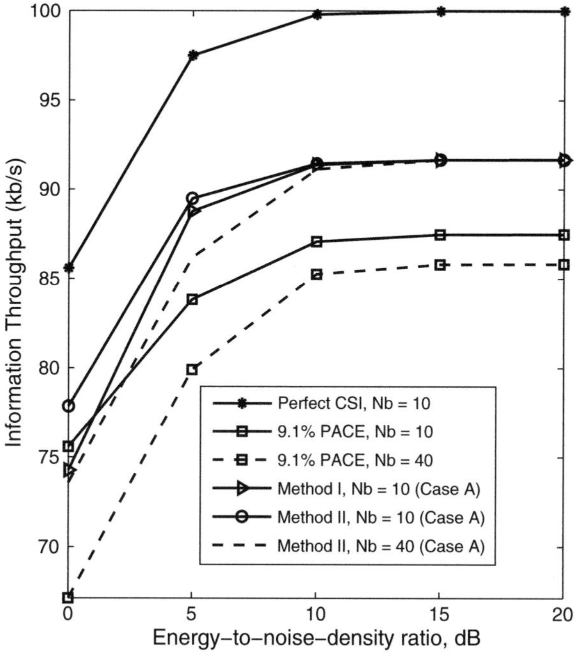

Fig. 9.8 Information throughput versus $\mathcal{E}_b/N_0$ for an IRA-coded iterative receiver in a single-user environment, varying N_b, unknown phase [106]

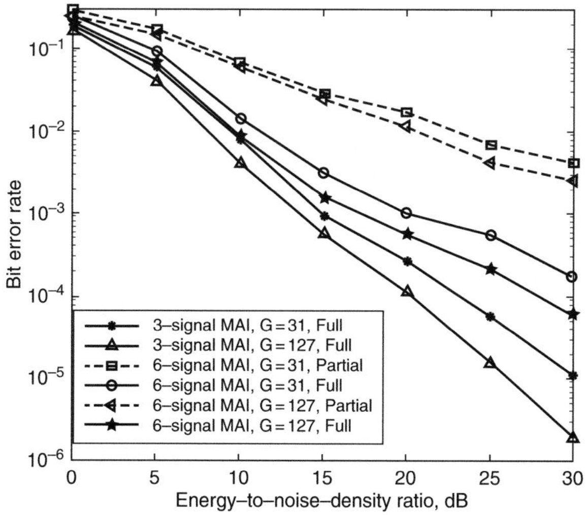

Fig. 9.9 BER versus $\mathcal{E}_b/N_0$ for an IRA-coded iterative receiver affected by an unknown phase and various component spreading factors, number of multiple-access interference (MAI) signals, and degrees of adaptation [106]

bit energies for all signals. The partially adaptive estimation is unable to cope with the interference caused by six multiple-access interference signals regardless of the spreading factor, whereas the fully adaptive estimation offers a substantial improvement in BER. The benefit of an increased component spreading factor ($G = 127$ versus $G = 31$) is more apparent at low bit error rates for fully adaptive estimation. For example, the fully adaptive estimation with three multiple-access interference signals improves by a factor of approximately 5 dB at $BER = 10^{-5}$, despite nonorthogonal spreading sequences and imperfect CSI.

Multipath Channel

A DS-CDMA system can exploit a frequency-selective fading channel by using a rake demodulator (Section 6.10). As an example, we assume a channel with three resolvable multipath components (with known delays) of the desired signal and a rake demodulator with three corresponding fingers. The multipath components undergo independent fading across the fingers, but follow the correlated fading model of (9.63) over time. The magnitudes of the fading amplitudes of the components follow an exponentially decaying power profile across the fingers:

$$E\left[\alpha_l^2\right] = e^{-(l-1)}, \quad l = 1, 2, 3. \tag{9.65}$$

Each interference signal has the same power level in each finger and undergoes independent correlated fading. Because of the independent multipath fading amplitudes for the desired signal, the EM-based channel estimation is performed separately in each finger. The rake demodulator performs maximal-ratio combining (Section 6.4) of the received symbol copies based on channel estimates computed at all fingers. The symbol metric obtained from the rake demodulator is then passed to the QPSK demodulator metric generator, which generates soft inputs for the common decoder. The soft outputs of the decoder are fed back to the three channel estimator blocks, which then recompute updated fading amplitudes.

Figure 9.10 displays the rake demodulator performance for three multiple-access interference signals with method II under case C, where all signals are spread by length-127 Gold sequences. It is observed that the additional diversity due to rake combining improves performance as expected, but the performance disparity between partially and fully adaptive estimation remains large.

Comparison of Options

The simulation results indicate that pilot symbols are not essential to the effectiveness of DS-CDMA receivers with coding, coherent detection, and channel estimation. If the pilot symbols are replaced by information symbols, the throughput increases relative to PACE whether or not interference is present. If the BER is the primary performance criterion, then replacing the pilot symbols by parity symbols

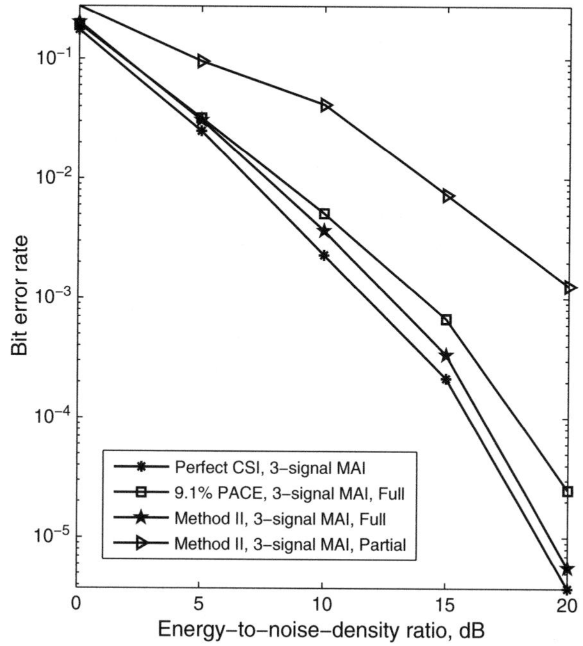

Fig. 9.10 BER versus $\mathcal{E}_b/N_0$ for an IRA-coded iterative rake receiver with three resolvable multipaths, three fingers, and three MAI signals [106]

gives a lower BER than PACE. If the spectral efficiency is of primary importance, then extending the symbol duration after the removal of the pilot symbols offers an improvement relative to PACE, albeit at the cost of a slight increase in the BER.

The simulation results indicate that the despreading and the subsequent estimation of the interference PSD enables the significant suppression of interference. This suppression is achieved without using the far more elaborate multiuser and interference-cancelation methods that could be implemented in a DS-CDMA receiver.

9.3 Guidance from Information Theory

Information theory (see Section 7.1) is renowned for establishing fundamental limits on what can be achieved by a communication system. The theory also provides insight into favorable choices of code rates and signal characteristics. The guidance provided by information theory is used in the next section to design robust frequency-hopping systems.

Let $\mathbf{X}$ and $\mathbf{Y}$ denote continuously distributed random vectors, which are vectors with components that are continuously distributed random variables. Let $f(\mathbf{x}, \mathbf{y})$ denote the joint density function of $\mathbf{X}$ and $\mathbf{Y}$, and let $f(\mathbf{x})$ and $f(\mathbf{y})$ denote the associated marginal density functions. If $\mathbf{X}$ is transmitted and $\mathbf{Y}$ is received, the *average mutual information* between $\mathbf{X}$ and $\mathbf{Y}$, in bits per channel use, is defined as [20, 69]

$$I(\mathbf{X};\mathbf{Y}) = \int_{R(\mathbf{y})}\int_{R(\mathbf{x})} f(\mathbf{x},\mathbf{y})\log_2 \frac{f(\mathbf{x},\mathbf{y})}{f(\mathbf{x})f(\mathbf{y})} d\mathbf{x}d\mathbf{y} \tag{9.66}$$

where $R(\mathbf{y})$ and $R(\mathbf{x})$ are the domains or regions of integration for $\mathbf{y}$ and $\mathbf{x}$, respectively. The *channel capacity* is defined as the maximum value of $I(\mathbf{X};\mathbf{Y})$ over all possible choices of the density function $f(\mathbf{x})$.

Digital communication systems transmit discrete-valued symbols and receive continuous-valued outputs. Let X denote a discrete random variable that is drawn from an input alphabet of q symbols and is applied to the input of a modulator. Let the continuously distributed random vector $\mathbf{Y}$ denote the channel outputs or matched-filter outputs. The *average mutual information between* X *and* $\mathbf{Y}$ is defined as

$$I(X,\mathbf{Y}) = \sum_{i=1}^{q} P[X=x_i]\int_{\mathsf{R}(\mathbf{y})} f(\mathbf{y}|x_i)\log_2 \frac{f(\mathbf{y}|x_i)}{f(\mathbf{y})} d\mathbf{y} \tag{9.67}$$

where $P[X=x_i]$ is the probability that $X = x_i, i = 1,2,\ldots,q$, and $f(\mathbf{y}|x_i)$ is the conditional density function of $\mathbf{Y}$ given that $X = x_i$. This equation can be obtained from (9.66) by making the replacements $f(\mathbf{x}) \to P[X=x_i]$ and $f(\mathbf{x},\mathbf{y}) \to f(\mathbf{y}\,|x_i)P[X=x_i]$ and replacing one of the integrals by a summation. The density function $f(\mathbf{y})$ may be expressed as

$$f(\mathbf{y}) = \sum_{i=1}^{q} P[X=x_i]f(\mathbf{y}|x_i). \tag{9.68}$$

If (9.67) is maximized with respect to $P[X=x_i]$, the average mutual information is called the *channel capacity of the discrete-input, continuous-output channel.*

Suppose that the channel symbols are selected to have equal probability so that $P[X=x_i] = 1/q, i = 1,2,\ldots,q$, in (9.67) and (9.68). Then the *symmetric channel capacity* is defined to be the average mutual information for equally likely symbols:

$$C = \log_2 q + \frac{1}{q}\sum_{i=1}^{q}\int_{\mathsf{R}(\mathbf{y})} f(\mathbf{y}|x_i)\log_2 \frac{f(\mathbf{y}|x_i)}{\sum_{i=1}^{q} f(\mathbf{y}|x_i)} d\mathbf{y}. \tag{9.69}$$

Consider a fading channel and a complex fading amplitude $\mathcal{A}$ during each symbol interval. The *ergodic channel capacity* is the channel capacity averaged over all possible channel states. If the channel symbols are equally likely, the *ergodic symmetric channel capacity* is

$$C = \log_2 q + \frac{1}{q}\sum_{i=1}^{q}\int_{\mathsf{R}(a)}\int_{\mathsf{R}(\mathbf{y})} g(a)f(\mathbf{y}|x_i,a)\log_2 \frac{f(\mathbf{y}|x_i,a)}{\sum_{i=1}^{q} f(\mathbf{y}|x_i,a)} d\mathbf{y}da \tag{9.70}$$

where $g(a)$ is the two-dimensional density function of the real and imaginary components of the complex fading amplitude, $\mathsf{R}(a)$ is the region of integration of the complex amplitude, and $f(\mathbf{y}\ |x_i, a)$ is the conditional density function of $\mathbf{Y}$ given that $X = x_i$ and the complex amplitude is $\mathcal{A} = a$.

9.4 Robust Frequency-Hopping Systems

This section describes and analyzes a robust frequency-hopping system with noncoherent detection, iterative turbo decoding and demodulation, and channel estimation [105]. The system is designed to be effective not only when operating over the AWGN and fading channels but also in environments with multiple-access interference and multitone jamming.

Noncoherent or differentially coherent demodulation has practical advantages and is often necessary because of the difficulty of phase estimation after every frequency hop. A plausible choice of modulation is orthogonal continuous-phase frequency-shift keying (CPFSK). With orthogonal CPFSK, the energy efficiency can be improved by increasing the alphabet size q, which is equal to the number of possible transmit frequencies in the signal set during each hop dwell interval. The problem is that a large bandwidth B_u of each frequency channel, although necessary to support a large number of transmit frequencies, reduces the number of frequency channels available when the hopping is over a spectral region with fixed bandwidth W. This reduction makes the system more vulnerable to both multiple-access frequency-hopping signals and multitone jamming. A reduction in B_u is obtained by using nonorthogonal CPFSK. As an example of the importance of B_u, consider multitone jamming of a frequency-hopping system with q-ary CPFSK in which the thermal noise is absent and each jamming tone has its carrier frequency within a distinct frequency channel. The uncoded symbol-error probability is given by (3.69), which explicitly indicates the significant benefit of a small bandwidth in reducing the effect of multitone jamming.

Robust system performance is provided by using nonorthogonal CPFSK, a turbo code, *bit-interleaved coded modulation* (BICM), iterative decoding and demodulation, and channel estimation. The bandwidth of q-ary CPFSK decreases with reductions in the modulation index h. Although the lack of orthogonality when $h < 1$ causes a performance loss for the AWGN and fading channels, the turbo decoder makes this loss minor compared with the gain against multiple-access interference and multitone jamming.

The system with noncoherent, nonorthogonal CPFSK has the following primary advantages relative to other systems with differential detection, coherent detection, or orthogonal modulation.

1. No extra reference symbol and no estimation of the phase offset in each dwell interval are required.
2. It is not necessary to assume that the phase offset is constant throughout a dwell interval.

3. The channel estimators are much more accurate and can estimate an arbitrary number of interference and noise spectral-density levels.
4. The compact spectrum during each dwell interval allows more frequency channels and, hence, enhances performance against multiple-access interference and multitone jamming.
5. Because noncoherent detection is used, system complexity is independent of the choice of h, and thus there is much more design flexibility than is possible in coherent CPFSK systems.

System Model

In the transmitter of the system, which uses bit-interleaved coded-modulation with iterative decoding (BICM-ID; Sections 1.7 and 6.9), encoded message bits are interleaved and then placed into a $1 \times N_d$ vector $\mathbf{d}$ with elements $d_i \in \{1, 2, \ldots, q\}$, each of which represents $m = \log_2 q$ bits. The vector $\mathbf{d}$ generates the sequence of tones that are frequency-translated by the carrier frequency of the frequency-hopping waveform. After the modulated signal passes through an AWGN or fading channel with partial-band or multiple-access interference, the receiver front-end dehops the signal, as shown in Figure 9.11. The dehopped signal passes through a bank of q matched filters, each of which is implemented as a quadrature pair (Section 2.5). The output of each matched filter is sampled at the symbol rate to produce a sequence of complex numbers. Assuming that symbol synchronization exists, the complex samples are then placed into a $q \times N_d$ matrix $\mathbf{Y}$ with an ith column that represents the outputs of the matched filters corresponding to the ith received symbol. The matrix $\mathbf{Y}$ is applied to the channel estimator and is used to produce an $m \times N_d$ matrix $\mathbf{Z}$ of demodulator bit metrics.

The demodulator exchanges information with both the turbo decoder and the channel estimators. After deinterleaving, the demodulator bit metrics are applied to the decoder. The decoder feeds a priori information (in the form of an $m \times N_d$ matrix $\mathbf{V}$ of decoder bit metrics) back to the demodulator and channel estimator, in accordance with the turbo principle. Frequency-selective fading changes the amplitude from hop to hop, and the partial-band and multiple-access interference

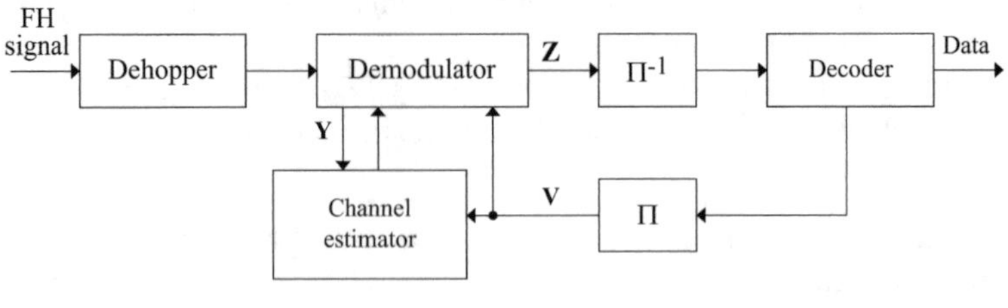

Fig. 9.11 Architecture of a receiver for a frequency-hopping system with a turbo code. Π = interleaver. Π^{-1} = deinterleaver

change the interference and noise during some hop dwell intervals. Consequently, estimates of the fading amplitude and the PSD of the interference and noise are computed for a block size N_b that is smaller than or equal to the number of symbols in the hop dwell interval. If there are N_b symbols per block, then there are $\lceil N_d/N_b \rceil$ blocks per codeword.

Demodulator Metrics

The complex envelope of a unit-energy q-ary CPFSK symbol waveform with zero initial phase offset is

$$s_l(t) = \frac{1}{\sqrt{T_s}} e^{j2\pi lht/T_s}, \quad 0 \le t \le T_s\,, \quad l = 1, 2, \ldots, q \tag{9.71}$$

where T_s is the symbol duration, h is the modulation index, and $j = \sqrt{-1}$. Because of the continuous-phase constraint, the initial phase of the CPFSK symbol i is $\phi_i = \phi_{i-1} + 2\pi lh$. The phase continuity ensures the compact spectrum of the CPFSK waveform. Suppose that symbol i of a codeword uses unit-energy waveform $s_{d_i}(t)$. If this codeword is transmitted over an AWGN channel with fading, the received signal for symbol i can be expressed in complex notation as

$$\begin{aligned} r_i(t) &= \mathrm{Re}\left[\alpha_i \sqrt{2\mathcal{E}_s} s_{d_i}(t) e^{j(2\pi f_c t + \theta_i)}\right] + n_i(t), \quad 0 \le t \le T_s \\ i &= 1, 2, \ldots, N_d \end{aligned} \tag{9.72}$$

where $n_i(t)$ is independent, zero-mean, white Gaussian noise with two-sided PSD $N_{0i}/2, f_c$ is the carrier frequency, $\mathcal{E}_s$ is the signal energy, and α_i is the magnitude of the complex fading amplitude. Without loss of generality, we assume $E[\alpha_i^2] = 1$ so that $\mathcal{E}_s$ is the average received symbol energy. The phase θ_i is the phase due to the contributions of the CPFSK constraint, the fading, and the frequency offset of the receiver.

Exploitation of the inherent memory in the CPFSK may be considered when computing the metric transferred from the demodulator to a decoder, but phase stability over several symbols is necessary, and the demodulator functions as a rate-one inner decoder. Furthermore, a trellis demodulator requires a rational h and the number of states depends on the denominator of h. More design flexibility exists if the demodulator metrics are computed on a symbol-by-symbol basis, and the memory in the turbo code is exploited rather than the memory in the modulation.

Matched-filter k, which is matched to $s_k(t)$, produces the output samples

$$y_{k,i} = \sqrt{2} \int_0^{T_s} r_i(t) e^{-j2\pi f_c t} s_k^*(t)dt\,, \quad i = 1, 2, \ldots, N_d, \ k = 1, 2, \ldots, q \tag{9.73}$$

where the $\sqrt{2}$ is inserted for mathematical convenience. The substitution of (9.71) and (9.72) into (9.73) and the approximation that each of the $\{s_k(t)\}$ has a spectrum confined to $|f| < f_c$ yields

$$y_{k,i} = \alpha_i \sqrt{\mathcal{E}_s} e^{j\theta_i} \rho_{d_i - k} + n_{k,i} \tag{9.74}$$

where

$$n_{k,i} = \sqrt{2} \int_0^{T_s} n_i(t) e^{-j2\pi f_c t} s_k^*(t) dt \tag{9.75}$$

and

$$\rho_l = \frac{\sin(\pi h l)}{\pi h l} e^{j\pi h l}. \tag{9.76}$$

As shown in Section 1.2, since $n_i(t)$ is zero-mean white noise and the spectra of the $\{s_k(t)\}$ are confined, it follows that each $n_{k,i}$ is zero-mean,

$$E[n_{k,i} n_{l,i}^*] = N_{0i} \rho_{l-k} \tag{9.77}$$

and that the $\{n_{k,i}\}$ have circular symmetry:

$$E[n_{k,i} n_{l,i}] = 0. \tag{9.78}$$

Since $n_i(t)$ is a Gaussian process, the real and imaginary components of $n_{k,i}$ are jointly Gaussian, and the set $\{n_{k,i}\}$ comprises *complex-valued jointly Gaussian random variables.*

Let $\mathbf{y}_i = [y_{1,i} \ldots y_{q,i}]^T$ denote the column vector of the matched-filter outputs corresponding to symbol i, and let $\mathbf{n} = [n_{1,i} \ldots n_{q,i}]^T$. Then given that the transmitted symbol is d_i, the symbol energy is $\mathcal{E}_s$, the fading amplitude is a_i, the noise PSD is $N_{0i}/2$, and the phase is θ_i, $\mathbf{y}_i = \overline{\mathbf{y}}_i + \mathbf{n}$, where $\overline{\mathbf{y}}_i = E[\mathbf{y}_i | d_i, \alpha_i \sqrt{\mathcal{E}_s}, N_{0i}, \theta_i]$. Equation (9.74) indicates that the kth component of $\overline{\mathbf{y}}_i$ is

$$\overline{y}_{k,i} = \alpha_i \sqrt{\mathcal{E}_s} e^{j\theta_i} \rho_{d_i - k}. \tag{9.79}$$

The covariance matrix of $\mathbf{y}_i$ is

$$\begin{aligned} \mathbf{R}_i &= E[(\mathbf{y}_i - \overline{\mathbf{y}}_i)(\mathbf{y}_i - \overline{\mathbf{y}}_i)^H \mid d_i, \alpha_i \sqrt{\mathcal{E}_s}, N_{0i}, \theta_i] \\ &= E[\mathbf{n}\mathbf{n}^H] \end{aligned} \tag{9.80}$$

and its elements are given by (9.77).

It is convenient to define the matrix $\mathbf{K} = \mathbf{R}_i / N_{0i}$ with components

$$K_{k,l} = \rho_{l-k}. \tag{9.81}$$

We can represent the conditional density function of $\mathbf{y}_i$ given that the transmitted symbol is d_i, the average symbol energy is $\mathcal{E}_s$, the fading amplitude is α_i, the noise PSD is $N_{0i}/2$, and the phase is θ_i as

$$g(\mathbf{y}_i|\boldsymbol{\psi}) = \frac{1}{\pi^q N_{0i}^q \det \mathbf{K}} \exp\left[-\frac{1}{N_{0i}}(\mathbf{y}_i - \overline{\mathbf{y}}_i)^H \mathbf{K}^{-1}(\mathbf{y}_i - \overline{\mathbf{y}}_i)\right] \tag{9.82}$$

where

$$\boldsymbol{\psi} = (d_i, \alpha_i\sqrt{\mathcal{E}_s}, N_{0i}, \theta_i) \tag{9.83}$$

and $\mathbf{K}$ is independent of $\boldsymbol{\psi}$.

An expansion of the quadratic in (9.82) yields

$$\begin{aligned} Q_i &= (\mathbf{y}_i - \overline{\mathbf{y}}_i)^H \mathbf{K}^{-1}(\mathbf{y}_i - \overline{\mathbf{y}}_i) \\ &= \mathbf{y}_i{}^H \mathbf{K}^{-1}\mathbf{y}_i + \overline{\mathbf{y}}_i{}^H \mathbf{K}^{-1}\overline{\mathbf{y}}_i - 2\,\mathrm{Re}(\mathbf{y}_i{}^H \mathbf{K}^{-1}\overline{\mathbf{y}}_i). \end{aligned} \tag{9.84}$$

Equations (9.79) and (9.81) indicate that $\overline{\mathbf{y}}_i$ is proportional to the $d_i th$ column of $\mathbf{K}$:

$$\overline{\mathbf{y}}_i = \alpha_i\sqrt{\mathcal{E}_s} e^{j\theta_i}\mathbf{K}_{:,d_i}. \tag{9.85}$$

Since $\mathbf{K}^{-1}\mathbf{K} = \mathbf{I}$, only the $d_i th$ component of the column vector $\mathbf{K}^{-1}\overline{\mathbf{y}}_i$ is nonzero, and

$$Q_i = \mathbf{y}_i{}^H \mathbf{K}^{-1}\mathbf{y}_i + \alpha_i^2\mathcal{E}_s - 2\alpha_i\sqrt{\mathcal{E}_s}\,\mathrm{Re}(y_{d_i,i}e^{-j\theta_i}). \tag{9.86}$$

For *noncoherent* signals, we assume that each θ_i is uniformly distributed over [0, 2π). Substituting (9.86) into (9.82), expressing $y_{d_i,i}$ in polar form, and using (H.16) of Appendix H.3 to integrate over θ_i, we obtain the density function

$$g(\mathbf{y}_i|\boldsymbol{\psi}) = \frac{\exp\left(-\frac{\mathbf{y}_i^H\mathbf{K}^{-1}\mathbf{y}_i + \alpha_i^2\mathcal{E}_s}{N_{0i}}\right)}{\pi^q N_{0i}^q \det \mathbf{K}} I_0\left(\frac{2\alpha_i\sqrt{\mathcal{E}_s}\,|y_{d_i,i}|}{N_{0i}}\right) \tag{9.87}$$

where $I_0(\cdot)$ is the modified Bessel function of the first kind and order zero. Since the white noise $n_i(t)$ is independent from symbol to symbol, $\mathbf{y}_i$ with the density function given by (9.87) is independent of $\mathbf{y}_l$, $i \neq l$.

Let $\widehat{A}$ and $\widehat{B}$ denote the estimates of $A = N_0$ and $B = 2\alpha\sqrt{\mathcal{E}_s}$, respectively, for a dwell interval of N_b symbols during which $\alpha_i = \alpha$ and $N_{0i} = N_0$ are constants. Let $b_k(i)$ denote bit k of symbol i and $\mathbf{Z}$ the $m \times N_d$ matrix with element $z_{k,i}$ equal to the log-likelihood ratio for $b_k(i)$ computed by the demodulator. The matrix $\mathbf{Z}$ is reshaped into a row vector and deinterleaved, and the resulting vector $\mathbf{z}'$ is fed into the turbo decoder. The extrinsic information $\mathbf{v}'$ at the output of the decoder is interleaved and reshaped into a $m \times N_d$ matrix $\mathbf{V}$ containing the a priori information:

$$v_{k,i} = \ln \frac{P\left[b_k(i) = 1|\mathbf{Z}\backslash z_{k,i}\right]}{P\left[b_k(i) = 0|\mathbf{Z}\backslash z_{k,i}\right]} \tag{9.88}$$

where conditioning on $\mathbf{Z}\backslash z_{k,i}$ means that the extrinsic information for bit $b_{k,i}$ is produced without using $z_{k,i}$.

Since $\mathbf{V}$ is fed back to the demodulator,

$$z_{k,i} = \ln \frac{P\left[b_k(i) = 1|\mathbf{y}_i, \gamma'_{\lceil i/N_b \rceil}, \mathbf{v}_i\backslash v_{k,i}\right]}{P\left[b_k(i) = 0|\mathbf{y}_i, \gamma'_{\lceil i/N_b \rceil}, \mathbf{v}_i\backslash v_{k,i}\right]} \tag{9.89}$$

where $\gamma' = \{\hat{A}, \hat{B}\}$, and $\lceil x \rceil$ denotes the smallest integer greater than or equal to x. Partition the set of symbols $\mathcal{D} = \{1, \ldots, q\}$ into two disjoint sets $\mathcal{D}_k^{(1)}$ and $\mathcal{D}_k^{(0)}$, where $\mathcal{D}_k^{(b)}$ contains all symbols labeled with $b_k = b$. As indicated by (1.215), the extrinsic information can then be expressed as

$$z_{k,i} = \ln \frac{\sum_{d \in \mathcal{D}_k^{(1)}} g(\mathbf{y}_i|d, \gamma'_{\lceil i/N_b \rceil}) \prod_{\substack{l=1 \\ l \neq k}}^{m} \exp\left(b_l(d) v_{l,i}\right)}{\sum_{d \in \mathcal{D}_k^{(0)}} g(\mathbf{y}_i|d, \gamma'_{\lceil i/N_b \rceil}) \prod_{\substack{l=1 \\ l \neq k}}^{m} \exp\left(b_l(d) v_{l,i}\right)} \tag{9.90}$$

where $b_l(d)$ is the value of the *lth* bit in the labeling of symbol d. Substituting (9.87) into (9.90) and canceling common factors, we obtain

$$z_{k,i} = \ln \frac{\sum_{d \in \mathcal{D}_k^{(1)}} I_0\left(\gamma_{\lceil i/N_b \rceil}|y_{d_i,i}|\right) \prod_{\substack{l=1 \\ l \neq k}}^{m} \exp\left(b_l(d) v_{l,i}\right)}{\sum_{d \in \mathcal{D}_k^{(0)}} I_0\left(\gamma_{\lceil i/N_b \rceil}|y_{d_i,i}|\right) \prod_{\substack{l=1 \\ l \neq k}}^{m} \exp\left(b_l(d) v_{l,i}\right)} \tag{9.91}$$

where only the ratio $\gamma = \hat{B}/\hat{A}$ is needed rather than the individual estimates.

Channel Estimators

Since the preceding and subsequent equations in this section refer to a specific receiver iteration, the superscript denoting the receiver-iteration number is omitted to simplify the notation.

Since under block fading and time-varying interference, A and B can change on a block-by-block basis; thus, each block is processed separately and in an identical fashion. To maintain robustness, the estimators make no assumptions regarding the distribution of the quantities to be estimated, nor do they make any assumptions regarding the correlation from block to block. The estimators directly use the channel observation for a single block while the observations of the other blocks are used indirectly through feedback of extrinsic information from the decoder. In this section, the matrix $\mathbf{Y}$ is a generic $q \times N_b$ received block, $\mathbf{y}_i$ is the ith column vector of $\mathbf{Y}$, the vector $\mathbf{D} = [D_1, \ldots, D_{N_b}]$ is the corresponding set of transmitted symbols, and $\{\hat{A}, \hat{B}\}$ is the corresponding set of channel estimators.

Rather than attempting to directly evaluate the maximum-likelihood estimates, the expectation-maximization (EM) algorithm can be used as an iterative approach to estimation. Let $\{\mathbf{Y}, \mathbf{D}\}$ denote the *complete* data set. Since $\ln h(\mathbf{d})$ is independent of A and B and hence does not affect the maximization, the log-likelihood of the complete data set is

$$\ln f(\mathbf{z}|A, B) = \ln\, g(\mathbf{y}|\mathbf{d}, A, B) + \ln h(\mathbf{d}) \sim \ln\, g(\mathbf{y}|\mathbf{d}, A, B). \tag{9.92}$$

Since $\mathbf{y}_i$ and $\mathbf{y}_l$ are independent for $i \neq l$, (9.87) implies that

$$g(\mathbf{y}|\mathbf{d}, A, B) = \frac{\exp\left[-\frac{H}{A} - \frac{N_b B^2}{4A} + \sum_{i=1}^{N_b} \ln I_0\left(\frac{B|y_{d_i,i}|}{A}\right)\right]}{(\pi^q A^q \det \mathbf{K})^{N_b}} \tag{9.93}$$

where

$$H = \sum_{i=1}^{N_b} \mathbf{y}_i^{\ H} \mathbf{K}^{-1} \mathbf{y}_i. \tag{9.94}$$

After dropping irrelevant constants, we obtain

$$\ln f(\mathbf{z}|A, B) \sim -qN_b \log A - \frac{H}{A} - \frac{N_b B^2}{4A} + \sum_{i=1}^{N_b} \ln I_0\left(\frac{B\,|y_{d_i,i}|}{A}\right). \tag{9.95}$$

The form of this equation indicates that the parameters A and B must both be estimated rather than just the ratio B/A.

Let r denote the EM iteration number, and $\hat{A}^{(r)}, \hat{B}^{(r)}$ the estimates of A, B during the rth iteration. The *expectation* step (E-step) requires the calculation of

$$Q(A, B) = E_{\mathbf{d}|\mathbf{y}, \hat{A}^{(r-1)}, \hat{B}^{(r-1)}}\left[\ln f(\mathbf{Z}|A, B)\right] \tag{9.96}$$

where the expectation is taken with respect to the unknown symbols $\mathbf{d}$ conditioned on $\mathbf{y}$ and the estimates $\hat{A}^{(r-1)}, \hat{B}^{(r-1)}$ from the previous EM iteration. Substituting (9.95) into (9.96), it is found that

$$Q(A, B) = -qN_b \ln A - \frac{H}{A} - \frac{N_b B^2}{4A} + \sum_{i=1}^{N_b} \sum_{k=1}^{q} p_{k,i}^{(r-1)} \ln I_0\left(\frac{B\,|y_{k,i}|}{A}\right) \tag{9.97}$$

where the fact that D_i is independent of $\hat{A}^{(r-1)}$ and $\hat{B}^{(r-1)}$ indicates that

$$p_{k,i}^{(r-1)} = P(D_i = k|\mathbf{y}_i, \hat{A}^{(r-1)}, \hat{B}^{(r-1)}) = \frac{g(\mathbf{y}_i|D_i = k, \hat{A}^{(r-1)}, \hat{B}^{(r-1)})P(D_i = k)}{g(\mathbf{y}_i|\hat{A}^{(r-1)}, \hat{B}^{(r-1)})} \tag{9.98}$$

and $P(D_i = k)$ is the probability that $D_i = k$, which is estimated by the decoder. Applying (9.87), we obtain

$$p_{k,i}^{(r-1)} = \alpha_i^{(r-1)} I_0\left(\frac{\hat{B}^{(r-1)}|y_{k,i}|}{\hat{A}^{(r-1)}}\right) P(D_i = k) \tag{9.99}$$

where $\alpha_i^{(r-1)}$ is the normalization factor forcing $\sum_{k=1}^{q} p_{k,i}^{(r-1)} = 1$; i.e.,

$$\alpha_i^{(r-1)} = \frac{1}{\sum_{k=1}^{q} I_0\left(\frac{\hat{B}^{(r-1)}|y_{k,i}|}{\hat{A}^{(r-1)}}\right) P(D_i = k)}. \tag{9.100}$$

The *maximization* step (M-step) is

$$\hat{A}^{(r)}, \hat{B}^{(r)} = \arg\max_{A,B} Q(A, B) \tag{9.101}$$

which can be found by setting the derivatives of the function $Q(A, B)$ with respect to A and B to zero. The solution to the corresponding system of equations is

$$\hat{A}^{(r)} = \frac{1}{qN_b}\left(H - \frac{N_b(\hat{B}^{(r)})^2}{4}\right) \tag{9.102}$$

$$\hat{B}^{(r)} = \frac{2}{N_b}\sum_{i=1}^{N_b}\sum_{k=1}^{q} p_{k,i}^{(r-1)} |y_{k,i}| F\left(\frac{4qN_b\hat{B}^{(r)}|y_{k,i}|}{4H - N_b(\hat{B}^{(r)})^2}\right) \tag{9.103}$$

where $F(x) = I_1(x)/I_0(x)$, and $I_1(\cdot)$ is the modified Bessel function of the first kind and order 1 defined by (H.13).

Although a closed-form solution to (9.103) is difficult to obtain, it can be found recursively by using the fixed-point iteration method of Section 9.1. The recursion involves initially replacing $\hat{B}^{(r)}$ on the right-hand side of (9.103) with $\hat{B}^{(r-1)}$ from the previous EM iteration. To select an initial estimate for B, consider what happens in the absence of noise. Without noise, (9.74) implies that either $|y_{k,i}| = a\sqrt{\mathcal{E}_s}$ (when $k = d_i$) or $|y_{k,i}| = 0$ (otherwise). Thus, an estimate for $a\sqrt{\mathcal{E}_s} = B/2$ can be achieved by taking the maximum $|y_{k,i}|$ over any column of $\mathbf{Y}$. To compensate for the noise, the average can be taken across all columns in the block, resulting in

$$\hat{B}^{(0)} = \frac{2}{N_b} \sum_{i=1}^{N_b} \max_k |y_{k,i}| \,. \tag{9.104}$$

The initial estimate of A is found from $\hat{B}^{(0)}$ by evaluating (9.102) for $r = 0$. After the initial values $\hat{A}^{(0)}$ and $\hat{B}^{(0)}$ are calculated, the initial probabilities $\{p_{k,i}^{(0)}\}$ are calculated from (9.99) and (9.100). The EM algorithm terminates when $\hat{B}^{(r)}$ converges to some fixed value, typically in fewer than 10 EM iterations.

The complexity of the channel estimation for each receiver iteration is as follows. The initial estimate of $\hat{B}$ calculated using (9.104) requires N_b maximizations over q values, $N_b - 1$ additions, and a single multiplication by $2/N_b$. The calculation of H in (9.94), which only needs to be computed once prior to the first EM iteration, requires $N_b q(q+1)$ multiplications and $N_b q^2 - 1$ additions. For each EM iteration, the calculation $\hat{A}^{(r)}$ using (9.102) requires only two multiplications and an addition. Calculating $p_{k,i}^{(r-1)}$ using (9.99) and (9.100) requires $3N_b q + 1$ multiplications, $N_b(q-1)$ additions, and $N_b q$ lookups of the $I_0(\cdot)$ function. Calculation of $\hat{B}^{(r)}$ by solving (9.103) is recursive, and complexity depends on the number of recursions for each value of r. Suppose that there are ξ recursions, then the calculation requires $N_b q + \xi(2N_b q + 4)$ multiplications, $\xi N_b q$ additions, and $\xi N_b q$ lookups of the $F(\cdot)$ function. A stopping criterion is used for the calculation of $\hat{B}$ such that the recursions stop once $\hat{B}$ is within 10% of its value during the previous recursion or a maximum number of 10 recursions is reached. With such a stopping criterion, an average of only 2 or 3 recursions are required.

Selection of Modulation Index

Let B_{max} denote the maximum bandwidth of the CPFSK modulation such that the hopping band accommodates enough frequency channels to ensure adequate performance against multiple-access interference and multitone jamming. We seek to determine the values of h, q, and code-rate R of the turbo code that provide good performance over the fading and AWGN channels in the presence of partial-band interference. For specific values of the modulation parameters h and q, the code rate is limited by the bandwidth requirement. Let $B_u T_b$ denote the normalized, 99% power bandwidth of the uncoded CPFSK modulation. This value can be found for nonorthogonal CPFSK by numerically integrating the power-spectrum equations of Section 3.4 or using (3.92) when the number of symbols per hop is large. When a code of rate R is used, the bandwidth becomes $B_c = B_u/R$. Since $B_c \leq B_{max}$ is required, the minimum code rate that achieves the bandwidth constraint is $R_{min} = B_u/B_{max}$.

Guidance in the selection of the best values of h, q, and $R \geq R_{\min}$ is provided by information theory. For specific values of h and q, we evaluate the capacity $C(\gamma)$ as a function of $\gamma = \mathcal{E}_s/N_0$ under a bandwidth constraint for both the Rayleigh and AWGN channels. Perfect channel-state information is assumed, and symbols are drawn from the signal set with equal probabilities. With these assumptions, a change of variables with $\mathbf{u} = \mathbf{y}_i/\sqrt{\mathcal{E}_s}$, (9.87), and (9.70), the ergodic symmetric capacity for the fading channel may be expressed as

$$C(\gamma) = \log_2 q - \frac{1}{q}\sum_{\nu=1}^{q}\int\int g(\alpha)f(\mathbf{u}|\nu,\alpha)\log_2\left[\frac{\sum_{k=1}^{q} I_0\left(2\alpha\gamma\left|u_k\right|\right)}{I_0\left(2\alpha\gamma\left|u_\nu\right|\right)}\right]d\mathbf{u}d\alpha \tag{9.105}$$

where $g(\alpha)$ is the density function of the magnitude of the fading amplitude, the $(2q+1)$-fold integration is over all values of α and the $2q$ real and imaginary components of $\mathbf{u}$, and

$$f(\mathbf{u}|\nu,\alpha) = \frac{\gamma^q \exp[-\gamma(\mathbf{u}^H\mathbf{K}^{-1}\mathbf{u} + \alpha^2)]}{\pi^q \det \mathbf{K}} I_0\left(2\alpha\gamma\left|u_\nu\right|\right). \tag{9.106}$$

Equation (9.105) is numerically integrated using the Monte Carlo method.

To determine the minimum $\mathcal{E}_b/N_0$ necessary to maintain $C(\gamma)$ above the code-rate R for specific values of q and h, we substitute $\mathcal{E}_s = R\mathcal{E}_b \log_2 q$ and solve the equation

$$R = C(R\mathcal{E}_b \log_2 q/N_0) \tag{9.107}$$

for all code rates such that $R_{min} \leq R \leq 1$. For noncoherent systems under severe bandwidth constraints, the R that minimizes $\mathcal{E}_b/N_0$ is typically $R = R_{\min}$, but under loose bandwidth constraints the R that minimizes $\mathcal{E}_b/N_0$ could be larger than R_{min} (in which case the actual bandwidth is less than B_{max}).

Figures 9.12 and 9.13 show plots of the minimum $\mathcal{E}_b/N_0$ versus h for $2 \leq q \leq 32$, $B_{\max}T_b = 2$, and $B_{\max}T_b = \infty$. Figure 9.12 is for the AWGN channel, and Figure 9.13 is for the Rayleigh fading channel. When $B_{\max}T_b = 2$, the curves are truncated because there is a maximum value of h beyond which no code exists that satisfies the bandwidth constraint. For each value of q, in each figure there is an optimal value of h that gives the smallest value of the minimum $\mathcal{E}_b/N_0$. This smallest value decreases with q, but there are diminishing returns and the implementation complexity increases rapidly for $q > 8$.

Let f_e denote the offset in the estimated carrier frequency at the receiver due to the Doppler shift and the frequency-synthesizer inaccuracy. The separation between adjacent frequencies in a CPFSK symbol is $hf_b/R\log_2 q$, where f_b denotes the

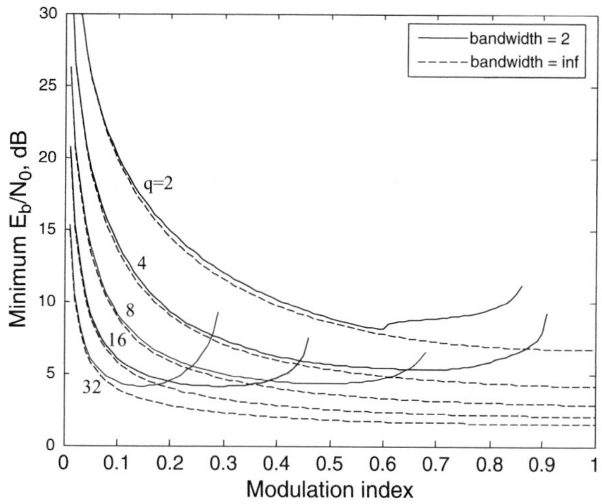

Fig. 9.12 Minimum $\mathcal{E}_b/N_0$ versus h for the AWGN channel, $2 \leq q \leq 32$, $B_{\max}T_b = 2$, and $B_{\max}T_b = \infty$ [105]

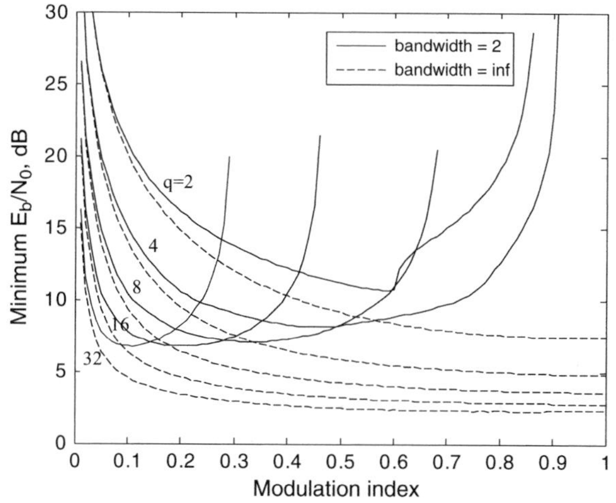

Fig. 9.13 Minimum $\mathcal{E}_b/N_0$ versus h for the Rayleigh channel, $2 \leq q \leq 32$, $B_{\max}T_b = 2$, and $B_{\max}T_b = \infty$ [105]

information-bit rate. Since this separation must be much larger than f_e if the latter is to be negligible as assumed in (9.73),

$$f_e << \frac{hf_b}{R\log_2 q} \tag{9.108}$$

is required. Since the optimal h decreases while $R\log_2 q$ increases with q, (9.108) is another reason to choose $q \leq 8$.

For $q = 4$ in Figure 9.13, $h = 0.46$ is the approximate optimal value when $B_{\max}T_b = 2$, and the corresponding code rate is approximately $R = 16/27$. For $q = 8$, $h = 0.32$ is the approximate optimal value when $B_{\max}T_b = 2$, and the corresponding code rate is approximately $R = 8/15$. For both $q = 8$ and $q = 4$, (9.108) is satisfied if $f_e << 0.2f_b$. At the optimal values of h, the plots indicate that the loss is less than 1 dB for the AWGN channel and less than 2 dB for the Rayleigh channel relative to what could be attained with the same value of q, $h = 1$ (orthogonal CPFSK), and an unlimited bandwidth.

Partial-Band Interference

Simulation experiments were conducted to assess the benefits and tradeoffs of using the nonorthogonal CPFSK coded modulation and accompanying channel estimator in a frequency-hopping system that suppresses partial-band interference. Interference is modeled as AWGN within a fraction μ of the hopping band. The PSD of the interference is I_{t0}/μ, where I_{t0} is the PSD over the receiver passband when $\mu = 1$ and the total interference power is conserved as μ varies. The parameter A represents the PSD due to the noise and the interference during a dwell interval. The bandwidth is assumed to be sufficiently small that the fading is flat within each frequency channel; hence, the symbols of a dwell interval undergo the same fading amplitude. The fading amplitudes are independent from hop to hop, which models the frequency-selective fading that varies after each hop. A fading block coincides with a dwell interval, and is thus suitable for the estimation of a single fading amplitude. Three alphabet sizes are considered: binary ($q = 2$), quaternary ($q = 4$), and octal ($q = 8$).

The simulated system uses the widely deployed turbo code from the Universal Mobile Telecommunications System (UMTS) specification, which has a constraint length of 4, a specified code-rate matching algorithm, and an optimized variable-length interleaver that is set to 2048. If the dwell time is fixed but the codeword length is extended beyond the specified 2048 bits, then the number of hops per codeword increases. As a result, the diversity order increases, but the benefit of increased diversity order obeys a law of diminishing returns, and the benefit is minor at bit error rates of approximately 10^{-3} or less. The values of modulation index h and code rate R are selected to be close to the information-theoretic optimal values given previously for Rayleigh fading under the bandwidth constraint $B_{\max}T_b = 2$. In particular, a system with $q = 2$ uses $h = 0.6$ and $R = 2048/3200$, one with $q = 4$ uses $h = 0.46$ and $R = 2048/3456$, and one with $q = 8$ uses $h = 0.32$ and $R = 2048/3840$.

A receiver iteration comprises the steps of channel estimation, demapping and demodulation, and one full turbo-decoder iteration. Up to 20 receiver iterations are executed. An early halting routine stops the iterations once the data is correctly decoded (which can be determined, for instance, by using the cyclic redundancy check specified in the UMTS standard). The number of hops per codeword may vary, and results below show the impact of changing this value.

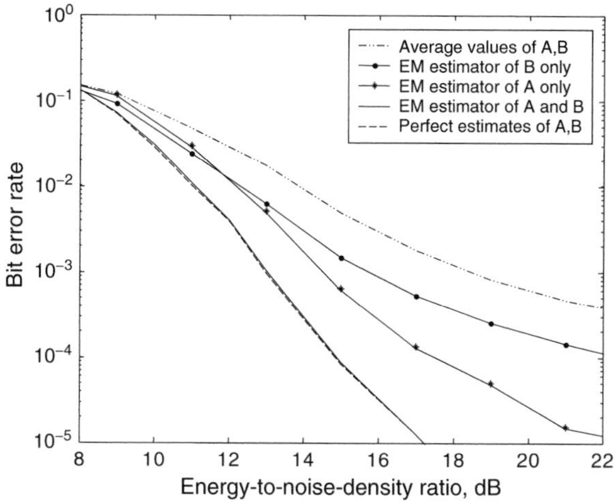

Fig. 9.14 Bit error rates of frequency-hopping systems with various estimators in Rayleigh block fading with partial-band interference, $\mu = 0.6$, $\mathcal{E}_b/I_{t0} = 13\,\text{dB}$, turbo-coded octal CPFSK, $h = 0.32$, code rate $2048/3840$, and 32 hops per codeword [105]

Figure 9.14 illustrates the influence of channel estimation on the BER of the system. For this figure, $\mu = 0.6$, $\mathcal{E}_b/I_{t0} = 13\,\text{dB}$, the channel undergoes block-by-block Rayleigh fading, and octal CPFSK is used with 32 hops per codeword. The uppermost curve in the figure shows the performance of a simple system that does not attempt to estimate A or B. Instead, the system sets these values to their statistical averages: $A = N_0 + I_{t0}$, and $B = 2E[\alpha]\sqrt{\mathcal{E}_s} = \sqrt{\pi\mathcal{E}_s}$, as implied by (E.31) of Appendix E.4. It can be seen that the performance of such a system is rather poor, as it has no knowledge of which hops have experienced interference and which have not. The system can be improved by estimating A and/or B on a block-by-block basis using the EM estimator. The second curve from the top shows the performance when only B is estimated on a block-by-block basis (and $A = N_0 + I_{t0}$), whereas the next curve down shows the performance when only A is estimated on a block-by-block basis (and $B = \sqrt{\pi\mathcal{E}_s}$). The second lowest curve shows the performance when both A and B are estimated on a block-by-block basis with the EM estimator, whereas the lowest curve shows the performance with perfect CSI, i.e., when A and B are known perfectly. As can be seen, there is a large gap between perfect CSI and simply using the average values of A and B. This gap can be partially closed by estimating either A and B independently on a block-by-block basis, and the gap closes almost completely by estimating them jointly.

Figure 9.15 illustrates the robustness of the estimator as a function of the alphabet size, channel type, and fraction of partial-band interference μ. The figure shows the value of $\mathcal{E}_b/N_0$ required to achieve a BER of 10^{-3} as a function of μ for several systems with $\mathcal{E}_b/I_{t0} = 13\,\text{dB}$. For each of the three alphabet sizes, both AWGN and block Rayleigh fading (again, 32 hops per codeword) are considered. For each of these six cases, the performance using perfect CSI and the performance with the EM

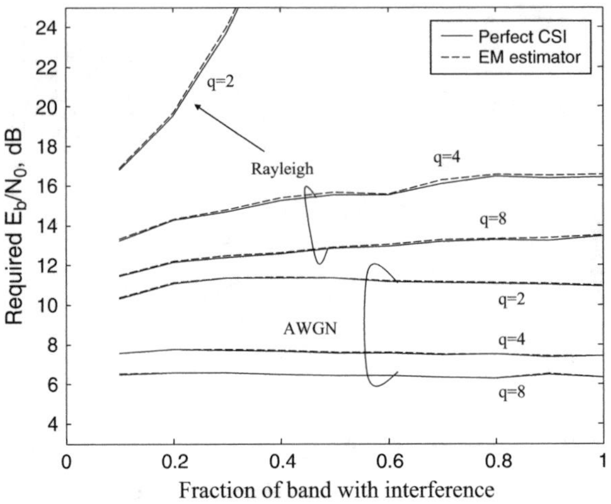

Fig. 9.15 Required $\mathcal{E}_b/N_0$ to achieve $BER = 10^{-3}$ as a function of μ for binary, quaternary, and octal turbo-coded FH-CPFSK systems in both AWGN and Rayleigh fading channels with $\mathcal{E}_b/I_{t0} = 13\,\text{dB}$ and $B_{\max}T_b = 2$ [105]

estimator are shown. Across the entire range of tested parameters, the estimator's performance nearly matches that of perfect CSI. The benefit of increasing the alphabet size is apparent. For instance, in AWGN, increasing q from 2 to 4 improves performance by about 4 dB, whereas increasing it again from 4 to 8 yields another 1.2 dB gain. The gains in Rayleigh fading are even more dramatic. Although the performance in AWGN is relatively insensitive to the value of μ, the performance in Rayleigh fading degrades as μ increases, and when $q = 2$, this degradation is quite severe.

If the hop rate increases, the increase in the number of independently fading dwell intervals per codeword implies that more diversity is available in the processing of a codeword. However, the shortening of the dwell interval makes the channel estimation less reliable by providing the estimator with fewer samples per block. The influence of the number of hops per codeword is shown in Figure 9.16 as a function of μ for quaternary and octal CPFSK using the EM estimator, Rayleigh block fading, and partial-band interference with $\mathcal{E}_b/I_{t0} = 13\,\text{dB}$. Since the codeword length is fixed for each q, increasing the number of hops per codeword, results in shorter blocks. For $q = 4$, there are 108, 54, or 27 symbols per hop when there are 16, 32, or 64 hops per codeword, respectively. For $q = 8$, there are 80, 40, or 20 symbols per hop when there are 16, 32, or 64 hops per codeword respectively. Despite the slow decline in the accuracy of the EM channel estimates, the diversity improvement is sufficient to produce improved performance as the number of code symbols per hop decreases. However, decreasing to fewer than 20 code symbols per hop begins to broaden the spectrum significantly, as indicated in Section 3.4, unless the parameter values are changed.

Frequency-hopping systems, such as GSM, Bluetooth, and combat net radios, use binary minimum-shift keying (MSK) with $h = 0.5$ or binary Gaussian FSK, and do

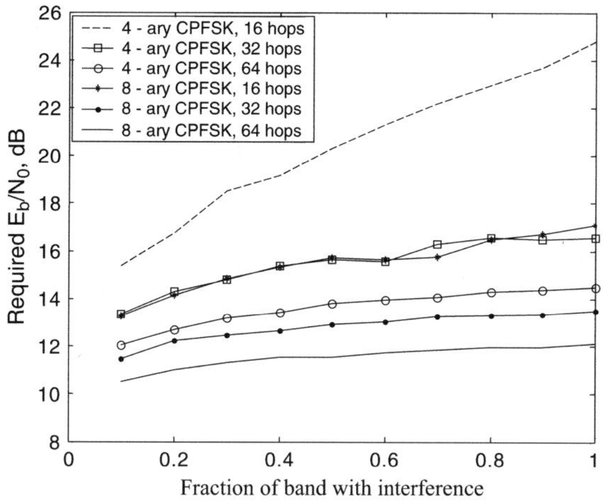

Fig. 9.16 Required $\mathcal{E}_b/N_0$ to achieve $BER = 10^{-3}$ as a function of μ for quaternary and octal FH-CPFSK systems in Rayleigh block fading with partial-band interference, EM estimation, $\mathcal{E}_b/I_{t0} = 13$ dB, and various hops per codeword [105]

not have fading-amplitude estimators. Figures 9.15 and 9.14 illustrate the substantial performance penalties resulting from the use of a binary modulation and the absence of fading-amplitude estimation respectively. The cost of the superior performance of the robust system is primarily the increased computational requirements. Other nonbinary frequency-hopping systems use channel estimators to achieve excellent performance against partial-band interference and AWGN. However, they are not resistant to multiple-access interference because the transmitted symbols are not spectrally compact, and the channel estimators are not designed to estimate multiple interference and noise spectral-density levels. As described subsequently, the robust system accommodates substantial multiple-access interference.

Asynchronous Multiple-Access Interference

Multiple-access interference may occur when two or more frequency-hopping signals share the same physical medium or network, but the hopping patterns are not coordinated. A collision occurs when two or more signals using the same frequency channel are received simultaneously (Sections 7.6 and 8.6). Since the probability of a collision in a network is decreased by increasing the number of frequency channels in the hopset, a spectrally compact modulation is highly desirable when the hopping band is fixed.

Simulation experiments were conducted to compare the effect of the number of mobiles of an ad hoc network on systems with different values of q and h. All network mobiles have asynchronous, statistically independent, randomly generated hopping patterns with negligible switching times between dwell intervals. Let T_i denote the random variable representing the relative transition time of frequency-hopping interference signal i or the start of its new dwell interval relative to that of the desired signal. The ratio T_i/T_s is uniformly distributed over the integers in $[0, N_h - 1]$, where N_h is the number of symbols per dwell interval.

Let M denote the number of frequency channels in the hopset shared by all mobiles. Since the interference and desired signal are not synchronized, parts of two dwell intervals of the interference signal coincide with a single dwell interval of the desired signal. Thus, two carrier frequencies are randomly generated by each interference signal during each dwell interval of the desired signal. Therefore, the probability is $1/M$ that the interference signal collides with the desired signal *before* T_i, and the probability is $1/M$ that the interference signal collides with the desired signal *after* T_i. Each interference signal transmits a particular symbol with probability $1/q$ (common values of q and h are used throughout the network). The response of each matched filter to an interference symbol is given by the same equations used for the desired signal. The soft-decision metrics sent to the decoder are generated in the usual manner but are degraded by the multiple-access interference.

The transmitted powers of the interference and the desired signals are the same. All the interference sources are randomly located at a distance from the receiver within four times the distance of the desired-signal source. All signals experience a path loss with an attenuation power law equal to 4 and independent Rayleigh fading. The interference signals also experience independent shadowing (Section 6.1) with a shadow factor equal to 8 dB.

The simulations consider CPFSK alphabet sizes from the set $q = \{2, 4, 8\}$. The hopping band has the normalized bandwidth $WT_b = 2000$. Both orthogonal and nonorthogonal modulation are considered. For the orthogonal case, the code rate is chosen to be $2048/6144$, which is close to the information-theoretic optimal value in Rayleigh fading when $h = 1$. Taking into account the 99% power bandwidth of the resulting signal, there are $M = 312$, 315, and 244 frequency channels for binary, quaternary, and octal orthogonal CPFSK, respectively. For the nonorthogonal case, a bandwidth constraint $B_{\max}T_b = 2$ is assumed; thus, there are $M = 1000$ frequency channels. As in the previous examples, values of h and R that are close to the information-theoretic optimal values for this bandwidth constraint are selected (i.e., $h = 0.6$ and $R = 2048/3200$ for $q = 2$, $h = 0.46$ and $R = 2048/3456$ for $q = 4$, and $h = 0.32$ and $R = 2048/3840$ for $q = 8$). In all cases, there are 32 hops per codeword.

In the presence of multiple-access interference, it is important to accurately estimate the values of A and B. The impact of the channel-estimation technique is illustrated in Figure 9.17 for a system with 30 mobiles, all transmitting nonorthogonal octal CPFSK. The uppermost curve shows what happens when the receiver ignores the presence of interference. In this case, B is set to its actual value (perfect CSI for B), whereas A is set to N_0, its value without interference. Performance can be improved by using the EM-based estimators for jointly estimating A and B on a block-by-block basis, as illustrated by the second curve from the top.

Unlike the partial-band interference case, where the interference power is constant for the entire duration of the hop, the interference power in the presence of multiple-access interference is not generally constant due to the asynchronous

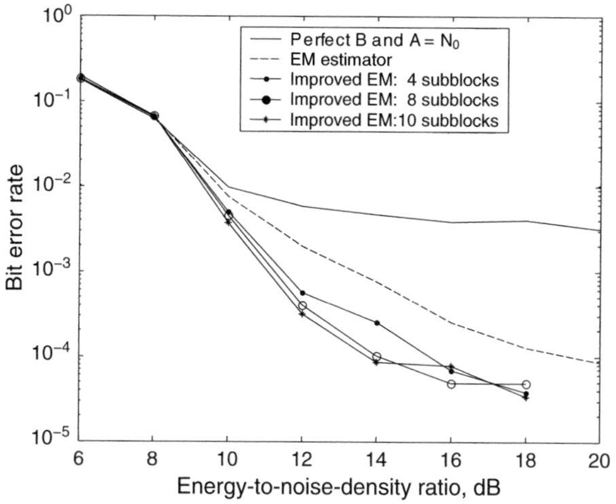

Fig. 9.17 Bit error rates of frequency-hopping systems with various estimators in Rayleigh block fading with multiple-access interference created by 30 mobiles. Turbo-coded octal CPFSK is used with $h = 0.32$, code rate 2048/3840, and 32 hops per codeword [105]

hopping. This fact suggests that performance can be improved by partitioning the block into multiple sub-blocks, and obtaining a separate estimate of A for each sub-block. Such estimates can be found from a simple modification of the EM-based estimator. The estimator first finds the value of B for the entire block from (9.103), as before. Next, it finds the value of A for each sub-block from (9.102) using the value of B found for the entire block and the value of D for just that sub-block (with N_b set to the size of the sub-block). The bottom three curves in Figure 9.17 show the performance when four, eight, or ten sub-blocks are used to estimate A. Although it is beneficial to use more sub-blocks at low $\mathcal{E}_b/N_0$, at higher $\mathcal{E}_b/N_0$ even using only four sub-blocks is sufficient to give significantly improved performance. This method of sub-block estimation essentially entails no additional complexity.

Figures 9.18 and 9.19 show the energy efficiency as a function of the number of mobiles. In Figure 9.18, the performance using the block EM estimator (no sub-block estimation) is shown. In particular, the minimum required value of $\mathcal{E}_b/N_0$ to achieve a bit error rate equal to 10^{-4} is given as a function of the number of mobiles. The performance is shown for both orthogonal and nonorthogonal modulation and the three values of q. For a lightly loaded system (fewer than five mobiles), the orthogonal systems outperform the nonorthogonal ones because orthogonal modulation is more energy efficient in the absence of interference. However, as the number of mobiles increases beyond about five, the nonorthogonal systems offer superior performance. The reason is that the improved spectral efficiency of a nonorthogonal modulation allows more frequency channels, thereby decreasing the probability of a collision. With orthogonal modulation, performance as a function of the number of mobiles degrades more rapidly as q increases, because larger values of q require larger bandwidths. In contrast, with nonorthogonal

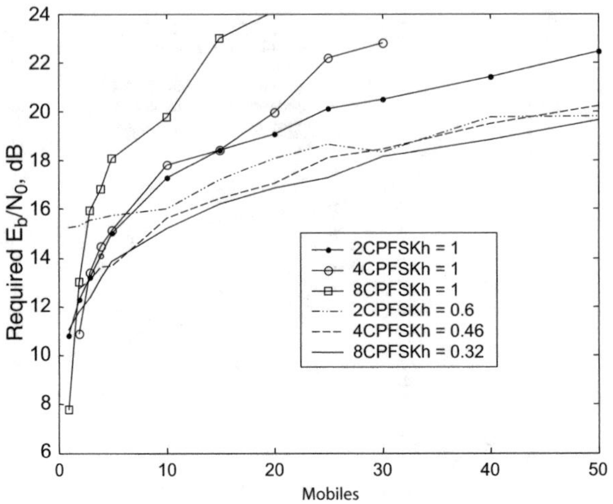

Fig. 9.18 Required $\mathcal{E}_b/N_0$ to achieve $BER = 10^{-4}$ as a function of the number of mobiles for turbo-coded FH-CPFSK systems in Rayleigh block fading. $B_{\max}T_b = 2$ for nonorthogonal signals [105]

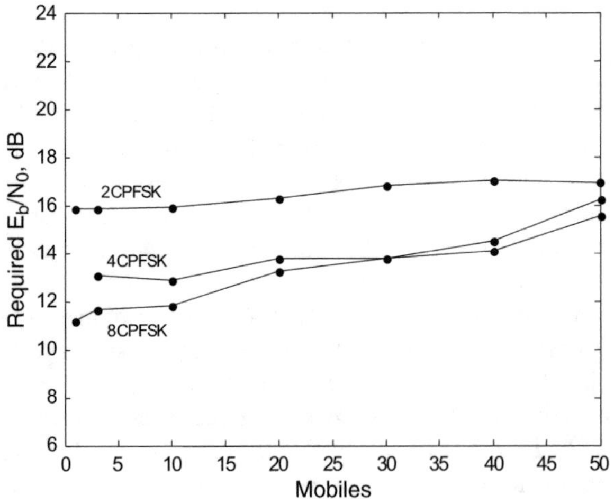

Fig. 9.19 Required $\mathcal{E}_b/N_o$ to achieve $BER = 10^{-4}$ as a function of the number of mobiles for improved EM estimation and nonorthogonal binary, quaternary, and octal turbo-coded FH-CPFSK systems in the presence of Rayleigh block fading [105]

modulation, the best performance is achieved with the largest value of q, although performance with $q = 2$ or $q = 4$ is only about 1-2 dB worse than with $q = 8$. When there are 50 mobiles, nonorthogonal CPFSK with $q = 8$ is about 3 dB more energy efficient than the more conventional orthogonal CPFSK with $q = 2$.

Figure 9.19 shows the performance with nonorthogonal CPFSK when sub-block estimation is used instead of block estimation, but the system parameters remain the same. For $q = 8$ there are ten sub-blocks of $40/10 = 4$ symbols, for $q = 4$ there are nine sub-blocks of $54/9 = 6$ symbols, and for $q = 2$ there are ten sub-blocks of $100/10 = 10$ symbols. A comparison with Figure 9.18 indicates that for $q = 8$ and 50 mobiles, there is a 4-dB gain in energy efficiency relative to the block estimator. It is also observed that when using the sub-block estimator, the performance is less sensitive to the number of mobiles, and that in a very lightly loaded system, the block estimator offers better performance (because then, A is likely to be constant for the entire hop).

The noncoherent frequency-hopping system with nonorthogonal CPFSK is highly robust in environments including frequency-selective fading, partial-band interference, multitone jamming, and multiple-access interference. The robustness is due to the iterative turbo decoding and demodulation, the channel estimator based on the expectation-maximization algorithm, and the spectrally compact modulation.

9.5 Problems

1 Consider a complete data vector that has a conditional density function of the form

$$f(\mathbf{z}|\boldsymbol{\theta}) = \alpha(\mathbf{z}) \exp\left[g(\boldsymbol{\theta}) + \boldsymbol{\theta}^T \boldsymbol{\beta}(\mathbf{z})\right]$$

where $\alpha(\mathbf{z})$ and $g(\boldsymbol{\theta})$ are scalar-valued functions, and $\boldsymbol{\beta}(\mathbf{z})$ is a vector-valued function. If $\mathbf{y}$ is an incomplete data vector, show that the EM estimator has the form

$$\nabla_{\boldsymbol{\theta}} g(\boldsymbol{\theta})_{\boldsymbol{\theta}=\hat{\boldsymbol{\theta}}_{i+1}} = h\left(\hat{\boldsymbol{\theta}}_i, \mathbf{y}\right)$$

and determine the function $h\left(\hat{\boldsymbol{\theta}}_i, \mathbf{y}\right)$.

2 Let X_{1i} and X_{2i} denote independent, zero-mean, Gaussian random variables with variances λ_1 and λ_2, respectively, for $i = 1, \ldots, N$. The received random vector is $\mathbf{Y} = [Y_1 \ldots Y_N]^T$, where $Y_i = aX_{1i} + bX_{2i}$, and a and b are known constants. It is desired to derive maximum-likelihood estimates of the variances. (a) Evaluate the likelihood equations and show that they provide a single equation that does not have a unique solution even when $N \to \infty$. Interpret this result in terms of the variances. (b) Define a complete data vector. Apply the EM algorithm to obtain estimates of the variances.

3 Consider two modifications of Example 1. (a) Assume that v is known and derive the EM estimator of A. (b) Assume that A is known and derive the EM estimator of v.

4 Let $\mathbf{Z}$ denote a complete data vector. Each component Z_i is an independent, identically distributed Gaussian random variable with a mean μ and variance v. The observed data vector is $\mathbf{Y}$, where each component is $Y_i = |Z_i|$. Apply the EM algorithm to obtain estimates of $\boldsymbol{\theta} = [\mu, v]^T$. Let $f(z|\boldsymbol{\theta})$ denote the Gaussian density with parameters equal to $\boldsymbol{\theta}$. Show that the estimates are

$$\widehat{\mu}_{k+1} = \frac{1}{n}\sum_{i=1}^{n} \frac{y_i f(y_i|\hat{\boldsymbol{\theta}}_k) - y_i f(-y_i|\hat{\boldsymbol{\theta}}_k)}{f(y_i|\hat{\boldsymbol{\theta}}_k) + f(-y_i|\hat{\boldsymbol{\theta}}_k)}$$

$$\widehat{v}_{k+1} = \frac{1}{n}\sum_{i=1}^{n} y_i^2 - \widehat{\mu}_{k+1}^2.$$

5 Consider the equation $f(x) = x^2 - 3x + 1 = 0$. Find two different representations of this equation that have the form $x = g(x)$, and find sufficient conditions for convergence of the fixed-point iterations for both representations. Show that the initial value $x_0 = 2$ does not satisfy one of these conditions for an interval that includes it. Show by numerical computation of a few iterations that convergence to the smaller solution of $f(x) = 0$ occurs nevertheless.

6 Consider the equation $f(x) = x^3 + x - 1 = 0$, which has a solution near $x = 0.68$. Find a representation of this equation having the form $x = g(x)$ and satisfying the sufficient condition for convergence of the fixed-point iteration for any initial x_0. Show that $x = 1 - x^3$ is not the required representation.

7 Derive (9.47) to (9.49) from (9.42).

8 Use calculations similar to those in the text to derive (9.59).

9 Show how the spectrum confinement leads to (9.74).

10 Derive (9.102) and (9.103) from (9.97).

Chapter 10
Detection of Spread-Spectrum Signals

The ability to detect the presence of spread-spectrum signals is often required by cognitive radio, ultra-wideband, and military systems. This chapter presents an analysis of the detection of spread-spectrum signals when the spreading sequence or the frequency-hopping pattern is unknown and cannot be accurately estimated by the detector. Thus, the detector cannot mimic the intended receiver, and alternative procedures are required. The goal is limited in that only detection is sought, not demodulation or decoding. Nevertheless, detection theory leads to impractical devices for the detection of spread-spectrum signals. An alternative procedure is to use a radiometer or energy detector, which relies solely on energy measurements to determine the presence of unknown signals. The radiometer has applications not only as a detector of spread-spectrum signals, but also as a general sensing method in cognitive radio and ultra-wideband systems.

10.1 Detection of Direct-Sequence Signals

A spectrum analyzer usually cannot detect a signal with a power spectral density (PSD) below that of the background noise, which has two-sided PSD $N_0/2$. If the spreading factor $G >> 1$, (2.61) indicates that the two-sided PSD of a direct-sequence signal with a maximal spreading sequence is upper-bounded by

$$\bar{S}_s(f) < \frac{\mathcal{E}_c}{2} \tag{10.1}$$

where $\mathcal{E}_c = A^2T_c/2$ is the energy per chip. Thus, a received $\mathcal{E}_c/N_0 > 1$ is an approximate necessary, but not sufficient, condition for a spectrum analyzer to detect a direct-sequence signal. If $\mathcal{E}_c/N_0 < 1$, detection may still be probable by other means. If not, the direct-sequence signal is said to have a *low probability of interception*.

D. Torrieri, *Principles of Spread-Spectrum Communication Systems*,
https://doi.org/10.1007/978-3-319-70569-9_10

Detection of a Known Signal in Additional White Gaussian Noise

Consider the detection problem of deciding on the presence or absence of a known signal $s(t, \boldsymbol{\theta})$ over an observation interval in the presence of additional white Gaussian noise (AWGN), where $\boldsymbol{\theta}$ denotes a vector of parameters. Based on the observation of the received signal $r(t)$, classical detection theory [33, 47] requires a choice to be made between the hypothesis H_1 that the signal is present and the hypothesis H_0 that the signal is absent. For the observation interval $[0, T]$, the hypotheses are

$$\begin{aligned} H_1 &: r(t) = s(t, \boldsymbol{\theta}) + n(t) \\ H_0 &: \quad\quad r(t) = n(t) \end{aligned} \tag{10.2}$$

where $n(t)$ is zero-mean, white Gaussian noise. The real-valued signal $s(t,\boldsymbol{\theta})$ belongs to the signal space $L^2[0, T]$ of complex-valued functions f such that $|f|^2$is integrable over $[0, T]$. As explained in Appendix F.1, the inner product of functions $f(t)$ and $g(t)$ in $L^2[0, T]$ is

$$\langle f(t), g(t)\rangle = \int_0^T f(t)g^*(t)dt. \tag{10.3}$$

Since both $s(t, \boldsymbol{\theta})$ and $n(t)$ are real-valued, we establish a suitable set of real-valued orthonormal basis functions (Appendix F.1) that simplify the detection problem. We first choose a complete set of real-valued orthonormal basis functions $\{\phi_i'(t)\}$ for $L^2[0, T]$. A new complete set of real-valued orthonormal basis functions $\{\phi_i(t)\}$ for $L^2[0, T]$ is generated by selecting the first basis function as

$$\phi_1(t) = \frac{s(t, \boldsymbol{\theta})}{\sqrt{\mathcal{E}}} \tag{10.4}$$

where the signal energy is

$$\mathcal{E} = \|s(t, \boldsymbol{\theta})\|^2 = \langle s(t, \boldsymbol{\theta}), s(t, \boldsymbol{\theta})\rangle = \int_0^T |s(t, \boldsymbol{\theta})|^2\, dt \tag{10.5}$$

and is assumed to be the same for all $\boldsymbol{\theta}$. The remaining basis functions are constructed by the Gram-Schmidt orthonormalization process (Appendix F.1) using the functions $\phi_1(t)$ and $\{\phi_i'(t)\}$.

In terms of the $\{\phi_i(t)\}$, the orthonormal expansions of $r(t), n(t), s(t, \boldsymbol{\theta})$ are

$$r(t) = \sum_{i=1}^{\infty} r_i\phi_i(t),\ n(t) = \sum_{i=1}^{\infty} n_i\phi_i(t),\ s(t, \boldsymbol{\theta}) = \sqrt{\mathcal{E}}\phi_1(t) \tag{10.6}$$

where the coefficients are

$$r_i = \langle r(t), \phi_i(t) \rangle, \; n_i = \langle n(t), \phi_i(t) \rangle, \; i \geq 1 \tag{10.7}$$

and

$$r_1 = \frac{1}{\sqrt{\mathcal{E}}} \int_0^T r(t) s(t, \boldsymbol{\theta}) \, dt \tag{10.8}$$

$$n_1 = \frac{1}{\sqrt{\mathcal{E}}} \int_0^T n(t) s(t, \boldsymbol{\theta}) \, dt. \tag{10.9}$$

Since an orthonormal expansion contains the same information that resides in $r(t)$, it follows that detection can be based on the coefficients of $r(t)$ with respect to the orthonormal expansion. Therefore, after substituting (10.2) and (10.5) into (10.8), the detection problem becomes the decision between the hypotheses

$$\begin{aligned} H_1 &: r_1 = \sqrt{\mathcal{E}} + n_1, \; r_i = n_i, \; i \geq 2 \\ H_0 &: \qquad\qquad\qquad r_i = n_i, \; i \geq 1. \end{aligned} \tag{10.10}$$

As shown in Appendix F.2, since $n(t)$ is white Gaussian noise, coefficients n_i with $i \geq 2$ are statistically independent of n_1 and hence contain no information about r_1. Therefore, the detection problem reduces to a decision between the hypotheses

$$\begin{aligned} H_1 &: r_1 = \sqrt{\mathcal{E}} + n_1 \\ H_0 &: \quad r_1 = n_1 \end{aligned} \tag{10.11}$$

which indicates that r_1 is a *sufficient statistic* that retains all the information in $r(t)$ relevant to the detection. This sufficient statistic provides the same correlation metric that is obtained by projecting $r(t)$ onto the one-dimensional subspace spanned by $s(t, \boldsymbol{\theta})$. If $s(t, \boldsymbol{\theta})$ is known, the sufficient statistic may be obtained by applying $r(t)$ to a filter matched to $s(t, \boldsymbol{\theta})/\sqrt{\mathcal{E}}$ and then sampling the output (Section 1.2). If $s(t, \boldsymbol{\theta})$ is known and the sufficient statistic exceeds or is equal to a threshold, then the detector decides in favor of hypothesis H_1; if the sufficient statistic is less than the threshold, then the detector decides in favor of H_0. The threshold may be set to ensure a tolerable false-alarm probability when $s(t, \boldsymbol{\theta})$ is absent.

An equivalent procedure when $s(t, \boldsymbol{\theta})$ is known is to compare the likelihood ratio of the sufficient statistic with a threshold. Let $f(r_1|H_1, \boldsymbol{\theta})$ denote the conditional density function or likelihood function of the sufficient statistic r_1 of (10.8) given hypothesis H_1 and the value of $\boldsymbol{\theta}$. Let $f(r_1|H_0, \boldsymbol{\theta})$ denote the conditional density function or likelihood function of r_1 given hypothesis H_0. The *likelihood ratio* is defined as $f(r_1|H_1, \boldsymbol{\theta})/f(r_1|H_0, \boldsymbol{\theta})$. If $\boldsymbol{\theta}$ is known and the likelihood ratio is larger than or equal to a threshold, then H_1 is accepted; if the likelihood ratio is less than the threshold, then H_0 is accepted. However, if the parameters are modeled as random variables with known probability distribution functions, then we compute an *average likelihood ratio* by averaging over the distribution functions. The average likelihood ratio of the received signal $r(t)$, which is compared with a threshold for a detection decision, is

$$\Lambda[r(t)] = E_{\boldsymbol{\theta}}\left[\frac{f(r_1|H_1,\boldsymbol{\theta})}{f(r_1|H_0,\boldsymbol{\theta})}\right] \tag{10.12}$$

where $E_{\boldsymbol{\theta}}[\cdot]$ is the average over the distribution functions of the parameter components of $\boldsymbol{\theta}$.

Since $n(t)$ is zero-mean white noise, n_1 is a zero-mean Gaussian random variable (Appendix F.2). The expected values of n_1 under the two hypotheses are

$$E[n_1 \mid H_1] = E[n_1 \mid H_0] = 0. \tag{10.13}$$

The autocorrelation of $n(t)$ is

$$R_n(\tau) = \frac{N_0}{2}\delta(\tau) \tag{10.14}$$

where $\delta(\tau)$ is the Dirac delta function. Therefore, (10.9) yields

$$E\left[n_1^2\right] = \frac{N_0}{2}. \tag{10.15}$$

The conditional density functions of r_1 under the two hypotheses are

$$f(r_1|H_1,\boldsymbol{\theta}) = \frac{1}{\sqrt{\pi N_0}}\exp\left[-\frac{(r_1-\sqrt{\mathcal{E}})^2}{N_0}\right] \tag{10.16}$$

$$f(r_1|H_0,\boldsymbol{\theta}) = \frac{1}{\sqrt{\pi N_0}}\exp\left(-\frac{r_1^2}{N_0}\right). \tag{10.17}$$

Substituting these equations into (10.12), we obtain

$$\Lambda[r(t)] = E_{\boldsymbol{\theta}}\left[\exp\frac{2\sqrt{\mathcal{E}}r_1-\mathcal{E}}{N_0}\right]. \tag{10.18}$$

Direct-Sequence Signals

Detection theory leads to various detection receivers depending on precisely what is assumed to be known about the direct-sequence signal to be detected. To account for uncertainty in the chip timing, several parallel detectors may be used, each of which implements a different chip timing. Assuming that this procedure ultimately provides accurate chip-timing information, we consider the design of a single detector with perfect chip-timing information. Another assumption, which greatly simplifies the mathematical analysis, is that whenever the signal is present, it is present during the entire observation interval. Even with the latter assumptions, we find that the application of detection theory leads to a very complicated receiver.

Consider the detection of a direct-sequence signal with BPSK modulation:

$$s(t, \boldsymbol{\theta}) = \sqrt{2\mathcal{E}_s}p(t)\cos(2\pi f_c t + \theta), \quad 0 \leq t \leq T \tag{10.19}$$

where $\mathcal{E}_s$ is the energy per symbol, f_c is the known carrier frequency, and θ is the carrier phase assumed to be constant over the *observation interval* $0 \leq t \leq T$. The spreading waveform $p(t)$, which subsumes the random data modulation and has unit energy over a symbol interval of duration T_s, is

$$p(t) = \sum_{i=-\infty}^{\infty} p_i \psi(t - iT_c) \tag{10.20}$$

where $p_i = \pm 1$ is a data-modulated chip, and $\psi(t)$ is the chip waveform of known duration T_c. The energy of $s(t, \boldsymbol{\theta})$ over the observation interval is

$$\mathcal{E} = \frac{\mathcal{E}_s T}{T_s}. \tag{10.21}$$

We wish to detect the direct-sequence signal with unknown random parameters $\boldsymbol{\theta} = (\mathcal{E}_s, \theta, \mathbf{p})$, where the vector $\mathbf{p}$ denotes the spreading sequence. The known chip timing allows the boundary of the observation interval of the received signal $r(t)$ to coincide with a chip transition.

By substituting (10.8), (10.19), and (10.21) into (10.18), the average likelihood ratio is expressed in terms of the signal waveforms as

$$\Lambda[r(t)] \equiv E_{\mathcal{E},\theta,\mathbf{p}}\left\{\exp\left[\frac{2\mathcal{E}\sqrt{2T_s/T}}{N_0}\int_0^T r(t)p(t)\cos(2\pi f_c t + \theta)\,dt - \frac{\mathcal{E}}{N_0}\right]\right\} \tag{10.22}$$

where $E_{\mathcal{E},\theta,\mathbf{p}}\{\cdot\}$ is the average over the distribution functions of $\mathcal{E}, \theta$, and $\mathbf{p}$.

For *coherent detection*, we make the assumption that θ is somehow accurately estimated so that it can be removed from consideration. Mathematically, we set $\theta = 0$ into (10.22). There are 2^{N_c} equally likely patterns of the spreading sequence of N_c chips when $T = N_c T_c$. Thus, the average likelihood ratio for the ideal coherent detector is

$$\Lambda[r(t)] \equiv E_{\mathcal{E}} \left\{ \sum_{j=1}^{2^{N_c}} \exp\left[\frac{2\mathcal{E}\sqrt{2T_s/T}}{N_0} \sum_{i=0}^{N_c-1} p_i^{(j)} r_{ci} - \frac{\mathcal{E}}{N_0} \right] \right\} \quad \text{(coherent)} \tag{10.23}$$

where $p_i^{(j)}$ is chip i of pattern j and

$$r_{ci} = \int_{iT_c}^{(i+1)T_c} r(t)\psi(t - iT_c)\cos(2\pi f_c t)\,dt. \tag{10.24}$$

For the far more realistic *noncoherent detection* of a direct-sequence signal, the received carrier phase is assumed to be uniformly distributed over $[0, 2\pi)$. Using a trigonometric expansion and then evaluating the expectation over the random spreading sequence, we obtain the average likelihood ratio:

$$\Lambda(r(t)) = E_{\mathcal{E},\theta} \left\{ \sum_{j=1}^{2^{N_c}} \exp\left[\frac{2\mathcal{E}\sqrt{2T_s/T}}{N_0} \sum_{i=0}^{N_c-1} p_i^{(j)} (r_{ci}\cos\theta - r_{si}\sin\theta) - \frac{\mathcal{E}}{N_0} \right] \right\} \tag{10.25}$$

where r_{ci} is defined by (10.24), and

$$r_{si} = \int_{iT_c}^{(i+1)T_c} r(t)\psi(t - iT_c)\sin(2\pi f_c t)\,dt. \tag{10.26}$$

Using (H.15) of Appendix H.3, the uniform distribution of θ, and trigonometry, and dropping an irrelevant factor, we obtain the average likelihood ratio:

$$\Lambda[r(t)] = E_{\mathcal{E}} \left[\sum_{j=1}^{2^{N_c}} I_0\left(\frac{2\mathcal{E}\sqrt{2R_j T_s/T}}{N_0} \right) \exp\left(-\frac{\mathcal{E}}{N_0} \right) \right] \quad \text{(noncoherent)} \tag{10.27}$$

where $I_0(\cdot)$ is the modified Bessel function of the first kind and order zero, and

$$R_j = \left[\sum_{i=0}^{N_c-1} p_i^{(j)} r_{ci} \right]^2 + \left[\sum_{i=0}^{N_c-1} p_i^{(j)} r_{si} \right]^2. \tag{10.28}$$

These equations define the theoretically optimal noncoherent detector for a direct-sequence signal when the signal parameters have known probability distribution functions. The presence of the desired signal is declared if (10.27) exceeds a threshold level.

Even if the signal energy is assumed to be known so that the expectations in (10.23) and (10.27) do not need to be evaluated, the implementation of either the coherent or noncoherent optimal detector would be very complicated. The computational complexity would grow exponentially with N_c, the number of chips in the observation interval. Additional complexity would be required because of the parallel processors necessary to account for the unknown chip timing. Calculations [67] indicate that the ideal coherent and noncoherent detectors typically provide advantages of 3 and 1.5 dB, respectively, over the far more practical wideband radiometer, which is analyzed in the next section. The use of several wideband radiometers can compensate for these advantages. Furthermore, implementation losses and imperfections in the optimal detectors are likely to be significant.

10.2 Radiometer

A *radiometer* or *energy detector* is a device that uses energy measurements to determine the presence of unknown signals [101]. The radiometer is notable for its extreme simplicity and its meager requirement of virtually no information about a target signal other than its rough spectral location. The radiometer has applications not only as a detector of spread-spectrum signals, but also as a sensing method in cognitive radio systems and as a detector in ultra-wideband systems.

Suppose that the signal to be detected is approximated by a zero-mean, white Gaussian process. Consider two hypotheses that both assume the presence of a zero-mean, bandlimited white Gaussian process over an observation interval $0 \leq t \leq T$. Under H_0, only noise is present, and the two-sided PSD over the signal band is $N_0/2$, whereas under H_1, both signal and noise are present, and the PSD is $N_1/2$ over this band. Using orthonormal basis functions as in the derivation of (10.16) and (10.17), we find that the conditional density functions are approximated by

$$f(\mathbf{r}|H_i) = \prod_{k=1}^{\infty} \frac{1}{\sqrt{\pi N_i}} \exp\left(-\frac{r_k^2}{N_i}\right), \quad i = 0, 1. \tag{10.29}$$

Calculating the likelihood ratio, taking the logarithm, and merging constants with the threshold, we find that the decision rule is to compare

$$V = \sum_{k=1}^{\infty} r_k^2 \tag{10.30}$$

with a threshold. If we use the properties of orthonormal basis functions, then we find that the test statistic is

$$V = \int_0^T r^2(t)dt \tag{10.31}$$

where the assumption of bandlimited processes is necessary to ensure the finiteness of the statistic. A device that implements this test statistic is called an *energy detector* or *radiometer*. Although it was derived for a bandlimited white Gaussian signal, the radiometer is a reasonable configuration for determining the presence of unknown deterministic signals. Although a single radiometer is incapable of determining whether one or more than one signal has been detected, narrowband interference can be rejected by the methods of Section 5.3.

Theoretically superior devices require much more information about the target signals than does the radiometer and have other practical limitations. These devices have greatly increased computational complexity compared with the radiometer, and they provide improved performance, primarily when the noise-estimation errors of the radiometer are substantial. As discussed subsequently, appropriate methods ensure that these errors and their impact remain small.

An ideal radiometer has the form shown in Figure 10.1. The input signal $r(t)$ is filtered, squared, and integrated to produce an output that is compared with a threshold. The receiver decides that the target signal has been detected if and only if the threshold is exceeded. Since accurate analog integrators are difficult to implement and very power consuming, much more practical radiometers use either baseband sampling, as illustrated in Figure 10.2, or bandpass sampling, as illustrated in Figure 10.3. The mathematical analyses of all three forms of the radiometer lead to the same performance equations, but the analyses of the two practical radiometers entail significantly fewer approximations and hence provide more reliable results. The analysis of the radiometer with bandpass sampling is provided subsequently.

The bandpass filter in Figure 10.3 is assumed to approximate an ideal rectangular filter that passes the target signal $s(t)$ with negligible distortion while eliminating interference and limiting the noise. If the target signal has an unknown arrival time, the observation interval during which samples are taken is a sliding window with the oldest sample discarded as soon as a new sample is taken. The filter has center frequency f_c, bandwidth W, and produces the output

$$y(t) = s(t) + n(t) \tag{10.32}$$

where $n(t)$ is bandlimited white Gaussian noise with a two-sided PSD equal to $N_0/2$.

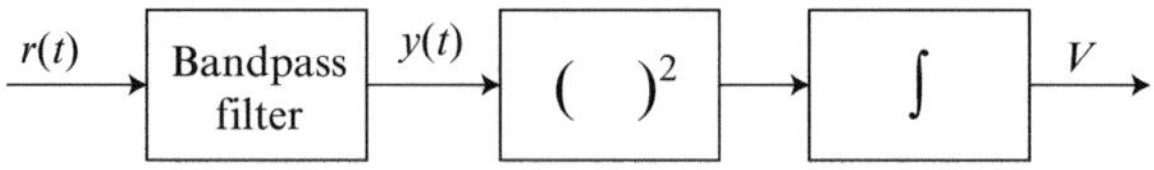

Fig. 10.1 Ideal radiometer

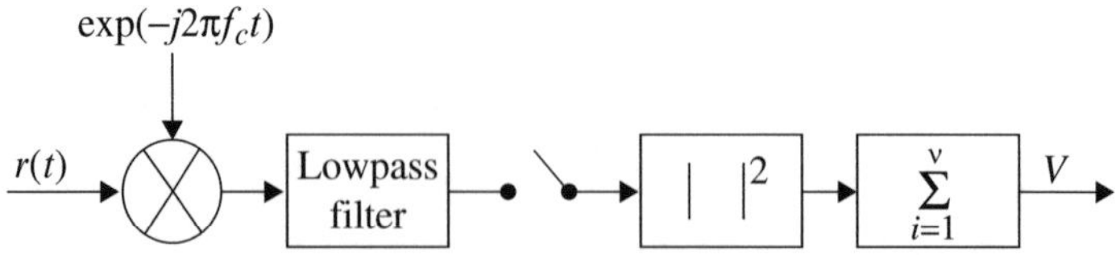

Fig. 10.2 Radiometer with baseband sampling

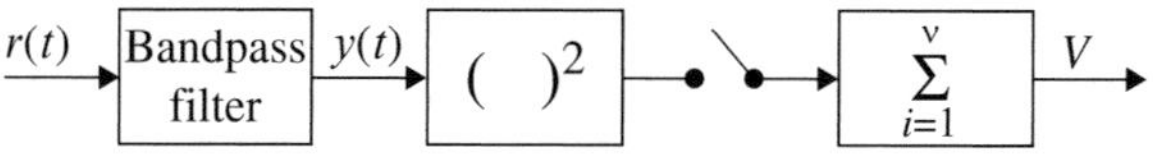

Fig. 10.3 Radiometer with bandpass sampling

As indicated by (D.16) of Appendix D.1, a bandlimited deterministic signal can be represented as

$$s(t) = s_c(t)\cos 2\pi f_c t - s_s(t)\sin 2\pi f_c t. \tag{10.33}$$

Since the spectrum of $s(t)$ is confined within the filter passband, $s_c(t)$ and $s_s(t)$ have frequency components confined to the band $|f| \leq W/2$. As indicated by (D.32) and (D.45) of Appendix D.2, the bandlimited Gaussian noise emerging from the bandpass filter can be represented in terms of quadrature components as

$$n(t) = n_c(t)\cos 2\pi f_c t - n_s(t)\sin 2\pi f_c t \tag{10.34}$$

where the PSDs of $n_c(t)$ and $n_s(t)$ are

$$S_c(f) = S_s(f) = \begin{cases} N_0, & |f| \leq W/2 \\ 0, & |f| > W/2. \end{cases} \tag{10.35}$$

The associated autocorrelation functions are

$$R_c(\tau) = R_s(\tau) = \sigma^2 \frac{\sin \pi W\tau}{\pi W\tau} \tag{10.36}$$

where $\sigma^2 = N_0 W$ is the noise power.

Let ν denote the number of samples collected by the radiometer of Figure 10.2. At sampling rate W, the duration of the observation interval of the target signal is $T = \nu/W$. Substituting (10.33) and (10.34) into (10.32), squaring, and sampling at rate W, we obtain the output

$$\begin{aligned} V = {} & \frac{1}{2}\sum_{i=1}^{\nu}\left\{(s_c[i] + n_c[i])^2 + (s_s[i] + n_s[i])^2\right\} \\ & + \frac{1}{2}\sum_{i=1}^{\nu}\left\{(s_c[i] + n_c[i])^2\, c[i] + (s_s[i] + n_s[i])^2\, s[i]\right\} \\ & - \sum_{i=1}^{\nu}\{(s_c[i] + n_c[i])\,(s_s[i] + n_s[i])\, s[i]\} \end{aligned} \tag{10.37}$$

where $s_c[i] = s_c(i/W), n_c[i] = n_c(i/W), s_s[i] = s_s(i/W), n_s[i] = n_s(i/W), c[i] = \cos(4\pi f_c i/W)$, and $s[i] = \sin(4\pi f_c i/W)$. This output provides a test statistic that is compared with a threshold.

If $\nu \gg 1$ and $f_c \gg W$, the fluctuations of $c[i]$ and $s[i]$ cause the final two summations in (10.37) to be negligible compared with the first summation. Equation (10.36) indicates that the zero-mean random variable $n_c[i]$ has variance σ^2 and is statistically independent of the zero-mean random variable $n_c[j]$ if $i \neq j$. Similarly, the zero-mean random variables $n_s[i]$ and $n_s[j]$ have variance σ^2 and are statistically independent if $i \neq j$. Since $n(t)$ is a zero-mean Gaussian process and has a PSD that is symmetrical about f_c, $n_c(t)$ and $n_s(t)$ are zero-mean, independent Gaussian processes (Appendix D.2); hence, $n_c[i]$ is statistically independent of $n_s[j]$. Therefore, for the AWGN channel in which $s_c[i]$ and $s_s[i]$ are deterministic, the test statistic may be expressed as

$$V = \frac{\sigma^2}{2} \sum_{i=1}^{\nu} (A_i^2 + B_i^2) \tag{10.38}$$

where the $\{A_i\}$ and the $\{B_i\}$ are statistically independent Gaussian random variables with unit variances and means

$$m_{1i} = E[A_i] = \frac{s_c[i]}{\sigma} \tag{10.39}$$

$$m_{2i} = E[B_i] = \frac{s_s[i]}{\sigma}. \tag{10.40}$$

The random variable $2V/\sigma^2$ has a *noncentral chi-squared distribution* (Appendix E.1) with 2ν degrees of freedom and noncentral parameter

$$\begin{aligned} \lambda &= \sum_{i=1}^{\nu} (m_{1i}^2 + m_{2i}^2) = \frac{1}{\sigma^2} \sum_{i=1}^{\nu} (s_c^2[i] + s_s^2[i]) \\ &\approx \frac{1}{N_0} \int_0^T [s_c^2(t) + s_s^2(t)]dt \approx \frac{2}{N_0} \int_0^T s^2(t)dt \end{aligned} \tag{10.41}$$

where the first approximation is obtained by dividing the integration interval into ν parts, each of duration $1/W$, and the second approximation uses (10.33) and $f_c \gg W$. Let $\mathcal{E}$ denote the energy of the target signal, and γ denote the *energy-to-noise-density ratio* $\mathcal{E}/N_0$. Assuming that the approximations in (10.41) cause a negligible error,

$$\lambda = 2\gamma \tag{10.42}$$

and the statistics of Gaussian random variables indicates that

$$E[V] = \sigma^2(\nu + \gamma). \tag{10.43}$$

Applying (10.41), (10.42), and the statistics of Gaussian variables to (10.38), and using (A.7) of Appendix A.1, which indicates that a zero-mean Gaussian random variable x has $E\left[x^4\right] = 3E\left[x^2\right]$, we obtain

$$var(V) = \sigma^4(\nu + 2\gamma). \tag{10.44}$$

From the noncentral chi-squared distribution, the density function of V is determined:

$$f_V(x) = \frac{1}{\sigma^2}\left(\frac{x}{\sigma^2\gamma}\right)^{(\nu-1)/2} \exp\left(-\frac{x}{\sigma^2} - \gamma\right) I_{\nu-1}\left(\frac{2\sqrt{x\gamma}}{\sigma}\right) u(x) \tag{10.45}$$

where $u(x)$ is the unit step function: $u(x) = 1, x \geq 0$, and $u(x) = 0,\ x < 0$, and $I_n(\cdot)$ is the modified Bessel function of the first kind and order n (Appendix H.3). Substituting (H.13) into (10.45) and then setting $\gamma = 0$, we obtain the density function in the absence of a signal:

$$f_V(x) = \frac{1}{\sigma^2(\nu - 1)!}\left(\frac{x}{\sigma^2}\right)^{\nu-1} \exp\left(-\frac{x}{\sigma^2}\right) u(x), \quad \gamma = 0. \tag{10.46}$$

Let V_t denote the threshold such that if $V > V_t$, the receiver decides that the target signal is present. Therefore, a false alarm occurs if $V > V_t$ when the target signal is absent. Integration of (10.46) over (V_t, ∞), a change of variables, and the application of (H.5) and (H.8) from Appendix H.1 gives the false-alarm probability

$$P_F = \frac{\Gamma(\nu, V_t/\sigma^2)}{\Gamma(\nu)} = \exp\left(-\frac{V_t}{\sigma^2}\right)\sum_{i=0}^{\nu-1}\frac{1}{i!}\left(\frac{V_t}{\sigma^2}\right)^i \tag{10.47}$$

where $\Gamma(a, x)$ is the incomplete gamma function, and $\Gamma(a) = \Gamma(a, 0)$ is the gamma function.

The threshold V_t is usually set to a value that ensures a specified P_F. Thus, if the estimate of σ^2 is σ_e^2, then

$$V_t = \sigma_e^2 G_\nu^{-1}(P_F) \tag{10.48}$$

where $G_\nu^{-1}(\cdot)$ is the inverse function of $P_F\left(V_t/\sigma^2\right)$. Since the series in (10.47) is finite, this inverse can be numerically computed by applying Newton's method. If the exact value of the noise power is known, then $\sigma_e^2 = \sigma^2$ in (10.48).

The target signal is detected if $V > V_t$ when the target signal is present during the observation interval. For the AWGN channel, the integration of (10.45) and a change of variables yields the detection probability

$$P_D = Q_\nu(\sqrt{2\gamma}, \sqrt{2V_t/\sigma^2}) \tag{10.49}$$

where $Q_m(\alpha, \beta)$ is the *generalized Marcum Q-function defined by* (H.26) of Appendix H.4.

For $\nu > 100$, the generalized Marcum Q-function in (10.49) is difficult to compute and to invert; thus, we seek an approximation. As indicated by (10.38), the test statistic V is the sum of independent random variables with finite means. Since the $\{A_i\}$ and the $\{B_i\}$ are Gaussian random variables, the $\{A_i^2\}$ and the $\{B_i^2\}$ have bounded fourth central moments. Equation (10.44) indicates that $[var(V)]^{3/2}/\nu \to \infty$ as $\nu \to \infty$. Therefore, if $\nu \gg 1$, the central limit theorem (Corollary A3, Appendix A.2) is applicable. The theorem indicates that the distribution of V is approximately Gaussian if $\nu \gg 1$. Using (10.43) and (10.44) and the Gaussian distribution, we obtain

$$P_D \simeq Q\left[\frac{V_t/\sigma^2 - \nu - \gamma}{\sqrt{\nu + 2\gamma}}\right], \quad \nu \gg 1 \tag{10.50}$$

where the Gaussian Q-function is defined as

$$Q(x) = \frac{1}{\sqrt{2\pi}} \int_x^\infty \exp(-\frac{y^2}{2})dy. \tag{10.51}$$

The error in the approximation (10.50) has been shown [37] to be less than 0.02 when $\nu > 100$.

Setting $\gamma = 0$ in (10.50) gives

$$P_F \simeq Q\left[\frac{V_t/\sigma^2 - \nu}{\sqrt{\nu + 2\gamma}}\right], \quad \nu \gg 1. \tag{10.52}$$

The approximations provided by (10.50) and (10.52) are very accurate if $\nu \geq 100$. Inverting (10.52), we obtain an approximate equation for V_t in terms of P_F, σ^2, and ν. Accordingly, if a required P_F is specified and the estimate of σ^2 is σ_e^2, then the threshold setting is

$$V_t \simeq \sigma_e^2[\sqrt{\nu}Q^{-1}(P_F) + \nu], \quad \nu \gg 1 \tag{10.53}$$

where $Q^{-1}(\cdot)$ denotes the inverse of the Gaussian Q-function, and ideally $\sigma_e^2 = \sigma^2$. In most applications, there is a required false-alarm rate F, which is the expected number of false alarms per unit time. If successive observation intervals do not overlap each other except possibly at end points, then the required probability of false alarm is $P_F = FT$.

Figure 10.4 depicts P_D versus the energy-to-noise-density ratio γ for a radiometer operating over the AWGN channel with $\sigma_e^2 = \sigma^2$ and $P_F = 10^{-3}$. Equation (10.48) is used to calculate V_t. Equation (10.49) is used to calculate P_D for $\nu = 10$ and $\nu = 100$. For $\nu \geq 1000$, (10.50) is used to calculate P_D. The figure illustrates that the signal energy required to achieve a desired P_D increases as ν increases because additional energy is needed to overcome the noise in each additional sample. However, since the average signal power is equal to $\gamma\sigma^2/\nu$, the average signal power required decreases as ν increases.

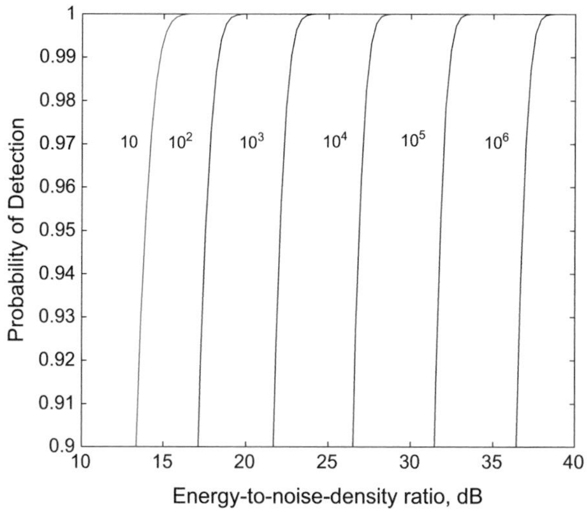

Fig. 10.4 Probability of detection for the AWGN channel, $\sigma_e^2 = \sigma^2$, and $P_F = 10^{-3}$. Plots are labeled with ν

Rayleigh Fading

If the coherence time of Rayleigh fading exceeds the observation interval, then the energy of a target signal in Rayleigh fading is a random variable with an exponential distribution (Appendix E.5) and average value $E[\mathcal{E}] = \bar{\mathcal{E}}$. Thus, the average probability of detection is

$$\bar{P}_D = \int_0^\infty \frac{1}{\bar{\gamma}} \exp\left(-\frac{\gamma}{\bar{\gamma}}\right) P_D(\gamma) d\gamma \tag{10.54}$$

where $\bar{\gamma} = \bar{\mathcal{E}}/N_0$ and $P_D(\gamma)$ is given by (10.49).

The substitution of (H.26) and (H.13) of Appendix H into (10.49), the interchange of the summation and integration, a change of variables, and the evaluation of the integral using (H.5) gives

$$P_D(\gamma) = \sum_{i=0}^{\infty} \frac{\Gamma(i+\nu, V_t/\sigma^2)\gamma^i e^{-\gamma}}{\Gamma(i+\nu)\Gamma(i+1)} \,. \tag{10.55}$$

The interchange of the summation and integration can be justified by first observing that $Q_m(\alpha, 0) - Q_m(\alpha, \beta) = 1 - Q_m(\alpha, \beta)$ is an integral over a finite interval. Since the series for the Bessel function is uniformly convergent and continuous over this interval, the series can be integrated term-by-term.

The substitution of (10.55) into (10.54), the application of the dominated convergence theorem to justify the interchange of the summation and integration, a change of variables, the evaluation of the integral using (H.3), and the substitution of (H.8) yields

$$\bar{P}_D = \sum_{i=0}^{\infty} \left(\frac{\bar{\gamma}}{\bar{\gamma}+1}\right)^i \frac{\exp\left(-\frac{V_t}{\sigma^2}\right)}{\bar{\gamma}+1} \sum_{k=0}^{\nu-1+i} \frac{\left(V_t/\sigma^2\right)^k}{k!}. \tag{10.56}$$

Dividing the inner series into two series, using (10.47), evaluating a geometric series, and then rearranging a double series, we obtain

$$\begin{aligned} \bar{P}_D &= \sum_{i=0}^{\infty} \left(\frac{\bar{\gamma}}{\bar{\gamma}+1}\right)^i \frac{\exp\left(-\frac{V_t}{\sigma^2}\right)}{\bar{\gamma}+1} \left[\sum_{k=0}^{\nu-1} \frac{\left(V_t/\sigma^2\right)^k}{k!} + \sum_{k=\nu}^{\nu-1+i} \frac{\left(V_t/\sigma^2\right)^k}{k!}\right] \\ &= P_F + \sum_{i=0}^{\infty} \sum_{k=\nu}^{\nu-1+i} \left(\frac{\bar{\gamma}}{\bar{\gamma}+1}\right)^i \frac{\exp\left(-\frac{V_t}{\sigma^2}\right)\left(V_t/\sigma^2\right)^k}{(\bar{\gamma}+1)\,k!} \\ &= P_F + \sum_{k=\nu}^{\infty} \frac{\exp\left(-\frac{V_t}{\sigma^2}\right)\left(V_t/\sigma^2\right)^k}{(\bar{\gamma}+1)\,k!} \sum_{i=k-\nu+1}^{\infty} \left(\frac{\bar{\gamma}}{\bar{\gamma}+1}\right)^i. \end{aligned} \tag{10.57}$$

Evaluation of the inner geometric series, use of the series for the exponential function to express the remaining infinite series for $\bar{P}_D$ as an exponential minus a finite series, and the application of (H.8) yields

$$\begin{aligned} \bar{P}_D &= P_F + \exp\left(-\frac{V_t}{\sigma^2}\right) \sum_{k=\nu}^{\infty} \frac{\left(V_t/\sigma^2\right)^k}{k!} \left(\frac{\bar{\gamma}}{\bar{\gamma}+1}\right)^{k-\nu+1} \\ &= P_F + \left(\frac{\bar{\gamma}+1}{\bar{\gamma}}\right)^{\nu-1} \exp\left(-\frac{V_t}{\sigma^2}\right) \left[\exp\left(\frac{\bar{\gamma}V_t/\sigma^2}{\bar{\gamma}+1}\right) - \sum_{k=0}^{\nu-1} \left(\frac{\bar{\gamma}V_t/\sigma^2}{\bar{\gamma}+1}\right)^k\right] \\ &= P_F + \left(\frac{\bar{\gamma}+1}{\bar{\gamma}}\right)^{\nu-1} \exp\left(-\frac{V_t/\sigma^2}{\bar{\gamma}+1}\right) \left[1 - \frac{\Gamma(\nu, \frac{\bar{\gamma}V_t/\sigma^2}{\bar{\gamma}+1})}{\Gamma(\nu)}\right] \end{aligned} \tag{10.58}$$

for $\bar{\gamma} > 0$. This equation expresses $\bar{P}_D$ in a closed form that is easily computed. Equations for $\bar{P}_D$ in the presence of Nakagami and Ricean fading may be found in [36] and [21].

The outputs of L radiometers, each processing ν samples, can be combined to provide diversity reception for fading channels. In a selection-combining diversity scheme, the largest radiometer output is compared with the threshold V_t to determine the presence of the target signal. Thus, if each radiometer processes independent noise, the probability of false alarm is

$$P_F = 1 - \prod_{i=1}^{L} [1 - P_{F0}(\sigma_i^2)] \tag{10.59}$$

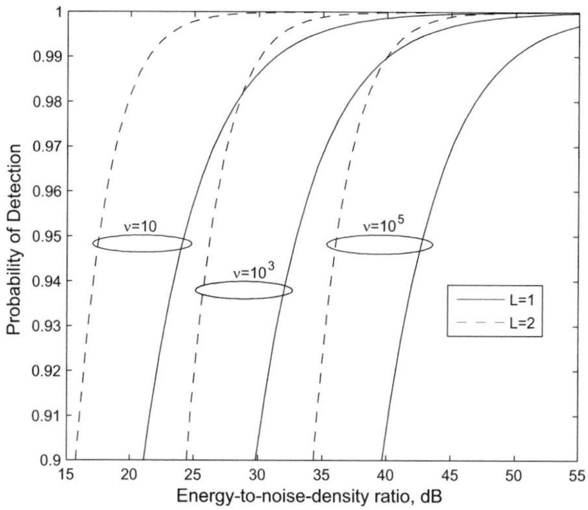

Fig. 10.5 Average probability of detection for Rayleigh fading, $\sigma_i^2 = \sigma_e^2 = \sigma^2, \bar{\gamma}_i = \bar{\gamma}, P_F = 10^{-3}$, and $L = 1$ and $L = 2$

where σ_i^2 is the noise power of radiometer i and $P_{F0}(\sigma_i^2)$ is given by the right-hand side of (10.47) with $\sigma^2 = \sigma_i^2$. Let σ_e^2 denote an estimated noise power that is known to exceed $\sigma_i^2, i = 1, 2, \ldots, L$, with high probability. Using this estimate for each σ_i^2 and solving (10.59) for V_t, we ensure that a specified P_F is almost always achieved. We obtain

$$V_t \simeq \sigma_e^2 G_\nu^{-1}(1 - (1 - P_F)^{1/L}) \tag{10.60}$$

which indicates that V_t must be increased with increases in L, the amount of diversity, if a specified P_F is to be achieved. If each radiometer receives a target signal that experiences independent Rayleigh fading, then the average probability of detection is

$$\bar{P}_D = 1 - \prod_{i=1}^{L}[1 - \bar{P}_{D0}(\bar{\gamma}_i, \sigma_i^2)] \tag{10.61}$$

where $\bar{\gamma}_i$ is the expected value of γ in radiometer i and $\bar{P}_{D0}(\bar{\gamma}_i, \sigma_i^2)$ is given by the right-hand side of (10.58) with $\sigma^2 = \sigma_i^2$ and $\bar{\gamma} = \bar{\gamma}_i$.

Figure 10.5 shows $\bar{P}_D$ versus the average energy-to-noise-density ratio $\bar{\gamma}$ for $\sigma_i^2 = \sigma_e^2 = \sigma^2, \bar{\gamma}_i = \bar{\gamma}, P_F = 10^{-3}$, and $L = 1$ and $L = 2$. A comparison of this figure with the preceding one indicates that the impact of the fading is pronounced at high values of $\bar{P}_D$. When $L = 1$, the required signal energy to achieve a specified $\bar{P}_D$ over the Rayleigh channel is increased by 8 dB or more relative to the required signal energy over the AWGN channel. When selection diversity with $L = 2$ is used, this increase is reduced to roughly 3 dB. However, further increases in L produce only minor and diminishing gains. An alternative diversity scheme is square-law combining, which is analyzed in [21] and provides a performance similar to that of selection combining.

Estimation of Noise Power

The greatest obstacle to the implementation and efficient operation of the radiometer is its extreme sensitivity to imperfect knowledge of the noise power. The sensitivity is due to the typical extensive overlapping of the density functions under the two hypotheses involving the presence or absence of the target signal. This overlap causes a small change in the threshold because of the imperfect estimation of the noise power to have a large impact on the probabilities of false alarm and detection.

To ensure that the threshold is large enough that the required P_F is achieved regardless of the true value of noise power σ^2, the power estimate must be set to the upper end of the measurement uncertainty interval. If measurements indicate that the noise power is between a lower bound σ_l^2 and an upper bound σ_u^2, then the threshold should be calculated using σ_u^2 as the power estimate σ_e^2.

The noise power is nonstationary primarily because of temperature variations, vibrations, aging, and nonstationary environmental noise. Laboratory measurements indicate that the noise power can change by 0.1% in a few minutes [35]. Thus, if measurements made shortly before the radiometer is used to detect a target signal indicate that the noise power σ^2 has an upper-bound of σ_u^2, and the noise power can change by no more than 0.1%, then using $\sigma_e^2 = 1.001\sigma_u^2$ to calculate the threshold ensures a specified P_F with high probability and limits the degradation to P_D or $\bar{P}_D$ caused by the fact that $\sigma_e^2 > \sigma^2$.

The noise power can be estimated by measuring the output V of a radiometer over ν_e samples when no target or interfering signals are received. An estimate of the noise power is $N_e = V/\nu_e$, and (10.43) and (10.44) indicate that its mean is σ^2 and its standard deviation is $\sigma^2/\sqrt{\nu_e}$. Since Chebyshev's inequality (Section 4.3) implies that it is highly unlikely that the measurement error exceeds five standard deviations, N_e is highly likely to lie between $\sigma_l^2 = \sigma^2(1-5/\sqrt{\nu})$ and $\sigma_u^2 = \sigma^2(1+5/\sqrt{\nu})$. Thus, setting $\sigma_e^2 = N_e(1+5.1/\sqrt{\nu_e})$ makes it highly likely that $\sigma_e^2 > \sigma^2$ and the specified P_F is achieved. The *error factor*

$$h = \frac{\sigma_e^2}{\sigma^2} \tag{10.62}$$

can be made arbitrarily close to unity by using a sufficiently large ν_e. For example, if $T_e = 1$ s is the duration of the observation interval for the noise-power estimation and $W = 1$ MHz, then $\nu_e = 10^6$ and $h \in (1, 1.01)$.

An auxiliary radiometer simultaneously operating over an adjacent spectral region can potentially be used to estimate the noise power in parallel with a

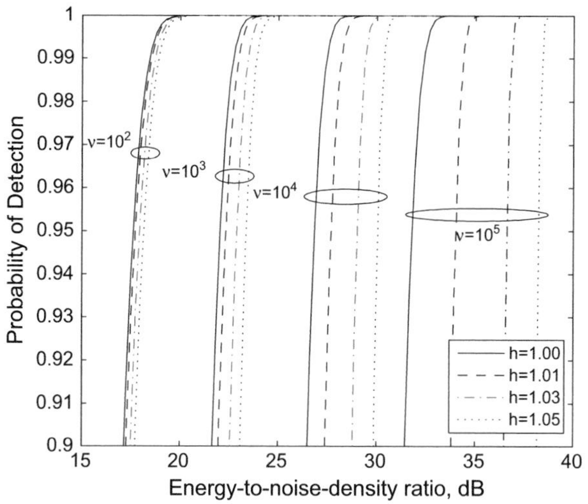

Fig. 10.6 Probability of detection for the AWGN channel, $P_F = 10^{-3}$, and various values of h

radiometer detecting a target signal. An advantage of this method is that time-dependent fluctuations in the noise power are generally negligible over a practical observation interval. The main requirements are that there is negligible target-signal power in the auxiliary radiometer and that the ratio of the two noise powers is an approximately known constant during the observation interval.

The impact of the imperfect estimation of the noise power is illustrated in Figures 10.6, 10.7, and 10.8 for $P_F = 10^{-3}$ and various values of the error factor. Figure 10.6 depicts P_D versus γ for a radiometer and the AWGN channel. As observed in this figure, the required signal energy for a specified P_D and ν increases with h. If $\nu \leq 10^4$ and $h \leq 1.01$, the increase is less than 0.90 dB. The increase in required signal energy when $h > 1.0$ is partially offset by the decrease in the actual P_F below the specified P_F because the threshold is increased by approximately the factor h, as observed from (10.53). Figures 10.7 and 10.8 depict $\bar{P}_D$ versus $\bar{\gamma}$ for a radiometer and the Rayleigh fading channel. In Figure 10.7, $L = 1$; in Figure 10.8, $L = 2, \sigma_i^2 = \sigma_e^2 = h\sigma^2$, and $\bar{\gamma}_i = \bar{\gamma}$. The effects of h and ν are fairly similar in Figures 10.6, 10.7, and 10.8. As ν increases, the frequency and accuracy of the noise-power estimates must increase if the performance degradation is to be minimized.

Other Implementation Issues

To avoid processing noise outside the spectral region occupied by the target signal, the bandpass filter should have as small a bandwidth as possible. If it is known that a target signal occupies a small band somewhere within the larger passband of the

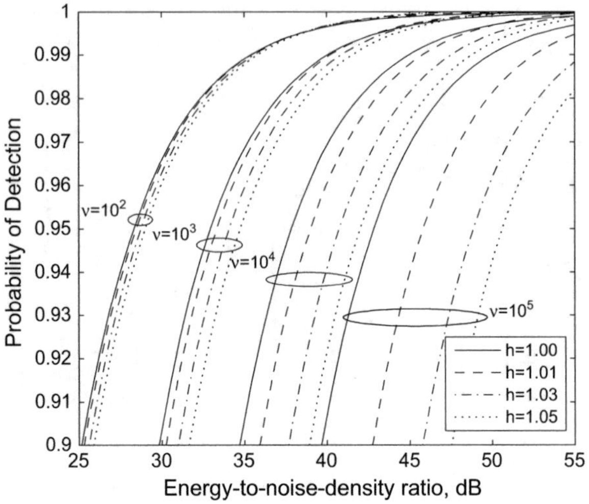

Fig. 10.7 Average probability of detection for Rayleigh fading, $P_F = 10^{-3}, L = 1$, and several values of h

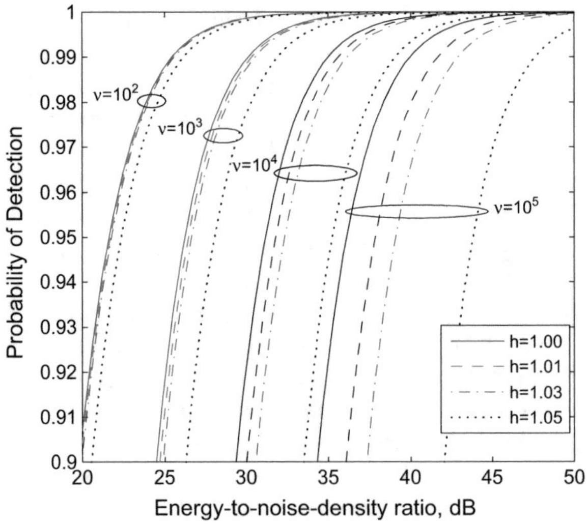

Fig. 10.8 Average probability of detection for Rayleigh fading, $P_F = 10^{-3}, L = 2$, and several values of h

bandpass filter, then the target signal can be isolated in its subband by inserting a fast Fourier transformer after the sampler in Figure 10.2. Parallel outputs of the transformer can be applied to separate radiometers, each one processing a distinct subband defined by the transformer. With this architecture, multiple signals can be simultaneously detected over several subbands.

Wideband signals, such as direct-sequence and ultra-wideband signals, can be detected by a radiometer, but both the sampling rate and the noise power increase with the bandwidth of the radiometer. The radiometer can serve as a basic component of a channelized radiometer for the detection of frequency-hopping signals (Section 10.4).

It is desirable for the observation interval to be large enough to collect the energy of not only the main target signal but also its significant multipath components. Thus, the extent of the observation interval or, equivalently, the number of samples, is largely determined by the delay power spectrum or intensity profile of the multipath to the degree that it is known. A sample at the end of the interval should only be added if the increase in processed signal energy is sufficient to compensate for the increase in noise variance in the radiometer output.

Let

$$\beta = Q^{-1}(P_F), \;\; \xi = Q^{-1}(P_D). \tag{10.63}$$

In practical applications, $P_F < 1/2 < P_D$ is required. For this requirement to be met, (10.50), (10.52), and $Q(0) = 1/2$ imply that

$$\beta > 0, \; \xi < 0, \; \gamma > \beta h\sqrt{\nu} + (h-1)\nu > 0 \tag{10.64}$$

are required. The required value of γ to achieve specified values of P_F and P_D over the AWGN channel may be obtained by inverting (10.49), which is computationally difficult, but can be closely approximated by inverting (10.50) if $\nu \gg 1$. The inversion of (10.50) entails the solution of a quadratic equation. The use of $\xi < 0$ in choosing the root, and the substitution of (10.53) and (10.62) yield the required value:

$$\gamma_r(\nu) \simeq \beta h\sqrt{\nu} + (h-1)\nu + \psi(\beta, \xi, \nu, h), \quad \nu \gg 1 \tag{10.65}$$

where

$$\psi(\beta, \xi, \nu, h) = \xi^2 - \xi\sqrt{\xi^2 + 2\beta h\sqrt{\nu} + (2h-1)\nu} \,. \tag{10.66}$$

Figure 10.9 shows the *required energy-to-noise-density ratio* $\gamma_r(\nu)$ versus ν for $P_D = 0.999, P_F = 10^{-3}$, and various values of h.

To increase the probability of detection, it is desirable to continue collecting additional samples if the increase in $\gamma_r(\nu)$, which is proportional to the required signal energy, is less than the increase in $\gamma(\nu)$, which is proportional to the signal energy processed by the radiometer. If we approximate the integer ν by treating it as a continuous variable, then collecting additional samples beyond ν_0 samples already collected is potentially useful if

$$\frac{\partial \gamma_r(\nu_0)}{\partial \nu} < \frac{\partial \gamma(\nu_0)}{\partial \nu} \,. \tag{10.67}$$

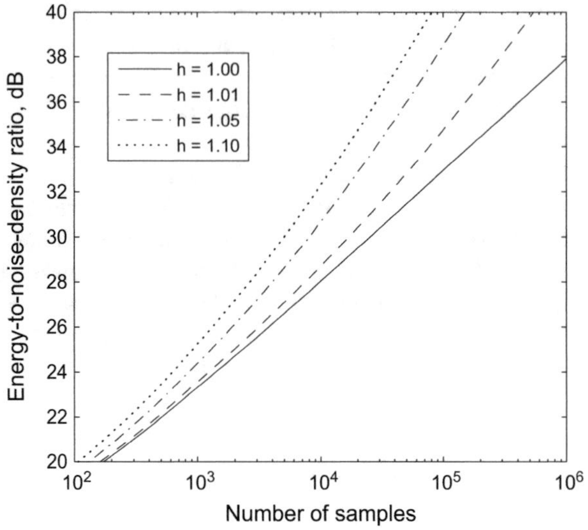

Fig. 10.9 Required energy-to-noise-density ratio for the AWGN channel, $P_D = 0.999, P_F = 10^{-3}$, and various values of h

The derivative $\partial\gamma_r(\nu_0)/\partial\nu$ can be determined from the curves of Figure 10.9 or by using (10.65). The derivative $\partial\gamma(\nu_0)/\partial\nu$ can be calculated from knowledge of the waveform of the target signal or from the intensity profile of the target signal's multipath, if the latter is known.

Example 1 Suppose that $P_D = 0.999$ and $P_F = 10^{-3}$ are required when the radiometer operates over the AWGN channel, $h = 1.05$, and $\nu_0 = 800$. Figure 10.9 indicates that $\gamma_r(\nu_0) \simeq 24$ dB and $\partial \log_{10} \gamma_r(\nu_0)/\partial \log_{10} \nu \approx 0.6$. Therefore, $\partial\gamma_r(\nu_0)/\partial\nu \approx 0.6\gamma_r(\nu_0)/\nu_0$. Suppose that the target signal has constant power over a time interval exceeding the observation interval of the radiometer so that $\partial\gamma(\nu_0)/\partial\nu \approx \gamma(\nu_0)/\nu_0$. Collecting more samples is useful if $\gamma(\nu_0) > 0.6\gamma_r(\nu_0)$ or $\gamma(\nu_0) > 21.8$ dB. If $\gamma(\nu_0) > 24$ dB, then P_D already exceeds its required value if $P_F = 10^{-3}$, and collecting more samples increases P_D further. If 21.8 dB $< \gamma(\nu_0) < 24$ dB, then collecting sufficient additional samples potentially allows $P_D = 0.999$ and $P_F = 10^{-3}$ to be achieved. □

A different perspective is gained by examining the value of the signal-to-noise ratio (SNR) required to achieve specified values of P_F and P_D over the AWGN channel. The *required SNR*, defined as

$$S_r(\nu) = \frac{\gamma_r(\nu)}{\nu} \tag{10.68}$$

may be computed using (10.65) and is shown in Figure 10.10 for the AWGN channel, $P_D = 0.999, P_F = 10^{-3}$, and various values of h. It is observed that for $S_r(\nu) = -14$ dB, roughly twice as many samples are needed when $h = 1.01$ as are needed when $h = 1.0$.

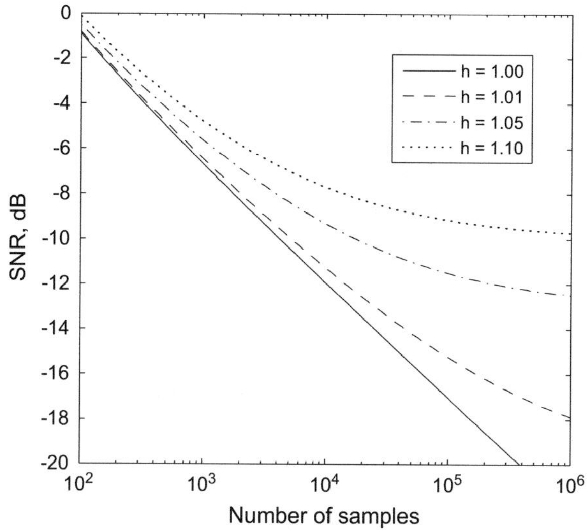

Fig. 10.10 Required signal-to-noise ratio for the AWGN channel, $P_D = 0.999, P_F = 10^{-3}$, and various values of h

Equations (10.68) and (10.65) indicate that if h is a constant, then

$$\lim_{\nu \to \infty} S_r(\nu) = h - 1. \tag{10.69}$$

The significance of this limit is that if the SNR of the target signal is below $h - 1$, then specified values of P_F and P_D cannot be achieved, no matter how many samples are collected.

Example 2 Suppose that the SNR of the target signal is approximately −12 dB over a long observation interval when it is present, and that $P_D = 0.999$ and $P_F = 10^{-3}$ are desired when the radiometer operates over the AWGN channel. Figure 10.10 indicates that the desired P_D and P_F cannot be achieved if $h \geq 1.10$. However, if the noise-power estimation is timely and accurate enough that $h \leq 1.05$, then the desired P_D and P_F can be achieved with the collection of $\nu \approx 3 \cdot 10^5$ or fewer samples; only $\nu \approx 2 \cdot 10^4$ or fewer are needed if $h \leq 1.01$. □

When the limit in (10.69) exceeds 0, it is called the *SNR wall*. Its existence is due to the assumption that h is a constant as the number of samples ν used to detect a signal increases. As shown previously, if h is estimated by using an energy detector, then it is highly probable that

$$1 - \frac{5}{\sqrt{\nu_e}} < h < 1 + \frac{5}{\sqrt{\nu_e}} \tag{10.70}$$

where ν_e is the number of samples used to determine σ_e^2. If $\nu_e \to \infty$ as $\nu \to \infty$, then $h \to 1$ and there is no SNR wall. Thus, the SNR wall exists only when the detection interval has a much longer duration than that of the interval used in the estimation of the noise power. An alternative analysis [51] that models σ_e^2 as a random variable generated by an energy detector confirms this result.

Because of its large bandwidth and low PSD, a direct-sequence signal is difficult to detect by any device that cannot despread it. The effectiveness of a radiometer in detecting a direct-sequence signal depends on the collection of enough samples and sufficient received signal energy.

10.3 Detection of Frequency-Hopping Signals

An interception receiver intended for the detection of frequency-hopping signals may be designed according to the principles of classical detection theory or according to more intuitive ideas. The former approach is useful in setting limits on what is possible, but the latter approach is more practical and flexible and less dependent on knowledge of the characteristics of the frequency-hopping signals.

To enable a tractable analysis according to classical detection theory, the idealized assumptions are made that the hopset and the *hop epoch timing* are known. The hop epoch timing comprises the hop duration T_h, the number of hops N_h, and the hop-transition times. We further assume that whenever the frequency-hopping signal is present, it occupies the entire observation interval. Consider slow frequency-hopping signals with continuous-phase modulation (FH/CPM) or continuous-phase frequency-shift keying (FH/CPFSK) that have negligible switching times. The frequency-hopping signal over the ith hop interval or dwell time is

$$s(t,\boldsymbol{\theta}) = \sqrt{2\mathcal{S}}\cos\left[2\pi f_{ci}t + \phi(\mathbf{d}_n,t) + \phi_i\right]\ ,\quad (i-1)T_h \le t < iT_h \tag{10.71}$$

where $\mathcal{S}$ is the average signal power, f_{ci} is the carrier frequency associated with the ith hop, $\phi(\mathbf{d}_n,t)$ is the CPM component that depends on the data sequence $\mathbf{d}_n$, and ϕ_i is the phase associated with the ith hop. The vector $\boldsymbol{\theta}$ denotes the parameters $\mathcal{S}, \{f_{ci}\}, \{\phi_i\}, i \in [1, N_h]$, and the components of $\mathbf{d}_n$, which are modeled as random variables. For the AWGN channel and observation interval $[0, N_hT_h]$, (10.18) and (10.71) indicate that the average likelihood ratio is

$$\Lambda[r(t)] = E_{\Theta}\left\{\exp\left[\frac{2\sqrt{2\mathcal{S}}}{N_0}\int_0^{N_hT_h} r(t)\cos\left[2\pi f_{ci}t + \phi(\mathbf{d}_n,t) + \phi_i\right]dt - \frac{\mathcal{S}N_hT_h}{N_0}\right]\right\} \tag{10.72}$$

where N_0 is known. The average likelihood ratio $\Lambda[r(t)]$ is compared with a threshold to determine whether a signal is present. The threshold may be set to ensure a tolerable false-alarm probability when the signal is absent.

The M carrier frequencies $\{f_j\}$ in the hopset are assumed to be equally likely over a given hop and statistically independent from hop to hop. Each of the N_d data sequences that can occur during a hop is assumed to be equally likely. Dividing the integration interval in (10.72) into N_h parts, averaging over the M frequencies, averaging over the N_d data sequences, and dropping factors that can be merged with the threshold, we obtain

$$\Lambda[r(t)] = E_{\mathcal{S}}\left\{\exp\left(-\frac{\mathcal{S}N_hT_h}{N_0}\right)\prod_{i=1}^{N_h}\sum_{j=1}^{M}\Lambda_{i,j}[r(t)|f_j]\right\} \tag{10.73}$$

$$\Lambda_{i,j}[r(t)|f_j] = E_{\phi_i}\left\{\sum_{n=1}^{N_d}\exp\left[\frac{2\sqrt{2\mathcal{S}}}{N_0}\int_{(i-1)T_h}^{iT_h} r(t)\cos\left[\chi_{j,n}(t)+\phi_i\right]\right]\right\} \tag{10.74}$$

where the expectations are over the distribution functions of the remaining random parameters $\mathcal{S}$ and ϕ_i, and

$$\chi_{j,n}(t) = 2\pi f_j t + \phi(\mathbf{d}_n, t). \tag{10.75}$$

These equations indicate the general structure of the theoretically optimal detector when the signal parameters are modeled as random variables with known probability distribution functions. When $\mathcal{S}$ is known, the optimal detector has the form illustrated in Figure 10.11.

For *coherent detection* of FH/CPM [8], we assume that the $\{\phi_i\}$ are somehow accurately estimated. Thus, we set $\phi_i = 0$ in (10.74) to obtain

$$\Lambda_{i,j}[r(t)|f_j] = \sum_{n=1}^{N_d}\exp\left\{\frac{2\sqrt{2\mathcal{S}}}{N_0}\int_{(i-1)T_h}^{iT_h} r(t)\cos\left[\chi_{j,n}(t)\right]\right\} \quad \text{(coherent).} \tag{10.76}$$

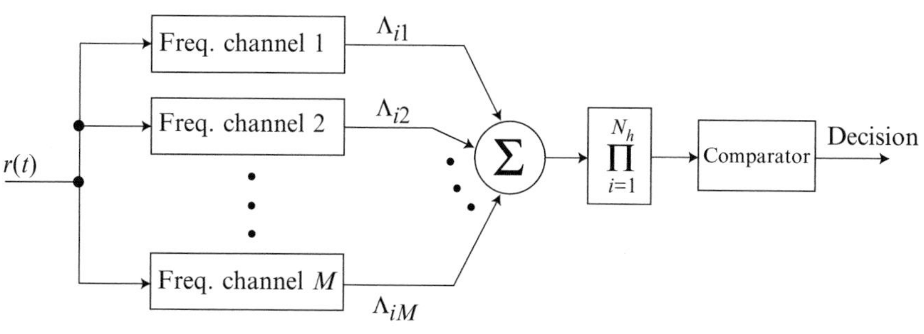

Fig. 10.11 General structure of optimal detector for frequency-hopping signal with N_h hops and M frequency channels

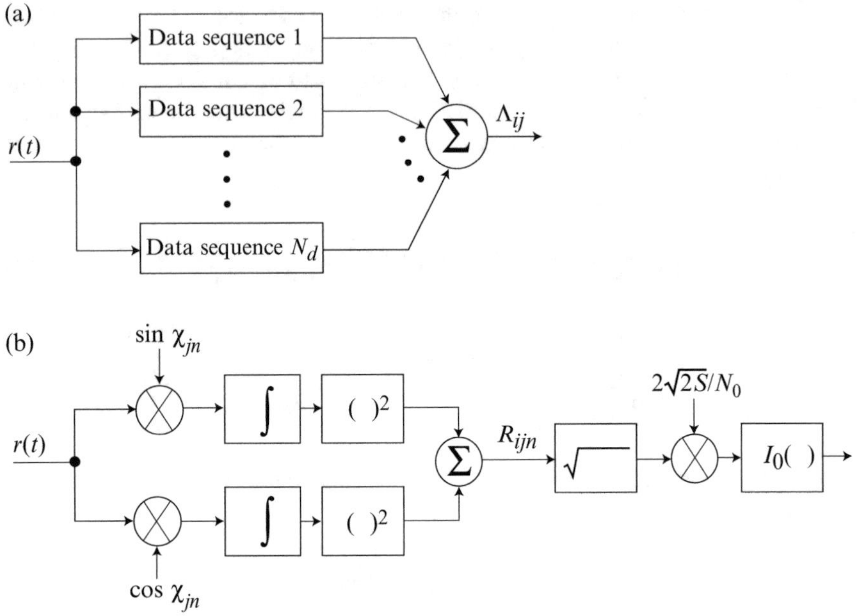

Fig. 10.12 Optimal noncoherent detector for slow frequency hopping with CPM: (**a**) basic structure of frequency channel j for hop i with parallel cells for N_d candidate data sequences, and (**b**) cell for data sequence n

This equation indicates how $\Lambda_{i,j}$ in Figure 10.11 is to be calculated for each hop i and each frequency channel j corresponding to carrier frequency f_j. Equations (10.73) and (10.76) define the optimal coherent detector for any slow frequency-hopping signal with CPM.

For noncoherent detection of FH/CPM [46], the received carrier phase ϕ_i is assumed to be uniformly distributed over $[0, 2\pi)$ during a given hop and statistically independent from hop to hop. Averaging over the random phase and dropping factors that can be merged with the threshold yields

$$\Lambda_{i,j}[r(t)|f_j] = \sum_{n=1}^{N_d} I_0\left(\frac{2\sqrt{2\mathcal{S}R_{i,j,n}}}{N_0}\right) \qquad \text{(noncoherent)} \tag{10.77}$$

where

$$R_{i,j,n} = \left\{\int_{(i-1)T_h}^{iT_h} r(t)\cos\left[\chi_{j,n}(t)\right]dt\right\}^2 + \left\{\int_{(i-1)T_h}^{iT_h} r(t)\sin\left[\chi_{j,n}(t)\right]dt\right\}^2. \tag{10.78}$$

Equations (10.73), (10.77), (10.78), and (10.75) define the optimal noncoherent detector for any slow frequency-hopping signal with CPM. When $\mathcal{S}$ is known, the means of producing (10.77) is diagrammed in Figure 10.12.

A major contributor to the huge computational complexity of the optimal detectors is the fact that with N_s data symbols per hop and an alphabet size q, there may be $N_d = q^{N_s}$ data sequences per hop. Consequently, the computational burden grows exponentially with N_s. However, if it is known that the data modulation is CPFSK with a modulation index $h = 1/n$, where n is a positive integer, the computational burden has a linear dependence on N_s [46]. Even then, the optimal detectors are extremely complex when the number of frequency channels is large.

Consider fast frequency hopping with one hop per orthogonal FSK channel symbol. Since the information is embedded in the sequence of carrier frequencies, the optimal coherent and noncoherent detectors are defined by (10.75), (10.76), and (10.77) with $N_d = 1$ and $\phi(\mathbf{d}_n, t) = 0$. Although the optimal detectors are simplified relative to those required for slow frequency hopping, they are still very complex when M is large.

Instead of basing detector design on the average likelihood ratio, a composite hypothesis test may be applied in which the presence of the signal is detected while simultaneously one or more of the unknown parameters under hypothesis H_1 are estimated. If we view a parameter as unknown but nonrandom, then maximum-likelihood estimation is applicable. A maximum-likelihood estimate of f_j is

$$\widehat{f}_j = \arg\max_{f_j} \Lambda_{ij}[r(t)|f_j]. \tag{10.79}$$

To simultaneously detect the signal while determining the frequency-hopping pattern, (10.73) is replaced by the *generalized likelihood ratio*:

$$\Lambda\left[r(t)\right] = E_S\left\{\exp\left(-\frac{SN_hT_h}{N_0}\right)\prod_{i=1}^{N_h}\max_{i\leq j\leq M}\left\{\Lambda_{i,j}\left[r(t)|\widehat{f}_j\right]\right\}\right\}. \tag{10.80}$$

Although the detection performance is suboptimal when the generalized likelihood ratio is used to design a detector, this detector is slightly easier to implement and analyze than the optimal one [8, 46]. However, the implementation complexity is still formidable.

10.4 Channelized Radiometer

Among the many alternatives to the optimal detector, two of the most useful are the *wideband radiometer* and the *channelized radiometer*. The wideband radiometer is notable in that it requires virtually no detailed information about the parameters of the frequency-hopping signals to be detected other than their rough spectral location. The price paid for this robustness is much worse performance than that of more sophisticated detectors that exploit additional information about the signal [46]. The channelized radiometer is designed to explicitly exploit the spectral

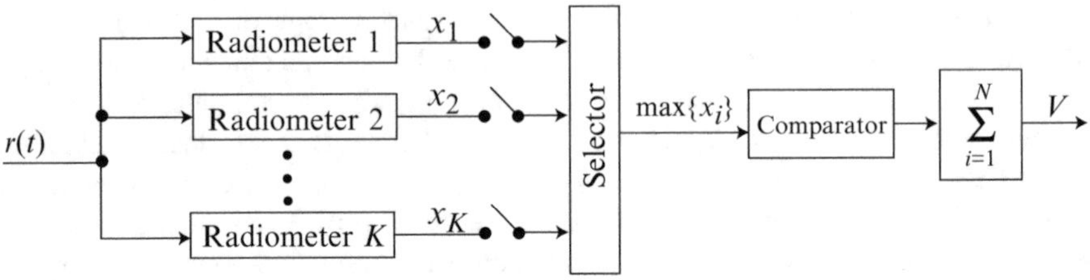

Fig. 10.13 Channelized radiometer

characteristics of frequency-hopping signals. In its optimal form, the channelized radiometer performs nearly as well as the ideal detector. In its suboptimal form, the channelized radiometer trades performance for practicality and the easing of the required a priori information about the signal to be detected.

A *channelized radiometer* comprises K parallel radiometers, each of which has the form of Figure 10.1 and monitors a disjoint portion of the hopping band of a frequency-hopping signal, as depicted in Figure 10.13. The largest of the sampled radiometer outputs is compared with a threshold V_t stored in a comparator. If the threshold is exceeded, the comparator sends a 1 to the summer; otherwise, it sends a 0. If the hop dwell epochs are at least approximately known, the channelized radiometer may improve its detection reliability by adding the 1's produced by N consecutive comparator outputs corresponding to multiple frequency hops of the signal to be detected. A signal is declared to be present if the sum V equals or exceeds the integer r, which serves as a second threshold. The two thresholds V_t and r are jointly optimized for the best system performance.

Ideally, $K = M$, the number of frequency channels in a hopset, but many fewer radiometers may be a practical or economic necessity; if so, each radiometer may monitor M_r frequency channels, where $1 \leq M_r \leq M$. Because of insertion losses and the degradation caused by a power divider, it is unlikely that many more than 30 parallel radiometers are practical. An advantage of each radiometer covering many frequency channels is the reduced sensitivity to imprecise knowledge of the spectral boundaries of frequency channels. Since it is highly desirable to implement the parallel radiometers with similar circuitry, their bandwidths are henceforth assumed to be identical.

To prevent steady interference in a single radiometer from causing false alarms, the channelized radiometer must be able to recognize when one of its constituent radiometers produces an output above the threshold for too many consecutive samples. The channelized system may then delete that constituent radiometer's output from the detection algorithm or it may reassign the radiometer to another spectral location.

In the subsequent analysis of the channelized radiometer of Figure 10.13, the observation interval of the parallel radiometers, which is equal to the sampling interval, is assumed to have a duration equal to the hop duration T_h. The effective observation time of the channelized radiometer, $T = NT_h$, should be less than the minimum expected message duration, to avoid processing extraneous noise. Let B denote the bandwidth of each of the M_r frequency channels encompassed by a radiometer passband. The sampling rate is B, and the number of samples per hop dwell time is $\nu = \lfloor T_h M_r B \rfloor$. Let P_{F1} denote the probability that a particular radiometer output at the end of a hop dwell time exceeds the comparator threshold V_t when no signal is present. For the AWGN channel, it follows in analogy with (10.47) that

$$P_{F1} = \frac{\Gamma(\nu, V_t/\sigma^2)}{\Gamma(\nu)} = \exp\left(-\frac{V_t}{\sigma^2}\right)\sum_{i=0}^{\nu-1}\frac{1}{i!}\left(\frac{V_t}{\sigma^2}\right)^i \tag{10.81}$$

where the noise power is $\sigma^2 = N_0 M_r B$. Thus, in analogy with (10.53),

$$\begin{aligned} V_t &= \sigma_e^2 G_\nu^{-1}(P_{F1}) \\ &\simeq \sigma_e^2[\sqrt{\nu}Q^{-1}(P_{F1}) + \nu], \quad \nu \gg 1 \end{aligned} \tag{10.82}$$

where $G_\nu^{-1}(\cdot)$ is the inverse function of $P_{F1}\left(V_t/\sigma^2\right)$, and σ_e^2 is an estimate σ^2 that can be obtained in the same manner as in Section 10.2. The probability that at least one of the K parallel radiometer outputs exceeds V_t is

$$P_{F2} = 1 - (1 - P_{F1})^K \tag{10.83}$$

assuming that the channel noises are statistically independent because the radiometer passbands are disjoint.

It is convenient to define the function

$$F(x, r, N) = \sum_{i=r}^{N}\binom{N}{i}x^i(1-x)^{N-i}. \tag{10.84}$$

If $y = F(x, r, N)$, then the inverse function is denoted by $x = F^{-1}(y, r, N)$, which may be easily computed using Newton's method. The probability of a false alarm of the channelized radiometer is the probability that the output V is equal to or exceeds a threshold r:

$$P_F = F(P_{F2}, r, N). \tag{10.85}$$

Therefore, if $\nu >> 1$, (10.82), (10.83), and (10.85) may be combined to determine the approximate threshold necessary to achieve a specified P_F:

$$V_t \simeq \sigma_e^2[\sqrt{\nu}Q^{-1}\left\{1 - \left[1 - F^{-1}\left(P_F, r, N\right)\right]^{1/K}\right\} + \nu], \quad \nu \gg 1 \tag{10.86}$$

where we assume that σ^2 does not vary across the hopping band; hence, there is one σ_e^2 and one V_t for all the parallel radiometers.

We assume that at most a single radiometer receives significant signal energy during each hop dwell time. Let P_{D1} denote the probability that a particular radiometer output exceeds the threshold when a signal is present in that radiometer. By analogy with (10.49) and (10.50),

$$P_{D1} = Q_{\nu}\left(\sqrt{2\mathcal{E}_h/N_0}\,,\ \sqrt{2V_t/\sigma^2}\right)$$
$$\simeq Q\left[\frac{V_t/\sigma^2 - \nu - \mathcal{E}_h/N_0}{\sqrt{\nu + 2\mathcal{E}_h/N_0}}\right], \quad \nu >> 1 \tag{10.87}$$

where $\mathcal{E}_h$ is the energy per hop dwell time. Let P_{D2} denote the probability that the threshold is exceeded by the sampled maximum of the parallel radiometer outputs. We assume that when a signal is present it occupies any one of M frequency channels with equal probability, and that all radiometer passbands are within the hopping band. Consequently, the signal has probability M_r/M of being in the passband of a particular radiometer, and the *monitored fraction* $\mu = KM_r/M$ is the probability that the signal is in the passband of any radiometer. Since a detection may be declared in response to a radiometer that does not receive the signal,

$$P_{D2} = \mu\left[1 - (1 - P_{D1})\,(1 - P_{F1})^{K-1}\right] + (1 - \mu)\,P_{F2}. \tag{10.88}$$

The number of hop dwell times during which the signal is actually present is $N_1 \le N$. The second threshold is exceeded if the comparator produces j 1's in response to these N_1 dwell times, $i-j$ 1's in response to the remaining $N - N_1$ dwell times that are observed, and $i \ge r$. Thus, the probability of detection when the signal is actually present during $N_1 \le N$ of the observed hop dwell times is

$$P_D = \sum_{i=r}^{N}\sum_{j=0}^{i}\binom{N_1}{j}\binom{N-N_1}{i-j}P_{D2}^{j}\,(1 - P_{D2})^{N_1-j}\,P_{F2}^{i-j}\,(1 - P_{F2})^{N-N_1-i+j}\,. \tag{10.89}$$

If at least the minimum duration of a frequency-hopping signal is known, the overestimation of N might be avoided so that $N_1 = N$. The detection probability then becomes

$$P_D = \sum_{i=r}^{N}\binom{N}{i}P_{D2}^{i}\,(1 - P_{D2})^{N-i}$$
$$= F(P_{D2}, r, N). \tag{10.90}$$

A reasonably good, but not optimal, choice for the second threshold is $r = \lfloor N/2 \rfloor$ when the full hopping band is monitored by the channelized radiometer, where $\lfloor x \rfloor$ denotes the largest integer less than or equal to x. In general, numerical results indicate that

$$r = \left\lfloor \mu\frac{N}{2} \right\rfloor \tag{10.91}$$

is a good choice for partial-band monitoring.

If a detection decision is made by summing N comparator outputs derived from a block of N hop dwell intervals of duration $T = NT_h$, and successive detection decisions are derived from successive blocks that do not overlap except possibly at end points, then the false-alarm rate F in units of false alarms per second is an appropriate design parameter. This type of detection is called *block detection,* and

$$P_F = FNT_h. \tag{10.92}$$

To prevent the risk of major timing misalignment of the observation interval with the time at which the signal is being transmitted, either block detection must be supplemented with hardware for arrival-time estimation, or the duration of successive block observation intervals should be less than roughly half the anticipated signal duration.

A different approach to mitigating the effect of a misalignment, called *binary moving-window detection*, is for a new block to comprise the preceding block with the oldest hop dwell interval discarded and the most recent hop dwell interval added. A false alarm is considered to be a new detection declaration at the end of a block when no signal is actually present. Thus, a false alarm occurs only if the comparator input for an added hop dwell time exceeds the threshold, the comparator input for the discarded hop dwell time did not, and the count for the intermediate hop dwell times was $r-1$. Therefore, the probability of a false alarm is

$$P_{F0} = C(0,1)C(r-1, N-1)C(1,1) \tag{10.93}$$

where

$$C(i,N) = \binom{N}{i} P_{F2}^{i}\left(1-P_{F2}\right)^{N-i}, \quad i \leq N. \tag{10.94}$$

Since a false alarm may occur after every hop dwell interval, the false-alarm rate is

$$F_0 = \frac{P_{F0}}{T_h} = \frac{r}{NT_h}\binom{N}{r} P_{F2}^{r}\left(1-P_{F2}\right)^{N+1-r}. \tag{10.95}$$

To compare the block and binary moving-window detectors, assume that P_{F2} is the same for both detectors. Since the right-hand side of (10.95) is proportional to the first term of the series representation of (10.85) and $F = P_F/NT_h$ for the block detector, the false-alarm rate F_0 of the binary moving-window detector has the upper bound given by

$$F_0 \leq rF. \tag{10.96}$$

If $P_{F2} << 1/N$, the upper bound is tight, which implies that the false-alarm rate is nearly r times as large for moving-window detection as it is for block detection.

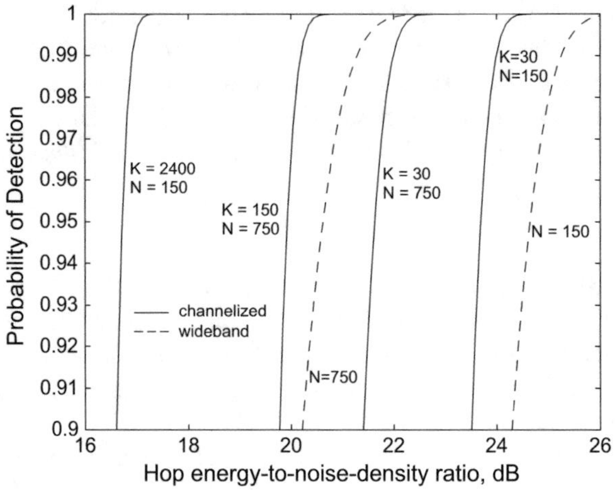

Fig. 10.14 Probability of detection versus $\mathcal{E}_h/N_0$ for channelized and wideband radiometers with full coverage, $N_1 = N$, $h = 1$, $M = 2400$, $F = 10^{-7}/T_h$, and $B = 250/T_h$

Thus, moving-window detection usually requires a higher comparator threshold for the same false-alarm rate and hence more signal power to detect a frequency-hopping signal. However, moving-window detection with $N \approx N_1 >> 1$ inherently limits the misalignment between the occurrence of the intercepted signal and some block observation interval. If the signal occurs during two successive blocks, then for one of the block observation intervals, the misalignment is not more than $T_h/2$.

As examples of the performance of the channelized radiometer, Figures 10.14 and 10.15 plot P_D versus $\mathcal{E}_h/N_0$. We assume that there are $M = 2400$ frequency channels, block detection is used, $F = 10^{-5}/T_h$, $B = 250/T_h$, and $\nu = T_h\mu MB/K = 6 \cdot 10^5\mu/K >> 1$. The signal duration is known and there is no misalignment so that $N_1 = N$. In Figure 10.14, the full hopping band is monitored so that $\mu = 1$, $h = \sigma_e^2/\sigma^2 = 1$, and P_D is plotted for several values of K and N. The figure also shows the results for a wideband radiometer with $\nu = NT_hMB = 6 \cdot 10^5 \cdot N$, and $N = 150$ or 750. The substantial advantage of the channelized radiometer with $K = M$ and $M_r = 1$ is apparent. The channelized radiometer with $K = 30$ is much better than the wideband radiometer when $N = 150$, but $K = 150$ is needed for the advantage of the channelized radiometer to be preserved when $N = 750$. As N increases, the channelized radiometer can retain its advantage over the wideband radiometer by increasing K accordingly.

In Figure 10.15, $N = 150$ and $K = 30$, but M_r and $h = \sigma_e^2/\sigma^2$ are variable. It is observed that when $h > 1$, the performance loss depends on the value of μ, which directly affects ν. The figure illustrates the tradeoff when K and M are fixed and the monitored fraction μ decreases. Since $M_r = \mu M/K = 80\mu$ decreases, fewer frequency channels are monitored, ν decreases, the sensitivity to $h > 1$ decreases, and less noise enters a radiometer. The net result is beneficial when

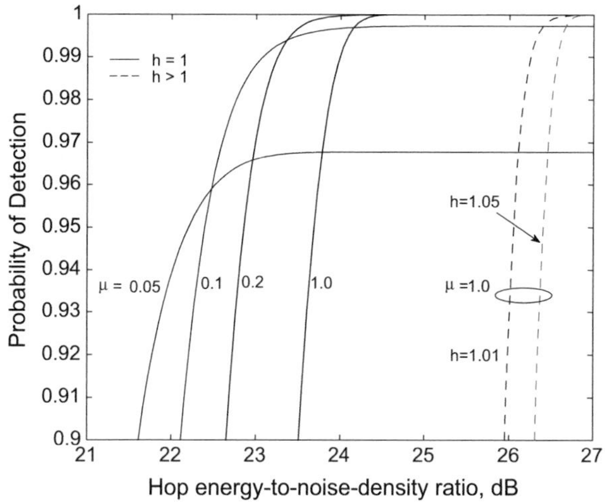

Fig. 10.15 Probability of detection for a channelized radiometer with different monitored fractions, $N_1 = N = 150$, $K = 30$, $M = 2400$, $F = 10^{-7}/T_h$, $B = 250/T_h$, and $h = 1.0, 1.01$, and 1.05

$\mu = 0.2$. However, the figure indicates that for $\mu = 0.1$ or 0.05, the hopping-band coverage becomes inadequate to enable a P_D greater than 0.998 and 0.968, respectively, regardless of $\mathcal{E}_h/N_0$. Thus, there is a minimum fraction $\mu_{\min}$ of the hopping band that must be monitored to ensure a specified P_D.

As $\mathcal{E}_h/N_0 \to \infty$, (10.87) indicates that $P_{D1} \to 1$. Therefore, (10.88) implies that $P_{D2} \to \mu + (1-\mu)P_{F2}$. Suppose that $\mu = \mu_{\min}$ for a specified P_D. The threshold V_t is raised to a sufficiently high level that $P_{F2} \ll \mu_{\min}$; hence, $P_{D2} \approx \mu_{\min}$. If detection is to be accomplished for the minimum monitored fraction, then $r = 1$ is the best choice for the second threshold. For $r = 1$ and $N_1 = N$, (10.90) yields

$$P_D = 1 - (1 - P_{D2})^N. \tag{10.97}$$

Since $P_{D2} \approx \mu_{\min}$, (10.97) implies that even if $\mathcal{E}_h/N_0 \to \infty$, the realization of a specified P_D requires the minimum monitored fraction

$$\mu_{\min} \approx 1 - (1 - P_D)^{1/N}. \tag{10.98}$$

Thus, if $P_D = 0.99$ and $N = N_1 = 150$, then $\mu_{\min} \approx 0.03$.

The channelized radiometer requires estimation of the arrival time of the frequency-hopping signal, the hop duration, and the minimum duration of the entire frequency-hopping signal. The channelized radiometer has a performance degradation due to inaccurate estimates or any timing misalignment. However, when the SNR is very high, the opportunity to intercept two partial pulses in one observation period due to a timing misalignment actually improves the detection probability of the channelized radiometer [55].

10.5 Problems

1 If the parameters of a signal $s(t)$ to be detected are known, then the sufficient statistic r_1 is compared with a threshold V_t to determine whether the target signal is present. Suppose that it is present and coincides with the observation interval. Assume that the AWGN $n(t)$ has two-sided PSD $N_0/2$ and that the signal energy is $\mathcal{E}$. (a) What is the probability of detection P_D? What is the probability of false alarm P_F? (b) Express P_D in terms of a required P_F. (c) What is the value of $\mathcal{E}/N_0$ necessary to ensure specified values of P_F and P_D?

2 Use (A.6) and (A.7) of Appendix A.1 for a standard Gaussian random variable to derive (10.44).

3 The *receiver operating characteristic* (ROC) is a traditional plot depicting P_D versus P_F for various values of ν or $\mathcal{E}/N_0$. For the AWGN channel, the ROC may be calculated from (10.50) and (10.47). Plot the ROC for the wideband radiometer with $\mathcal{E}/N_0 = 20$ dB and no noise-measurement error. Let $\nu = 10^4$ and 10^5.

4 (a) Derive (10.55) using the method described in the text. (b) Derive (10.56) using the method described in the text.

5 Derive (10.65) using the method described in the text.

6 Find conditions under which (10.65) indicates that a negative energy is required. What is the physical implication of this result?

7 Consider a channelized radiometer that is to detect a single hop of a frequency-hopping signal. Assume that $N_1 = N = 1$ and $r = 1$. (a) Find V_t in terms of P_F that does not require $F^{-1}(\cdot)$. (b) If $KM_r = M$ and $\mathcal{E}_h/N_0 >> \nu + V_t/\sigma^2$, show that the $\mathcal{E}_h/N_0$ required for specified values of P_D and P_F is

$$\mathcal{E}_h/N_0 \approx 2\left\{Q^{-1}\left[1 - \frac{1-P_D}{(1-P_F)^{(K-1)/K}}\right]\right\}^2 .$$

Appendix A
Gaussian Random Variables and Processes

A.1 General Characteristics

Let X denote a random variable. Then, X is a *standard Gaussian or normal random variable* if it has a mean $E[X] = 0$, variance $\sigma^2 = E\left[X^2\right] > 0$, and density function

$$f(x) = \frac{1}{\sqrt{2\pi}\sigma} \exp\left(-\frac{x^2}{2\sigma^2}\right). \tag{A.1}$$

To verify that $f_X(x)$ is a legitimate density function, we prove that

$$\int_{-\infty}^{\infty} f(x)\, dx = \frac{1}{\sqrt{\pi}} \int_{-\infty}^{\infty} \exp(-x^2)\, dx = 1. \tag{A.2}$$

The first equality follows from the substitution of (A.1) and a change of variable. Squaring the second integral, applying Fubini's theorem (Section C.1) to equate the result to a double integral, changing variables, and applying Fubini's theorem to perform successive integrations, we obtain

$$\begin{aligned}\left[\int_{-\infty}^{\infty} \exp(-x^2)\, dx\right]^2 &= \int_{-\infty}^{\infty}\int_{-\infty}^{\infty} \exp\left[-(x^2+y^2)\right] dxdy \\ &= \int_{0}^{2\pi}\int_{0}^{\infty} \exp(-\rho^2)\rho d\rho d\theta \\ &= \pi \end{aligned} \tag{A.3}$$

which proves the second equality of (A.2). The standard Gaussian distribution function is

D. Torrieri, *Principles of Spread-Spectrum Communication Systems*,
https://doi.org/10.1007/978-3-319-70569-9

$$F(x) = \frac{1}{\sqrt{2\pi}\sigma} \int_{-\infty}^{x} \exp\left(-\frac{y^2}{2}\right) dy. \tag{A.4}$$

Finite moments of the standard Gaussian density function exist for any non-negative integer k. An integration by parts indicates that

$$\begin{aligned} E\left[X^k\right] &= \frac{1}{\sqrt{2\pi}} \int_{-\infty}^{\infty} x^k \exp\left(-\frac{x^2}{2}\right) dx \\ &= -\frac{1}{\sqrt{2\pi}} \int_{-\infty}^{\infty} x^{k-1} \frac{d}{dx} \exp\left(-\frac{x^2}{2}\right) dx \\ &= \frac{k-1}{\sqrt{2\pi}} \int_{-\infty}^{\infty} x^{k-2} \exp\left(-\frac{x^2}{2}\right) dx \\ &= (k-1)\, E\left[X^{k-2}\right], \quad k \geq 2. \end{aligned} \tag{A.5}$$

Since $E[X] = 0$ and $E\left[X^2\right] = 1$, it follows by induction that

$$E\left[X^{2k+1}\right] = 0, \quad k \geq 0. \tag{A.6}$$

$$E\left[X^{2k}\right] = (2k-1)\,(2k-3)\ldots 1, \quad k \geq 1. \tag{A.7}$$

The *characteristic function* (Appendix C.2) of a random variable X is defined as $E\left[e^{juX}\right]$, where $j = \sqrt{-1}$ and $-\infty < u < \infty$. Therefore, the *characteristic function of the standard Gaussian random variable* is

$$h_s(u) = \frac{1}{\sqrt{2\pi}} \int_{-\infty}^{\infty} \exp\left(-\frac{x^2}{2}\right) \exp\left(jux\right) dx. \tag{A.8}$$

To evaluate this integral, we apply Cauchy's integral theorem to a contour integral in the complex plane over the rectangle with vertices $(-x_0, 0)$, $(x_0, 0)$, $(x_0, x_0 + ju)$, and $(-x_0, -x_0 + ju)$. The complex integration variable is z, and the integrals over the vertical sides of the rectangle become negligible as $x_0 \to \infty$. The integral from $(-x_0, 0)$ to $(x_0, 0)$ approaches $h_s(u)$ as $x_0 \to \infty$. Since there are no singularities within the transformed rectangle,

$$\begin{aligned} 0 &= \lim_{x_0 \to \infty} \oint \frac{1}{\sqrt{2\pi}} \exp\left(-\frac{z^2}{2}\right) \exp\left(juz\right) dz \\ &= h_s(u) - \frac{1}{\sqrt{2\pi}} \exp\left(-\frac{u^2}{2}\right) \int_{-\infty}^{\infty} \exp\left(-\frac{x^2}{2}\right) dx \\ &= h_s(u) - \exp\left(-\frac{u^2}{2}\right) \end{aligned} \tag{A.9}$$

where the final equality is obtained by applying (A.2). Thus, the characteristic function of the standard Gaussian random variable is $\exp\left(-u^2/2\right)$, and

$$\int_{-\infty}^{\infty} \exp\left(-\frac{x^2}{2}\right) \exp\left(jux\right) dx = \sqrt{2\pi} \exp\left(-\frac{u^2}{2}\right). \tag{A.10}$$

If X is a standard Gaussian random variable, then $Y = \mu + \sigma X$ is a Gaussian or normal random variable with mean μ and variance σ^2. By including the constant random variable as a standard Gaussian random variable with $\sigma = 0$, we have $\sigma^2 \geq 0$. The characteristic function of Y is $h(u) = E\left[e^{juY}\right] = E\left[e^{ju(\mu+\sigma X)}\right]$. Using (A.10), we obtain the *characteristic function of a Gaussian random variable*:

$$h(u) = \exp\left(ju\mu - \frac{\sigma^2 u^2}{2}\right). \tag{A.11}$$

Since the distribution function of a random variable is uniquely determined by the characteristic function (Appendix C.2), *a necessary and sufficient condition that a random variable is Gaussian is that its characteristic function takes the form of* (A.11).

The *joint characteristic function* of the random variables $X_1, \ldots, X_n$ is defined as

$$h(\mathbf{u}) = E\left[\exp\left(j\mathbf{u}^T\mathbf{X}\right)\right] = E\left[\exp\left(j\sum_{k=1}^{n} u_k X_k\right)\right] \tag{A.12}$$

where $\mathbf{u} = [u_1\, u_2\, \ldots\, u_n]^T$. If the random variables are independent, zero-mean Gaussian random variables with $var\,(X_i) = \sigma_i^2$, then

$$\begin{aligned} h(\mathbf{u}) &= \prod_{i=1}^{n} E\left[\exp\left(ju_i X_i\right)\right] \\ &= \prod_{k=1}^{n} \exp\left(-\frac{\sigma_k^2 u^2}{2}\right). \end{aligned} \tag{A.13}$$

An $n \times 1$ random column vector $\mathbf{X} = [X_1\, \ldots\, X_n]^T$ has components that are random variables. Let $\boldsymbol{\mu} = E\left[\mathbf{X}\right]$ denote the mean vector, and let $\mathbf{K}$ denote the $n \times n$ *covariance matrix*

$$\mathbf{K} = E\left[(\mathbf{X}-\boldsymbol{\mu})(\mathbf{X}-\boldsymbol{\mu})^T\right] \tag{A.14}$$

which is symmetric. Since $\mathbf{x}^T\mathbf{K}\mathbf{x} = E\left[\left[(\mathbf{X}-\boldsymbol{\mu})^T\mathbf{x}\right]^2\right]$, $\mathbf{K}$ is nonnegative definite. *A Gaussian random vector* X is defined as one with a characteristic function of the form

$$h(\mathbf{u}) = E\left[\exp\left(j\mathbf{u}^T\mathbf{X}\right)\right] = \exp\left(j\mathbf{u}^T\boldsymbol{\mu} - \frac{1}{2}\mathbf{u}^T\mathbf{K}\mathbf{u}\right) \tag{A.15}$$

where $\boldsymbol{\mu}$ is the mean vector, and $\mathbf{K}$ is the symmetric, nonnegative-definite covariance matrix. The components of a Gaussian random vector $X_1, \ldots, X_n$ are called *jointly Gaussian random variables.*

Theorem A1 *The random vector* $\mathbf{X} = \mathbf{AY} + \boldsymbol{\mu}$ *is a Gaussian random vector if* $\mathbf{Y}$ *has components that are independent zero-mean Gaussian random variables.*

Proof If $\mathbf{Y}$ is an $n \times 1$ Gaussian random vector with components that are independent zero-mean Gaussian random variables, then it has a mean $\mathbf{0}$ with an $n \times n$ diagonal covariance matrix $\mathbf{D}$. If $\mathbf{X} = \mathbf{AY} + \boldsymbol{\mu}$, then the characteristic function of $\mathbf{X}$ is

$$\begin{aligned} E\left[\exp\left(j\mathbf{u}^T\mathbf{X}\right)\right] &= \exp\left(j\mathbf{u}^T\boldsymbol{\mu}\right) E\left[\exp\left(j\mathbf{u}^T\mathbf{AY}\right)\right] \\ &= \exp\left(j\mathbf{u}^T\boldsymbol{\mu} - \frac{1}{2}\mathbf{u}^T\mathbf{K}\mathbf{u}\right), \quad \mathbf{K} = \mathbf{ADA}^T. \end{aligned} \tag{A.16}$$

Since the characteristic function of $\mathbf{X}$ has the form of (A.15), $\mathbf{X}$ is a Gaussian random vector. □

Theorem A2 *An* $n \times 1$ *Gaussian random vector* $\mathbf{X}$ *with an* $n \times 1$ *mean vector* μ *and an* $n \times n$ *covariance matrix* $\mathbf{K}$ *can be expressed as* $\mathbf{X} = \mathbf{AY} + \mu$*, where the components of* $\mathbf{Y}$ *are independent Gaussian random variables with zero mean, and* $\mathbf{A}$ *is an* $n \times n$ *orthogonal matrix.*

Proof Since the $n \times n$ matrix $\mathbf{K}$ is a symmetric nonnegative definite, it can be diagonalized by an orthogonal matrix (Appendix G). Let $\mathbf{A}$ denote an $n \times n$ orthogonal matrix, such that $\mathbf{A}^T\mathbf{KA} = \mathbf{D}$, where $\mathbf{D}$ is an $n \times n$ diagonal matrix with diagonal elements equal to the $\{\lambda_k\}$, which are the eigenvalues of $\mathbf{K}$. Define $\mathbf{Y} = \mathbf{A}^T(\mathbf{X} - \boldsymbol{\mu})$. Then, $\mathbf{Y}$ has a mean $\mathbf{0}$, covariance $\mathbf{A}^T\mathbf{KA} = \mathbf{D}$, and characteristic function

$$\begin{aligned} E\left[\exp\left(j\mathbf{u}^T\mathbf{Y}\right)\right] &= \exp\left(-\frac{1}{2}\mathbf{u}^T\mathbf{D}\mathbf{u}\right) = \exp\left(-\frac{1}{2}\sum_{k=1}^{n}\lambda_k u_k^2\right) \\ &= \prod_{k=1}^{n}\exp\left(-\frac{1}{2}\lambda_k u_k^2\right). \end{aligned} \tag{A.17}$$

By the uniqueness of the characteristic function, (A.17), and (A.13), the $\{Y_k\}$ are independent, and Y_k is a zero-mean Gaussian random variable with variance λ_k. The orthogonality of $\mathbf{A}$ implies that $\mathbf{A}^T = \mathbf{A}^{-1}$, and hence $\mathbf{X} = \mathbf{AY} + \boldsymbol{\mu}$. □

Theorem A3 *If* $\mathbf{K}$ *is invertible, an* $n \times 1$ *Gaussian random vector* $\mathbf{X}$ *with an* $n \times 1$ *mean vector* $\boldsymbol{\mu}$ *and an* $n \times n$ *covariance matrix* $\mathbf{K}$ *has a density function*

$$f_{\mathbf{X}}(\mathbf{x}) = (2\pi)^{-n/2}(\det \mathbf{K})^{-1/2} \exp\left[-\frac{1}{2}(\mathbf{X}-\boldsymbol{\mu})^T\mathbf{K}^{-1}(\mathbf{X}-\boldsymbol{\mu})\right] \tag{A.18}$$

which indicates that a Gaussian density function is completely determined by its mean vector and its covariance matrix.

Proof If the symmetric nonnegative-definite matrix $\mathbf{K}$ is invertible, it is positive definite and every eigenvalue λ_k of $\mathbf{K}$ is positive (Appendix G). Let $\mathbf{A}$ denote an $n \times n$ orthogonal matrix, such that $\mathbf{A}^T\mathbf{K}\mathbf{A} = \mathbf{D}$, where $\mathbf{D}$ is an $n \times n$ diagonal matrix with positive diagonal elements equal to the $\{\lambda_k\}$. Then, Theorem A2, (A.17), and (A.13) imply that $\mathbf{Y} = \mathbf{A}^T(\mathbf{X}-\boldsymbol{\mu})$ has a density function

$$\begin{aligned} f_{\mathbf{Y}}(\mathbf{y}) &= \prod_{k=1}^{n} (2\pi\lambda_k)^{-1/2} \exp\left(-\frac{y_k^2}{2\lambda_k}\right) \\ &= (2\pi)^{-n/2}(\det \mathbf{D})^{-1/2} \exp\left(-\frac{\mathbf{y}^T\mathbf{D}^{-1}\mathbf{y}}{2}\right). \end{aligned} \tag{A.19}$$

Since $\mathbf{A}$ is orthogonal, the Jacobian of the transformation $\mathbf{X} = \mathbf{A}\mathbf{Y}+\boldsymbol{\mu}$ is $\left|\det \mathbf{A}^T\right| = \left|\det \mathbf{A}^{-1}\right| = 1$. Since $\mathbf{A}^T\mathbf{K}\mathbf{A} = \mathbf{D}$ implies that $\mathbf{A}\mathbf{D}^{-1}\mathbf{A}^T = \mathbf{K}^{-1}$ and $\det \mathbf{D} = \det \mathbf{K}$, the density function of $\mathbf{X}$ is given by (A.18). □

If $\mathbf{K}$ is singular, then $\mathbf{X}$ does not have a density function.

Theorem A4 *If $\mathbf{X}$ is a Gaussian random vector, $\mathbf{B}$ is an arbitrary $n \times n$ matrix, and $\mathbf{Z} = \mathbf{B}\mathbf{X}$, then $\mathbf{Z}$ is a Gaussian random vector. Thus, the linear transformation of a Gaussian random vector is itself a Gaussian random vector.*

Proof According to Theorem A1, $\mathbf{X}$ can be expressed as $\mathbf{X} = \mathbf{A}\mathbf{Y}+\boldsymbol{\mu}$, where $\mathbf{Y}$ is a vector with components that are independent zero-mean Gaussian random variables. Then $\mathbf{Z} = \mathbf{B}\mathbf{A}\mathbf{Y} + \mathbf{B}\boldsymbol{\mu}$, which is a Gaussian random vector by Theorem A1. □

It is important to note that *even if $\mathbf{X}$ has Gaussian components, $\mathbf{X}$ may not be a Gaussian random vector*. Thus, $\mathbf{Z} = \mathbf{B}\mathbf{X}$ may not be a Gaussian random vector if $\mathbf{X}$ has Gaussian components but is not a Gaussian random vector.

Theorem A5 *The components of a Gaussian random vector $\mathbf{X}$ are independent random variables if and only if they are uncorrelated and have positive variances.*

Proof If the component random variables of a Gaussian random vector $\mathbf{X}$ are uncorrelated, then $\mathbf{K}$ is diagonal. If all variances are positive, then the diagonal elements of $\mathbf{K}$ are positive. Since $\mathbf{K}$ is invertible, Theorem A3 indicates that the density function of $\mathbf{X}$ is the product of the density functions of its components, and hence the components are independent. Conversely, if the components of a Gaussian random vector $\mathbf{X}$ are independent random variables, then (A.14) implies that $\mathbf{K}$ is diagonal, and hence the components are uncorrelated. □

A *complex $n \times 1$ random vector* has the form $\mathbf{X} = \mathbf{X}_1 + j\mathbf{X}_2$, where $\mathbf{X}_1$ and $\mathbf{X}_2$ are real-valued random vectors with means $\boldsymbol{\mu}_1$ and $\boldsymbol{\mu}_2$ respectively. Thus, $E[\mathbf{X}] = \boldsymbol{\mu} = \boldsymbol{\mu}_1 + j\boldsymbol{\mu}_2$. Let $\mathbf{K}_i$ denote the $n \times n$ covariance matrix of $\mathbf{X}_i$, $i = 1, 2$. We define the $n \times n$ cross-covariance matrices of $\mathbf{X}_1$ and $\mathbf{X}_2$ as

$$\mathbf{K}_{12} = E\left[(\mathbf{X}_1-\boldsymbol{\mu}_1)(\mathbf{X}_2-\boldsymbol{\mu}_2)^T\right],\ \mathbf{K}_{21} = E\left[(\mathbf{X}_2-\boldsymbol{\mu}_2)(\mathbf{X}_1-\boldsymbol{\mu}_1)^T\right]. \quad \text{(A.20)}$$

The $n \times n$ covariance matrix of $\mathbf{X}$ is

$$\begin{aligned}\mathbf{K} &= E\left[(\mathbf{X}-\boldsymbol{\mu})(\mathbf{X}-\boldsymbol{\mu})^H\right] \\ &= \mathbf{K}_1 + \mathbf{K}_2 + j(\mathbf{K}_{21} - \mathbf{K}_{12}).\end{aligned} \quad \text{(A.21)}$$

The density function of $\mathbf{X}$ is defined as the density function of the $2n \times 1$ real-valued vector $\mathbf{X}_c = [\mathbf{X}_1\ \mathbf{X}_2]^T$. The $2n \times 2n$ covariance matrix of $\mathbf{X}_c$ is

$$\mathbf{K}_c = \begin{bmatrix} \mathbf{K}_1 & \mathbf{K}_{12} \\ \mathbf{K}_{21} & \mathbf{K}_2 \end{bmatrix}. \quad \text{(A.22)}$$

A complex $n \times 1$ random vector $\mathbf{X}$ is *circularly symmetric* if

$$E\left[(\mathbf{X}-\boldsymbol{\mu})(\mathbf{X}-\boldsymbol{\mu})^T\right] = \mathbf{0}. \quad \text{(A.23)}$$

Expanding this equation in terms of its real and imaginary parts, we find that circular symmetry implies that

$$\mathbf{K}_1 = \mathbf{K}_2,\ \mathbf{K}_{21} = -\mathbf{K}_{12}. \quad \text{(A.24)}$$

A complex $n \times 1$ random vector $\mathbf{X}$ is a *complex Gaussian random vector* if the $2n \times 1$ real-valued vector $\mathbf{X}_c = [\mathbf{X}_1\mathbf{X}_2]^T$ is a Gaussian random vector. Let $\mathbf{D}$ denote a real-valued diagonal matrix with positive elements. If a complex Gaussian random vector $\mathbf{X}$ has covariance matrix $\mathbf{K} = \mathbf{D}$, then

$$\mathbf{D} = \mathbf{K}_1 + \mathbf{K}_2,\ \mathbf{K}_{21} = \mathbf{K}_{12}. \quad \text{(A.25)}$$

If $\mathbf{X}$ is also circularly symmetric, then (A.22), (A.24), and (A.25) imply that

$$\mathbf{K}_c = \begin{bmatrix} \frac{1}{2}\mathbf{D} & \mathbf{0} \\ \mathbf{0} & \frac{1}{2}\mathbf{D} \end{bmatrix} \quad \text{(A.26)}$$

which indicates that the components of $\mathbf{X}_c$ are uncorrelated and hence are independent Gaussian random variables.

A stochastic process $X(t) = \{X_t,\ t \in T\}$ is called a *Gaussian process* if every finite linear combination of the form

$$Y = \sum_{i=1}^{N} a_i X_{t_i} \quad \text{(A.27)}$$

is a Gaussian random variable.

A zero-mean stochastic process $n(t)$ is called a *white process* if its autocorrelation is

$$E[n(t)n(t+\tau)] = \frac{N_0}{2}\delta(\tau) \tag{A.28}$$

where $\delta(\tau)$ is the Dirac delta function, and $N_0/2$ is the two-sided power spectral density of this process. A white process is an idealization of a physical process because it requires an infinite bandwidth and power and zero correlation time. However, consider noise that has a flat spectrum across the passband of a band-limiting filter in a receiver. Since the filter blocks the noise spectrum beyond the filter passband, the hypothetical existence of the blocked spectrum does not affect the noise in the filter output. Consequently, a white process provides the standard mathematical model for thermal, shot, and environmental noise.

A.2 Central Limit Theorem

The *central limit theorem* establishes conditions under which the sum of many random variables has an approximately normal or Gaussian distribution. The proof exploits the following fundamental theorem. *Let $F_n(x)$ and $F(x)$ denote distribution functions with characteristic functions $h_n(u)$ and $h(u)$* respectively. *A necessary and sufficient condition for $F_n(x) \to F(x)$ is that $h_n(u) \to h(u)$ for each u* [6, 9].

In the proof of the central limit theorem, Taylor series are needed. Let $f^{(n)}(x)$ denote the nth derivative of $f(x)$.

Taylor's Theorem for Complex-Valued Functions Let $f(x)$ denote a complex-valued function of a real variable x with $n+1$ continuous derivatives on an open interval, including the origin. Then, for all x in the interval,

$$f(x) = \sum_{k=0}^{n} \frac{f^{(k)}(0)\,x^k}{k!} + R_n(x) \tag{A.29}$$

where the remainder is

$$R_n(x) = \frac{x^{n+1}}{n!}\int_0^1 f^{(n+1)}(xy)\,(1-y)^n\,dy. \tag{A.30}$$

If $\left|f^{(n+1)}(u)\right| \le c$ for all u on the open interval, then

$$|R_n(x)| \le \frac{c\,|x|^{n+1}}{(n+1)!}. \tag{A.31}$$

Proof Integrating by parts the integral in (A.30), we obtain

$$R_n(x) = -\frac{f^{(n)}(0)\,x^n}{n!} + R_{n-1}(x).$$

Repeated substitutions into the right-hand side of this equation and the final substitution of $R_0(x) = f(x) - f(0)$ proves (A.29). Substituting $\left|f^{(n+1)}(u)\right| \leq c$ and

$$\int_0^1 (1-y)^n\,dy = \frac{1}{n+1}$$

into (A.30) proves (A.31). □

Applying Taylor's theorem to a series expansion of $\exp(jx)$ about the origin $x = 0$, we obtain

$$\exp(jx) = \sum_{k=0}^{n} \frac{(jx)^k}{k!} + \frac{\theta\,|x|^{n+1}}{(n+1)!}, \quad |\theta| \leq 1 \tag{A.32}$$

where $j = \sqrt{-1}$ and y is real-valued.

Taylor's Theorem for Analytic Functions Let $f(z)$ denote an analytic function over an open disk $\mathcal{D}$ including the origin in the complex plane. Then, for every $z \in \mathcal{D}$, we have the Taylor series

$$f(z) = \sum_{k=1}^{\infty} \frac{f^{(k)}(0)\,z^k}{k!}. \tag{A.33}$$

Proof Let C denote a circle centered at the origin and within $\mathcal{D}$. Using Cauchy's integral formula,

$$f(z) = \frac{1}{2\pi j}\oint_C \frac{f(z_1)}{z_1 - z}dz_1$$

where z lies within C, z_1 is the integration variable, and the integration path is counterclockwise around C. From the definition of a derivative in the complex plane and Cauchy's integral formula, we find that

$$f^{(k)}(z) = \frac{k!}{2\pi j}\oint_C \frac{f(z_1)}{(z_1 - z)^{k+1}}dz_1. \tag{A.34}$$

From the formula for a finite geometric sum, we obtain

$$\frac{1}{1-t} = \sum_{k=0}^{\infty} t^k, \quad |t| < 1$$

where the convergence is uniform. Since $|z/z_1| < 1$, (A.34) and the geometric sum imply that

$$f(z) = \frac{1}{2\pi j} \oint_C \frac{f(z_1)}{z_1} \sum_{n=0}^{\infty} \left(\frac{z}{z_1}\right)^k dz_1$$

$$= \frac{1}{2\pi j} \sum_{n=0}^{\infty} z^k \oint_C \frac{f(z_1)}{z_1^{k+1}} dz_1$$

where the interchange of the order of integration and summation is valid because of the uniform convergence. Substitution of (A.34) with $z = 0$ into this equation yields (A.33). □

A Taylor series for the principal branch of $\ln(1+z)$ is

$$\ln(1+z) = \sum_{k=1}^{\infty} \frac{(-1)^{k+1} z^k}{k}$$

$$= z + z^2 \sum_{k=2}^{\infty} \frac{(-1)^{k+1} z^{k-2}}{k}, \quad |z| < 1. \tag{A.35}$$

We define

$$\zeta = \frac{z^2}{|z|^2} \sum_{k=2}^{\infty} \frac{(-1)^{k+1} z^{k-2}}{k}. \tag{A.36}$$

If $|z| \le 1/2$, then

$$|\zeta| \le \sum_{k=2}^{\infty} \frac{|z|^{k-2}}{k} \le \frac{1}{2} \sum_{k=2}^{\infty} |z|^{k-2} \le 1, \quad |z| \le 1/2. \tag{A.37}$$

Therefore,

$$\ln(1+z) = z + \zeta |z|^2, \quad |z| \le 1/2, \ |\zeta| \le 1. \tag{A.38}$$

Central Limit Theorem *Suppose that for each n, the sequence* $X_1, X_2, \ldots, X_n$ *is independent, and that each* X_k *has finite mean* m_k, *finite variance* σ_k^2, *and distribution function* $F_k(x)$. *Let* $S_n = X_1 + X_2 + \ldots + X_n$ *and* $T_n = (S_n - E[S_n])/s_n$, *where* $s_n^2 = var(S_n) = \sum_{k=1}^n \sigma_k^2$. *If for every positive* ϵ,

$$\sum_{k=1}^{n} \frac{1}{s_n^2} \int_{|x-m_k| \ge \epsilon s_n} (x - m_k)^2 \, dF_k(x) \to 0 \ \ as \ \ n \to \infty$$

then T_n converges in distribution to a standard Gaussian random variable with distribution given by (A.4).

Proof We assume in the proof that $m_k = 0$ with no loss of generality because $S_n - E[S_n] = \sum_{k=1}^{\infty}(X_k - m_k)$, and $E[X_k - m_k] = 0$.

Let h_k and ϕ_n denote the characteristic functions of X_k and T_n, respectively. The independence of each X_k implies that

$$\phi_n(u) = E\left[e^{juT_n}\right] = \prod_{k=1}^{n} E\left[e^{juX_k/s_n}\right] = \prod_{k=1}^{n} h_k\left(\frac{u}{s_n}\right).$$

Since the convergence of the characteristic functions determines the convergence of the distribution functions, the theorem is proved if it is shown that $\phi_n(u) \to \exp\left(-u^2/2\right)$, which is equivalent to showing that

$$\ln(\phi_n(u)) = \sum_{k=1}^{n} \ln\left(h_k\left(\frac{u}{s_n}\right)\right) \to -u^2/2 \text{ as } n \to \infty. \tag{A.39}$$

The partitioning of the defining integral gives

$$h_k\left(\frac{u}{s_n}\right) = \int_{|x|<\epsilon s_n} e^{jux/s_n} dF_k(x) + \int_{|x|\geq \epsilon s_n} e^{jux/s_n} dF_k(x)$$

for each positive ϵ. Substituting (A.32) with $n = 2$ and $n = 1$ into the first and second integrals, respectively, and using $m_k = E[X_k] = 0$, we obtain

$$h_k\left(\frac{u}{s_n}\right) = 1 + \frac{u^2}{2}\theta_1\alpha_{nk} - \frac{u^2}{2}\beta_{nk} + \frac{|u|^3}{6s_n^3}\theta_2 \int_{|x|<\epsilon s_n} |x|^3 dF_k(x) \tag{A.40}$$

where $|\theta_1|, |\theta_2| \leq 1$ and

$$\alpha_{nk} = \frac{1}{s_n^2}\int_{|x|\geq \epsilon s_n} x^2 dF_k(x), \quad \beta_{nk} = \frac{1}{s_n^2}\int_{|x|<\epsilon s_n} x^2 dF_k(x). \tag{A.41}$$

Since $|x|^3 < \epsilon s_n x^2$ when $|x| < \epsilon s_n$, (A.40) may be expressed as

$$h_k\left(\frac{u}{s_n}\right) = 1 + \gamma_{nk} \tag{A.42}$$

where

$$\gamma_{nk} = \frac{u^2}{2}\theta_1\alpha_{nk} - \frac{u^2}{2}\beta_{nk} + \frac{|u|^3}{6}\epsilon\theta_3\beta_{nk} \tag{A.43}$$

and $|\theta_3| < 1$. If $|\gamma_{n,k}| \leq 1/2$, the application of (A.38) yields

$$\sum_{k=1}^{n} \ln\left(h_k\left(\frac{u}{s_n}\right)\right) = \sum_{k=1}^{n} \gamma_{nk} + \zeta \sum_{k=1}^{n} |\gamma_{nk}|^2 \tag{A.44}$$

where $|\zeta| \leq 1$.

From the hypothesis of the theorem, as $n \to \infty$,

$$\sum_{k=1}^{n} \alpha_{nk} \to 0, \ \ \alpha_{nk} \to 0. \tag{A.45}$$

It follows from (A.41) that

$$\sum_{k=1}^{\infty} (\alpha_{nk} + \beta_{nk}) = \frac{1}{s_n^2} \sum_{k=1}^{\infty} \sigma_k^2 = 1.$$

This equation and (A.45) imply that, as $n \to \infty$,

$$\sum_{k=1}^{n} \beta_{nk} \to 1 \tag{A.46}$$

and hence

$$\sum_{k=1}^{n} \gamma_{nk} \to -u^2/2 + \frac{|u|^3}{6} \theta_3 \epsilon \text{ as } n \to \infty. \tag{A.47}$$

Since (A.41) indicates that $0 \leq \beta_{nk} < \epsilon^2$, we obtain

$$\max_k |\gamma_{nk}| < \frac{u^2}{2} \epsilon^2 + \frac{|u|^3}{6} \epsilon^3 \tag{A.48}$$

for sufficiently large n. Thus, $|\gamma_{nk}| \leq 1/2$ for all k and u if n is sufficiently large and ϵ is sufficiently small. Using (A.48), (A.43), (A.45), and (A.46), we obtain

$$\sum_{k=1}^{n} |\gamma_{nk}|^2 \leq \max_k |\gamma_{nk}| \sum_{k=1}^{n} |\gamma_{nk}| < \left(\frac{u^2}{2} \epsilon^2 + \frac{|u|^3}{6} \epsilon^3\right) \left(\frac{u^2}{2} + \frac{|u|^3}{6} \epsilon\right) \tag{A.49}$$

for sufficiently large n. Thus, for any positive δ, (A.44), (A.47), and (A.49) imply that

$$\left| \sum_{k=1}^{n} \ln\left(h_k\left(\frac{u}{s_n}\right)\right) + u^2/2 \right| < \delta$$

if ϵ is chosen sufficiently small and n is sufficiently large, which proves (A.39). □

Corollary A1 *Suppose that the sequence $X_1, X_2, \ldots, X_n$ is independent for each n and identically distributed so that each X_k has a finite mean m, finite variance $\sigma^2 > 0$, and a distribution function $F(x)$. Let $S_n = X_1 + X_2 + \ldots + X_n$ and $T_n = (S_n - nm) / \sigma\sqrt{n}$. Then, T_n converges in distribution to a standard Gaussian random variable.*

Proof Since $s_n^2 = n\sigma^2$, $m_k = m$, and $F_k(x) = F(x)$,

$$\sum_{k=1}^{n} \frac{1}{s_n^2} \int_{|x-m_k|\geq \epsilon s_n} (x - m_k)^2 \, dF_k(x) = \frac{1}{\sigma^2} \int_{|x-m|\geq \epsilon\sigma\sqrt{n}} (x - m)^2 \, dF(x)$$

which converges to zero by Lebesgue's dominated convergence theorem, as σ^2 is finite and positive, and $\left\{|x - m| \geq \epsilon\sigma\sqrt{n}\right\}$ converges to the empty set as $n \to \infty$. □

To prove the next corollary, we use Chebyshev's inequality (Section 4.3).

Corollary A2 *Suppose that the sequence $X_1, X_2, \ldots, X_n$ is independent for each n, and that each X_k has a finite mean m_k, distribution function $F_k(x)$, and is uniformly bounded with $|X_k - m_k| < M$ for all k. Let $S_n = X_1 + X_2 + \ldots + X_n$ and $T_n = (S_n - E[S_n]) / s_n$, where $s_n^2 = var(S_n) = \sum_{k=1}^{n} \sigma_k^2$. If $s_n \to \infty$, then T_n converges in distribution to a standard Gaussian random variable.*

Proof Using $|X_k - m_k| < M$ for all k and Chebyshev's inequality, we obtain

$$\begin{aligned}
\sum_{k=1}^{n} \frac{1}{s_n^2} \int_{|x-m_k|\geq \epsilon s_n} (x - m_k)^2 \, dF_k(x) &\leq \sum_{k=1}^{n} \frac{4M^2}{s_n^2} P\{|x - m_k| \geq \epsilon s_n\} \\
&\leq \sum_{k=1}^{n} \frac{4M^2\sigma_k^2}{s_n^4\epsilon^2} = \frac{4M^2}{s_n^2\epsilon^2} \to 0. \ \square
\end{aligned}$$

To prove the next corollary, we use the indicator function and the Cauchy-Schwarz inequality for random variables. The *indicator function* I_A of a set A is the function on the sample space Ω that assumes the value 1 on A and 0 on the complement of A. Substituting $x = X/\sqrt{E[X^2]}$ and $y = Y/\sqrt{E[Y^2]}$ into the inequality $2xy \leq x^2 + y^2$, and then taking the expected value of both sides of the resulting equation, we obtain the *Cauchy-Schwarz inequality*:

$$E[XY] \leq \sqrt{E[X^2]\,E[Y^2]}. \tag{A.50}$$

Corollary A3 *Suppose that the sequence $X_1, X_2, \ldots, X_n$ is independent for each n, and that each X_k has finite mean m_k, finite variance σ_k^2, and distribution function $F_k(x)$. Let $S_n = X_1 + X_2 + \ldots + X_n$ and $T_n = (S_n - E[S_n])/s_n$, where $s_n^2 = var(S_n) = \sum_{k=1}^{n} \sigma_k^2$. If $s_n^3/n \to \infty$, and the fourth central moment are uniformly bounded so that $E\left[(X_k - m_k)^4\right] < M^2$ for all k, then T_n converges in distribution to a standard Gaussian random variable.*

Proof Applying the Cauchy-Schwarz inequality, we obtain

$$E\left[(X_k - m_k)^2\right] = \sigma_k^2 \leq M.$$

Applying the Cauchy-Schwarz inequality, Chebyshev's inequality, $\sigma_k^2 \leq M$, and $s_n^3/n \to \infty$, we obtain

$$\begin{aligned}
\sum_{k=1}^{n} \frac{1}{s_n^2} \int_{|x-m_k| \geq \epsilon s_n} (x - m_k)^2 \, dF_k(x) &= \sum_{k=1}^{n} \frac{1}{s_n^2} E\left[(X_k - m_k)^2 I_{\{|x-m_k| \geq \epsilon s_n\}}\right] \\
&\leq \sum_{k=1}^{n} \frac{M}{s_n^2} \sqrt{P[|x - m_k| \geq \epsilon s_n]} \\
&\leq \sum_{k=1}^{n} \frac{M^{3/2}}{s_n^3 \epsilon} = \frac{M^{3/2}/\epsilon}{s_n^3/n} \to 0. \quad \square
\end{aligned}$$

Appendix B
Moment-Generating Function and Laplace Transform

B.1 Moment-Generating Function

The *moment-generating function* of the random variable X with distribution function $F(x)$ is defined as

$$M(s) = E\left[e^{sX}\right] = \int_{-\infty}^{\infty} e^{sx} dF(x) \tag{B.1}$$

for all real-valued s for which the integral is finite. Thus, the moment-generating function is the two-sided Laplace transform restricted to real values of s. If $s_0 > 0$ and $M(s)$ is defined throughout $(-s_0, s_0)$, then $\exp(|sx|)$ is integrable for $|s| < s_0$. Since its series expansion indicates that $\exp(|sx|) \geq |sx|^k / k!, k \geq 0$, X has finite moments of all orders.

If $s_0 > 0$ and $M(s)$ are defined throughout $(-s_0, s_0)$, the difference quotient

$$\frac{M(s+\Delta s) - M(s)}{\Delta s} = \int_{-\infty}^{\infty} \frac{e^{(s+\Delta s)x} - e^{sx}}{\Delta s} dF(x), \ s \in (-s_0, s_0) \tag{B.2}$$

is finite when $s + \Delta s \in (-s_0, s_0)$ and $s_0 > 0$. Taking the limit of both sides of this equation as $\Delta s \to 0$ and applying the Lebesgue-dominated convergence theorem, we find that the derivative of $M(s)$ is

$$M'(s) = \int_{-\infty}^{\infty} x e^{sx} dF(x), \ s \in (-s_0, s_0). \tag{B.3}$$

Since finite moments of all orders exist, the preceding derivation can be extended successively to higher-order derivatives of $M(s)$. The kth derivative is

$$M^{(k)}(s) = \int_{-\infty}^{\infty} x^k e^{sx} dF(x), \ s \in (-s_0, s_0). \tag{B.4}$$

D. Torrieri, *Principles of Spread-Spectrum Communication Systems*,
https://doi.org/10.1007/978-3-319-70569-9

Thus, if $M(s)$ exists in some neighborhood of 0, then the kth moment of X is

$$E\left[X^k\right] = M^{(k)}(0) \tag{B.5}$$

which indicates that the moment-generating function is aptly named.

B.2 Laplace Transform

The *Laplace transform* of a nonnegative random variable X with a distribution function $F(x)$ concentrated on $[0, \infty)$ is defined as

$$\mathcal{L}(s) = E\left[e^{-sX}\right] = \int_0^\infty e^{-sx} dF(x),\ Re(s) \geq 0 \tag{B.6}$$

where s is a complex variable. The moment-generating function of a nonnegative random variable is obtained from its Laplace transform by replacing the complex variable s with the real variable $-s$. In the following exposition, we restrict attention to $\mathcal{L}(s)$ for the nonnegative real variable $s \geq 0$ because it simplifies the analysis.

Theorem B1 *If $F(x)$ is the distribution function of a nonnegative random variable X, then the kth derivative of the Laplace transform $\mathcal{L}(s)$ is*

$$\mathcal{L}^{(k)}(s) = (-1)^k \int_0^\infty x^k e^{-sx} dF(x),\ k \geq 0,\ s > 0. \tag{B.7}$$

If X has a kth moment, then

$$E\left(X^k\right) = (-1)^k \mathcal{L}^{(k)}\left(0^+\right). \tag{B.8}$$

Proof The proof is by mathematical induction. Equation (B.7) is true for $k = 0$ by definition (B.6). Assuming that it is true for $k = n$ and applying Taylor's theorem (Appendix A.2) to e^{-hx}, we find that for $s + h > 0$,

$$\begin{aligned}\frac{\mathcal{L}^{(n)}(s+h) - \mathcal{L}^{(n)}(s)}{h} &= (-1)^n \int_0^\infty x^n e^{-sx} \frac{e^{-hx} - 1}{h} dF(x)\\ &= (-1)^{n+1} \int_0^\infty x^{n+1} e^{-sx} dF(x) + R\end{aligned} \tag{B.9}$$

where

$$|R| \leq \frac{h}{2} \int_0^\infty x^{n+2} e^{-sx} dF(x). \tag{B.10}$$

The right-hand side of (B.10) is finite if $s > 0$. Therefore, taking the limit as $h \to 0$ in (B.9), we verify (B.7) for $k = n + 1$. If X has a kth moment, then taking the limit of (B.7) as $s \to 0$ from the right and applying the dominated convergence theorem, we obtain (B.8). □

Theorem B2 *The Laplace transform $\mathcal{L}(s)$ of a random variable X uniquely determines its distribution function $F(x)$.*

Proof For positive y, (B.7) implies that

$$\sum_{k=0}^{\lfloor sy \rfloor} \frac{(-1)^k}{k!} s^k \mathcal{L}^{(k)}(s) = \int_0^\infty H(s, y, x)\, dF(x), \; s > 0$$

where

$$H(s, y, x) = \sum_{k=0}^{\lfloor sy \rfloor} e^{-sx} \frac{(sx)^k}{k!}, \; s > 0.$$

Let Z denote a discrete random variable mean $\lambda > 0$, variance λ, and Poisson distribution

$$P[Z = k] = e^{-\lambda} \frac{\lambda^k}{k!}, \; k = 0, 1, \ldots$$

If $t > 0$,

$$P[Z \leq \lambda t] = \sum_{k=0}^{\lfloor \lambda t \rfloor} e^{-\lambda} \frac{\lambda^k}{k!} \tag{B.11}$$

which implies that

$$\lim_{\lambda \to \infty} \sum_{k=0}^{\lfloor \lambda t \rfloor} e^{-\lambda} \frac{\lambda^k}{k!} = 1, \; t > 1. \tag{B.12}$$

Chebyshev's inequality (Section 4.3) yields

$$P[|Z - \lambda| \geq \lambda \epsilon] \leq \frac{1}{\lambda \epsilon^2}$$

and hence

$$\lim_{\lambda \to \infty} P[|Z - \lambda| \geq \lambda \epsilon] = 0. \tag{B.13}$$

Therefore, Z has a value concentrated in $[\lambda(1-\epsilon), \lambda(1+\epsilon)]$ for $\epsilon > 0$ as $\lambda \to \infty$, and hence (B.11) implies that

$$\lim_{\lambda\to\infty} \sum_{k=0}^{\lfloor \lambda t \rfloor} e^{-\lambda} \frac{\lambda^k}{k!} = 0, \ t < 1. \tag{B.14}$$

Application of (B.12) and (B.14) indicates that if $y > x \geq 0$, then $H(s, y, x) \to 1$ as $s \to \infty$; if $0 \leq y < x$, then $H(s, y, x) \to 0$ as $s \to \infty$. Therefore, the dominated convergence theorem indicates that at all continuity points of $F(y)$, we have

$$\lim_{s\to\infty} \sum_{k=0}^{\lfloor sy \rfloor} \frac{(-1)^k}{k!} s^k \mathcal{L}^{(k)}(s) = \int_0^y dF(x) = F(y). \tag{B.15}$$

Since $F(y)$ is right continuous, (B.15) determines $F(y)$ as a function of $\mathcal{L}(s)$. If the Laplace transform is the same for distributions $F_1(y)$ and $F_2(y)$, then (B.15) indicates that $F_1(y) = F_2(y)$.□

Theorem B3 *The Laplace transform $\mathcal{L}_t(s)$ of a sum of independent nonnegative random variables X_1 and X_2 with Laplace transforms $\mathcal{L}_1(s)$ and $\mathcal{L}_2(s)$, respectively, is*

$$\mathcal{L}_t(s) = \mathcal{L}_1(s)\,\mathcal{L}_2(s). \tag{B.16}$$

Proof Independence of the random variables implies that

$$\mathcal{L}_t(s) = E\left[e^{-s(X_1+X_2)}\right] = E\left[e^{-sX_1}\right]E\left[e^{-sX_2}\right] = \mathcal{L}_1(s)\,\mathcal{L}_2(s). \ \square$$

Appendix C
Fourier Transform and Characteristic Function

C.1 Fourier Transform

The Fourier transform of a complex-valued, integrable function $g(x)$ is defined as

$$\mathcal{F}(g) = \int_{-\infty}^{\infty} e^{-j2\pi fx} g(x)\, dx \tag{C.1}$$

where $j = \sqrt{-1}$ and $-\infty < f < \infty$. Since integration is a linear operation,

$$\mathcal{F}(ag + bh) = a\mathcal{F}(g) + b\mathcal{F}(h) \tag{C.2}$$

for integrable functions $g(x)$ and $h(x)$ and constants a and b. The following is the inversion theorem most commonly used.

Theorem C1 *If $g(x)$ is a bounded, continuous, and integrable function, and its Fourier transform $\mathcal{F}(g) = \widehat{g}(f)$ is an integrable function, then*

$$g(x) = \int_{-\infty}^{\infty} e^{j2\pi fx} \widehat{g}(f)\, df. \tag{C.3}$$

Proof The substitution of the identity $\lim_{\sigma\to\infty}(e^{-2\pi^2 f^2/\sigma^2}) = 1$ and (C.1) into (C.3). Further evaluation yields

$$\begin{aligned}
\int_{-\infty}^{\infty} e^{j2\pi fx} \widehat{g}(f)\, df &= \lim_{\sigma\to\infty} \int_{-\infty}^{\infty} e^{j2\pi fx} e^{-2\pi^2 f^2/\sigma^2} \int_{-\infty}^{\infty} e^{-j2\pi fz} g(z)\, dz df \\
&= \lim_{\sigma\to\infty} \int_{-\infty}^{\infty} g(z) \left[\int_{-\infty}^{\infty} e^{j2\pi f(x-z)} e^{-2\pi^2 f^2/\sigma^2} df \right] dz
\end{aligned}$$

D. Torrieri, *Principles of Spread-Spectrum Communication Systems*,
https://doi.org/10.1007/978-3-319-70569-9

$$= \frac{1}{\sqrt{2\pi}} \lim_{\sigma\to\infty} \int_{-\infty}^{\infty} g(z)\, \sigma \exp\left[-\frac{\sigma^2 (z-x)^2}{2}\right] dz$$
$$= \frac{1}{\sqrt{2\pi}} \int_{-\infty}^{\infty} \lim_{\sigma\to\infty} [g\left(x + \frac{u}{\sigma}\right)] \exp\left(-\frac{u^2}{2}\right) du$$
$$= g(x).$$

In the first equality, the integrability of $\widehat{g}(f)$ and the dominated convergence theorem justify taking the limit outside the outer integral. In the second equality, the integrability of $g(x)$ and Fubini's theorem (see below) justify the interchange of the order of integration. The third equality follows from a change of integration variable and (A.10) of Appendix A.1. The fourth equality results from a change in the integration variable followed by application of the dominated convergence theorem to justify taking the limit inside the integral. The final equality is obtained by taking the limit and then applying (A.2). □

The convolution of functions g and h is the function $g \star h$ defined by

$$(g \star h)(x) = \int_{-\infty}^{\infty} g(x-y)\, h(y)\, dy. \tag{C.4}$$

Convolution Theorem

(a) If g and h are bounded and integrable with Fourier transforms $\mathcal{F}(g)$ and $\mathcal{F}(h)$, respectively, then

$$\mathcal{F}(g * h) = \mathcal{F}(g)\,\mathcal{F}(h).$$

(b) If $\mathcal{F}(g)$ and $\mathcal{F}(h)$ are bounded and integrable, then

$$\mathcal{F}(gh) = \mathcal{F}(g) * \mathcal{F}(h).$$

Proof

(a) Since g and h are bounded and integrable, $g*h$ is integrable. Therefore, Fubini's theorem justifies the following interchange of the order of integration, and we obtain

$$\mathcal{F}(g * h) = \int_{-\infty}^{\infty} \left[\int_{-\infty}^{\infty} g(x-y)\, h(y)\, dy\right] e^{-j2\pi fx} dx$$
$$= \int_{-\infty}^{\infty} e^{-j2\pi fy} h(y) \left[\int_{-\infty}^{\infty} e^{-j2\pi f(x-y)} g(x-y)\, dx\right] dy$$
$$= \mathcal{F}(g)\,\mathcal{F}(h)$$

where the final equality results from a change in the integration variable.

(b) If $\mathcal{F}(g)$ and $\mathcal{F}(h)$ are bounded and integrable, then $\mathcal{F}(g) * \mathcal{F}(h)$ is integrable. Let $\mathcal{F}^{-1}(\cdot)$ denote the inverse Fourier transform. A derivation almost identical to the preceding one yields

$$\begin{aligned}\mathcal{F}^{-1}\left[\mathcal{F}(g) * \mathcal{F}(h)\right] &= \mathcal{F}^{-1}\left[\mathcal{F}(g)\right]\mathcal{F}^{-1}\left[\mathcal{F}(h)\right] \\ &= gh\,.\end{aligned}$$

Taking the Fourier transform of both sides of this equation completes the proof. □

Parseval's Identities If g, $\widehat{g}(f)$, h, and $\widehat{h}(f)$ are bounded and integrable, then

$$\int_{-\infty}^{\infty} g(x)\, h^*(x)\, dx = \int_{-\infty}^{\infty} \widehat{g}(f)\, \widehat{h}^*(f)\, df$$

and

$$\int_{-\infty}^{\infty} |g(x)|^2\, dx = \int_{-\infty}^{\infty} |\widehat{g}(f)|^2\, df.$$

Proof Since $g(x)$ and $\widehat{h}(f)$ are bounded and integrable,

$$\int_{-\infty}^{\infty} \left| e^{-j2\pi fx} \widehat{h}^*(f)\, g(x) \right| df = |g(x)| \int_{-\infty}^{\infty} \left| \widehat{h}(f) \right| df < \infty.$$

Therefore, Fubini's theorem justifies the following interchange of the order of integration, and we obtain

$$\begin{aligned}\int_{-\infty}^{\infty} g(x)\, h^*(x)\, dx &= \int_{-\infty}^{\infty} g(x) \left[\int_{-\infty}^{\infty} e^{-j2\pi fx} \widehat{h}^*(f)\, df \right] dx \\ &= \int_{-\infty}^{\infty} \widehat{h}^*(f) \left[\int_{-\infty}^{\infty} e^{-j2\pi fx} g(x)\, dx \right] df \\ &= \int_{-\infty}^{\infty} \widehat{g}(f)\, \widehat{h}^*(f)\, df.\end{aligned}$$

The second identity of the theorem follows by setting $g = h$. □

Application of Fubini's Theorem

Fubini's theorem is applied several times in this section and elsewhere in this book. This theorem, which is proved using measure theory [6, 9, 114], states that under certain conditions, a double integral may be evaluated as either of two iterated integrals:

$$\int_{-\infty}^{\infty}\int_{-\infty}^{\infty} f(x,y)\,dxdy = \int_{-\infty}^{\infty}\left[\int_{-\infty}^{\infty} f(x,y)\,dx\right]dy$$
$$= \int_{-\infty}^{\infty}\left[\int_{-\infty}^{\infty} f(x,y)\,dy\right]dx. \tag{C.5}$$

To apply the theorem, we compute or bound one of the iterated integrals with $|f(x,y)|$ in place of $f(x,y)$. If the result is finite, then the double integral of $|f(x,y)|$ is finite, which implies that the double integral of $f(x,y)$ may be computed as either of the two iterated integrals.

C.2 Characteristic Function

The *characteristic function* of the random variable X with distribution function $F(x)$ is

$$h(u) = E\left[e^{juX}\right] = \int_{-\infty}^{\infty} e^{jux}dF(x) \tag{C.6}$$

where $j = \sqrt{-1}$ and $-\infty < u < \infty$. Since $|\exp(jux)| \le 1$, the characteristic function always exists, and $|h(u)| \le h(0) = 1$. Since

$$|h(u+h) - h(u)| \le \int_{-\infty}^{\infty} \left|e^{juh} - 1\right| dF(x) \tag{C.7}$$

application of the bounded convergence theorem indicates that $h(u)$ is continuous. If a continuous density function $f(x)$ exists, then $h(u)$ is the conjugate *Fourier transform* of the density function:

$$h(u) = \int_{-\infty}^{\infty} e^{jux} f(x)\,dx. \tag{C.8}$$

The principal advantage of the characteristic function relative to the Laplace transform is that the characteristic function is applicable to a random variable that may take negative values.

The usefulness of the characteristic function depends on the fact that it uniquely determines the distribution function from which it is derived. To prove this fact, we need to evaluate the integral

$$I_1 = P\int_{-\infty}^{\infty} \frac{e^{jx}}{x}dy = \lim_{\epsilon\to 0}\left(\int_{-\infty}^{-\epsilon} \frac{e^{jx}}{x}dy + \int_{+\epsilon}^{\infty} \frac{e^{jx}}{x}dy\right) \tag{C.9}$$

where P denotes the Cauchy principal value, which is defined to avoid the singularity at the origin when the integral is a Riemann integral. However, if the integral is considered a Lebesgue integral, then the singularity has measure 0, and does not have to be avoided. Since $\cos(x)/x$ is an odd function and $\sin(x)/x$ is an even function, I_1 reduces to

$$\begin{aligned} I_1 &= \int_{-\infty}^{\infty} \frac{e^{jx}}{x} dx \\ &= 2j \int_0^{\infty} \frac{\sin(x)}{x} dx. \end{aligned} \tag{C.10}$$

The integrand is uniformly bounded, and the Lebesgue integral exists because

$$\int_{(n-1)\pi}^{n\pi} \frac{\sin(x)}{x} dx \tag{C.11}$$

alternates in sign for each positive integer n, and its absolute value decreases to zero.

Theorem C2

$$P \int_{-\infty}^{\infty} \frac{e^{jx}}{x} dx = 2j \int_0^{\infty} \frac{\sin(x)}{x} dx = j\pi. \tag{C.12}$$

Proof We apply Cauchy's integral theorem to a contour integral of $\exp(-jz)/z$, where z is a complex variable. The contour includes a large semicircle C_1 with a radius c in the upper complex plane and a small semicircle C_2 with a radius ϵ traversed clockwise around the pole at the origin. Since there are no singularities within the contour,

$$\int_{-c}^{-\epsilon} \frac{e^{jx}}{x} dx + \int_{\epsilon}^{c} \frac{e^{jx}}{x} dx + \int_{C_1} \frac{e^{jz}}{z} dz + \int_{C_2} \frac{e^{jz}}{z} dz = 0. \tag{C.13}$$

After changing the integration variable in the third integral by substituting $z = ce^{j\theta}$ and then applying the dominated convergence theorem, we obtain

$$\begin{aligned} \lim_{c\to\infty} \int_{C_1} \frac{e^{jz}}{z} dz &= \lim_{c\to\infty} \int_0^{-\pi} je^{jc\cos\theta - c\sin\theta} d\theta \\ &= 0. \end{aligned}$$

After changing the integration variable in the fourth integral by substituting $z = \epsilon e^{j\theta}$ and then applying the dominated convergence theorem, we obtain

$$\begin{aligned} \lim_{\epsilon\to 0} \int_{C_2} \frac{e^{jz}}{z} dz &= \lim_{\epsilon\to 0} \int_{-\pi}^{0} je^{j\epsilon\cos\theta - \epsilon\sin\theta} d\theta \\ &= -j\pi. \end{aligned}$$

Using these results and taking the limit of (C.13) as $c \to \infty$ and $\epsilon \to 0$, we obtain (C.12). □

Theorem C3 *If the distribution function $F(x)$ has the characteristic function $h(u)$, then*

$$F(b) - F(a) = \lim_{L\to\infty} \int_{-L}^{L} \frac{e^{-jua} - e^{-jub}}{j2\pi u} h(u)\, du \tag{C.14}$$

for all points a and $b > a$ at which $F(x)$ is continuous, and the distribution function is uniquely determined. If the characteristic function is integrable, then the distribution function has a continuous density function given by

$$f(x) = \frac{1}{2\pi} \int_{-\infty}^{\infty} e^{-jux} h(u)\, du. \tag{C.15}$$

Proof Let

$$I_L = \int_{-L}^{L} \frac{e^{-jua} - e^{-jub}}{j2\pi u} h(u)\, du. \tag{C.16}$$

Substituting (C.6) into this integral gives

$$I_L = \int_{-L}^{L} \frac{e^{-jua} - e^{-jub}}{j2\pi u} \left[\int_{-\infty}^{\infty} e^{-jux} dF(x) \right] du. \tag{C.17}$$

To interchange the order of integration on the right-hand side of (C.17), we first observe that

$$\begin{aligned} \left| \frac{e^{-jua} - e^{-jub}}{j2\pi u} e^{jux} \right| &= \left| \frac{e^{-jua} - e^{-jub}}{j2\pi u} \right| = \left| \int_{a}^{b} \frac{e^{-juy}}{2\pi} dy \right| \\ &\leq \frac{b-a}{2\pi} \end{aligned} \tag{C.18}$$

and

$$\int_{-L}^{L} \left[\int_{-\infty}^{\infty} \frac{b-a}{2\pi} dF(x) \right] du = \frac{L(b-a)}{\pi} < \infty$$

which indicates that Fubini's theorem is applicable to (C.17). Interchanging the order of integration and then changing integration variables, we obtain

$$I_L = \int_{-\infty}^{\infty} G_L(x)\, dF(x)$$

where

$$G_L(x) = \int_{-L(x-a)}^{L(x-a)} \frac{e^{jy}}{j2\pi y} dy - \int_{-L(x-b)}^{L(x-b)} \frac{e^{jy}}{j2\pi y} dy.$$

By Theorem C2, each of these integrals is bounded by $1/2$, and hence $|G_L(x)| \leq 1$. Therefore, the bounded convergence theorem implies that

$$\lim_{L\to\infty} I_L = \int_{-\infty}^{\infty} \lim_{L\to\infty} G_L(x)\, dF(x). \tag{C.19}$$

Applying Theorem C2, we find that for $a < b$,

$$\lim_{L\to\infty} G_L(x) = \begin{cases} 0 & x < a \text{ or } x > b \\ 1/2 & x = a \text{ or } x = b \\ 1 & a < x < b. \end{cases}$$

Substituting this equation into (C.19), evaluating the integral, and then equating the result with the limit of (C.16), we obtain (C.14) for all points a and $b > a$ at which $F(x)$ is continuous. Since $F(x)$ is right continuous, $h(u)$ determines $F(x)$ everywhere. Thus, the characteristic function uniquely determines the distribution function.

If $h(u)$ is integrable, then (C.14) and (C.18) indicate that

$$F(b) - F(a) \leq \frac{(b-a)}{2\pi} \int_{-\infty}^{\infty} |h(u)|\, du$$

and hence $F(x)$ is continuous. If $f(x)$ is defined by (C.15), application of the dominated convergence theorem proves that $f(x)$ is continuous.

Applying Fubini's theorem to interchange the order of integration, we find that

$$\begin{aligned} \int_a^x f(y)\, dy &= \frac{1}{2\pi} \int_{-\infty}^{\infty} \left[\int_a^x e^{-juy} dy \right] h(u)\, du \\ &= \lim_{L\to\infty} \int_{-L}^{L} \frac{e^{-jua} - e^{-jux}}{j2\pi u} h(u)\, du \\ &= F(x) - F(a). \end{aligned}$$

By the continuity of $f(x)$, this equation implies that the derivative of $F(x)$ is $f(x)$. Since $F(x)$ is monotonically increasing, $f(x)$ is nonnegative everywhere; hence, $f(x)$ is the density function for $F(x)$. □

Let $h^{(k)}(u)$ denote the kth derivative of $h(u)$ with respect to u. The following theorem enables the calculation of the kth moment of a random variable without the restrictive condition in Theorem B2 of Appendix B.2.

Theorem C4 *If the random variable X has a distribution function $F(x)$ and a characteristic function $h(u)$, and if $E\left[|X|^k\right] < \infty$ for a positive integer k, then*

$$h^{(k)}(u) = \int_{-\infty}^{\infty} (jx)^k e^{jux} dF(x) \tag{C.20}$$

and

$$E\left[X^k\right] = j^{-k} h^{(k)}(0). \tag{C.21}$$

Proof Since $\left|(jx)^k e^{jux}\right| = |x|^k$ and $E\left[|X|^k\right] < \infty$, we can differentiate the right-hand side of (C.20) k times under the integral sign, which proves (C.20). Using the dominated convergence theorem and taking the limit in (C.6) proves (C.21). □

From definition (C.6) and an evaluation similar to that in Theorem B3 of Appendix B.2, it follows that *the characteristic function $h_t(u)$ of the sum of independent random variables X_1 and X_2 with characteristic functions $h_1(u)$ and $h_2(u)$*, respectively, *is*

$$h_t(u) = h_1(u)\, h_2(u). \tag{C.22}$$

Appendix D
Signal Characteristics

D.1 Bandpass Signals

A *bandpass signal* has its power spectrum in a spectral band surrounding a carrier frequency, which is usually at the center of the band. The *Hilbert transform* provides the basis for signal representations that facilitate the analysis of bandpass signals and systems. Let P denote the Cauchy principal value of an integral. The Hilbert transform of a function $g(t)$ is defined as

$$\begin{aligned} H[g(t)] = \hat{g}(t) &= \frac{1}{\pi} P \int_{-\infty}^{\infty} \frac{g(u)}{t-u}\, du \\ &= \frac{1}{\pi} \lim_{\epsilon \to 0} \left[\int_{-\infty}^{t-\epsilon} \frac{g(u)}{t-u}\, du + \int_{t+\epsilon}^{\infty} \frac{g(u)}{t-u}\, du \right] \end{aligned} \tag{D.1}$$

provided that the integral exists as a principal value.

Since (D.1) has the form of the convolution of $g(t)$ with $1/\pi t$, $\hat{g}(t)$ results from passing $g(t)$ through a linear filter with an impulse response equal to $1/\pi t$. The transfer function of the filter is given by the Fourier transform of $1/\pi t$. For this function, the Fourier transform is

$$\mathcal{F}\left[\frac{1}{\pi t}\right] = P \int_{-\infty}^{\infty} \frac{e^{-j2\pi f t}}{\pi t}\, dt \tag{D.2}$$

where $j = \sqrt{-1}$. Changing variables and applying Theorem C2 of Appendix C.2, we obtain

$$\begin{aligned} \mathcal{F}\left[\frac{1}{\pi t}\right] &= \frac{-\text{sgn}(f)}{\pi} P \int_{-\infty}^{\infty} \frac{e^{jx}}{x}\, dx \\ &= -j\, \text{sgn}(f) \end{aligned} \tag{D.3}$$

D. Torrieri, *Principles of Spread-Spectrum Communication Systems*,
https://doi.org/10.1007/978-3-319-70569-9

where *sgn(f)* is the *signum function* defined by

$$\operatorname{sgn}(f) = \begin{cases} 1, & f > 0 \\ 0, & f = 0 \\ -1, & f < 0. \end{cases} \tag{D.4}$$

Let $G(f) = \mathcal{F}[g(t)]$ and $\hat{G}(f) = \mathcal{F}[\hat{g}(t)]$. Applying the convolution theorem of Fourier analysis (Appendix C.1) to (D.1) and then substituting (D.3), we obtain

$$\hat{G}(f) = -j \operatorname{sgn}(f) G(f). \tag{D.5}$$

Because $H[\hat{g}(t)]$ results from passing $g(t)$ through two successive filters, each with a transfer function $-j \operatorname{sgn}(f)$,

$$H[\hat{g}(t)] = -g(t) \tag{D.6}$$

provided that $G(0) = 0$.

Equation (D.5) indicates that taking the Hilbert transform corresponds to introducing a phase sift of $-\pi/2$ radians for all positive frequencies and $+\pi/2$ radians for all negative frequencies. Consequently,

$$H[\cos 2\pi f_c t] = \sin 2\pi f_c t \tag{D.7}$$

$$H[\sin 2\pi f_c t] = -\cos 2\pi f_c t. \tag{D.8}$$

These relations can be formally verified by taking the Fourier transform of the left-hand side of (D.7) or (D.8), applying (D.5), and then taking the inverse Fourier transform of the result. If $G(f) = 0$ for $|f| > W$ and $f_c > W$, the same method yields

$$H[g(t) \cos 2\pi f_c t] = g(t) \sin 2\pi f_c t \tag{D.9}$$

$$H[g(t) \sin 2\pi f_c t] = -g(t) \cos 2\pi f_c t. \tag{D.10}$$

A *bandpass signal* is a signal with a Fourier transform that is negligible except for $f_c - W/2 \leq |f| \leq f_c + W/2$, where $0 \leq W < 2f_c$ and f_c is the center frequency. If $W << f_c$, the bandpass signal is often called a *narrowband signal*. A complex-valued signal with a Fourier transform that is nonzero only for $f > 0$ is called an *analytic signal*.

Consider a bandpass signal $g(t)$ with Fourier transform $G(f)$. The analytic signal $g_a(t)$ associated with $g(t)$ is defined to be the signal with Fourier transform

$$G_a(f) = [1 + \operatorname{sgn}(f)]G(f) \tag{D.11}$$

which is zero for $f \leq 0$ and is confined to the band $|f - f_c| \leq W/2$ when $f > 0$. The inverse Fourier transform of $G_a(f)$ and (D.5) imply that

$$g_a(t) = g(t) + j\hat{g}(t). \tag{D.12}$$

The *complex envelope* of $g(t)$ is defined by

$$g_l(t) = g_a(t)e^{-j2\pi f_c t} \tag{D.13}$$

where f_c is the center frequency if $g(t)$ is a bandpass signal. Since the Fourier transform of $g_l(t)$ is $G_a(f + f_c)$, which occupies the band $|f| \leq W/2$, the complex envelope is a baseband signal that may be regarded as an *equivalent lowpass representation* of $g(t)$. Equations (D.12) and (D.13) imply that $g(t)$ can be expressed in terms of its complex envelope as

$$g(t) = \text{Re}[g_l(t)e^{j2\pi f_c t}]. \tag{D.14}$$

The complex envelope can be decomposed as

$$g_l(t) = g_c(t) + jg_s(t) \tag{D.15}$$

where $g_c(t)$ and $g_s(t)$ are real-valued functions. Therefore, (D.14) yields

$$g(t) = g_c(t)\cos(2\pi f_c t) - g_s(t)\sin(2\pi f_c t). \tag{D.16}$$

Since the two sinusoidal carriers are in phase quadrature, $g_c(t)$ and $g_s(t)$ are called the *in-phase* and *quadrature* components of $g(t)$, respectively. These components are lowpass signals confined to $|f| \leq W/2$.

Applying Parseval's identity from Fourier analysis (Appendix C.1) and then (D.5), we obtain

$$\int_{-\infty}^{\infty} \hat{g}^2(t)\, dt = \int_{-\infty}^{\infty} |\hat{G}(f)|^2\, df = \int_{-\infty}^{\infty} |G(f)|^2\, df = \int_{-\infty}^{\infty} g^2(t)\, dt. \tag{D.17}$$

Therefore,

$$\begin{aligned} \int_{-\infty}^{\infty} |g_l(t)|^2\, dt &= \int_{-\infty}^{\infty} |g_a(t)|^2\, dt = \int_{-\infty}^{\infty} g^2(t)\, dt + \int_{-\infty}^{\infty} \hat{g}^2(t)\, dt \\ &= 2\int_{-\infty}^{\infty} g^2(t)\, dt = 2\mathcal{E} \end{aligned} \tag{D.18}$$

where $\mathcal{E}$ denotes the energy of the bandpass signal $g(t)$.

D.2 Stationary Stochastic Processes

A stochastic process is called *wide-sense stationary* if its mean is independent of the sampling time, and its autocorrelation depends only on the time difference between samples. Consider a real-valued, wide-sense-stationary stochastic process $n(t)$ that is a zero-mean with autocorrelation

$$R_n(\tau) = E[n(t)n(t+\tau)] \tag{D.19}$$

where $E[x]$ denotes the expected value of x. The Hilbert transform of this process is the real-valued stochastic process defined by

$$\hat{n}(t) = \frac{1}{\pi}\int_{-\infty}^{\infty} \frac{n(u)}{t-u} du \tag{D.20}$$

where we assume that the Cauchy principal value of the integral exists for almost every sample function of $n(t)$. This equation indicates that $\hat{n}(t)$ is a zero-mean stochastic process. The zero-mean processes $n(t)$ and $\hat{n}(t)$ are *jointly wide-sense stationary* if their correlation and cross-correlation functions are not functions of t.

An application of (D.20) and (D.19) gives the cross-correlation of $n(t)$ and $\hat{n}(t)$:

$$\begin{aligned} R_{n\hat{n}}(\tau) = E[n(t)\hat{n}(t+\tau)] &= \frac{1}{\pi}\int_{-\infty}^{\infty} \frac{R_n(u)}{\tau-u} du \\ &= \hat{R}_n(\tau). \end{aligned} \tag{D.21}$$

An application of this result and (D.6) yields the autocorrelation

$$\begin{aligned} R_{\hat{n}}(\tau) &= E[\hat{n}(t)\hat{n}(t+\tau)] \\ &= \frac{1}{\pi}\int_{-\infty}^{\infty} \frac{\hat{R}_n(t+\tau-u)}{t-u} du = -\frac{1}{\pi}\int_{-\infty}^{\infty} \frac{\hat{R}_n(x)}{\tau-x} dx \\ &= R_n(\tau). \end{aligned} \tag{D.22}$$

Equations (D.19), (D.21), and (D.22) indicate that $n(t)$ and $\hat{n}(t)$ are jointly wide-sense stationary. Since $n(t)$ is wide-sense stationary, $R_n(\tau)$ is an even function. It then follows from (D.22) that $R_{\hat{n}}(\tau)$ is an even function. Equation (D.21) and a change in the integration variable indicate that $\widehat{R}_n(\tau)$ and $R_{n\hat{n}}(\tau)$ are odd functions. Thus,

$$\begin{aligned} &R_n(-\tau) = R_n(\tau), \; R_{\hat{n}}(-\tau) = R_{\hat{n}}(\tau) \\ &\hat{R}_n(-\tau) = -\hat{R}_n(\tau), \; R_{n\hat{n}}(-\tau) = -R_{n\hat{n}}(\tau). \end{aligned} \tag{D.23}$$

The *analytic signal* associated with $n(t)$ is the complex-valued zero-mean process defined by

$$n_a(t) = n(t) + j\hat{n}(t). \tag{D.24}$$

The autocorrelation of the analytic signal is defined as

$$R_a(\tau) = E[n_a^*(t)n_a(t+\tau)] \tag{D.25}$$

where the asterisk denotes the complex conjugate. Using (D.19) and (D.21) to (D.25), we obtain

$$R_a(\tau) = 2R_n(\tau) + 2j\hat{R}_n(\tau) \tag{D.26}$$

which establishes the wide-sense stationarity of the analytic signal.

Since (D.19) indicates that $R_n(\tau)$ is an even function, (D.21) yields

$$R_{n\hat{n}}(0) = \hat{R}_n(0) = 0 \tag{D.27}$$

which indicates that $n(t)$ and $\hat{n}(t)$ are uncorrelated. Equations (D.22), (D.26), and (D.27) yield

$$R_{\hat{n}}(0) = R_n(0) = 1/2R_a(0). \tag{D.28}$$

The *complex envelope* of $n(t)$ or the *equivalent lowpass representation* of $n(t)$ is the zero-mean stochastic process defined by

$$n_l(t) = n_a(t)e^{-j2\pi f_c t} \tag{D.29}$$

where f_c is an arbitrary frequency usually chosen as the center or carrier frequency of $n(t)$. The complex envelope can be decomposed as

$$n_l(t) = n_c(t) + jn_s(t) \tag{D.30}$$

where $n_c(t)$ and $n_s(t)$ are real-valued, zero-mean stochastic processes.

Equations (D.24) and (D.29) imply that

$$n(t) = Re[n_l(t)e^{j2\pi f_c t)}]. \tag{D.31}$$

The substitution of (D.30) into (D.31) gives an in-phase and quadrature representation:

$$n(t) = n_c(t)\cos(2\pi f_c t) - n_s(t)\sin(2\pi f_c t). \tag{D.32}$$

Substituting (D.24) and (D.30) into (D.29) we find that

$$n_c(t) = n(t)\cos(2\pi f_c t) + \hat{n}(t)\sin(2\pi f_c t) \tag{D.33}$$

$$n_s(t) = \hat{n}(t)\cos(2\pi f_c t) - n(t)\sin(2\pi f_c t). \tag{D.34}$$

Using (D.33), (D.34), (D.19), (D.21), (D.22), (D.23), and trigonometric identities, we obtain the autocorrelations of $n_c(t)$ and $n_s(t)$, which are

$$R_c(\tau) = E[n_c(t)n_c(t+\tau)] = R_n(\tau)\cos(2\pi f_c\tau) + \hat{R}_n(\tau)\sin(2\pi f_c\tau) \tag{D.35}$$

$$R_s(\tau) = E[n_s(t)n_s(t+\tau)] = R_c(\tau) \tag{D.36}$$

and the cross-correlations

$$R_{cs}(\tau) = E[n_c(t)n_s(t+\tau)] = \hat{R}_n(\tau)\cos(2\pi f_c\tau) - R_n(\tau)\sin(2\pi f_c\tau) \tag{D.37}$$

$$R_{sc}(\tau) = E[n_s(t)n_c(t+\tau)] = -R_{cs}(\tau). \tag{D.38}$$

These equations show explicitly that if $n(t)$ is wide-sense stationary, then $n_c(t)$ and $n_s(t)$ are jointly wide-sense stationary with identical autocorrelation functions. From (D.23), we obtain

$$R_c(-\tau) = R_c(\tau),\ R_s(-\tau) = R_s(\tau)$$
$$R_{cs}(-\tau) = -R_{cs}(\tau),\ R_{sc}(-\tau) = -R_{sc}(\tau). \tag{D.39}$$

Since

$$R_c(0) = R_s(0) = R_n(0),\ R_{cs}(0) = 0 \tag{D.40}$$

the variances of $n(t)$, $n_c(t)$, and $n_s(t)$ are all equal, and $n_c(t)$ and $n_s(t)$ are uncorrelated.

Equations (D.30) and (D.39) imply that

$$E[n_l(t)n_l(t+\tau)] = 0. \tag{D.41}$$

A complex-valued, zero-mean stochastic process that satisfies this equation is called a *circularly symmetric* process. Thus, *the complex envelope of a zero-mean, wide-sense stationary process is a circularly symmetric process.* The autocorrelation of a complex envelope is defined as

$$R_l(\tau) = E[n_l^*(t)n_l(t+\tau)]. \tag{D.42}$$

Substituting (D.29) and (D.26) into (D.42), we obtain

$$R_l(\tau) = 2e^{-j2\pi f_c \tau}\left[R_n(\tau) + j\hat{R}_n(\tau)\right] \tag{D.43}$$

which shows that $n_l(t)$ is a zero-mean, wide-sense-stationary process. Since $R_n(\tau)$ and $\hat{R}_n(\tau)$ are real-valued,

$$R_n(\tau) = \frac{1}{2}Re\left[R_l(\tau)e^{j2\pi f_c \tau}\right]. \tag{D.44}$$

Power Spectral Density

The *power spectral density* (PSD) of a signal is the Fourier transform of its autocorrelation. Let $S_n(f)$, $S_c(f)$, and $S_s(f)$ denote the PSDs of $n(t)$, $n_c(t)$, and $n_s(t)$, respectively. We assume that $S_n(f)$ occupies the band $f_c - W/2 \le |f| \le f_c + W/2$ and that $f_c > W/2 \ge 0$. Taking the Fourier transform of (D.35), using (D.5), and simplifying, we obtain

$$S_c(f) = S_s(f) = \begin{cases} S_n(f - f_c) + S_n(f + f_c), & |f| \le W/2 \\ 0, & |f| > W/2. \end{cases} \tag{D.45}$$

Similarly, the cross-spectral density of $n_c(t)$ and $n_s(t)$ can be derived by taking the Fourier transform of (D.37) and using (D.5). After simplification, the result is

$$S_{cs}(f) = \begin{cases} j[S_n(f - f_c) - S_n(f + f_c)], & |f| \le W/2 \\ 0, & |f| > W/2. \end{cases} \tag{D.46}$$

Since $R_n(\tau)$ is an even function, $S_n(f)$ is a real-valued, even function. If $S_n(f)$ is *locally symmetric* about f_c so that

$$\begin{aligned} S_n(f_c + f) &= S_n(f_c - f), \\ &= S_n(f - f_c), \quad |f| \le W/2 \end{aligned} \tag{D.47}$$

then (D.46) indicates that $S_{cs}(f) = 0$, which implies that

$$R_{cs}(\tau) = 0 \tag{D.48}$$

for all τ. Thus, $n_c(t)$ and $n_s(t + \tau)$ are uncorrelated for all τ when $S_n(f)$ is locally symmetric.

The PSD of $n_l(t)$, which we denote as $S_l(f)$, can be derived by calculating the Fourier transform of (D.43), and using (D.5). If $S_n(f)$ occupies the band $f_c - W/2 \leq |f| \leq f_c + W/2$ and $f_c > W/2 \geq 0$, then

$$S_l(f) = \begin{cases} 4S_n(f + f_c), & |f| \leq W/2 \\ 0, & |f| > W/2 \end{cases} \tag{D.49}$$

which indicates that $S_l(f)$ is a real-valued function. Therefore, expanding the right-hand side of (D.44) by using $Re[z] = (z + z^*)/2$ and then taking the Fourier transform yields

$$S_n(f) = \frac{1}{4}S_l(f - f_c) + \frac{1}{4}S_l(-f - f_c). \tag{D.50}$$

White Gaussian Noise

The communication channel is often modeled as an *additive white Gaussian noise* (AWGN) *channel* for which the noise in the receiver is a zero-mean, white Gaussian process with autocorrelation

$$R_n(\tau) = E[n(t)n(t+\tau)] = \frac{N_0}{2}\delta(\tau) \tag{D.51}$$

and the *two-sided noise PSD*

$$S_n(f) = \frac{N_0}{2}. \tag{D.52}$$

The Hilbert transform $\hat{n}(t)$ is the limit of Riemann sums that are linear combinations of $n(t)$. Therefore, Theorem A1 of Appendix A.1 implies that if $n(t)$ is a zero-mean, white Gaussian process, then $\hat{n}(t)$ and $n(t)$ are zero-mean jointly Gaussian processes. Equations (D.33) and (D.34) and Theorem A4 of Appendix A.1 then imply that $n_c(t)$ and $n_s(t)$ are zero-mean jointly Gaussian processes. Since (D.40) shows that they are uncorrelated for a specific value of t, $n_c(t)$ and $n_s(t)$ are statistically independent, zero-mean Gaussian random variables with equal variances.

D.3 Downconverter

Most modern digital receivers use a *downconverter* to convert the received signal into a filtered baseband signal. The main components of a downconverter for the desired signal $s(t)$ with carrier or center frequency f_c are shown in Figure D.1 (a).

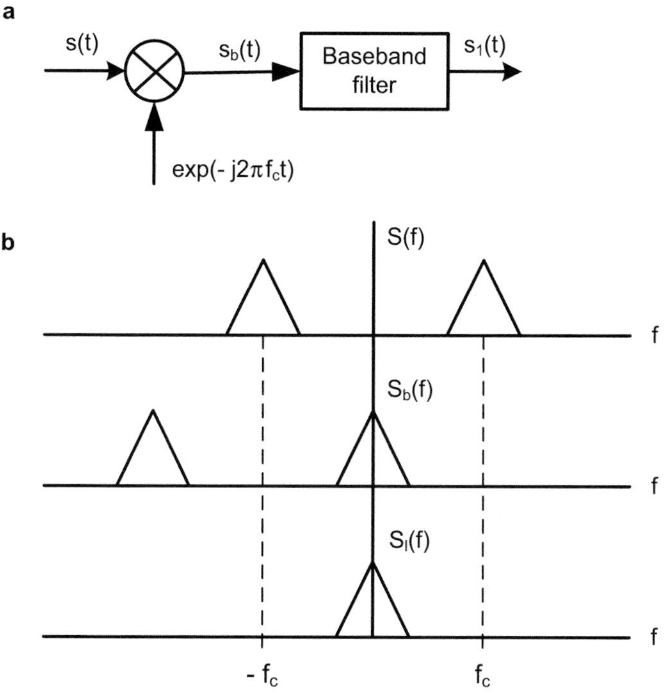

Fig. D.1 Envelope extraction: (**a**) downconverter and (**b**) associated spectra

Consider the received signal

$$r(t) = s(t) + n(t) \tag{D.53}$$

where $n(t)$ is the noise. Let $h(t)$ denote the impulse response of the baseband filter. The output of the baseband filter due to $s(t)$ is

$$s_1(t) = \int_{-\infty}^{\infty} s(\tau)e^{-j2\pi f_c \tau} h(t-\tau)\, d\tau. \tag{D.54}$$

Using (D.14) to express $s(t)$ in terms of its complex envelope $s_l(\tau)$ and then substituting $Re(x) = (x + x^*)/2$, where x^* denotes the complex conjugate of x, we obtain

$$s_1(t) = \frac{1}{2}\int_{-\infty}^{\infty} s_l(\tau)h(t-\tau)\, d\tau + \frac{1}{2}\int_{-\infty}^{\infty} s_l(\tau)h(t-\tau)e^{-j4\pi f_c \tau} d\tau. \tag{D.55}$$

The second term is the Fourier transform of $s_l(\tau)h(t-\tau)$ evaluated at frequency $-2f_c$. Assuming that $s_l(\tau)$ and $h(t-\tau)$ have transforms confined to $|f| < f_c$, their product has a transform confined to $|f| < 2f_c$, and the second term in (D.55) vanishes, leaving

$$s_1(t) = \frac{1}{2}\int_{-\infty}^{\infty} s_l(\tau)h(t-\tau)\,d\tau. \tag{D.56}$$

This convolution is equivalent to the multiplication of the corresponding Fourier transforms. Therefore, if the Fourier transform of $h(t)$ is a constant over the passband of $s_l(t)$, then $s_1(t)$ is proportional to $s_l(t)$, which indicates that the complex envelope of the received signal can be extracted by the downconverter. The spectra of the received signal $s(t)$, the input to the baseband filter $s_b(t) = s(t)\exp(-j2\pi f_c t)$, and the complex envelope $s_l(t)$ are depicted in Figure D.1 (b).

The downconverter alters the character of the noise $n(t)$ entering it. The complex-valued noise at the output of the downconverter is

$$z(t) = \int_{-\infty}^{\infty} n(u)e^{-j2\pi f_c u}h(t-u)du. \tag{D.57}$$

We assume that the noise is a zero-mean, white Gaussian process. The approximating Riemann sums of the real and imaginary parts of the integral in (D.57) are sums of independent zero-mean Gaussian random variables. Therefore, the real and imaginary components of $z(t)$ are *jointly zero-mean Gaussian random variables* (Appendix A.1), and $z(t)$ is zero-mean. The autocorrelation of a wide-sense-stationary, complex-valued process $z(t)$ is defined as

$$R_z(\tau) = E[z^*(t)z(t+\tau)]. \tag{D.58}$$

Substituting (D.57), interchanging the expectation and integration operations, using (D.51) to evaluate one of the integrals, and then changing variables, we obtain

$$R_z(\tau) = \frac{N_0}{2}\int_{-\infty}^{\infty} h^*(u)h(u+\tau)du. \tag{D.59}$$

Equations (D.57) and (D.51) imply that

$$E[z(t)z(t+\tau)] = \frac{N_0}{2}e^{-j4\pi f_c t}\int_{-\infty}^{\infty} e^{j4\pi f_c u}h(u+\tau)h(u)\,du. \tag{D.60}$$

The integrand is the Fourier transform of $h(u+\tau)h(u)$ evaluated at frequency $-2f_c$. Assuming that $h(u)$ has a Fourier transform confined to $|f| < f_c$, $h(u+\tau)h(u)$ has a Fourier transform confined to $|f| < 2f_c$, and

$$E[z(t)z(t+\tau)] = 0. \tag{D.61}$$

A complex-valued stochastic process $z(t)$ that satisfies (D.61) is called a *circularly symmetric* process.

Let $z_R(t)$ and $z_I(t)$ denote the real and imaginary parts of $z(t)$, respectively. Since $z(t)$ is zero-mean, $z_R(t)$ and $z_I(t)$ are zero-mean. Setting $\tau = 0$ in (D.61), equating (D.58) and (D.59) for $\tau = 0$, and then solving the preceding equations, we obtain

$$E[(z_R(t))^2] = E[(z_I(t))^2] = \frac{1}{2}E[|z(t)|^2] = \frac{N_0}{4}\int_{-\infty}^{\infty} |h(u)|^2 \, du \qquad \text{(D.62)}$$

$$E[(z_R(t)z_I(t))] = 0 \qquad \text{(D.63)}$$

which proves that $z_R(t)$ and $z_I(t)$ are zero-mean, independent Gaussian processes with the same variance.

D.4 Sampling Theorem

The Fourier transform and the inverse Fourier transform of a bounded, complex-valued, continuous-time signal $x(t)$ are

$$X(f) = \int_{-\infty}^{\infty} x(t)e^{-j2\pi ft}dt \qquad \text{(D.64)}$$

$$x(t) = \int_{-\infty}^{\infty} X(f)e^{j2\pi ft}df. \qquad \text{(D.65)}$$

A sample of this signal is $x_n = x(nT)$, where $1/T$ is the sampling rate and n is an integer. The *discrete-time Fourier transform (DTFT) is defined as*

$$X(e^{j\theta}) = \sum_{n=-\infty}^{\infty} x_n e^{-jn\theta} \qquad \text{(D.66)}$$

if the infinite sum converges to a finite value for all values of θ. A sufficient condition for convergence is

$$\sum_{n=-\infty}^{\infty} |x_n| < \infty \qquad \text{(D.67)}$$

which implies that $X(e^{j\theta})$ is uniformly convergent over $[-\pi, \pi]$. Equation (D.66) indicates that $X(e^{j\theta})$ is a periodic function of θ with period 2π.

Let $\mathcal{Z}$ denote the set of all integers. The *inverse DTFT* of $X(e^{j\theta})$ is

$$x_n = \frac{1}{2\pi}\int_{-\pi}^{\pi} X(e^{j\theta})e^{jn\theta}d\theta, \; n \in \mathcal{Z} \qquad \text{(D.68)}$$

which is verified by the direct substitution of (D.66) into the integrand and an evaluation. Since the complex exponentials constitute a complete set of orthonormal basis functions over $[-\pi, \pi]$, a uniqueness property can be shown [6]: if $X_1(\theta)$ and $X_2(\theta)$ have the same inverse DTFT, then $X_1(\theta) = X_2(\theta)$ for almost all θ.

The *sampling theorem* relates $x(t)$ and x_n in the frequency domain, indicating that $X(e^{j2\pi fT})$ is the sum of shifted versions of $X(f)$.

Sampling Theorem *The discrete-time and continuous-time Fourier transforms of a bounded continuous-time signal* $x(t)$ *are related by*

$$X(e^{j2\pi fT}) = \frac{1}{T}\sum_{i=-\infty}^{\infty} X\left(f - \frac{i}{T}\right) \tag{D.69}$$

if the series converges uniformly over the interval $f \in [-1/2T, +1/2T]$.

Proof Consider a bounded continuous-time signal $x(t)$ with Fourier transform $X(f)$ and $x_n = x(nT)$. According to the uniqueness property, the theorem can be proved by showing that the inverse DTFT of the right side of (D.69) for all $n \in \mathcal{Z}$ is x_n. Let I_n denote this inverse DTFT for $n \in \mathcal{Z}$. Using the uniform convergence to interchange the summation and integration, we obtain

$$\begin{aligned} I_n &= \int_{-1/2T}^{1/2T}\left[\sum_{i=-\infty}^{\infty} X\left(f - \frac{i}{T}\right)\right] e^{j2\pi nfT}\,df \\ &= \sum_{i=-\infty}^{\infty}\int_{-1/2T}^{1/2T} X\left(f - \frac{i}{T}\right) e^{j2\pi nfT}\,df, \quad n \in \mathcal{Z}. \end{aligned} \tag{D.70}$$

Changing the integration variable in this equation and using $e^{j2\pi ni} = 1$ when n and i are integers, we obtain

$$I_n = \sum_{i=-\infty}^{\infty}\int_{(2i-1)/2T}^{(2i+1)/2T} X(f)e^{j2\pi nfT}\,df \tag{D.71}$$

$$= \int_{-\infty}^{\infty} X(f)e^{j2\pi nfT}\,df \tag{D.72}$$

$$= x(nT) = x_n, \quad n \in \mathcal{Z} \tag{D.73}$$

which completes the proof. □

If a continuous-time signal $x(t)$ with Fourier transform $X(f)$ is *bandlimited* so that $X(f) = 0$ for $|f| > 1/2T$, then (D.69) indicates that

$$X(e^{j2\pi fT}) = \frac{1}{T}X(f), \quad -\frac{1}{2T} \le f \le \frac{1}{2T}. \tag{D.74}$$

Substitution of this equation into (D.65) and the use of (D.66) yield

$$\begin{aligned} x(t) &= \int_{-1/2T}^{1/2T} X(f)e^{j2\pi ft}df = T\int_{-1/2T}^{1/2T} X(e^{j2\pi fT})e^{j2\pi ft}df \\ &= T\int_{-1/2T}^{1/2T} \sum_{n=-\infty}^{\infty} x_n e^{-j2\pi fnT}e^{j2\pi ft}df \\ &= T\sum_{n=-\infty}^{\infty} x_n \int_{-1/2T}^{1/2T} e^{j2\pi f(t-nT)}df. \end{aligned} \tag{D.75}$$

Evaluating the final integral, we obtain the formula for the exact reconstruction of a bandlimited $x(t)$ from its samples:

$$x(t) = \sum_{n=-\infty}^{\infty} x_n \, sinc\left(\frac{t-nT}{T}\right), \quad X(f) = 0 \text{ for } |f| > 1/2T \tag{D.76}$$

where $sinc\,(x) = \sin(\pi x)/\pi x$. This equation implies that a signal with a one-sided bandwidth W that is sampled at the rate $1/T > 2W$ is uniquely defined by its samples. If the sampling rate is not high enough that $X(f) = 0$ for $|f| > 1/2T$, then the terms in the sum in (D.69) overlap, which is called *aliasing,* and the samples may not specify a unique continuous-time signal.

Appendix E
Probability Distribution Functions

E.1 Chi-Squared Distribution

Consider the random variable

$$Z = \sum_{i=1}^{N} X_i^2 \tag{E.1}$$

where the $\{X_i\}$ are independent Gaussian random variables with means $\{m_i\}$ and common variance σ^2. The random variable Z is said to have a *noncentral chi-squared (also chi-square) distribution* with N degrees of freedom and a *noncentral parameter*

$$\lambda = \sum_{i=1}^{N} m_i^2. \tag{E.2}$$

To derive the density function of Z, we first note that each X_i has the density function

$$f_{X_i}(x) = \frac{1}{\sqrt{2\pi}\sigma} \exp\left[-\frac{(x-m_i)^2}{2\sigma^2}\right]. \tag{E.3}$$

From elementary probability, the density function of $Y_i = X_i^2$ is

$$f_{Y_i}(x) = \frac{1}{2\sqrt{x}}[f_{X_i}(\sqrt{x}) + f_{X_i}(-\sqrt{x})]\, u(x) \tag{E.4}$$

where $u(x)$ is the unit step function: $u(x) = 1,\ x \geq 0$, and $u(x) = 0, x < 0$. Substituting (E.3) into (E.4), expanding the exponentials, and simplifying, we obtain

D. Torrieri, *Principles of Spread-Spectrum Communication Systems*,
https://doi.org/10.1007/978-3-319-70569-9

the density function

$$f_{Y_i}(x) = \frac{1}{\sqrt{2\pi x}\sigma} \exp\left(-\frac{x+m_i^2}{2\sigma^2}\right) \cosh\left(\frac{m_i\sqrt{x}}{\sigma^2}\right) u(x). \tag{E.5}$$

Characteristic functions, moment-generating functions, and Laplace transforms all uniquely determine distribution functions. Since Y_i assumes only nonnegative values, it is convenient to use the Laplace transform. The *Laplace transform* of a continuous nonnegative random variable Y is defined as (Appendix B.2)

$$\mathcal{L}(s) = E[e^{-sY}] = \int_0^\infty f_Y(x)e^{-sx}dx \tag{E.6}$$

where $f_Y(x)$ is the density function of Y. To evaluate the Laplace transform of $Y_i = X_i^2$, we substitute (E.5) into (E.6), change the integration variable by setting $y = \sqrt{(1+2\sigma^2 s)x/2\sigma^2}$, expand the range of integration because $\cosh(\cdot)$ is an even function, and obtain

$$\mathcal{L}_i(s) = \frac{1}{\sqrt{(1+2\sigma^2 s)\pi}} \int_{-\infty}^\infty \exp\left(-\frac{m_i^2}{2\sigma^2} - y^2\right) \cosh\left(\frac{ym_i\sqrt{2}}{\sigma\sqrt{1+2\sigma^2 s}}\right) dy. \tag{E.7}$$

Using $\cosh(z) = (e^z + e^{-z})/2$, separating the integral into two integrals, completing the squares in the arguments of the exponentials, and then observing that a Gaussian density function must integrate to unity, we obtain the Laplace transform of Y_i:

$$\mathcal{L}_i(s) = \frac{\exp\left(\frac{-sm_i^2}{1+2\sigma^2 s}\right)}{(1+2\sigma^2 s)^{1/2}}, \quad Re(s) > -\frac{1}{2\sigma^2}. \tag{E.8}$$

The Laplace transform of a sum of independent random variables is equal to the product of the individual Laplace transforms. Because Z is the sum of the $\{Y_i\}$, the Laplace transform of Z is

$$\mathcal{L}_Z(s) = \frac{\exp\left(\frac{-s\lambda}{1+2\sigma^2 s}\right)}{(1+2\sigma^2 s)^{N/2}}, \quad Re(s) > -\frac{1}{2\sigma^2} \tag{E.9}$$

where we have used (E.2).

The noncentral chi-squared density with N degrees of freedom and noncentral parameter λ is the unique density (see Section B.2) with the Laplace transform given by (E.9). As shown subsequently, this density is

$$f_Z(x) = \frac{1}{2\sigma^2}\left(\frac{x}{\lambda}\right)^{(N-2)/4} \exp\left[-\left(\frac{x+\lambda}{2\sigma^2}\right)\right] I_{N/2-1}\left(\frac{\sqrt{x\lambda}}{\sigma^2}\right) u(x) \tag{E.10}$$

where $I_n(\cdot)$ is the modified Bessel function of the first kind and order n (Appendix H.3). Thus, the noncentral chi-squared distribution is

$$F_Z(x) = \int_0^x \frac{1}{2\sigma^2}\left(\frac{y}{\lambda}\right)^{(N-2)/4} \exp\left(-\frac{y+\lambda}{2\sigma^2}\right) I_{N/2-1}\left(\frac{\sqrt{y\lambda}}{\sigma^2}\right) dy\,, \quad x \geq 0. \tag{E.11}$$

To prove that $F_Z(\infty) = 1$, and hence that $F_Z(x)$ is a legitimate distribution function, we first need to establish an integral identity:

$$\int_0^\infty x^{(N-2)/4} \exp(-x)\, I_{N/2-1}\left(\sqrt{4xa}\right) dx = a^{(N-2)/4} \exp(a)\,, \quad a \geq 0. \tag{E.12}$$

To prove (E.12), we substitute (H.13) of Appendix H.3 into the integral, apply the monotone convergence theorem to interchange the integral and infinite summation, apply (H.1), simplify, and then recognize the power series expansion of $\exp(a)$. A change of variables in (E.12) then proves that $F_Z(\infty) = 1$.

To prove that $f_Z(x)$ is given by (E.10), we substitute (E.10) into (E.6), change variables, and use (E.12) to obtain (E.9).

If N is even so that $N/2$ is an integer, then a change of variables in (E.11) yields

$$F_Z(x) = 1 - Q_{N/2}\left(\frac{\sqrt{\lambda}}{\sigma}, \frac{\sqrt{x}}{\sigma}\right), \quad x \geq 0 \tag{E.13}$$

where $Q_m(\alpha, \beta)$ is the *generalized Marcum Q-function* (Appendix H.4). The moments of Z can be obtained by using (E.1) and the properties of independent Gaussian random variables. The mean and variance of Z are

$$E[Z] = N\sigma^2 + \lambda, \quad \sigma_z^2 = 2N\sigma^4 + 4\lambda\sigma^2 \tag{E.14}$$

where σ^2 is the common variance of the $\{X_i\}$. Alternatively, the moments of Z can be obtained by applying Theorem B1 of Appendix B.2 and (E.9).

From (E.9), it follows that the sum of two independent noncentral chi-squared random variables with N_1 and N_2 degrees of freedom, noncentral parameters λ_1 and λ_2, respectively, and the same parameter σ^2 is a noncentral chi-squared random variable with $N_1 + N_2$ degrees of freedom and noncentral parameter $\lambda_1 + \lambda_2$.

E.2 Central Chi-Squared Distribution

To determine the density function of Z when the $\{X_i\}$ have zero-means, we substitute (H.13) of Appendix H.3 into (E.10) and then take the limit as $\lambda \to 0$.

We obtain the *central chi-squared density* with N degrees of freedom:

$$f_Z(x) = \frac{1}{(2\sigma^2)^{N/2}\Gamma(N/2)} x^{N/2-1} \exp\left(-\frac{x}{2\sigma^2}\right) u(x). \tag{E.15}$$

The central chi-squared distribution with N degrees of freedom, which is concentrated on the positive x-axis, is

$$F_Z(x) = \frac{1}{(2\sigma^2)^{N/2}\Gamma(N/2)} \int_0^x y^{N/2-1} \exp\left(-\frac{y}{2\sigma^2}\right) dy, \ x \geq 0. \tag{E.16}$$

In terms of the incomplete gamma function $\gamma(a, b)$ defined by (H.6), we have

$$F_Z(x) = \frac{\gamma\left(\frac{N}{2}, \frac{x}{2\sigma^2}\right)}{\Gamma(N/2)} u(x). \tag{E.17}$$

The moments of Z may be obtained by direct integration using (E.15) and (H.1) of Appendix H.1. The mean and variance of Z are

$$E[Z] = N\sigma^2, \ \ \sigma_z^2 = 2N\sigma^4. \tag{E.18}$$

If N is even, so that $N/2$ is an integer, then integrating (E.16) by parts $N/2 - 1$ times yields

$$F_Z(x) = 1 - \exp\left(-\frac{x}{2\sigma^2}\right) \sum_{i=0}^{N/2-1} \frac{1}{i!}\left(\frac{x}{2\sigma^2}\right)^i, \ \ x \geq 0. \tag{E.19}$$

If $N = 1$, then changing the integration variable to $z = \sqrt{y}$ in (E.16) and using (H.20), we obtain

$$F_Z(x) = \left[1 - 2Q\left(\frac{\sqrt{x}}{\sigma}\right)\right] u(x), \ N = 1. \tag{E.20}$$

E.3 Rice Distribution

Consider the random variable

$$R = \sqrt{X_1^2 + X_2^2} \tag{E.21}$$

where X_1 and X_2 are independent Gaussian random variables with means m_1 and m_2, respectively, and a common variance σ^2. The distribution function of R must satisfy $F_R(r) = F_Z(r^2)$, where $Z = X_1^2 + X_2^2$ has a chi-squared distribution with two degrees of freedom. Therefore, (E.13) with $N = 2$ implies that R has a Rice distribution:

$$F_R(r) = 1 - Q_1\left(\frac{\sqrt{\lambda}}{\sigma}, \frac{r}{\sigma}\right), \quad r \geq 0 \tag{E.22}$$

where $\lambda = m_1^2 + m_2^2$. The *Rice density*, which may be obtained by differentiation of (E.22), is

$$f_R(r) = \frac{r}{\sigma^2}\exp\left(-\frac{r^2+\lambda}{2\sigma^2}\right) I_0\left(\frac{r\sqrt{\lambda}}{\sigma^2}\right) u(r). \tag{E.23}$$

The moments of even order can be derived from (E.21) and the moments of the independent Gaussian random variables. The second moment is

$$E[R^2] = 2\sigma^2 + \lambda. \tag{E.24}$$

In general, moments of the Rice distribution are given by an integration over the density function in (E.23). Substituting (H.13) of Appendix H.3 into the integrand, interchanging the summation and integration, changing the integration variable, and using (H.1) of Appendix H.1, we obtain a series that is recognized as a special case of the confluent hypergeometric function. Thus,

$$E[R^n] = (2\sigma^2)^{n/2}\exp\left(-\frac{\lambda}{2\sigma^2}\right)\Gamma\left(1+\frac{n}{2}\right)\, {}_1F_1\left(1+\frac{n}{2}, 1; \frac{\lambda}{2\sigma^2}\right), \quad n \geq 0 \tag{E.25}$$

where ${}_1F_1(\alpha, \beta; x)$ is the *confluent hypergeometric function defined by* (H.27) of Appendix H.5.

The Rice density often arises in the context of a transformation of variables. Let X_1 and X_2 represent independent Gaussian random variables with common variance σ^2 and means λ and zero respectively. Let R and Θ be implicitly defined by $X_1 = R\cos\Theta$ and $X_2 = R\sin\Theta$. Then (E.21) and $\Theta = \tan^{-1}(X_2/X_1)$ describe a transformation of variables. Therefore, the joint density function of R and Θ is

$$f_{R,\Theta}(r,\theta) = \frac{r}{2\pi\sigma^2}\exp\left(-\frac{r^2 - 2r\lambda\cos\theta + \lambda^2}{2\sigma^2}\right), \quad r \geq 0, \quad |\theta| \leq \pi. \tag{E.26}$$

The density function of R is obtained by integration over θ. Using (H.15) of Appendix H.3, this density function reduces to the Rice density (E.23). The density function of the angle Θ is obtained by integrating (E.26) over r. Completing the square of the argument in (E.26), changing variables, and using the definition of the Gaussian Q-function or complementary error function (Appendix H.4), we obtain

$$f_{\Theta}(\theta) = \frac{1}{2\pi} \exp\left(-\frac{\lambda^2}{2\sigma^2}\right) + \frac{\lambda \cos\theta}{\sqrt{2\pi}\sigma} \exp\left(-\frac{\lambda^2 \sin^2\theta}{2\sigma^2}\right)\left[1 - Q\left(\frac{\lambda\cos\theta}{\sigma}\right)\right],$$
$$|\theta| \le \pi. \tag{E.27}$$

Since (E.26) cannot be written as the product of (E.23) and (E.27), the random variables R and Θ are not independent.

E.4 Rayleigh Distribution

A Rayleigh-distributed random variable is defined by (E.21) when X_1 and X_2 are independent Gaussian random variables with zero-means and a common variance σ^2. Since $F_R(r) = F_Z(r^2)$, where Z has a central chi-squared distribution with two degrees of freedom, (E.19) with $N = 2$ implies that the *Rayleigh distribution* is

$$F_R(r) = 1 - \exp\left(-\frac{r^2}{2\sigma^2}\right), \quad r \ge 0. \tag{E.28}$$

The *Rayleigh density*, which may be obtained by differentiation of (E.28), is

$$f_R(r) = \frac{r}{\sigma^2} \exp\left(-\frac{r^2}{2\sigma^2}\right) u(r). \tag{E.29}$$

By a change of the variable in the defining integral, any moment of R can be expressed in terms of the gamma function:

$$E[R^n] = (2\sigma^2)^{n/2}\Gamma\left(1 + \frac{n}{2}\right). \tag{E.30}$$

Using the properties of the gamma function (Appendix H.1), we obtain the mean and the variance of a Rayleigh-distributed random variable:

$$E[R] = \sqrt{\frac{\pi}{2}}\sigma, \quad \sigma_R^2 = \left(2 - \frac{\pi}{2}\right)\sigma^2. \tag{E.31}$$

Since X_1 and X_2 have zero means, the joint density function of the random variables $R = \sqrt{X_1^2 + X_2^2}$ and $\Theta = \tan^{-1}(X_2/X_1)$ is given by (E.26) with $\lambda = 0$. Therefore,

$$f_{R,\Theta}(r, \theta) = \frac{r}{2\pi\sigma^2} \exp\left(-\frac{r^2}{2\sigma^2}\right), \quad r \ge 0, \quad |\theta| \le \pi. \tag{E.32}$$

Integration over θ yields (E.29), and integration over r yields the uniform density function:

$$f_{\Theta}(\theta) = \frac{1}{2\pi}, \quad |\theta| \leq \pi. \tag{E.33}$$

Since (E.32) equals the product of (E.29) and (E.33), the random variables R and Θ are independent. In terms of these random variables, $X_1 = R\cos\Theta$ and $X_2 = R\sin\Theta$. A straightforward calculation using the independence and density functions of R and Θ verifies that X_1 and X_2 are zero-mean, independent, Gaussian random variables with common variance σ^2.

E.5 Exponential Distribution

A random variable X has the *exponential distribution* with parameter $\alpha > 0$ if it has the density function

$$f_X(x) = \alpha e^{-\alpha x} u(x). \tag{E.34}$$

The corresponding distribution function is

$$F_X(x) = 1 - e^{-\alpha x}, \ x \geq 0. \tag{E.35}$$

Integrations using (H.3) and (H.2) yield

$$E[X] = \frac{1}{\alpha}, \quad var(X) = \frac{1}{\alpha^2}. \tag{E.36}$$

Since the square of a Rayleigh-distributed random variable may be expressed as $R^2 = X_1^2 + X_2^2$, where X_1 and X_2 are zero-mean, independent, Gaussian random variables with common variance σ^2, R^2 has a central chi-squared distribution with two degrees of freedom. Therefore, (E.15) with $N = 2$ indicates that the square of a Rayleigh-distributed random variable has an exponential density with mean $2\sigma^2$.

E.6 Gamma Distribution

A random variable X has the *gamma distribution* with parameters $\alpha, \beta > 0$ if it has the density

$$f(x; \alpha, \beta) = \frac{\alpha^\beta}{\Gamma(\beta)} x^{\beta-1} e^{-\alpha x} u(x). \tag{E.37}$$

Successive integrations by parts yield the corresponding distribution function:

$$F(x;\alpha,\beta) = [1 - \Gamma(\beta,\alpha x)]\, u(x) \tag{E.38}$$

where the incomplete gamma function is defined by (H.5). Integrations of (E.37) using (H.3) and (H.2) determine the moments of X. The mean and variance are

$$E[X] = \frac{\beta}{\alpha}, \quad var\,(X) = \frac{\beta}{\alpha^2}\,. \tag{E.39}$$

Let $\{f\,(x;\alpha,\beta)\,,\beta > 0\}$ denote the family of gamma densities with the same parameter $\alpha > 0$ but different values of $\beta > 0$. Let $Z = X_1 + X_2$ denote the sum of two independent random variables with density functions $f\,(x;\alpha,\beta_1)$ and $f\,(x;\alpha,\beta_2)$ respectively. From elementary probability theory, it follows that the density function of Z is determined by the convolution of these two density functions:

$$\begin{aligned}[f\,(\cdot;\alpha,\beta_1) * f\,(\cdot;\alpha,\beta_2)]\,(x) &= \int_0^x f\,(y;\alpha,\beta_1) f\,(x-y;\alpha,\beta_2)\,dy \\ &= \frac{\alpha^{\beta_1+\beta_2} e^{-\alpha x}}{\Gamma\,(\beta_1)\,\Gamma\,(\beta_2)} \int_0^x y^{\beta_1-1}\,(x-y)^{\beta_2-1}\,dy \\ &= \frac{\alpha^{\beta_1+\beta_2} x^{\beta_1+\beta_2-1} e^{-\alpha x}}{\Gamma\,(\beta_1)\,\Gamma\,(\beta_2)} B\,(\beta_1,\beta_2)\end{aligned} \tag{E.40}$$

where the star $\star$ denotes the convolution operation, and the beta function $B\,(\cdot,\cdot)$ is defined by (H.10) of Appendix H.2. Substituting (H.11) into (E.40) and using (E.37), we obtain

$$[f\,(\cdot;\alpha,\beta_1) * f\,(\cdot;\alpha,\beta_2)]\,(x) = f\,(x;\alpha,\beta_1+\beta_2) \tag{E.41}$$

which indicates that the family of gamma densities $\{f\,(x;\alpha,\beta)\,,\beta > 0\}$ is closed under convolution. By mathematical induction, if

$$Z = \sum_{i=1}^{N} X_i \tag{E.42}$$

is the sum of N independent random variables, and X_i has a gamma density with parameters α and β_i, then Z has a gamma density with parameters α and $\sum_{i=1}^{N} \beta_i$.

Let Z in (E.42) denote the sum of N independent, exponentially distributed random variables with the same parameter $\alpha > 0$. Since the exponential density is equal to $f(x;\alpha,1)$, (E.41) implies that the density function of Z is the gamma density $f(x;\alpha,N)$.

Equations (E.5) and (H.4) indicate that a central chi-squared density with *one* degree of freedom is equal to $f\left(x; 1/2\sigma^2, 1/2\right)$. Therefore, if Z in (E.42) denotes the sum of N independent central chi-squared random variables with the same parameter σ^2, then the application of (E.41) provides another proof that the density function of Z is the gamma density $f\left(x; 1/2\sigma^2, N/2\right)$, as indicated by (E.15).

Appendix F
Orthonormal Functions and Parameter Estimation

F.1 Deterministic Functions

A complete normed vector space is one in which every Cauchy sequence converges at a member of the vector space. A *Hilbert space* is a complete vector space with an inner product and a norm defined by the inner product. The *signal space* $L^2[0,T]$ is the Hilbert space of complex-valued functions $f(t)$ such that $|f(t)|^2$ is integrable over $[0,T]$. The inner product of functions $f(t)$ and $g(t)$ in $L^2[0,T]$ is

$$\langle f(t), g(t)\rangle = \int_0^T f(t)g^*(t)dt \tag{F.1}$$

where the asterisk denotes the complex conjugate. The norm of $f(t) \in L^2[0,T]$ is denoted by $\|f(t)\|$, and its square is

$$\|f(t)\|^2 = \langle f(t), f(t)\rangle = \int_0^T |f(t)|^2 dt < \infty. \tag{F.2}$$

These definitions indicate that the norm is nonnegative,

$$\langle f(t), g(t)\rangle = \langle g(t), f(t)\rangle^* \tag{F.3}$$

and

$$\langle f(t) + h(t), g(t)\rangle = \langle f(t), g(t)\rangle + \langle h(t), g(t)\rangle . \tag{F.4}$$

Functions that are equal almost everywhere are considered equivalent, and $\|f(t)\| = 0$ implies that $f(t) = 0$.

D. Torrieri, *Principles of Spread-Spectrum Communication Systems*,
https://doi.org/10.1007/978-3-319-70569-9

Let $\lambda = -\langle f(t), g(t)\rangle / \|f(t)\|$, where $\|f(t)\| \neq 0$. Then

$$\begin{aligned} 0 &\le \|\lambda f(t) + g(t)\|^2 \\ &= \left|\lambda \|f(t)\| + \frac{\langle f(t), g(t)\rangle}{\|f(t)\|}\right|^2 - \frac{|\langle f(t), g(t)\rangle|^2}{\|f(t)\|^2} + \|g(t)\|^2 \\ &= -\frac{|\langle f(t), g(t)\rangle|^2}{\|f(t)\|^2} + \|g(t)\|^2 \end{aligned} \tag{F.5}$$

which implies the *Cauchy-Schwarz inequality for an inner product*:

$$|\langle f(t), g(t)\rangle| \le \|f(t)\| \, \|g(t)\| . \tag{F.6}$$

This inequality remains valid when $\|f(t)\| = 0$.

Let $\{\phi_i(t)\}$ denote a countable set of functions $\phi_i(t), \phi_i(t), \ldots$ in $L^2[0, T]$. The set $\{\phi_i(t)\}$ is orthonormal if

$$\langle \phi_i(t), \phi_k(t)\rangle = \delta_{ik}, \quad i \neq k \tag{F.7}$$

where $\delta_{ii} = 1$, and $\delta_{ik} = 0$, $k \neq i$. An orthonormal subset B of $L^2[0, T]$ is a *basis* for $L^2[0, T]$ if B is not a proper subset of any other orthonormal subset of $L^2[0, T]$. The subspace $S(B)$ *spanned* by B is the smallest closed subspace of $L^2[0, T]$ containing all the orthonormal basis functions of B. An orthonormal basis B is *complete* if $S(B) = L^2[0, T]$.

Using functional analysis [60], many specific complete orthonormal bases can be constructed. An example of an orthonormal basis in $L^2[0, T]$ comprises the complex exponential functions in a Fourier series representation of a function; that is, the basis is

$$\left\{\sqrt{1/T}\exp(j2\pi kt/T), \ k \ge 1\right\}.$$

An example of a real-valued orthonormal basis in $L^2[0, T]$ comprises the sine and cosine functions in a Fourier series representation of a function; that is, the basis is

$$\left\{\sqrt{2/T}\sin(2\pi kt/T), \ \sqrt{2/T}\cos(2\pi kt/T), \ k \ge 1\right\}.$$

A countable sequence of functions $\{f_n(t), n \ge 1\}$ in $L^2[0, T]$ converges to a function $f(t)$ in $L^2[0, T]$ if $\|f_n(t) - f(t)\| \to 0$ as $n \to \infty$. Consider a complete set of *orthonormal basis functions* $\{\phi_i(t)\}$ for $L^2[0, T]$. We define the finite expansion of $f(t) \in L^2[0, T]$ in terms of the first N basis functions as

$$S(f,N) = \sum_{i=1}^{N} f_i \phi_i(t) \tag{F.8}$$

where the expansion coefficients are defined as

$$f_i = \langle f(t), \phi_i(t) \rangle, \; i \geq 1. \tag{F.9}$$

It can be shown [6, 60] that $\|f(t) - S(f,N)\| \to 0$ as $N \to \infty$, and hence

$$f(t) = \lim_{N\to\infty} \sum_{i=1}^{N} f_i \phi_i(t) = \sum_{i=1}^{\infty} f_i \phi_i(t). \tag{F.10}$$

Suppose that $f(t)$, $g(t) \in L^2[0,T]$, $\|f(t) - S(f,N)\| \to 0$ as $N \to \infty$, and $\|g(t) - S(g,N)\| \to 0$ as $N \to \infty$. The linearity of the inner product gives

$$\begin{aligned}\langle f(t), g(t)\rangle &= \langle f(t) - S(f,N), g(t)\rangle - \langle f(t) - S(f,N), g(t) - S(g,N)\rangle \\ &\quad + \langle f(t), g(t) - S(g,N)\rangle + \langle S(f,N), S(g,N)\rangle . \end{aligned} \tag{F.11}$$

Applying Cauchy-Schwarz inequality successively to the first three terms on the right and using $\|f(t) - S(f,N)\| \to 0$ and $\|g(t) - S(g,N)\| \to 0$ as $N \to \infty$, it follows that each of these terms approaches zero as $N \to \infty$. Therefore, taking $N \to \infty$ and using (F.8), we obtain

$$\begin{aligned}\langle f(t), g(t)\rangle &= \lim_{N\to\infty} \left\langle \sum_{i=1}^{N} f_i \phi_i(t), \sum_{i=1}^{N} g_i \phi_i(t) \right\rangle \\ &= \lim_{N\to\infty} \sum_{i=1}^{N} f_i g_i^* \\ &= \sum_{i=1}^{\infty} f_i g_i^* \end{aligned} \tag{F.12}$$

and hence

$$\|f(t)\|^2 = \sum_{i=1}^{\infty} |f_i|^2 . \tag{F.13}$$

The *Gram-Schmidt orthonormalization* procedure starts with a countable set of linearly independent functions $\{g_i(t)\}$ and then uses them in the construction of a complete set of orthonormal basis functions $\{\phi_i(t)\}$, $i = 1, 2, \ldots$, in $L^2[0,T]$. In the procedure, each new function $\phi_i(t)$, $i \leq N$, is a linear combination of $g_i(t)$ and the previously constructed $\{\phi_k(t)\}$, $k < i \leq N$. The linear combinations are selected to ensure orthonormality. The defining equations of the procedure are

$$\phi_i(t) = \alpha_i \left[g_i(t) - \sum_{k=1}^{N} \langle g_i(t), \phi_k(t) \rangle \phi_k(t) \right] \tag{F.14}$$

where

$$\alpha_i = \left[\| g_i(t) \|^2 - \sum_{k=1}^{N} [\langle g_i(t), \phi_k(t) \rangle]^2 \right]^{-1/2}. \tag{F.15}$$

F.2 White Gaussian Noise

The *Dirac delta function* $\delta(t, s)$ is defined as the nonstandard function such that for any continuous function $h(s)$, we have

$$\int_{-\infty}^{\infty} h(s)\, \delta(t-s)\, ds = h(t). \tag{F.16}$$

For each $t, s \in [0, T]$, we expand $\delta(t-s)$ in terms of a complete set of orthonormal basis functions as

$$\delta(t-s) = \sum_{i=1}^{\infty} \delta_i(t)\, \phi_i(s) \tag{F.17}$$

where

$$\delta_i(t) = \langle \delta(t-s), \phi_i(s) \rangle = \int_0^T \delta(t-s)\, \phi_i^*(s) ds. \tag{F.18}$$

Applying (F.16) to evaluate the integral and substituting the result into (F.17), we obtain the identity

$$\delta(t-s) = \sum_{i=1}^{\infty} \phi_i^*(t) \phi_i(s). \tag{F.19}$$

Consider zero-mean, white Gaussian noise $n(t)$ over $[0, T]$. Its autocorrelation function is

$$E[n(t)n(s)] = \frac{N_0}{2} \delta(t-s). \tag{F.20}$$

Given a complete set of orthonormal basis functions $\{\phi_i(t)\}$ in $L^2[0,T]$, we define the projection of $n(t)$ onto the subspace spanned by the first N basis functions as

$$S(n,N,t) = \sum_{i=1}^{N} n_i\phi_i(t) \tag{F.21}$$

where

$$n_i = \langle n(t), \phi_i(t)\rangle = \int_0^T n(t)\phi_i^*(t)dt,\ i \geq 1. \tag{F.22}$$

Using $E[n(t)] = 0$, (F.16), (F.20), and (F.7) in (F.22), we obtain

$$E[n_i] = 0,\ E\left[|n_i|^2\right] = \frac{N_0}{2},\ i \geq 1 \tag{F.23}$$

$$E\left[n_i n_k^*\right] = 0,\ i \neq k. \tag{F.24}$$

The white Gaussian noise may be represented by the expansion

$$n(t) = \sum_{i=1}^{\infty} n_i\phi_i(t) \tag{F.25}$$

in the sense that $E[S^*(n,N,t)\,S(n,N,s)] \to \delta(t-s)$ as $N \to \infty$. To prove this result, we calculate

$$\begin{aligned}\lim_{N\to\infty} E[S^*(n,N,t)\,S(n,N,s)] &= \lim_{N\to\infty} E[\sum_{i=1}^{N} n_i^*\phi_i^*(t)\sum_{i=1}^{N} n_k\phi_k(s)]\\ &= \frac{N_0}{2}\sum_{i=1}^{\infty}\phi_i^*(t)\phi_i(s)\\ &= \delta(t-s)\end{aligned} \tag{F.26}$$

where (F.24) and (F.23) are used in the second equality, and (F.20) is used in the third equality.

The approximating Lebesgue or Riemann sums of the real and imaginary parts of the integral in (F.22) are sums of independent, zero-mean Gaussian random variables because the noise is white. Therefore, the real and imaginary components of the $\{n_i\}$ are *jointly zero-mean Gaussian random variables* (Theorem A4, Appendix A.1). If the basis functions are real-valued, then the $\{n_i\}$ are real-valued, (F.24) implies that they are uncorrelated, and hence they are statistically independent of each other.

F.3 Estimation of Waveform Parameters

Consider the estimation of waveform parameters that are components of the vector $\boldsymbol{\theta}$. The observed signal is

$$r(t) = s\left(t, \boldsymbol{\theta}\right) + n\left(t\right), \; 0 \le t \le T \tag{F.27}$$

where the real-valued signal $s\left(t, \boldsymbol{\theta}\right)$ has a known waveform except for $\boldsymbol{\theta}$, and $n\left(t\right)$ is zero-mean, white Gaussian noise with autocorrelation function given by (F.20). For all values of the waveform parameters, $s(t, \boldsymbol{\theta})$ belongs to the signal space $L^2\left[0, T\right]$ of complex-valued functions f such that $|f|^2$is integrable over $[0, T]$.

Let $\{\phi_i(t)\}$ denote a complete set of real-valued orthonormal basis functions in $L^2\left[0, T\right]$. The orthonormal expansions of the functions in (F.27) are

$$r\left(t\right) = \sum_{i=1}^{\infty} r_i\phi_i(t), \; s\left(t, \boldsymbol{\theta}\right) = \sum_{i=1}^{\infty} s_i\left(\boldsymbol{\theta}\right)\phi_i(t), \; n\left(t\right) = \sum_{i=1}^{\infty} n_i\phi_i(t) \tag{F.28}$$

with coefficients

$$r_i = \langle r\left(t\right), \phi_i\left(t\right)\rangle = \int_0^T r(t)\phi_i(t)dt = s_i\left(\boldsymbol{\theta}\right) + n_i \tag{F.29}$$

$$s_i\left(\boldsymbol{\theta}\right) = \langle s\left(t, \boldsymbol{\theta}\right), \phi_i\left(t\right)\rangle = \int_0^T s\left(t, \boldsymbol{\theta}\right)\phi_i(t)dt \tag{F.30}$$

$$n_i = \langle n\left(t\right), \phi_i\left(t\right)\rangle = \int_0^T n(t)\phi_i(t)dt. \tag{F.31}$$

Each n_i is a statistically independent, real-valued, Gaussian random variable, and (F.23) indicates that the conditional density function of r_i is

$$f(r_i|\boldsymbol{\theta}) = \frac{1}{\sqrt{\pi N_0}}\exp\left[-\frac{\left(r_i - s_i\left(\boldsymbol{\theta}\right)\right)^2}{N_0}\right]. \tag{F.32}$$

To avoid convergence problems arising from the assumption of white noise, we initially consider the $N \times 1$ vector $\mathbf{r} = [r_1 \, r_2 \, \ldots \, r_N]^T$. Applying (F.32) and the statistical independence of the r_i, we obtain the log-likelihood function

$$\ln f\left(\mathbf{r}|\boldsymbol{\theta},N\right) = -\ln\sqrt{\pi N_0} - \frac{1}{N_0}\sum_{i=1}^{N} r_i^2 + \frac{2}{N_0}\sum_{i=1}^{N} r_i s_i\left(\boldsymbol{\theta}\right) - \frac{1}{N_0}\sum_{i=1}^{N} s_i^2\left(\boldsymbol{\theta}\right). \tag{F.33}$$

The first sum may not converge, but it is irrelevant to the maximum-likelihood estimation and may be dropped. The factor N_0 then also becomes irrelevant and may be similarly dropped. Taking $N \to \infty$, and applying (F.12), we obtain the *sufficient statistic for maximum-likelihood estimation of* $\boldsymbol{\theta}$:

$$\Lambda_s[r(t)] = 2\int_0^T r(t)s(t,\boldsymbol{\theta})dt - \int_0^T s^2(t,\boldsymbol{\theta})dt. \tag{F.34}$$

If $s(t,\boldsymbol{\theta})$ depends on a random vector ϕ, we base the maximum-likelihood estimation on the *average log-likelihood ratio* defined as $E_\phi[\ln f(\mathbf{r}|\boldsymbol{\theta},N)]$, where $E_\phi[\cdot]$ denotes the expected value with respect to ϕ. Therefore, the sufficient statistic becomes

$$\Lambda_a[r(t)] = E_\phi\left[2\int_0^T r(t)s(t,\boldsymbol{\theta})dt - \int_0^T s^2(t,\boldsymbol{\theta})dt\right] \tag{F.35}$$

and the maximum-likelihood estimator is

$$\widehat{\theta} = \arg\max_{\boldsymbol{\theta}} \Lambda_a[r(t)]. \tag{F.36}$$

F.4 Cramer-Rao Inequality

The Cramer-Rao inequality provides a lower bound on the variance of an unbiased estimator. Consider a random vector $\mathbf{X}$ and a single unknown parameter θ. The conditional density function of $\mathbf{X}$ is $f(\mathbf{x}|\theta)$, and the partial derivative $f^{(1)}(\mathbf{x}|\theta)$ with respect to θ exists for all $(\mathbf{x},\theta)$. Assume that there is an integrable function $g(\mathbf{x})$ such that $\left|\widehat{\theta}(\mathbf{x})f^{(1)}(\mathbf{x}|\theta)\right| \le g(\mathbf{x})$ for all $(\mathbf{x},\theta)$ and that $\int_{-\infty}^{\infty} g(\mathbf{x})\,d\mathbf{x} < \infty$. Let $\widehat{\theta}(\mathbf{X})$ denote an estimator of θ. This estimator is *unbiased* if

$$E\left[\widehat{\theta}(\mathbf{X})\right] = \theta = \int_{-\infty}^{\infty} \widehat{\theta}(\mathbf{x})f(\mathbf{x}|\theta)\,d\mathbf{x}. \tag{F.37}$$

Differentiating both sides of the second equality with respect to θ, we obtain

$$1 = \int_{-\infty}^{\infty} \widehat{\theta}(\mathbf{x})\frac{\partial}{\partial\theta}f(\mathbf{x}|\theta)\,d\mathbf{x} = \int_{-\infty}^{\infty} \widehat{\theta}(\mathbf{x})\frac{\partial \ln f(\mathbf{x}|\theta)}{\partial\theta}f(\mathbf{x}|\theta)\,d\mathbf{x} \tag{F.38}$$

where the interchange of the derivative and integration is valid because of the assumption about $g(\mathbf{x})$. Since $f(\mathbf{x}|\theta)$ integrates to unity,

$$\begin{aligned} 0 &= \frac{\partial}{\partial\theta}\int_{-\infty}^{\infty} f(\mathbf{x}|\theta)\,d\mathbf{x} \\ &= \int_{-\infty}^{\infty} \frac{\partial}{\partial\theta}f(\mathbf{x}|\theta)\,d\mathbf{x} \\ &= \int_{-\infty}^{\infty} \frac{\partial \ln f(\mathbf{x}|\theta)}{\partial\theta}f(\mathbf{x}|\theta)\,d\mathbf{x}. \end{aligned} \tag{F.39}$$

Combining (F.38) and (F.39) and using the Cauchy-Schwarz inequality for random variables (Appendix A.2) yields

$$\begin{aligned} 1 &= \int_{-\infty}^{\infty} \left[\widehat{\theta}(\mathbf{x}) - \theta\right] \frac{\partial \ln f(\mathbf{x}|\theta)}{\partial \theta} f(\mathbf{x}|\theta)\, d\mathbf{x} \\ &= E\left\{\left[\widehat{\theta}(\mathbf{X}) - \theta\right] \frac{\partial \ln f(\mathbf{X}|\theta)}{\partial \theta}\right\} \\ &= \left[var\left(\widehat{\theta}\right)\right]^{2} E\left\{\left[\frac{\partial \ln f(\mathbf{X}|\theta)}{\partial \theta}\right]^{2}\right\}^{1/2} \end{aligned} \tag{F.40}$$

which implies *Cramer-Rao inequality for the variance of an unbiased estimator*:

$$var\left(\widehat{\theta}\right) \geq \left(E\left\{\left[\frac{\partial \ln f(\mathbf{X}|\theta)}{\partial \theta}\right]^{2}\right\}\right)^{-1}. \tag{F.41}$$

If the second partial derivative $f^{(2)}(\mathbf{x}|\theta)$ with respect to θ exists for all $(\mathbf{x}, \theta)$, there is an integrable function $h(\mathbf{x})$ such that $\left|f^{(2)}(\mathbf{x}|\theta)\right| \leq h(\mathbf{x})$ for all $(\mathbf{x}, \theta)$, and $\int_{-\infty}^{\infty} h(\mathbf{x})\, d\mathbf{x} < \infty$, then differentiating the final integral in (F.39) gives

$$0 = \int_{-\infty}^{\infty} \left\{\frac{\partial^2 \ln f(\mathbf{x}|\theta)}{\partial \theta^2} + \left[\frac{\partial \ln f(\mathbf{x}|\theta)}{\partial \theta}\right]^{2}\right\} f(\mathbf{x}|\theta)\, d\mathbf{x}. \tag{F.42}$$

Combining this equation with the lower bound in (F.41), we obtain an alternative version of Cramer-Rao inequality:

$$var\left(\widehat{\theta}\right) \geq -\left(E\left[\frac{\partial^2 \ln f(\mathbf{X}|\theta)}{\partial \theta^2}\right]\right)^{-1}. \tag{F.43}$$

We calculate the Cramer-Rao inequality for the observed signal of (F.27), assuming that $\partial s(t,\theta)/\partial\theta$ is in $L^2[0,T]$ and that each coefficient $s_i(\theta)$ of $s(t,\theta)$ is differentiable for all θ. Differentiating (F.33), we obtain

$$\frac{\partial \ln f(\mathbf{r}|\theta, N)}{\partial \theta} = \frac{2}{N_0}\left[\sum_{i=1}^{N} r_i \frac{\partial s_i(\theta)}{\partial \theta} - \sum_{i=1}^{N} s_i(\theta) \frac{\partial s_i(\theta)}{\partial \theta}\right]. \tag{F.44}$$

Let $S(s, N, \theta)$ and $S\left(s^{(1)}, N, \theta\right)$ denote the finite expansions of $s(t,\theta)$ and $\partial s(t,\theta)/\partial\theta$, respectively, in terms of the first N basis functions. Then

$$\frac{\partial S(s, N, \theta)}{\partial \theta} = \sum_{i=1}^{N} \frac{\partial s_i(\theta)}{\partial \theta} \phi_i(t) = S\left(s^{(1)}, N, \theta\right) \tag{F.45}$$

where the orthonormality of the basis functions is used to prove the second equality. Therefore,

$$\frac{\partial s(t,\theta)}{\partial\theta} = \lim_{N\to\infty} S\left(s^{(1)}, N, \theta\right) = \sum_{i=1}^{\infty} \frac{\partial s_i(\theta)}{\partial\theta}\phi_i(t). \tag{F.46}$$

Taking the limit of (F.44) as $N \to \infty$, applying (F.46) and (F.12), and then substituting (F.27) gives

$$\begin{aligned} \frac{\partial \ln f(\mathbf{r}|\theta)}{\partial\theta} &= \lim_{N\to\infty} \frac{\partial \ln f(\mathbf{r}|\theta,N)}{\partial\theta} \\ &= \frac{2}{N_0}\left[\int_0^T r(t)\frac{\partial s(t,\theta)}{\partial\theta}dt - \int_0^T s(t,\theta)\frac{\partial s(t,\theta)}{\partial\theta}dt\right] \\ &= \frac{2}{N_0}\int_0^T n(t)\frac{\partial s(t,\theta)}{\partial\theta}dt. \end{aligned} \tag{F.47}$$

Applying (F.41) and (F.20), we find that the Cramer-Rao inequality for a waveform parameter θ is

$$var\left(\widehat{\theta}\right) \geq \frac{N_0}{2}\left\{\int_0^T \left[\frac{\partial s(t,\theta)}{\partial\theta}\right]^2 dt\right\}^{-1}. \tag{F.48}$$

Appendix G
Hermitian Positive-Definite Matrices

An $n \times n$ matrix $\mathbf{A}$ has an eigenvector $\mathbf{u}$ and an eigenvalue λ if $\mathbf{Au} = \lambda\mathbf{u}$. A set of vectors $\mathbf{u}_1, \ldots, \mathbf{u}_n$ are *orthonormal* if

$$\mathbf{u}_i^H \mathbf{u}_k = 0, \; i \neq k, \; \text{and} \; \|\mathbf{u}_i\|^2 = \mathbf{u}_i^H \mathbf{u}_i = 1 \tag{G.1}$$

where $\|\cdot\|$ denotes the *Euclidean norm* of a vector, and the superscript H denotes the conjugate transpose. A *unitary matrix* $\mathbf{U}$ is an $n \times n$ matrix with orthonormal column vectors. Therefore, $\mathbf{U}$ has rank n, $\mathbf{U}^H\mathbf{U} = \mathbf{I}$, and

$$\mathbf{U}^{-1} = \mathbf{I}\mathbf{U}^{-1} = \mathbf{U}^H\mathbf{U}\mathbf{U}^{-1} = \mathbf{U}^H \tag{G.2}$$

where $\mathbf{I}$ denotes the identity matrix. A unitary matrix with real-valued elements is called an *orthogonal matrix*.

The $n \times n$ matrix $\mathbf{A}$ has of a *complete set of n orthonormal eigenvectors* $\mathbf{u}_1, \ldots, \mathbf{u}_n$ with corresponding eigenvalues $\lambda_1, \ldots, \lambda_n$ if there are n orthonormal eigenvectors satisfying

$$\mathbf{A}\mathbf{u}_i = \lambda_i \mathbf{u}_i, \; 1 \leq i \leq n. \tag{G.3}$$

If an $n \times n$ matrix $\mathbf{A}$ has of a complete set of n orthonormal eigenvectors, and $\mathbf{U}$ is a unitary matrix with column vectors equal to $\mathbf{u}_1, \ldots, \mathbf{u}_n$, then $\mathbf{U}^H\mathbf{A}\mathbf{u}_i = \lambda_i \mathbf{U}^H \mathbf{u}_i$, $1 \leq i \leq n$, and hence the diagonal matrix of eigenvalues is

$$\Lambda = \mathbf{U}^H\mathbf{A}\mathbf{U}. \tag{G.4}$$

Conversely, if an $n \times n$ matrix $\mathbf{A}$ and an $n \times n$ unitary matrix $\mathbf{U}$ satisfy (G.4) for some diagonal matrix $\boldsymbol{\Lambda}$ with diagonal elements $\lambda_1, \ldots, \lambda_n$, then an application of (G.2) proves that $\mathbf{A}$ has of a *complete set of n orthonormal eigenvectors* $\mathbf{u}_1, \ldots, \mathbf{u}_n$ with

D. Torrieri, *Principles of Spread-Spectrum Communication Systems*,
https://doi.org/10.1007/978-3-319-70569-9

corresponding eigenvalues $\lambda_1, \ldots, \lambda_n$. Applying (G.2) to (G.4) twice, we obtain the *spectral decomposition*

$$\mathbf{A} = \mathbf{U}\boldsymbol{\Lambda}\mathbf{U}^H = \mathbf{U}\boldsymbol{\Lambda}\mathbf{U}^{-1}. \tag{G.5}$$

An $n \times n$ matrix $\mathbf{A}$ is *diagonalizable* if $\mathbf{B}^{-1}\mathbf{A}\mathbf{B} = \mathbf{D}$ for some nonsingular matrix $\mathbf{B}$ and diagonal matrix $\mathbf{D}$. Thus, if an $n \times n$ matrix $\mathbf{A}$ and an $n \times n$ unitary matrix $\mathbf{U}$ satisfy (G.5) for some diagonal matrix $\boldsymbol{\Lambda}$ with diagonal elements $\lambda_1, \ldots, \lambda_n$, then $\mathbf{A}$ is diagonalizable.

An $n \times n$ *Hermitian matrix* $\mathbf{A}$ is a matrix satisfying $\mathbf{A}^H = \mathbf{A}$. A Hermitian matrix with real-valued elements is called a *symmetric matrix.*

Theorem *An $n \times n$ Hermitian matrix $\mathbf{A}$ has a complete set of orthonormal eigenvectors and can be diagonalized.*

Proof The proof is by mathematical induction. The result is true if $n = 1$. Assume that the hypothesis is true for $k \times k$ Hermitian matrices, and let $\mathbf{A}$ denote a $(k + 1) \times (k + 1)$ Hermitian matrix. This matrix has at least one eigenvector $\mathbf{u}_1$ with a unit norm. Let λ_1 denote the corresponding eigenvalue. Using the Gram-Schmidt process, we determine orthonormal vectors $\mathbf{u}_1, \ldots, \mathbf{u}_{k+1}$ that constitute the basis for a complex $(k + 1)$-dimensional vector space. Let $\mathbf{Z}$ denote the $(k + 1) \times (k + 1)$ unitary matrix with these orthonormal vectors as its columns. The $(k + 1) \times (k + 1)$ matrix $\mathbf{Z}^H\mathbf{A}\mathbf{Z}$ is Hermitian, and its first column is

$$\mathbf{Z}^H\mathbf{A}\mathbf{u}_1 = \lambda_1\mathbf{Z}^H\mathbf{u}_1 = \lambda_1\mathbf{e}_1 \tag{G.6}$$

where $\mathbf{e}_1 = [1\ 0\ \ldots\ 0]^H$. Therefore, as $\mathbf{Z}^H\mathbf{A}\mathbf{Z}$ is a Hermitian matrix, it must have the form

$$\mathbf{Z}^H\mathbf{A}\mathbf{Z} = \begin{bmatrix} \lambda_1 & \mathbf{0} \\ \mathbf{0} & \mathbf{Y} \end{bmatrix} \tag{G.7}$$

where $\mathbf{Y}$ is a $k \times k$ Hermitian matrix. According to the induction hypothesis, a $k \times k$ unitary matrix $\mathbf{V}_1$ exists such that $\mathbf{V}_1^H\mathbf{Y}\mathbf{V}_1 = \mathbf{D}$, where $\mathbf{D}$ is a $k \times k$ diagonal matrix with its ith diagonal component equal to λ_{i+1}. Let $\mathbf{V}$ denote the $(k + 1) \times (k + 1)$ unitary matrix

$$\mathbf{V} = \begin{bmatrix} \lambda_1 & \mathbf{0} \\ \mathbf{0} & \mathbf{V}_1 \end{bmatrix}. \tag{G.8}$$

The $(k + 1) \times (k + 1)$ matrix $\mathbf{U} = \mathbf{Z}\mathbf{V}$ is unitary because $\mathbf{U}^H\mathbf{U} = \mathbf{V}^H\mathbf{Z}^H\mathbf{Z}\mathbf{V} = \mathbf{I}$, and

$$\mathbf{U}^H\mathbf{A}\mathbf{U} = \mathbf{V}^H\mathbf{Z}^H\mathbf{A}\mathbf{Z}\mathbf{V} = \begin{bmatrix} \lambda_1 & \mathbf{0} \\ \mathbf{0} & \mathbf{V}_1^H\mathbf{Y}\mathbf{V}_1 \end{bmatrix}$$

$$= \begin{bmatrix} \lambda_1 & \mathbf{0} \\ \mathbf{0} & \mathbf{D} \end{bmatrix} = \boldsymbol{\Lambda} \tag{G.9}$$

where $\mathbf{\Lambda}$ is a $(k+1) \times (k+1)$ diagonal matrix with its ith diagonal component equal to λ_i. Therefore, the columns of $\mathbf{U}$ comprise a complete set of orthonormal eigenvectors, and $\mathbf{A}$ is diagonalizable. □

For an $n \times n$ Hermitian matrix $\mathbf{A}$, implies that

$$\mathbf{u}_i^H \mathbf{A} \mathbf{u}_i = \lambda_i \|\mathbf{u}_i\|^2, \; 1 \le i \le n. \tag{G.10}$$

The left-hand side of this equation is real as it is identical to its conjugate transpose, and hence each λ_i is real. Therefore, *the eigenvalues of a Hermitian matrix are real.*

An $n \times n$ Hermitian matrix is *positive semidefinite* if for all $n \times 1$ column vectors $\mathbf{x}$,

$$\mathbf{x}^H \mathbf{A} \mathbf{x} \ge \mathbf{0} \tag{G.11}$$

and it is *positive definite* if

$$\mathbf{x}^H \mathbf{A} \mathbf{x} > \mathbf{0}, \; \mathbf{x} \ne \mathbf{0}. \tag{G.12}$$

If $\mathbf{Y}$ is an $n \times 1$ random vector, then the $n \times n$ matrix $E\left[\mathbf{Y}\mathbf{Y}^H\right]$ is positive semidefinite because $\mathbf{x}^H E\left[\mathbf{Y}\mathbf{Y}^H\right]\mathbf{x} = E\left[\left|\mathbf{Y}^H\mathbf{x}\right|^2\right] \ge 0$ for all $n \times 1$ column vectors $\mathbf{x}$.

Equation (G.10) indicates that a *Hermitian positive-semidefinite matrix has non-negative eigenvalues,* and that a *Hermitian positive-definite matrix has positive eigenvalues.* If a Hermitian positive-definite matrix were singular, then $\lambda = 0$ would be one or more of its eigenvalues. Therefore, since all its eigenvalues are positive, *a Hermitian positive-definite matrix is invertible.*

If $\mathbf{B}$ is an $n \times n$ Hermitian positive-definite matrix and $\mathbf{A}$ is an $n \times n$ invertible matrix, then $\mathbf{x}^H\mathbf{A}^H\mathbf{B}\mathbf{A}\mathbf{x} = (\mathbf{A}\mathbf{x})^H\mathbf{B}\mathbf{A}\mathbf{x} \ne \mathbf{0}$ because $\mathbf{A}\mathbf{x} \ne \mathbf{0}$. Therefore, $\mathbf{A}^H\mathbf{B}\mathbf{A}$ *is an $n \times n$ Hermitian positive-definite matrix.*

The *trace of a matrix* is equal to the sum of its diagonal elements. From the definitions of matrix multiplication and the trace, it follows that $tr\,(\mathbf{AB}) = tr\,(\mathbf{BA})$ for compatible matrices $\mathbf{A}$ and $\mathbf{B}$. If $\mathbf{A}$ is a Hermitian positive-definite matrix, then its spectral decomposition and positive eigenvalues imply that

$$\begin{aligned} tr\,(\mathbf{A}) &= tr\left(\mathbf{U}\Lambda\mathbf{U}^H\right) = tr\left(\Lambda\mathbf{U}^H\mathbf{U}\right) \\ &= tr\,(\mathbf{\Lambda}) > 0. \end{aligned} \tag{G.13}$$

Therefore, if $\mathbf{A}$ is a Hermitian positive-definite matrix, then $tr\,(\mathbf{A}) > 0$. Similarly, if $\mathbf{A}$ is a Hermitian positive-semidefinite matrix, then $tr\,(\mathbf{A}) \ge 0$.

Appendix H
Special Functions

H.1 Gamma Functions

The *gamma function* is defined as

$$\Gamma(x) = \int_0^\infty y^{x-1}e^{-y}dy, \quad \text{Re}(x) > 0. \tag{H.1}$$

An integration by parts indicates that

$$\Gamma(1+x) = x\Gamma(x). \tag{H.2}$$

A direct integration yields $\Gamma(1) = 1$. Therefore, when n is a positive integer,

$$\Gamma(n) = \int_0^\infty y^{n-1}e^{-y}dy = (n-1)!\,, \quad n \text{ is positive integer.} \tag{H.3}$$

Changing the integration variable by substituting $y = z^2$ in (H.1), observing that the integrand is an even function, and using (A.2), it is found that

$$\Gamma(1/2) = \sqrt{\pi}. \tag{H.4}$$

The *incomplete gamma functions* are defined as

$$\Gamma(a,x) = \int_x^\infty e^{-t}t^{a-1}dt, \quad Re\,(a) > 0 \tag{H.5}$$

and

$$\gamma(a,x) = \int_a^x e^{-t}t^{a-1}dt, \quad Re\,(a) > 0. \tag{H.6}$$

D. Torrieri, *Principles of Spread-Spectrum Communication Systems*,
https://doi.org/10.1007/978-3-319-70569-9

Therefore,

$$\Gamma(a) = \Gamma(a, x) + \gamma(a, x). \tag{H.7}$$

When $a = n$ is a positive integer, the integration of $\Gamma(n, x)$ by parts $n - 1$ times yields

$$\Gamma(n, x) = (n-1)! e^{-x} \sum_{i=0}^{n-1} \frac{x^i}{i!} \tag{H.8}$$

and hence

$$\gamma(n, x) = (n-1)! \left(1 - e^{-x} \sum_{i=0}^{n-1} \frac{x^i}{i!}\right). \tag{H.9}$$

H.2 Beta Function

The *beta function* is defined as

$$B(x, y) = \int_0^1 t^{x-1}(1-t)^{y-1}\, dt, \quad x > 0, \;\; y > 0. \tag{H.10}$$

The identity

$$B(x, y) = B(y, x) = \frac{\Gamma(x)\Gamma(y)}{\Gamma(x+y)} \tag{H.11}$$

is proved by substituting $y = z^2$ in the integrand of (H.1), expressing the product $\Gamma(a)\Gamma(b)$ as a double integral, changing to polar coordinates, integrating over the radius to obtain a result proportional to $\Gamma(a + b)$, and then changing the variable in the remaining integral to obtain $B(a, b)\Gamma(a + b)$.

In (H.10), set $x = (k + 1)/2$ and $y = 1/2$, and then apply (H.11) . The change of variable $t = cos^2\theta$, $0 \leq \theta \leq \pi/2$, in the integrand of (H.10), and similarly the change of variable $t = \sin^2\theta$, $0 \leq \theta \leq \pi/2$, yield

$$\int_0^{\pi/2} \cos^k\theta\, d\theta = \int_0^{\pi/2} \sin^k\theta\, d\theta = \frac{\sqrt{\pi}\,\Gamma\left(\frac{k+1}{2}\right)}{2\Gamma\left(\frac{k+2}{2}\right)}, \quad k \geq 0. \tag{H.12}$$

H.3 Bessel Functions of the First Kind

The *modified Bessel function* of the first kind and order n is defined as

$$I_n(x) = \sum_{i=0}^{\infty} \frac{(x/2)^{n+2i}}{i!\Gamma(n+i+1)} \tag{H.13}$$

where the gamma function may be replaced by a factorial if n is an integer. Therefore, the modified Bessel function of the first kind and order zero is defined as

$$I_0(x) = \sum_{i=0}^{\infty} \frac{1}{i!i!} \left(\frac{x}{2}\right)^{2i}. \tag{H.14}$$

A substitution of the series expansion of the exponential function and a term-by-term integration using (H.12) verifies the representation

$$I_0(x) = \frac{1}{2\pi} \int_0^{2\pi} e^{x\cos u} du. \tag{H.15}$$

Since the cosine is a periodic function and the integration is over the same period, we may replace $\cos u$ with $\cos(u+\theta)$ for any θ in (H.15). A trigonometric expansion with $x_1 = |x| \cos\theta$ and $x_2 = |x| \sin\theta$ then yields

$$\begin{aligned} I_0(|x|) &= \frac{1}{2\pi} \int_0^{2\pi} \exp\{\mathrm{Re}[|x|\, e^{j(u+\theta)}]\} du \\ &= \frac{1}{2\pi} \int_0^{2\pi} \exp\left(x_1 \cos u - x_2 \sin u\right) du, \quad |x| = \sqrt{x_1^2 + x_2^2}\,. \end{aligned} \tag{H.16}$$

A term-by-term differentiation of (H.14) yields

$$I_1(x) = \frac{d}{dx} I_0(x). \tag{H.17}$$

The *Bessel function* of the first kind and order n is defined as

$$J_n(x) = \sum_{i=0}^{\infty} \frac{(-1)^i \,(x/2)^{n+2i}}{i!\,\Gamma(n+i+1)}. \tag{H.18}$$

A substitution of the series expansion of the exponential function and a term-by-term integration using (H.12) verifies the representation

$$J_0(x) = \frac{1}{2\pi} \int_0^{2\pi} e^{jx\sin u} du. \tag{H.19}$$

H.4 Q-Functions

The Gaussian *Q-function* is defined as

$$Q(x) = \frac{1}{\sqrt{2\pi}} \int_x^\infty \exp\left(-\frac{y^2}{2}\right) dy = \frac{1}{2} erfc\left(\frac{x}{\sqrt{2}}\right) \tag{H.20}$$

where erfc($\cdot$) is the *complementary error function*. The results in Appendix A.1 imply that $Q(-\infty) = 1$, $Q(0) = 1/2$, and $Q(\infty) = 0$.

The Gaussian Q-function can be recast into a form in which the limits of the integral are not only finite but also independent of the argument of the function. This form facilitates computation and enables simplified analyses. To derive this form, we use (H.20) and $Q(-\infty) = 1$. For $x \geq 0$,

$$\begin{aligned} Q(x) &= \frac{1}{\sqrt{2\pi}} \int_x^\infty \exp\left(-\frac{y^2}{2}\right) dy \left[\frac{1}{\sqrt{2\pi}} \int_{-\infty}^\infty \exp\left(-\frac{z^2}{2}\right) dz\right] \\ &= \frac{1}{2\pi} \iint\limits_{y \geq x \geq 0,\, z > -\infty} \exp\left(-\frac{y^2 + z^2}{2}\right) dydz \\ &= \frac{1}{2\pi} \int_0^\pi \int_{x/\sin\theta}^\infty \exp\left(-\frac{r^2}{2}\right) rdrd\theta. \end{aligned} \tag{H.21}$$

In the second equality, Fubini's theorem justifies expressing the successive integrations as a double integral over a region in the plane. The third equality is the result of changing the Cartesian coordinates to polar coordinates. After integrating over the radius and using the periodic character of the angular coordinate, we obtain the desired form:

$$Q(x) = \frac{1}{\pi} \int_0^{\pi/2} \exp\left(-\frac{x^2}{2\sin^2\theta}\right) d\theta, \; x \geq 0. \tag{H.22}$$

If $x, y \geq 0$, then

$$Q(x+y) = \frac{1}{\pi} \int_0^{\pi/2} \exp\left(-\frac{x^2 + y^2 + 2xy}{2\sin^2\theta}\right) d\theta. \tag{H.23}$$

Since $\left(y^2 + 2xy\right) / \sin^2\theta \geq y^2$ when $\theta \in [0, \pi/2]$,

$$Q(x+y) \leq \exp\left(-\frac{y^2}{2}\right) Q(x), \; x, y \geq 0. \tag{H.24}$$

This inequality provides a generalization of the Chernoff bound on the standard Gaussian Q-function. Similarly,

$$Q(\sqrt{x+y}) \leq \exp\left(-\frac{y}{2}\right) Q(\sqrt{x}), \; x, y \geq 0. \tag{H.25}$$

The *generalized Marcum Q-function* is defined as

$$Q_m(\alpha, \beta) = \int_{\beta}^{\infty} x \left(\frac{x}{\alpha}\right)^{m-1} \exp\left(-\frac{x^2+\alpha^2}{2}\right) I_{m-1}(\alpha x)\, dx \tag{H.26}$$

and m is an integer. Since $Q_m(\alpha, 0) - Q_m(\alpha, \beta) = 1 - Q_m(\alpha, \beta)$ is an integral over a finite interval, it can be numerically integrated, and hence $Q_m(\alpha, \beta)$ numerically evaluated.

H.5 Hypergeometric Function

The *confluent hypergeometric function* is defined as

$${}_1F_1(\alpha, \beta; x) = \sum_{i=0}^{\infty} \frac{\Gamma(\alpha+i)\Gamma(\beta)x^i}{\Gamma(\alpha)\Gamma(\beta+i)i!}, \quad \beta \neq 0, -1, -2, \ldots \tag{H.27}$$

and the series converges for all finite x.

References

1. F. Adachi, D. Garg, S. Takaoka, and K. Takeda, "Broadband CDMA Techniques," İ *IEEE Wireless Commun.*, vol. 44, pp. 8–18, April 2005.
2. F. Adachi and K. Takeda, "Bit Error Rate Analysis of DS-CDMA with Joint Frequency-Domain Equalization and Antenna Diversity Combining," *IEICE Trans. Commun.*, vol. E87-B, pp. 2291–3001, Oct. 2004.
3. F. Adachi, M. Sawahashi, and K. Okawa, "Tree-structured Generation of Orthogonal Spreading Codes with Different Lengths for Forward Link of DS-CDMA Mobile Radio," *IEE Electronics Letters*, vol. 33, pp. 27–28, Jan. 1997.
4. S. Ahmed, L.-L. Yang, and L. Hanzo, "Erasure Insertion in RS-Coded SFH FSK Subjected to Tone Jamming and Rayleigh Fading," *IEEE Trans. Commun.*, vol. 56, pp. 3563–3571, Nov. 2007.
5. R. B. Ash and W. P. Novinger, *Complex Variables, 2nd ed.* Dover, 2004.
6. R. B. Ash and C. A. Doleans-Dade, *Probability and Measure Theory, 2nd ed.,* Academic Press, 2000.
7. J. R. Barry, E. A. Lee, and D. G. Messerschmitt, *Digital Communication, 3rd ed.,* Kluwer Academic, 2004.
8. N. C. Beaulieu, W. L. Hopkins, and P. J. McLane, "Interception of Frequency-Hopped Spread-Spectrum Signals," *IEEE. J. Select. Areas Commun.*, vol. 8, pp. 853–870, June 1990.
9. P. Billingsley, *Probability and Measure, 3rd ed., Wiley,* 1995.
10. N. Bonello, S. Chen, and L. Hanzo, "Low-Density Parity-Check Codes and Their Rateless Relatives," *IEEE Commun. Surveys Tut.*, vol. 13, pp. 3–26, first quarter, 2011.
11. W. R. Braun, "PN Acquisition and Tracking Performance in DS/CDMA Systems with Symbol-Length Spreading Sequence," *IEEE Trans. Commun.*, vol. 45, pp. 1595–1601, Dec. 1997.
12. H. Cai, Y. Yang, Z. Zhou, and X. Tang, "Strictly Optimal Frequency-Hopping Sequence Sets With Optimal Family Sizes," *IEEE Trans. Inf. Theory, vol. 62,* pp. 1087–1093, Feb. 2016.
13. G. Caire, G. Taricco, and E. Biglieri, "Bit-interleaved coded modulation," *IEEE Trans. Inform. Theory*, vol. 44, pp. 927–946, May 1998.
14. J.-J. Chang, D.-J. Hwang, and M.-C. Lin, "Some Extended Results on the Search for Good Convolutional Codes," *IEEE Trans. Inform. Theory*, vol. 43, pp. 1682–1697, Sept. 1997.
15. H. Chen, R. G. Maunder, and L. Hanzo, "A Survey and Tutorial on Low-Complexity Turbo Coding Techniques and a Holistic Hybrid ARQ Design Example," *IEEE Commun. Surveys Tut.*, vol. 15, pp. 1546–1566, fourth quarter, 2013.

D. Torrieri, *Principles of Spread-Spectrum Communication Systems*,
https://doi.org/10.1007/978-3-319-70569-9

16. A. Chockalingam, P. Dietrich, L. B. Milstein, and R. R. Rao, "Performance of Closed-Loop Power Control in DS-CDMA Cellular Systems," *IEEE Trans. Veh. Technol.*, vol. 47, pp. 774–789, Aug. 1998.
17. J. Choi, *Adaptive and Iterative Signal Processing in Communications,* Cambridge Univ. Press, 2006.
18. K. Choi, K. Cheun, and T. Jung, "Adaptive PN Code Acquisition Using Instantaneous Power-Scaled Detection Threshold under Rayleigh Fading and Pulsed Gaussian Noise Jamming," *IEEE Trans. Commun.*, vol. 50, pp. 1232–1235, Aug. 2002.
19. E. K. P. Chong and S. H. Zak, *An Introduction to Optimization, 4th ed.*, Wiley, 2013.
20. T. M. Cover and J. M. Thomas, *Elements of Information Theory, 2nd ed.*, Wiley, 2006.
21. F. F. Digham, M. S. Alouini, and M. K. Simon, "On the Energy Detection of Unknown Signals over Fading Channels," *IEEE Trans. Commun.*, vol. 55, pp. 21–24, Jan. 2007.
22. P. S. R. Diniz, *Adaptive Filtering: Algorithms and Practical Implementation, 4th ed.,* Springer, 2012.
23. S. Emami, "UWB Communication Systems: Conventional and 60 GHz: Principles, Design, and Standards," Springer, 2013.
24. Y. Fang, G. Bi, Y. L. Guan, and F. C. M. Lau, "A Survey on Protograph LDPC Codes and Their Applications," *IEEE Commun. Surveys Tut.,* vol. 17, pp. 1989–2016, fourth quarter, 2015.
25. Y. Fang, G. Han, P. Chen, F. C. M. Lau, G. Chen, and L. Wang, "A Survey on DCSK-based Communication Systems and Their Application to UWB Scenarios," *IEEE Commun. Surveys Tut.,* vol. 18, pp. 1804–1837, fourth quarter, 2016.
26. B. Farhang-Boroujeny, *Adaptive Filters: Theory and Applications, 2nd ed.,* Wiley, 2013.
27. S. Gezici, "Mean Acquisition Time Analysis of Fixed-Step Serial Search Algorithms," *IEEE Trans. Commun.*, vol. 8, pp. 1096–1101, March 2009.
28. M. Goresky and A. Klapper, *Algebraic Shift Register Sequences*, Cambridge Univ. Press, 2012.
29. T. T. Ha, *Theory and Design of Digital Communication Systems*, Cambridge Univ. Press, 2011.
30. L. Hanzo, T. H. Liew, B. L. Yeap, R. Y. S. Lee, and S. X. Ng, *Turbo Coding, Turbo Equalisation and Space-Time Coding, 2nd ed.,* Wiley, 2011.
31. A. R. Hammons and P. V. Kumar, "On a Recent 4-Phase Sequence Design for CDMA," *IEICE Trans. Commun.*, vol. E76-B, pp. 804–813, Aug. 1993.
32. S. Haykin, *Adaptive Filter Theory, 5th ed.,* Prentice-Hall, 2013.
33. C. W. Helstrom, *Elements of Signal Detection and Estimation*, Prentice Hall, 1995.
34. K. Higuchi et al., "Experimental Evaluation of Combined Effect of Coherent Rake Combining and SIR-Based Fast Transmit Power Control for Reverse Link of DS-CDMA Mobile Radio," *IEEE J. Select. Areas Commun.*, vol. 18, pp. 1526–1535, Aug. 2000.
35. D. A. Hill and E. B. Felstead, "Laboratory Performance of Spread Spectrum Detectors," *IEE Proc.-Commun.*, vol. 142, pp. 243–249, Aug. 1995.
36. D. Horgan and C. C. Murphy, "Fast and Accurate Approximations for the Analysis of Energy Detection in Nakagami-m Channels," *IEEE Commun. Lett.*, vol. 17, pp. 83–86, Jan. 2013.
37. D. Horgan and C. C. Murphy, "On the Convergence of the Chi Square and Noncentral Chi Square Distributions to the Normal Distribution," *IEEE Commun. Lett.*, vol. 17, pp. 2233–2236, Dec. 2013.
38. V. M. Jovanovic, "Analysis of Strategies for Serial-Search Spread-Spectrum Code Acquisition—Direct Approach," *IEEE Trans. Commun.*, vol. 36, pp. 1208–1220, Nov. 1988.
39. E. Kreyszig, *Advanced Engineering Mathematics, 10th ed.,* Wiley, 2011.
40. V. Kuhn, *Wireless Communications over MIMO Channels*, Wiley, 2006.
41. Y. M. Lam and P. H. Wittke, "Frequency-Hopped Spread-Spectrum Transmission with Band-Efficient Modulations and Simplified Noncoherent Sequence Estimation," *IEEE Trans. Commun.*, vol. 38, pp. 2184–2196, Dec. 1990.
42. K. Lange, *Optimization*, 2nd ed., Springer, 2013.

43. E. G. Larsson and P. Stoica, *Space-Time Block Coding for Wireless Communications*, Cambridge Univ. Press, 2003.
44. J. S. Lee, L. E. Miller, and Y. K. Kim, "Probability of Error Analyses of a BFSK Frequency-Hopping System with Diversity under Partial-Band Jamming Interference—Part II: Performance of Square-Law Nonlinear Combining Soft Decision Receivers," *IEEE Trans. Commun.*, vol. 32, pp. 1243–1250, Dec. 1984.
45. S. J. Leon, *Linear Algebra with Applications, 9th ed.,* Pearson, 2014.
46. B. K. Levitt, U. Cheng, A. Polydoros, and M. K. Simon, "optimal Detection of Slow Frequency-Hopped Signals," *IEEE Trans. Commun.*, vol. 42, pp. 1990–2000, Feb./March/April 1994.
47. B. Levy, *Principles of Signal Detection and Parameter Estimation*, Springer, 2008.
48. X. Li, A. Chindapol, and J. A. Ritcey, "Bit-interleaved Coded Modulation with Iterative Decoding and 8PSK Modulation," *IEEE Trans. Commun.*, vol. 50, pp. 1250–1257, Aug. 2002.
49. C-F. Li, Y-S Chu, J-S Ho, and W-H Sheen, "Cell Search in WCDMA Under Large-Frequency and Clock Errors: Algorithms to Hardware Implementation," *IEEE Trans. Circuits and Syst.*, vol. 55, pp. 659–671, March 2008.
50. T. G. Macdonald and M. B. Pursley, "The Performance of Direct-Sequence Spread Spectrum with Complex Processing and Quaternary Data Modulation," *IEEE J. Select. Areas Commun.*, vol. 18, pp. 1408–1417, Aug. 2000.
51. A. Mariani, A. Giorgetti, and M. Chiani, "Effects of Noise Power Estimation on Energy Detection for Cognitive Radio Applications," *IEEE Trans. Commun.*, vol. 59, pp. 3410–3420, Dec. 2011.
52. M. Medley, G. Saulnier, and P. Das, "The Application of Wavelet-Domain Adaptive Filtering to Spread-Spectrum Communications," *Proc. SPIE Wavelet Applications for Dual-Use*, vol. 2491, pp. 233–247, April 1995.
53. H. Meyr and G. Polzer, "Performance Analysis for General PN Spread-spectrum Acquisition Techniques," *IEEE Trans. Commun.*, vol. 31, pp. 1317–1319, Dec. 1983.
54. L. E. Miller, J. S. Lee, R. H. French, and D. J. Torrieri, "Analysis of an Anti-jam FH Acquisition Scheme," *IEEE Trans. Commun.*, vol. 40, pp. 160–170, Jan. 1992.
55. L. E. Miller, J. S. Lee, and D. J. Torrieri, "Frequency-Hopping Signal Detection Using Partial Band Coverage," *IEEE Trans. Aerosp. Electron. Syst.*, vol. 29, pp. 540–553, April 1993.
56. L. B. Milstein, "Interference Rejection Techniques in Spread Spectrum Communications," *Proc. IEEE*, vol. 76, pp. 657–671, June 1988.
57. T. K. Moon, *Error Correction Coding*, Wiley, 2005.
58. D. Morgan, *Surface Acoustic Wave Filters with Applications to Electronic Communications and Signal Processing, 2nd Ed.*, Academic Press, 2007.
59. H. Mukhtar, A. Al-Dweik, and A. Shami, "Turbo Product Codes: Applications, Challenges and Future Directions," *IEEE Trans. Commun. Surveys Tut.*, vol. 18, pp. 3052–3069, fourth quarter, 2016.
60. J. T. Oden and L. Demkowicz, *Applied Functional Analysis, 2nd. ed.*, CRC Press, 2010.
61. S.-M. Pan, D. E. Dodds, and S. Kumar, "Acquisition Time Distribution for Spread-Spectrum Receivers," *IEEE J. Select. Areas Commun.*, vol. 8, pp. 800–808, June 1990.
62. D. R. Pauluzzi and N. C. Beaulieu, "A Comparison of SNR Estimation Techniques for the AWGN Channel," *IEEE Trans. Commun.*, vol. 48, pp. 1681–1691, Oct. 2000.
63. D. Peng and P. Fan, "Lower bounds on the Hamming auto- and cross correlations of frequency-hopping sequences," *IEEE Trans. Inf.* Theory, vol. 50, pp. 2149–2154, Sept. 2004.
64. R. L. Peterson, R. E. Ziemer, and D. E. Borth, *Introduction to Spread Spectrum Communications,* Prentice Hall, 1995.
65. C. Phillips, D. Sicker, and D. Grunwald, "A Survey of Wireless Path Loss Prediction and Coverage Mapping Methods," *IEEE Trans. Commun. Surveys Tut.*, vol. 15, pp. 255–270, first quarter, 2013.
66. A. Polydoros and C. L. Weber, "A Unified Approach to Serial-Search Spread Spectrum Code Acquisition," *IEEE Trans. Commun.*, vol. 32, pp. 542–560, May 1984.

67. A. Polydoros and C. L. Weber, "Detection Performance Considerations for Direct-Sequence and Time-Hopping LPI Waveforms," *IEEE J. Select. Areas Commun.*, vol. 3, pp. 727–744, Sept. 1985.
68. B. Porat, *A Course in Digital Signal Processing*, Wiley, 1997.
69. J. G. Proakis and M. Salehi, *Digital Communications, 5th ed.,* McGraw-Hill, 2008.
70. M. B. Pursley, "Spread-Spectrum Multiple-Access Communications," in *Multi-User Communications Systems*, G. Longo, ed., Springer-Verlag, 1981.
71. M. B. Pursley, D. V. Sarwate, and W. E. Stark, "Error Probability for Direct-Sequence Spread-Spectrum Multiple-Access Communications—Part 1: Upper and Lower Bounds," *IEEE Trans. Commun.*, vol. 30, pp. 975–984, May 1982.
72. H. Puska, H. Saarnisaari, J. Iinatti, and P. Lilja, "Serial Search Code Acquisition Using Smart Antennas with Single Correlator or Matched Filter," *IEEE Trans. Commun.*, vol. 56, pp. 299–308, Feb. 2008.
73. C. A. Putman, S. S. Rappaport, and D. L. Schilling, "Comparison of Strategies for Serial Acquisition of Frequency-Hopped Spread-Spectrum Signals," *IEE Proc.*, vol. 133, pt. F, pp. 129–137, April 1986.
74. C. A. Putman, S. S. Rappaport, and D. L. Schilling, "Tracking of Frequency-Hopped Spread-Spectrum Signals in Adverse Environments," *IEEE Trans. Commun.*, vol. 31, pp. 955–963, Aug. 1983.
75. R. Pyndiah, "Near-optimal decoding of product codes: block turbo codes," *IEEE Trans. Commun.*, vol. 46, pp. 1003–1010, Aug. 1998.
76. Y. Rahmatallah and S. Mohan, Member, "Peak-To-Average Power Ratio Reduction in OFDM Systems: A Survey And Taxonomy," *IEEE Commun. Surveys Tut.*, vol. 15, pp. 1567–1592, fourth quarter, 2013.
77. B. Razavi, *RF Microelectronics, 2nd ed.,* Pearson Prentice Hall, 2012.
78. M. Rice, *Digital Communications: A Discrete-Time Approach,* Pearson Prentice Hall, 2009.
79. U. L. Rohde and M. Rudolph, *RF/Microwave Circuit Design for Wireless Applications, 2nd ed.,* Wiley, 2013.
80. W. E. Ryan and S. Lin, *Channel Codes: Classical and Modern,* Cambridge Univ. Press, 2009.
81. D. S. Saini and M. Upadhyay, "Multiple Rake Combiners and Performance Improvement in 3G and Beyond WCDMA Systems," *IEEE Trans. Veh. Technol.*, vol. 58, pp. 3361–3370, Sept. 2009.
82. M. Sawahashi, K. Higuchi, H. Andoh, and F. Adachi, "Experiments on Pilot Symbol-Assisted Coherent Multistage Interference canceler for DS-CDMA Mobile Radio," *IEEE J. Select. Areas Commun.*, vol. 20, pp. 433–449, Feb. 2002.
83. A. H. Sayed, *Adaptive Filters,* Wiley, 2008.
84. C. Schlegel and A. Grant, *Coordinated Multiuser Communications.* Springer, 2006.
85. M. K. Simon and M.-S. Alouini, *Digital Communication over Fading Channels, 2nd ed.,* Wiley, 2004.
86. J. R. Smith, *Modern Communication Circuits, 2nd ed.*, McGraw-Hill, 1998.
87. R. A. Soni and R. M. Buehrer, "On the Performance of Open-Loop Transmit Diversity Techniques for IS-2000 Systems: A Comparative Study," *IEEE Trans. Wireless Commun.*, vol. 3, pp. 1602–1615, Sept. 2004.
88. G. Stuber, *Principles of Mobile Communication, 4th ed.,* Springer, 2017.
89. S-Y. Sun, H.-H. Chen, and W.-X. Meng, "A Survey on Complementary-Coded MIMO CDMA Wireless Communications," *IEEE Commun. Surveys Tut.*, vol. 17, pp. 52–69, first quarter, 2015.
90. W. Suwansantisuk and M. Z. Win, "Multipath Aided Rapid Acquisition: Optimal Search Strategies," *IEEE Trans. Inform. Theory*, vol. 53, pp. 174–193, Jan. 2007.
91. S. Talarico, M. C. Valenti, and D. Torrieri, "Optimization of an Adaptive Frequency-Hopping Network," *IEEE Military Commun. Conf.*, Oct. 2015.
92. W. M. Tam, F. C. M. Lau, and C. K. Tse, *Digital Communications with Chaos: Multiple Access Techniques and Performance Evaluation,* Oxford, U. K.: Elsevier, 2007.

93. B. S. Tan, K. H. Li, and K. C. Teh, "Transmit Antenna Selection Systems," *IEEE Vehicular Technol. Mag.*, vol. 8, pp. 104–112, Sept. 2013.
94. X. Tan and J. M. Shea, "An EM Approach to Multiple-Access interference Mitigation in Asynchronous Slow FHSS Systems," *IEEE Trans. Wireless Commun.*, vol. 7, pp. 2661–2670, July 2008.
95. S. Tanaka, A. Harada, and F. Adachi, "Experiments on Coherent Adaptive Antenna Array Diversity for Wideband DS-CDMA Mobile Radio," *IEEE J. Select. Areas Commun.*, vol. 18, pp. 1495–1504, Aug. 2000.
96. S. Tantaratana, A. W. Lam, and P. J. Vincent, "Noncoherent Sequential Acquisition of PN Sequences for DS/SS Communications with/without Channel Fading," *IEEE Trans. Commun.*, vol. 43, pp. 1738–1745, Feb./March/April 1995.
97. D. J. Torrieri, "The Performance of Five Different Metrics Against Pulsed Jamming," *IEEE Trans. Commun.*, vol. 34, pp. 200–207, Feb. 1986.
98. D. J. Torrieri, "Fundamental Limitations on Repeater Jamming of Frequency-Hopping Communications," *IEEE J. Select. Areas Commun.*, vol. 7, pp. 569–578, May 1989.
99. D. Torrieri, *Principles of Secure Communication Systems*, 2nd ed., Artech House, 1992.
100. D. Torrieri, "Performance of Direct-Sequence Systems with Long Pseudonoise Sequences," *IEEE J. Select. Areas Commun.*, vol. 10, pp. 770–781, May 1992.
101. D. Torrieri, "The Radiometer and Its Practical Implementation," *IEEE Military Commun. Conf.*, Oct. 2010.
102. D. Torrieri and K. Bakhru, "Anticipative Maximin Adaptive-Array Algorithm for Frequency-Hopping Systems," *IEEE Military Commun. Conf.*, Nov. 2006.
103. D. Torrieri and K. Bakhru, "The Maximin Adaptive-Array Algorithm for Direct-Sequence Systems," *IEEE Trans. Signal Processing*, vol. 55, pp.1853–1861, May 2007.
104. D. Torrieri and K. Bakhru, "Adaptive-Array Algorithm for Interference Suppression Prior to Acquisition of Direct-Sequence Signal," *IEEE Trans. Wireless Commun.*, vol. 7, pp. 3341–3346, Sept. 2008.
105. D. Torrieri, S. Cheng, and M. C. Valenti, "Robust Frequency Hopping for Interference and Fading Channels," IEEE Trans. Communications, vol. 56, pp. 1343–1351, Aug. 2008.
106. D. Torrieri, A. Mukherjee, and H. M. Kwon, "Coded DS-CDMA systems with Iterative Channel Estimation and No Pilot Symbols," *IEEE Trans. Wireless Commun.*, vol. 9, pp. 2012–2021, June 2010.
107. D. Torrieri and M. C. Valenti, "The Outage Probability of a Finite Ad Hoc Network in Nakagami Fading," *IEEE Trans. Commun.*, vol. 60, pp. 3509–3518, Nov. 2012.
108. D. Torrieri and M. C. Valenti, "Exclusion and Guard Zones in DS-CDMA Ad Hoc Networks," *IEEE Trans. Commun.*, vol. 61, pp. 2468–2476, June 2013.
109. D. Torrieri, M. C. Valenti, and S. Talarico, "An Analysis of the DS-CDMA Cellular Uplink for Arbitrary and Constrained Topologies," *IEEE Trans. Commun.*, vol. 61, pp. 3318–3326, Aug. 2013.
110. M. C. Valenti, S. Cheng, and D. Torrieri, "Iterative Multisymbol Noncoherent Reception of Coded CPFSK," *IEEE Trans. Commun.*, vol. 58, pp. 2046–2054, July 2010.
111. M. C. Valenti and S. Cheng, "Iterative Demodulation and Decoding of Turbo Coded M-ary Noncoherent Orthogonal Modulation," *IEEE J. Select. Areas Commun.*, vol. 23, pp. 1738–1747, Sept. 2005.
112. M.C. Valenti and M. Fanaei, "The interplay between modulation and channel coding," Chapter 5 of *Transmission Techniques for Digital Communications*, Elsevier, 2016.
113. M. C. Valenti, D. Torrieri, and S. Talarico, "A New Analysis of the DS-CDMA Cellular Downlink Under Spatial Constraints," *Intern. Conf. Computing, Networking, Commun.*, Jan. 2013.
114. S. S. Venkatesh, *The Theory of Probability*, Cambridge Univ. Press, 2013.
115. S. Verdu, *Multiuser Detection,* Cambridge Univ. Press, 1998.
116. X. Wang and H. V. Poor, *Wireless Communication Systems*, Prentice Hall, 2004.

117. S. P. Weber, J. G. Andrews, X. Yang, and G. de Veciana, "Transmission Capacity of Wireless Ad Hoc Networks with Successive Interference Cancellation," *IEEE Trans. Inf. Theory, vol. 53,* pp. 2799–2812, Aug. 2007.
118. P. H. Wittke, Y. M. Lam, and M. J. Schefter, "The Performance of Trellis-Coded Nonorthogonal Noncoherent FSK in Noise and Jamming," *IEEE Trans. Commun.*, vol. 43, pp. 635–645, Feb./March/April 1995.
119. S. Won and L. Hanzo, "Initial Synchronisation of Wideband and UWB Direct Sequence Systems: Single- and Multiple-Antenna Aided Solutions," *IEEE Commun. Surveys Tut.*, vol. 14, pp. 87–108, first quarter, 2012.
120. S. Xie and S. Rahardja, "Performance Evaluation for Quaternary DS-SSMA Communications with Complex Signature Sequences over Rayleigh-fading Channels," *IEEE Trans. Wireless Commun.*, vol. 4, pp. 266–77, Jan. 2005.
121. L. L. Yang and L. Hanzo, "Serial Acquisition Performance of Single-Carrier and Multicarrier DS-CDMA over Nakagami-m Fading Channels," *IEEE Trans. Wireless Commun.,* vol. 1, pp. 692–702, Oct. 2002.
122. Z. Zhou, X. Tang, X. Niu, and U. Parampalli, "New Classes of Frequency-Hopping Sequences With Optimal Partial Correlation," *IEEE Trans. Inf. Theory,* , vol. 58, pp.453–458, Jan. 2012.

Index

D. Torrieri, *Principles of Spread-Spectrum Communication Systems*,
https://doi.org/10.1007/978-3-319-70569-9

D

N

O

P